BOILING HEAT TRANSFER AND TWO-PHASE FLOW

Second Edition

Series in Chemical and Mechanical Engineering

G. F. Hewitt and C. L. Tien, *Editors*

Carey, Liquid–Vapor Phase-Change Phenomena: An Introduction to the Thermophysics of Vaporization and Condensation Processes in Heat Transfer Equipment

Diwekar, Batch Distillation: Simulation, Optimal Design and Control

FORTHCOMING TITLES

Tong and Tang, Boiling Heat Transfer and Two-Phase Flow, Second Edition

BOILING HEAT TRANSFER AND TWO-PHASE FLOW

Second Edition

L. S. Tong, Ph.D.

Y. S. Tang, Ph.D.

USA	Publishing Office	Taylor & Francis 1101 Vermont Avenue, N.W., Suite 200 Washington, D.C. 20005–3521 Tel: (202) 289-2174 Fax: (202) 289-3665
	Distribution Center:	Taylor & Francis 1900 Frost Road, Suite 101 Bristol, PA 19007-1598 Tel: (215) 785-5800 Fax: (215) 785-5515
UK		Taylor & Francis Ltd. 1 Gunpowder Square London EC4A 3DE Tel: 171 583 0490 Fax: 171 583 0581

BOILING HEAT TRANSFER AND TWO-PHASE FLOW, Second Edition

Copyright © 1997 Taylor & Francis. All rights reserved. Printed in the United States of America. Except as permitted under the United States Copyright Act of 1976, no part of this publication may be reproduced or distributed in any form or by any means, or stored in a database or retrieval system, without the prior written permission of the publisher.

1 2 3 4 5 6 7 8 9 0   B R B R   9 8 7

The editors were Lynne Lackenbach and Holly Seltzer. Cover design by Michelle Fleitz. Prepress supervisor was Miriam Gonzalez.

A CIP catalog record for this book is available from the British Library.
⊗ The paper in this publication meets the requirements of the ANSI Standard Z39.48-1984 (Permanence of Paper)

Library of Congress Cataloging-in-Publication Data

Tong, L. S. (Long-sun)
  Boiling heat transfer and two-phase flow/L. S. Tong and Y. S. Tang.
    p. cm.
  Includes bibliographical references.

  1. Heat—Transmission. 2. Ebullition. 3. Two-phase flow.
  I. Tang, Y. S. (Yu S.). II. Title.
QC320.T65 1997
536'.2—dc20   96-34009
                                                                                                  CIP

ISBN 1-56032-485-6 (case)

In Memory of Our Parents

CONTENTS

	Preface	xv
	Preface to the First Edition	xvii
	Symbols	xix
	Unit Conversions	xxix

1 INTRODUCTION   1

1.1 Regimes of boiling   1
1.2 Two-Phase Flow   3
1.3 Flow Boiling Crisis   4
1.4 Flow Instability   4

2 POOL BOILING   7

2.1 Introduction   7
2.2 Nucleation and Dynamics of Single Bubbles   7
    2.2.1 Nucleation   8
        2.2.1.1 Nucleation in a Pure Liquid   8
        2.2.1.2 Nucleation at Surfaces   10
    2.2.2 Waiting Period   19
    2.2.3 Isothermal Bubble Dynamics   23
    2.2.4 Isobaric Bubble Dynamics   26
    2.2.5 Bubble Departure from a Heated Surface   37

		2.2.5.1 Bubble Size at Departure	37
		2.2.5.2 Departure Frequency	40
		2.2.5.3 Boiling Sound	44
		2.2.5.4 Latent Heat Transport and Microconvection by Departing Bubbles	45
		2.2.5.5 Evaporation-of-Microlayer Theory	45
2.3	Hydrodynamics of Pool Boiling Process		50
	2.3.1 The Helmholtz Instability		50
	2.3.2 The Taylor Instability		52
2.4	Pool Boiling Heat Transfer		54
	2.4.1 Dimensional Analysis		55
		2.4.1.1 Commonly Used Nondimensional Groups	55
		2.4.1.2 Boiling Models	58
	2.4.2 Correlation of Nucleate Boiling Data		60
		2.4.2.1 Nucleate Pool Boiling of Ordinary Liquids	60
		2.4.2.2 Nucleate Pool Boiling with Liquid Metals	71
	2.4.3 Pool Boiling Crisis		80
		2.4.3.1 Pool Boiling Crisis in Ordinary Liquids	81
		2.4.3.2 Boiling Crisis with Liquid Metals	97
	2.4.4 Film Boiling in a Pool		102
		2.4.4.1 Film Boiling in Ordinary Liquids	103
		2.4.4.2 Film Boiling in Liquid Metals	109
2.5	Additional References for Further Study		116

3 HYDRODYNAMICS OF TWO-PHASE FLOW 119

3.1	Introduction		119
3.2	Flow Patterns in Adiabatic and Diabatic Flows		120
	3.2.1 Flow Patterns in Adiabatic Flow		120
	3.2.2 Flow Pattern Transitions in Adiabatic Flow		128
		3.2.2.1 Pattern Transition in Horizontal Adiabatic Flow	130
		3.2.2.2 Pattern Transition in Vertical Adiabatic Flow	133
		3.2.2.3 Adiabatic Flow in Rod Bundles	136
		3.2.2.4 Liquid Metal–Gas Two-Phase Systems	140
	3.2.3 Flow Patterns in Diabatic Flow		140
3.3	Void Fraction and Slip Ratio in Diabatic Flow		147
	3.3.1 Void Fraction in Subcooled Boiling Flow		152
	3.3.2 Void Fraction in Saturated Boiling Flow		155
	3.3.3 Diabatic Liquid Metal–Gas Two-Phase Flow		159
	3.3.4 Instrumentation		161
		3.3.4.1 Void Distribution Measurement	161
		3.3.4.2 Interfacial Area Measurement	163
		3.3.4.3 Measurement of the Velocity of a Large Particle	164
		3.3.4.4 Measurement of Liquid Film Thickness	166

3.4	Modeling of Two-Phase Flow		168
	3.4.1 Homogeneous Model/Drift Flux Model		168
	3.4.2 Separate-Phase Model (Two-Fluid Model)		170
	3.4.3 Models for Flow Pattern Transition		172
	3.4.4 Models for Bubbly Flow		173
	3.4.5 Models for Slug Flow (Taitel and Barnea, 1990)		174
	3.4.6 Models for Annular Flow		177
		3.4.6.1 Falling Film Flow	177
		3.4.6.2 Countercurrent Two-Phase Annular Flow	180
		3.4.6.3 Inverted Annular and Dispersed Flow	180
	3.4.7 Models for Stratified Flow (Horizontal Pipes)		182
	3.4.8 Models for Transient Two-Phase Flow		183
		3.4.8.1 Transient Two-Phase Flow in Horizontal Pipes	185
		3.4.8.2 Transient Slug Flow	186
		3.4.8.3 Transient Two-Phase Flow in Rod Bundles	186
3.5	Pressure Drop in Two-Phase Flow		187
	3.5.1 Local Pressure Drop		187
	3.5.2 Analytical Models for Pressure Drop Prediction		188
		3.5.2.1 Bubbly Flow	188
		3.5.2.2 Slug Flow	190
		3.5.2.3 Annular Flow	191
		3.5.2.4 Stratified Flow	191
	3.5.3 Empirical Correlations		194
		3.5.3.1 Bubbly Flow in Horizontal Pipes	196
		3.5.3.2 Slug Flow	200
		3.5.3.3 Annular Flow	201
		3.5.3.4 Correlations for Liquid Metal and Other Fluid Systems	202
	3.5.4 Pressure Drop in Rod Bundles		207
		3.5.4.1 Steady Two-Phase Flow	207
		3.5.4.2 Pressure Drop in Transient Flow	209
	3.5.5 Pressure Drop in Flow Restriction		210
		3.5.5.1 Steady-State, Two-Phase-Flow Pressure Drop	210
		3.5.5.2 Transient Two-Phase-Flow Pressure Drop	217
3.6	Critical Flow and Unsteady Flow		219
	3.6.1 Critical Flow in Long Pipes		220
	3.6.2 Critical Flow in Short Pipes, Nozzles, and Orifices		225
	3.6.3 Blowdown Experiments		228
		3.6.3.1 Experiments with Tubes	228
		3.6.3.2 Vessel Blowdown	230
	3.6.4 Propagation of Pressure Pulses and Waves		231
		3.6.4.1 Pressure Pulse Propagation	231
		3.6.4.2 Sonic Wave Propagation	236
		3.6.4.3 Relationship Among Critical Discharge Rate, Pressure Propagation Rate, and Sonic Velocity	239
3.7	Additional References for Further Study		242

4 FLOW BOILING — 245

4.1	Introducton	245
4.2	Nucleate Boiling in Flow	248
	4.2.1 Subcooled Nucleate Flow Boiling	248
	4.2.1.1 Partial Nucleate Flow Boiling	248
	4.2.1.2 Fully Developed Nucleate Flow Boiling	257
	4.2.2 Saturated Nucleate Flow Boiling	258
	4.2.2.1 Saturated Nucleate Flow Boiling of Ordinary Liquids	259
	4.2.2.2 Saturated Nucleate Flow Boiling of Liquid Metals	265
4.3	Forced-Convection Vaporization	265
	4.3.1 Correlations for Forced-Convection Vaporization	266
	4.3.2 Effect of Fouling Boiling Surface	268
	4.3.3 Correlations for Liquid Metals	268
4.4	Film Boiling and Heat Transfer in Liquid-Deficient Regions	274
	4.4.1 Partial Film Boiling (Transition Boiling)	275
	4.4.2 Stable Film Boiling	276
	4.4.2.1 Film Boiling in Rod Bundles	277
	4.4.3 Mist Heat Transfer in Dispersed Flow	277
	4.4.3.1 Dispersed Flow Model	279
	4.4.3.2 Dryout Droplet Diameter Calculation	281
	4.4.4 Transient Cooling	283
	4.4.4.1 Blowdown Heat Transfer	283
	4.4.4.2 Heat Transfer in Emergency Core Cooling Systems	287
	4.4.4.3 Loss-of-Coolant Accident (LOCA) Analysis	288
	4.4.5 Liquid-Metal Channel Voiding and Expulsion Models	297
4.5	Additional References for Further Study	299

5 FLOW BOILING CRISIS — 303

5.1	Introduction	303
5.2	Physical Mechanisms of Flow Boiling Crisis in Visual Observations	304
	5.2.1 Photographs of Flow Boiling Crisis	304
	5.2.2 Evidence of Surface Dryout in Annular Flow	309
	5.2.3 Summary of Observed Results	309
5.3	Microscopic Analysis of CHF Mechanisms	317
	5.3.1 Liquid Core Convection and Boundary-Layer Effects	318
	5.3.1.1 Liquid Core Temperature and Velocity Distribution Analysis	319
	5.3.1.2 Boundary-Layer Separation and Reynolds Flux	320
	5.3.1.3 Subcooled Core Liquid Exchange and Interface Condensation	323
	5.3.2 Bubble-Layer Thermal Shielding Analysis	328
	5.3.2.1 Critical Enthalpy in the Bubble Layer (Tong et al., 1996a)	329

		5.3.2.2 Interface Mixing	336
		5.3.2.3 Mass and Energy Balance in the Bubble Layer	342
	5.3.3	Liquid Droplet Entrainment and Deposition in High-Quality Flow	343
	5.3.4	CHF Scaling Criteria and Correlations for Various Fluids	351
		5.3.4.1 Scaling Criteria	351
		5.3.4.2 CHF Correlations for Organic Coolants and Refrigerants	357
		5.3.4.3 CHF Correlations for Liquid Metals	360
5.4	Parameter Effects on CHF in Experiments		366
	5.4.1	Pressure Effects	367
	5.4.2	Mass Flux Effects	369
		5.4.2.1 Inverse Mass Flux Effects	369
		5.4.2.2 Downward Flow Effects	373
	5.4.3	Local Enthalpy Effects	377
	5.4.4	CHF Table of p-G-X Effects	378
	5.4.5	Channel Size and Cold Wall Effects	378
		5.4.5.1 Channel Size Effect	378
		5.4.5.2 Effect of Unheated Wall in Proximity to the CHF Point	379
		5.4.5.3 Effect of Dissolved Gas and Volatile Additives	382
	5.4.6	Channel Length and Inlet Enthalpy Effects and Orientation Effects	383
		5.4.6.1 Channel Length and Inlet Enthalpy Effects	383
		5.4.6.2 Critical Heat Flux in Horizontal Tubes	387
	5.4.7	Local Flow Obstruction and Surface Property Effects	391
		5.4.7.1 Flow Obstruction Effects	391
		5.4.7.2 Effect of Surface Roughness	391
		5.4.7.3 Wall Thermal Capacitance Effects	392
		5.4.7.4 Effects of Ribs or Spacers	393
		5.4.7.5 Hot-Patch Length Effects	394
		5.4.7.6 Effects of Rod Bowing	395
		5.4.7.7 Effects of Rod Spacing	395
		5.4.7.8 Coolant Property (D_2O and H_2O) Effects on CHF	396
		5.4.7.9 Effects of Nuclear Heating	397
	5.4.8	Flow Instability Effects	398
	5.4.9	Reactor Transient Effects	399
5.5	Operating Parameter Correlations for CHF Predictions in Reactor Design		401
	5.5.1	W-3 CHF Correlation and THINC-II Subchannel Codes	405
		5.5.1.1 W-3 CHF Correlation	405
		5.5.1.2 THINC II Code Verification	410
	5.5.2	B & W-2 CHF Correlation (Gellerstedt et al., 1969)	415
		5.5.2.1 Correlation for Uniform Heat Flux	415
		5.5.2.2 Correlation for Nonuniform Heat Flux	416
	5.5.3	CE-1 CHF Correlation (C-E Report, 1975, 1976)	416
	5.5.4	WSC-2 CHF Correlation and HAMBO Code	417

		5.5.4.1 Bowring CHF Correlation for Uniform Heat Flux (Bowring, 1972)	417
		5.5.4.2 WSC-2 Correlation and HAMBO Code Verification (Bowring, 1979)	418
	5.5.5	Columbia CHF Correlation and Verification	423
		5.5.5.1 CHF Correlation for Uniform Heat Flux	423
		5.5.5.2 COBRA IIIC Verification (Reddy and Fighetti, 1983)	425
		5.5.5.3 Russian Data Correlation of Ryzhov and Arkhipow (1985)	426
	5.5.6	Cincinnati CHF Correlation and Modified Model	427
		5.5.6.1 Cincinnati CHF Correlation and COBRA IIIC Verification	427
		5.5.6.2 An Improved CHF Model for Low-Quality Flow	428
	5.5.7	A.R.S. CHF Correlation	429
		5.5.7.1 CHF Correlation with Uniform Heating	429
		5.5.7.2 Extension A.R.S. CHF Correlation to Nonuniform Heating	431
		5.5.7.3 Comparison of A.R.S. Correlation with Experimental Data	432
	5.5.8	Effects of Boiling Length: CISE-1 and CISE-3 CHF Correlations	433
		5.5.8.1 CISE-1 Correlation	433
		5.5.8.2 CISE-3 Correlation for Rod Bundles (Bertoletti et al., 1965)	439
	5.5.9	GE Lower-Envelope CHF Correlation and CISE-GE Correlation	441
		5.5.9.1 GE Lower-Envelope CHF Correlation	441
		5.5.9.2 GE Approximate Dryout Correlation (GE Report, 1975)	443
	5.5.10	Whalley Dryout Predictions in a Round Tube (Whalley et al., 1973)	447
	5.5.11	Levy's Dryout Prediction with Entrainment Parameter	449
	5.5.12	Recommendations on Evaluation of CHF Margin in Reactor Design	453
5.6	Additional References for Further Study		454

6 INSTABILITY OF TWO-PHASE FLOW 457

6.1	Introduction		457
	6.1.1	Classification of Flow Instabilities	458
6.2	Physical Mechanisms and Observations of Flow Instabilities		458
	6.2.1	Static Instabilities	460
		6.2.1.1 Simple Static Instability	460
		6.2.1.2 Simple (Fundamental) Relaxation Instability	461
		6.2.1.3 Compound Relaxation Instability	462

	6.2.2	Dynamic Instabilities		463
		6.2.2.1	Simple Dynamic Instability	463
		6.2.2.2	Compound Dynamic Instability	465
		6.2.2.3	Compound Dynamic Instabilities as Secondary Phenomena	466
6.3	Observed Parametric Effects on Flow Instability			468
	6.3.1	Effect of Pressure on Flow Instability		469
	6.3.2	Effect of Inlet and Exit Restrictions on Flow Instability		470
	6.3.3	Effect of Inlet Subcooling on Flow Instability		470
	6.3.4	Effect of Channel Length on Flow Instability		471
	6.3.5	Effects of Bypass Ratio of Parallel Channels		471
	6.3.6	Effects of Mass Flux and Power		471
	6.3.7	Effect of Nonuniform Heat Flux		471
6.4	Theoretical Analysis			473
	6.4.1	Analysis of Static Instabilities		473
		6.4.1.1	Analysis of Simple (Fundamental) Static Instabilities	473
		6.4.1.2	Analysis of Simple Relaxation Instabilities	473
		6.4.1.3	Analysis of Compound Relaxation Instabilities	473
	6.4.2	Analysis of Dynamic Instabilities		474
		6.4.2.1	Analysis of Simple Dynamic Instabilities	476
		6.4.2.2	Analysis of Compound Dynamic Instabilities	478
		6.4.2.3	Analysis of Compound Dynamic Instabilities as Secondary Phenomena (Pressure Drop Oscillations)	478
6.5	Flow Instability Predictions and Additional References for Further Study			479
	6.5.1	Recommended Steps for Instability Predictions		479
	6.5.2	Additional References for Further Study		480

APPENDIX	Subchannel Analysis (Tong and Weisman, 1979)	481
A.1	Mathematical Representation	481
A.2	Computer Solutions	484
	REFERENCES	491
	INDEX	533

PREFACE

Since the original publication of *Boiling Heat Transfer and Two-Phase Flow* by L. S. Tong almost three decades ago, studies of boiling heat transfer and two-phase flow have gone from the stage of blooming literature to near maturity. Progress undoubtedly has been made in many aspects, such as the modeling of two-phase flow, the evaluation of and experimentation on the forced-convection boiling crisis as well as heat transfer beyond the critical heat flux conditions, and extended research in liquid-metal boiling. This book reexamines the accuracy of existing, generally available correlations by comparing them with updated data and thereby providing designers with more reliable information for predicting the thermal hydraulic behavior of boiling devices. The objectives of this edition are twofold:

1. To provide engineering students with up-to-date knowledge about boiling heat transfer and two-phase flow from which a consistent and thorough understanding may be formed.
2. To provide designers with formulas for predicting real or potential boiling heat transfer behavior, in both steady and transient states.

The chapter structure remains close to that of the first edition, although significant expansion in scope has been made, reflecting the extensive progress advanced during this period. At the end of each chapter (except Chapter 1), additional, recent references are given for researchers' outside study.

Emphasis is on applications, so some judgments based on our respective experiences have been applied in the treatment of these subjects. Various workers from international resources are contributing to the advancement of this complicated field. To them we would like to express our sincere congratulations for their valuable contributions. We are much indebted to Professors C. L. Tien and G. F. Hewitt for their review of the preliminary manuscript. Gratitude is also due to the editor Lynne Lachenbach as well as Holly Seltzer, Carolyn Ormes, and Lisa Ehmer for their tireless editing.

L. S. Tong
Gaithersburg, Maryland

Y. S. Tang
Bethel Park, Pennsylvania

PREFACE TO THE FIRST EDITION

In recent years, boiling heat transfer and two-phase flow have achieved worldwide interest, primarily because of their application in nuclear reactors and rockets. Many papers have been published and many ideas have been introduced in this field, but some of them are inconsistent with others. This book assembles information concerning boiling by presenting the original opinions and then investigating their individual areas of agreement and also of disagreement, since disagreements generally provide future investigators with a basis for the verification of truth.

The objectives of this book are

1. To provide colleges and universities with a textbook that describes the present state of knowledge about boiling heat transfer and two-phase flow.
2. To provide research workers with a concise handbook that summarizes literature surveys in this field.
3. To provide designers with useful correlations by comparing such correlations with existing data and presenting correlation uncertainties whenever possible.

This is an engineering textbook, and it aims to improve the performance of boiling equipment. Hence, it emphasizes the boiling crisis and flow instability. The first five chapters, besides being important in their own right, serve as preparation for understanding boiling crisis and flow instability.

Portions of this text were taken from lecture notes of an evening graduate course conducted by me at the Carnegie Institute of Technology, Pittsburgh, during 1961–1964.

Of the many valuable papers and reports on boiling heat transfer and two-phase flow that have been published, these general references are recommended:

"Boiling of Liquid," by J. W. Westwater, in *Advances in Chemical Engineering* **1**

(1956) and **2** (1958), edited by T. B. Drew and J. W. Hoopes, Jr., Academic Press, New York.

"Heat Transfer with Boiling," by W. M. Rohsenow, in *Modern Development in Heat Transfer,* edited by W. Ibele, Academic Press (1963).

"Boiling," by G. Leppert and C. C. Pitts, and "Two-Phase Annular-Dispersed Flow," by Mario Silvestri, in *Advances in Heat Transfer* **1,** edited by T. F. Irvine, Jr., and J. H. Hartnett, Academic Press (1964).

"Two-Phase (Gas-Liquid) System: Heat Transfer and Hydraulics, An Annotated Bibliography," by R. R. Kepple and T. V. Tung, ANL-6734, USAEC Report (1963).

I sincerely thank Dr. Poul S. Larsen and Messrs. Hunter B. Currin, James N. Kilpatrick, and Oliver A. Nelson and Miss Mary Vasilakis for their careful review of this manuscript and suggestions for many revisions; the late Prof. Charles P. Costello, my classmate, and Dr. Y. S. Tang, my brother, for their helpful criticisms, suggestions, and encouragement in the preparation of this manuscript. I am also grateful to Mrs. Eldona Busch for her help in typing the manuscript.

L. S. TONG

SYMBOLS*

A	constant in Eq. (2-10), or in Eq. (4-27)
A_c	cross-sectional area for flow, ft²
A_h	heat transfer area, ft²
A_{vc}	vena contracta area ratio
a	acceleration, ft/hr²
a	gap between rods, ft
a	void volume per area, Eq. (3-40), ft
B	constant in Eq. (2-10)
B	dispersion coefficient
b	thickness of a layer, ft
C	slip constant ($= \alpha/\beta$)
C	constant, or accommodation coefficient
C	crossflow resistance coefficient
C	concentration, lb/ft³
C, c_p	specific heat at constant pressure, Btu/lb °F
C_c	contraction coefficient
C_{fg}	friction factor
C_i	concentration of entrained droplets in gas core of subchannel i
c_o	empirical constant, Eq. (5-16)
D	diffusion constant
D	damping coefficient
D_b	bubble diameter, ft
D_e	equivalent diameter of flow channel, ft
D_h	equivalent diameter based on heated perimeter, ft
DNBR	predicted over observed power at DNB, Eq. (5-123)
d	wire or rod diameter, ft, or subchannel equivalent diameter, in

* Unless otherwise specified, British units are shown to indicate the dimension used in the book.

E	energy, ft-lb
E	free flow area fraction in rod bundles, used in Eq. (4-31)
E	(wall-drop) heat transfer effectiveness
E_{bow}	bowing effect on CHF
E_L	liquid holdup
e	emissivity of heating surface
e	$e = 2.718$
e	constant
F	force, such as surface tension force, F_s, and tangential inertia force, F_t
F	a parameter (forced convection factor) Eq. (4-15), $F = \text{Re}_{tp}/\text{Re}_L)^{0.8}$
F	free energy, ft-lb
F	friction factor based on D_e (Weisbach), or frictional pressure gradient
F	shape factor applied to non-uniform heat flux case, or empirical rod-bundle spaces factor
F'	activation energy, ft-lb
F_e	view factor including surface conditions
F_k	a fluid-dependent factor in Kandlikare's Eq. (4-25)
$\mathbf{F}$	force vector
f	friction factor based on r_h (Fanning, $F = 4f$), as f_f, f_G, f_i are friction factors between the liquid and wall, the gas and the wall, and the gas–liquid interface, respectively
f	frequency, hr^{-1}
$f_m(z)$	a mixing factor in subchannel analysis, Eq. (5-132)
G	mass flux, lb/hr ft^2
G	volumetric flow rate, ft^3/hr
G_o	empirical parameter for gas partial pressure in cavity, ft-lb/°R
G', G^*	effective mixing mass flux in and out the bubble layer, lb/hr ft^2
g	acceleration due to gravity, ft/hr^2
g_c	conversion ratio, lb ft/lb hr^2
g(mH)	difference in axial pressure gradient caused by the cross flow
H	enthalpy, Btu/lb
H_{fg}	latent heat of evaporation, Btu/lb
H_{in}	inlet enthalpy, Btu/lb
ΔH_{sub}	subcooling enthalpy ($H_{sat} - H_{local}$), Btu/lb
h	heat transfer coefficient, Btu/hr ft^2 °F
h	mixture specific enthalpy, Btu/lb
h	height of liquid level, ft
I	flow inertia ($\rho L/A$), lb/ft^4
i_b	turbulent intensity at the bubble layer–core interface
J	volumetric flux, ft/hr
J	mechanical–thermal conversion ratio, $J = 778$ ft-lb/Btu

J	mixture average superficial velocity, ft/hr
J_G	crossflow of gas per unit length of bundle, ft/hr ft
K	a gas constant, or scaling factors
K	inlet orifice pressure coefficient
K	grid loss coefficient
k	a parameter, Eq. (3-39), or mass transfer coefficient
k	thermal conductivity, Btu/hr ft²
k	ratio of transverse and axial liquid flow rates per unit length in Eq. (5-51)
L	length of heated channel, ft
ℓ	length in different zones, as ℓ_s = length of liquid slug zone and ℓ_f = length of film zone
ℓ	Prandtl mixing length, ft
ln	logarithm to the base e
M	mass, lb
M	molecular weight
M_k	mass transfer per unit time and volume to phase k, lb/hr ft³
m	constant exponent in Eq. (2-78)
m	mass per pipe volume, lb/ft³
m	wave number ($= 2\pi/\lambda$)
N	number of nuclei or molecules
N_{AV}	Avogadro's constant
N_f	dimensionless inverse viscosity, Eq. (3-93)
n	number of nuclei
n	number of rods
n	bubble density or nucleus density, ft⁻²
n	droplet flux, ft⁻²
n	wave angular velocity, hr⁻¹
n	constant exponent, Eq. (2-78)
$\mathbf{n}_G$	normal vector in gas phase direction
P	power, Btu/hr
$\tilde{P}$	perimeter for gas or liquid phase
p	pressure, lb/ft² or psi
Δp	pressure drop. psi
Q	volumetric flow rate, ft³/hr
Q_k	heat transferred per unit time and volume to phase k, Btu/hr ft³
q	heat transfer rate, Btu/hr
q'	linear power, Btu/hr ft
q''	heat flux, Btu/hr ft²
$\overline{q''}$	average heat flux, Btu/hr ft²
q'''	power density, Btu/hr ft³
$\mathbf{q}$	heat flux vector
R	resistance, hr °F/Btu

xxii SYMBOLS

R	radius of bubble, ft
R	liquid holdup, or liquid fraction
R'	dimensionless heater radius, $R' = R[g_c\sigma/g(\rho_L - \rho_G)]^{-1/2}$
R_{eff}	effective radius, $[= R(1 + 0.02\theta/R')]$, ft
R_f	ratio of rough-pipe friction factor to smooth-pipe friction factor
R_g	gas constant
r	radius, ft
r_h	hydraulic radius, $D_e = 4r_h$, ft
S	slip ratio, or boiling suppression factor
S	periphery on which the stress acts, ft
s	width, or thickness, ft
s	entropy, Btu/lb °F
T	temperature, °F
T'	temperature deviations, °F
T_∞	temperature in superheated liquid layer, °F
ΔT_{FDB}	ΔT_{sat} at the beginning of fully developed boiling, °F
$\Delta T_{\text{J\&L}}$	Lens and Lottes temperature difference, °F
T_{LB}	bulk temperature of coolant at start of local boiling, °F
$[T]$	$n \times n$ matrix with elements $\partial P_j^c/\partial V_k^i$
ΔT_{sat}	$(T_{\text{wall}} - T_{\text{sat}})$, °F
ΔT_{sub}	subcooling $(T_{\text{sat}} - T_{\text{local}})$, °F
t	time, hr
t	average film thickness, in
U	internal energy
U_b	velocity of vapor blanket in the turbulent stream [Eq. (5-45)], ft/hr
U_{bl}	velocity of liquid at $y = \delta_m + (D_b/2)$ (Fig. 5.21), ft/hr
U_o	relative velocity (or rise velocity), ft/hr
U_s	velocity of sound in the vapor, ft/hr
$\underline{\underline{U}}$	metric tensor of the space
u	velocity in the axial direction, or radial liquid velocity, ft/hr
u'	local velocity deviation (in the axial direction)
u^*	friction velocity in Eq. (3-124)
u_{GJ}	drift velocity in Eq. (3-58), ft/hr
u_{GM}	gas velocity relative to the velocity of the center of mass, ft/hr
$\mathbf{u}$	velocity vector
$\overline{u'v'}$	Reynolds stress, time average of the product of the velocity deviations in the axial and radial direction
V	volume, ft³
V	velocity, ft/hr
V_∞	terminal velocity, ft/hr
v	velocity in the normal direction, ft/hr
v	specific volume, ft³/lb

v_{fg}	specific volume change during evaporation, ft³/lb
v'	local velocity deviation, ft/hr
W	weight, lb
W	mass flux, lb/hr ft²
W_B	critical power over boiling length
w	flow rate, lb/hr
w	frequency
w'	flow exchange rate per unit length by mixing, lb/hr ft
X	quality, weight percent of steam
X_{tt}	Lockhart and Martinelli parameter, $$X_{tt} = \left(\frac{1-X}{X}\right)^{0.9}\left(\frac{\rho_G}{\rho_L}\right)^{0.5}\left(\frac{\mu_f}{\mu_G}\right)^{0.1}$$
X	group of parameters, Eq. (3-7)
X'	static quality defined by Eq. (3-38)
x	length in x direction, ft
Y	axial heat flux profile parameter in Eq. (5-122)
Y	group of parameters in Eq. (3-8)
Y_r	a parameter for wall effects on vapor blanket circulation
Y'	subchannel imbalance factor in Eq. (5-122)
y	length in y direction, ft
y	a parameter [$= \ln(p)$]
z	axial length, ft
z_b	distance from the inlet to the bulk boiling, ft
z_d	distance from the inlet to the void detachment, ft
z_{LB}	distance from the inlet to the start of local boiling, ft
z^*	distance from the inlet to the merging point of the Bowring void curve and the Martinelli-Nelson void curve, ft
α	thermal diffusivity ($= k/\rho c$) ft²/hr
α	absorptivity of liquid
α	void fraction
α'	dimensionless thermal diffusion coefficient ($= \varepsilon/Vb$)
$\langle\alpha\rangle$	average void fraction
α_0	steady-state sonic velocity, ft/hr
β	vapor volumetric rate ratio, or an entrainment parameter
β	volumetric compressibility of two-phase flow
β	bubble contact angle between liquid and solid surfaces
β_1, β_2	Parameters in wall-drop effectiveness calculation, Eq. (3-95)
Γ	volumetric interfacial area, Eq. (3-56)
Γ	volumetric flow per unit width of parallel-plate channel
γ	constant, or angle
γ'	isentropic exponent for vapor compression (c_p/c_v)
δ	boundary-layer or thermal-layer thickness, ft

δ_c	wave crest amplitude
ε	eddy viscosity, ft²/hr
ε	parameter for void fraction correlation
ε	ratio of liquid convective heat transfer to bubble latent heat transport
ε_H	eddy thermal conductivity, ft²/hr
ζ	constant
η	amplitude of a wave, ft
η	a function related to the critical distance, Eq. (2-112)
θ	angle, deg
θ	time, hr
θ	temperature difference, °F
κ	a constant
λ	wavelength, ft, or a scalar quantity
λ	ratio of superficial velocities, Eq. (3-104)
μ	viscosity, lb/ft hr
ν	kinematic viscosity, ft²/hr
ν_s	slug frequency
ξ	constant, a measure of inert gas in cavity at start of boiling, Eq. (2-20)
π	$\pi = 3.1416$
ρ	density, lb/ft³
σ	surface tension, lb/ft
σ	area ratio (A_1/A_2)
$\sigma_{S\text{-}B}$	Stefan-Boltzmann constant ($= 17.3 \times 10^{-10}$ Btu/hr ft² °R⁴
τ	nondimensional time, τ_D, drag relaxation time; τ_t, thermal relaxaton time
τ	shear force, lb/ft²
$\bar{\bar{\tau}}$	stress tensor
ϕ	a function, or heat flux, Btu/hr ft²
ϕ	contact angle, or angle from the vertical line
ϕ	average chemical function
ϕ_i	mass flux across the interface
ϕ_{LO}	$(\Delta p_{TPF}/\Delta p_{LO})^{1/2}$
ψ	a function
ψ	apex angle (Fig. 2.3)
ω	angular velocity, hr⁻¹
ω	frequency of oscillation, hr⁻¹

Superscripts

+	refers to nondimensional parameter
−	refers to time average or mean value

*	refers to critical value, or nondimensional parameter
i, o	refers to inlet and outlet values, respectively, Eq. (A-11)

Subscripts

A, B	refers to phase A and phase B, respectively
a	refers to apparent property, such as $\rho_a =$ apparent density, Eq. (3-42)
B	refers to boiling condition
b	refers to bubble property or bulk flow condition
c	refers to crud, or cavity
c	refers to core condition
c, crit	refers to critical condition
c_{ij}	refers to turbulent interchange of entrained drops between sub-channels of types i and j
D	refers to drag
D, d	refers to droplet or deposition
d	refers to bubble departure condition or droplet condition
E, e	refers to liquid entrainment
e	refers to exit condition
e_1, e_2, e_3	refers to dry patch due to evaporation at respective stages
F, f	refers to liquid film condition, such as pressure, p_F, and temperature, T_f
f	refers to saturated liquid
fg	refers to phase change from liquid to vapor
f.c.	refers to forced convection
g, G	refers to gas, or vapor, condition
g	refers to grid spacer
i	refers to inner diameter
i	refers to interfacial value
i	refers to subchannel type i
j	refers to vapor jets
j	refers to number of subchannels
ℓ, L	refers to saturated liquid condition
ℓ'	refers to local subcooled liquid condition
m	refers to matrix channel equivalent
m	refers to mixture property
m	refers to bubble collapse time (maximum)
o	refers to initial condition or outer diameter
o	refers to quantities at center, such as α_o is void fraction at the center
r	refers to reduced properties, such as p_r, T_r
r	refers to size r of the nucleation site

r	refers to bubble resonance
S	refers to superficial value, such as superficial velocity, V_S
s	refers to suspension
s	refers to slug
t	refers to thermal
u	refers to slug unit in slug flow geometry
v	refers to saturated vapor condition
w	refers to wall condition
w	refers to waiting period
atn	refers to attenuation coefficient
bulk	refers to bulk flow condition
conv	refers to forced-convection component
crit	refers to critical condition
DFB	refers to departure from film boiling
DNB	refers to departure from nucleate boiling
do	refers to dryout condition
eff	refers to effective value
elev.	refers to elevation
FB	refers to film boiling
FDB	refers to fully developed nucleate boiling
fric	refers to friction
GPF	refers to the friction of a flow with gas mass velocity component
HT	refers to homogeneous, isothermal conditions
hor	refers to horizontal flow
IB	refers to incipient boiling
LB	refers to local boiling condition
LDF	refers to Leidenfrost state
LE	refers to entrained liquid
LO	refers to the friction of a liquid flow with total mass flux
LPF	refers to the friction of a flow with liquid mass flux
LS	refers to liquid slug
max	refers to maximum value
mom	refers to momentum
NB	refers to nucleate boiling
Ob	refers to obstructions
rel	refers to relative value
sat	refers to saturated condition
SM	refers to Sauter mean, as in d_{SM}, Sauter mean diameter
sub	refers to subcooled condition
sup	refers to superheated condition
TB	refers to transition boiling, or Taylor bubble
td	refers to crossflow due to droplet deposition
TH	refers to a group of thermodynamic similitude

te	refers to liquid crossflow due to reentrainment
tot	refers to total condition
TP	refers to two-phase
TPF	refers to two-phase friction
vert	refers to vertical flow
ups	refers to upstream
(W-3)	refers to W-3 CHF correlation, Eq. (5-113)

Nondimensional Groups

Bo	boiling number $(= q''/H_{fg}\rho_G V)$
Co	convection number $\{= [(1 - X)/X]^{0.8} (\rho_G/\rho_L)^{0.5}\}$
Fr	Froude number $(= V^2/gD_e)$
Gr	Grashof number $(= L^3\rho^2\beta g \, \Delta T/\mu^2)$
Ja	Jacob number $[= c_p\rho_L(T_w - T_b)/H_{fg}\rho_G]$
Nu	Nusselt number $(= D_b q''/\Delta T_w k_L)$
Pr	Prandtl number $(= c\mu/k)$
Re	Reynolds number $(= D_b G/\mu)$
We	Weber number $(= D_b \rho V^2/\sigma g_c)$

UNIT CONVERSIONS

Acceleration, $1 \text{ft/s}^2 = 0.305 \text{ m/s}^2$
Area, $1 \text{ ft}^2 = 9.29 \times 10^{-2} \text{ m}^2$
Density, $1 \text{ lb}_m/\text{ft}^3 = 16.02 \text{ kg/m}^3$
Force, $1 \text{ lb}_f = 4.448 \text{ N}$
Heat flow, $1 \text{ Btu/h ft}^2 = 3.152 \text{ W/m}^2$
Heat transfer coefficient, $1 \text{ Btu/h ft}^2 \text{ °F} = 5.678 \text{ W/m}^2 \text{ °C}$
Length, $1 \text{ ft} = 0.305 \text{ m}$
Mass, $1 \text{ lb}_m = 0.454 \text{ kg}$
Mass flow rate, $1 \text{ lb}_m/\text{h} = 1.26 \times 10^{-4} \text{ kg/s}$
Mass flux, $1 \text{ lb}_m/\text{ft}^2 \text{ h} = 1.356 \times 10^{-3} \text{ kg/m}^2 \text{ s}$
Power, $1 \text{ Btu/h} = 0.293 \text{ W}$
Pressure, $1 \text{ psi} = 6.895 \times 10^3 \text{ Pa}$; $1 \text{ atm} = 1.013 \times 10^5 \text{ Pa}$
Specific heat, $1 \text{ Btu/lb}_m \text{ °F} = 4.184 \times 10^3 \text{ J/kg °C}$
Surface tension, $1 \text{ lb}_f/\text{ft} = 14.59 \text{ N/m}$
Thermal energy, $1 \text{ Btu} = 1.055 \times 10^3 \text{ J}$
Thermal conductivity, $1 \text{ Btu/h ft °F} = 1.73 \text{ W/m °C}$
Thermal diffusivity, $1 \text{ ft}^2/\text{h} = 2.581 \times 10^{-5} \text{ m}^2/\text{s}$
Thermal resistance, $1 \text{ °F h ft}^2 \text{ Btu} = 0.176 \text{°C m}^2/\text{W}$
Velocity, $1 \text{ ft/s} = 0.305 \text{ m/s}$
Viscosity, $1 \text{ lb}_m/\text{ft s} = 1.488 \text{ Pa s}$
Volumetric flow rate, $1 \text{ ft}^3/\text{s} = 2.832 \times 10^{-2} \text{ m}^3/\text{s}$
Volumetric heat generation, $1 \text{ Btu/h ft}^3 = 10.343 \text{ W/m}^3$

CHAPTER
ONE

INTRODUCTION

Boiling heat transfer is defined as a mode of heat transfer that occurs with a change in phase from liquid to vapor. There are two basic types of boiling: pool boiling and flow boiling. *Pool boiling* is boiling on a heating surface submerged in a pool of initially quiescent liquid. *Flow boiling* is boiling in a flowing stream of fluid, where the heating surface may be the channel wall confining the flow. A boiling flow is composed of a mixture of liquid and vapor and is the type of *two-phase flow* that will be discussed in this book. Because of the very high heat transfer rate in boiling, it has been used to cool devices requiring high heat transfer rates, such as rocket motors and nuclear reactors. Its applications in modern industry are so important that large amounts of research in many countries have been devoted to understanding its mechanisms and behavior, especially since the publication of the first edition of this book. The results have not yet been entirely satisfactory in clarifying boiling phenomena and in correlating experimental data on heat transfer in nucleate boiling. This is largely because of the complexity and irreproducibility of the phenomena, caused by the fact that the surface conditions (i.e., the surface roughness, the deposition of foreign materials, or the absorption of gas on the surface) become inherent factors that influence bubble generation (Nishikawa and Fujita, 1990).

1.1 REGIMES OF BOILING

There are several boiling regimes in pool boiling as well as in flow boiling. The only difference lies in the influence of flow effect. The buoyancy effect is significant in a pool boiling system, while the flow forced-convection effect is significant in flow boiling inside a channel.

The various regimes of boiling in a typical case of pool boiling in water at atmospheric pressure are shown in Figure 1.1, which is the conventional log–log representation of heat flux versus wall superheat. These boiling regimes were observed by previous researchers, namely, Leidenfrost (1756), Lang (1888), McAdams et al. (1941), Nukiyama (1934), and Farber and Scorah (1948). In the range A–B (Fig. 1.1), the water is heated by natural convection. With the mechanism of single-phase natural convection, the heat transfer rate q'' is proportional to $(\Delta T_{sat})^{5/4}$. In the range B–C, the liquid near the wall is superheated and tends to evaporate, forming bubbles wherever there are nucleation sites such as tiny pits or scratches on the surface. The bubbles transport the latent heat of the phase change and also increase the convective heat transfer by agitating the liquid near the heating surface. The mechanism in this range is called *nucleate boiling* and is characterized by a very high heat transfer rate for only a small temperature difference. There are two subregimes in nucleate boiling: local boiling and bulk boiling. Local boiling is nucleate boiling in a subcooled liquid, where the bubbles formed at the heating surface tend to condense locally. Bulk boiling is nucleate boiling in a saturated liquid; in this case, the bubbles do not collapse. In the nucleate boiling range, q'' varies as $(\Delta T_{sat})^n$, where n generally ranges from 2 to 5. However, the heat flux in nucleate boiling cannot be increased indefinitely. When the population of bubbles becomes too high at some high heat flux point C, the outgoing bubbles may ob-

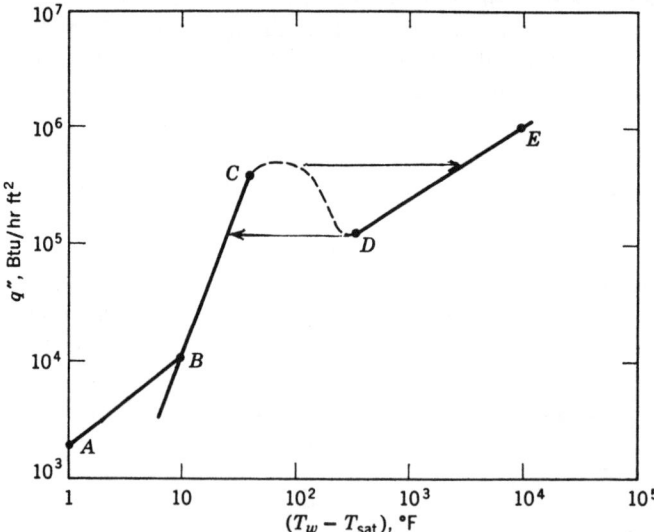

Figure 1.1 Pool boiling regimes: A–B, natural convection; B–C, nucleate boiling; C–D, partial film boiling; D–E, stable film boiling.

struct the path of the incoming liquid. The vapor thus forms an insulating blanket covering the heating surface and thereby raises the surface temperature. This is called the *boiling crisis,* and the maximum heat flux just before reaching crisis is *critical heat flux,* which can occur in pool boiling or in various flow patterns of flow boiling (see Sec. 1.3). In the past, the terminology of the boiling crisis was not universal. The pool boiling crisis with constant heat flux supply, or crisis occurring in an annular flow is sometimes called *burnout,* and that occurring in bubbly flow is sometimes called *departure from nucleate boiling* (DNB). In this book, all three terms are used interchangeably.

In the range *C–D,* immediately after the critical heat flux has been reached, boiling becomes unstable and the mechanism is then called *partial film boiling* or *transition boiling.* The surface is alternately covered with a vapor blanket and a liquid layer, resulting in oscillating surface temperatures. If the power input to the heater is maintained, the surface temperature increases rapidly to point *D* while the heat flux steadily decreases. In the range *D–E,* a stable vapor film is formed on the heating surface and the heat transfer rate reaches a minimum. This is called *stable film boiling.* By further increasing the wall temperature, the heat transfer rate also is increased by thermal radiation. However, too high a temperature would damage the wall. Hence, for practical purposes, the temperature is limited by the material properties.

The boiling regimes mentioned above also exist in flow boiling. The mechanisms are more complicated, however, owing to the fact that two-phase flow plays an important role in the boiling process. For instance, the flow shear may cut off the bubbles from the wall so that the average bubble size is reduced and the frequency is increased. Other interactions between the boiling process and two-phase flow are discussed in the next section. As in pool boiling, the range of the boiling curve of interest for most practical applications is that of nucleate boiling (*B–C*), where very high heat fluxes can be attained at relatively low surface temperatures.

1.2 TWO-PHASE FLOW

Two-phase flows are classified by the void (bubble) distributions. Basic modes of void distribution are bubbles suspended in the liquid stream; liquid droplets suspended in the vapor stream; and liquid and vapor existing intermittently. The typical combinations of these modes as they develop in flow channels are called *flow patterns.* The various flow patterns exert different effects on the hydrodynamic conditions near the heated wall; thus they produce different frictional pressure drops and different modes of heat transfer and boiling crises. Significant progress has been made in determining flow-pattern transition and modeling.

The microscopic picture of the flow in the proximity of the heated wall can be described in terms of two-phase boundary-layer flow. The macroscopic effect of a two-phase flow on the frictional pressure drop still relies on empirical correlations.

1.3 FLOW BOILING CRISIS

Boiling crisis is a combined phenomenon of hydrodynamics and heat transfer. Owing to excessively high wall temperature, the boiling crisis usually results in damage to the heating surface in a constant-energy-input system. It is imperative, therefore, to predict and prevent the occurrence of the crisis in boiling equipment.

As the flow boiling crisis occurs at a very high heat flux, the prediction of such a crisis has to be closely related with the flow boiling heat transfer, and the appropriate model should also be related with the two-phase flow pattern existing at the CHF conditions. There are two types of parameters by which a flow boiling crisis can be described. One type is the operational parameters of a boiling system, such as system pressure, mass flux, and channel geometry, which are set a priori. An engineering correlation of flow boiling crisis for design purposes can be developed from these parameters, and the parameter effects can be evaluated without revealing the mechanism of the crisis. The other type comprises microscopic parameters such as flow velocity near the wall, local voids, coolant properties, and surface conditions. The latter parameters can be used in calculating the principal forces acting on a bubble or on a control volume. Such data can be useful in modeling the flow pattern, or can be used in organizing significant nondimensional groups to give a phenomenologically meaningful correlation that may reveal the mechanism of the boiling crisis. Early attempts to obtain generalized predictions of flow boiling crises were often based on the assumption that the underlying mechanism was essentially the same as for pool boiling. It has been generally agreed that this is not the case (Tong, 1972; Tong and Hewitt, 1972; Weisman, 1992).

1.4 FLOW INSTABILITY

Flow instability is a phenomenon of combined hydrodynamic and thermodynamic nature and is caused by the large momentum change introduced by boiling of a two-phase flow. It may start with small, constant-amplitude flow oscillations in a channel at low power input. If the power input is increased, the amplitude increases as the flow becomes unstable. Such flow instabilities occurring in boiling equipment can be in the form of instability between parallel channels, flow instability in a natural-circulation loop, or flow instability caused by the difference in pressure drops of interchanging flow patterns. Bouré et al. (1973) classified flow instability phenomena in two categories: static instability and dynamic instability. Within each category there are fundamental (or simple) and compound instabilities. Flow excursion or Ledinegg instability and flow pattern transition instability are examples of fundamental static instabilities, while bumping, geysering, or chugging is a compound relaxation instability. Acoustic oscillations are examples of fundamental dynamic instability, while parallel channel instability is a compound dynamic instability.

Ruddick (1953) and Lowdermilk et al. (1958) found that flow oscillation can induce a premature boiling crisis. Moreover, in a boiling water reactor the flow oscillation may induce a nuclear instability. Thus, in designing a boiling system, it is imperative to predict and prevent those operational conditions that might create flow oscillation.

CHAPTER
TWO

POOL BOILING

2.1 INTRODUCTION

Pool boiling occurs when a heater is submerged in a pool of initially stagnant liquid. When the surface temperature of the heater exceeds the saturation temperature of the liquid by a sufficient amount, vapor bubbles nucleate on the heater surface. The bubbles grow rapidly in the superheated liquid layer next to the surface until they depart and move out into the bulk liquid. While rising as the result of buoyancy, they either collapse or continue their growth, depending on whether the liquid is locally subcooled or superheated. Thus, in pool boiling, a complex fluid motion around the heater is initiated and maintained by the nucleation, growth, departure, and collapse of bubbles, and by natural convection. A thorough understanding of the process of boiling heat transfer in both pool boiling and flow boiling requires investigation of, first, the thermodynamics of the single bubble and, second, the hydrodynamics of the flow pattern resulting from many bubbles departing from a heated surface. Later in this chapter, the correlation of heat transfer data will be developed with water and liquid metals, both of which are used as coolants in nuclear reactors.

2.2 NUCLEATION AND DYNAMICS OF SINGLE BUBBLES

The life of a single bubble may be summarized as occurring in the following phases: nucleation, initial growth, intermediate growth, asymptotic growth, possible collapse. In the ebullition cycle, however, a waiting period occurs in a bubble site just after the departure of a bubble and before a new bubble is formed, which was shown by a shadowgraph and Schlieren technique (Hsu and Graham, 1961). This waiting period between two consecutive appearances of bubbles can be described meaningfully only in the lower heat flux range, where bubbles are discrete. Nucleation is a molecular-scale process in which a small bubble (nucleus) of a size just in excess of the thermodynamic equilibrium [see Eq. (2-6)] is formed. Initial growth from the nucleation size is controlled by inertia and surface tension effects. The growth rate is small at first but increases with bubble size as the surface tension effects become less significant. In the intermediate stage of accelerated growth, heat transfer becomes increasingly important, while inertia effects begin to lose significance. When the growth process reaches the asymptotic stage, it is controlled by the rate of heat transferred from the surrounding liquid to facilitate the evaporation at the bubble interface. If the bubble, during its growth, contacts the subcooled liquid, it may collapse. The controlling phenomena for the collapse process are much the same as for the growth process but are encountered in reverse order.

2.2.1 Nucleation

The primary requirement for nucleation to occur or for a nucleus to subsist in a liquid is that the liquid be superheated. There are two types of nuclei. One type is formed in a pure liquid; it can be either a high-energy molecular group, resulting from thermal fluctuations of liquid molecules; or a cavity, resulting from a local pressure reduction such as occurs in accelerated flow. The other type, formed on a foreign object, can be either a cavity on the heating wall or suspended foreign material with a nonwetted surface. The latter type is obviously of importance for boiling heat transfer equipment.

2.2.1.1 Nucleation in a pure liquid. According to the kinetic theory for pure gases and liquids, there are local fluctuations of densities, which are clusters of molecules in a gas and holes (or vapor clusters) in a liquid. Frenkel (1955) established the population distribution of such holes of phase B in a liquid of continuum phase A by Boltzmann's formula,

$$N_r = C \exp\left[\frac{-\Delta F(r)}{KT}\right] \tag{2-1}$$

where r is the size of the hole, N_r is the population of a hole of size r, ΔF is the difference of free energy between two phases, K is the Boltzmann constant or the gas constant per molecule, and T is the absolute temperature. In the case of vapor clusters (phase B) contained in a liquid phase (phase A), the free-energy difference can be expressed as

$$\Delta F = \left[\frac{-(\phi_A - \phi_B)}{V_B}\right]\left(\frac{4r^3}{3}\right) + 4\pi\sigma r^2$$

The term ($4\pi\sigma r^2$) is to account for the additional surface energy due to the presence of an interface, assuming that the surface tension, σ, for a flat surface is applicable to the spherical vapor phase hole. ϕ_A and ϕ_B are the average chemical potentials of each molecule in phases A and B, respectively. V_B is the molecular volume for phase B. If $\phi_A < \phi_B$, then phase A is thermodynamically stable or, in this case the liquid is subcooled, and ΔF increases monotonically with r. On the other hand, in a superheated liquid, $\phi_B < \phi_A$, the ΔF has two terms with opposing signs. Increasing r will first increase ΔF, until a maximum is attained, and then the function will decrease to a negative value. The maximum ΔF is located at $r^* = 2\sigma V_B/(\phi_A - \phi_B)$. The corresponding Boltzmann distribution will have a minimum at r^*, but the population will increase rapidly for values of $r > r^*$. Therefore, in a superheated liquid, a large population of bubbles will exist with $r > r^*$ (Hsu and Graham, 1976). In addition, the bubble population can be raised by increasing the temperature. The rate of nucleation can be shown to be

$$\frac{dn}{dt} = C_n \exp\left(\frac{\Delta F'}{KT}\right) \qquad (2\text{-}2)$$

where n is the number of molecules, $\Delta F'$ is the activation energy, and C_n is a coefficient. Many theories have been proposed to determine C_n and $\Delta F'$ (Cole, 1970). Thus the nucleation work is equivalent to overcoming an energy barrier. If the liquid superheat is increased, more liquid molecules carry enough kinetic energy to be converted to this energy of activation. Consequently, there is a higher probability of the vapor cluster growing. When the vapor cluster is large enough, a critical size is eventually achieved at which the free energy drops due to the rapid decrease of surface energy with further increase of size. From then on the nucleation becomes a spontaneous process.

For a nucleus to become useful as a seed for subsequent bubble growth, the size of the nucleus must exceed that of thermodynamic equilibrium corresponding to the state of the liquid. The condition for thermodynamic equilibrium at a vapor–liquid interface in a pure substance can be written as

$$p_G - p_L = \sigma\left(\frac{1}{R_1} + \frac{1}{R_2}\right)$$

where R_1 and R_2 are the principal radii of curvature of the interface. For a spherical nucleus of radius R, the above equation becomes the Lapalace equation,

$$p_G - p_L = 2\sigma/R \tag{2-3}$$

For a bulk liquid at pressure p_L, the vapor pressure p_G of the superheated liquid near the wall can be related to the amount of superheat, $(T_G - T_{sat})$, by the Clausius-Clapeyron equation,

$$\frac{dp}{dT} = \frac{H_{fg} J}{v_{fg} T} \tag{2-4}$$

which, in finite-difference form and for $v_G \gg v_L$, that is, the specific volume change during evaporation, $v_{fg} = v_G$, gives

$$p_G - p_L = \frac{(T_G - T_{sat})H_{fg} J \rho_G}{T_{sat}} \tag{2-5}$$

Combining Eqs. (2-3) and (2-5) yields

$$R = \left(\frac{2\sigma}{JH_{fg}\rho_G}\right)\left(\frac{T_{sat}}{T_G - T_{sat}}\right) \tag{2-6}$$

for the equilibrium bubble size. Hence, for increasing superheat, the nucleation size (cavity) can be smaller, and by Eq. (2-2) the number of nuclei formed per unit time increases. Another implication of Eq. (2-6) is that only a nucleus of the equilibrium size is stable. A smaller nucleus will collapse, and a larger nucleus will grow. In other words, Eq. (2-6) represents the minimum R corresponding to a given liquid superheat that will grow, or the minimum superheat corresponding to the nucleus's radius R.

2.2.1.2 Nucleation at surfaces. Typical nucleation sites at the cavities of heating surface are shown in Figure 2.1. The angle ϕ is called the contact angle. A large angle ϕ may have a better chance to trap gas inside the cavity by a capillary effect. The wall superheat required for bubble growth to occur from a nucleation site of

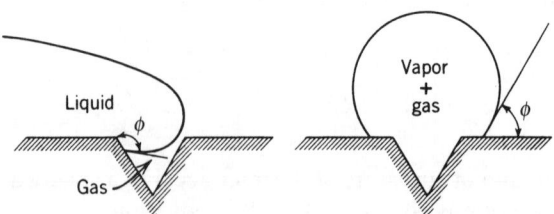

Figure 2.1 Nucleation from cavities.

a solid surface was thought to be calculable from Eq. (2-6), where the critical radius may be taken as equal to the cavity radius. Such predictions, however, do not agree with observed data when a solid is used as the heating surface. Hsu (1962) suggested that the criterion for the formation of a bubble on a solid surface, Eq. (2-6), might be invalid when the solid surface alone is hot, and the difference must be related to the nature of the temperature field in the liquid immediately adjacent to the solid. The liquid temperature can be represented by the temperature profile in a thermal layer. Because of turbulence in the bulk of the liquid, the thermal layer cannot grow beyond a limiting thickness δ. If the liquid in the thermal layer is renewed by some disturbance, the temperature profile will reestablish itself by means of transient conduction and will ultimately grow into a linear profile as time approaches infinity (Marcus and Dropkin, 1965). The model is depicted in Figure 2.2. Note that there is one range of cavity size for which the bubble temperature is lower than the liquid temperature at the bubble cap. This is the size range in which the bubble embryo will grow to make a cavity into an active site. The maximum and minimum sizes can be determined by solving two equations,

$$\frac{T_b - T_\infty}{T_w - T_\infty} = \frac{\delta - C_1 r}{\delta} \tag{2-7}$$

and

$$T_b - T_{sat} = \frac{2\sigma T_{sat}}{C_2 r_c H_{fg} \rho_G} \tag{2-8}$$

The above nucleation criterion for boiling agrees with the experimental results of Clark et al. (1959) and of Griffith and Wallis (1960). It was also verified qualitatively by Bergles and Rohsenow (1964), who showed that nucleation in a forced-convective channel is suppressed when the limiting thermal boundary-layer thickness is thinned by increases in the bulk velocity. They further proposed a modified nucleation criterion for the case of a forced-convective channel by assuming that (1) vapor density can be calculated from the ideal gas law, (2) the bubble height is equal to cavity radius, and (3) the relationship $-k(dt/dy) = h(T_w - T_\infty)$ holds for the liquid temperature profile rather than using a limiting thermal layer thickness of δ. Their resulting incipience criterion, however, cannot be expressed in a simple analytical form (Hsu and Graham, 1976).

Shai (1967) followed Bergles and Rohsenow's approach and calculated the required liquid superheat as

$$(T_G - T_{sat}) = \frac{T_G T_{sat}}{B} \log_{10}\left(1 + \frac{2\sigma}{r p_{sat}}\right) \tag{2-9}$$

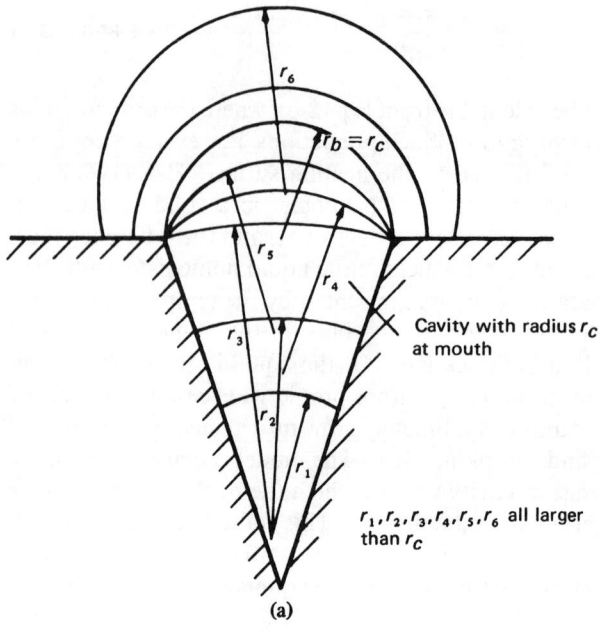

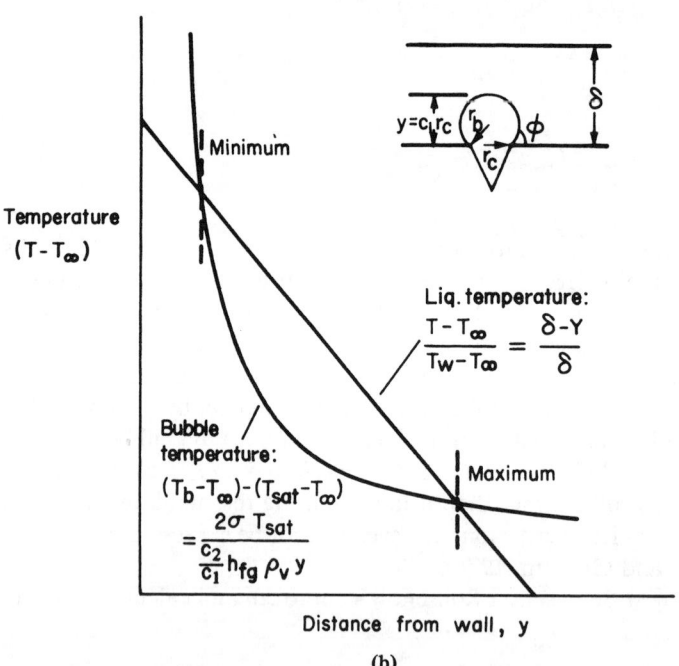

Figure 2.2 Bubble activation: (*a*) change of radius of curvature of a bubble as it grows out from a cavity; (*b*) criterion for activation of ebullition site. (From Hsu and Graham, 1976. Copyright © 1976 by Hemisphere Publishing Corp., New York. Reprinted with permission.)

where B is the constant in the conventional form of the vapor pressure–temperature relation,

$$\log_{10} p = A - \frac{B}{T} \tag{2-10}$$

By comparison of Eq. (2-9) and the Clausius-Clapeyron equation with the perfect gas approximation,

$$T_G - T_{sat} = \left(\frac{T_G T_{sat} R_G}{H_{fg}}\right) \ln\left(1 + \frac{2\sigma}{rp_{sat}}\right) \tag{2-11}$$

it can be shown

$$B = \frac{H_{fg}}{R_G \ln 10} \tag{2-12}$$

As the value of B is very nearly a constant for a wide range of pressures in Eq. (2-10), Eq. (2-9) is more useful. Thus,

$$T_G = \frac{T_{sat}}{1 - (T_{sat}/B) \log_{10}(1 + 2\sigma/rp_{sat})} \tag{2-13}$$

For all practical values of $(2\sigma/rp_{sat})$, the term

$$\left(\frac{T_{sat}}{B}\right) \log_{10}\left(1 + \frac{2\sigma}{rp_{sat}}\right) \ll 1$$

can be simplified by expanding into a Taylor series, and solve for $(T_G - T_{sat})$; thus

$$T_G - T_{sat} = \frac{(T_{sat})^2}{B} \log_{10}\left(1 + \frac{2\sigma}{rp_{sat}}\right) \tag{2-14}$$

Equation (2-14) provides a way to calculate the liquid temperature in equilibrium with the ready-to-grow bubble if the saturation pressure or temperature, the value of B, and the cavity radius are known (Shai, 1967). Several modified versions of nucleation criteria have since been advanced. An example is the model proposed by Lorenta et al. (1974), which takes into account both the geometric shape of the cavity and the wettability of the surface (in terms of contact angle ϕ). Consider an idealized conical cavity with apex angle ψ, and a liquid with a flat front penetrating into it (Fig. 2.3a). Assume that once the vapor is trapped in by the liquid front, the interface readjusts to form a cap with radius of curvature r_n. Conservation of vapor

14 BOILING HEAT TRANSFER AND TWO-PHASE FLOW

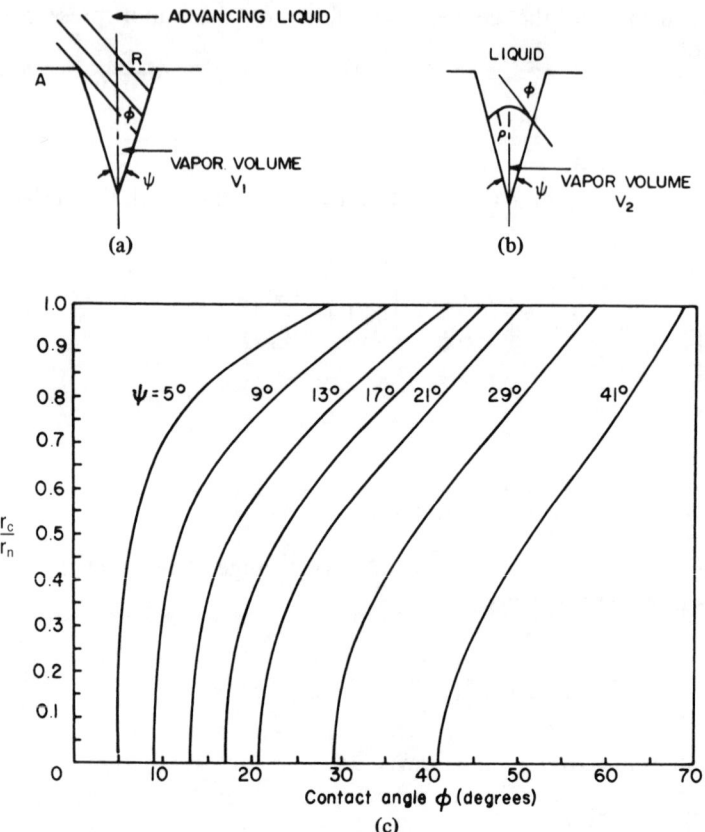

Figure 2.3 Vapor trapping model and resulting nucleation radius as functions of ψ and ϕ. (*a*) Vapor trapping process; (*b*) Formation of radius of curvature; (*c*) Relation between nucleation radius, contact angle, and apex angle. (From Lorentz et al., 1974. Copyright © 1974 by Hemisphere Publishing Corp., New York. Reprinted with permission.)

volume demands that the radius of curvature r_n be a function of ψ and ϕ. Figure 2.3c shows such a relationship. Thus, if the activation of a bubble site on a surface is known for one liquid, that information can be used to determine the incipience of boiling for other liquids of different contact angles (Hsu and Graham, 1976). Singh et al. (1974) extended Lorentz's work to include the effects of cavity depth, wettability, and liquid entering velocity on the stability of nucleation size. According to their thesis, only those cavities that are stable vapor trays can later be sites for bubble growth, and all the thermal criteria apply only to stable cavities.

For the nucleate boiling of a liquid metal, Shai made the following remarks (Shai, 1967). He gave values of B for different metals, as shown in Table 2.1. For comparison, the value of B for water in the range of pressures between 30 psi (0.2 MPa) and 960 psi (6.5 MPa) is calculated from steam tables to be 3732.83. As

Table 2.1 Values of B (in °R) in Eq. (2-14) for liquid metals and sources of data

Sodium	Potassium	Rubidium	Cesium
9396.8 (Bonilla, 1962)	7797.6 (Lemmon, 1964)	6994.7 (Achener, 1964)	6880.2 (Achener, 1964)
	7707.3 (Achener, 1965)	7062.5 (Tepper, 1964)	6631.7 (Achener, 1964)
		6983.0 (Bonilla, 1962)	6665.3 (Tepper, 1964)
			(Bonilla, 1962)

was assumed by Bergles and Rohsenow (1964), the bubble will grow if the liquid temperature at a distance $y = r$ from the wall is equal to or greater than the vapor temperature in the hemispherical bubble, T_G. Since r is much less than the thickness of the thermal layer, δ, a linear temperature profile in the liquid layer can be assumed,

$$T_w - T_L = \frac{q_o r}{k_L}$$

Thus, the maximum surface temperature difference before the bubble grows is

$$T_w(0, 0) - T_{sat} = (T_G - T_{sat}) + \frac{q_o r}{k_L}$$

For liquid metals having high thermal conductivity, the second term on the right-hand side can be safely neglected, and the maximum surface temperature difference is simply

$$T_w(0, 0) - T_{sat} = T_G - T_{sat}$$
$$= \left(\frac{T_{sat}^2}{B}\right) \log_{10}\left(1 + \frac{2\sigma}{r p_{sat}}\right) \quad (2\text{-}15)$$

Thus it can be seen that the degree of superheat is much greater in liquid metals than in water for the same pressure and cavity size, because of their much higher values of $(T_{sat})^2$. Also, for the same cavity size, pressure, and heat flux, the time required to build the thermal layer as well as its thickness will be much greater in liquid metals than in other liquids (see Sec. 2.2.2).

The role of thermal fluctuations in bubble nucleation of pool boiling was shown experimentally by Dougall and Lippert (1967); see Figure 2.4. Their experiments were conducted using water, at atmospheric pressure, boiled from a 2-in. (5 cm)-diameter copper surface that was located in the bottom plate of a 2-gal (7.6-liter) aluminum container. The copper boiling surface was prepared by pol-

16 BOILING HEAT TRANSFER AND TWO-PHASE FLOW

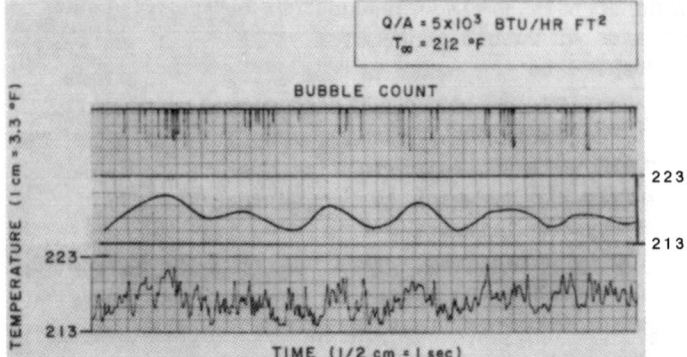

Figure 2.4 Saturated boiling liquid temperature variations within thermal sublayer, and bubble counts departing from heating surface. (From Dougall and Lippert, 1967. Copyright © 1967 by American Society of Mechanical Engineers, New York. Reprinted with permission.)

ishing with different Durite papers and finally with fine crocus cloth and was reconditioned by polishing with the crocus cloth and washing with alcohol prior to each test. Test data were essentially divided into two series. One was obtained from thermocouples located approximately 0.004 in. (0.10 mm) above the boiling surface, and the other was obtained from thermocouples positioned outside the sublayer at a height of 0.100 in. (2.5 mm) above the surface. In presenting the data, only a small portion of the complete test run is shown in the figure. Two sets of data are shown in the figure. One set represents the bubble departing from the surface as is seen as sharp blips interrupting an otherwise straight line (at the top of the figure). The second set (shown at the bottom of the figure) records the measured temperature fluctuations near the boiling site, which have two different characteristics: the rapid fluctuations exemplified by jagged peaks, which vary in magnitude but appear consistently throughout the tests; and a rambling or gross variation on which the jagged peaks are superimposed. This gross behavior is shown by a heavy curve in the middle of the figure. Dougall and Lippert believed this gross temperature variation to be indicative of the fluid eddies near the nucleation site at the boiling surface. The figure represents typical data obtained by these authors and is similar to the results obtained by Staniszewski (1959) and by Han and Griffith (1965a).

It is difficult to obtain accurate and reproducible data for boiling liquid metal, particularly when measuring incipient boiling superheats. The temperature and pressure fluctuations that immediately follow an incipient boiling event for a liquid metal are usually much greater than those that follow a similar event for an ordinary liquid, primarily because of the large superheats generally observed with liquid metals and the relatively low pressure under which the event is caused to occur, but also to some extent because of their high thermal conductivities. At the instant of inception, the pressure rises very rapidly as a result of the explosive growth rate of the bubble. Then, after a fraction of a second, it falls sharply, often to below the original value, and then quickly rises again. These fluctuations continue, although their magnitude decreases with time and they tend to damp out in a few tenths of a second; they may soon be followed by another series if the generated bubble rises to a subcooled region and there collapses. At the instant of boiling inception, the temperature begins to drop sharply because of the very rapid growth of the bubble and the high local latent heat demand. The temperature generally drops to about the saturation value, and then, if stable nucleate boiling ensues, rises to some mildly fluctuating level between T_{sat} and the incipient boiling temperature, T_L.

Deane and Rohsenow (1969) boiled sodium on a smooth nickel plate containing a single 0.0135-in. (0.34-mm)-diameter × 0.01-in. (0.25-mm)-deep cylindrical cavity. At three different boiling pressures, their measured stable-boiling superheats, based on the average of the T_w maxima (as T_w fluctuated with the cyclic ebullition process), showed excellent agreement with those predicted by Eq. (2-3) using 0.00675 in. (0.17 mm) for r. Excellent agreement between the experimental incipient boiling results and those predicted by Eq. (2-3) was also reported by Schultheiss (1970) from boiling sodium tests on polished stainless steel surfaces, where each surface contained a single 0.017-in. (0.4-mm)-diameter artificial cylindrical cavity whose depth ranged from 0.017 in. (0.4 mm) to 0.085 in. (2.0 mm). Although all the above experimental results were obtained with cylindrical cavities, conical cavities would probably give the same results, as long as $\psi/2 < \phi \leq 90°$ (Dwyer, 1976).

Considerable attention has been given to the effect of preboiling conditions on the incipient boiling wall superheat of liquid metals. As indicated before, an active cavity is one that is either partially or wholly filled with vapor or a mixture of vapor and inert gas. Liquid metals, particularly the alkali metals, tend to fill or quench the larger cavities in a heating surface because of their strong wetting characteristics, thus liquid metals penetrate the cavities under subcooled conditions prior to the heating/boiling process. The extent of such penetration depends on many factors, among which are pressure, temperature, the inert gas content of the cavity, the inert gas concentration in the liquid metal, the thickness of the oxide coating on the cavity walls, as well as the oxide concentration in the liquid metal. Maximum penetration occurs with the greatest subcooling experienced by the system, and it corresponds to the minimum cavity radius to which the liquid enters, assuming that the equivalent cavity radius decreases with depth. The greater the extent to which this occurs, the greater will be the required superheat for boiling inception. Maximum penetration therefore corresponds to maximum boiling suppression. This is the basis of the "equivalent-cavity" model proposed by Holtz (1966), reflecting the effect of preboiling conditions. It allows the determination of a value for r for use in the bubble equilibrium equation, Eq. (2-3), when working with smooth heating surfaces where cavities are too small to measure directly (Dwyer, 1976). Figure 2.5 shows schematically an idealized active cavity where a force balance on the interface gives

18 BOILING HEAT TRANSFER AND TWO-PHASE FLOW

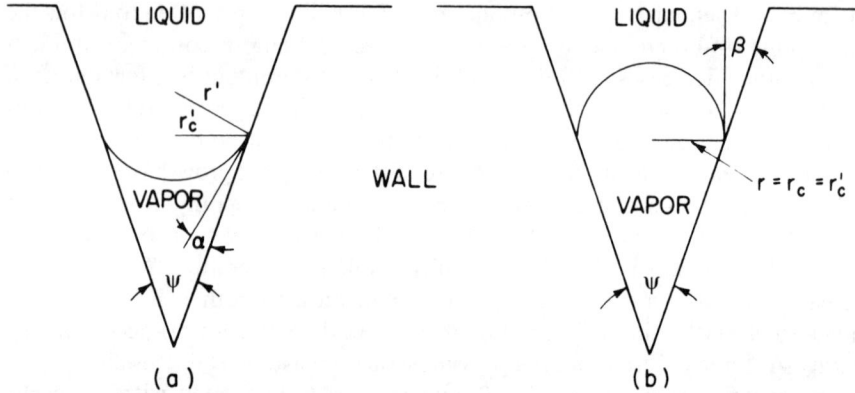

Figure 2.5 Idealized representation of an active cavity: (*a*) situation at maximum boiling suppression; (*b*) situation at incipient boiling. (From Dwyer, 1976. Copyright © 1976 by American Nuclear Society, LaGrange Park, IL. Reprinted with permission.)

$$p_G + p_{i.g.} = p_L + \frac{2\sigma}{r} \tag{2-16}$$

Under the conditions for deepest liquid penetration, Eq. (2-16) becomes

$$p'_L - p'_G - p'_{i.g.} = \frac{2\sigma'}{r'} \tag{2-17}$$

where the primes represent the conditions of maximum boiling suppression. Dwyer (1969) extended Holtz's concept and not only related r at incipient boiling to r' at maximum boiling suppression but also related $p_{i.g.}$ to $p'_{i.g.}$. As a result, he was able to achieve simultaneous solutions of Eqs. (2-3) and (2-17) on a more realistic basis. Assuming the perfect gas law,

$$p'_{i.g.} = \frac{n'_G R T'_L}{V'} \tag{2-18}$$

where n'_G is the number of moles of inert gas in the cavity, T'_L is the absolute temperature of the cavity environment, and V' is the volume of the cavity at maximum penetration conditions, Eq. (2-17) can be written as (Dwyer, 1976)

$$p'_L - p'_G - \frac{\xi' T'_L}{(r')^3} = \frac{2\sigma'}{r'} \tag{2-19}$$

where

$$\xi' = \frac{3n'_G R(\tan \psi/2)}{\pi[\cos(\psi/2 + \alpha)]^3}$$

The value of ξ' depends on the characteristics of the particular surface liquid system, the amount of inert gas trapped in the cavity, and the maximum penetration conditions. Similarly, Eq. (2-16) can be converted to

$$p_G + \frac{\xi T_L}{r^3} - p_L = \frac{2\sigma}{r} \qquad (2\text{-}20)$$

where

$$\xi = \frac{3n_G R(\tan \psi/2)}{\pi[\cos(\psi/2 + \alpha)]^3}$$

and is a measure of the amount of inert gas in the cavity at the instant of boiling inception (Dwyer, 1976).

Experimental results from Chen (1968), Holtz (1971), and Holtz and Singer (1968, 1969) more or less confirmed the validity of Holtz's maximum boiling suppression theory, while some other experimental studies by Deane and Rohsenow (1969), Schultheiss and Smidt (1969) and Kottowski and Grass (1970) did not confirm. This disagreement of experimental results was explained by Dwyer (1976), along with the details of all the experiments.

2.2.2 Waiting Period

As stated in Section 2.1, there is a "waiting period" between the time of release of one bubble and the time of nucleation of the next at a given nucleation site. This is the period when the thermal boundary layer is reestablished and when the surface temperature of the heater is reheated to that required for nucleation of the next bubble. To predict the waiting period, Hsu and Graham (1961) proposed a model using an active nucleus cavity of radius r_c which has just produced a bubble that eventually departs from the surface and has trapped some residual vapor or gas that serves as a nucleus for a new bubble. When heating the liquid, the temperature of the gas in the nucleus also increases. Thus the bubble embryo is not activated until the surrounding liquid is hotter than the bubble interior, which is at

$$T_G = T_{sat} + \frac{2\sigma T_{sat}}{Cr_c h_{fg} \rho_G}$$

The heating of the liquid can be approximated by the transient conduction of heat to a slab of finite thickness δ_t with the conventional differential equation,

$$\frac{\partial^2 T}{\partial y^2} = \frac{1}{\alpha}\frac{\partial T}{\partial t}$$

and the boundary conditions,

$$t = 0 \quad T = T_\infty$$
$$y = 0 \quad T = T_w \quad y = \delta \quad T = T_\infty$$

The solution to this equation yields the transient temperature profile in the thermal layer, δ. When the curve of the superheat of the bubble nucleus versus distance (y/δ) is superposed on such a transient temperature profile at various values of dimensionless time ($\alpha t/\delta^2$), the intersection of the two curves gives the waiting time for a given cavity size and a given wall temperature. For a given cavity size, increasing wall temperature shortens the waiting time; however, for a given wall temperature, there is a range of cavity sizes for which the waiting time is finite (Hsu and Graham, 1976).

Han and Griffith (1965a) developed a modified version of an ebullition cycle and, using a simplification in the temperature profile equation, they obtained an explicit form for the waiting period:

$$t_w = \frac{\delta^2}{\pi\alpha} = \left(\frac{9}{4\pi\alpha}\right)\left[\frac{(T_w - T_\infty)r_c}{T_w - T_{sat}(1 - 2\sigma/r_c\rho_G h_{fg})}\right]^2 \qquad (2\text{-}21)$$

Hatton and Hall (1966) further improve the model by taking into consideration the heat capacity of the heating element, since during the waiting period the heat surface has to be heated to a sufficiently high temperature to initiate the new bubble. The heat balance used by Hatton and Hall is

$$q''' \delta_w = \rho_w C_w \delta_w \left(\frac{dT_w}{dt}\right) + k_L\left(\frac{dT}{dy}\right) \qquad (2\text{-}22)$$

where q''' is the heat generation rate per unit volume per unit time, and δ_w, C_w, and ρ_w are the thickness, specific heat, and density of the heating surface, respectively. However, they obtained the expression for the waiting period by representing (dT/dy) by ($T_w - T_\infty$)/δ_L, or a linear liquid temperature profile. Unfortunately, a model involving both liquid and solid wall heat capacity yields quite a complicated analysis. Deane and Rohsenow (1969), in their thermal analysis of the waiting and bubble growth periods, did consider both capacities. Their analysis was influenced a good deal by the results of their pool boiling experiments with sodium. They assumed that no dry patch appeared in the center of the microlayer (a thin layer of liquid lies between the bubble and the heating surface, the existence of which

POOL BOILING 21

during the bubble's early growth was first suggested by Snyder and Edwards (1956)). At the instant of bubble departure, the residual microlayer is swept off the heating surface and replaced by liquid at the saturation temperature corresponding to the system pressure (Fig. 2.6). Assuming one-dimensional heat transfer in the solid, the governing energy equation for the solid is

$$\frac{\partial^2 [T_w(x, \theta) - T_{sat}]}{\partial x^2} = \frac{1}{\alpha_w} \frac{\partial [T_w(x, \theta) - T_{sat}]}{\partial \theta} \tag{2-23}$$

where x is the distance from the surface in the solid, as shown in Figure 2.6. The equation for the liquid is

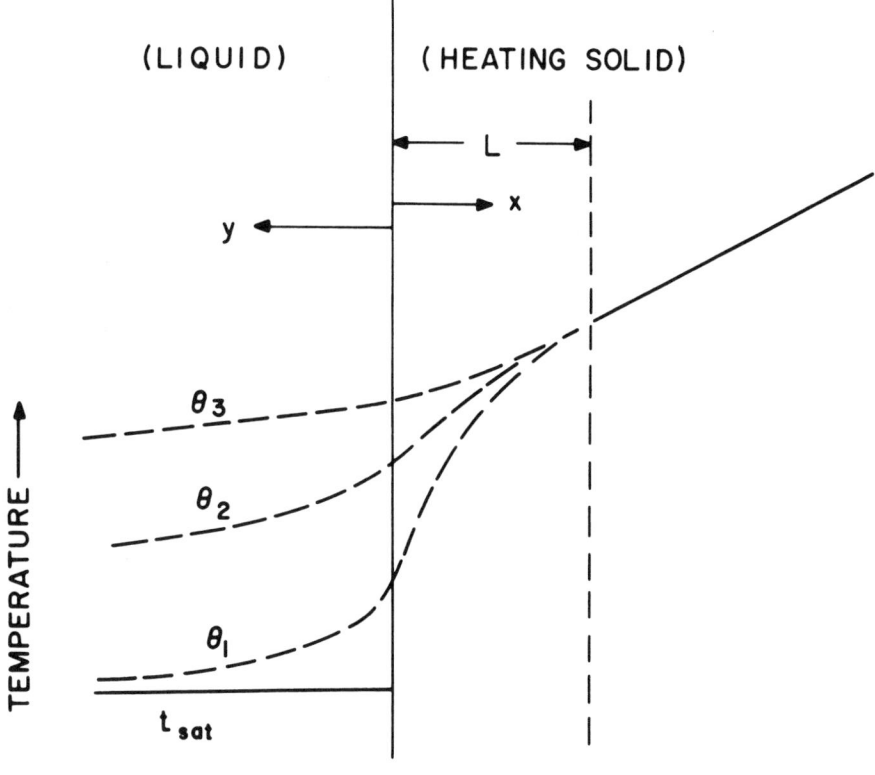

Figure 2.6 Temperature profiles in liquid and heating solid during waiting period, according to mathematical model of Deane and Rohsenow. (From Deane & Rohsenow, 1969. Copyright © 1969 by American Society of Mechanical Engineers, New York. Reprinted with permission.)

$$\frac{\partial^2 [T_L(y, \theta) - T_{sat}]}{\partial y^2} = \frac{1}{\alpha_L} \frac{\partial [T_L(y, \theta) - T_{sat}]}{\partial \theta} \qquad (2\text{-}24)$$

with the following initial and boundary conditions.

Initial conditions: (1) The temperature distribution in the solid at the beginning of the waiting period is the same as that at the end of the bubble growth period. (2) The temperature of the liquid at the beginning of the waiting period is assumed to be uniform at T_{sat}.

Boundary conditions:

$$T_w(0, \theta) - T_{sat} = T_L(0, \theta) - T_{sat}$$

$$k_w \frac{\partial [T_w(0, \theta) - T_{sat}]}{\partial x} = -k_L \frac{\partial [T_L(0, \theta) - T_{sat}]}{\partial y}$$

$$= h_{eq}[T_w(0, \theta) - T_{sat}]$$

$$T_w(0, \theta_b) - T_{sat} = T_w(0, 0) - T_{sat}$$

where θ_b is the ebullition cycle time. Deane and Rohsenow solved the above system of equations numerically for the same conditions as those under which they ran one of their nucleate pool boiling experiments with sodium. The value of h_{eq} was varied until the ebullition cycle time obtained from the analysis,* θ_b, equaled that obtained experimentally. Equations (2-23) and (2-24) were solved by an iterative procedure until the temperature distribution used in the solid at the beginning of the bubble growth period was the same as that calculated for the end of the waiting period (Dwyer, 1976). As will be discussed in the next section, there is evidence that, in general, the length of the bubble growth period for liquid metals is shorter than that for ordinary liquids and probably seldom exceeds a few tens of milliseconds. Experimental results show that the total ebullition cycle time, θ_b, for liquid metals is typically of the order of 1 sec. Thus the waiting period represents a very high fraction of the time θ_b. In their experiments with sodium, Deane and Rohsenow (1969) estimated it to be 98%. The reasons for such longer waiting periods than for ordinary liquids are (Dwyer, 1976)

1. The nucleation superheats are generally higher.
2. The thermal boundary-layer thicknesses in the liquid before bubble nucleation are much greater.
3. The heat capacity (or thermal inertia) of the heating wall is generally higher.

* To calculate the total ebullition cycle time, the length of the bubble growth period, θ_d, is also needed, which can be estimated by first estimating the bubble diameter at the time of departure and then estimating the time for the bubble to grow to that size, as will be discussed in later sections.

2.2.3 Isothermal Bubble Dynamics

When the size of a bubble nucleus formed in a liquid exceeds that of thermodynamic equilibrium as given by Eq. (2-3), the bubble will grow because of the excess vapor pressure that is no longer balanced by the surface tension forces. During the initial stage of growth, the inertia of the surrounding liquid and the surface tension forces control the process. Considering a single spherical bubble in an incompressible inviscous liquid of infinite extent (a homogeneous medium), the equation of motion for the spherically symmetrical domain of the liquid is

$$\frac{\rho_L}{g_c}\left(\frac{\partial u}{\partial t} + u\frac{\partial u}{\partial r}\right) = -\frac{\partial p}{\partial r} \qquad (2\text{-}25)$$

where u is the radial liquid velocity. Integrating the equation of continuity,

$$\frac{1}{r^2}\frac{\partial (r^2 u)}{\partial r} = 0 \qquad (2\text{-}26)$$

from the radius of the bubble interface R to r, where $r > R$, gives the radial liquid velocity u in terms of the interface velocity (if mass transfer due to evaporation is neglected):

$$u = \dot{R}\left(\frac{R}{r}\right)^2 \qquad (2\text{-}27)$$

where $\dot{R} = dR/dt$ and $\ddot{R} = d^2R/dt^2$.

Substituting Eq. (2-27) into Eq. (2-25) gives

$$\left(\frac{\rho_L}{g_c}\right)\left[\frac{(2R\dot{R}^2 + R^2\ddot{R})}{r^2} - \frac{2\dot{R}^2 R^4}{r^5}\right] = -\frac{\partial p}{\partial r}$$

which, upon integration from the bubble interface R to infinity, gives

$$\left(\frac{\rho_L}{g_c}\right)\left(R\ddot{R} + \frac{3\dot{R}^2}{2}\right) = p_L(R) - p_L(\infty) \qquad (2\text{-}28)$$

The vapor pressure in the bubble is related to the liquid pressure at the bubble interface and the surface tension force by Eq. (2-3). Introducing this result into Eq. (2-28), the *Rayleigh equation* (Rayleigh, 1917) for isothermal bubble dynamics is obtained as

$$\left(\frac{\rho_L}{g_c}\right)\left(R\ddot{R} + \frac{3\dot{R}^2}{2}\right) = \Delta p - \frac{2\sigma}{R} \qquad (2\text{-}29)$$

where Δp can be related to the liquid superheat by the Clausius-Clapeyron equation, as before, to give the initial value of the driving pressure difference for bubble growth. The initial bubble size, which must exceed the critical size given by Eq. (2-3), is generally not known. Idealized models have thus been postulated to describe bubble growth with Eq. (2-29). In completely inertia-controlled bubble growth, p_G can be replaced by p_o^*, (as shown later by point a in Fig. 2.9), and if surface tension and acceleration terms are dropped as soon as the bubble has grown significantly beyond its initial radius, which can be obtained rather quickly, Eq. (2-29) becomes (Dwyer, 1976)

$$\frac{dR}{d\theta} = \left[\frac{(2/3)g_c(p_\infty^* - p_\infty)}{\rho_L}\right]^{1/2} \qquad (2\text{-}29a)$$

or

$$R = \left[\frac{(2/3)g_c(p_\infty^* - p_\infty)}{\rho_L}\right]^{1/2} \theta \qquad (2\text{-}29b)$$

At a later stage of bubble growth, heat diffusion effects are controlling (as point c in Fig. 2.9), and the solution to the coupled momentum and heat transfer equations leads to the asymptotic solutions and is closely approximated by the leading term of the Plesset-Zwick (1954) solution,

$$\frac{dR}{d\theta} = \frac{k_L(T_\infty - T_\infty^*)}{[H_{fg}\,\rho_G(\pi\alpha_L\theta/3)^{1/2}]} \qquad (2\text{-}29c)$$

which, after integration, can be expressed in the simple form

$$R = \frac{2k_L(T_\infty - T_\infty^*)}{[H_{fg}\,\rho_G(\pi\alpha_L/3)^{1/2}]}\,\theta^{1/2} \qquad (2\text{-}29d)$$

Note that high superheats, large liquid thermal conductivities, low pressures, and low bubble frequencies, all of which are more typical of liquid metals, tend to give bubble dynamics that approach the inertia-controlled case as the bubble growth rates are high. On the other hand, low superheats, low conductivities, high pressures, and high bubble frequencies tend to give bubble dynamics that approach the heat diffusion-controlled case as the bubble growth rates are relatively low (as is typical of water).

One solution that was considered by Rayleigh (Lamb, 1945) for the determination of bubble collapse time, t_m, used the model of a bubble with initial size R_m, suddenly subjected to a constant excess liquid pressure p_L. Neglecting the surface tension and the gas pressure in the bubble, Eq. (2-29) may be rearranged to

$$\frac{d(R^3 \dot{R}^2)}{dt} = \left(\frac{-2 g_c p_L}{\rho_L}\right) R^2 \dot{R} \tag{2-30}$$

with initial conditions

$$R(0) = R_m \qquad\qquad W$$
$$\dot{R}(0) = 0$$

By integration,

$$R^3 \dot{R}^2 = \left(\frac{2 g_c p_L}{3 \rho_L}\right)(R_m^3 - R^3) \tag{2-31}$$

Introducing the dimensionless radius, $R^+ = R/R_m$, and rearranging, Eq. (2-31) leads to the integral form

$$\frac{t}{R_m} = \left(\frac{3 \rho_L}{2 g_c p_L}\right)^{1/2} \int_1^{R^+} \frac{(R^+)^{3/2} \, dR^+}{[1 - (R^+)^3]^{1/2}} \tag{2-32}$$

from which the collapse time t_m is evaluated as

$$\frac{t_m}{R_m} \simeq 0.915 \left(\frac{\rho_L}{6 p_L g_c}\right)^{1/2} \tag{2-33}$$

For the time and size relationship for bubble growth to the maximum size, R_m, Bankoff and Mikesell (1959) arrived at essentially the same results. From observation of growth and collapse of bubbles in highly subcooled boiling, they postulated a model such that, after incipience of very rapid bubble growth, the difference between the pressure at the bubble interface and the pressure at a great distance from the bubble, $\Delta p = (p_G - p_\infty) - (2\sigma/R)$, remains essentially constant at some negative value. Using the conditions at the maximum bubble size, $R = R_m$ at $t = t_m$ and where $\dot{R}_m = 0$, this model circumvents the difficulties of specifying initial conditions. The solution, similar to Eq. (2-32) for bubble size versus time, may be written as

$$\psi = \left(\frac{t}{R_m}\right)\left(\frac{-2\,\Delta p}{3\rho_L}\right)^{1/2} = \int_0^{R^+} \frac{(R^+)^{3/2}\,d(R^+)}{(1-R^+)^{1/2}} \qquad (2\text{-}34)$$

Bankoff and Mikesell (1959) compared this solution with the data of Ellion (1954) and that of Gunther and Kreith (1950), as shown in Figure 2.7.

2.2.4 Isobaric Bubble Dynamics

In the later part of bubble life, i.e., a few milliseconds after growth has begun, the inertia of the surrounding liquid and the surface tension forces can be neglected. The pressure can be considered to be uniform (isobaric) throughout the bubble and the liquid. At this stage, bubble growth is governed by the rate at which heat can be supplied from the superheated liquid to the bubble interface to facilitate the vapor formulation associated with growth. Because the importance of inertia effects relative to the effects of heat transfer can be expressed in the group,

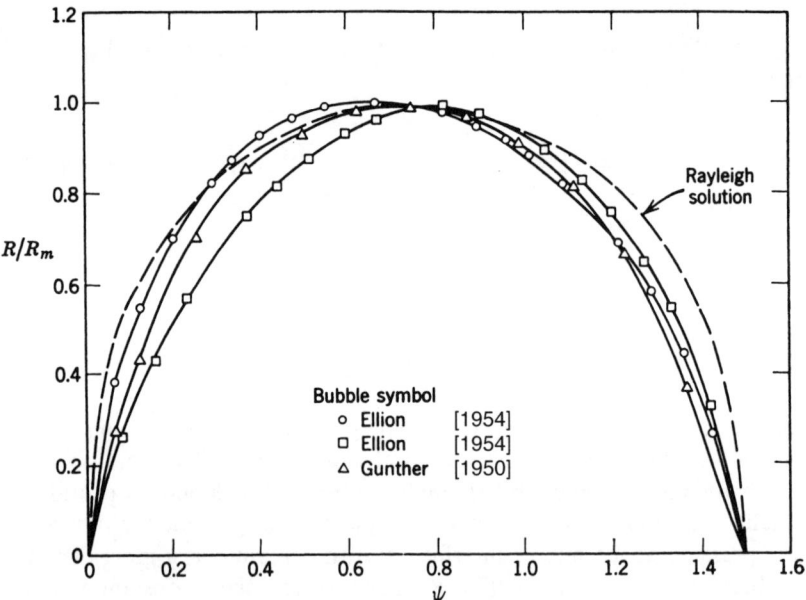

Figure 2.7 Bubble size versus time function ψ. (From Bankoff and Mikesell, 1959. Copyright © 1959 New York. Reprinted with permission.)

$$\left(\frac{E_i}{E_h}\right) = \frac{g_c D_b^2 \Delta p}{4\alpha_L^2 \rho_L}$$

as the bubble grows older, the ratio becomes smaller. If we consider first the growth of a single spherical isobaric bubble in a uniformly superheated liquid of infinite extent, the heat conduction and convection problem of spherical symmetry is given by

$$\frac{\partial T}{\partial t} + u\frac{\partial T}{\partial r} = \alpha_L \left(\frac{1}{r^2}\right) \frac{\partial\left[r^2\left(\frac{\partial T}{\partial r}\right)\right]}{\partial r} \tag{2-35}$$

with boundary conditions

$$T(r, 0) = T_{sup} \quad r > R$$
$$T(R, t) = T_{sat}$$
$$T(\infty, t) = T_{sup}$$

where T_{sup} denotes the superheated temperature of the liquid and where the radial liquid velocity u for an incompressible liquid is subject to the continuity equation. Accounting for the difference in density between the liquid and the vapor phase, an interface mass balance gives Eq. (2-36) by integration of Eq. (2-26):

$$u(r) = \left(\frac{\rho_L - \rho_G}{\rho_L}\right) \dot{R} \left(\frac{R}{r}\right)^2 \tag{2-36}$$

A heat balance at the bubble interface gives

$$k\frac{\partial T(R, t)}{\partial r} = \rho_G H_{fg} \dot{R} \tag{2-37}$$

Scriven (1959) and Bankoff and Mikesell (1959) have derived the exact solution to Eq. (2-35) subject to Eqs. (2-36) and (2-37). Characteristic of similarity solutions for bubble growth, the solution is in the form

$$R(t) = 2C_1(\alpha_L t)^{1/2} \tag{2-38}$$

where the growth constant, C_1, is an implicit function of the *Jakob number,* Ja, and the density ratio (ρ_L/ρ_G) given by

$$\text{Ja} = \left(\frac{\rho_L c_L}{\rho_g H_{fg}}\right)(T_{\text{sup}} - T_{\text{sat}})$$

$$= 2C_1^2 \exp\left[C_1^2\left(3 + \frac{\rho_G}{\rho_L}\right)\right]\int_{c_1}^{\infty} \frac{1}{x}\exp\left[\frac{-x^3 - 2(1 - \rho_G/\rho_L)C_1^3}{x}\right]dx \tag{2-39}$$

where the liquid superheat is $(T_{\text{sup}} - T_{\text{sat}})$. Solutions to Eq. (2-39) as obtained by Scriven (1959) are shown in Figure 2.8. It is interesting to note that within the validity of this model the bubble growth rate becomes unbounded as the pressure of the system approaches critical. Thus, the maximum superheat the liquid can sustain under normal bubble growth is H_{fg}/c_p. In most cases not close to the critical state, $\rho_G/\rho_L \ll 1$, and for large values of Jakob number, the growth constant C_1 may be approximated, as shown by Scriven, by

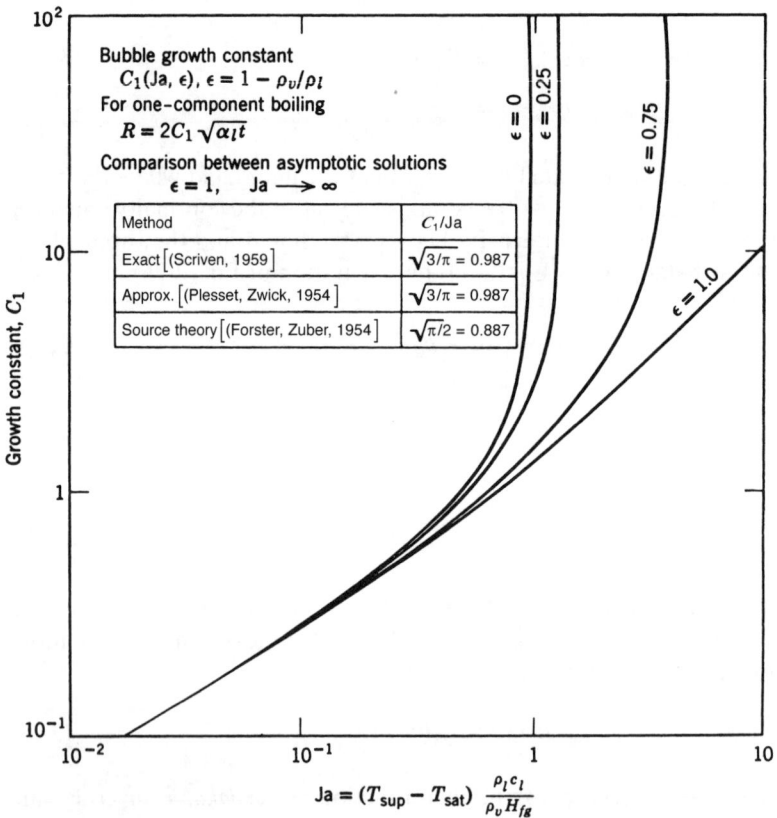

Figure 2.8 Bubble growth in one-component superheated liquid. (From Scriven, 1959. Copyright © 1959 by Elsevier Science Ltd, Kidlington, UK. Reprinted with permission.)

$$C_1 = \left(\frac{3}{\pi}\right)^{1/2} \text{Ja} = \left(\frac{3}{\pi}\right)^{1/2} (T_{\text{sup}} - T_{\text{sat}}) \frac{\rho_L c_L}{\rho_G H_{fg}} \qquad (2\text{-}40)$$

which coincides with the approximate solution obtained earlier by Plesset and Zwick (1954). Forster and Zuber (1954) formulated and solved the coupled problem of the intermediate bubble growth (where inertia and surface tension forces are still important) by combining the Rayleigh equation (2-29) with the Clausius-Clapeyron equation and the appropriate Green's function for the heat conduction problem. Their solution for the asymptotic stage, where heat transfer predominates, gives a growth constant of $(\pi)^{1/2}/2$ times the Jakob number. A remarkably good approximation to these results is obtainable by neglecting the radial liquid motion around the bubble and treating the thermal boundary layer as a slab. Measuring x from the bubble interface, the formulation becomes

$$\frac{\partial T}{\partial t} = \alpha_L \frac{\partial^2 T}{\partial x^2} \qquad (2\text{-}41)$$

with boundary conditions

$$T(x, 0) = T_{\text{sup}}$$

$$T(0, t) = T_{\text{sat}}$$

$$T(\infty, t) = T_{\text{sup}}$$

The solution to Eq. (2-41) can be shown as

$$T - T_{\text{sat}} = (T_{\text{sup}} - T_{\text{sat}}) \, \text{erf}\left[\frac{x}{2(\alpha_L t)^{1/2}}\right] \qquad (2\text{-}42)$$

The interface heat balance, similar to Eq. (2-37), becomes

$$k \frac{\partial T(0, t)}{\partial x} = \rho_G H_{fg} \dot{R} \qquad (2\text{-}43)$$

where, from Eq. (2-42),

$$\frac{\partial T(0, t)}{\partial x} = \frac{T_{\text{sup}} - T_{\text{sat}}}{(\pi \alpha_L t)^{1/2}} \qquad (2\text{-}44)$$

Combining Eqs. (2-43) and (2-44), the solution for the bubble size becomes of the form of Eq. (2-38), where the growth constant C_1 now is $1/(\pi)^{1/2}$ times the Jakob number.

Bubble growth in a uniformly superheated liquid according to Eq. (2-38) has been verified experimentally by Fritz and Ende (1936) and by Dergarabedian (1960). Theofanous et al. (1969) suggested a more rigorous analysis of the dynamics of bubble growth in a uniformly superheated liquid of infinite extent, taking into account not only inertia, viscous and surface tension forces, and heat transfer, but also allowing for nonequilibrium at the vapor–liquid interface and for time-dependent variation of the vapor density in the bubble. Their model starts with a minute vapor bubble in mechanical and thermal equilibrium with the surrounding superheated liquid, Eq. (2-3). Theofanous et al. set up five governing differential equations (Dwyer, 1976):

1. Continuity for the vapor,

$$\left(\frac{1}{R}\right)\left(\frac{dR}{d\theta}\right) + \left(\frac{1}{3p_G}\right)\left(\frac{dp_G}{d\theta}\right) - \left(\frac{1}{3T_G}\right)\left(\frac{dT_G}{d\theta}\right) = \frac{R_G W T_G}{M R p_G} \qquad (2\text{-}45)$$

where R_G = gas constant
W = net mass flux of vapor into bubble
M = molecular weight of vapor
v = kinematic viscosity of liquid

2. Equation of motion for the liquid,

$$\frac{d^2R}{d\theta^2} + \frac{3}{2R}\left(\frac{dR}{d\theta}\right)^2 + \left(\frac{4v}{R^2}\right)\left(\frac{dR}{d\theta}\right) + \frac{2g_c\sigma}{\rho_L R^2} + \frac{g_c(p_\infty - p_G)}{\rho_L R} = 0 \qquad (2\text{-}46)$$

3. Vapor energy balance,

$$J(c_v)_G\left(\frac{M}{R_G}\right)\left(\frac{R}{T_G}\right)\left(\frac{dT_G}{d\theta}\right) = -3\left(\frac{dR}{d\theta}\right)$$

$$+ 3C\left(\frac{Mg_c}{2\pi R_G}\right)^{1/2} J(c_v)_G \left(\frac{p_L^*}{p_G}\right)\frac{(T_L - T_G)}{\sqrt{T_L}} \qquad (2\text{-}47)$$

4. Liquid energy balance,

$$(T_L - T_\infty)(3r_L^2 + 3Rr_L + 3R^2)\frac{dR}{d\theta} + 2(T_L - T_\infty)(r_L^2 + 3r_L R - 9R^2)\frac{dR}{d\theta}$$

$$+ (r_L - R)(r_L^2 + 3r_L R + 6R^2)\frac{dT_L}{d\theta} = \frac{60\alpha_L R^2(T_L - T_\infty)}{r_L - R} \qquad (2\text{-}48)$$

5. Boundary condition involving net mass transfer,

$$\frac{(T_L - T_\infty)}{(r_L - R)^2}\left(\frac{dR}{d\theta}\right) - \frac{(T_L - T_\infty)}{(r_L - R)^2}\left(\frac{dr_L}{d\theta}\right) - C_A\left(\frac{Mg_c}{2\pi R_G}\right)^{1/2}\left(\frac{H_{fg}}{k_L(T_G)^{1/2}}\right)\left(\frac{dp_G}{d\theta}\right)$$

$$+ C_A\left(\frac{Mg_c}{2\pi R_G}\right)^{1/2}\left(\frac{H_{fg}}{4k_L}\right)\left[\frac{p_G}{(T_G)^{3/2}}\right]\left(\frac{dT_G}{d\theta}\right) + \left\{\frac{1}{r_L - R} + C_A\left(\frac{Mg_c}{2\pi R_G}\right)^{1/2}\right.$$

$$\left. \times \left(\frac{H_{fg}}{2k_L}\right)\left(\frac{dp_L^*}{dT_L}\right)\left(\frac{1}{\sqrt{T_L}}\right) - C_A\left(\frac{Mg_c}{2\pi R_G}\right)^{1/2}\left(\frac{H_{fg}}{4k_L}\right)\frac{p_L^*}{(T_L)^{3/2}}\right\}\frac{dT_L}{d\theta} = 0 \qquad (2\text{-}49)$$

where C_A is the accommodation coefficient. The determination of C_A in a given situation presents a problem, as there is no clear concensus on the method of estimating it. Theofanous et al. (1969) treated it as a parameter.

The above analysis requires the simultaneous solution of five nonlinear differential equations. Mikic et al. (1970) and Vohr (1970) independently developed alternative methods, which avoid a numerical technique requiring the use of a digital computer. Their method consisted of closed-form solutions to sets of simpler equations that rest on the same basic principles but with certain simplifying assumptions. Shortly after these publications, Board and Duffey (1971) presented another simplified solution that did not contain as many simplifying assumptions as had been used before, which required the numerical solution of only the coupled equations of motion and heat transfer [much simpler than the solution of Theofanous et al. (1969)] (Dwyer, 1976). Like Mikic et al. (1970), they assumed that the vapor in the bubble was in thermodynamic equilibrium with the liquid of the bubble wall (that is, $p_G = p_L$ and $T_G = T_L$, and points d and e in Figure 2.9 coincide to become point b on the saturation curve). On the other hand, they allowed for variation in ρ_G and for both acceleration and surface tension effects. However, the last two affect bubble growth only in the very early stages (see Fig. 2.11). Only those equations they developed and simultaneously solved are given here; for the detailed derivation the reader is referred to the description by Dwyer (1976). Board and Duffey employed the Rayleigh momentum equation, (2-29), to take care of the inertia and surface tension effects, using the following form of the Clausius-Clapeyron equation:

$$\ln p_G = A - H_{fg}\big|_{T_\infty} \frac{MJ}{R_G T_G} - \left(\frac{MJ}{R_G}\right)\left(\frac{dH_{fg}}{dT}\right)\ln T_G \qquad (2\text{-}50)$$

where A is a constant of integration. They handled the heat transfer problems via

$$\frac{dQ}{d\theta} = 4\pi R^2 k_L \left(\frac{\partial T}{\partial r}\right)_{R,\theta} \qquad (2\text{-}51)$$

32 BOILING HEAT TRANSFER AND TWO-PHASE FLOW

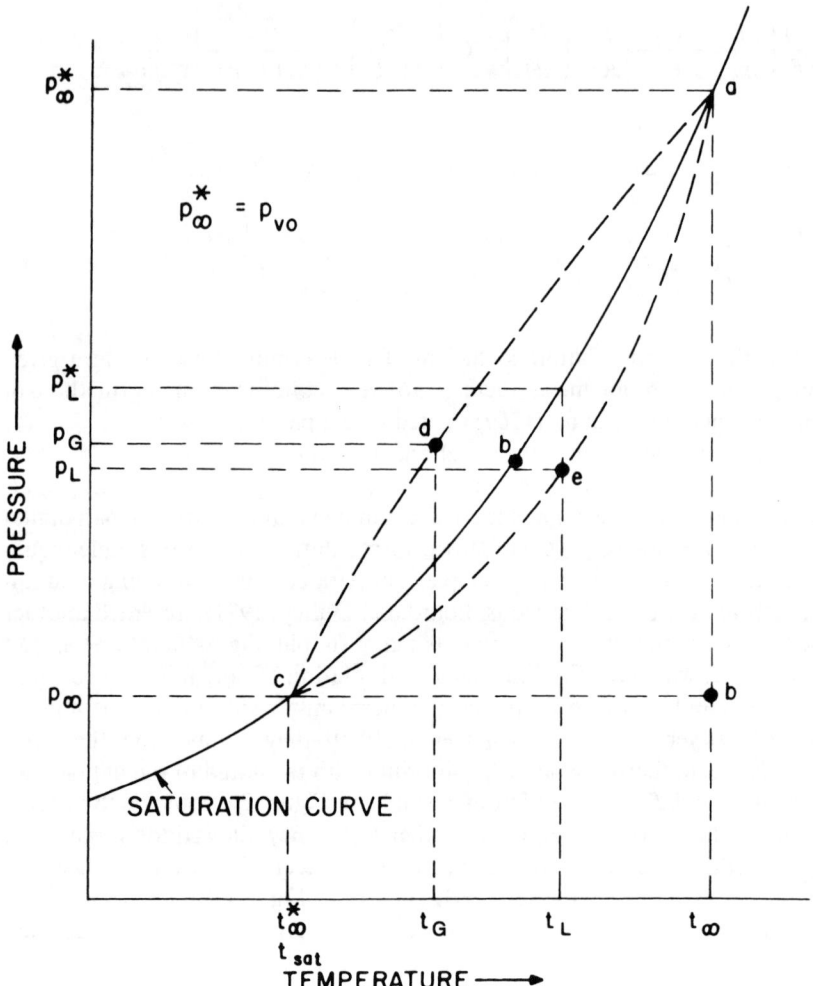

Figure 2.9 Representation of pressure–temperature relationship that exists during the growth period of a spherical bubble in a superheated liquid of infinite extent. (From Dwyer, 1976. Copyright © 1976 by American Nuclear Society, LaGrange Park, IL. Reprinted with permission.)

where

$$\left(\frac{\partial T}{\partial r}\right)_{R,\theta} = \frac{T_\infty - T_G}{[\pi k_L \theta/3(c_p)_L \rho_L]^{1/2}} \quad (2\text{-}52)$$

and

$$\frac{1}{4\pi R^2}\frac{dQ}{d\theta} = \rho_G \left\{ H_{fg}\Big|_{T_\infty} + \left[\frac{dH_{fg}}{dT} + (c_p)_L\right](T_\infty - T_G)\right\}\frac{dR}{d\theta}$$

$$+ \frac{R\rho_G}{3}\left[(c_p)_L + \left(\frac{T_\infty}{T_G}\right)\left(\frac{dH_{fg}}{dT}\right) - \frac{H_{fg}\Big|_{T_\infty}}{T_G}\right]\frac{dT_G}{d\theta}$$

$$+ \frac{R}{3}\left\{H_{fg}\Big|_{T_\infty} + \left[\frac{dH_{fg}}{dT} + (c_p)_L\right](T_\infty - T_G)\right\}\frac{d\rho_G}{d\theta} \quad (2\text{-}53)$$

The vapor density ρ_G was expressed in terms of p_G and T_G through the perfect gas law:

$$\rho_G = \frac{Mp_G}{R_G T_G} \quad (2\text{-}54)$$

Equations (2-29), (2-50), (2-51), and (2-54) along with Eqs. (2-52) and (2-53) were solved simultaneously by Board and Duffey to obtain the four unknowns R, p_G, T_G, and ρ_G. The starting conditions for their numerical solution were those represented by the equilibrium relation, Eq. (2-2), at the instant that growth begins ($\theta = 0$), and growth was initiated by increasing $r(\theta = 0)$ by 0.05%. Some of their results are shown in Figures 2.10 and 2.11. In Figure 2.10, the Board-Duffey curve falls only slightly below that of Theofanous et al. for $C_A = 1$. Figure 2.11, from Board and Duffey's article, shows very good agreement between their curve and that given by Mikic et al. for bubble growth in sodium superheated 100°C (212°F) at 1.7 atm (0.2 MPa) except at the very early stages ($\theta < 10^{-6}$ sec). This is because the Mikic et al. solution, not intended for the very early stages, is based on the assumptions that $R = 0$ at $\theta = 0$ and that acceleration and surface tension forces are negligible (Dwyer, 1976). Board and Duffey found that within the limits of experimental error, their solution agreed with their own data on the growth of vapor bubbles in water at constant and uniform superheats up to 20°C (68°F).

The application of the above-mentioned solutions of the bubble growth problem to liquid metals has been shown by Figures 2.10 and 2.11. Based on the information presented here, some of Dwyer's conclusions are cited (Dwyer, 1976):

1. In the early stages of bubble growth, the growth rate for all liquids is controlled by inertia effects. This phase is extended to larger bubble sizes in the case of liquid metals because of the higher growth rate associated with these liquids. The inertia-controlled phase lasts for (R/r_o) to 1,000, or for bubble diameters up to 1 mm (0.04 in.).

34 BOILING HEAT TRANSFER AND TWO-PHASE FLOW

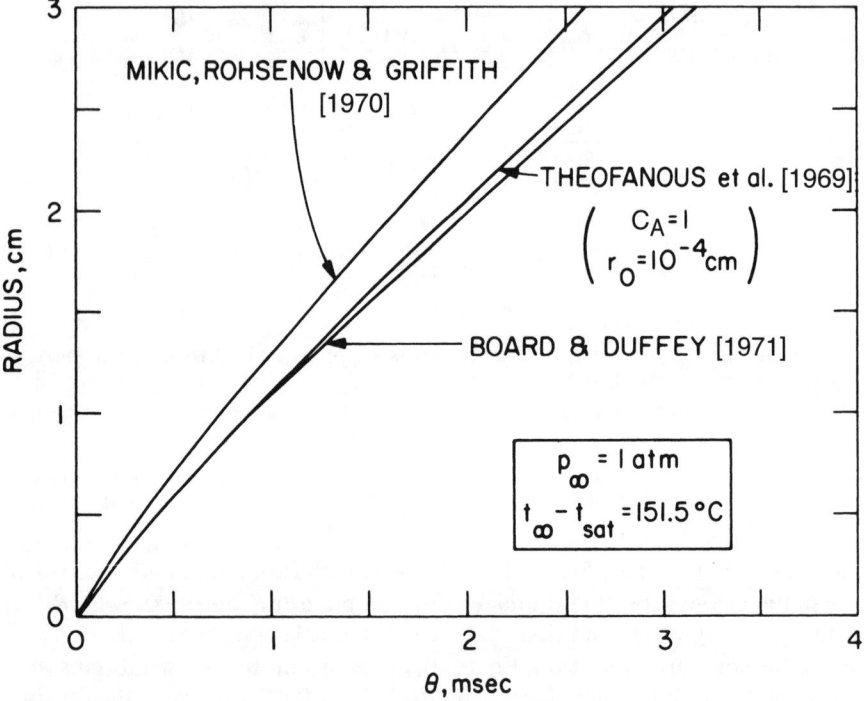

Figure 2.10 Calculated growth rates for a spherical vapor bubble in a uniformly heated, large volume of sodium under highly superheated conditions. (From Dwyer, 1976. Copyright © 1976 by American Nuclear Society, LaGrange Park, IL. Reprinted with permission.)

2. In the later stages, bubble growth is controlled more and more by heat transfer to the bubble wall, although for a high-conductivity liquid such as sodium, inertia effects are dominant throughout most of the growth period.
3. If the accommodation coefficient C_A is equal or close to unity for liquid metals, as appears most likely for "clean" systems, then bubble growth in such liquids is little affected by mass transfer effects. It has been illustrated that the growth rate curves for $C_A = 1$ and $C_A = \infty$ are not very far apart.
4. The method of Theofanous et al. (1969) should be the most accurate for predicting bubble growth rates in large volumes of liquid metals at uniform superheats, although there has been no experimental data against which to test it directly.

A number of investigators have studied the problem of isobaric bubble growth in an initially nonuniformly superheated liquid, such as occurs when a bubble grows in a thin superheated liquid layer on a heater surface. Considering such a bubble surrounded by this nonuniformly superheated liquid layer during its growth, relations can be derived among bubble growth rate, maximum bubble size,

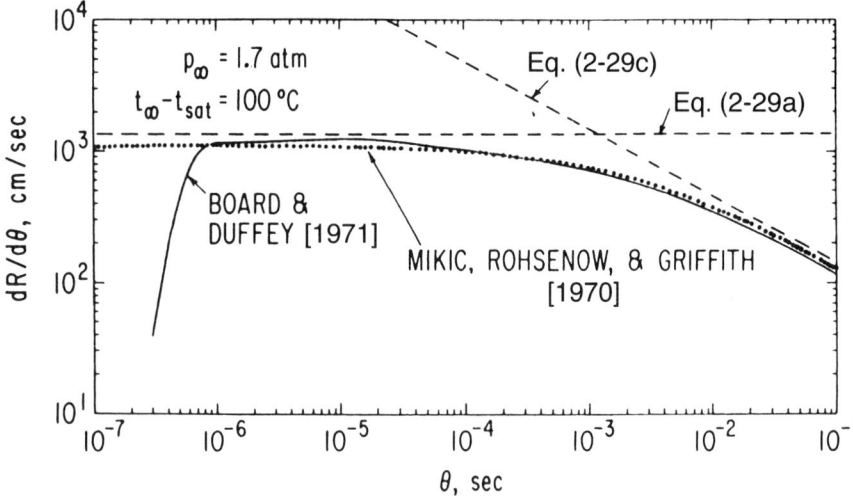

Figure 2.11 Comparison of calculated growth rates for a spherical bubble in a uniformly heated, large volume of superheated sodium. (From Board and Duffey, 1971. Copyright © 1971 by Elsevier Science Ltd., Kidlington, UK. Reprinted with permission.)

and time required to attain maximum size. Using the one-dimensional analysis, Eq. (2-43), the heat balance for the bubble may be written (Zuber, 1961) as

$$\rho_G H_{fg} \dot{R} = A' \left[\frac{k(T_{s.h.} - T_{sat})}{(\pi \alpha_L t)^{1/2}} - q_b'' \right] \quad (2\text{-}55)$$

where the first term in brackets accounts for the heat flux conducted to the bubble from the superheated liquid layer, and the second term, q_b'', is the heat flux from the bubble to the bulk liquid. Then q_b'' may be approximated by the actual heat flux from the heating surface, and the constant A' has a value between 1.0 and $\sqrt{3}$, depending on whether the form of the heat conduction equation is for a slab or a sphere. Integrating Eq. (2-55), we obtain

$$R = A' \left(\frac{2}{\sqrt{\pi}} \right) \text{Ja} \left[1 - \left(\frac{q_b'' \sqrt{\pi \alpha_L t}}{2k \, \Delta T} \right) \right] \sqrt{\alpha_L t}$$

The maximum bubble radius occurs when $dR/dt = 0$ at time $t = t_m$, where

$$\sqrt{\pi \alpha_L t_m} = \frac{k(T_{s.h.} - T_{sat})}{q_b''}$$

Thus the equation for the transient bubble size becomes

$$R = A'\left(\frac{2}{\sqrt{\pi}}\right) \mathrm{Ja}\left(1 - \frac{1}{2}\sqrt{\frac{t}{t_m}}\right)\sqrt{\alpha_L t} \qquad (2\text{-}56)$$

and

$$R_{max} = A'\left(\frac{1}{\sqrt{\pi}}\right) \mathrm{Ja}\sqrt{\alpha_L t_m}$$

Consequently,

$$\frac{R}{R_{max}} = \sqrt{\frac{t}{t_{max}}}\left(2 - \sqrt{\frac{t}{t_m}}\right) \qquad (2\text{-}57)$$

This equation agrees with Ellion's data (1954).

Hsu and Graham (1961) took into consideration the bubble shape and incorporated the thermal boundary-layer thickness, δ, into their equation, thus making the bubble growth rate a function of δ. Han and Griffith (1965b) took an approach similar to that of Hsu and Graham with more elaboration, and dealt with the constant-wall-temperature case. Their equation is

$$\phi_v H_{fg} \rho_G \left[4\pi R^2\left(\frac{dR}{dt}\right)\right]$$

$$= \phi_c \phi_s (4\pi R^2) k_L c_L \rho_L \left[\frac{d(T - T_{sat})}{dx}\right]_{x=0} + \phi_{base} 4\pi R^2 h_G (T_w - T_{sat}) \qquad (2\text{-}58)$$

where ϕ_c = curvature factor, $1 < \phi_c < \sqrt{3}$

ϕ_s = surface factor = $\dfrac{1 + \cos\phi}{2}$

ϕ_{base} = base factor = $\dfrac{\sin^2\phi}{4}$

ϕ_v = volume factor = $\dfrac{1}{4}[2 + \cos\phi(2 + \sin^2\phi)]$

h_G = heat transfer coefficient from heating surface to vapor

and

$$\left|\frac{d(T - T_{sat})}{dx}\right|_{x=0} = \frac{T - T_{sat}}{\sqrt{\pi \alpha t}} - \left(\frac{T_w - T_\infty}{\delta}\right) \text{erf}\left(\frac{\delta}{\sqrt{\pi \alpha t}}\right)$$

with $\delta = \sqrt{\pi \alpha t_w}$ and t_w = waiting period. Thus they tied the waiting period explicitly to the superheated thermal layer of the growth period.

A longer waiting period means a larger growth rate; such a trend has been observed experimentally (Hatton and Hall, 1966). There are three limiting conditions for ϕ_c to satisfy:

1. For a bubble to grow in an infinite fluid field with a superheat of $(T_w - T_{sat})$, $\phi_s = 1$, $\phi_v = 1$, $\phi_{base} = 0$, and $\delta >> R$,

$$\frac{dR}{dt} = \frac{\phi_c}{\sqrt{\pi}} \left[\frac{(T_w - T_{sat})\rho_L c_L}{\rho_G H_{fg}}\right] \sqrt{\frac{\alpha}{t}}$$

   If it is to reduce to Scriven's equation, the curvature factor ϕ_c must be $\sqrt{3}$.
2. For contact angle $\phi = 0$ and $\delta << R$, Eq. (2-58) should be reduced to Plesset-Zwick's equation based on the thin boundary layer assumption; thus, $\phi_c = \pi/2$.
3. For $\phi = \pi$ the equation should be reduced to the one-dimensional case; thus, $\phi_c = 1$.

These conditions were the basis for constructing the form for ϕ_c. Equation (2-58) is considered the most general form for bubble growth rate in an initially *nonuniformly* superheated liquid whereby the effects of contact angle, waiting period, surface heat flux, and sucooling are all considered (Hsu and Graham, 1976).

2.2.5 Bubble Departure from a Heated Surface

The following phenomena pertaining to bubble departure from a heated surface are discussed in this section: bubble size at departure, departure frequency, boiling sound, and heat transfer effects by departing bubbles.

2.2.5.1 Bubble size at departure. At departure from a heated surface, the bubble size may theoretically be obtained from a dynamic force balance on the bubble. This should include allowance for surface forces, buoyancy, liquid inertia due to bubble growth, viscous forces, and forces due to the liquid convection around the bubble. For a horizontally heated surface, the maximum static bubble size can be determined analytically as a function of contact angle, surface tension, and liquid–vapor density difference. From such analyses, Fritz (1935) and Wark (1933) found the bubble diameter D_b just breaking off from a surface to be

$$D_b = C_d \phi \left[\frac{2 g_c \sigma}{g(\rho_L - \rho_G)} \right]^{1/2} \quad (2\text{-}59)$$

where the constant $C_d = 0.0148$ for bubbles of hydrogen and water vapor in water and the contact angle ϕ is in degrees.

Staniszewski (1959) conducted experiments on bubble departure sites for boiling water and alcohol under various pressures and found the bubble departure diameter to be linearly proportional to the bubble growth rate at the last stage.

Cole and Shulman (1966) later proposed

$$D_d = 0.0208 \phi \left[\frac{g_c \sigma}{g(\rho_L - \rho_G)} \right]^{1/2} \left[1 + 0.0025 \left(\frac{dD}{dt} \right)^{3/2} \right] \quad (2\text{-}60)$$

based on data of six fluids, with dD/dt expressed in millimeters per second.

In a detailed study of bubble departure, Hatton and Hall (1966) concluded that bubble departure is relatively independent of heat flux but strongly dependent on cavity size D_c and pressure. They presented a bubble departure criterion including the effect of cavity size:

$$D_d^3 \frac{(\rho_L - \rho_G)}{6} = \frac{64 X^4}{g} \left(\frac{C_D \rho_L}{8} - \frac{\rho_L}{12} + \frac{\rho_G}{6} \right) \\ - \frac{g_c \sigma}{g} \left(\frac{D_c^2}{D_d} - D_c \sin \phi \right) \quad (2\text{-}61)$$

where

$$X = \frac{\sqrt{3} K \Delta T}{H_{fg} \rho_G \sqrt{\pi \alpha}} = \left(\frac{3\alpha}{\pi} \right)^{1/2} (\text{Ja})$$

with K = Boltzmann constant and $\Delta T = (4\sigma T_{sat})/H_{fg}\rho_G D_c$, so that the growth rate is

$$\frac{dD}{dt} = 2X(t^{-1/2})$$

The terms in the equation of Hatton and Hall represent:

Buoyancy force = (drag force + liquid inertia + vapor inertia)
− (surface tension − excess pressure)

To take into account that the constant C_d in Eq. (2-59) is really dependent to a significant extent on the boiling pressure, Cole and Rohsenow (1969) proposed a modified correlation that also appears to work quite well for a large variety of ordinary liquids for saturate boiling:

$$\left[\frac{g(\rho_L - \rho_G)D_b^2}{g_c \sigma}\right]^{1/2} = C_d' \left(\frac{\rho_L C_{pL} T_{sat}}{\rho_G H_{fg}}\right)^{5/4} \qquad (2\text{-}62)$$

where the proportionality constant C_d' is equal to 1.5×10^{-4} for water and 4.65×10^{-4} for other (mostly organic) liquids. It can be seen that the term on the left, the dimensionless departure diameter, is equal to $\sqrt{2C_d}\,\phi$ from Eq. (2-59), and the quantity in parentheses on the right-hand side is a modified Jakob number. Primarily through the vapor density term in Ja, it takes care of the variation in pressure. As indicated by Deane and Rohsenow (1969), this expression can be used for liquid metals, as the value of C_d' for sodium should be the same as that for water. Their observation was based on the fact that the value of the modified Jakob number is the same for both sodium and water at the same boiling pressure. A value of 4.65×10^{-4} is recommended for potassium, as its modified Jakob number is closer to those of organics than to that of water. Equation (2-62) is presumably restricted to commercially smooth heating surfaces having more or less normal cavity size distribution and to reduced pressures of less than 0.2, which is usually in the range of interest for liquid metals. There is no ϕ term in Eq. (2-62), because for most situations, the value of ϕ under dynamic conditions appears to be fairly uniform at 45°. Thus it would not be expected that the values of C_d' given above would hold for either very poorly wetting or very strongly wetting systems (Dwyer, 1976). This means that values of D_b for the poorly wetting mercury–steel system and for the strongly wetting mercury–nickel system could not be estimated from the above C_d' values. Such a restriction, however, should not hold for the normally wetting alkali metal–stainless steel systems (as used in liquid metal-cooled reactors). Like Eq. (2-59), this equation, Eq. (2-62), is based on the equality of buoyancy and surface tension forces at time of liftoff for a spherically shaped growing bubble. Liquid metals apparently produce hemispherical bubbles for a large fraction of their growth and tend to grow bigger. Nevertheless, the bubble growth times predicted by Eq. (2-62) appear reasonable, and on that basis one can assume that the calculated D_b values are at least qualitatively correct (Dwyer, 1976).

Experimental results on the size of departing vapor bubbles during nucleate boiling of a liquid metal have been reported by Bobrovich et al. (1967). They boiled potassium on a horizontal stainless steel rod heater in a vessel of a rectangular parallelepiped shape. Two rigs were employed, with the following dimensions:

	Heater			Vessel	
Rig	Diameter	Length	Width	Length	Height
First	14 mm (0.55 in.)	80 mm (3.2 in.)	100 mm (4 in.)	180 mm (7.1 in.)	280 mm (11 in.)
Second	9 mm (0.35 in.)	80 mm (3.2 in.)	70 mm (2.8 in.)	180 mm (7.1 in.)	280 mm (11 in.)

and with heat fluxes in the range 70,000–95,000 Btu/hr ft² or 220–299 kW/m² for heater 1, and 162,000–216,000 Btu/hr ft² or 509–679 kW/m² for heater 2. Bubble growth times, departure diameters, and bubble rise velocities were measured by means of high-speed X-ray photography. The data points, although few and scattered, show very little dependence on pressure, which is in contrast to the predictions of Eq. (2-62), as well as to experimental results on ordinary liquids. Thus the degree of applicability of the equation to liquid metals has not been confirmed.

2.2.5.2 Departure frequency. When a bubble starts to grow on a heating surface, a time interval t_d is required for it to depart from the surface. Griffith (1958) suggested that the inertia of the liquid helps to detach the bubble from the surface and carry it away. Cold liquid then rushes in behind the departing bubble and touches the heating surface. As mentioned before, a time interval t_w, a "waiting period," is required to heat the new liquid layer so that nucleation, which is a prerequisite for growth of the next bubble from the same site, can occur. If t_d represents the time of departure and t_w the waiting period, the bubble frequency f_b is then defined as $1/(t_w + t_d)$.

The increase of heat flux activates more nucleation sites, thus increasing the bubble population. The increase of heat flux also reduces the generation waiting period and the time of bubble departure. The rapid growth of a bubble induces a high inertia effect on the bubble; acting away from the surface, it thus reduces the departure bubble size. An increase of pressure raises the saturation temperature, which in turn reduces the surface tension, σ. A decrease of surface tension results in a smaller bubble size or in less required superheat. A correlation between departure size and departure frequency may be obtained as follows. Peebles and Garber (1953) observed the velocity of bubble rise in a gravitational field as

$$V_b = 1.18 \left[\frac{\sigma g_c g(\rho_L - \rho_G)}{\rho_L^2} \right]^{1/4} \tag{2-63}$$

According to Jakob and Linke (1933),

$$V_b = D_b f_b \left(\frac{t_w + t_d}{t_d}\right)$$

where D_b is the bubble departure size. Hence,

$$D_b f_b = \frac{t_d}{t_d + t_w} (1.18) \left[\frac{\sigma g_c g(\rho_L - \rho_G)}{\rho_L^2}\right]^{1/4} \tag{2-64}$$

By assuming that $t_d = t_w$, Jakob (1949) obtained

$$D_b f_b = \frac{1.18}{2} \left[\frac{\sigma g_c g(\rho_L - \rho_G)}{\rho_L^2}\right]^{1/4}$$

This assumption was not confirmed by Hsu and Graham (1961) or by Westwater and Kirby (1963). The latter observed that $D_b f_b$ for carbon tetrachloride is approximately 1,200 ft/hr (366 m/hr), which is different from the value of 920 ft/hr (280 m/hr) for liquids that Jakob tested. Westwater and Kirby also found that the product was not constant at high heat fluxes. At higher heat fluxes (such as $q'' > 0.2 q_{cr}$), t_w is usually small compared with t_d. If the heat flux is sufficiently high to make $t_w \ll t_d$, the maximum rate of bubble generation is reached because the vertical distance between successive bubbles is essentially zero. We then have the simple relationship

$$V_b = D_b f_b \tag{2-65}$$

Based on his own experimental results with water as well as those of others with water and methanol, Ivey (1967) showed that Eq. (2-65) is approximately correct at higher heat fluxes and larger bubble sizes. These are the conditions under which both the departure diameter and the frequency are controlled by hydrodynamic factors. Ivey also showed that a single relationship between f_b and D_b does not hold over the entire range of D_b, as he found that experimental results of different fluids fall into three different regions, depending on both bubble diameter and heat flux, and that a different D_b–f_b relationship exists for each region. These regions are (1) a hydrodynamic region in which the major forces acting on the bubble are those of buoyancy and drag, (2) a thermodynamic region in which the frequency of bubble formation is governed largely by thermodynamic conditions during growth, and (3) a transition region between the above two, in which buoyancy, drag, and surface tension forces are of the same order of magnitude.

1. The hydrodynamic region has received considerable attention over the years. Equations (2-63) and (2-64) follow the buoyancy–drag force balance theory. If we

combine Eqs. (2-64) and (2-59) through the elimination of σ, and if we assume that $\rho_L \gg \rho_G$, we have

$$D_b^{1/2} f_b = \left(\frac{1.18}{2^{1/4} C_d^{1/2} \phi^{1/2}}\right)\left(\frac{t_d}{t_w + t_d}\right) g^{1/2} \tag{2-66}$$

At atmospheric pressure, C_d for ordinary liquids is usually taken at 0.0148 where ϕ is expressed in degrees, of which the dynamic value is usually taken as 45°, and in the hydrodynamic-controlled region $t_d/(t_d + t_w)$ is ~1. By substituting these values in Eq. (2-66), we obtain

$$D_b^{1/2} f_b = 1.2 g^{1/2}$$

where the coefficient is approximate and is for ordinary liquids only.

A similar equation was derived by Cole (1960) for ordinary liquids with $\rho_G \ll \rho_L$ and assuming $V_b = D_b f_b$:

$$D_b^{1/2} f_b = 1.15 g^{1/2}$$

Using two sets of data on water and one on methanol, which apparently fell in the hydrodynamic region, Ivey (1967) obtained the relation

$$D_b^{1/2} f_b = 0.9 g^{1/2}$$

with considerable scatter in the data, as is usually found in such experiments. Ivey recommended that his equation be applied to large bubbles ($D_b > 0.5$ cm, or 0.2 in.), where the heat flux ranges from medium to high ($q'' > 0.20 q_{cr}$), or to medium-size bubbles ($0.1 < D_b < 0.5$ cm, or $0.04 < D_b < 0.2$ in.) at high heat fluxes ($q'' > 0.8 q_{cr}$), where drag and buoyancy are the dominant forces. It may be concluded that the three simplified equations for ordinary fluids agree as well as can be expected (Dwyer, 1976).

2. In the thermodynamic region, Ivey found only one set of data on water and one on nitrogen, which indicated that

$$D_b^2 f_b = \text{constant} \tag{2-66a}$$

where the constants are quite different for these two liquids. He concluded that equations of this type apply generally in the case of small bubbles ($D_b < 0.05$ cm or 0.02 in.) and in the case of medium-size bubbles ($0.05 < D_b < 0.5$ cm, or $0.02 < D_b < 0.2$ in.) at very low heat fluxes.

3. On the basis of six sets of data on water, two on methanol, and one each on isopropanol and carbon tetrachloride, all falling in the transition region, Ivey obtained the correlation

$$(D_b)^{3/4} f_b = 0.44 g^{1/2} \tag{2-66b}$$

where the coefficient 0.44 has the units of cm$^{1/4}$. He found this equation to be applicable to situations where bubble diameters range from 0.05 cm (0.02 in.) at high heat fluxes to 1 cm (0.4 in.) at low fluxes.

Malenkov (1971) recommended a single equation for all regions:

$$f_b D_b = \frac{1}{\pi(1-\bar{\alpha})} \left[\frac{gD_b(\rho_L - \rho_G)}{2(\rho_L + \rho_G)} + \frac{2g_c \sigma}{D_b(\rho_L + \rho_G)} \right]^{1/2} \quad (2\text{-}67)$$

where $\bar{\alpha}$, the dimensionless vapor content of the boundary layer above the heating surface, is defined by the equation

$$\bar{\alpha} = \frac{G_G}{G_G + V_b}$$

and G_G is the volumetric flow rate of vapor per unit area of heating surface. The bracketed term on the right-hand side of Eq. (2-67) represents the bubble rise velocity, V_b. Equation (2-67) is based on the premise that the bubble detachment frequency is determined by the oscillation frequency of the liquid surrounding the chain of rising bubbles and that the relationship between this frequency and bubble size is given by

$$f = \frac{V_b}{\pi D_b (1-\bar{\alpha})}$$

Malenkov claimed that Eq. (2-67) effectively correlated five sets of water data, three sets of methanol data, and one each of ethanol, *n*-pentane, and carbon tetrachloride—all obtained at 1 atm—in 11 different investigations.

For liquid metal boiling, however, Eq. (2-67) showed poor agreement with the experimental results of Bobrovich et al. (1967). In Ivey's correlations, it is expected that the thermodynamic region will not normally be applicable to liquid metals, because their bubble growth is very rarely thermally controlled. Consequently, in this case, Eq. (2-64) for the hydrodynamic region is applicable and is combined with Eq. (2-62), leaving the ratio $[t_d/(t_w + t_d)]$ as a variable:

$$(D_b)^{1/2} f_b = (1.18/C_d'^{1/2}) \left(\frac{t_d}{t_w + t_d} \right) \frac{g^{1/2}}{(\text{Ja}^*)^{5/8}} \quad (2\text{-}68)$$

where the modified Jakob number is defined by

$$\text{Ja}^* = \frac{\rho_L C_{pL} T_{\text{sat}}}{\rho_G H_{fg}}$$

Equation (2-68) should theoretically be more applicable to liquid metals and for pressures well beyond atmospheric, if reliable means of estimating C'_d and $[t_d/(t_w + t_d)]$ become available.

As shown in Section 2.2.5.1, a value of C'_d of 1.5×10^{-4} is recommended for sodium and a value of 4.65×10^{-4} for potassium (because of their respective modified Jakob numbers). Suffice it to say that the relationship between bubble size and detachment frequency in nucleate boiling of liquid metals is not yet well established, even though it is fundamental to a good understanding of such boiling process.

2.2.5.3 Boiling sound. The audible sound differs in different types of boiling. It is not clear whether the boiling sound is caused by the formation and collapse of bubbles or by vibration of the heating surface. The sound emitted from methanol pool boiling on a copper tube at atmospheric pressure has been measured by Westwater and his co-workers (1955). The results for a frequency range of 25–7,500 cps are plotted in Figure 2.12. They reported that nucleate boiling is the most quiet, with transition boiling next, and film boiling the noisiest of the three types of boiling studied. The sound of subcooled flow boiling was reported by Goldmann (1953), who noted that the noise level increases as the heat flux increases toward the burnout condition. A "whistle" was detected in the heating of a supercritical fluid.

Boiling sound was also detected by Taylor and Steinhaus (1958). The sound was audible at a heat flux of 3.9×10^6 Btu/hr ft² (10.6×10^6 kcal/h m²) and a plate temperature of 350°F (177°C), but the audible sound disappeared at a heat flux of

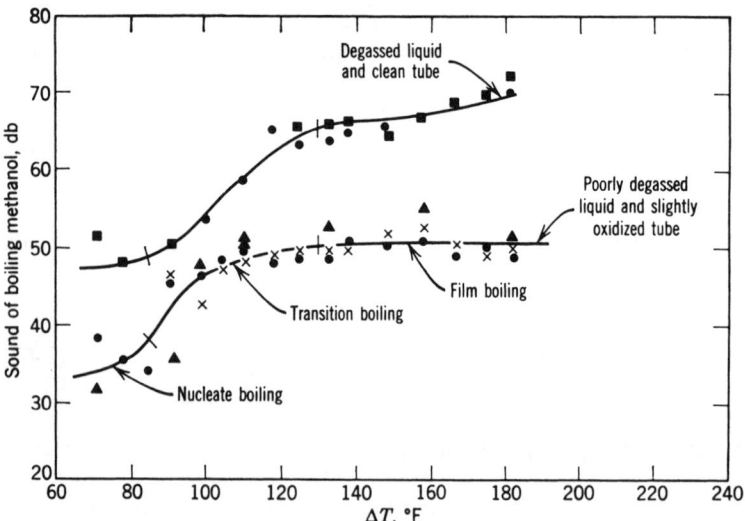

Figure 2.12 The sound of boiling methanol on a case ³⁄₈-in., horizontal, steam-heated copper tube at 1 atm. (From Westwater et al., 1955. Copyright © 1955 by American Association for the Advancement of Science, Washington, DC. Reprinted with permission.)

7.6 × 10⁶ Btu/hr ft² (2.1 × 10⁷ kcal/h m²) and a plate temperature of 500°F (315°C) when a fraction of the surface was covered by a vapor film. The sound frequency was about 10 kc for 3.4 × 10⁶ Btu/hr ft² (9.2 × 10⁶ kcal/h m²) and 4 kc for 5.4 × 10⁶ Btu/hr ft² (1.5 × 10⁷ kcal/h m²). These results are contradictory to the trend of bubble frequency. The fundamental mode of vibration for a copper plate, $\tfrac{3}{16}$ in. (0.48 cm) thick and $6\tfrac{3}{16}$ in. (15.7 cm) in diameter, clamped around the edge, is about 12 kc, and it is expected to decrease when the heat flux or plate temperature increases.

2.2.5.4 Latent heat transport and microconvection by departing bubbles. The amount of heat transferred by latent heat transport of bubbles can be calculated from the bubble density, D_b, and departure frequency data. Its order of magnitude can be demonstrated by one typical set of the data obtained by Gunther and Kreith (1950):

Observed heat flux	1.04 × 10⁶ Btu/hr ft²	(2.8 × 10⁶ kcal/h m²)
Liquid temperature	98°F	(37°C)
Wall temperature	270°F	(132°C)
Avg. departure, D_b	0.030 in.	(0.076 cm)
Departure frequency	1,000 cps	
Number of bubbles per unit surface area	280 per in.²	(43 per cm²)

Thus, the amount of latent heat per bubble can be obtained,

$$h_{fg}\rho_G V_b = (970\ \text{Btu/lb})(2 \times 10^{-5}\ \text{lb/in.}^3)(7 \times 10^{-6}\ \text{in.}^3)$$
$$= 1.4 \times 10^{-7}\ \text{Btu/bubble} \quad \text{or} \quad 3.53 \times 10^{-5}\ \text{cal/bubble}$$

and the heat transfer rate due to latent heat transport is

$$q''_{\text{latent}} = 1.4 \times 10^{-7} \times 280 \times 1{,}000 \times 144 \times 3{,}600$$
$$= 20{,}000\ \text{Btu/hr ft}^2 \quad \text{or} \quad 54{,}200\ \text{kcal/hr m}^2$$

The latent heat transport accounts for only 2% of the total heat flux in this case. However, it was observed by several investigators that the total heat transfer rate is proportional to this value, q''_{latent}, because it is proportional to the bubble volume and the number of bubbles that cause intense agitation of the liquid layer close to the surface. This agitation, termed *microconvection,* together with the liquid–vapor exchange, were considered to be the key to excellent characteristics of boiling heat transfer (Forster and Greif, 1959).

2.2.5.5 Evaporation-of-microlayer theory. A later hypothesis for the mechanism of nucleate boiling considers the vaporization of a microlayer of water underneath the bubble. This was first suggested by Moore and Mesler (1961), who measured the surface temperature during nucleate boiling of water at atmospheric pressure and found that the wall temperature occasionally drops 20–30°F (11–17°C) in about 2 msec. Their calculation indicated that this rapid removal of heat was possibly caused by vaporization of a thin water layer under a bubble. This hypothesis was further verified by Sharp (1964), whose experiments demonstrated the existence of an evaporating liquid film at the base of bubbles during nucleate boiling by two optical techniques; by Cooper and Lloyd (1966, 1969) with boiling toluene; and by Jawurek (1969), who measured the radial variation in the microlayer thickness directly during pool boiling experiments with ethanol and methanol. Table 2.2 shows a summary of microlayer thickness measurements from different boiling fluids. Although agreement among different tests was hardly evident, the range of the magnitude is nevertheless revealing. The extent of contribution of microlayer evaporation to the enhanced heat transfer in nucleate boiling is still not well defined, ranging from less than 20% to nearly 100% of the energy required for bubble growth (Voutsinos and Judd, 1975; Judd and Hwang, 1976; Fath and Judd, 1978; Koffman and Plasset, 1983; Lee and Nydahl, 1989.)

An analytical model for the evaporation of the liquid microlayer was proposed by Dzakowic (1967) which used a nonlinear heat flux boundary condition derived from kinetic theory considerations and the transient one-dimensional heat conduction equation (Fig. 2.13). In this study, a combination of four fluids (water, nitrogen, n-pentane, and ammonia) and two heater materials (type 302 stainless steel and copper) were used. Dzakowic observed that liquid microlayer evaporation cooled the stainless steel surface (low thermal conductivity) more rapidly than the copper surface (high thermal property values); and for a given heater material, the rate of surface cooling for the different fluids increased in the order n-pentane, water, ammonia, nitrogen. This order of different evaporating rates coincides with the ordering of the superheat temperatures required for typical nucleate boiling.

Table 2.2 Summary of microlayer measurements

Investigations	Boiling Fluid	Heating surface	Heat flux 10^3/Btu/h ft^2	Pressure (in. Hg)	r (mm)	δ_o (mm)
Sharp (1964)	Water	Glass (tiny scratched)	10–15	2–3	~0	0.0004
					1.7	0.0004
Cooper and Lloyd (1966, 1969)	Toluene	glass	7.2–15	52–103	0.4	0.005
					5.0	0.030
Katto and Yokoya (1966)	Water	Copper (with interference pl.)		30	5.0	0.011
Jawurek (1969)	Methanol	Glass (stannic oxide film)	19.7	180	~0	0.0005

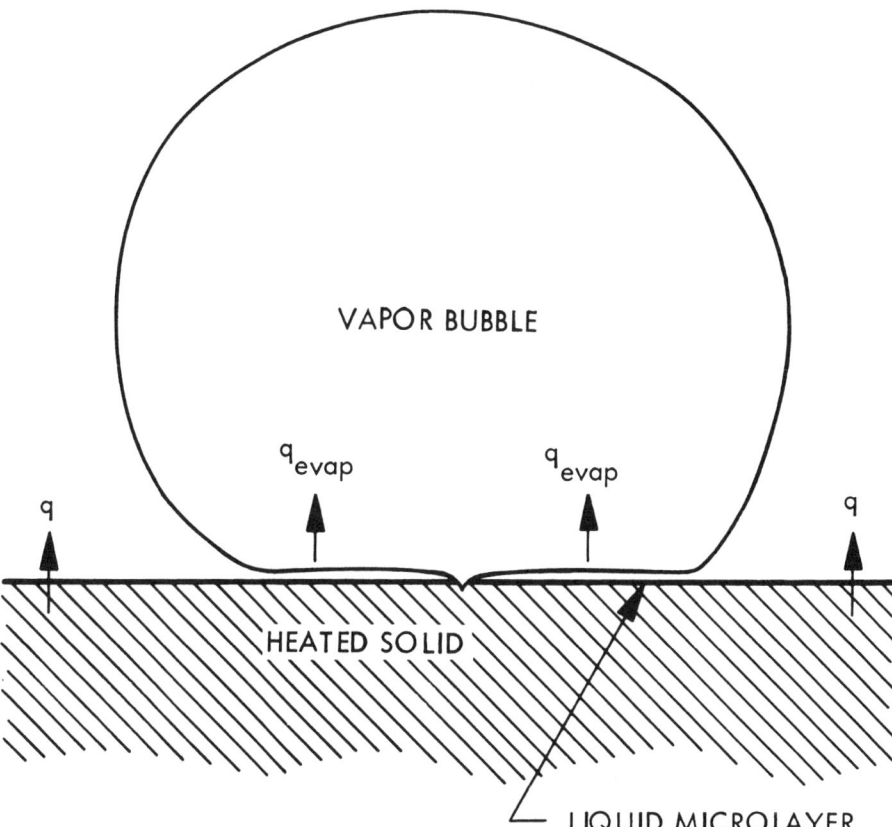

Figure 2.13 A newly formed vapor bubble illustrating an evaporating liquid microlayer.

This follows the view that larger surface temperature drops must be associated with high rates of microlayer evaporation or high local heat flux. The evaporation-of-microlayer theory appeared to explain why the nucleate boiling heat transfer coefficients are much higher than, for example, forced-convection liquid-phase heat transfer coefficients. According to this theory, it is due not so much to the increase in microconvection adjacent to the heating surface as to the high-flux microlayer evaporation phase, and to the high-flux, cold liquid heating phase of the ebullition cycle. However, this theory is also subject to objections (see Sec. 2.4.1.2).

For liquid metals the superiority of nucleate boiling heat transfer coefficients over those for forced-convection liquid-phase heat transfer is not as great as for ordinary liquids, primarily because the liquid-phase coefficients for liquid metals are already high, and the bubble growth period for liquid metals is a relatively short fraction of the total ebullition cycle compared with that for ordinary fluids. In the case of liquid metals, the initial shape of the bubbles is hemispheric, and it becomes spherical before leaving the heating surface. This is because of very rapid inertial force-controlled growth deriving from high wall temperatures, high liquid temperatures, low boiling pressures, high heat fluxes, and long waiting periods. Figure 2.14 is a schematic drawing showing three stages in the bubble growth period as postulated for a liquid metal (Dwyer, 1976), with Figure 2.14a representing stages 1 and 2, and Figure 2.14b representing stages 2 and 3. Dimensions r_{e1}, r_{e2}, and r_{e3} are radii of the dry patch (due to evaporation) at the respective stages. A number of changes are shown from stage 1 to stage 2 (Fig. 2.14a), such as the radius of the dry patch and the thickness of the microlayer between r_{e2} and R_1 and between R_1 and R_2. These changes are brought about by a decrease in the rate of bubble growth concomitant with a decrease in relative importance of the dominant inertia forces compared with that of surface tension, which tend to make the bubble shape more spherical. For radii greater than r_x, the rate of liquid evaporation from the surface of the microlayer is less than the rate of liquid inflow from the periphery of the bubble. Between the times of stage 2 and stage 3 (Fig. 2.14b), the bubble is approaching the limit of its growth, inertial forces have become negligible, and the surface tension force is being overcome by the force of buoyance. This results in bubble growth rate that is controlled by the heat transfer through the liquid–vapor interface, with the bubble becoming spherical with a smaller dry patch and thicker microlayer at all points. From the experimental data presented in Table 2.2, even the largest value of δ_o is less than 1% of the bubble radius. Dwyer and Hsu (1975) predicted a maximum initial microlayer thickness of 0.021 mm (or 0.0008 in.) at $r = 20.9$ mm (0.82 in.), or only 0.05% of the bubble diameter, which indicates that in the boiling of liquid metals the microlayer should have a very low heat capacity compared with that of the typical plate heater.

Dwyer and Hsu (1976) further analyzed theoretically the size of a dry patch during nucleate boiling of a liquid metal; they concluded that in hemispherical bubble growth on a smooth metallic surface, the dry patch is expected to be negligibly small compared to the base area of the bubble. For example, Dwyer and Hsu (1976) showed by calculation that with sodium boiling on a type 316 stainless steel surface under a pressure of 100 mm Hg (1.9 psia) and at a cavity mouth radius corresponding to $r_c = 5 \times 10^{-4}$ in. (0.013 mm), the volume of liquid evaporated from the microlayer by the time the bubble grows to a diameter of 2 cm (0.8 in.) is extremely small (only 2.5% of the original microlayer formation). Note that the base diameter at the end of bubble growth in this case is estimated to be 4.2 cm (1.6 in.). Also from that calculation, the average value of local heat flux from microlayer to the bubble was shown to be roughly twice that of heat flux from the curved interface (hemisphere) to the bubble. The bubble growth rate (similar to the heat transfer-controlled mode) can be obtained from an energy balance,

$$Q_{bb} = Q_{micro} + Q_{curved}$$

where Q_{bb} = rate of energy gain by the bubble
Q_{micro} = heat rate from the microlayer to the bubble
Q_{curved} = heat rate from the curved interface to the bubble

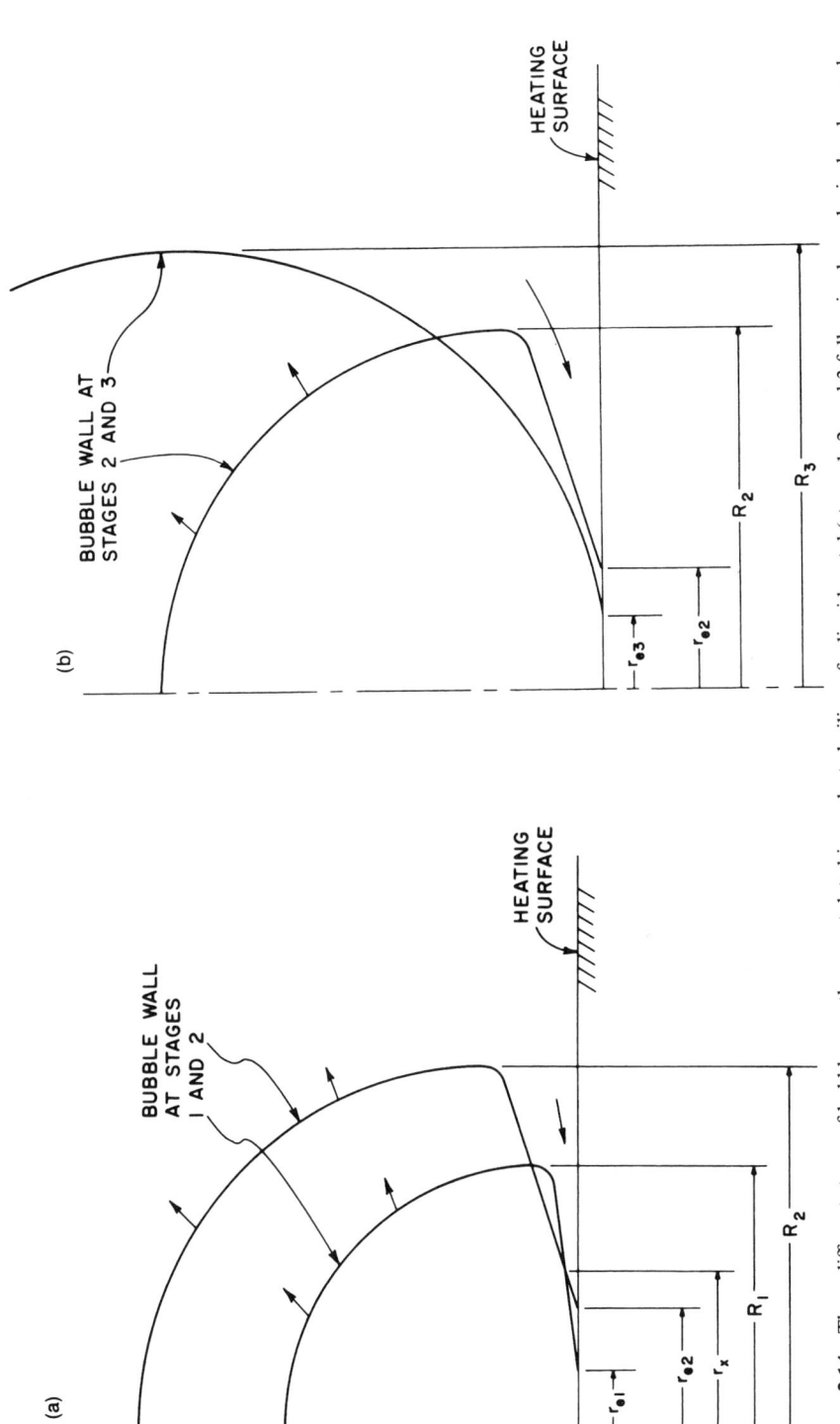

Figure 2.14 Three different stages of bubble growth as postulated in nucleate boiling of a liquid metal (stages 1, 2, and 3 follow in chronological order, and stage 2 in (*a*) and (*b*) are the same). (From Dwyer, 1976. Copyright © 1976 by American Nuclear Society, LaGrange Park, IL. Reprinted with permission.)

2.3 HYDRODYNAMICS OF POOL BOILING PROCESS

As mentioned in the previous section, the heat transfer rate in nucleate pool boiling is usually very high. This high flux could be attributed to vapor–liquid exchange during the growth and movement of bubbles (Forster and Grief, 1959), and to the vaporization of (Moore and Mesler, 1961) or transient conduction to the microlayer and liquid–vapor interface (Han and Griffith, 1965b) before the bubble's departure. The fact that the boiling heat flux is insensitive to the degree of liquid subcooling can be explained by the process of a quantity of hot liquid being pushed from the heating surface into the main stream and, at the same time, a stream of cold liquid being pulled down to the heating surface (Fig. 2.15). Since increased subcooling reduces bubble size and thus reduces the intensity of agitation, this effect nearly cancels the subcooling effect on turbulent liquid convection. This vapor–liquid exchange effect is expected to become smaller at higher pressures because of the smaller bubble size. The effect is also smaller because the value of the Jakob number, $c_p \rho_L (T_w - T_b)/H_{fg} \rho_G$, decreases with increasing pressures up to about 2,000 psia (14 MPa) for water, the Jakob number being the ratio of the amount of sensible heat absorbed by a unit volume of cooling liquid to the latent heat transported by the same volume of bubbles.

Because the bubble population increases with heat flux, a point of peak flux may be reached in nucleate boiling where the outgoing bubbles jam the path of the incoming liquid. This phenomenon can be analyzed by the criterion of a Hemholtz instability (Zuber, 1958) and thus serves to predict the incipience of the boiling crisis (to be discussed in Sec. 2.4.4). Another hydrodynamic aspect of the boiling crisis, the incipience of stable film boiling, may be analyzed from the criterion for a Taylor instability (Zuber, 1961).

2.3.1 The Helmholtz Instability

When two immiscible fluids flow relative to each other along an interface of separation, there is a maximum relative velocity above which a small disturbance of the interface will amplify and grow and thereby distort the flow. This phenomenon is

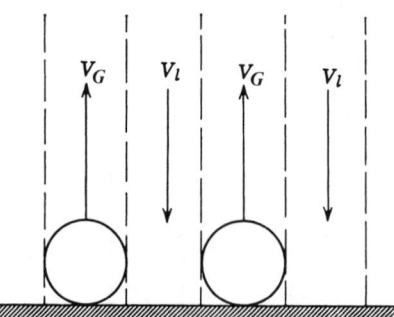

Figure 2.15 Vapor–liquid counterflow in pool boiling.

known as the *Helmholtz instability*. According to Lamb (1945) and Zuber (1958), the velocity of propagation, c, of a surface wave along a vertical vapor jet of upward velocity V_G, with an adjacent downward liquid jet of velocity V_L (of opposite sign), may be expressed as

$$c^2 = \left(\frac{n}{m}\right)^2 = \frac{mg_c\sigma}{\rho_L + \rho_G} - \frac{\rho_L\rho_G}{(\rho_L + \rho_G)^2}(V_G - V_L)^2 \qquad (2\text{-}69)$$

where the wave number m is $(2\pi/\lambda)$, the wave angular velocity n is

$$n = \left(\frac{2\pi}{\lambda}\right)c = mc$$

and, for a harmonic wave form, the amplitude is

$$\eta = \eta_o e^{-int}\cos(mx)$$

Equation (2-69) is developed by assuming that the fluids are of infinite depth and the force of gravity is not included. The condition for a stable jet is that the wave angular velocity n is real, i.e.,

$$\left(\frac{n}{m}\right)^2 > 0$$

Equation (2-69), subject to this condition, gives

$$\left(\frac{mg_c\sigma}{\rho_L + \rho_G}\right) > \frac{\rho_L\rho_G}{(\rho_L + \rho_G)^2}(V_G - V_L)^2 \qquad (2\text{-}70)$$

In the steady state, the incoming liquid flow equals the vapor flow; hence, by continuity,

$$-V_L = \left(\frac{\rho_G}{\rho_L}\right)V_G \qquad (2\text{-}71)$$

Substituting Eq. (2-71) into Eq. (2-70) and rearranging gives

$$\frac{\rho_L\sigma mg_c}{\rho_G(\rho_L + \rho_G)} > V_G^2$$

Thus, the maximum vapor velocity for a stable vapor stream from the surface is

$$V_G = \left[\frac{\rho_L \sigma m g_c}{\rho_G (\rho_L + \rho_G)}\right]^{1/2} \tag{2-72}$$

If the vapor stream velocity exceeds this value, vapor cannot easily get away and thus a partial vapor blanketing (film boiling) may occur. This result is used to predict the maximum heat flux by relating the heat flux to the vapor velocity (see Sec. 2.4.4).

2.3.2 The Taylor Instability

The criterion for stable film boiling on a horizontal surface facing upward can be developed from the *Taylor instability*. This says that the stability of an interface of (capillary) wave form between two fluids of different densities depends on the balance of the surface tension energy and the sum of the kinetic and potential energy of the wave. Whenever the former is greater than the latter, a lighter fluid can remain underneath the heavier fluid. This is the condition of stable film boiling from a horizontal surface, as shown in Figure 2.16. The total energy per wavelength of a progressing wave due to kinetic and potential energy is (Lamb, 1945)

$$\left(\frac{E}{\lambda}\right)_{tot} = \frac{g(\rho_L - \rho_G)\eta_o^2}{2g_c} \tag{2-73}$$

where η_o is again the maximum amplitude of the wave. The energy of the wave due to surface tension is

$$\left(\frac{E}{\lambda}\right)_\sigma = \left(\frac{1}{\lambda}\right) \int_0^\lambda \Delta P \, \eta \, dx \tag{2-74}$$

where the pressure differential for a curved surface of the wave form

$$\eta = \eta_o \sin mx = \eta_o \sin\left(\frac{2\pi x}{\lambda}\right)$$

can be evaluated as follows from a force balance on a differential element, ds, of the surface (Fig. 2.17):

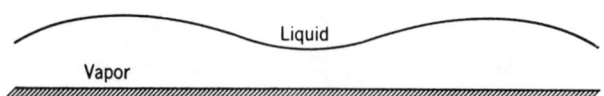

Figure 2.16 Stable film boiling from a horizontal surface.

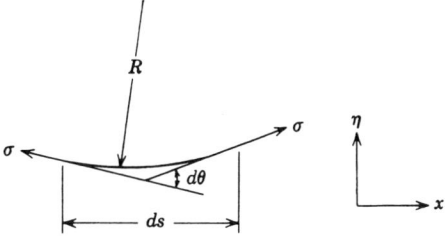

Figure 2.17 Force balance on a differential element of a curved surface.

$$\Delta p \, ds = \sigma \sin(d\theta) \cong \sigma \, d\theta \qquad (2\text{-}74a)$$

or

$$\Delta p = \sigma\left(\frac{d\theta}{ds}\right)$$

From the geometry of the curve interface,

$$\frac{d\theta}{ds} = \frac{1}{R} \cong \frac{d^2\eta}{dx^2}$$

Then

$$\Delta p = \sigma\left(\frac{d^2\eta}{dx^2}\right) = \left(\frac{2\pi}{\lambda}\right)^2 \sigma \eta_c \sin mx \qquad (2\text{-}74b)$$

Substituting this value of ΔP into Eq. (2-74) results in

$$\left(\frac{E}{\lambda}\right)_\sigma = \left(\frac{\sigma \eta_o^2}{\lambda}\right) \int_0^{2\pi} \frac{2\pi}{\lambda} \sin^2(mx) \, d(mx)$$

$$= \frac{\sigma \eta_o^2}{\lambda}\left(\frac{2\pi}{\lambda}\right)\pi \qquad (2\text{-}75)$$

To satisfy the condition of a stable wave, $(E/\lambda)_{tot} < (E/\lambda)_\sigma$, the wavelength must be smaller than a certain critical value, λ_c:

$$\lambda_o < \lambda_c = 2\pi\left[\frac{g_c \sigma}{g(\rho_L - \rho_G)}\right]^{1/2} \qquad (2\text{-}76)$$

In a physical system, the disturbance wavelength can be interpreted as the distance between nucleation sites or as the departure bubble size. This criterion was used by Zuber (1959) to predict the incipience of the critical heat flux assuming that the vapor rises from the heating surface in cylindrical columns, whose average diameter is $\lambda_o/2$ and whose centers are spaced λ_o units apart (Fig. 2.18), where $\lambda_c \leq \lambda_o \leq \lambda_d$. On the basis of experimental observations, Lienhard and Dhir (1973b) suggested that for flat horizontal heaters whose dimensions are large compared with the value of λ_d, a good estimate of the length of the Helmholtz critical wavelength is λ_d, the "most dangerous" wavelength, which is defined by Bellman and Pennington (1954) as

$$\lambda_d = 2\pi \left[\frac{3 g_c \sigma}{g(\rho_L - \rho_G)} \right]^{1/2} \tag{2-77}$$

2.4 POOL BOILING HEAT TRANSFER

Boiling at a heated surface, as has been shown, is a very complicated process, and it is consequently not possible to write and solve the usual differential equations of motion and energy with their appropriate boundary conditions. No adequate description of the fluid dynamics and thermal processes that occur during such a process is available, and more than two mechanisms are responsible for the high

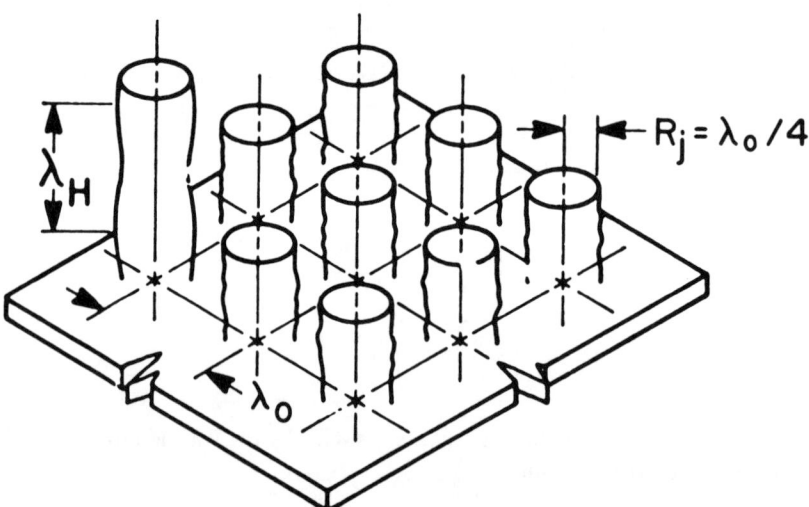

Figure 2.18 Vapor jet configuration for boiling on a horizontal flat-plate heater, as postulated by Zuber (1959). Adapted from Leinhard and Dhir, 1973. Reprinted with permission of U.S. Department of Energy.)

heat flux in nucleate pool boiling (Sec. 2.3). Over the years, therefore, theoretical analyses have been for the most part empirical, and have leaned on the group parameter approach. Since the high liquid turbulence in the vicinity of the heating surface is considered to be dominant, at least in a portion of the ebullition cycle, it is natural to correlate the boiling heat transfer rates in a similar fashion as in single-phase turbulent-flow heat transfer phenomena by an equation of the type

$$\mathrm{Nu} = f(\mathrm{Re}, \mathrm{Pr})$$

Thus many theoretical correlations start with the form

$$\mathrm{Nu}_b = a(\mathrm{Re}_b)^m (\mathrm{Pr}_L)^n \tag{2-78}$$

where a = a constant coefficient
m, n = constant exponents
Nu_b = boiling Nusselt number
Pr_L = liquid Prandtl number

To give a qualitative description of various boiling mechanisms and facilitate the empirical correlation of data, it is necessary to employ dimensional analysis.

2.4.1 Dimensional Analysis

2.4.1.1 Commonly used nondimensional groups. The commonly used nondimensional groups in boiling heat transfer and two-phase flow are summarized as follows. Some are used more frequently than others, but all represent the boiling mechanisms in some fashion.

The *boiling number* (Bo) is the ratio of vapor velocity away from the heating surface to flow velocity parallel to the surface, V. The vapor velocity is evaluated on the basis of heat transfer by latent heat transport.

$$\mathrm{Bo} = \frac{q''}{H_{fg} \rho_G V} \tag{2-79}$$

The *buoyancy modulus* (Bu) is defined as the ratio of the density difference to the liquid density:

$$\mathrm{Bu} = \frac{\rho_L - \rho_G}{\rho_L} \tag{2-80}$$

The *Euler number* (Eu) is defined as the ratio of the pressure force to the inertial force, as in the form

$$\text{Eu} = \frac{g_c \Delta p}{\rho V^2} \qquad (2\text{-}81)$$

where ρ can be the density of either the mixture or a single-phase component, and Δp can be the frictional pressure drop of flow or the pressure difference across the boundary of a bubble.

The *Froude number* (Fr) is the ratio of the inertial force to the gravitational force of the liquid:

$$\text{Fr} = \frac{V^2}{gD_b} \qquad (2\text{-}82)$$

The *Jakob number* (Ja) is the ratio of the sensible heat carried by a liquid to the latent heat of a bubble with the same volume,

$$\text{Ja} = \frac{c_p \rho_L (T_w - T_b)}{H_{fg} \rho_G} \qquad (2\text{-}83)$$

which indicates the relative effectiveness of liquid–vapor exchange.

The *Kutateladze number* (B) is the coefficient in the correlation for the pool boiling crisis, Eq. (2-128), or

$$B = \frac{q''_{\text{crit}}}{H_{fg}\, \rho_G^{1/2} [g_c g \sigma (\rho_L - \rho_G)]^{1/4}} \qquad (2\text{-}84)$$

The *boiling Nusselt number* (Nu_b), or Nusselt number for bubbles, is defined as the ratio of the boiling heat transfer rate to the conduction heat transfer rate through the liquid film,

$$\text{Nu}_b = \frac{\delta q''}{k_L (T_w - T_b)} \qquad (2\text{-}85)$$

where δ = the thickness of liquid film; it can be of the same order of magnitude as a bubble diameter, or it may be chosen as some other dimension, depending on the physical model used.

The *Prandtl number* of a liquid (Pr_L) is defined as the ratio of the kinematic viscosity to the thermal diffusivity of the liquid:

$$\text{Pr}_L = \frac{c_p \mu}{k_L} \qquad (2\text{-}86)$$

The *boiling Reynolds number* or *bubble Reynolds number* (Re$_b$) is defined as the ratio of the bubble inertial force to the liquid viscous force, which indicates the intensity of liquid agitation induced by the bubble motion:

$$\text{Re}_b = \frac{\rho_G V_b D_b}{\mu_L} \quad (2\text{-}87)$$

Substituting the bubble departure diameter from Eq. (2-59) and identifying V_b with the liquid velocity in Eq. (2-79), it becomes

$$\text{Bo Re}_b = \frac{q''\beta}{H_{fg}\mu_L}\left[\frac{g_c\sigma}{g(\rho_L-\rho_G)}\right]^{1/2} \quad (2\text{-}87a)$$

The *spheroidal modulus* (So) is defined as the ratio of conduction heat flux through the vapor film to the evaporation heat flux:

$$\text{So} = \frac{k_G(T_w - T_{\text{sat}})/\delta}{H_{fg}\rho_G V_G} \quad (2\text{-}88)$$

Combining (So) with the Reynolds number based on the film thickness, δ,

$$(\text{So})(\text{Re}_\delta) = \frac{k_G(T_w-T_{\text{sat}})}{\delta H_{fg}\rho_G V_G}\left(\frac{\rho_G V_G \delta}{\mu_G}\right)$$

$$= \frac{k_G(T_w-T_{\text{sat}})}{H_{fg}\mu_G} \quad (2\text{-}89)$$

This nondimensional group describes the spheroidal state of film boiling.

The *superheat ratio* (Sr) is defined as the ratio of liquid superheat at the heating surface to the heat of evaporation:

$$\text{Sr} = \frac{c_L(T_w-T_{\text{sat}})}{H_{fg}} \quad (2\text{-}90)$$

It is the product of the bubble Reynolds and the liquid Prandtl number divided by the boiling Nusselt number (Nu$_b$), which is equivalent to the Stanton number in single-phase convective heat transfer.

The *Weber-Reynolds number* (Re/We) is defined as the ratio of surface tension of a bubble to viscous shear on the bubble surface due to bubble motion:

$$\frac{\text{Re}}{\text{We}} = \frac{g_c \sigma / D_b}{\mu_L V_b / D_b} = \frac{g_c \sigma}{\mu_L V_b} \tag{2-91}$$

2.4.1.2 Boiling models. As the result obtained by applying dimensional analysis is limited by the validity and completeness of the assumptions made prior to the analysis, experiments are the only safe basis for determining the correctness and adequacy of the assumptions. Several suggested models are reviewed briefly here.

Bubble agitation mechanism **(Fig. 2.19a)** An appreciable degree of fluid mixing occurs near the heater surface during boiling, as evidenced by Schlieren and shadowgraph high-speed motion pictures (Hsu and Graham, 1961, 1976). While this mechanism can be an important contributor to the effective nucleate boiling coefficient, it does not appear to be a singular cause of the large heat transfer coefficient noted in nucleate boiling (Kast, 1964; Graham et al., 1965).

Vapor–liquid exchange mechanism **(Fig. 2.19b)** This model (Forster and Greif, 1959) is in some respects similar to the bubble agitation model, but it avoids certain objections of the latter. This mechanism visualizes a means of pumping a slug of hot liquid away from the wall and replacing it with a cooler slug, with a growing and departing bubble acting as the piston to do the pumping. The fact that the Jakob number (Sec. 2.4.1.1) attains values as high as 100 justifies the dominance of the liquid-exchange mechanism, and Forster and Greif claim that this vapor–liquid exchange mechanism explains the boiling mechanism for both saturated and subcooled liquids (Sec. 2.3). There is some controversy, however, about their assumption on the volume of the hot liquid slug participating in the exchange (Bankoff and Mason, 1962), which may leave the potential heat flux contribution of this mechanism as calculated by Forster and Greif (1959) questionable.

Microlayer evaporative mechanism **(Fig. 2.19c)** Section 2.2.5.5 gives the model for this mechanism. A theoretical heat transfer rate is possible through evaporation of a fluid into a receiver:

$$q = (aH_{fg}) \left(\frac{2\pi R_g T_{\text{sat}}}{g_c M} \right)^{-1/2} (p_L - p_G) \tag{2-92}$$

where a = coefficient of evaporation (Wyllie, 1965)
R_g = gas constant
M = molecular weight

In a highly idealized case representing the case of steady-state evaporation into a vacuum ($p_G = 0$), Hsu and Graham (1976) use Eq. (2-92) to compute the rates of

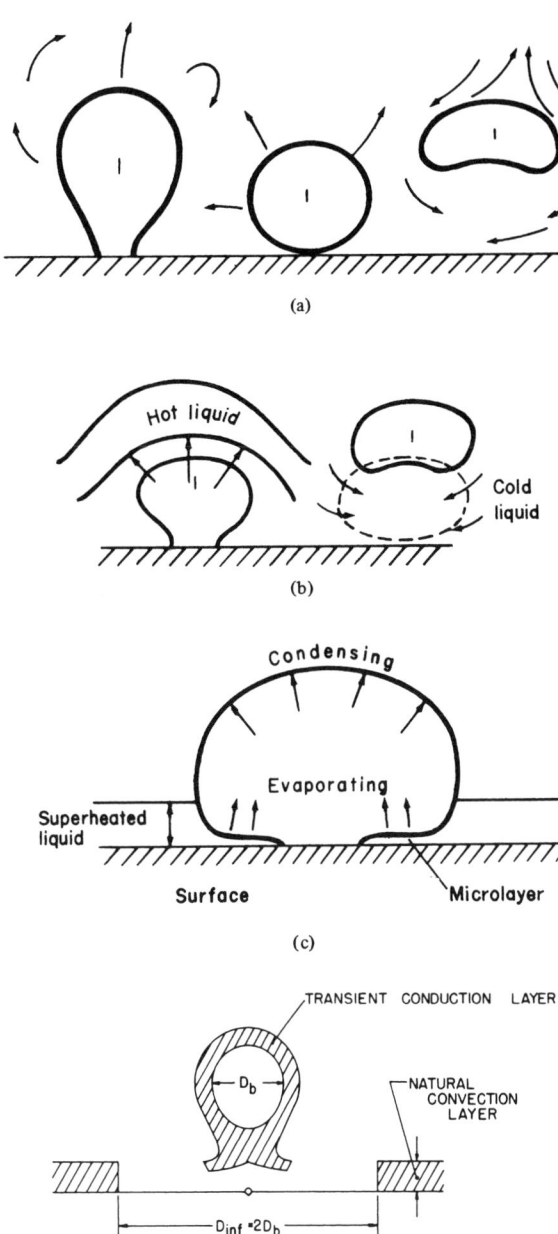

Figure 2.19 Schematic diagrams of boiling models: (*a*) bubble agitation model; (*b*) vapor–liquid exchange mechanism; (*c*) microlayer evaporation mechanism. (From Hsu and Granham, 1976. Copyright © 1976 by Hemisphere Publishing Corp., New York. Reprinted with permission.) (*d*) Transient conduction to, and subsequent replacement of, superheated liquid layer. (From Mikic and Rohsenow, 1969. Copyright © 1969 by American Society of Mechanical Engineers, New York. Reprinted with permission.)

heat transfer for mercury, water, and other materials at a saturation pressure of 1 atm. Values for water and mercury are as follows:

Water: $T_{sat} = 672°R$, $H_{fg} = 970$ Btu/lb, $M = 18$, $a = 0.04$, $q'' = 2.78 \times 10^6$ Btu/hr ft², or 8.8×10^6 W/m²

Mercury: $T_{sat} = 1135°R$, $H_{fg} = 126$ Btu/lb, $M = 200.6$, $a = 1.0$, $q'' = 44.4 \times 10^6$ Btu/hr ft², or 140×10^6 W/m²

Although the above idealized case cannot represent nucleate boiling, the highly effective heat transfer mechanism due to evaporation over even a portion of the heat transfer surface at any one time is evident.

Transient conduction to, and subsequent replacement of, superheated liquid layer (Fig. 2.19d) Mikic and Rohsenow (1969) considered this model, which was first suggested by Han and Griffith (1965a), to be the single most important contributor to the heat transfer in nucleate boiling. Based on an implicit reasoning, they disregarded the evaporation of the microlayer as a dominant factor in most practical cases. With the basic mechanism for a single active cavity site, a departing bubble from the heated surface will remove with it (by action of a vortex ring created in its wake) a part of the superheated layer. The area from which the superheated layer is removed, known as the area of influence, can be approximately related to the bubble diameter at departure as (Han and Griffith, 1965a; Fig. 2.19d)

$$D_{inf} = 2D_b$$

Following the departure of the bubble and the superheated layer from the area of influence, the liquid at T_{sat} from the main body of the fluid comes in contact with the heating surface at T_w. This bears some similarity to the vapor–liquid exchange mechanism, except for the quantity of liquid involved.

2.4.2 Correlation of Nucleate Boiling Data

2.4.2.1 Nucleate pool boiling of ordinary liquids.

Rohsenow's early correlation Using the form of correlation shown in Eq. (2-78), and assuming that the convection mechanism associated with bubble liftoff is of prime importance, Rohsenow (1952) found empirically, with data obtained from a 0.024-in. (0.6-mm) platinum wire in degassed distilled water, that

$$\frac{Re_b Pr_L}{Nu_b} = C(Re_b)^{m'}(Pr_L)^{n'}$$

which is an expression for the superheat ratio, Sr [Eq. (2-90)]. Thus

$$\frac{c_L(T_w - T_{sat})}{H_{fg}} = 0.013 \left[\frac{q''}{\mu_L H_{fg}} \sqrt{\frac{g_c \sigma}{g(\rho_L - \rho_G)}} \right]^{0.33} \left(\frac{c_L \mu_L}{k_L} \right)^{1.7} \quad (2\text{-}93)$$

This equation was shown to correlate well not only with boiling data for water in a pressure range of 14.7 to 2,465 psia (0.1 to 16.8 Mpa), but also with other data of different surface–fluid combinations: water–nickel and stainless steel, carbon tetrachloride–copper, isopropyl alcohol–copper, and potassium carbonate–copper when the constant C takes values ranging from 0.0027 to 0.015. The exponent on the Prandtl number of Eq. (2-93) being greater than 1 yields a negative sign on the Prandtl number exponent in the conventional expression of the Nusselt number, Eq. (2.78). This has appeared "illogical" to many scientists (Westwater, 1956). On the basis of additional experimental data, Rohsenow later recommended a value of 1 for water for this exponent in Eq. (2-93) (Dwyer, 1976).

Vapor–liquid exchange correlation Forster and Greif (1959), based on the vapor–liquid exchange mechanism, also proposed a correlation in the form of Eq. (2.78), employing the following definitions of Re_b and Nu_b:

$$Re_b = \frac{\rho_L}{\mu_L} \left[\frac{(T_w - T_{sat}) c_L \rho_L (\pi \alpha_L)^{1/2}}{H_{fg} \rho_G} \right]^2 \quad (2\text{-}94)$$

and

$$Nu_b = \frac{q[2\sigma/(\rho_G - \rho_L)]}{(T_w - T_{sat}) k_L} \quad (2\text{-}95)$$

Thus their final equation became

$$\frac{2q\sigma/(\rho_G - \rho_L)}{(T_w - T_{sat}) k_L} = C_1 \left\{ \frac{\rho_L}{\mu_L} \left[\frac{(\rho_G - \rho_L) T_{sat} c_L \rho_L (\pi \alpha_L)^{1/2}}{J(H_{fg} \rho_G)^2} \right]^2 \right\}^{1/5} (Pr_L)^{1/3} \quad (2\text{-}96)$$

where J is the work–heat conversion factor. As mentioned previously, the implication of a thermal-layer thickness of the order of the maximum bubble radius has been criticized by others.

Estimation of microlayer evaporation The model, incorporating the evaporation from a microlayer surface underneath a bubble attached to the heater surface, was used by Hendricks and Sharp (1964). With water as the fluid, at somewhat subcooled conditions, the heat transfer rates were as high as 500,000 Btu/hr ft², or

1,580,000 W/m². This maximum value approaches the magnitude of the idealized heat flux predicted from the kinetic theory of evaporation, Eq. (2-92). Bankoff and Mason (1962) reported values of the heat transfer coefficient at a surface where rapidly growing and collapsing bubbles were present. The heat transfer coefficients ranged in magnitude from 13,000 to 300,000 Btu/hr ft² °F, or 74,000 to 1,700,000 W/m² °C. Despite the experimental evidence of the existence of an evaporative microlayer, Mikic and Rohsenow (1969) questioned that if the evaporative model is indeed the governing mechanism in nucleate boiling, why the theories for bubble growth on a heated surface (Han and Griffith, 1965a; Mikic et al., 1970; Sec. 2.2.4), which neglected the microlayer evaporation and which were essentially based on the extension of the model used in the derivation of a bubble growth in a uniformly superheated liquid, gave very good agreement with experimental results.

Mikic-Rohsenow's correlation Mikic and Rohsenow (1969) therefore proposed a new correlation, based on the transient conduction to the superheated layer mechanism described in Section 2.4.1.2:

$$q''_{boil} = 2(\pi f)^{1/2}(k_L \rho_L c_L)^{1/2}(D_b)^2 n(T_w - T_{sat}) \tag{2-97}$$

where q''_{boil} is the average heat flux over the whole heating surface, A_T, due to boiling; f is the frequency of bubble departures from the particularly active site considered; and D_b is the bubble diameter at departure. The above equation assumes that the area of influence is proportional to D_b^2 such that $A_i = \pi D_b^2$ (Han and Griffith, 1965a), and also that these areas of influence of neighboring bubbles do not overlap. That is,

$$q''_{boil} = q''_{Ai} \pi D_b^2 n \tag{2-98}$$

where n is the number of active sites per unit area of heating surface (N/A_T). Further, assuming that

1. (Contact area)/(area of influence, A_i) $\ll 1$
2. The circulation of liquid in the vicinity of a growing bubble due to thermocapilarity effects on vapor–liquid bubble interface is negligible
3. $\dfrac{\text{(Contact area at departure)} \times q''_{micro}}{A_i \times q''_{Ai}} \ll 1$

   where q''_{micro} is the average heat flux through the microlayer at the base of a growing bubble

we can express the average heat flux over the area of influence of a single bubble,

$$q''_{Ai} = f\int_0^{1/f} q''_{tran} \, dt = 2\left(\frac{f}{\pi\alpha}\right)^{1/2} k(T_w - T_{sat}) \tag{2-99}$$

where q''_{tran} is transient heat conduction flux from the heated surface to the liquid in contact with it following the departure of the bubble and the superheated layer, which can be expressed as

$$q''_{tran} = \frac{k(T_w - T_{sat})}{(\pi \alpha t)^{-1/2}}$$

Thus Eq. (2-97) can be obtained by substituting Eq. (2-99) into Eq. (2-98).

Mikic and Rohsenow (1969) used the following correlations for n, D_b, and f:

$$n = C_1 r_s^m \left(\frac{H_{fg} \rho_G}{2 T_{sat} \sigma} \right)^m (T_w - T_{sat})^m \tag{2-100}$$

where C_1 is a dimensional constant (1/unit area), r_s is a radius for which n would be one per unit area, and m is an empirical exponent in the power law for the cumulative number of active sites as reported by Brown (1967);

$$D_b = C_2 \left[\frac{\sigma g_c}{g(\rho_L - \rho_G)} \right]^{1/2} (\text{Ja}^*)^{5/4} \tag{2-101}$$

where $C_2 = 1.5 \times 10^{-4}$ for water and 4.65×10^{-4} for other fluids (Cole, 1970), and Ja* is a modified Jacob number,

$$\text{Ja}^* = \frac{\rho_L c_L T_{sat}}{\rho_G H_{fg}}$$

and

$$f D_b = \phi(\rho_G, \rho_L, \sigma, g) \tag{2-102}$$

According to Ivey (1967), the above equation may be in different forms in three different regions [Sec. 2.2.5.2, Eqs. (2-66), (2-66a), and (2-66b)]. Since f in Eq. (2-97) is interpreted as an average f over the whole heating surface, and since heat flux is not a strong function of the frequency, a best single approximation applicable in the whole range of interest is used (Cole, 1967):

$$f D_b = C_3 \left[\frac{\sigma g_c g(\rho_L - \rho_G)}{\rho_L^2} \right]^{1/4} \tag{2-103}$$

where $C_3 = 0.6$ was chosen as an average value that best accommodates the experimental results of Cole (1967). Substituting Eqs. (2-100), (2-101), and (2-103) in Eq.

(2-97) results in the following expression for the heat flux due to the boiling alone from the heated surface:

$$\frac{q''_{boil}}{M_L H_{fg}} \left[\frac{g_c \sigma}{g(\rho_L - \rho_G)} \right]^{1/2} = B_1 \phi^{m+1} (T_w - T_{sat})^{m+1} \quad (2\text{-}104)$$

where B_1 and ϕ, dimensional quantities that depend partly on heating surface characteristics, are given by

$$B_1 = \frac{C_1 C_2^{5/3} C_3^{1/2} (r_s J/2)^m (2/\sqrt{\pi})(g_c)^{11/8}}{(g)^{9/8}} \quad (2\text{-}105)$$

and

$$\phi^{m+1} = \frac{k_L^{1/2} \rho_L^{17/8} c_L^{19/8} H_{fg}^{(m-23/8)} \rho_G^{(m-15/8)}}{\mu_L (\rho_L - \rho_G)^{9/8} \sigma^{(m-11/8)} T_{sat}^{(m-15/8)}} \quad (2\text{-}106)$$

The total heat flux from the entire boiling surface can be expressed as

$$q'' = \left(\frac{A_{n.c.}}{A_{tot}} \right) (q''_{n.c.}) + q''_{boil}$$

where $A_{n.c.}$ and $q''_{n.c.}$ are the area and heat flux of the natural-convection component, respectively.

$q''_{n.c.}$ can be correlated as suggested by Han and Griffith (1965b) and by Judd and Hwang (1976):

$$q''_{n.c.} = 0.14 k_L \left[\left(\frac{g\beta}{v_L^2} \right) \left(\frac{\mu_L c_L}{k_L} \right) \right]^{1/3} (T_w - T_{bulk})^{4/3}$$

According to this model, for a given surface and a given fluid (i.e., constant B), the value of

$$N_R = (q''_{boil}) \left[\frac{\sigma g_c}{g(\rho_L - \rho_G)} \right]^{1/2} / (H_{fg} \mu_L)$$

can be expressed as a function of $\phi(T_w - T_{sat})$. Both B and ϕ, the functional relation between the heat flux and $(T_w - T_{sat})$ or ΔT, depend on the cavity size distribution of the boiling surface. However, the relation between N_R and $\phi \Delta T$ is independent of pressure. Consequently, if the boiling properties of the surface are unknown, one may determine m as well as B from data for q'' versus ΔT, and use this value of m to predict the boiling heat transfer at any pressure (Mikic and Rohsenow, 1969). This procedure was applied to the experimental data of Addoms (1948) for pool boiling of water on a platinum wire of two different diameters (Fig. 2.20). The value for m was taken to be 2.5, resulting in (Mikic and Rohsenow, 1969)

$$q'' \propto (T_w - T_{sat})^{3.5}$$

Similarly, in Figure 2.21, N_R is plotted versus $\phi \Delta T$ for boiling n-pentate, benzene, and ethyl alcohol on a flat chromium surface (Cichelli and Bonilla, 1945). The value for m in all three cases was chosen to be 3. In the cases considered, the measured q'' from experimental data was interpreted as q''_{boil}, neglecting the natural-convection contributions because the flux levels in the experiments were

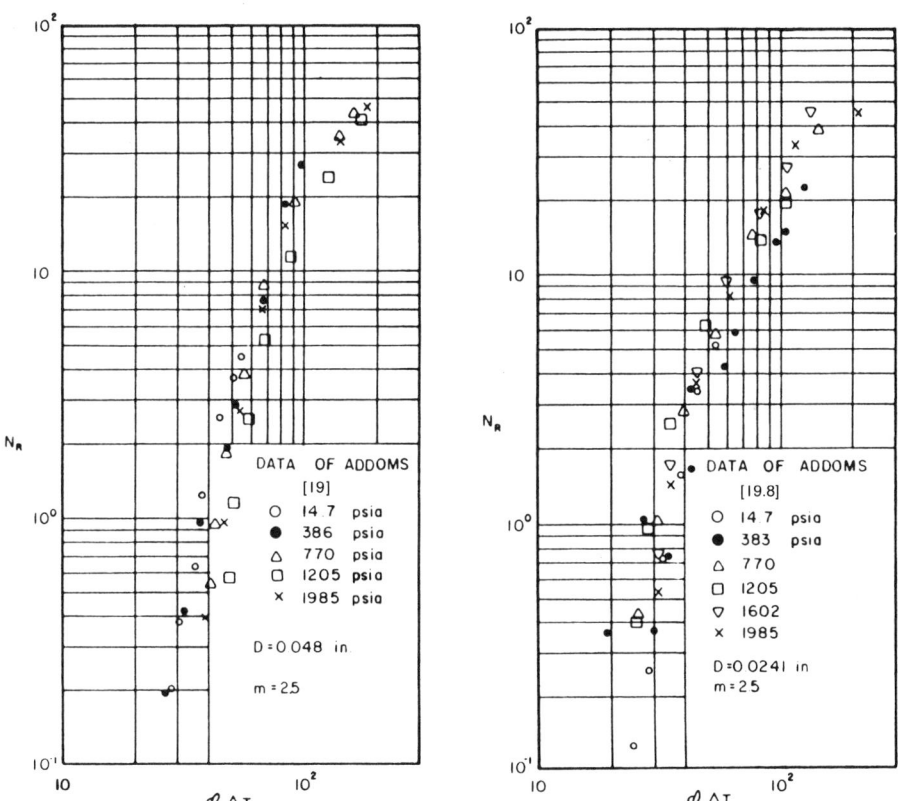

Figure 2.20 Nucleate pool boiling of water on a platinum wire at different pressures, data of Addoms, (1948): (*a*) wire diameter = 0.0241 in. (0.061 cm); (*b*) wire diameter = 0.048 in. (0.122 cm). (From Mikic and Rohsenow, 1969. Copyright © 1969 by American Society of Mechanical Engineers, New York. Reprinted with permission.)

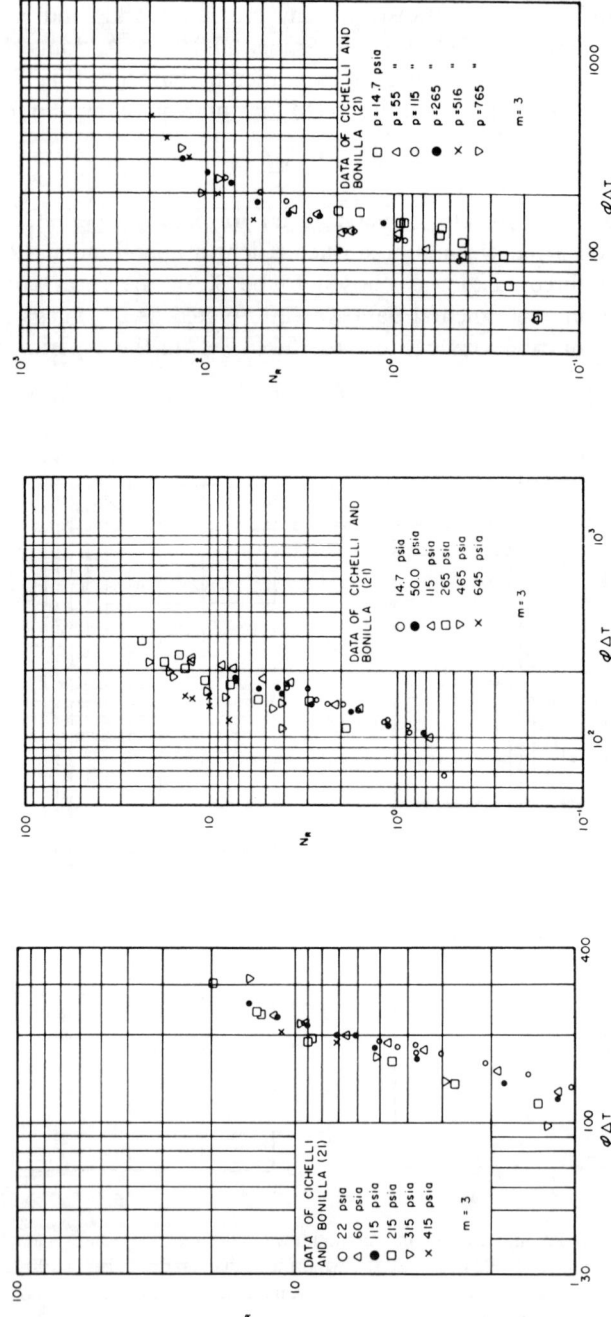

Figure 2.21 Nucleate pool boiling, on a flat chromium surface, at different pressures, of (*a*) *n*-pentane; (*b*) benzene; (*c*) ethyl alcohol. Data of Cichelli and Bonilla (1945). (From Mikic and Rohsenow, 1969. Copyright © 1969 by American Society of Mechanical Engineers, New York. Reprinted with permission.)

sufficiently high. It appeared that Eq. (2-104) correlated experimental results for all four liquids satisfactorily, and this equation is therefore recommended for ordinary liquids.

It is interesting to note that in a photographic investigation of saturated nucleate boiling by Anderson et al. (1970) with five different fluids boiling on a transparent oxide-coated glass surface, the active nucleation site density data obtained was correlated with surface temperature using the Gaertner site activation theory (Gaertner, 1963b, 1965):

$$\frac{N}{A} = N_0 \exp\left\{\left[\frac{-16\pi\sigma^3 M^2 N_{Av}}{3\rho_L^2 R_g^3 [\ln(p_\infty/p_G)]^2}\right](\phi)\left(\frac{1}{T_w}\right)^3\right\} \quad (2\text{-}100a)$$

where N_o = constant, sites/ft^2
N_{Av} = Avogadro's constant
R_g = universal gas constant
M = molecular weight
ϕ = constant

Anderson et al. found that the product $\{16\pi\sigma^3 M^2 N_{Av}/3\rho_L^2 R_g^3[\ln(p_\infty/p_G)]^2\}(\phi)$ is constant for a particular surface, regardless of the fluid properties, and even for some other surface properties reported in the literature. Thus it was suggested that the relationship Eq. (2–100a) can be simplified to

$$\frac{N}{A} = N_0 \exp\left[\frac{-3.305 \times 10^9}{T_w^3(°K)}\right] \quad (2\text{-}100b)$$

Model of composite mechanism In low-heat-flux cases, to model nucleate boiling of composite processes (mechanisms), Graham and Hendericks (1967) proposed an overall model by weighing the various heat transfer process with the area fractions occupied by each process. The area function is determined by bubble population, ebullition cycle, surface wettability, bubble departure diameters, etc. In the nucleate boiling regime, as the heat flux or wall superheat is raised, more and more nucleate centers are activated and the frequency of bubbles at a given site increases while the waiting period decreases. The net result is the coalescence of bubbles in both the vertical and lateral directions. Figure 2.22 shows the various types of merging bubbles (Kirby and Westwater, 1965). Under normal gravity, the type IIa bubble, which coalesces in the vertical direction, is not encountered as frequently as those shown as type IIb. Type IIc is the situation when the merging of type IIb becomes more frequent and involves more than two roots. Gaertner (1961), in his study of bubble distribution, showed that bubble sites are randomly distributed and can be represented by a Poisson distribution:

68 BOILING HEAT TRANSFER AND TWO-PHASE FLOW

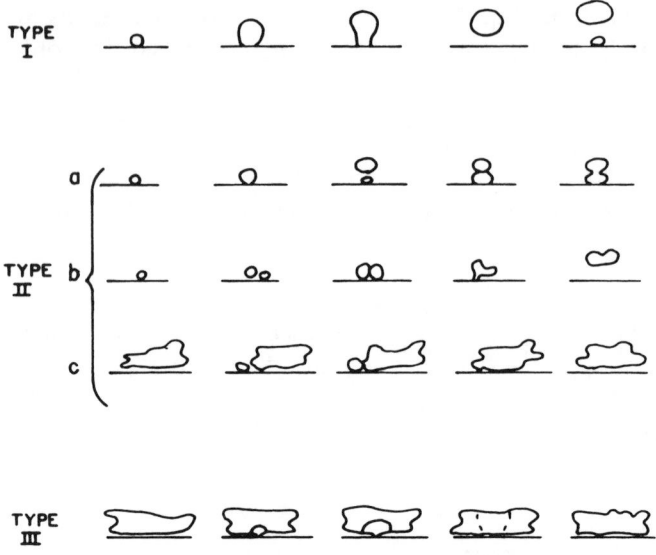

Figure 2.22 Sketches of imagined profile views of three types of discrete growing bubbles. (From Kirby and Westwater, 1965. Copyright © 1965 by American Institute of Chemical Engineers, New York. Reprinted with permission.)

$$P(\mathrm{Na}) = \frac{[e^{-\overline{\mathrm{Na}}}(\overline{\mathrm{Na}})^{\mathrm{Na}}]}{(\mathrm{Na})!} \tag{2-107}$$

where $P(\mathrm{Na})$ is the probability of finding the Na number of sites in a cell of area a when the mean site population density is $\overline{N}$. Graham and Hendricks divided the boiling heat transfer surface into three parts: (1) a projected surface underneath a growing bubble, consisting of a possible dry spot due to evaporation with negligibly small heat transfer contribution after it is formed, and a wetted contact area that can be related to a contact angle and is considered to be an evaporative microlayer, supplying of vapor into the bubble; (2) a surface where the thermal layer is being prepared for ebullition; and (3) a surface where no boiling activity takes place (heat is transferred by natural convection) (Fig. 2.23). When the Mikic and Rohsenow model is used, the evaporation and enhanced turbulent convection processes are replaced by the transient conduction and pumping process, Eq. (2-96), in the area (1). This argument was settled by Judd (1989), who suggested that the microlayer evaporation heat transfer be included to represent the periodic removal of energy that is extracted from the surface during the growth period in order to evaporate the microlayer. The q''_{boil} represents the periodic removal of the energy that accumulates in the liquid that replaced the previous bubble during the waiting period by entrainment in the wake of the subsequent bubble, over area (2). The two effects are complementary, and while they both occur in the vicinity of the nucleation site,

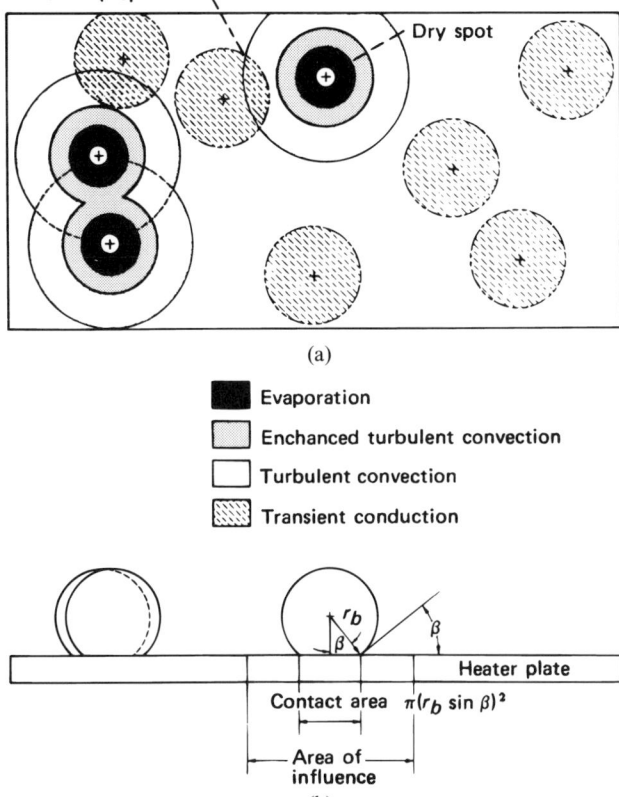

Figure 2.23 Instantaneous representation of nucleate boiling surface showing distribution of heat transfer mechanisms: (a) plan view; (b) profile view. (From Hsu and Graham, 1976. Copyright © 1976 by Hemisphere Publishing Corp., New York. Reprinted with permission.)

they occur at different times and do not conflict. Judd compared experimental results (Ibrahim and Judd, 1985) with predictions of a model involving nucleate boiling only, a model involving nucleate boiling and natural convection, and a model involving nucleate boiling, natural convection, and microlayer evaporation (Judd, 1989). These figures are reproduced in Figure 2.24. Good agreement is seen in Figure 2.24c between the experimental results and the theoretical predictions at $N/A_T = 14{,}000$ sites/m² over the entire range of subcooling values. While the rough agreement between the experimental bubble period and the predictions at $N/A_T = 19{,}000$ sites/m² in Figure 2.24a confirms that the nucleate boiling heat transfer mechanism is the dominant factor in the prediction of heat transfer, the agreement cannot support this as the only mechanism. Although better agreement is shown in Figure 2.24b at the higher levels of subcooling, where a model involving nucleate

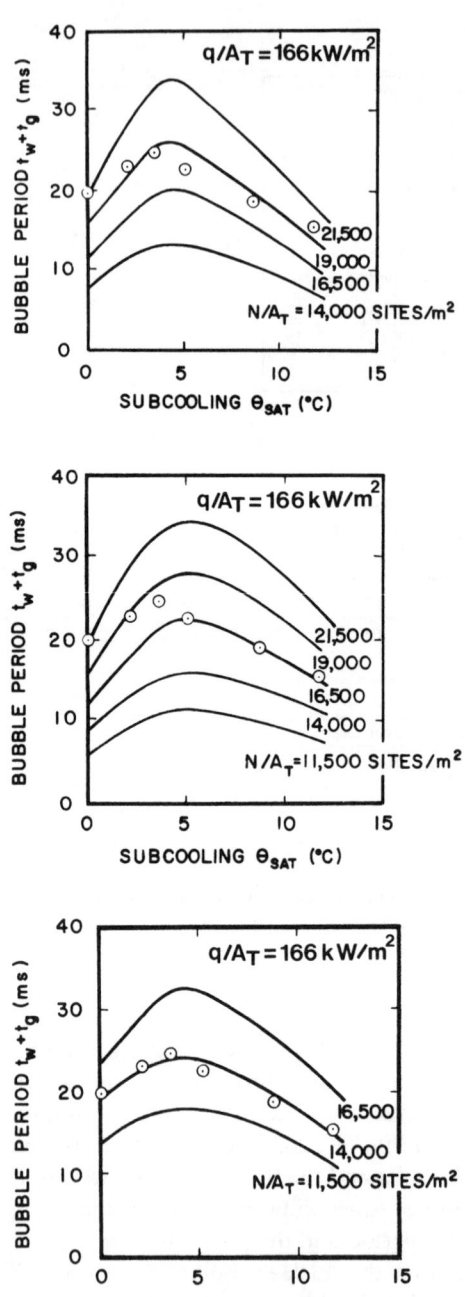

Figure 2.24 Comparison of experimental results of bubble period with predictions of a model involving different mechanisms: (a) nucleate boiling only; (b) nucleate boiling and natural convection; (c) nucleate boiling, natural convection, and microlayer evaporation. (From Judd, 1989. Copyright © 1989 by American Society of Mechanical Engineers, New York. Reprinted with permission.)

boiling and natural convection was used, the overall agreement still does not sufficiently support the said model. These model predictions were given by Judd (1989).

A boiling heat transfer model incorporating nucleate boiling, natural convection, and microlayer evaporation was formulated as

$$\frac{q}{A_T} = q''_{boil} + \left(\frac{q_{n.c.}}{A_{n.c.}}\right)\left[1 - \left(\frac{A_{n.b.}}{A_T}\right)\right] + \left(\frac{q_{micro}}{A_T}\right) \quad (2\text{-}108)$$

wherein the relationship

$$\frac{q''_{micro}}{A_T} = \rho_L H_{fg} V_{micro} f\left(\frac{N}{A_T}\right) \quad (2\text{-}109)$$

was introduced to predict the contribution of microlayer evaporation heat transfer. The volume of the microlayer that had evaporated at $t = t_g$ was computed from the instantaneous microlayer profile $\delta(r, t_g)$ as determined by Judd (1989).

2.4.2.2 Nucleate pool boiling with liquid metals. Because of the special properties of liquid metals, it has been shown in previous sections that the nucleation superheats for liquid metals tend to be higher than for ordinary liquids, the bubble growth rates tend to be higher, and the waiting times are longer for liquid metals. The blind use of the theoretical correlations for predicting liquid-metal nucleate boiling heat transfer rates thus becomes risky. Under certain conditions, however, some of the correlations may be used with confidence if the proper value of the coefficient α for the particular system is known, as long as they are not used at low pressures (say, $p_L/p_{cr} < 0.001$), and if the value of m [in Eq. (2-78)] for the particular heating surface is also known (which for a commercially smooth metallic heating surface is about $\frac{2}{3}$). A judicious choice of a theoretical equation for use in such an instance can be made (Dwyer, 1976). Kutateladze's correlation (Kutateladze, 1952), as modified by Minchenko (1960),

$$\frac{q''}{(t_w - t_{sat})k_L}\left(\frac{g_c\sigma}{g(\rho_L - \rho_G)}\right)^{1/2} = \alpha\left(\frac{q''g_c p_L}{H_{fg}\rho_G v_L g(\rho_L - \rho_G)}\right)^{0.7}(c_L\mu_L/k_L)^{0.7} \quad (2\text{-}110)$$

appears to be the most generally dependable for predicting the influences of q'', p_L, and Pr on h when boiling alkali metals on commercially smooth stainless steel surfaces and at reduced pressures greater than 0.001. The value of α, which normally varies with the liquid metal/heating surface system, must be estimated from experimental data. The reduced pressure at 0.001 has a special meaning in that the boiling data points taken with three different liquid metals (sodium, potassium, and cesium) on a smooth flat plate over the pressure range 0.015 to 0.5 atm (1.5 × 10^3 to 0.5 × 10^5 Pa) exhibited a sharp break when correlated by the empirical equation (Subbotin et al., 1970)

$$h = C\left[\frac{kH_{fg}\rho_L J}{\sigma T_{sat}^2}\right]^{1/3} q''^{2/3} \left(\frac{p_L}{p_{cr}}\right)^s \quad (2\text{-}111)$$

where $C = 8.0$ and $s = 0.45$ for $p_L/p_{cr} < 0.001$, and where $C = 1.0$ and $s = 0.15$ for $p_L/p_{cr} > 0.001$ (Fig. 2.25). Here $kH_{fg}\rho_L J/\sigma T_{sat}^2$ is a dimensional physical property parameter used to bring the results on the three different alkali metals into a single relationship.

Because of the heterogeneous nucleation from active cavities in the solid surface, some active cavities may become deactive (i.e., all the trapped vapor in the cavities condenses) during the various stages of boiling in a pool. If under certain conditions all the active cavities in the surface become deactivated, the boil will stop, which causes a temperature rise in the heating solid. In so doing, the liquid superheat is increased and might in turn activate some smaller cavities to resume boiling, which will then reactivate even larger cavities. For a fixed heat flux, this phenomenon causes fluctuation of temperature between the boiling point and the natural-convection point (when the boiling stops) and is a state of unstable boiling, or bumping. Several investigators (Madsen and Bonilla, 1959; Marto and Rohsenow, 1965) have reported unstable behavior of liquid alkali metals during boiling while stable boiling of ordinary fluids almost always exists. Shai (1967) established a criterion for stable boiling of alkali metals by considering a cylindrical cavity of length-to-radius ratio, $l/r > 10$. The cycling behavior requires that the temperatures and the temperature gradients be the same after each full cycle, and that the

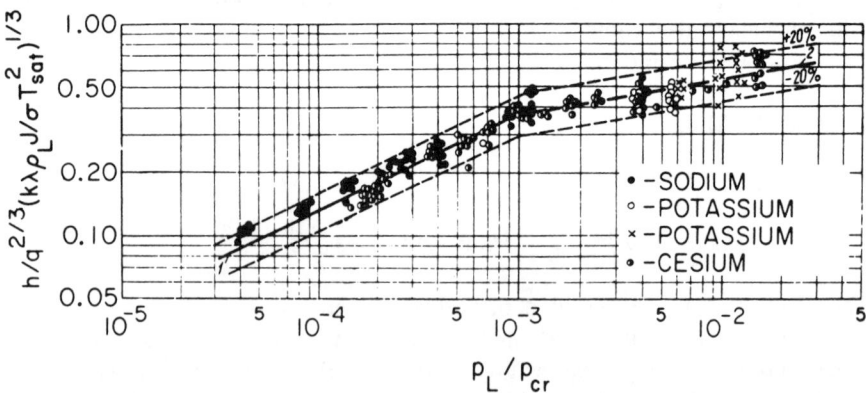

Figure 2.25 Stable nucleate pool boiling data of alkali metals on a smooth, flat, stainless steel surface. Data of Subbotin et. al. (1970) and Cover and Balzhiser (1964). (From Dwyer, 1976. Copyright © 1976 by American Nuclear Society, LaGrange Park, IL. Reprinted with permission.)

average heat flux at the surface be equal to the constant heat flux far away from the surface, where no variations of temperature occur during the cycle. Shai found the distance, X_{cr}, that the liquid travels from the solid surface into the cavity at which condensation stops if, during the waiting time t_w, the minimum temperature at that point reaches the vapor temperature, T_G:

$$X_{cr} = \eta(\beta)L \qquad (2\text{-}112)$$

where $\eta(\beta)$ is a dimensionless critical distance that is a function of the liquid–solid properties, and L is the relaxation length from the surface, where the temperature gradient becomes equal to the average gradient during the full cycle. Beyond this point the gradient is constant with time and position. Thus if a cavity has a depth of $l > X_{cr}$, it will always contain vapor; that is, it will remain an active cavity. The dimensionless critical distance $\eta(\beta)$ is defined by Shai (1967) as

$$\eta(\beta) = \frac{2(\pi\beta + 4)Z_o^2}{(\pi)^{1/2}(2+\beta)J}$$

where $\beta = (k_w/k_L)(\alpha_L/\alpha_w)^{1/2}$ and $Z_o = X_{cr}/[2(\alpha_w t_{cr})^{1/2}]$. The relaxation length, L, can be approximated by

$$L \approx 1.55(\alpha_w t_w)^{1/2}$$

For alkali metals with small cavities at low pressures, the value of l for a given heat flux may not be achievable. Since t_w can be expressed as a function of average heat transfer rate per unit area, q_o'', and liquid properties, Eq. (2-112) can be rearranged and solve for the heat flux:

$$q_o'' > 1.55(T_w - T_{sat})_{t=0}(k_L \rho_L c_L \alpha_w)^{1/2} \eta(\beta) I(\beta)/l \qquad (2\text{-}113)$$

where

$$I(\beta) = \frac{(6\pi + 6\pi\beta^2 + 7\pi\beta + 16\beta)}{3\sqrt{\pi}(1+\beta)(\pi\beta + 4)}$$

Equation (2-113) means that any cylindrical cavity for any liquid–solid combination under a given pressure has a minimum heat flux below which boiling will not be stable, and a transition between natural convection and stable nucleate boiling (bumping) is always observed.

Effect of surface roughness Just as in the case of ordinary liquids, the heating surface roughness can have a large effect on the nucleate boiling heat transfer. The most extensive and systematic investigations in the area of such effects are those by Rohsenow and co-workers at the Massachusetts Institute of Technology. Figure

2.26 shows some nucleate boiling heat transfer results obtained by Deane and Rohsenow (1969) for sodium boiling on a polished nickel surface containing a single artificial cylindrical cavity 0.0135 in. (0.343 mm) in diameter × 0.100 in. (2.54 mm) deep, located in the center of the plate. The plate was 2.5 in. (6.35 cm) in diameter and constituted the bottom of a vertical cylindrical vessel. The abscissa contains $\overline{T}_w''$, which is the average of the maximum values of the surface temperature as taken from the recording chart to provide a valid comparison. The three vertical lines in Figure 2.26 represent the simple force balance of a bubble having a radius equal to that of the artificial cavity [Laplace bubble equilibrium equation, Eq. (2-3)] for the three system pressures indicated. The agreement of predicted superheat at all values of the heat flux with the stable boiling data is excellent, as shown in the figure. Because their experiments were carried out by increasing the heat flux in increments until the mode of heat transfer switched from natural convection to nucleate, considerable unstable boiling is also shown at the lowest pressure and lower fluxes. The picture is different at the highest pressure employed, 93 mm Hg (1.8 psia), where the nucleate boiling was completely stable at all heat fluxes.

Marto and Rohsenow (1965, 1966) boiled sodium on a horizontal 2.56-in. (6.5-cm)-diameter disk welded to the bottom of either a 1-ft (0.3-m) length of nominal $2\frac{1}{2}$-in. (6.35-cm) stainless steel pipe or a same size A-nickel pipe with no thermocouple wells and was polished on the inside to a mirror finish. The test disks were made of either A-nickel or type 316 stainless steel, had various surface characteristics, and were heated from below by tantalum radiation heaters. Each disk was $\frac{3}{4}$ in. (1.9 cm) thick and held six Inconel-sheathed (0.0635-in., or 1.6 mm in diameter) thermocouples that were radially penetrated with four different centerline distances from the heating surface for determining the heat flux and the surface temperature. The six different heating surfaces tested were (1) mirror finish; (2) lapped finish (with abrasive material suspended in oil:LAP-A with grade A compound of 280 grit and LAP-F with grade F compound of 100 grit); (3) artificial porous weld (porous weld placed on mirror-finished disk); (4) porous coating (sintering a 0.03-in.-thick disk of porous A-nickel onto a normal nickel disk); and (5) artificial cavities (twelve distributed doubly reentrant cavities). Photographs of these surfaces are shown in Figure 2.27. Figure 2.28 shows nucleate boiling curves for these surfaces, indicating the great influence of heating surface roughness, or structure, on the heat transfer behavior. Generally, the rougher the surface, the lower the required superheat for a given heat flux and the greater the slope of the curve. The slopes of the lines in the plot vary from 1.1 for the porous surface to 6.3 for the surface with the doubly reentrant cavities, while the line for the LAP F surface shows a curvature (the LAP-A surface results not shown in Fig. 2.27 were of similar shape), presumably due to the rather unique cavity size distribution produced in the lapping process (Dwyer, 1976).

Bonilla et al. (1965) studied the effect of parallel scratches on a polished stainless steel plate when boiling mercury with a small portion of sodium as wetting agent. The mirror-finished stainless steel plate was scored by a tempered steel nee-

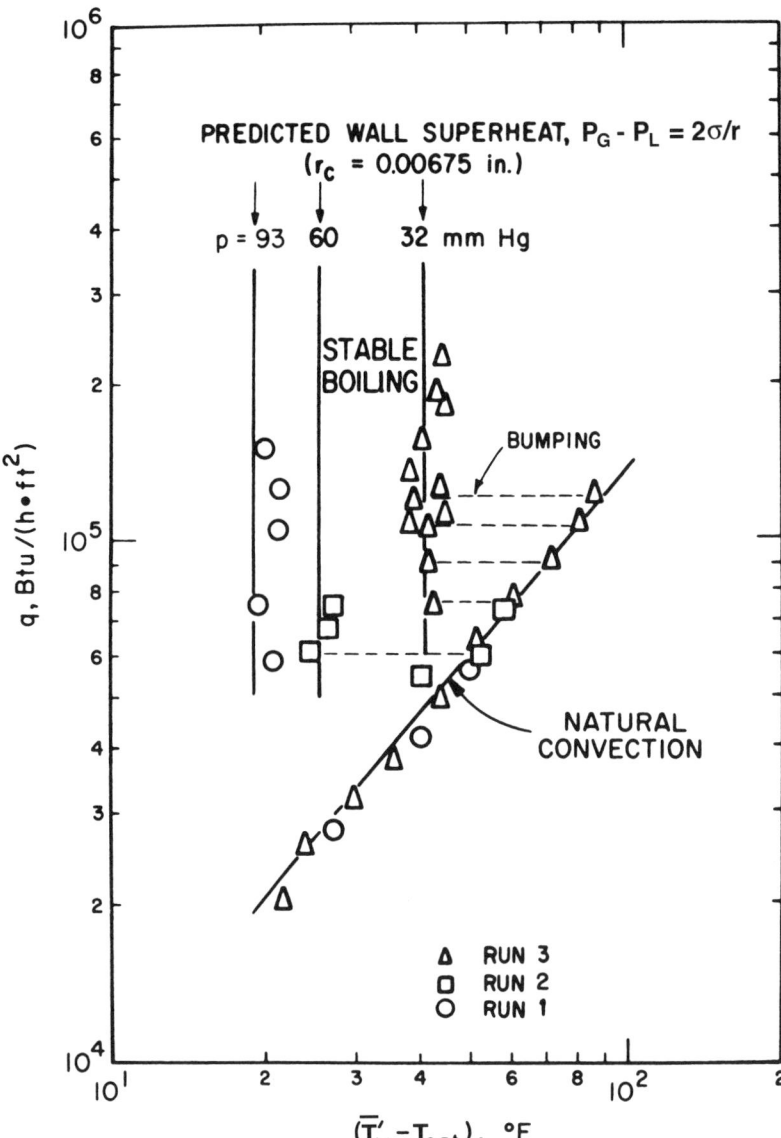

Figure 2.26 Heat transfer characteristics for stable nucleate boiling of sodium on a polished, flat, horizontal, nickel plate containing a single cylindrical artificial cavity. (From Deane & Rohsenow, 1969. Copyright © 1969 by American Society of Mechanical Engineers, New York. Reprinted with permission.)

76 BOILING HEAT TRANSFER AND TWO-PHASE FLOW

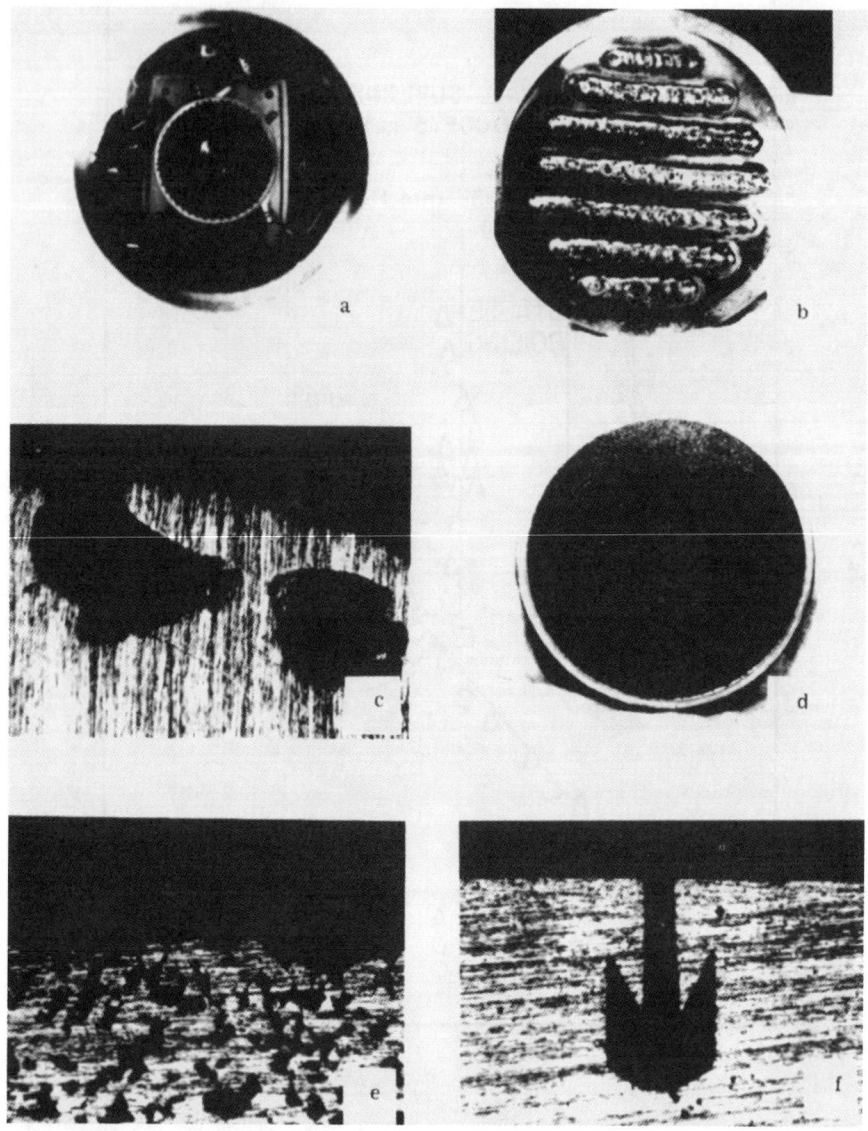

Figure 2.27 Photographs and photomicrographs of nickel heat transfer test surfaces: (*a*) mirror finish no. 1; (*b*) porous welds; (*c*) photomicrograph of porous welds (100×); (*d*) porous coating; (*e*) photomicrograph of porous coating (100×); (*f*) photomicrograph of double-reentrant cavity cross section (65×). (From Marto and Rohsenow, 1966. Copyright © 1966 by American Society of Mechanical Engineers, New York. Reprinted with permission.)

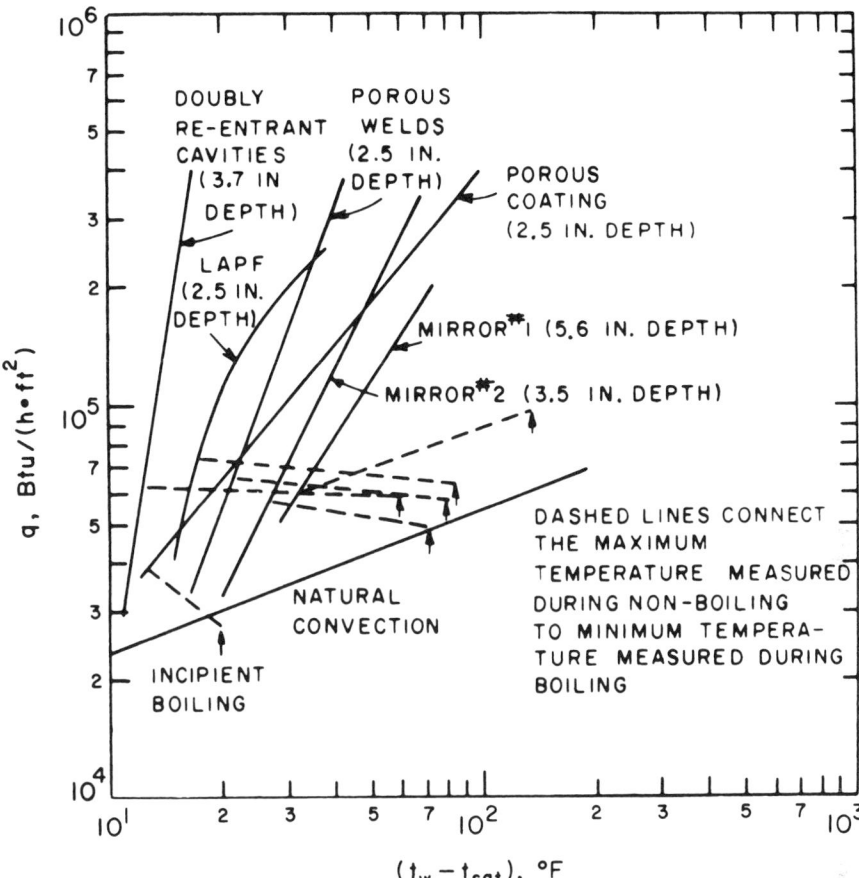

Figure 2.28 Experimental heat transfer results for saturated nucleate boiling of sodium from nickel disks at average pressure of 65 mm Hg. (From Marto and Rohsenow, 1966. Copyright © 1966 by American Society of Mechanical Engineers, New York. Reprinted with permission.)

dle with scratches about 0.003-in. (0.76 mm) wide and 0.004-in. (0.1 mm) deep. Runs were made with smooth plates, with the scratches $\frac{3}{8}$ in. (9.53 mm) apart, and again with the scratches $\frac{1}{8}$ in. (3.18 mm) apart. The results are shown in Figure 2.29 and indicate that the greater is the number of scratches, the steeper is the q-versus-$(T_W - T_b)$ line. Thus, for the scratched surfaces, the average active cavity size tends to decrease less as the heat flux is increased because the average active cavity size is larger—in other words, the condition represented by the vertical lines in Figure 2.26 (Dwyer, 1976). It was also noted that, with scratched surfaces, much less audible bumping for the same boiling flux was observed, which was due to better distribution of active nucleation sites and to lower wall superheats compared with

78 BOILING HEAT TRANSFER AND TWO-PHASE FLOW

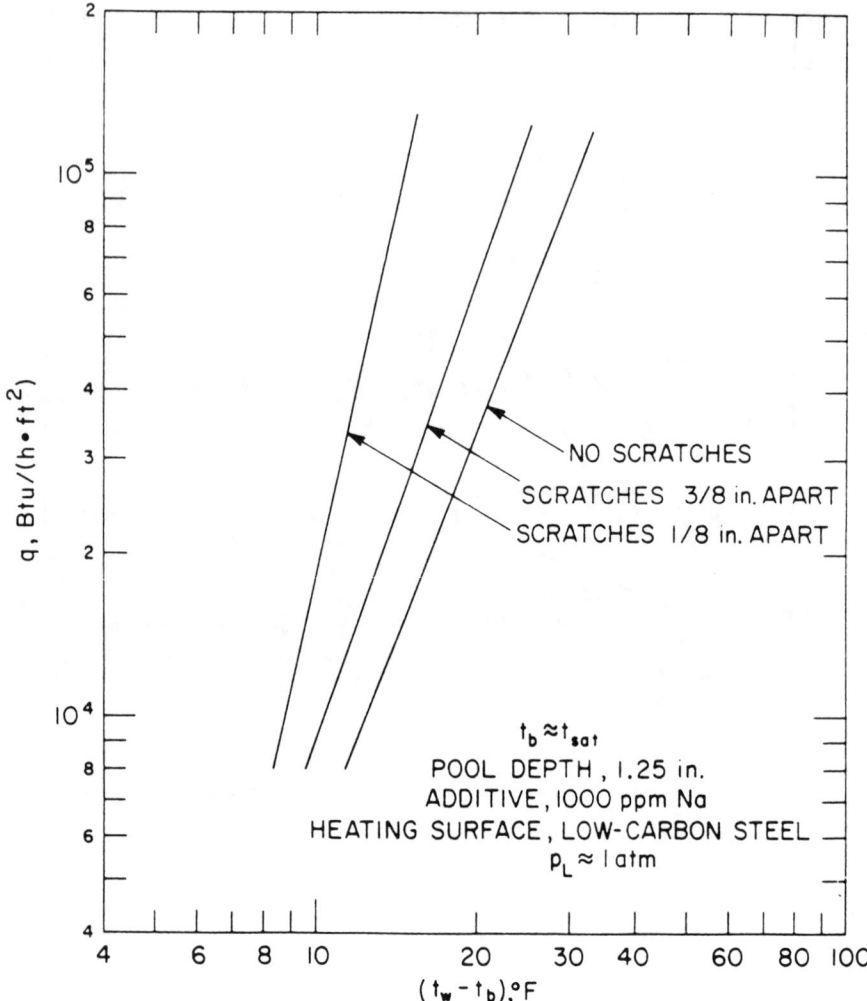

Figure 2.29 Experimental results for nucleate pool boiling of mercury (with wetting agents) above a horizontal plate with parallel scratches. (From Bonilla et al., 1965. Copyright © 1965 by American Institute of Chemical Engineers, New York. Reprinted with permission.)

those for the unscratched plate. Investigations on the effect of heater geometry and spatial orientation were reported by Korneev (1955), Clark and Parkman (1964), and Borishansky et al. (1965). It may be fairly safe to conclude that in stable nucleate boiling of liquid metals, heat transfer coefficients for the top and side of horizontal rod heaters and those for vertical rod heaters are, for all practical purposes, the same. The heat transfer coefficients for the bottom of horizontal rod heaters are, however, appreciably lower than those of the side and top. Comparing experimental results of some investigators using rod heaters with those of others using flat horizontal plates, heat transfer coefficients for the plates are not appreciably different from those for the top and sides of horizontal rod heaters and those for the side of vertical rod heaters. It is apparent that, even for liquid metals, the effect of spatial orientation of the heating surface on stable nucleate boiling heat transfer is not great, as long as the growth and departure of the vapor bubbles on the surface are not inhibited significantly (Dwyer, 1976).

Effect of degree of wetting The phenomenon of wetting is very different depending on whether the liquids are heavy metals such as mercury, or alkali metals. The former, being quite unreactive chemically, do not reduce oxides on the surfaces of most structural materials and alloys. The latter, on the other hand, are quite reactive and can reduce surface oxides from most iron-based alloys, including all the stainless steels. Thus the alkali metals usually wet common heating surfaces while under normal conditions mercury does not. However, with mercury, heating surface wetting can vary from zero (as in the Hg/Cr/O system) to 100% (as in the Hg/Cu system). By the addition of a small amount of a strong reducing metal such as Mg to the mercury (Korneev, 1955), the wetting of stainless steels and other iron-based alloys is greatly enhanced. Alkali metals wet stainless steels so well that the larger surface cavities tend to fill up and thereby become inactive. This, in turn, raises the boiling superheat, and lowers the heat transfer coefficient. Nevertheless, because heat is transferred from the heating surface to the liquid in nucleate boiling, good thermal contact between the two is of primary importance. A discussion of the effect of wetting on nucleation sites can be found in Section 2.2.1.2.

Effect of oxygen concentration in alkali metals An increase of oxygen concentration in sodium has generally been found to cause a significant decrease in the incipient-boiling superheat and thus is probably also true for stable nucleate boiling superheats. However, one would not expect a large effect as long as the oxygen concentration is well below its solubility limit, other things being equal. (This last would not be true if significant concentrations of oxygen, after a period of time, change the microstructure of heating surfaces exposed to alkali metals.) This is because the effects of subsaturation concentration of oxygen on the physical properties of alkali metals (such as sodium) have been found to be very slight, with the possible exception of surface tension, which does not affect the heat transfer coefficient significantly (Dwyer, 1976). As shown by Kudryavtsev et al.'s (1967) experiments with sodium boiling on a smooth, flat stainless steel plate, there was no effect on the heat transfer when the oxygen concentration in the sodium is increased from 10 to 1,000 ppm at the lowest saturation temperature tested of 700°C (1292°F). (The solubility of oxygen in sodium at this temperature is 5,300 ppm, which is far above the 1,000 ppm used in the experiments.)

Schultheiss (1970) also investigated experimentally the incipient boiling superheat of sodium in wall cavities with sodium oxide concentration as one of the parameters. The results were discussed with respect to the physicochemical conditions of the system solid wall material/oxygen/liquid metal. This change of microstructure on the heating surface contributed to the effect of oxide concentration on incipient boiling superheat or boiling behavior. Contrary to the pool boiling case, definite oxide-level effects on convective boiling initiation superheat values were reported by Logan et al. (1970).

Effect of aging Experimenters have found that the heat transfer behavior of a nucleate pool boiling liquid metal system nearly always changes from the time of the first run; if the system is run long enough, however, the behavior eventually attains an unchanging pattern. Whereas good wetting can generally be achieved with alkali metals in a matter of a few hours, it may take several weeks to achieve the same with mercury unless wetting agents are used. After wetting is achieved, the inert gas evolved from the cavity sites in the boiling heating surface raises the superheat at an increasingly slower rate to approach steady state. This was confirmed by Marto and Rohsenow (1966) who reported that for boiling sodium on a polished nickel disk containing porous weldments, only a few hours of degassing time was necessary to remove the noncondensable gases from their particular heating surface based on the observed heat transfer coefficients. Long-range corrosion effects may appear in the form of pitted surfaces, which occur relatively slowly and are only observed after periods of months. The time involved is almost too long to be observed in pool boiling equipments.

Based on what has been discussed so far, it is recommended that one use an empirical correlation for pool boiling of liquid metals based on data of experimental conditions that match or closely simulate the conditions in question.

2.4.3 Pool Boiling Crisis

As shown in Figure 1.1, during nucleate pool boiling (B–C) a large increase in heat flux is achieved at the expense of a relatively small increase in ($T_w - T_{sat}$) until the vertical chains of discrete bubbles begin to coalesce into vapor streams or jets, and a further increase in flux causes the vapor–liquid interfaces surrounding these jets to become unstable. When this occurs, the flow of liquid toward the wall begins to be obstructed, and a point (point C) is reached where the vapor bubbles begin to spread over the heating surface. This marks the beginning of the departure from nucleate boiling (DNB). As the temperature is increased further, the heat flux goes through a maximum called the *critical heat flux* (CHF), which corresponds to a sudden drop of heat transfer coefficient and in turn causes a surface temperature surge. This phenomenon leads to the name "boiling crisis." Beyond this point, depending on whether q'' or T_w is the controlled independent variable, the curve may follow the dotted line C–D, which is the transition boiling region, or T_w may jump from its value at point C to a point on the D–E line. When the latter phenomenon occurs (when q'' is controlled and differentially increased beyond C), some heating surfaces cannot withstand the large increase in temperature and thus melt. For this reason, the CHF is also known as the *burnout* heat flux. If q'', as the independent variable, is decreased slightly below its value at point D, then T_w will drop suddenly to a point on the line B–C. Thus, unless the wall temperature is controlled, large boiling instability can result if the heat flux exceeds the CHF in nucleate boiling or falls below the minimum heat flux in film boiling. Line D–E represents the stable film boiling regime, which will be discussed in Section 2.4.4.

2.4.3.1 Pool Boiling Crisis in Ordinary Liquids.

Theoretical considerations As far as these authors are aware, Chang (1957) was the first to propose a wave model for boiling and introduce some basic ideas about boiling heat transfer. Zuber, on the other hand, formulated the wave theory for boiling crisis (Zuber, 1958; Zuber et al., 1961; Sec. 2.3.1). Using wave motion theory and the Helmholtz stability requirement, Chang (1962) derived a general equation for the CHF both with and without forced convection and subcooling. As commonly accepted, the pool boiling crisis was considered to be limited by the maximum rate of bubble generation from a unit area of the heating surface. Chang treated this as one stability problem for a bubble growing or moving in an intensively turbulent field while the surface tension force gave rise to a stabilizing effect, but the dynamic force tends to destabilize the motion. Thus, the CHF condition is governed by a critical Weber number. Chang assumed (1) the existence of statistically mean values of the final bubble size, bubble frequency, and number of bubble sites per unit area of the heating surface; (2) that bubbles are spherical or equivalent to a sphere; and (3) at CHF, a bubble on the heating surface has developed to its departure size under hydrodynamic and thermodynamic equilibrium (the last assumption is not required for saturated boiling). Chang's model was based on a force balance of a bubble in contact with the wall, as shown in Figure 2.30. The acting forces are

Buoyancy force : $\quad F_B = \dfrac{C_B (r_b)^3 (\rho_L - \rho_G) g}{g_c}$ (2-114)

Surface tension force : $\quad F_S = C_s r_b \sigma$ (2-115)

Tangential inertia force : $\quad F_t = \dfrac{C_t (r_b)^2 \rho_L (u_{\text{rel}})^2}{g_c}$ (2-116)

Normal inertia force : $\quad F_n = \dfrac{C_n (r_b)^2 \rho_L (v_{\text{rel}})^2}{g_c}$ (2-117)

where u_{rel} and v_{rel} are components of the relative velocity between the liquid and the bubble, parallel and normal to the wall, respectively. C_B, C_s, C_t, and C_n are

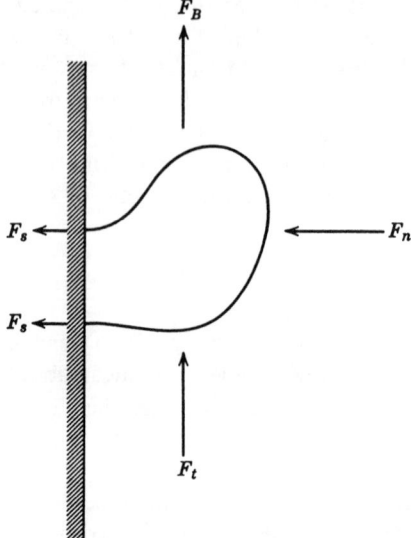

Figure 2.30 Equilibrium condition of a bubble.

constants. By neglecting the normal inertia force, Chang obtained a force balance for a bubble on a vertical wall as

$$C_B \left(\frac{g}{g_c}\right)(\rho_L - \rho_G)r_b^2 + C_t \rho_L (u_{rel})^2 \frac{r_b}{g_c} - C_s \sigma = 0 \qquad (2\text{-}118)$$

from which the final departure size of the bubble is obtained as

$$r_b = \left(\frac{g_c C_t}{2 C_B}\right)\left[\frac{\sigma(u_{rel})^2}{\rho_L (V_b)^4}\right]\left\{\left[1 + 4 \frac{C_s C_B}{(C_t)^2}\left(\frac{V_b}{u_{rel}}\right)^4\right]^{1/2} - 1\right\} \qquad (2\text{-}119)$$

where V_b is a dimensionless group that is proportional to the bubble detaching velocity as given by Peebles and Garber (1953), Eq. (2-63):

$$V_b = \left[\frac{gg_c \sigma(\rho_L - \rho_G)}{\rho_L^2}\right]^{1/4}$$

For saturated pool boiling on vertical surfaces, the Weber number (a ratio of dynamic force to the stabilizing force) alone determines the stability of a rising bubble:

$$\text{We}_b = \frac{\rho_L (v_{\text{rel}})^2 r_b}{g_c \sigma}$$

The critical resulting velocity of the liquid relative to the bubble, v_{rel}^*, can be evaluated from the critical Weber number, We*:

$$v_{\text{rel}}^* = \left[\frac{\text{We}_b^* g_c \sigma}{\rho_L r_b} \right]^{1/2} \tag{2-120}$$

By further simplification, Chang (1961) obtained the critical flux for vertical surfaces,

$$q_{\text{sat vert}}'' = 0.098(\rho_G)^{1/2} H_{fg} [\sigma g (\rho_L - \rho_G)]^{1/4} \tag{2-121}$$

and for boiling from horizontal surfaces, a constant ratio of $(q_{\text{sat vert}}''/q_{\text{sat hor}}'') = 0.75$ was adopted based on Bernath's (1960) comparison of a great number of CHFs from vertical heaters with those from horizontal ones. Thus,

$$q_{\text{sat hor}}'' = 0.13(\rho_G)^{1/2} H_{fg} [\sigma g (\rho_L - \rho_G)]^{1/4} \tag{2-122}$$

The above equation agrees with Kutateladze's correlation, which was derived through dimensional analysis (Kutateladze, 1952):

$$q_{\text{crit}}'' = K^{1/2} \{ H_{fg} (\rho_G)^{1/2} [\sigma g_c (\rho_L - \rho_G) g]^{1/4} \} \tag{2-123}$$

where K is a product of dimensionless groups,

$$K = \frac{V_b^2}{Lg} \left[\frac{\rho_G}{(\rho_L - \rho_G)} \right] \left(\frac{q''}{H_{fg} \rho_G V_b} \right)^2 \left[\frac{g(\rho_L - \rho_G) L^2}{g_c \sigma} \right]^{1/2} \tag{2-124}$$

and L is a characteristic length (e.g., the cavity diameter).

Kutateladze's data on various surface conditions of horizontal wires and disks indicate that the average value of $K^{1/2}$ is 0.16, in a range from 0.13 to 0.19. This equation agrees well with the pool boiling critical fluxes obtained by Cichelli and Bonilla (1945) for a number of organic liquids.

For subcooled pool boiling from vertical surface, Eq. (2-122) becomes

$$q_{\text{sub pool}}'' = (C_3)^{1/4} [0.0206(\rho_G \rho_L)^{1/2} H_{fg} + C_2 \rho_L C_p \Delta T_L] V_b \tag{2-125}$$

where C_2 represents the proportionality constant in the equation

$$q''_{\text{crit}} = C_2 \rho_G H_{fg} v^*_{\text{rel}}$$

Values of C_2 and $(C_3)^{1/4}$ are as follows (Chang, 1961):

Type/liquid	Heater	C_1	C_2	$(C_3)^{1/4}$
Subcooled pool boiling water	Vertical	0.0206	0.0106	6.62
Subcooled pool boiling ethanol	Horizontal	0.0206	0.0065	6.30

Ivey and Morris (1962) reported the ratio of subcooled critical flux to saturated critical flux of pool boiling in water, ethyl alcohol, ammonia, carbon tetrachloride, and isooctane for pressures from 4.5 to 500 psia (0.3 to 34 × 10⁵ Pa) as

$$\frac{q''_{\text{crit, sub}}}{q''_{\text{crit, sat}}} = 1 + 0.1\left(\frac{\rho_G}{\rho_L}\right)^{1/4}\left[\frac{C_P \rho_L (T_{\text{sat}} - T_b)}{H_{fg} \rho_G}\right] \quad (2\text{-}126)$$

Zuber (1958, 1959) and with his colleagues (Zuber et al., 1961), approached the CHF from the transition boiling side because the hydrodynamic processes are more ordered and better defined for this side than for the nucleate boiling side. Zuber (1959) assumed that in transition boiling a vapor film separates the heating surface from the boiling liquid (Fig. 2.16), but because of Taylor instability, the liquid–vapor interface is in the form of two-dimensional waves, and vapor bubbles burst through the interface in time-and-space regularity at the nodal points of the waves (Sec. 2.3.2). The heat flux is assumed to be proportional to the frequency of the interfacial waves. As the flux is increased, the velocity of the rising vapor relative to the descending liquid reaches the point where Helmholtz instability sets in, obstructing the liquid flow toward the heating surface. Thus the CHF in transition boiling is determined by both Taylor and Helmholtz instabilities,

$$q''_{\text{crit}} = H_{fg} \rho_G V_b \left(\frac{A_G}{A}\right)$$

Substituting V_b from Eq. (2-72), $V_b = (g_c \sigma m/\rho_G)^{1/2}$, and from Figure 2.18, $A_G/A = \pi(\lambda_o/4)^2/(\lambda_o)^2 = \pi/16$,

$$m = \text{Helmholtz critical wave number} = \frac{2\pi}{\lambda_c} = \frac{2\pi}{2\pi R_j} = 4/\lambda_o$$

Since $\lambda_c \leq \lambda_o \leq \lambda_d$, Zuber (1959) recommended the use of a reasonable average between λ_c and λ_d, and for pressures much less than the critical pressure,

$$q''_{crit} = 0.13 H_{fg}\, \rho_G \left[\frac{g_c g \sigma(\rho_L - \rho_G)}{\rho_G^2}\right]^{1/4} \quad (2\text{-}127)$$

which is same as the formulations of Kutateladze (1952) and Chang (1961).

Moissis and Berenson (1962) also derived the pool boiling CHF on horizontal surfaces by means of hydrodynamic transitions. Instead of taking λ_o between values of λ_c and λ_d, they used the most unstable wavelength as proportional to the jet diameter, D_j,

$$\lambda = \frac{2\pi}{m} = 6.48 D_j$$

and

$$D_j = 4.7 \left[\frac{g_c \sigma}{g(\rho_L - \rho_G)}\right]^{1/2}$$

Instead of Eq. (2-72), the value of V_G was determined from

$$(V_G - V_L) \leq C_1 \left[\frac{g_c \sigma m(\rho_L + \rho_G)}{\rho_L \rho_G}\right]^{1/2}$$

where C_1 is a geometric factor that takes into account the three-dimensionality and the finite thickness of the vapor columns. Using values of the experimental constant obtained from available CHF data for boiling from horizontal surfaces for $\rho_G \ll \rho_L$, Moissis and Berenson obtained the following final equation:

$$q''_{crit} = 0.18 H_{fg}\, \rho_G^{1/2} [g_c g \sigma(\rho_L - \rho_G)]^{1/4} \quad (2\text{-}128)$$

One more expression for q''_{crit} is that due to Lienhard and Dhir (1973b), who suggested $\lambda_c = \lambda_d$ (Sec. 2.3.2) and found the proportionality constant in the above equation to be 0.15, or

$$q''_{crit} = 0.15 H_{fg}\, \rho_G^{1/2} [g_c g \sigma(\rho_L - \rho_G)]^{1/4} \quad (2\text{-}128a)$$

The coefficient B (the Kutateladze number), when determined experimentally, was found to vary significantly from one liquid to another and also with pressure. Kutateladze and Malenkov (1974) published an expression for B in correlating "boiling" and "bubbling" data,

86 BOILING HEAT TRANSFER AND TWO-PHASE FLOW

$$B = f\left\{\left(\frac{1}{U_s}\right)\left[\frac{g_c g\sigma}{\rho_L - \rho_G}\right]^{1/4}\right\} \quad (2\text{-}129)$$

$$= \frac{q''_{\text{crit}}}{H_{fg}\,\rho_G^{1/2}\,[g_c g\sigma(\rho_L - \rho_G)]^{1/4}}$$

where U_s represents the velocity of sound in the vapor. Based on a total of 22 data points in boiling experiments with five different ordinary liquids at different pressures and in bubbling experiments with six different gas–liquid combinations, the value of B varied from 0.06 to 0.19 over a range of the dimensionless velocity-of-sound parameter, $(1/U_s)[g_c g\sigma/(\rho_L - \rho_G)]^{1/4}$, from 15 to 70 × 10^{-5}. Note that the velocity of sound for an ideal gas is proportional to $(\rho_L/\rho_G)^{1/2}$.

Effects of experimental parameters on pool boiling crisis with ordinary liquids In the following paragraphs the influences on the CHF in pool boiling of a number of experimental parameters are examined.

Effect of surface tension and wettability The effect of surface tension is represented as $q''_{\text{crit}} \simeq \sigma^{1/4}$ in the theoretical equation shown above. However, the effect of surface nonwettability on the maximum and transition boiling heat fluxes studied by Gaertner (1963a) and Stock (1960) gave conflicting observations. Liaw and Dhir (1986) observed both steady-state and transient nucleate boiling water data on vertical copper surfaces having contact angles of 38° and 107°, whereas their film boiling data were steady state (Fig. 2.31). For a given wall superheat, the transition boiling heat fluxes for the cooling curve are much lower than those for

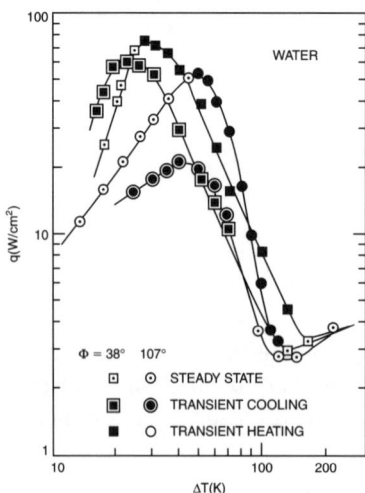

Figure 2.31 Boiling curve of water at contact angles of 38 and 107°. (From Liaw and Dhir, 1986. Copyright © 1986 by Hemisphere Publishing Corp., Washington, DC. Reprinted with permission.)

the heating curve. The CHFs obtained in transient cooling experiments are also lower than the steady-state heat fluxes. Further, the difference between the steady-state and transient CHFs and between the heating and cooling transition boiling heat fluxes are observed to increase with increase in the contact angle or decrease in the wettability of the test surface. Figure 2.32 shows the dimensionless CHF, which was defined previously as the Kutateladze number (B),

$$\frac{q''_{crit}}{\rho_G^{1/2} H_{fg} [\sigma g(\rho_L - \rho_G)]^{1/4}}$$

as a function of the contact angle. The CHF decreases with contact angle, whereas the difference between the steady-state and transient CHFs increases. At a contact angle of 107°, these critical heat fluxes are only about 50% and 20% respectively, of the value predicted from the hydrodynamic theory (Liaw and Dhir, 1986). A few experiments have also been conducted with Freon-113, for which the contact angle with polished copper was found to be near zero. The steady-state nucleate and film boiling data were obtained, as well as transition boiling data under transient heating and cooling modes. In these cases, the heating and cooling transition boiling curves nearly overlapped, and the steady-state and transient CHFs were within 10% of the values predicted from the hydrodynamic theory.

Effect of bubble density at elevated saturation pressures Gorenflo et al. (1986) interpreted the burnout event as locally coalescing bubbles, and suggested an ap-

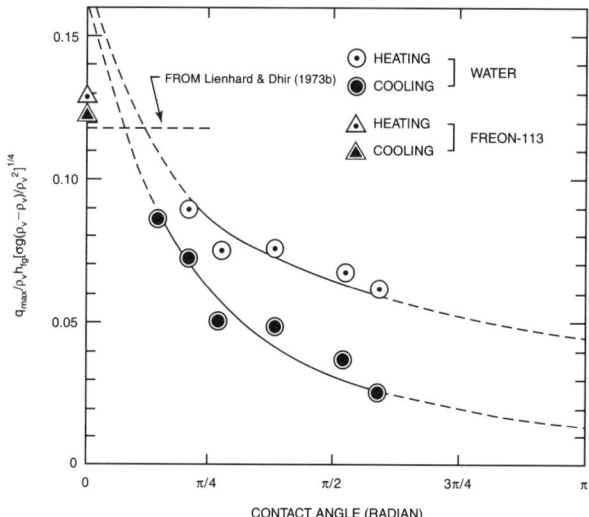

Figure 2.32 Dependence of the dimensionless critical heat flux on the contact angle during steady-state heating and transient cooling. (From Liaw and Dhir, 1986. Copyright © 1986 by Hemisphere Publishing Corp., Washington, DC. Reprinted with permission.)

proach using the transient conduction model for the heat transfer at active nucleation centers with assumptions about the distribution of the centers at high bubble density. It was shown that the relative pressure dependence of the CHF, q'' at burnout, and the heat transfer coefficient at single bubbles with heat flux-controlled bubble growth are similar, so they combined this technique with experimental data for the heat transfer coefficient in nucleate pool boiling. By assuming that burnout always occurs at a certain bubble density on the heated surface, the relative pressure dependence of q''_{crit} can be described quite well by the following equations:

$$\frac{q''_{crit}}{q''_{crit,o}} = 2.8(p_R)^{0.4}(1 - p_R) \qquad \text{for } p_R \geq 0.1 \qquad (2\text{-}130)$$

$$\frac{q''_{crit}}{q''_{crit,o}} = 1.05[(p_R)^{0.2} + (p_R)^{0.5}] \qquad \text{for } p_R \leq 0.1 \qquad (2\text{-}131)$$

where $q''_{crit,o}$ is the experimental value of the burnout heat flux at $p_R = 0.1$ (Gorenflo, 1984). These equations demonstrate the same relative pressure dependence of critical fluxes as do the correlations of Kutateladze (1959) and Noyes (1963).

Effect of surface conditions In his pioneering study of transition boiling heat transfer, Berenson (1960) used a copper block, heated from below by the condensation of high-pressure steam and cooled on top by the boiling of a low-boiling-point fluid. He found that while the nucleate boiling heat flux was extremely dependent on surface finish, the burnout heat flux in pool boiling was only slightly dependent on the surface condition of the heater. He obtained about a 15% total variation of q''_{crit} over the full range of surface finishes, with the roughest surfaces giving the highest values. The film boiling heat flux, however, was independent of the surface condition of the heater.

Ramilison and Lienhard (1987) re-created Berenson's flat-plate transition boiling experiment with a reduced thermal resistance in the heater, and improved access to certain portions of the transition boiling regime. Tests were made on Freon-113, acetone, benzene, and n-pentane boiling on horizontal flat copper heaters that had been mirror-polished, "roughened," or Teflon-coated. The resulting data reproduced and clarified certain features observed by Berenson (1960): a modest, or nonserious, surface-finish dependence of boiling CHF, and the influence of surface chemistry on both the minimum heat flux and the mode of transition boiling (Ramilison and Lienhard, 1987). The complete heat transfer data for acetone and Freon-113 are reproduced in Figures 2.33 and 2.34, respectively. These data, along with those of Berenson, were compared with the hydrodynamic CHF prediction of Lienhard et al. (1973c) as cited in Section 2.4.3.1. The "rough" surface data were consistently between 93% and 98% of the prediction. The highly polished surfaces

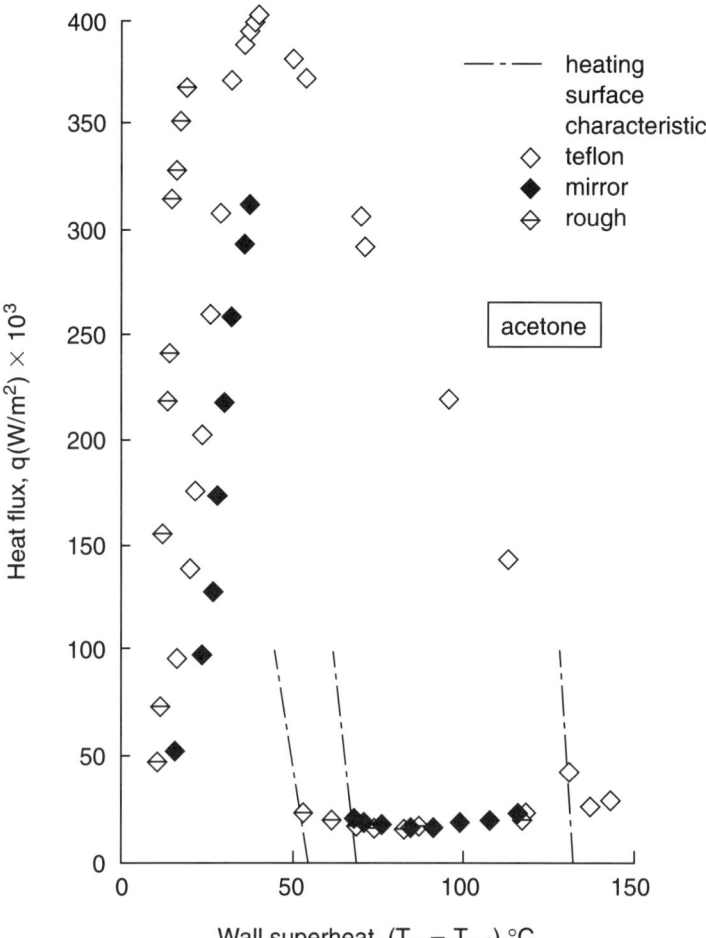

Figure 2.33 Boiling curves for acetone boiling on Teflon-coated, mirror-finished, and "rough" surfaces. (From Ramilison and Lienhard, 1987. Copyright © 1987 by American Society of Mechanical Engineers, New York. Reprinted with permission.)

consistently gave CHFs between 81% and 87% of the prediction. The Teflon-coated surfaces, on the other hand, gave values that exceeded the prediction by 4% to 10%.

That oxidized surfaces appear to yield higher CHFs than clean metallic surfaces was also reported by Ivey and Morris (1965). They found little difference in the CHF for wires that are not prone to severe oxidation. Results of tests with 0.020-in. wires of platinum, Chromel, silver, stainless steel, and nickel yielded CHFs of 350,000 + 20% BTU/hr ft² (1.1×10^6 + 20% W/m²). The scatter in the data for a given wire, as well as for different materials, was within the 20% band.

90 BOILING HEAT TRANSFER AND TWO-PHASE FLOW

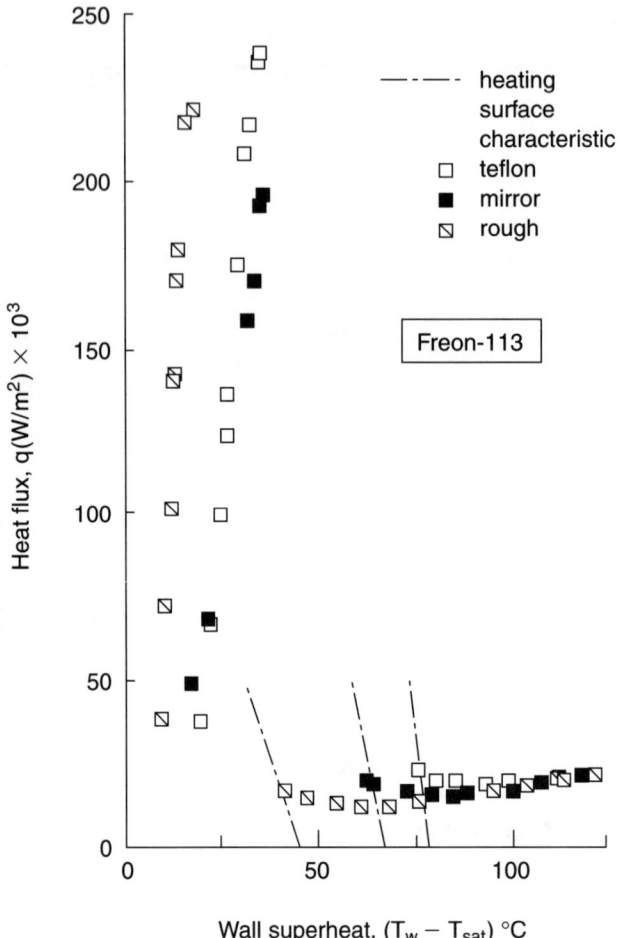

Figure 2.34 Boiling curves for Freon-113 boiling on Teflon-coated, mirror-finished, and "rough" surfaces. (From Ramilison and Lienhard, 1987. Copyright © 1987 by American Society of Mechanical Engineers, New York. Reprinted with permission.)

Effect of diameter, size, and orientation of heater Using data available then, Bernath (1960) studied the diameter effect of a horizontal cylindrical heater on the critical heat flux in pools of saturated water at atmospheric pressure. The results indicated that CHF increases as the heater diameter increases up to about 0.1 in. (2.5 mm), then levels off.

Sun and Lienhard (1970) found, through extensive experimentation with various liquids, that the vapor removal configuration in the region of the CHF depends on the diameter of the heater. Their analysis divided horizontal cylinders into "small" and "large" cylinders based on the dimensionless heater radius R' as defined by

$$R' = \frac{R}{[g_c \sigma/g(\rho_L - \rho_G)]^{1/2}}$$

For small cylinders, $0.2 < R' < 2.4$,

$$q''_{\text{crit}} = 0.123 H_{fg}\, \rho_G^{1/2} \left[\frac{(g_c)^3 \sigma^3 g(\rho_L - \rho_G)}{R^2} \right]^{1/8} \qquad (2\text{-}132)$$

Lienhard and Dhir (1973b) confirmed this equation by correlating with good accuracy about 900 data points obtained on several liquids over the dimensionless radius range $0.15 < R' < 1.2$ and over appreciable pressure and acceleration ranges. For large cylinders, $R' > 2.4$, Sun and Lienhard observed that the "most dangerous" Taylor unstable wavelength, λ_d [Eq. (2-77)], of the horizontal liquid–vapor interface was much smaller than the jet diameter and also was smaller than the normal Raleigh unstable wavelength $2\pi(R + \delta)$, observed with small cylinders. Thus,

$$q''_{\text{crit}} = 0.118 H_{fg}\, (\rho_G)^{1/2} [g_c g \sigma(\rho_L - \rho_G)]^{1/4} \qquad (2\text{-}133)$$

which indicates that for large cylinders, q''_{crit} is independent of R, and as in the case of large, flat surfaces, varies as the one-fourth power of g.

Bernath (1960) also studied the orientation effect on the CHF and found that the critical flux from a vertical heater is only about 75% of that of a horizontal heater under the same conditions. Further discussion of the latter effect is given in the section on the related effect of acceleration.

Effect of agitation The CHF of pool boiling can be increased considerably by introducing agitation, as shown by Pramuk and Westwater (1956) in experiments with boiling methanol at 1 atm (Fig. 2.35).

Effect of acceleration The effect of acceleration implied in Eq. (2-123) is that $q''_{\text{crit}}(a/g)^{0.25}$. This was confirmed by Merte and Clark (1961). Various other exponents for a/g were found by other experimenters at different pressures and heaters, e.g., Adams (1962) and Beasant and Jones (1963). Costello and Adams (1961) indicated that the exponent is dictated by surface characteristics and is not purely hydrodynamic.

Costello et al. (1965) also found that a flat ribbon heater, mounted on a slightly wider block, induced strong side flows. This induced convection effect on the CHF was identified by Lienhard and Keeling (1970) in their study of the gravity effect on pool boiling CHF. They developed a method for correlating such an effect under conditions of variable gravity, pressure, and size, as well as for various boiling liquids (e.g., methanol, isopropanol, acetone, and benzene). The effect was illus-

92 BOILING HEAT TRANSFER AND TWO-PHASE FLOW

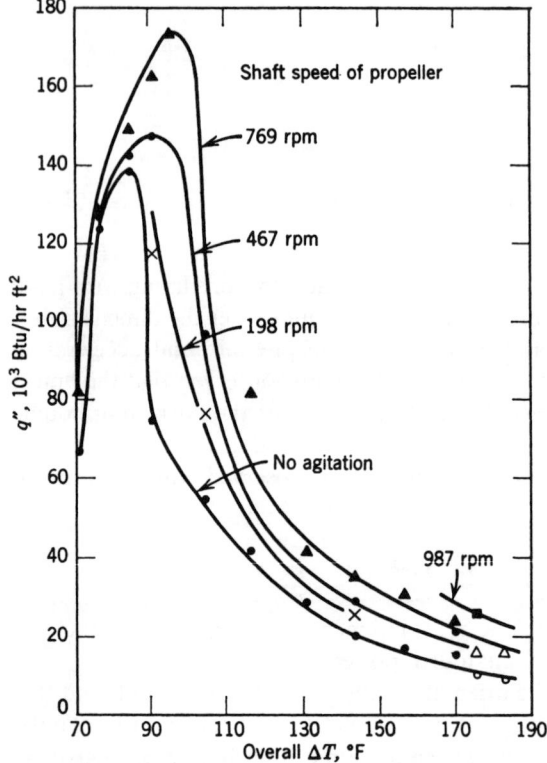

Figure 2.35 Effect of agitation in methanol boiling at 1 atm. (From Pramuk and Westwater, 1956. Copyright © 1956 by American Institute of Chemical Engineers, New York. Reprinted with permission.)

trated, and the correlation verified, with a large amount of CHF data obtained on a horizontal ribbon heater in a centrifuge, embracing an 87-fold range of gravity, a 22-fold range of width, and a 15-fold variation of reduced pressure. By means of dimensional analysis, they obtained a correlating relationship,

$$\frac{q''_{crit}}{q''_{critF}} = f\left[L', I, \sqrt{1 + \left(\frac{\rho_G}{\rho_L}\right)}\right] = f\left[L', N, \sqrt{1 + \left(\frac{\rho_G}{\rho_L}\right)}\right] \quad (2\text{-}134)$$

where $q''_{critF} = q''_{crit}$ for an infinite flat plate [e.g., Eq. (2-123)]

L = characteristic length

L' = dimensionless size = $L\sqrt{\dfrac{g(\rho_L - \rho_G)}{\sigma}}$

I = induced-convection scale parameter

$$= \frac{[(\text{inertia force})(\text{surface tension force})]^{1/2}}{(\text{viscous force})} = \sqrt{\frac{\rho_G L \sigma}{\mu^2}}$$

N = induced-convection buoyancy parameter

$$= \frac{[(\text{inertia force})(\text{surface tension force})^{3/2}]}{(\text{viscous force})^2 (\text{buoyant force})^{1/2}} = \left(\frac{I^2}{L'}\right)$$

For a horizontal ribbon heater, as used by Lienhard and Keeling (1970), $L = W$, the width of ribbon heater,

$$L' = W' = W\left[\frac{g(\rho_L - \rho_G)}{\sigma}\right]^{1/2}$$

Figures 2.36a and 2.36b show the resulting correlating surfaces in sets of contours. The correlation function $f(I, W')$ is presented in Figure 2.36a, which gives $q''_{\text{crit}}/q''_{\text{critF}}$-versus-$I$ contours as obtained in a number of cross plots. Further articulation of Eq. (2-134) was made for other geometries, including ribbons and finite plates (Lienhard and Dhir, 1973b), spheres (Ded and Lienhard, 1972). By collecting burnout curves for various geometries, Lienhard and Dhir (1973b) found that the function in Eq. (2-134) becomes a constant when R' (or L') is large (> 2) and so q''_{crit} depends on $g^{1/4}$.

For smaller cylinders (or for lower gravity), the gravity dependence is more complicated. Lienhard (1985) concluded that q''_{crit} for large heaters varies as $g^{1/4}$, but q must be gravity-independent in the region of the slugs-and-columns regime (Fig. 2.37), because the standing jets provide escape paths for the vapor. He based his conclusion on the experimental data of Nishikawa et al. (1983) with a plate oriented at an angle θ, which varied between 0 and 175° from a horizontal, upward-facing position. In this reference, photographs were included in this article that showed how the jets bend over, as the plate is tilted beyond 90°, and slide up the plate as large amorphous slugs. This means that a modified hydrodynamic theory would be required to predict q''_{crit} in this region, as suggested by Katto (1983) at the same conference. Katto proposed that q''_{crit} might occur as the result of Helmholtz instability, not in the large jet but in small feeder jets (below the obvious jets) formed by the merging bubbles. It is possible that such a hydrodynamic mechanism would dictate a higher q''_{crit} than is given by the conventional prediction, and it can only come into play when the large jets are eliminated, as they are in subcooled boiling or when the heater is tilted beyond 90° (Lienhard, 1985).

Effect of subcooling The correlation of Ivey and Morris, Eq. (2-126), correlated their own data with that of Kutateladze (Kutateladze and Shneiderman,

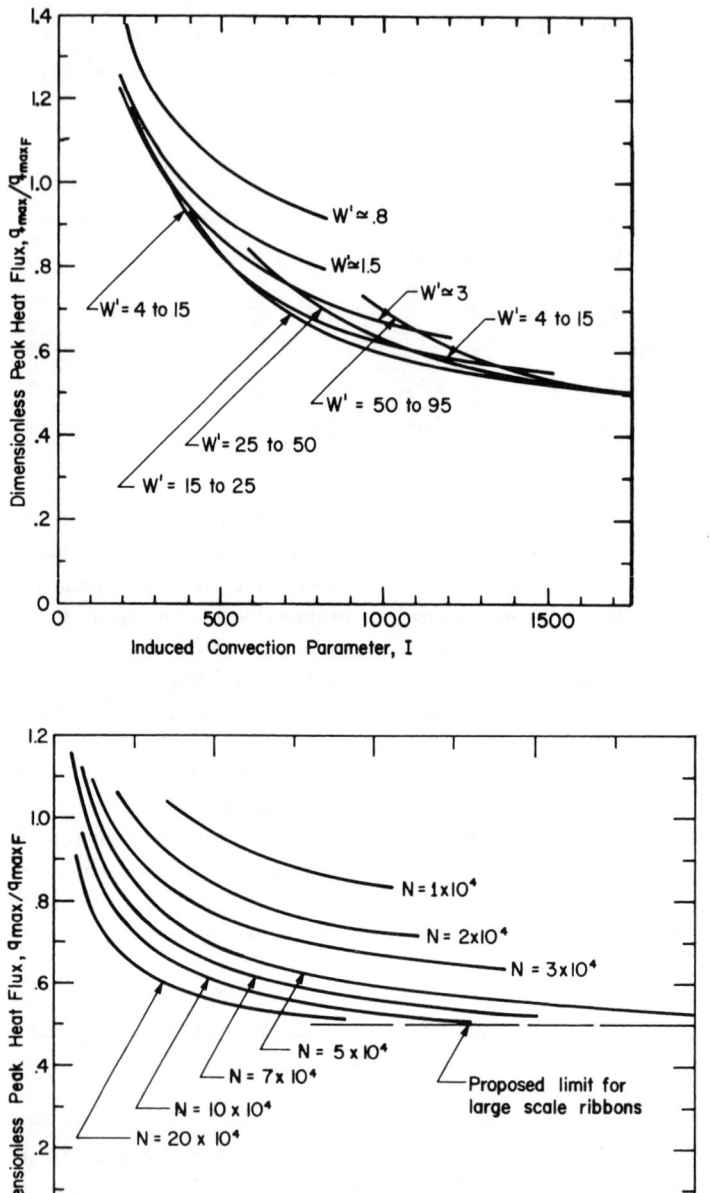

Figure 2.36 (a) $q''_{crit}/q''_{crit,\,F}$ versus I contours for eight values of W'. (b) $q''_{crit}/q''_{crit,\,F}$ versus W'' contours for seven values of N. (From Lienhard and Keeling, 1970. Copyright © 1970 by American Society of Mechanical Engineers, New York. Reprinted with permission.)

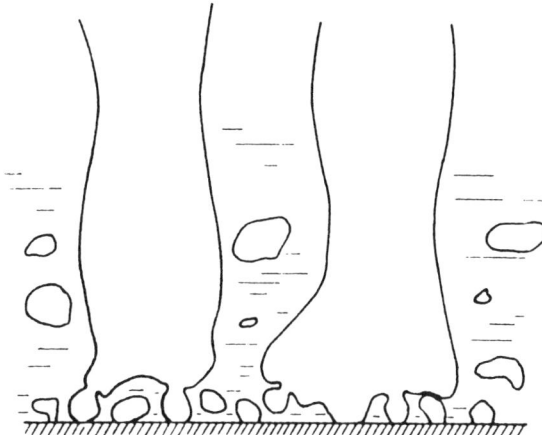

Figure 2.37 Vapor removal in the slugs-and-columns regime of nucleate pool boiling. (From Lienhard, 1985. Copyright © 1985 by American Society of Mechanical Engineers. Reprinted with permission.)

1953) for horizontal wires 1.22 to 2.67 mm (0.05 to 0.105 in.) in diameter in water, in the range $0 < \Delta T_{sub} < 72°C$ (130°F). The correlation was accurate only within ±25% and failed to represent data for other geometries.

Elkassabgi and Lienhard (1988) provided an extensive subcooled pool boiling CHF data set of 631 observations on cylindical electric resistance heaters ranging from 0.80 to 1.54 mm (0.03 to 0.06 in.) in diameter with four liquids (isopropanol, acetone, methanol, and Freon-113) at atmospheric pressure and up to 140°C (252°F) subcooling. They normalized $q''_{\text{crit sub}}$ data by Sun and Lienhard's (1970) saturated $q''_{\text{crit SL}}$ prediction in terms of B,

$$B = \frac{q''_{\text{crit}}}{(\rho_G)^{1/2} H_{fg} [\sigma g(\rho_L - \rho_G)]^{1/4}} = 0.1164 + 0.297 \exp\left(-3.44\sqrt{R'}\right) \quad (2\text{-}135)$$

which made it possible to see that $q''_{\text{crit sub}}$ passed through three identifiable regimes of boiling behavior as ΔT_{sub} was increased (Fig. 2.38). These three regimes of subcooled burnout behavior were more sharply evident in photographs (Lienhard, 1988) as low subcooling, intermediate subcooling, and highly subcooled regions.

For low subcooling,

$$\frac{q''_{\text{crit sub}}}{q''_{\text{crit}}} = 1 + f(R')(\text{Ja})(\text{Pe})^{-1/4} = 1 + 4.28(\text{Ja})(\text{Pe})^{-1/4} \quad (2\text{-}136)$$

where Pe (Peclet number) $= \sigma^{3/4}/\{\alpha[g(\rho_L - \rho_G)]^{1/4} \rho_G^{1/2}\}$. The values of q''_{crit} were based on experimentally determined values. Equation (2-136) represented the data within a root-mean-square (rms) error of ±5.95%.

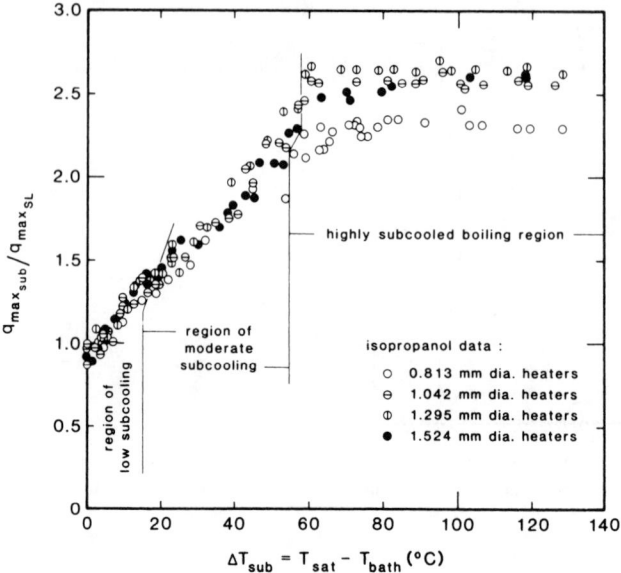

Figure 2.38 Effect of liquid subcooling on peak heat flux, for heaters of various sizes in isopropanol. (From Elkssabgi and Lienhard, 1988. Copyright © 1988 by American Society of Mechanical Engineers, New York. Reprinted with permission.)

For intermediate subcooling,

$$\text{Nu} = 28 + 1.50(\text{Ra})^{1/4}(\beta \Delta T_{\text{sub}})^{7/8} \tag{2-137}$$

where $\text{Nu} = \dfrac{q''_{\text{crit sub}}(2R_{\text{eff}})}{k \Delta T_{\text{sub}}}$

$\text{Ra} = \dfrac{g\beta\Delta T_{\text{sub}}(2R_{\text{eff}})^3}{\alpha \nu}$

α, ν = thermal diffusivity and kinetic viscosity, respectively

β = volumetric coefficient of thermal expansion

$R_{\text{eff}} = R\left(1 + \dfrac{0.02\theta}{R'}\right)$

θ = retreating contact angle

$R' = R\left[\dfrac{\sigma}{g(\rho_L - \rho_G)}\right]^{-1/2}$

Equation (2-137) represents the data accurately within a rms error of ±7.06%.

For high subcooling, the ratio of $q''_{\text{crit sub}}$ to the upper limit of boiling heat fluxes (an effusion limit to burnout) becomes

$$\phi = \frac{q''_{\text{crit sub}}}{\rho_G H_{fg} \sqrt{R_g T_{\text{sat}}/2\pi}} = 0.01 + 0.0047 \exp(-1.11 \times 10^{-6} X) \quad (2\text{-}138)$$

where R_g = ideal gas constant

$$X = \frac{R(R_g T_{\text{sat}})^{1/2}}{\alpha}$$

Equation (2-138) represents the data with a rms deviation of 6.82%.

Elkassabgi and Lienhard (1988) discovered that most of the existing subcooled boiling data for horizontal cylinders are data for which R' is too small, a significant liquid cross flow had been imposed, or for which the system parameters had been incompletely reported. A serious problem that is yet unresolved with the three new q''_{crit} predictions is determining which of them is appropriate for a heater of a given size in a given liquid at a given ΔT_{sub}. As of now, this determination must often be made after the fact.

2.4.3.2 Boiling crisis with liquid metals.

Theoretical considerations Thermal conductivity and wetting properties are two major differences between liquid metals and ordinary liquids, of which both influence the magnitude of the CHF. When the theoretical correlations were presented in previous sections they were derived solely on the basis of hydrodynamic considerations, without consideration of thermal conductivity or any other thermal transport property. While this is apparently quite permissible for ordinary liquids, it will be shown that this does not hold true for liquid metals. As was shown before, the theoretical CHF equation for boiling ordinary liquid on a flat plate allows only for the heat carried away from the heating surface by the departing columns of vapor, which can be referred to by q''_{evap}. An addition contribution to the CHF of boiling liquid metals is the heat transfer from the surface due to the temperature gradient in the liquid, which can be referred to by $q''_{c\text{-}c}$, the conduction-convection contribution. The latter term is relatively large in liquid metals (because of their high thermal conductivities). This leads to the theoretical equation for critical saturated stable nucleate boiling heat flux (Noyes and Lurie, 1966):

$$q''_{\text{crit}} = q''_{\text{evap}} + q''_{c\text{-}c} \quad (2\text{-}139)$$

While there is no reason to assume that basic features of the hydrodynamic vapor-removal mechanism, q''_{evap}, do not hold for liquid metals, the term $q''_{c\text{-}c}$ for liquid metals is much larger than that of ordinary liquids, and even greater than q''_{evap}. As yet, no theoretical method of estimating q''_{c-c} for liquid metals has been found, and one must rely on empirical correlations. The fact that q''_{c-c} is not significantly affected by variation in pressure makes q''_{crit} for liquid metals much less dependent on pressure than that for ordinary liquids. Subbotin et al. (1968) developed a CHF correlation via the form of expression of Eq. (2-139), using the empirically determined ratio (q''_{c-c}/q''_{evap}), based on experimental data for sodium, potassium, rubidium, and cesium from six different sources obtained with both horizontal flat surfaces and cylinders:

$$\left(\frac{q''_{c-c}}{q''_{crit}}\right) = \left(\frac{45}{p_{cr}}\right)\left(\frac{p_{cr}}{p_L}\right)^{0.4} \quad (2\text{-}140)$$

where p_{cr} in the ratio ($45/p_{cr}$) is in atmospheres, but k_L is notably missing even though Eq. (2-140) does represent a liquid-phase conduction-convection contribution to the CHF. The final correlation of Subbotin et al. (1968) is therefore

$$q''_{crit} = 0.14 H_{fg} (\rho_G)^{1/2} \left[1 + \left(\frac{45}{p_{cr}}\right)\left(\frac{p_{cr}}{p_L}\right)^{0.4}\right] [g_c g \sigma(\rho_L - \rho_G)]^{1/4} \quad (2\text{-}141)$$

Predictions of the effect of subcooling With experimental results of subcooled nucleate boiling CHF for liquid metals lacking, the $q''_{\text{crit sub}}$ can only be estimated by Kutateladze's equation (1952),

$$q''_{\text{crit sub}} = q''_{\text{crit}} + q''_{\text{sub}} \quad (2\text{-}142)$$

A combination of two separate energy-transfer mechanisms were considered for q''_{sub}: a conduction mechanism as in saturation boiling, and a sensible heat transport mechanism.

Predictions of the effect of pressure The boiling pressures for liquid metals are relatively low and their critical pressures are high, thus the CHF data are usually obtained in the reduced pressure range of 10^{-4} to 10^{-2} and are generally correlated by a simple power function of the form (Dwyer, 1976)

$$q''_{\text{crit}} = a(p_L)^b \quad (2\text{-}143)$$

where a and b are empirical constants. Dwyer compared a number of theoretical correlations [e.g., Eqs. (2-127), (2–128a), (2-129), and (2-133)] and showed the dependence of the CHF on the boiling pressure (or value of the exponent b) at three different pressures for such liquid metals as Hg, Na, K, and Cs. Part of Dwyer's

Table 2.3 Values of b in Eq. (2-143) from various theoretical CHF correlations for ordinary liquids in pool boiling

Type of heater	L/M	$p_L =$ 1 psia	$p_L =$ 10 psia	$p_L =$ 50 psia
For horizontal plates	Hg	0.45	0.45	0.45
From Eqs. (2-127) and (2-128a)	Na	0.43	0.41	0.38
($L/\lambda_d \geq 3$)	K	0.43	0.41	0.35
	Cs	0.42	0.40	0.33
From Eq. (2-129)	Hg	0.43	0.42	0.41
($L/\lambda_d \geq 3$)	Na	0.40	0.37	0.32
	K	0.42	0.36	0.31
	Cs	0.41	0.36	0.29
For horizontal cylinders	Hg	0.45	0.45	0.45
From Eq. (2-133)	Na	0.43	0.41	0.38
($R' \geq 2.4$)	K	0.43	0.41	0.35
	Cs	0.42	0.40	0.33

Source: Dwyer (1976). Copyright © 1976 by American Nuclear Society, LaGrange Park, IL. Reprinted with permission.

results are shown in Table 2.3. As in the case of ordinary liquids, theoretical correlations for predicting the CHF of boiling on horizontal plates and about horizontal cylinders are given, because these two types of heaters have been used almost exclusively in pool boiling.

The results show a tendency for b to decrease with an increase in pressure, and the higher the critical pressure of a given liquid metal, the less this tendency will be. Thus, in the case of mercury, some correlations show no decrease or a very slight decrease in b over the pressure range compared. However, this theoretical dependence is about three times greater than that observed experimentally with liquid metals, which suggests strongly that these theoretical correlations for CHF require revisions to achieve applicability to liquid metals.

Effects of experimental parameters of pool boiling crisis with liquid metals

Effect of geometric factors As shown in Section 2.4.3.1, Lienhard and Dhir (1973b) expressed the minimum dimension, L, of a horizontal flat-plate heater in terms of the dimensionless ratio L/λ_d. For ordinary liquids they found that the CHF is constant as long as $L/\lambda_d > 3$ (Eq. 2–128a). Otherwise, the CHF depends on the actual number of vapor jets (Lienhard and Dhir, 1973b),

$$q''_{crit} = \left[\frac{N_j (\lambda_d)^2}{A} \right] [q''_{crit} \text{ from Eq. (2-128}a)] \quad (2\text{-}144)$$

where N_j is the number of vapor jets on the heating surface area A with vertical side walls. For liquid metals, q''_{crit} in Eq. (2-144) becomes q''_{evap}, as shown in Eq. (2-139). Since q''_{c-c} is presumably far less sensitive to plate size in the lower L/λ_d range, it can be concluded that the effect of reducing L/λ_d below 3 on q''_{crit} is apparently much less for liquid metals than for ordinary liquids.

For horizontal cylinders, for all practical purposes, we are not concerned with D' values less than 0.5, and for convenience, we follow Lienhard and Dhir's (1973b) suggestion of using diametral size corresponding to $R' = 1$ as the dividing line between "small" and "large" heaters. For Na, K, and Hg boiling under 1 atm pressure and a normal gravitational force, this value of R' corresponds to heater diameters of 0.32 in. (8.1 mm), 0.25 in. (6.4 mm), and 0.14 in. (3.6 mm), respectively. For the value of q''_{evap}, Eq. (2-132) is recommended for small cylinders and Eq. (2-133) for large cylinders, as described in Section 2.4.3.1. For estimating q''_{crit} in saturated stable nucleate boiling of liquid metals on either large horizontal plates or horizontal cylinders, the generalized empirical correlation, Eq. (2-145), is recommended, as suggested by Kirillov (1968), who invoked the law of corresponding states (Borishansky, 1961),

$$q''_{crit} = 3.12 \times 10^5 (k_L)^{0.6} \left(\frac{p_L}{p_{cr}} \right)^{1/6} \tag{2-145}$$

Another generalized empirical correlation by Subbotin et al. (1968), Eq. (2-141), is also recommended for this purpose.

Effect of surface conditions While the value of the CHF is assumed not to be significantly affected by variation in heating surface roughness for ordinary liquids, some experiments with boiling liquid metals (cesium) on horizontal 0.43-in. (11-mm)-diameter stainless steel-clad cylindrical heaters of three different surface types (Kutateladze et al., 1973; Avksentyuk and Mamontova, 1973) showed different magnitudes and kinds of crisis. These experimenters tested three types of surfaces:

1. Smooth (formerly USSR Roughness Class No. 6)
2. Smooth, with artificial reentrant cavities [with diameters of outer circular openings of 0.15–0.20 mm (0.006–0.008 in.)]
3. Corrugated [0.1-mm (0.004-in.)-wide grooves over the entire heating surface]

Their results showed the following. Surface 1 gave direct transition from liquid-phase natural-convection heat transfer to film boiling with CHF values of 160,000 Btu/hr ft² (503 kW/m²), independent of the pressure. Surface 2 gave stable nucleate boiling with CHF values much greater than those obtained with surface 1, and which increased as the 0.1 power of the pressure. Surface 3, on the other hand, gave nucleate boiling that was less stable than that obtained with surface 2 and lower CHF values. Thus it appears that the magnitude and kind of boiling crisis with liquid metals depends on the topographical characteristics of the heating surface. For clean surfaces with the same microstructure and the same degree of wetting, the type of surface material is believed to have no significant effect on the magnitude of the CHF (Dwyer, 1976).

Effect of physical properties Physical properties of liquid metals that have significant effects on CHF values are thermal conductivity, latent heat of vaporization, and surface tension.

The thermal conductivity of the liquid, although it has no effect on q''_{evap}, does have a significant positive effect on both q''_{c-c} and q''_{sub}, which accounts for the larger values of CHF for saturated boiling liquid metals. It has an even greater effect on the CHF for subcooled boiling, depending on the pressure and the amount of subcooling (see the later section on the effect of subcooling).

Latent heat of vaporization has, as shown before, a strong positive effect on q''_{evap} and appears to have the same effect on q''_{c-c}, but it may have a negative or negligible effect on q''_{sub}. Thus it has an important effect on the CHF for saturated boiling but a gradually diminishing one as the amount of subcooling is increased (Dwyer, 1976). The specific heat of liquid, on the other hand, has little effect on $q''_{crit\ sat}$ but can have an appreciable effect on $q''_{crit\ sub}$, as will again be shown later.

Surface tension, σ, as shown in previous theoretical equations for q''_{evap} of boiling liquid metals or for q''_{crit} of ordinary liquids, is included in the form $(\sigma)^{1/4}$, or as high as $(\sigma)^{3/8}$. A comparable effect of σ is found on q''_{c-c}. However, its effect on q''_{sub} is not so certain, as will also be discussed later under the effect of subcooling.

Effect of oxygen concentration Oxygen concentrations below the solubility limits in alkali metals are believed to promote boiling stability and therefore increase the CHF. If the oxygen solubility limit is exceeded, the surface is coated by the oxygen concentration, and the alkali metal oxide is chemically reduced such that the degree of wetting is affected (reduced), the boiling will obviously become less stable and the CHF will be reduced.

Effect of acceleration As for ordinary liquids, q''_{evap} for boiling liquid metals varies as $g^{1/4}$. According to Eq. (2-140), the degree of acceleration has the same effect on q''_{c-c} as it does on q''_{evap}. It can be expected that the acceleration has a significant positive effect on q''_{crit} of liquid metals.

Effect of subcooling As shown in Eq. (2-142), the effect of subcooling is expressed by the term q''_{sub}. For ordinary liquids, Ivey and Morris (1962) recommended a modified Kutateladze equation:

$$q''_{\text{crit sub}} = q''_{\text{crit sat}} \left[0.1(\rho_L - \rho_G)^{3/4} \left(\frac{c_{pL}}{H_{fg}} \right)(T_{\text{sat}} - T_{\text{bulk}}) \right] \qquad (2\text{-}146)$$

This equation was derived based on a sensible heat transport mechanism. When applied to boiling liquid metals, $q''_{\text{crit sat}}$ in this equation becomes q''_{evap}, which can be obtained by Eq. (2–128a). Substituting this into Eq. (2-142),

$q''_{\text{crit sub}} = q''_{\text{crit}}$ [from Eq. (2-141)]

$$+ q''_{\text{crit}}[\text{from Eq. (2-128}a\text{)}] \left[0.1 \left(\frac{\rho_L}{\rho_G} \right)^{3/4} \left(\frac{c_{pL}}{H_{fg}} \right)(T_{\text{sat}} - T_{\text{bulk}}) \right] \qquad (2\text{-}147)$$

It should be noted that without experimental data on the subcooled pool boiling crisis in liquid metals, the above equation cannot be verified. Another mechanism for estimating the subcooling contribution to the CHF was used for boiling with ordinary liquids (i.e., a conduction mechanism). The two mechanisms may operate simultaneously, along with the hydrodynamics and conduction-convection mechanisms (Dwyer, 1976).

2.4.4 Film Boiling in a Pool

Immediately after critical heat flux is reached, the boiling mechanism becomes unstable. This regime is called *partial film boiling* or *transition boiling*. The vapor is formed in transition boiling by explosive bursts that occur at random locations and can be observed in high-speed motion pictures (Westwater and Santangelo, 1955). The frequency of the vapor bursts is very high. For an overall ΔT of 133°F (74°C) in the transition boiling of methanol on a $\frac{3}{8}$-in. (0.95-cm) copper tube, each inch of the photographed side of the tube exhibited 84 bursts per second. The heat flux of transition boiling is between that of nucleate boiling and stable film boiling, as shown by curve *C–D* of Figure 1.1.

As the wall temperature increases further, the boiling again becomes stable and is called *stable film boiling*. Under this condition the surface is so hot that the momentum of the rapidly evaporating vapor between the liquid and the hot surface forms a vapor cushion that prevents the liquid from wetting the surface. This is termed the *spheroidal state,* or the *Leidenfrost (1756) point*. It thus represents, for a given pressure, the minimum value of T_w at which the liquid will not wet the heating surface in a normal wetting liquid to solid system. Heat transfer in stable film boiling is normally accomplished by conduction through and radiation across the vapor film. Depending on the boiling pressure and the heating surface temperature, the relative radiant contribution can vary widely. Thus the total heat transfer coefficient, h, is generally expressed as the sum of a convective coefficient, h_c, and an effective radiation coefficient, (fh_r), where f is a constant.

2.4.4.1 Film Boiling in Ordinary Liquids.

Theoretical considerations

Convective heat transfer correlations for film boiling
1. Film boiling on horizontal plates. Following Chang's (1957, 1959) and Zuber's (1958, 1959) application of the Taylor instability theory to transition and film boiling heat transfer from a flat horizontal plate, various wave theory calculational models were developed that are distinguishable from each other by whether bubble release was assumed to be regular or random, and whether the vapor flow in the film was assumed to be laminar or inertia-dominated (Frederking et al., 1966). Of these four possible representative models, Sciance et al. (1967) found those based on the assumptions of regular vapor release and laminar vapor flow to be superior when compared with experimental results on four organic liquids. This set of assumptions was made by Berenson (1960). Employing the "most dangerous" wavelength, λ_d [Eq. (2-77)], and the bubble diameter relationship

$$D_b = 4.7 \left[\frac{g_c \sigma}{g(\rho_L - \rho_G)} \right]^{1/2} \qquad (2\text{-}148)$$

which was based on data obtained during film boiling of water, benzene, ethyl alcohol, and carbon tetrachloride on horizontal surfaces (Borishansky, 1953), Berenson derived an equation for the minimum heat flux:

$$q''_{\min} = 0.09 \rho_G H'_{fg} \left[\frac{g(\rho_L - \rho_G)}{(\rho_L + \rho_G)} \right]^{1/2} \left[\frac{g_c \sigma}{g(\rho_L - \rho_G)} \right]^{1/4} \qquad (2\text{-}149)$$

where H'_{fg} is "effective" heat of vaporization, first defined by Bromley (1950), as the difference between the heat content of the vapor at T_f and that of liquid at T_{sat}.
The equation for the convective heat transfer coefficient then becomes

$$h_c = 0.425 \left\{ \frac{(k_G)^3 H'_{fg} \rho_G g(\rho_L - \rho_G)}{\mu_G (T_w - T_{\text{sat}})[g_c \sigma/g(\rho_L - \rho_G)]^{1/2}} \right\}^{1/4} \qquad (2\text{-}150)$$

for fluxes close to $q''_{\min}$. This equation can be expressed in generalized form as

$$(\mathrm{Nu}_B)_f = 0.425(\mathrm{Ra}_B^*)_f^{1/4}\left[\frac{H'_{fg}}{c_{pG}(T_w - T_{sat})}\right]_f^{1/4} \quad (2\text{-}151)$$

where $(\mathrm{Nu}_B) = \dfrac{h_c B}{k_G}$

$$(\mathrm{Ra}_B^*) = \mathrm{Gr}_B^* \, \mathrm{Pr}_G$$

$$= \left[\frac{B^3 \rho_G (\rho_L - \rho_G) g}{\mu_G^2}\right]\left(\frac{c_{pG}\mu_G}{k_G}\right)$$

$$B = \text{Laplace reference length} = \left[\frac{g_c \sigma}{g(\rho_L - \rho_G)}\right]^{1/2}$$

and f (subscript) means that the physical properties of the vapor are evaluated at pressure p_L and temperature T_f.

A similar equation was derived by Hamill and Baumeister (1967), except for the definition of the effective heat of vaporization, H'_{fg}.

2. *Film boiling on horizontal cylinders.* In his pioneering work on film boiling on a horizontal cylinder, Bromley proposed a classical theoretical equation in dimensionless form:

$$(\mathrm{Nu}_D)_f = 0.62(\mathrm{Ra}_D^*)_f^{1/4}\left[\frac{H'_{fg}}{c_{pG}(T_w - T_{sat})}\right]_f^{1/4} \quad (2\text{-}152)$$

where $(\mathrm{Nu}_D) = \dfrac{h_c D}{k_G}$

$$(\mathrm{Ra}_D^*) = (\mathrm{Gr}_D^*)(\mathrm{Pr}_G) = \left[\frac{D^3 \rho_G (\rho_L - \rho_G) g}{\mu_G^2}\right]\left(\frac{c_{pG}\mu_G}{k_G}\right)$$

By modifying Berenson's equation (2-151) for film boiling on a flat plate, Sciance et al. (1967) suggested the correlation

$$(\mathrm{Nu}_B)_f = 0.369\left[\frac{\mathrm{Ra}_B^*}{(T_r)^2}\right]_f^{0.267}\left[\frac{H'_{fg}}{c_{pG}(T_w - T_{sat})}\right]_f^{0.267} \quad (2\text{-}153)$$

which was based on their study of methane, ethane, propane, and *n*-butane on the surface of a horizontal gold-plated cylinder 0.81 in. (2.06 cm) in diameter by 4 in. (10 cm) long. This equation was found to be roughly on a par with Eq. (2-152) when compared with experimental results on a wide variety of nonmetallic liquids over wide pressure and diameter ranges (Clements and Colver, 1964).

3. *Film boiling on a vertical surface.* The thermal and hydrodynamic behaviors are essentially the same for both a cylinder and a flat plate, as long as the diameter of the cylinder is considerably larger than the thickness of the vapor film. Consequently, Bromley (1950) recommended an equation very similar to Eq. (2-152), with a change in the characteristic length D to L, where L is the vertical distance from the bottom of the plate. Figure 2.39 illustrates how the vapor starts to rise in laminar motion from the bottom edge of the vertical plate, and the vapor–liquid interface is smooth. The vapor film thickness increases with height until, a short distance up the plate, the interface begins to show capillary waves. The laminar sublayer reaches a critical thickness δ_c at some height L_c, at which point, transition from laminar to turbulent flow occurs (Fig. 2.39). Hsu and Westwater (1960) calculated the coefficient in Eq. (2-152) to be 0.943, and used a new effective latent heat of vaporization, H''_{fg}, so that the correlation becomes

$$(Nu_L)_f = 0.943[(Ra_L^*)_f]^{1/2} \left[\frac{H''_{fg}}{c_{pG}(T_w - T_{sat})} \right]^{1/2}$$

where $L \leq L_c$, $T_f = (T_w + T_{sat})/2$, and $H''_{fg} = H_{fg}[1 + 0.34 c_{pG}(T_w - T_{sat})/H_{fg}]^2$. Thus,

$$(Nu_L)_f = 0.943(Ra_L^*)_f^{1/2} \left\{ \frac{H_{fg}[1 + 0.34 c_{pG}(T_w - T_{sat})/H_{fg}]^2}{c_{pG}(T_w - T_{sat})} \right\}^{1/2} \quad (2\text{-}154)$$

The values of L_c and δ_c are calculated from standard theoretical relations for the velocity profile in the laminar film (parabolic relationship) and for the dependence of film thickness on vertical distance (one-fourth power relationship), respectively. Hsu and Westwater obtained the following expression for the local heat transfer coefficient in the turbulent region:

$$(h_{c,x})_{turb} = k_G \left\{ \frac{2}{3} \left[\frac{3A}{(3B' + 1)} \right] (x - L_c) + \left(\frac{1}{\delta_c} \right)^2 \right\}^{1/2} \quad (2\text{-}155)$$

where x is the distance up from the bottom of the heated surface, and

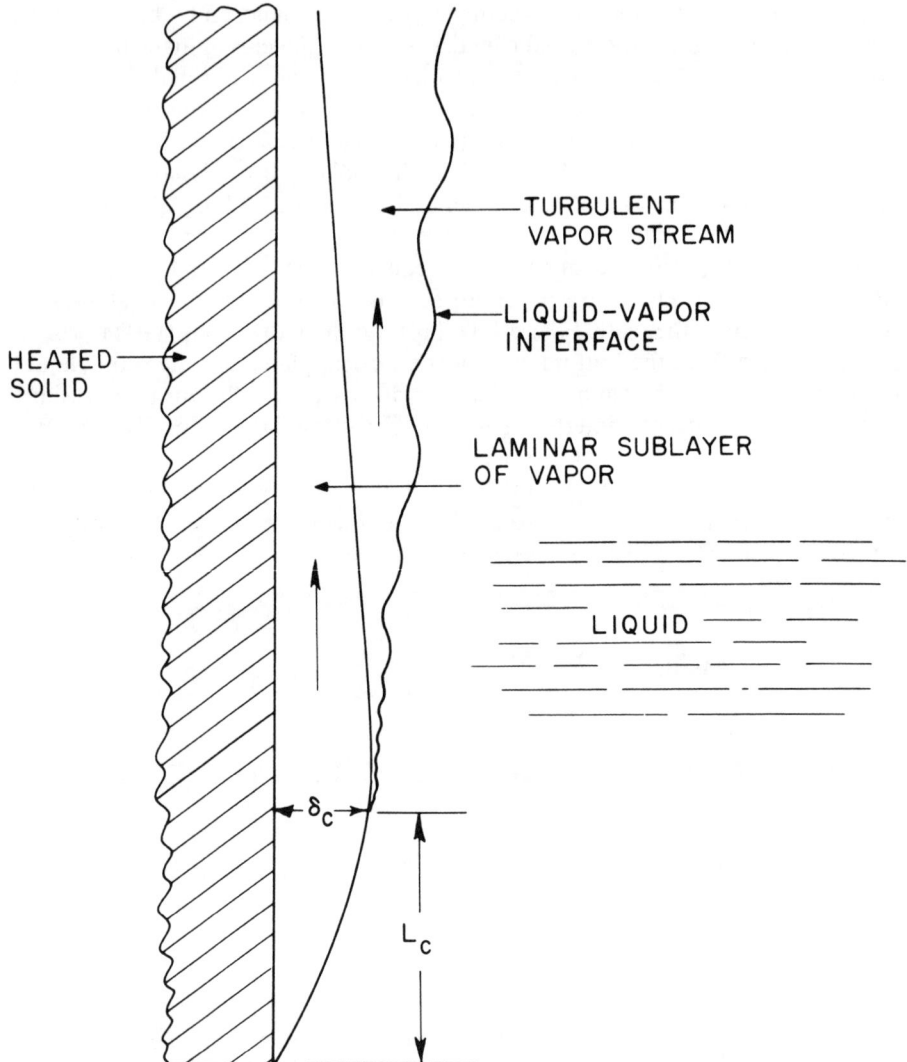

Figure 2.39 Schematic drawing of the growth of the vapor film in film boiling on a vertical surface. (From Dwyer, 1976. Copyright © 1976 by American Nuclear Society, LaGrange Park, IL. Reprinted with permission.)

$$A = \left[\frac{g(\rho_L - \rho_{Gt})}{\rho_{Gt}}\right]\left[\frac{\rho_G}{\mu_G \text{Re}^*}\right]^2$$

$$B' = \frac{\mu_G + (f\rho_{Gt}\mu_G \text{Re}^*/2\rho_G) + k_G(T_w - T_{\text{sat}})/H_{fg}}{k_G(T_w - T_{\text{sat}})/H_{fg}}$$

where ρ_G is the average density of the vapor in the laminar film, while ρ_{Gt} is the density of the vapor in the turbulent core. The friction factor, f, refers to the vapor–liquid interface (see Hsu and Westwater, 1960). The average value of the convective heat transfer coefficient over the vertical distance (L_c to L) is therefore

$$(h_{c,L})_{\text{turb}} = \int_{L_c}^{L} \frac{(h_{c,x})_{\text{turb}} \, dx}{L - L_c}$$

and the overall convective heat transfer coefficient from the bottom of the heater to the vertical distance $L (L > L_c)$ is

$$\bar{h}_c = \frac{h_c L_c + (h_{c,L})_{\text{turb}}(L - L_c)}{L} \tag{2-156}$$

where h_c is obtained from Eq. (2.154).

Radiation heat transfer in film boiling The relative radiant contribution can vary widely depending on the boiling pressure and the heating surface temperature. For cryogenic and low-boiling ordinary liquids, the radiation contribution often can be neglected, while for liquid metals this is rarely true. Radiation heat transfer can be calculated by using the expression

$$h_r = \frac{\sigma_{\text{S-B}} F_e (T_w^4 - T_{\text{sat}}^4)}{T_w - T_{\text{sat}}} \tag{2-157}$$

where F_e is the view factor, including surface conditions, and $\sigma_{\text{S-B}}$ is the Stefan-Boltzmann constant, 0.173×10^{-8} Btu/ft² hr °R⁴ (4.88×10^{-8} kg cal/m² h K⁴).

Using the view factor for radiation between infinite parallel plates,

$$\frac{1}{F_e} = \frac{1}{e} + \frac{1}{\alpha} - 1$$

Eq. (2-157) becomes (Bromley, 1950)

$$h_r = \left(\frac{\sigma_{\text{S-B}}}{1/e + 1/\alpha - 1}\right)\left(\frac{T_w^4 - T_{\text{sat}}^4}{T_w - T_{\text{sat}}}\right)$$

where e = emissivity of heating surface
α = absorptivity of liquid

Effects of experimental parameters on film boiling

Effect of reduced temperature T_r Clements and Colver (1964), based on data with many liquids over a wide range of diameters for boiling on horizontal cylinders, achieved a considerable improvement in Eq. (2-152) by modifying it empirically to

$$(\mathrm{Nu}_D)_f = 0.9 \left(\frac{\mathrm{Ra}_D^*}{T_r^2} \right)^{1/4} \left[\frac{H_{fg} + 0.5 c_{pG}(T_w - T_{\mathrm{sat}})}{c_{pG}(T_w - T_{\mathrm{sat}})} \right]^{1/4} \tag{2-158}$$

which is similar to Eq. (2-153).

Effect of diameter By analyzing results of several investigators, including their own, with isopropanol and Freon-113 in horizontal cylinders ranging from fine wires to large tubes, Breen and Westwater (1962) concluded that the critical wavelength, λ_c, is sometimes preferred over the diameter as the characteristic length in the Nusselt and modified Grashof numbers. Depending on the value of the dimensionless ratio λ_c/D, the boiling characteristics fall into three different regimes. Breen and Westwater succeeded in correlating the data in all three regimes by the equation

$$(\mathrm{Nu}_{\lambda_c})_f = \left(\frac{0.59 + 0.069\lambda_c}{D} \right) \left(\mathrm{Ra}_{\lambda_c}^* \right)_f^{1/4}$$

$$\times \left\{ \frac{H_{fg}[1 + 0.34 c_{pG}(\Delta T)/H_{fg}]^2}{c_{pG}\Delta T} \right\}_f^{1/4} \tag{2-159}$$

Effect of superheat $(T_w - T_{\mathrm{sat}})$ **on** h_c All correlations for h_c that have been presented so far, with the exception of Eq. (2-155), indicate that h_c should vary inversely as the one-fourth power of the wall superheat (ΔT), if all other things are equal. The last condition, however, cannot hold if ΔT is varied. The result is that h_c shows an appreciably larger dependence on the superheat (varies inversely as much as the one-third power).

Effect of acceleration Very little data are available for the g effects on film boiling from a horizontal or vertical surface. Some multi-g data for film boiling of liquid hydrogen was reported by Graham et al. (1965). The heat transfer coefficient was enhanced by an exponential ratio of the gravity raised to the one-fourth power. For large-diameter horizontal cylinders, h_c should be proportional to $g^{3/8}$. For small cylinders, the surface tension is more important, and the g effect is reduced. Taking data for film boiling of Freon-113 on a 0.188-in. (0.48-cm)-oustide-diameter tube for the acceleration ratio a/g ranging from 1 to 10, Pomerantz (1964) presented an empirical equation:

$$h_c = 0.62 \left[\frac{k^3 \rho_G (\rho_L - \rho_G) g H_{fg} (1 + 0.5 c_{pG} \Delta T / H_{fg})}{\mu_G (T_w - T_{sat}) D} \right]^{1/4} (D/\lambda)^{0.172} \quad (2\text{-}160)$$

The effect of sub-g acceleration, as discussed by Siegel (1967), diminishes as the zero-g condition is approached.

Effect of velocity and subcooling The relative motion of a heating surface to the Leidenfrost drop was studied by Baumeister and Schoessow (1968), who related the vaporization time of a moving drop to that for a stationary drop through an empirical correlation.

The combined effect of turbulent convection from liquid with high subcooling and radiation for film boiling on a flat surface was analyzed by Hamill and Baumeister (1967), resulting in the expression

$$h_T = h_{\text{sat film}} + 0.88 h_r + 0.12 h_{\text{turb conv}} \left(\frac{T_{\text{sat}} - T_L}{T_w - T_{\text{sat}}} \right) \quad (2\text{-}161)$$

When subcooling is very high, film boiling may not be sustained.

Summary of experimental data For film boiling of ordinary liquids, the following equations can be used:

Horizontal plate: Berenson, Eq. (2-150)
Horizontal wires, cylinders: Sciance et al., Eq. (2-153)
Vertical plates: Bromley, Eq. (2-152); and Hsu and Westwater, Eq. (2-154)

2.4.4.2 Film boiling in liquid metals.

Theoretical considerations

Effect of pressure on h_c Using the previously mentioned theoretical equations for h_c [i.e., Eqs. (2-150), (2-153), (2-154), and (2-158)], effect of pressure on the prediction, at constant $(T_w - T_{sat})$, can be calculated for Hg, K, and Na in the pressure range of 1–50 psia (0.1–5 MPa). The dependence of h_c on p_L in this pres-

Table 2.4 Dependency of h_c on p_L

Heater	Equation	Boiling Liquid	Effect of p_L
Horizontal cylinders	(2-153)	Hg	$\propto p_L^{0.33}$
		K	$\propto p_L^{0.25}$
		Na	$\propto p_L^{0.29}$
	(2-157)	Hg	$p_L^{0.33}$
		K	$p_L^{0.23}$
		Na	$p_L^{0.26}$
Horizontal plates	(2-150)	Hg	$p_L^{0.37}$
		K	$p_L^{0.29}$
		Na	$p_L^{0.32}$
Vertical surfaces	(2-154)	Hg	$p_L^{0.36}$
		K	$p_L^{0.28}$
		Na	$p_L^{0.31}$

sure range is given in Table 2.4 (Dwyer, 1976). The physical properties for the superheated vapor were estimated assuming a superheat of 700°F (389°C). It can be concluded that the dependency of h_c on p_L is quite significant and is shown to be essentially the same by all equations, but significantly greater for mercury than for the two alkali metals.

Effect of superheat ($T_w - T_{sat}$) Using the same equations, the dependency of h_c on ($T_w - T_{sat}$) in the range of 500–1500°F (278–833°C) for boiling potassium at $p_L = 1$ atm (Dwyer, 1976) are given in Table 2.5.

Sodium would show a similar result, and mercury, a slightly greater dependency.

Effect of molecular diffusion and vapor-phase chemical reactions Liquid metal vapors consist of molecules and gaseous atoms. Working with alkali metals, Ewing et al. (1967) found that the molecules are principally dimers and tetramers. The

Table 2.5 Dependency of h_c on superheat

Heater	Equation	Effect of superheat
Horizontal cylinders	Eq. (2-153)	$h_c \propto (T_w - T_{sat})^{-0.35}$
	Eq. (2-157)	$h_c \propto (T_w - T_{sat})^{-0.34}$
Horizontal plates	Eq. (2-150)	$h_c \propto (T_w - T_{sat})^{-0.33}$
Vertical surfaces	Eq. (2-154)	$h_c \propto (T_w - T_{sat})^{-0.34}$

equilibrium concentrations of monomer, dimer, and tetramer species are temperature- and pressure-dependent, and can be represented as

$$nM = M_n \tag{2-162}$$

where M represents the atomic species of the metal. It can be seen that at constant pressure, an increase in temperature causes the equilibrium to shift to the left; while at constant temperature, an increase in pressure causes it to shift to the right. At saturation conditions, temperature and pressure are coupled; and varied together, they exert opposite influences on the equilibrium. Since the pressure effect outweighs the temperature effect, an increase in temperature actually produces an increase in the dimer and tetramer concentrations. For instance, at its normal boiling point, K vapor contains about 16% (by weight) dimer and a negligibly small concentration of tetramer, under equilibrium conditions. At this constant pressure, the dimer concentration in the K vapor drop sharply with temperature increase across the film, falling to about 4% (by weight) at 1,993°F (1,089°C), corresponding to a superheat of 600°F (333°C). Considering only this dimerization reaction, which is significant in connection with film boiling of an alkali metal, the concentration of monomer decreases with distance from the heating surface, while that of the dimer increases. These concentration gradients promote diffusion of monomer toward the relatively cold vapor–liquid interface, and of the dimer molecules toward the relatively hot heating surface–vapor interface. As the monomers diffuse toward the cold liquid–vapor interface, they react *exothermally* to yield dimer molecules, which in turn diffuse toward the hot heating surface and, at the same time, decompose *endothermally* to produce monomer atoms. The net result is therefore an additional mode of heat transport from the heating surface to the boiling liquid, which is dependent on the particular liquid metal, the kinetics of the chemical reactions, the pressure, and the wall superheat. The easiest way to deal with this effect is to combine it with the conduction mode by employing an effective thermal conductivity that includes the polymerization effect. A problem, however, is estimating the degree of approach to chemical equilibrium; the assumption of complete equilibrium obviously maximizes the effect (Dwyer, 1976).

Effects of experimental parameters on pool film boiling with liquid metals

Effect of pressure Figure 2.40 shows the heat transfer coefficients for film boiling of potassium on a horizontal type 316 stainless steel surface (Padilla, 1966). Curve A shows the experimental results; curve B is curve A minus the radiant heat contribution (approximate because of appreciable uncertainties in the emissivities of the stainless steel and potassium surfaces). Curve C represents Eq. (2-150) with the proportionality constant arbitrarily increased to 0.68 and the use of the "equilibrium" value of k_G as given by Lee et al. (1969).

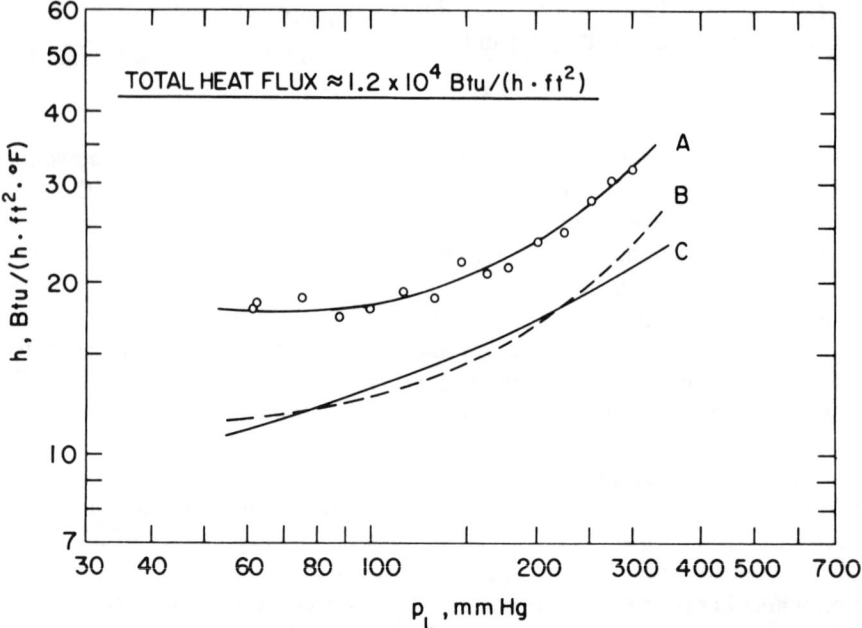

Figure 2.40 Heat transfer coefficient for film boiling of potassium on a horizontal type 316 stainless steel surface as a function of pressure. Curve A shows the experimental results of Padilla (1966). Curve B is curve A minus the radiant heat contribution. Curve C represents Eq. (2-150) with the proportionality constant arbitrarily increased to 0.68. (From Dwyer, 1976. Copyright © 1976 by American Nuclear Society, LaGrange Park, IL. Reprinted with permission.)

Effect of wetting The differences between the boiling curves of complete wetting, partial wetting, and nonwetting systems are illustrated in Figures 2.41a and 2.41b, the former being a plot of q'' versus $(T_w - T_{sat})$ and the latter, a plot of h vesus q''. The A–B–C–D–E–G–H curve in each figure represents a completely wetted surface, while the A–B–F–G–H curve represents a completely unwetted surface. For the first case, E–G represents the transition boiling regime and G–H represents the stable film boiling regime. For the second (unwetted) case, B–F represents the partial film boiling regime, and F–H the stable film boiling regime. The liquid-phase portion of the boiling curves, A–B, is meant to be the same for both wetted and unwetted surfaces, which is true only for perfectly "clean," gas-free systems (Dwyer, 1976). For contaminated (real) systems, the "unwetted" curve would be lower than the "wetted" curve.

Pure mercury does not easily wet steels and certain other structural alloys, thus it is an unwetted case. It causes the direct transition from liquid phase to film boiling heat transfer. The phenomenon has also, on occasion, been observed with alkali metals (Noyes and Lurie, 1966; Avksentyuk and Mamontova, 1973). Figure 2.42 shows experimental heat transfer results for pool boiling of pure mercury on

POOL BOILING 113

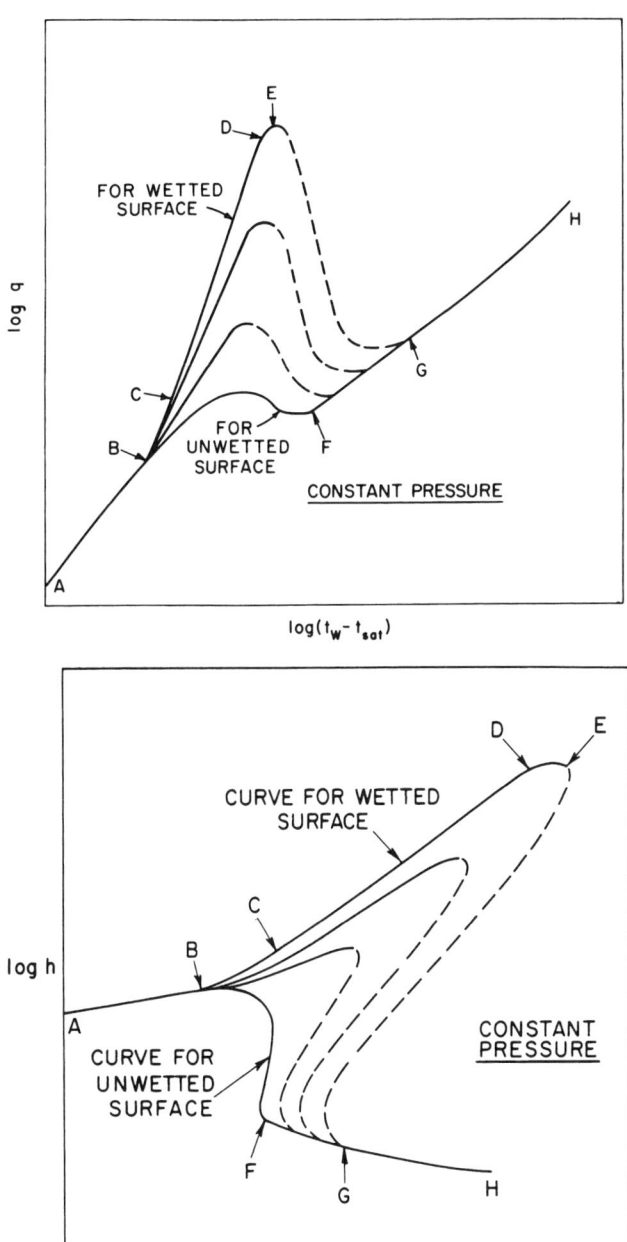

Figure 2.41 Curves illustrating the effect of surface wetting on pool boiling heat transfer: (a) heat flux versus $(t_w - t_{sat})$; (b) heat flux versus heat transfer coefficient. (From Dwyer, 1976. Copyright © 1976 by American Nuclear Society, LaGrange Park, IL. Reprinted with permission.)

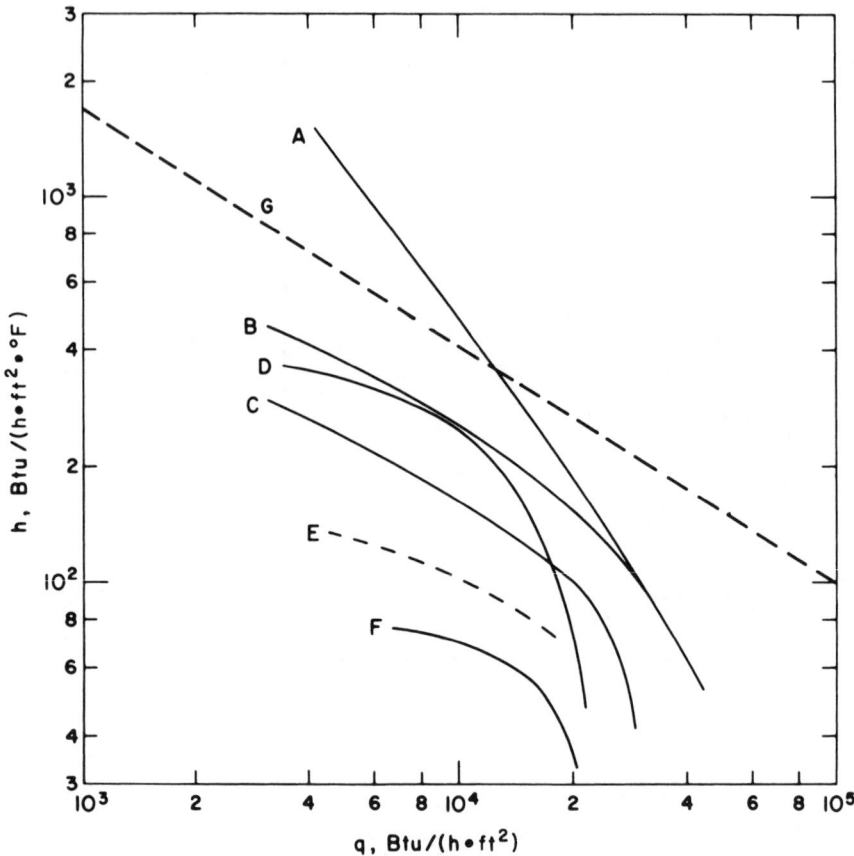

Figure 2.42 Experimental heat transfer results for pool boiling of pure mercury on the outside of an unwetted horizontal cylindrical heater at 1 atm.

Curve	Investigator(s)	Circum location	Heating surface	Diameter
A	Korneev (1955)	Top	Carbon steel	0.87 in. (2.2 cm)
B	Korneev (1955)	Bottom	Carbon steel	0.87 in. (2.2 cm)
C	Lyon et al. (1955)	Top side	Stainless steel	0.75 in. (1.9 cm)
D	Clark and Parkman (1964)	Side	Haynes 25	1.00 in. (2.5 cm)
E	Clark and Parkman (1964)	Side	Sicromo 9M	1.00 in. (2.5 cm)
F	Clark and Parkman (1964)	Top	Haynes 25 Sicromo 9M	1.00 in. (2.5 cm)

(From Dwyer, 1976. Copyright © 1976 by American Nuclear Society, LaGrange Park, IL. Reprinted with permission.)

the outside of unwetted horizontal cylindrical heaters at 1 atm. Differently marked curves are from different investigators, with varying diameters [for *A* and *B*, the diameter was 0.87 in. (2.2 cm), for *C* it was 0.75 in. (1.9 cm), and for *D*, *E*, and *F* it was 1.00 in. (2.54 cm)] and heating surfaces. Line *G* is placed on Figure 2.42 with a slope of -0.60 for reference, as predicted by Eq. (2-158) for mercury, and does not take into account any radiation effects. Note that the latter effect, if anything, would decrease the slope with an increase in heat flux. Curves *B–F* roughly maintain this same slope up to a heat flux of 20,000 Btu/hr ft^2 (62.9 kW/m^2), after which vapor blanketing apparently exerts a dominant effect. Other mercury boiling experiments on the outsides of horizontal cylindrical heaters have been reported (Lee, 1968; Turner and Colver, 1971), in which nonwetting of the heating surfaces existed. However, in both cases the results contribute little to the understanding of stable film boiling heat transfer with liquid metals. With the loss of interest in using mercury as a working medium in space power generation, and the difficulties of obtaining accurate results in film boiling of mercury—even from experiments that are seemingly well planned and conducted—we failed to find any new experimental data to shed light on this subject.

Effect of subcooling Because of the highthermal conductivity of liquid metals, the subcooling effect can be expected to be greater with such liquids than with ordinary liquids. In the past, subcooled film boiling heat transfer with liquid metals has not been studied, even though it may be relevant to accidental overheating problems in sodium-cooled power reactors. For instance, a major safety concern is the potential molten fuel–liquid metal coolant interaction. In such a case, it occurs in a highly transient state and is called vapor explosion. The vapor generation rates require large quantities of heat for latent heat of vaporization and expansion work; this heat either has to be stored in the coolant in a nonequilibrium state or transferred very rapidly between fluids. A number of physical mechanisms have been postulated to account for this heat storage or transfer (Anderson and Armstrong, 1973), which is different from stable film boiling.

Summary of experimental data Film boiling correlations have been quite successfully developed with ordinary liquids. Since the thermal properties of metal vapors are not markedly different from those of ordinary liquids, it can be expected that the accepted correlations are applicable to liquid metals with a possible change of proportionality constants. In addition, film boiling data for liquid metals generally show considerably higher heat transfer coefficients than is predicted by the available theoretical correlations for h_c. Radiant heat contribution obviously contributes to some of the difference (Fig. 2.40). There is a third mode of heat transfer that does not exist with ordinary liquids, namely, heat transport by the combined process of chemical dimerization and mass diffusion (Eq. 2–162).

2.5 ADDITIONAL REFERENCES FOR FURTHER STUDY

Additional references are given here for recent research work on the subject of this chapter, which are recommended for further study.

Cooper (1984) tried to correlate pool boiling heat flow with such parameters as reduced pressure, P_r, and molecular weight, M. The need for properties determination can thereby be bypassed and some estimates of boiling rates can be provided when experiments of the specific conditions are not available. Westwater et al. (1989) also showed that some simplification is possible in the prediction of the nucleate boiling curve and the transition curve for members of a homologous group such as Freons. From the hypothesis of thermodynamic similarity, modified by taking into account fluid-specific parameters, Leiner (1994) suggested a general correlation for nucleate boiling heat transfer for various fluids. The physical quantities are nondimensionalized in the equation by fluid-specific scaling units that are thermodynamic critical properties or power products of critical data of the fluids. This correlation is also supposed to allow for estimating nucleate boiling heat transfer coefficients in poorly known fluids.

A framework for a unified model for nucleate and transition boiling was suggested by Dhir and Liaw (1989). They developed an area- and time-averaged model for saturated pool boiling heat fluxes assuming the existence of stationary vapor stems at the wall. However, the maximum heat fluxes obtained from this model are valid only for surfaces that are not well wetted. Further discussion of nucleate and transition boiling heat transfer under pool and external flow conditions is provided by Dhir (1990). Recent advances that have been made toward a mechanistic understanding of nucleate and transition boiling are presented. Comments are also made regarding the theoretical and experimental studies that should be made in the future.

Judd (1989) interpreted experimental results of Ibrahim and Judd (1985), in which the bubble period first increased and then decreased as subcooling varied over the range $0 \leq (T_{sat} - T_{in}) \leq 15°C$ (27°F), by means of a comprehensive model incorporating the contributions of nucleate boiling, natural convection, and microlayer evaporation components. The mechanism responsible for the nucleation of bubbles at exactly the frequency required at each level of subcooling is the subject of their continuing research.

Models available to explain the CHF phenomenon are the hydrodynamic instability model and the macrolayer dryout model. The former postulates that the increase in vapor generation from the heater surface causes a limit of the steady-state vapor escape flow when CHF occurs. The latter postulates that a liquid sublayer (macrolayer)* formed on the heating surface (see Secs. 2.2.5.5 and 2.4.1.2)

---

* Readers should note that in Section 2.2.5.5, the term *microlayer* is used for the liquid sublayer beneath a single bubble. The macrolayer here includes a liquid sublayer and vapor stems. It is the same layer as shown in Figure 5.21.

with an initial thickness is evaporated away during a hovering period of the overlying vapor mass when the CHF appears (Katto, 1994b).

For the hydrodynamic instability model, Lienhard and Dhir (1973b) extended the Zuber model to the CHF on finite bodies of several kinds (see Sec. 2.3.1, Fig. 2.18). Lienhard and Hasan (1979) proposed a mechanical energy stability criterion: "The vapor-escape wake system in a boiling process remains stable as long as the net mechanical energy transfer to the system is negative." They concluded that there is no contradiction between this criterion and the hydrodynamic instability model.

For the macrolayer dryout model, Bhat et al. (1983) initially analyzed boiling heat transfer by assuming a macrolayer of invariable thickness, leading to the result that heat conduction across the macrolayer contributes a major portion of the heat transfer. Haramura and Katto (1983) presented a hydrodynamic model of the CHF that is applicable widely to both pool and forced-convection boiling on submerged bodies in saturated liquids. They imposed Helmholtz instability on the vapor–liquid interface of columnar vapor stems distributed in a thin liquid layer wetting a heating surface. Several improvements have been reported (Chyu, 1987, 1989; Pan and Lin, 1989; Jairajkuri and Saini, 1991). Recently, Bergles (1992) reviewed these two CHF models along with a somewhat classical "bubble packing" model, stating that the steady-state vapor escape flow from the heater surface is not necessarily in agreement with visual observations and that the simple hydrodynamic instability model is experiencing a challenge from the macrolayer dryout model. On the other hand, Lienhard (1985, 1988) and Dhir (1990, 1992) criticized the Haramura and Katto (1983) model, related to the macrolayer dryout model (Katto, 1994b). The controversy has not yet been settled.

Prediction of parametric effects on transition boiling under pool boiling conditions have also been presented by Pan and Lin (1991). The important parametric effects investigated include cavity size distribution, coating thickness, substrate thermal properties, system pressure, and liquid subcooling level. The model has the ability to predict a wide range of parametric effects on transition boiling critical heat fluxes. Cheng and Tichler (1991) presented a correlation for free-convection boiling in a special case of thin rectangular channels. Three mechanisms of burnout are called out, i.e., the pool boiling limit, the circulation limit, and the flooding limit associated with a transition in flow regime.

The basic mechanism of dryout almost invariably involves the rupture of a residual thin liquid film, either as a microlayer underneath the bubbles or as a thin annular layer in a high-quality burnout scenario. Bankoff (1994), in his brief review of significant progress in understanding the behavior of such thin films, discussed some significant questions that still remain to be answered.

CHAPTER
THREE

HYDRODYNAMICS OF TWO-PHASE FLOW

3.1 INTRODUCTION

From an engineering viewpoint, the final objective of studying two-phase flow is to determine the heat transfer and pressure drop characteristics of a given flow. In nuclear reactors, the flow may be in parallel flow channels as in the reactor core, or in a large pipe and connections as in the cooling loops under accidental conditions. One of the important boundary conditions is the presence or absence of heat transfer: Thus the adiabatic two-phase flow is differentiated from the diabatic (with heat addition) flow. In the latter case, the flow with heat addition is a coupled thermohydrodynamic problem. On the one hand, heat transfer causes phase change and hence a change of phase distribution and flow pattern; on the other hand, it causes a change in the hydrodynamics, such as a pressure drop along the flow path that affects the heat transfer characteristics. Furthermore, a single-component, two-phase flow in a channel (or conduit) can hardly become fully developed at low pressure because of the shape change in large bubbles and the inherent pressure change along the channel, which continually change the state of the fluid and thereby change the phase distribution and flow pattern. Both of these observations suggest high complexity in diabatic two-phase flow, where a local or point description is insufficient without knowledge of the previous "history" of the flow. Additional complexities are introduced by the hydrodynamic instabilities and the occasional departure from thermodynamic equilibrium between the phases. To avoid such complexities, many global analysis and experiments have so far been conducted, with reasonable success, that are based on the assumptions of fully developed flow patterns and without heat addition to the flow. Thus a great deal of information has become available on flow patterns, phase distribution, and pressure drop in adiabatic flows, often of two-component, gas–liquid mixtures (as in the oil and gas industries); see e.g., Govier and Aziz (1972), Taitel and Dukler (1976b; Taitel et al., 1980), Barnea et al. (1982a), Mishima and Ishii (1984), and McQuillan (1985). Such information provides basic mechanism of pattern transition and pressure drop characteristics. The application of adiabatic conditions to diabatic conditions with mass transfer in one-component gas–liquid system requires certain modifications.

In the following sections, the flow patterns, void fraction and slip ratio, and local phase, velocity, and shear distributions in various flow patterns, along with measuring instruments and available flow models, will be discussed. They will be followed by the pressure drop of two-phase flow in tubes, in rod bundles, and in flow restrictions. The final section deals with the critical flow and unsteady two-phase flow that are essential in reactor loss-of-coolant accident analyses.

3.2 FLOW PATTERNS IN ADIABATIC AND DIABATIC FLOWS

The hydrodynamic behavior of two-phase flows, such as pressure drop, void fraction, or velocity distribution, varies in a systematic way with the observed flow pattern (or regime), just as in the case of a single-phase flow, whose behavior depends on whether the flow is in the laminar or turbulent regime. However, in contrast to single-phase flow, there still is a lack of generalizing principles for gas–liquid flows that could serve as a framework for solving practical problems. For instance, for two-phase flow, we do not have such comfortable phenomenological principles as Prandtl's mixing-length theory, the methods of analogies such as Colburn's j factors, or the simplifications allowed by boundary-layer theory (Dukler, 1978). The identification of a flow regime automatically provides a picture of the phase boundaries. The location of the phase boundaries in turn allows one to make various order-of-magnitude calculations using integrated forms of the momentum and continuity equations. Such calculations suggest what variables might be worthwhile investigating and what kind of behavior is to be expected (Griffith, 1968). Although instruments such as hot wire probes, conductance probes, and other sampling probes have been developed, a visual or photographic definition of the flow regime is the best except in the case of metallic fluids. Both still and motion pictures are used to map flow regimes.

3.2.1 Flow Patterns in Adiabatic Flow

Figures 3.1a and 3.1b show the flow patterns in vertical and horizontal pipes, respectively. Obviously, stratified flow does not exist in vertical flow, because of the relative direction of the flow and gravitational force, and a more symmetrical flow pattern is possible in vertical flow than in horizontal flow. Flow patterns identified in the figure can be described as follows.

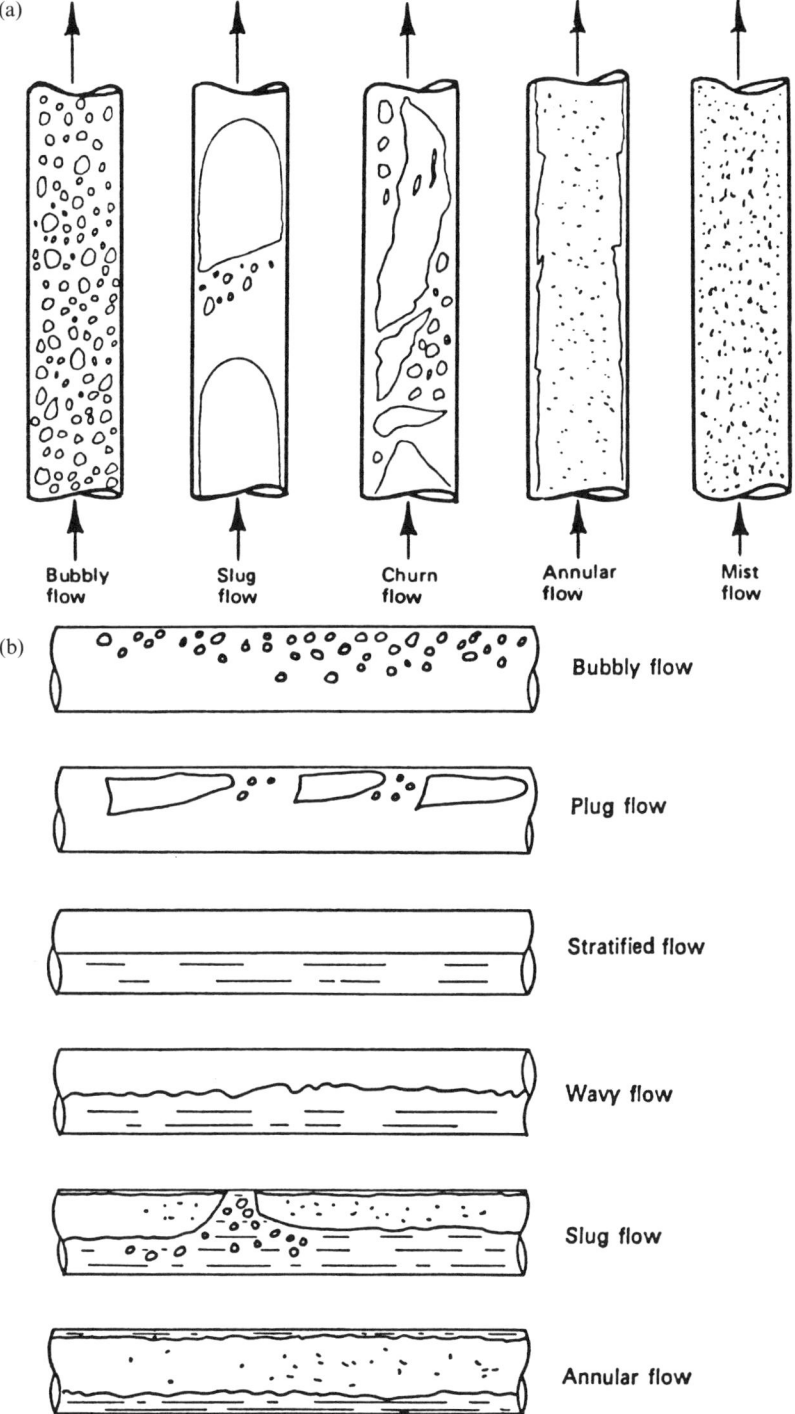

Figure 3.1 Typical two-phase flow patterns: (A) flow patterns in vertical flow; (B) flow patterns in horizontal flow.

Bubbly flow. In bubbly flow, the gas phase is moving as isolated bubbles in a liquid continuum. This flow pattern occurs at low void fractions.

Slug flow. In vertical slug flow, bubbles that have nearly the same diameter as the tube and that have a characteristic rounded front move along separated by liquid slugs within which may contain a dispersion of smaller bubbles. In horizontal slug flow, large liquid slugs move behind bubbles large enough to cover the entire diameter. This can give rise to chugging flow instabilities because of the differences in density and compressibility in different sections. Such flow occurs with moderate void fractions and relatively low flow velocity, and it can be considered a transition between bubbly and annular flow (described below). Such transitions may take place in more than one step, as shown for the next two flow regimes.

Plug flow. Plug in horizontal flow, as shown in Figure 3.1*b*, consists of enlongated gas bubbles. Although the name is sometimes used interchangeably with slug flow, it is differentiable in horizontal flow by the shape of the gas cavity or the appearance of "staircase" hydraulic jumps at the tails of air pockets (Ruder and Hanratty, 1990).

Churn flow. As the gas velocity is increased, the slug flow regime begins to break down and the gas bubbles become unstable, leading to an oscillating, "churning" flow (especially in air–water systems). Thus an alternative name for this region is *unstable slug flow.*

Wispy annular flow. In the wispy annular flow regime, there is a continuous, relatively slow-moving liquid film on the tube walls and a more rapidly moving entrained phase in the gas core (Griffith, 1968). This regime is different from annular flow by the nature of the entrained phase, which appears to flow in large agglomerates somewhat resembling ectoplasm.

Annular flow. In annular flow there is a continuous liquid in an annulus along the wall and a continuous gas/vapor phase in the core. The gas core may contain entrained droplets—dispersed mist—while the discontinuous gas phase appears as bubbles in the annulus. This flow pattern occurs at high void fractions and high flow velocities. A special case of annular flow is that where there is a gas/vapor film along the wall and a liquid core in the center. This type is called *inverse annular flow* and appears only in subcooled stable film boiling (see Sec. 3.4.6.3)

Among the earliest flow regime investigations in horizontal flow were Kosterin's (1949) in the USSR, using air and water in horizontal pipes from 2.5 cm (1 in.) to 10 cm (4 in.) in inside diameters, and Bergelin and Gazley's (1949) in the United States, also with air and water, in a 2.5-cm (1-in.) horizontal pipe. Early investigations in vertical pipes (also in 2.5-cm, or 1-in.-diameter pipes) with two-component systems were reported by Kozlov (1954). Griffith and Wallis (1961) correlated the transition boundaries using nondimensional groups. Experiments with a water–steam system at high pressures was first reported by Bennett et al. (1965b) under

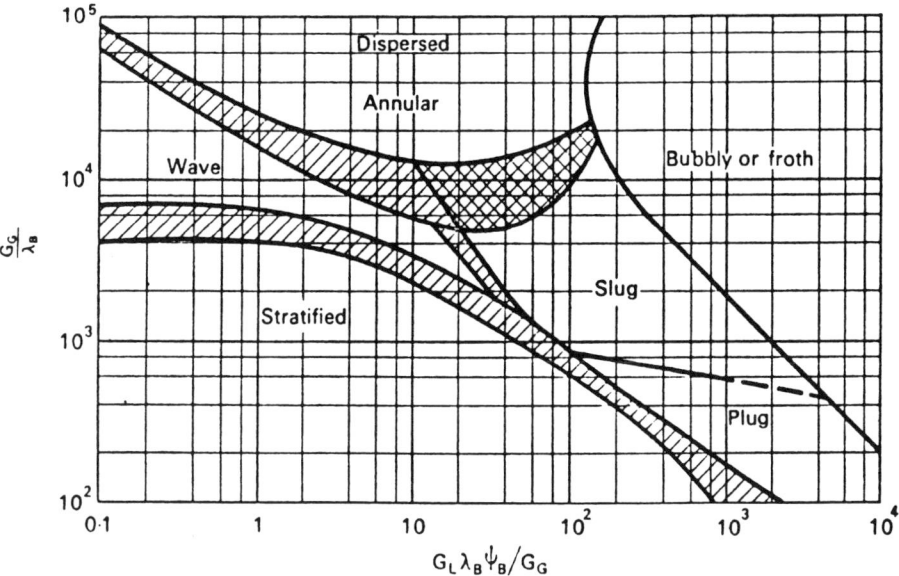

Figure 3.2 Flow pattern diagram for horizontal flow. (From Scott, 1963. Copyright © 1963 by Academic Press, New York. Reprinted with permission.)

steady-state and fully developed flow conditions. The flow regimes were observed in photographs, and two-phase flow was generated by heat addition to pure water immediately before the visual section. Baker (1954, 1960) looked at data from a variety of sources and came up with a flow regime map that has long been considered a representative plot for horizontal flow. This plot, as modified by Scott (1963) is shown in Figure 3.2. The parameters used are $G_G \lambda$ and $\lambda \psi G_L / G_G$, where G_G and G_L are the gas/vapor and liquid mass flux, respectively, based on total cross-sectional area of the pipe, and

$$\lambda = \left[\left(\frac{\rho_G}{\rho_{air}} \right) \left(\frac{\rho_L}{\rho_{water}} \right) \right]^{1/2} \tag{3-1}$$

$$\psi = \left(\frac{\sigma_{water}}{\sigma} \right) \left[\mu_L \left(\frac{\rho_{water}}{\rho_L} \right) \right]^{2/3} \tag{3-2}$$

The last two factors are used primarily for the translation from mostly air–water data to other gas–oil systems. This empirical mapping, like many others since then, such as those of Butterworth (1972) and Wallis and Dobson (1973), suffers from a lack of basis for the mechanisms that are responsible for the transition of flow regimes. An accurate analysis of regime transitions with flows in horizontal pipes

124 BOILING HEAT TRANSFER AND TWO-PHASE FLOW

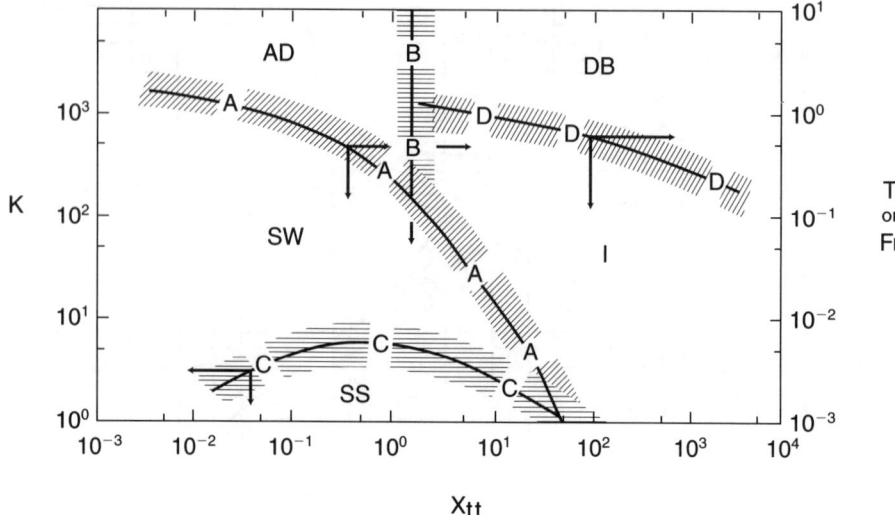

Figure 3.3 Horizontal flow regime map: curves A, B, (Fr) versus X_{tt}; curve C, K versus X_{tt}; curve D, T versus X_{tt}. (AD, annular dispersed; DB, dispersed bubble; SW, stratified wavy; I, intermittent; SS, stratified smooth.) (From Taitel and Dukler, 1976b. Copyright © 1976 by American Institution of Chemical Engineers, New York. Reprinted with permission.)

was first reported by Taitel and Dukler (1976b), using a physically based model for steady adiabatic flow without phase change. Their proposed map has been compared successfully with a large amount of experimental data [including 1,000 flow regime data points examined by Mandhane et al. (1974) for the air–water system in horizontal pipes ranging in size from 1.3 to 15 cm, or $\frac{1}{2}$ to 5.9 in. in diameter]. It is one of the best results actually available, because it considers the influence of the pipe diameter (Cumo and Naviglio, 1988). Figure 3.3 shows this generalized flow regime map for horizontal flow. Unlike Figure 3.2, different pairs of flow parameters are used for the flow regime maps. Thus the basic assumption, previously implied, that transitions between any two flow regimes can be dictated by the same pair of flow parameters (as coordinates) is removed. Figure 3.3 makes use of different coordinates for the different transitions represented as functions of the Martinelli parameter, X_{tt}. These functions are defined as (Taitel and Dukler, 1976b) follows:

$$Fr = \left(\frac{\rho_G}{\rho_L - \rho_G}\right)^{1/2} \left[\frac{V_{SG}}{(dg \cos \theta)^{1/2}}\right] \quad (3\text{-}3)$$

$$K = \left[\frac{\rho_G V_{SG}^2 V_{SL} \rho_L}{(\rho_L - \rho_G) g \mu_L \cos \theta}\right]^{1/2} \quad (3\text{-}4)$$

$$T = \left[\frac{(dp/dz)_L}{(\rho_L - \rho_G) g \cos \theta} \right]^{1/2} \qquad (3\text{-}5)$$

and

$$\begin{aligned} X_{tt} &= \frac{(dp/dz)_L}{(dp/dz)_G} \\ &= \left(\frac{\rho_G}{\rho_L}\right)^{0.5} \left(\frac{\mu_L}{\mu_G}\right)^{0.1} \left(\frac{1-X}{X}\right)^{0.9} \end{aligned} \qquad (3\text{-}6)$$

where $(dp/dz)_L$ and $(dp/dz)_G$ are pressure drops calculated as if the liquid or the gaseous flow were the sole flow in the duct. θ is the slope angle of the channel with respect to the horizontal ($\theta = 0$ for horizontal flow). It will be shown later (Sec. 3.4.7) that the liquid holdup and dimensionless pressure drop are truly a unique function of the Martinell parameter in the separate-flow model for the stratified flow.

For vertical two-phase flow, a differentiation of flow direction (upflow or downflow) is necessary because of the different relative direction of the gravity force with respect to the inertial force. This leads to a differentiation of the characteristics of various flow regimes and of the relative existence zones. Until recently, flow pattern maps for upward vertical flow were based largely on experimental observations for pipes ranging from 1 cm to 5 cm (0.4 to 2 in.) in diameter, as suggested by Duns and Ros (1963), Sternling (1965), Wallis (1969), and Govier and Aziz (1972). Recent studies on upward vertical flow (Taitel et al., 1980) resulted in the flow pattern map shown in Figure 3.4, where flow regimes are characterized as zones of prevalence of one of the forces acting on the system, and the transition lines, shown as A, B, C, D, and E curves, are obtained from the equilibrium among the forces (see next section). A particular feature of this map is the importance of the parameter L/d on the transition from slug to churn flow, where L is the distance traversed by the mixture to the observing point. Another particularity is related to smaller diameters (<5 cm or 2 in.), where the bubbly zone becomes a slug zone at low superficial velocities of the gaseous phase (Cumo and Navigilo, 1988).

This map has been checked by many researchers, indicating that it is applicable to a wide range of conditions. Also shown in Figure 3.4 are correlations derived by Mishima and Ishii (1984), which used similar basic principles except for the slug-to-churn transition.* These authors pointed out that, in view of the practical applications of the separate-fluid model to transient analysis, flow regime criteria based on the superficial velocities of the liquid and gas may not be consistent with the separate-flow model formulation. A direct geometric parameter such as the

---

* As indicated by Taitel et al. (1980), it is extremely difficult to discriminate visually between churn and slug flow at higher liquid velocities.

126 BOILING HEAT TRANSFER AND TWO-PHASE FLOW

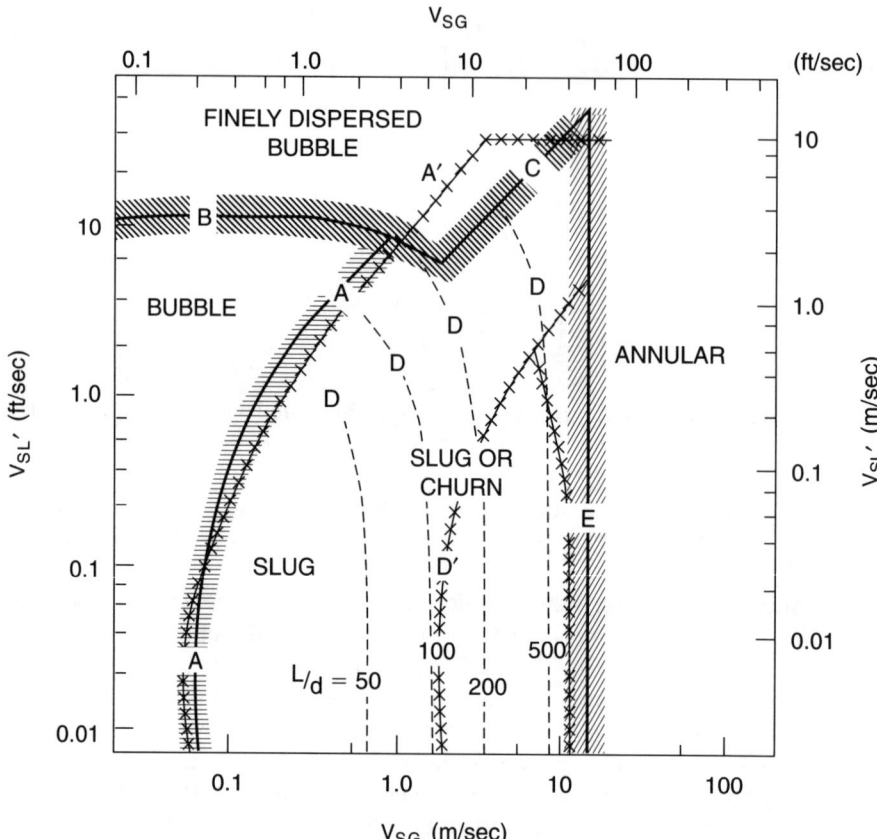

Figure 3.4 Vertical upflow regime map (d = 2.5 cm), air–water at 25°C and 0.1 MPa. A, B, C, D, E. (From Taitel et al., 1980. Copyright © 1980 by American Institute of Chemical Engineers, New York. Reprinted with permission.) A′, D′. D′ and D are the boundary between slug and churn flow. (From Mishima and Ishii, 1984. Copyright © 1984 by Elsevier Sci. Ltd., Kidlington, UK. Reprinted with permission.)

void fraction may be more suitable for use in flow regime criteria than the above-proposed traditional parameters.

The characteristics of various flow regimes and boundaries of vertical downflow regimes were defined by Oshimowo and Charles (1974) (Fig. 3.5). Their proposed map is shown in Figure 3.6, based on experimental data obtained with air and different liquids in a channel of 2.54 cm (1 in.) at a pressure of 172 kPa (1.7 bar). They pointed out that in downflow the structure of the bubbly flow shows a tendency to concentrate in the center of the channel. The bubble agglomeration, increasing with increase of the gaseous flow at a constant liquid flow rate, leads to the formation of plugs whose tops are rounded as in the upflow, while the base is flat (irregular) and has bubble trains. The bubbles may coalesce in oblong plugs

HYDRODYNAMICS OF TWO-PHASE FLOW 127

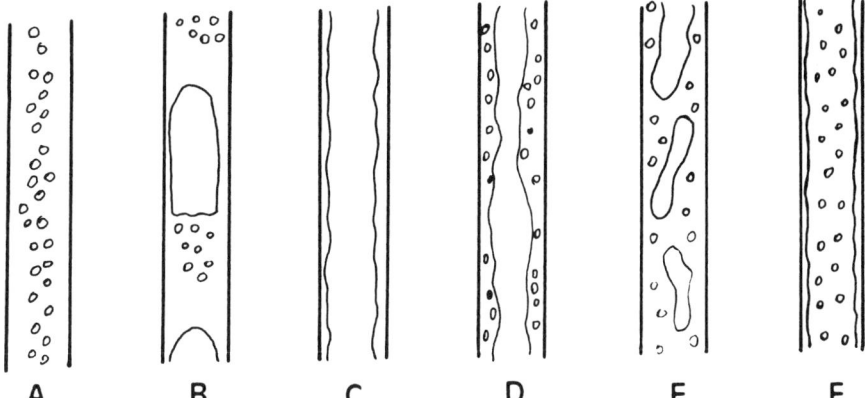

Figure 3.5 Flow regimes in vertical downflow: (A) bubbly flow; (B) slug flow; (C) falling film flow; (D) bubbly falling film flow; (E) churn flow; (F) dispersed annular flow. (From Oshimowo and Charles, 1974. Copyright © 1974 by Canadian Society of Chemical Engineers, Ottawa, Ont. Reprinted with permission.)

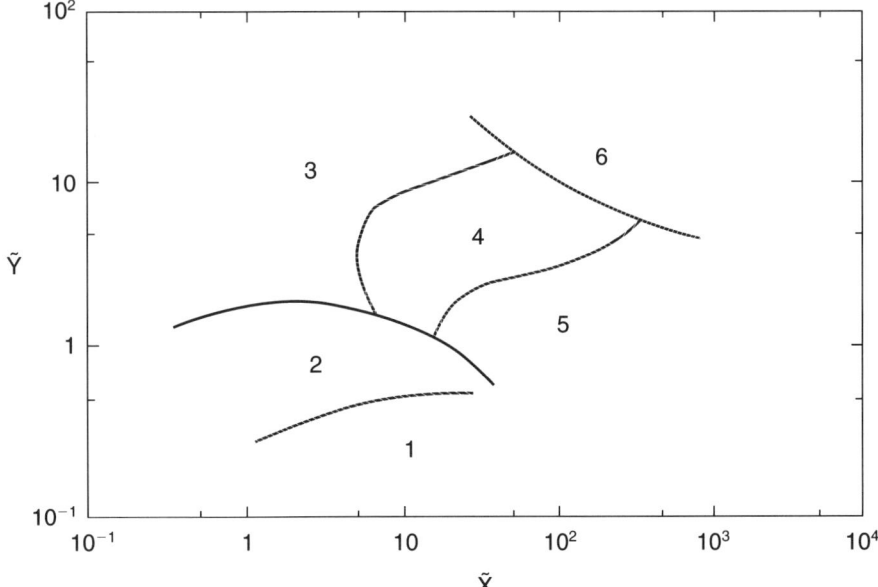

Figure 3.6 Vertical downflow regime map: (1) bubbles; (2) slugs; (3) falling film; (4) bubbly falling film; (5) churn; (6) dispersed annular. (From Oshimowo and Charles, 1974. Copyright © 1974 by Canadian Society of Chemical Engineers, Ottawa, Ont. Reprinted with permission.)

(churn flow), and may occupy the whole section that confines the liquid on the wall, leading to a falling film regime, a bubbly falling film, and a dispersed annular regime as the gaseous-phase flow increases (Oshimowo and Charles, 1974). Figure 3.6 uses two parameters:

$$\tilde{X} = \frac{\text{Fr}}{(\mu_L/\mu_G)^{1/2}\left[(\rho_L/\rho_G)(\sigma/\sigma_{\text{std}})^3\right]^{1/8}} \quad (3\text{-}7)$$

and

$$\tilde{Y} = \left(\frac{\beta}{1-\beta}\right)^{1/2} \quad (3\text{-}8)$$

where

$$\text{Fr} = \text{Froude number}$$

$$= \frac{(V_{SG} + V_{SL})^2}{gd} \quad (3\text{-}9)$$

$$= \frac{[\alpha V_G + (1-\alpha)V_L]^2}{gd}$$

$$\beta = \text{Volumetric vapor quality}$$

$$= \frac{Q_G}{Q} \quad (3\text{-}10)$$

$$= \left[1 + \left(\frac{1-X}{X}\right)\left(\frac{\rho_G}{\rho_L}\right)\right]^{-1}$$

Note that because of the numerous parameters which influence the spatial disposition of the two phases, the definition of the most suitable reference axes to be adopted for flow regime mapping has thus become a prerequisition problem. Other mappings of the downflow data based on empirical correlation of dimensionless groups were reported by Golan and Stenning (1969–1970), Martin (1973), and Spedding and Nguyen (1980).

3.2.2 Flow Pattern Transitions in Adiabatic Flow

The first attempt to analyze the flow pattern of an adiabatic gas–liquid two-phase flow in terms of the dominant physical forces acting on the system was made by Quandt (1965). The criteria for prediction of major flow patterns were developed in terms of dimensionless groups, such as the mixture Froude number, Fr_{mix}, and the mixture Weber number, We_{mix}:

$$Fr_{mix} = \frac{\overline{V}^2}{gd} \qquad (3\text{-}11)$$

$$We_{mix} = \frac{d\overline{\rho}\,\overline{V}^2}{\sigma g_c} \qquad (3\text{-}12)$$

In the one-dimensional analysis the important forces are

$$\text{Pressure gradient}: \quad F_p = \left(\frac{2f\overline{V}^2\overline{\rho}}{g_c d}\right) \qquad (3\text{-}13)$$

$$\text{Interfacial surface tension}: \quad F_\sigma = \frac{\sigma}{d^2} \qquad (3\text{-}14)$$

$$\text{Gravitational force}: \quad F_g = \frac{\rho_L g}{g_c} \qquad (3\text{-}15)$$

$$\text{Inertial force}: \quad F_i = \frac{\rho V^2}{g_c d} \qquad (3\text{-}16)$$

$$\text{Viscous force}: \quad F_v = \frac{\mu V}{d^2} \qquad (3\text{-}17)$$

Simplified expressions were developed for estimating the magnitude of each force. Comparison of the boundaries of the observed flow patterns with the analytical criteria derived by Quandt showed that the bubble, dispersed, and annular flow patterns are subclasses of a pressure gradient-controlled flow. Similarly, flow patterns identified as slug, wave, stratified, and falling film are subclasses of a gravity-controlled situation.

A variety of mechanisms have since been suggested to explain the physical basis for the observed transition between flow patterns (Ishii, 1975; Taitel and Dukler, 1976b; Taitel et al., 1980; Barnea, 1987; Taitel and Barnea, 1990). Dukler and Taitel (1991b) summarized the various mechanisms to explain such flow pattern transitions (Table 3.1), where the letters in Table 3.1 identify the theoretical curves shown in the accompanied graphs (Fig. 3.7) for different flow directions. A word of caution was given in the reference:

Table 3.1 Mechanisms for flow in tubes

Transition boundary	Patterns	Mechanism
A	Stratified to nonstratified	Kelvin Helmholtz instability
B	Intermittent to annular	$h/D \geq 0.35$ and A
C	Stratified smooth to wavy	Jeffreys, wind–wave interactions
D	Intermittent to dispersed bubble	Turbulent fluctuations versus buoyancy forces
E	Bubble to intermittent	$\alpha \geq 0.25$
F	Intermittent to dispersed bubble	Turbulent fluctuations versus surface forces
G	Dispersed bubble to slug or churn	$\alpha \geq 0.52$
H	Slug to churn	Entry length to develop stable slug
J	Annular to slug or churn	Minimum gas velocity to lift largest drop
K	Stratified smooth to wavy (incl.)	$Fr \geq 1.5$
L	Stratified wavy annular	Trajectory of drops torn from liquid film
	Minimum inclination angle to show bubble flow	Lift versus buoyant forces
M	Annular to slug churn	Matching of voids for slug and annular flow pattern

Source: Dukler and Taitel (1991b). Copyright © 1991 by University of Houston, Houston, TX. Reprinted with permission.

The matter of predicting flow pattern transitions is still in a developing state. In some cases the transition mechanisms are reliably understood, other mechanisms suggested are more in the nature of speculations.... If the ability to predict these transition conditions is to become more reliable and more generalized, it will require considerable additional research into mechanisms and into the basic fluid mechanical factors that enter into these mechanisms.

3.2.2.1 Pattern transition in horizontal adiabatic flow. An accurate analysis of pattern transitions on the basis of prevailing force(s) with flows in horizontal channels was performed and reported by Taitel and Dukler (1976b). In addition to the Froude and Weber numbers, other dimensionless groups used are

$$\text{Pressure coefficient}: \quad 2f = \frac{F_p}{F_i} \tag{3-18}$$

$$\text{Reynolds number}: \quad \text{Re} = \frac{F_i}{F_v} = \frac{dV\rho}{\mu g_c} \tag{3-19}$$

Each of the transitions shown in Figure 3.3 has been interpreted in the light of force balances. Thus, the transition between wavy (SW) flow and annular (AD) (or intermittent, I) flow is caused by the occurrence of instability conditions for the waves when the gaseous phase accelerates and produces a reduction of the local

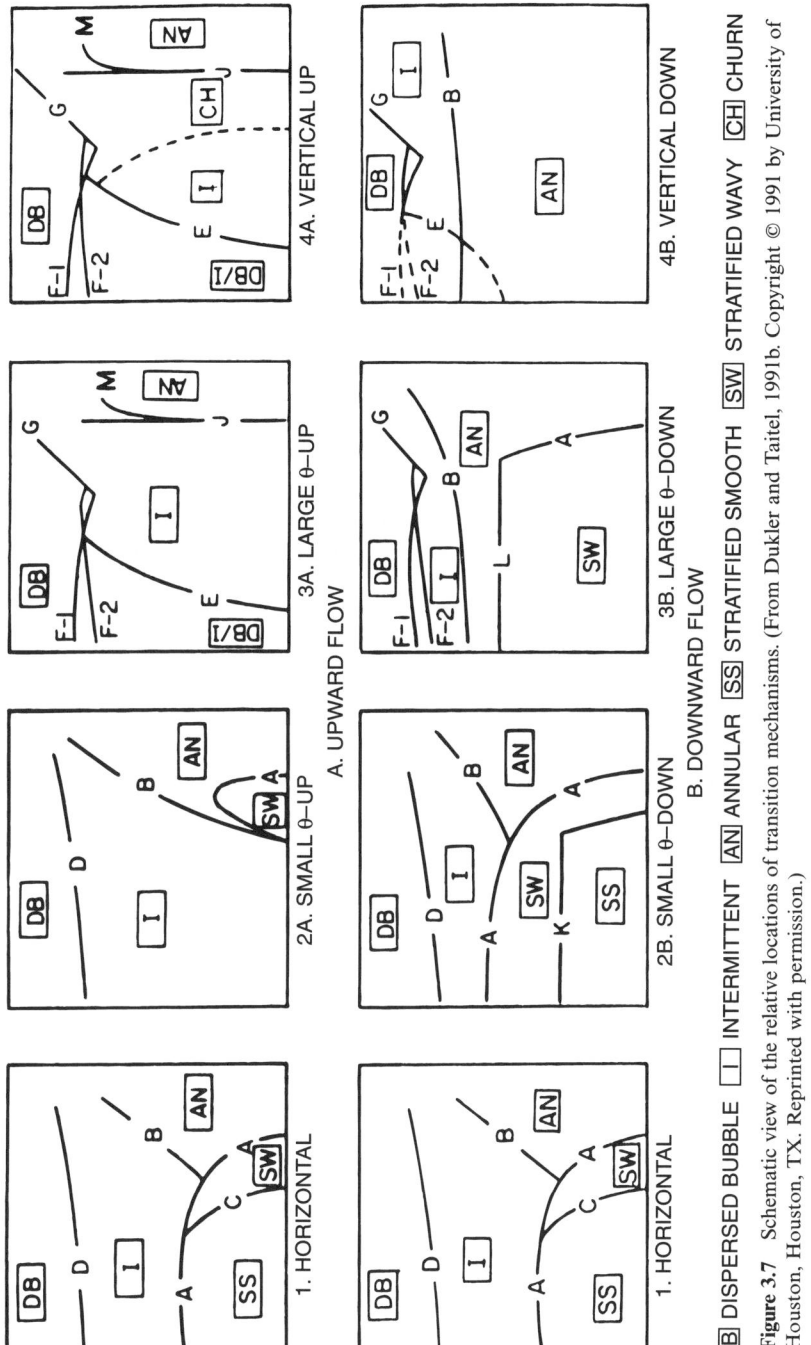

Figure 3.7 Schematic view of the relative locations of transition mechanisms. (From Dukler and Taitel, 1991b. Copyright © 1991 by University of Houston, Houston, TX. Reprinted with permission.)

static pressure leading to an upflow motion of the liquid below, which is against the gravity force. From the equilibrium of such forces, a criterion for the behavior of the waves may be obtained. The transition between intermittent (I) (plug flow) and the bubbly regime (dispersed bubble, DB) is identified as the condition in which, at high liquid and low gaseous flows, the high turbulence in the liquid, F_T, is able to prevail over the buoyancy forces, F_b, which tend to maintain the gaseous phase concentrated at the top of the duct ($F_T \geq F_b$). As a consequence, the breaking of gaseous pockets into small bubbles occurs, and these bubbles tend to mix with the liquid. The transition between intermittent (I) and annular dispersed liquid (AD) depends uniquely on the liquid level in a stratified equilibrium horizontal flow, as long as the criterion is satisfied under which finite waves that appear on the stratified liquid will grow. The transition between the stratified (SS) and wavy (SW) regimes is coincident with the occurrence of waves on the interface surface, due to the effect of a drag induced by the faster gaseous phase. From the theory, the transition conditions are predicted by the values of the dimensionless X_{tt}, and either F, T, or K. From the definition of these dimensionless groups, it is seen that once the fluid properties, pipe size, and inclination are fixed, the only variables that remain are superficial velocities, V_{SL} and V_{SG}, which are readily calculable from the known flow rates. Figure 3.8 compares the prediction of the theory for low-pressure, air–water flow in a 2.5-cm (1-in.)-I.D. pipe with the experimental data collected for that pipe size by Mandhane et al. (1974). The theoretically predicted transitions were arrived at without the use of any data, and the cross-hatched curves represent Mandhane's fit to experimental data with no theory. The agreement is quite satisfactory (Dukler, 1978). Not shown in this figure is a pseudo-slug flow in which the liquid slug does not block the entire flow cross section or only plugs the pipe momentarily, without keeping its identity (Lin and Hanratty, 1987).

In practical conditions, as in pressurized water reactors, cocurrent steam–water two-phase flow under pressure occurs in horizontal pipes of large size. Thus, Nakamura et al. (1991) reported such experiments under pressures ranging from 3 to 12 MPa (400 to 1,740 psia) with pipe inside diameters up to 180 mm (7 in.). Their results revealed that the conditions for flow regime transition between separated and intermittent flow depend strongly on system pressure, and the slug flow regime was hardly observed for pressures higher than 8.9 MPA (1,290 psia). A comparative analysis of data obtained from 180-mm (8-in.)-I.D. and 87.3-mm (4-in.)-I.D. was made. The 4-in.-pipe data showed that the flow regime transition to nonstratified (intermittent or dispersed) flow can be well represented by the Taitel and Dukler (1976b) model as modified by Nakamura et al. (1991). The modification is to replace the gas velocity, u_G, with the gas–liquid relative velocity, $u_G - u_L$, such that (Anoda et al., 1989)

$$u_G - u_L > \left(1 - \frac{h_L}{D}\right)\left[\frac{(\rho_L - \rho_G)g \cos \Theta A_G}{\rho_G (dA_L/dh_L)}\right]^{1/2}$$

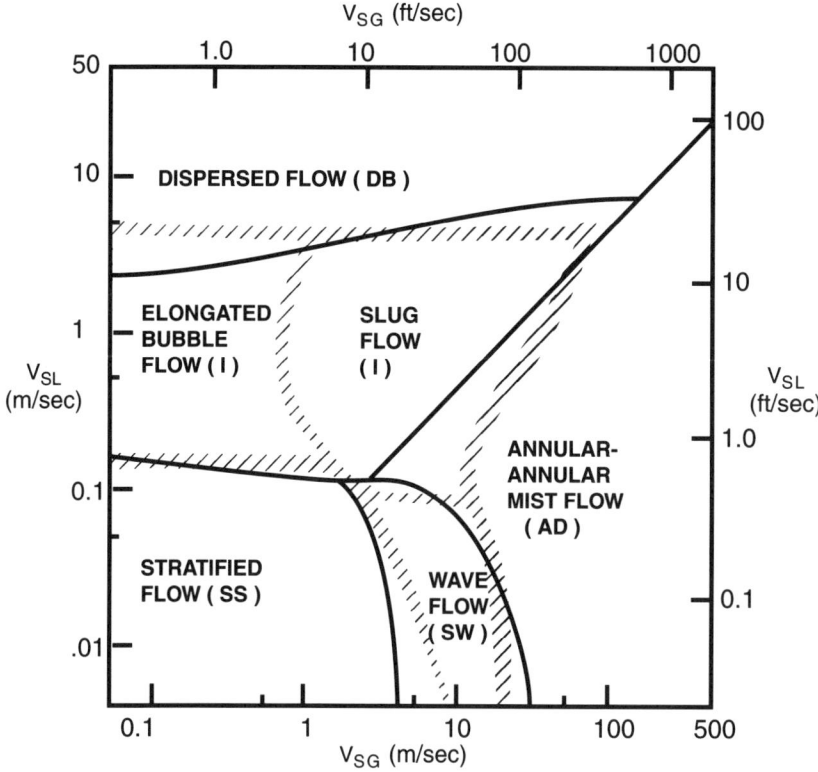

Figure 3.8 Comparison of theory and experiments (water–air horizontal flow at 25°C and 1 atm pressure with diameter of 2.5 cm). Solid lines: theory. (From Dukler, 1978. Copyright © 1978 by National Council of Canada. Reprinted with permission.) Fuzzy lines: experimental data. (From Mandhane et al., 1974. Copyright © 1974 by Elsevier Science Ltd., Kidlington, UK. Reprinted with permission.)

Figure 3.9 compares the separated-to-intermittent flow regime boundaries for both 8-in.- and 4-in.-pipe experiments. The conditions are presented in terms of nondimensional superficial velocities $J_k^* = J_k[\rho_k/(\rho_L - \rho_G)gD]^{1/2}$ (or the density-modified Froude numbers). For 4-in.-pipe experiments, data have been obtained only for 3 MPa, while for the 8-in.-pipe experiments, data have been obtained for pressures of 3.0, 5.0, and 7.3 MPa. Note the value of J_L^* at the flow regime transition increased with pressure for the 8-in.-pipe experiments.

3.2.2.2 Pattern transition in vertical adiabatic flow. *Upward vertical flow* has been studied intensively, both because of the simplicity of the geometric condition and the relevance in applications. The map shown in Figure 3.4 is the result of rather recent and relevant studies into the interpretation of regime transition mechanisms. In this figure, the transition between bubbly flow and slug flow occurs be-

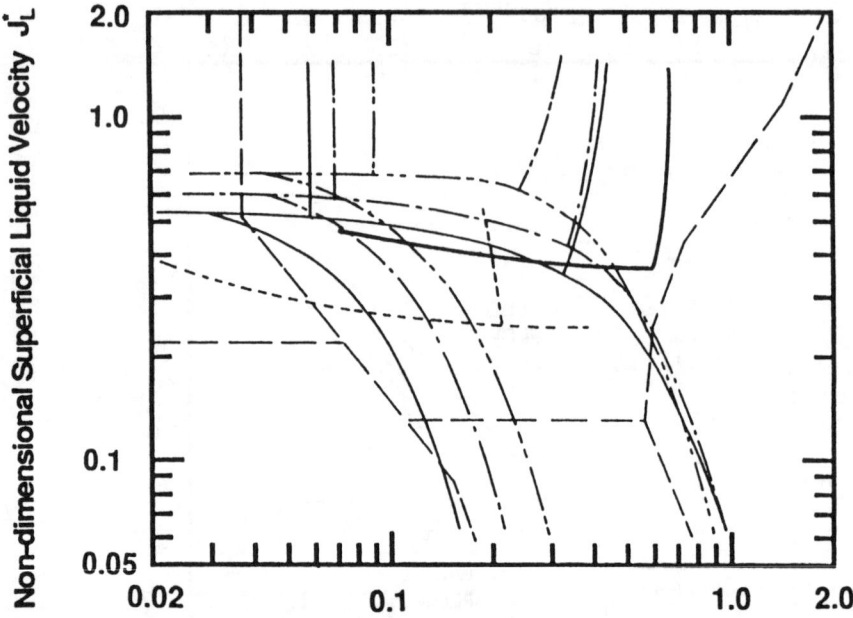

Figure 3.9 Comparison of flow regime maps at high pressures.

Steam-Water Data
(Nakamura et al., 1991)

——— 8-inch Pipe 3 MPa	- - - - Simpson et al. (1981) I.D. = 216 mm
— · — 5 MPa	
— - - — 7.3 MPa	— — — Mandhane et al. (1974) I.D. = 50 mm
——— 4-inch Pipe 3 MPa	

Air-Water Data

(From Nakamura et al., 1991. Copyright © 1991 by Amerian Nuclear Society, LaGrange Park, IL. Reprinted with permission.)

cause of the bubble coalescence process, but is obstructed by the turbulent fluctuations that increase with the flow rate and lead to a greater probability of destruction of the bubbles. At each flow rate and pressure, the transition condition represents the equilibrium between the two processes, as shown by curve A (Taitel et al., 1980; Cumo and Naviglio, 1988),

$$V_{SL} = 3.0 V_{SG} - 1.15 \left[\frac{(\rho_L - \rho_G) g \sigma}{(\rho_L)^2} \right]^{1/4} \tag{3-20}$$

This curve has no meaning for values of the diameter lower than a critical value, d_{cr}, expressed by

$$\left[\frac{\rho_L^2 g d_{cr}^2}{(\rho_L - \rho_G)\sigma}\right]^{1/4} = 4.36 \tag{3-21}$$

It may be assumed that within limited dragging action of the fluid, the critical value of the void fraction at which coalescence prevails is about 25% (Taitel et al., 1980; Cumo and Naviglio, 1988).

The transition to annular flow is defined as the minimum velocity of the gaseous phase needed to drag the droplets with the maximum admissible dimensions under specific conditions, curve E,

$$V_{SG} \rho_G^{1/2} = 3.1[(\rho_L - \rho_G)g\sigma]^{1/4} \tag{3-22}$$

At low values of V_{SG}, the coalescence of drops may take place and liquid bridges appear, leading to churn–slug flow. The transition from slug to churn depends on the parameter L/d, as mentioned before, according to the correlation, curve D,

$$\frac{L}{d} = 40.6\left[\frac{V_{SG} + V_{SL}}{gd^{1/2}} + 0.22\right] \tag{3-23}$$

The transition between finely dispersed bubbles and bubbly flow is represented by curve B,

$$V_{SL} + V_{SG} = 4.0\left\{\left[\frac{d^{0.429}(\sigma/\rho_L)^{0.089}}{(\mu_L/\rho_L)^{0.072}}\right]\left[\frac{g(\rho_L - \rho_G)}{\rho_L}\right]^{0.446}\right\} \tag{3-24}$$

and between finely dispersed bubbles and churn–slug flow, by curve C,

$$V_{SL} = V_{SG} - 0.765\left[\frac{(\rho_L - \rho_G)g\sigma}{\rho_L^2}\right]^{1/4} \tag{3-25}$$

For *vertical downflow*, the characteristics of the various flow regimes (Fig. 3.5) were mentioned in the previous section. Such characteristics have been defined by Oshimowo and Charles (1974). The transitions are shown in Figure 3.6, with different parameters based on the experimental data.

Particular conditions may occur for two-phase downflows in vertical or inclined channels (ducts) that are not completely described by the flow regime maps. Flooding occurs as the rising vapor completely blocks descending liquid. With lowering velocities of the vapor phase, this condition is preceded by the phenomenon of dragging liquid drops in a vapor upflow (carryover). When the vapor velocity is further lowered, combined with greater descending liquid mass flux, the phenomenon of dragging the vapor downflow (carryunder) in the liquid flow occurs. Carryunder and carryover are of particular interest in the study and design of moisture separator equipment, while flooding is a phenomenon of interest mainly in light water reactor safety (Cumo and Naviglio, 1988). The flooding model was originated by semiempirical studies, of which one correlation is applicable when gravity forces are much more dominant than the viscous forces (Wallis, 1969):

$$(V_{SG}^*)^{0.5} + m(V_{SL}^*)^{0.5} = C \qquad (3\text{-}26)$$

where $m = 0.8$–1.0, $C = 0.7$–1.0, and

$$V_{SG}^* = \text{Wallis parameter}$$

$$= \frac{V_{SG}\, \rho_G^{1/2}}{[gd(\rho_L - \rho_G)]^{1/2}} \qquad (3\text{-}27)$$

$$V_{SL}^* = \frac{V_{SL}\, \rho_L^{1/2}}{[gd(\rho_L - \rho_G)]^{1/2}} \qquad (3\text{-}28)$$

The transition of slug to annular or slug to stratified flow can be interpreted as a flooding condition. This approach describes the condition in which a wave, formed on the liquid film, may become unstable and increase indefinitely until it forms a slug (Bankoff and Lee, 1983).

3.2.2.3 Adiabatic flow in rod bundles. Reported experiments with flow pattern transitions in rod bundles are rather few in number, although such information is needed to predict such transitions in a pressurized water reactor during a loss of coolant accident (LOCA) or for the analysis and design of two-phase flow on the shell side of heat exchangers (steam generators). Early experiments with rod bundles for gas–liquid adiabatic systems were made by Cravarolo and Hassid (1965) using argon, water, and ethyl alcohol as the working fluids at pressures ranging from atmospheric to 2.76 MPa (400 psia). The bubble and dispersed-annular flow regimes were emphasized, and empirical correlation of the liquid volume fraction with mass flow rate, quality, gas density, and equivalent diameter was suggested (Cravarolo and Hassid, 1965). Bergles et al. (1968) studied the preheated two-phase upflow of steam–water at a pressure of 6.9×10^6 N/m² (1,000 psia) in a four-rod bundle arranged in a square array. Similar studies were undertaken by Williams and Peterson (1978) at 2.76×10^6 N/m² (400 psia), 8.27×10^6 N/m² (1,200 psia), and 13.79×10^6 N/m² (2,000 psia) in a four-rod bundle arranged in a linear array. In both of these studies, bubble, froth, slug, and annular flows were observed, except the latter study show that some of the flow patterns may be absent at certain

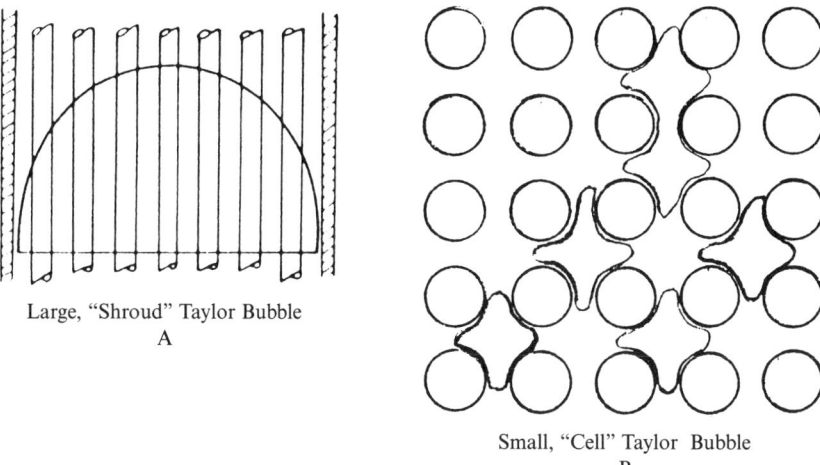

Figure 3.10 Two types of Taylor bubbles. From Venkateswararao et al., 1982. Copyright © 1982 by Elsevier Science Ltd., Kidlington, UK. Reprinted with permission.)

pressures and flow rates (Venkateswararao et al., 1982). Venkateswararao et al. reported their experiments with air–water upflow in a 24-rod bundle (on a square array) inside a cylindrical shell and suggested a mechanistic basis for the prediction of flow pattern transition. They also pointed out that two types of large bubbles were observed in the slug flow, as shown in Figure 3.10: (A) large, Taylor-type bubbles whose caps are penetrated by a number of rods, which in some cases are large enough to occupy almost the entire cross-sectional area of the shell and which are designated as shroud Taylor bubbles; (B) nearly spherical capped bubbles occupying the space in a four-rod cell, whose caps are not penetrated by the rods, and which are designated as cell Taylor bubbles (Venkateswararao et al., 1982).

The basic mechanism for transition from bubble to slug flow appears to be the same as in vertical pipe flow. That is, as the gas flow rate is increased for a given liquid flow rate, the bubble density increases, many collisions occur and cell-type Taylor bubbles are formed, and the transition to slug flow takes place. As shown in the case of vertical pipe upflow, Taitel et al. (1980) assumed that this transition takes place when $\alpha_c = 0.25$. This criterion is also applicable here. However, because of the preferable geometry in the rod bundle, where the bubbles are observed to exist, instead of in the space between any two rods, this void fraction of 0.25 applies to the local preferable area only, α_L. The local voids, α_L, can be related to the average void by (Venkateswararao et al., 1982)

$$\frac{\alpha}{\alpha_L} = \frac{[(2)^{1/2}(P/d) - 1]^2}{[(4/\pi)(P/d)^2 - 1]} \qquad (3\text{-}29)$$

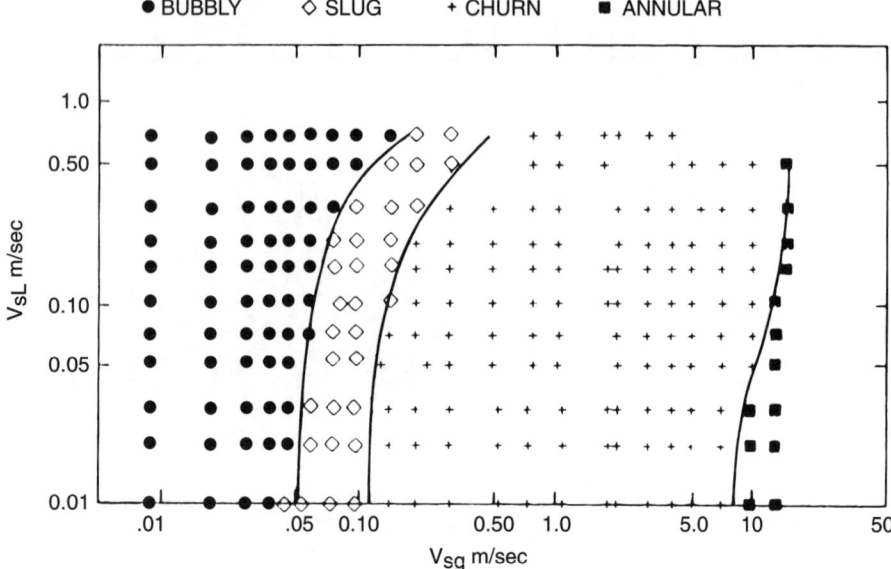

Figure 3.11 Experimentally observed flow patterns. (From Venkateswararao et al., 1982. Copyright © 1982 by Elsevier Science Ltd., Kidlington, UK. Reprinted with permission.)

The churn flow is characterized by irregular alternating motion of liquid and gas; i.e., the direction of the liquid flow changes in an erratic and irregular way from upflow to downflow and vice versa. Liquid flows downward not only as a film but also as units of liquid that occupy much of the cross-sectional area. This regime differs from slug flow in that the propagation velocities of the large bubbles and the liquid slugs in slug flow are always uniformly upward, while that of the film alongside the bubble is uniformly downward. Churn-to-annular transition takes place at high gas flow rates, as the liquid flows upward along all rod surfaces as thin annular film, while gas flows in the rest of the free area. The liquid interface is highly wavy, and the gas carries entrained liquid drops torn from the liquid film. Similar to the vertical flow in pipes, the mechanism for this transition is related to the minimum gas velocity necessary to transport the largest drop upward. At gas velocities less than the minimum, liquid begins to fall back, accumulates, and bridges the space, only to be thrust upward again, and thus the alternating motion characteristic of churn flow is observed (Venkateswararao et al., 1982). At low liquid flow rates, when the film is so thin that liquid entrainment can no longer be controlling, the mechanism of void matching comes into play. Flow patterns as observed over the operable flow rate range are shown in Figure 3.11.

The effect of flow obstructions on the flow pattern transitions in horizontal two-phase flow was studied by Salcudean and Chun (1983). The practical importance of the problem is related to the use of rod spacing devices in water-cooled nuclear reactors. In general, these devices are expected to affect the flow distribu-

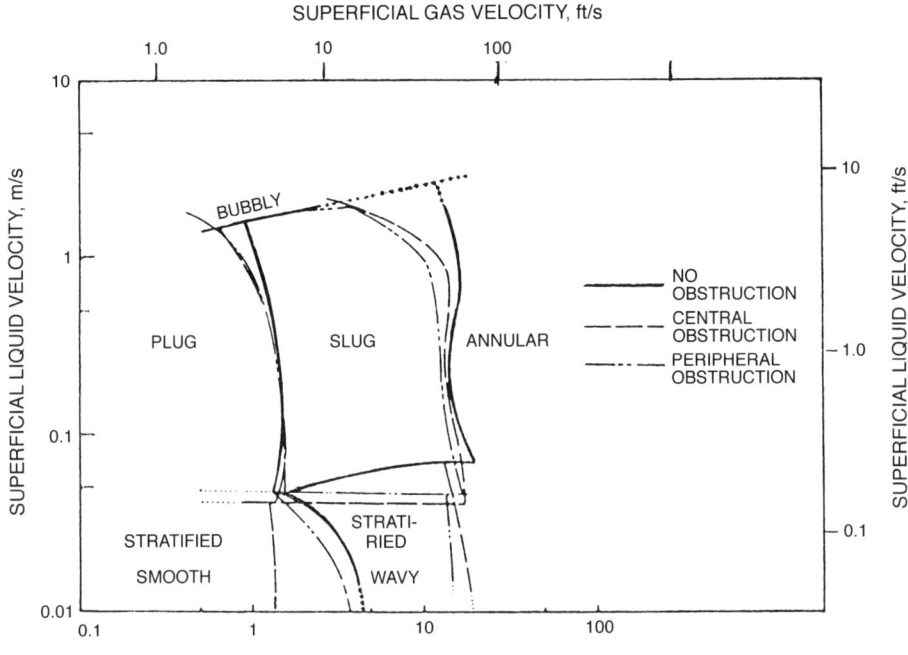

Figure 3.12 Comparison of flow pattern transition boundaries. (From Salcudean and Chun, 1983. Copyright © 1983 by Elsevier Science Ltd., Kidlington, UK. Reprinted with permission.)

tion, enhance flow homogenization, and frequently improve heat transfer. By using two types of obstruction, central and peripheral, within a 25.4-mm (1-in.)-I.D. horizontal Plexiglas tube, the flow pattern transition boundaries were compared with obstructionless flow as shown in Figure 3.12. Note that the central obstruction appears to have the strongest effect on the transition from stratified smooth to stratified wavy and from stratified wavy to intermittent flow. The peripheral obstruction has a stronger effect on the transition from intermittent to annular flow (Salcudean & Chun, 1983). One of the important results in these studies is the fact that a length-to-subchannel hydraulic diameter exceeding 30 is needed to assure the recovery of the flow from the effects of redevelopment. Measurements of local gas velocity and void fraction with vertical tube (rod) bundles on a support plate consisting of 3 × 4 tubes of 1.9 cm (0.75 in.) at a 2.69-cm (1.06-in.) square pitch were recently reported (Moujaes and Dougall, 1990). This configuration simulates the tube bundle's support plate in a pressurized water reactor (PWR) steam generator. In addition to 12 large holes of 2.027 cm (0.8 in.) for the tube array, a set of 6 smaller holes of 1.27 cm (0.5 in.) was drilled in between the rectangular grid to allow the majority of the two-phase flow through. The average volumetric fluxes ranged from a J_G of 1.52 m/s (5 ft/s) to 2.44 m/s (8 ft/s) and a J_L of 0.47 m/s (1.5 ft/s) to 0.68 m/s (2.2 ft/s). The flow regime observed for these ranges is

140 BOILING HEAT TRANSFER AND TWO-PHASE FLOW

"churn-turbulent," with an average void fraction of $0.3 < \alpha < 0.8$. At 10.68 cm (7.75 in.) downstream from the support plate (at about $7D_h$), no significant change in the void fraction from that at the flow holes was observed.

3.2.2.4 Liquid metal–gas two-phase systems. The use of liquid metals in nuclear power systems for terrestrial and space applications has made the prediction of hydraulics of liquid metal–gas two-phase systems important. The further development of conceptual design of fusion reactors and liquid–metal magnetohydrodynamic (MHD) power generators also demands accurate and reliable knowledge of liquid metal–gas two-phase flows under a magnetic field. The alkali liquid metals, as a group, exhibit some unusual physical properties. For example, in the application range of 650–980°C (1200–1800°F), these metals are several decades removed from their critical points, so liquid-to-gas density ratios of the order of several thousand are common (compared to those of subatmospheric-pressure steam) and presage low values of liquid fraction (Baroczy, 1968).

Not many liquid metal–gas two-phase flow regimes have been reported. Recently, a two-phase flow regimes map for the eutectic NaK-78 and nitrogen system was reported, as detected by the impedance probe method and statistically determined by a certain quantity representing time-varying characteristics of mixture density or interfacial configuration. The map is shown in Figure 3.13 (Michiyoshi et al., 1986). The solid lines in the figure show experimental flow regime boundaries determined by Michiyoshi et al., whereas the broken lines represent Mishima and Ishii's correlation (1984). The presence of bubbly, slug, churn, and annular flow regimes in adiabatic metal–gas flow is shown. The corresponding case under diabatic conditions will be covered in the next section.

3.2.3 Flow Patterns in Diabatic Flow

Figure 3.14 illustrates the transition of flow regimes in a horizontal diabatic flow (Becker, 1971). The asymmetric phase distribution and possible phase separation have a severe effect on heat transfer.

The application of the adiabatic flow pattern diagrams to the diabatic case should be regarded as somewhat uncertain (Hewitt, 1978). Although conflicting conclusions had previously been voiced by several investigators, Soliman and Azer (1971), Travis and Rohsenow (1973), and Palen et al. (1977, 1980) stated that flow pattern maps for no mass transfer were not suitable to predict their results on flow with mass transfer. Generalized coordinate maps prepared for horizontal tubes are presented in Figure 3.15 for boiling cases of three values of the dimensionless heat transfer parameter, Q, defined as (Dukler and Taitel, 1991b),

$$Q = \frac{q\pi V_{VS}}{H_{fg} D(dP/dz)_{VS}}$$

HYDRODYNAMICS OF TWO-PHASE FLOW **141**

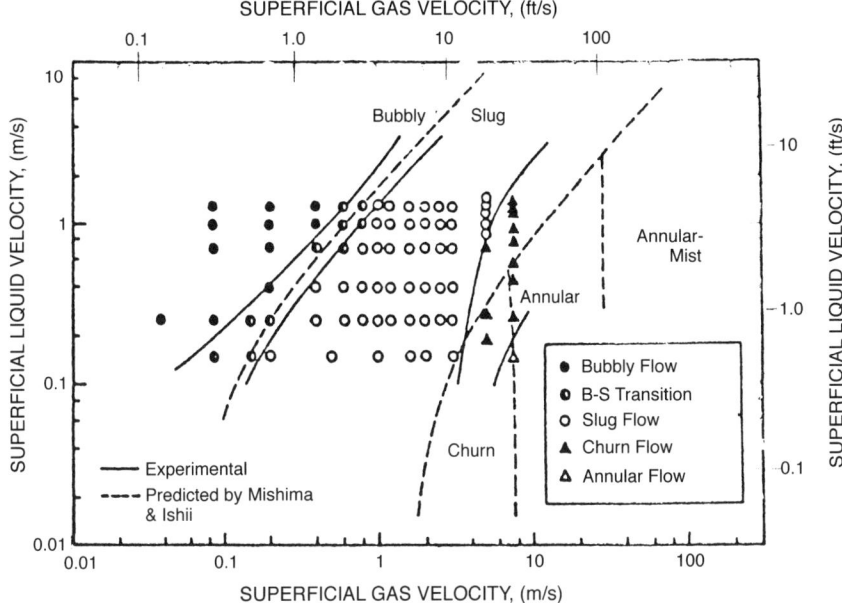

Figure 3.13 Liquid metal–gas two-phase flow regime map. [From Michiyoshi et al., 1986. Copyright © 1986 by Hemisphere Publishing Corp., New York. Reprinted with permission.)

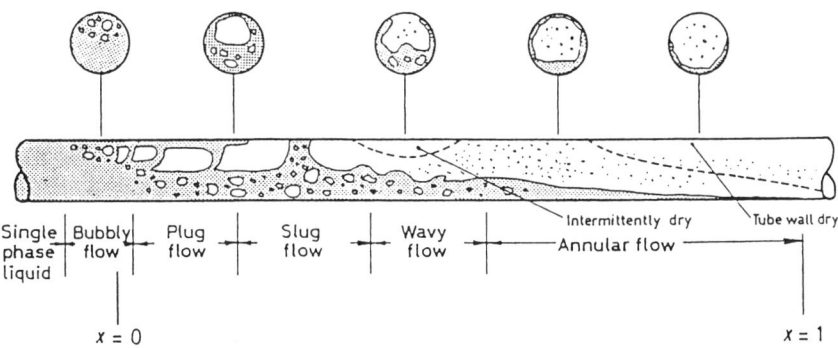

Figure 3.14 Flow regimes for horizontal diabatic flow. (From Becker, 1971. Copyright © 1971 by Studsvik AB, Nyköping, Sweden. Reprinted with permission.)

where V_{VS} is the superficial velocity of that portion of the gas that flows above the interface as additional gas flows as bubbles with the vapor in the boiling case; H_{fg} is latent heat of vaporization.

The curves for $Q = 0$ correspond to the results shown in Figure 3.3 for the adiabatic case. Note that the transition criteria F and T shown in Figure 3.15 are modified from Eqs. (3-3) and (3-5) by substituting ρ_M for ρ_L and V_{VS} for V_{SG}.

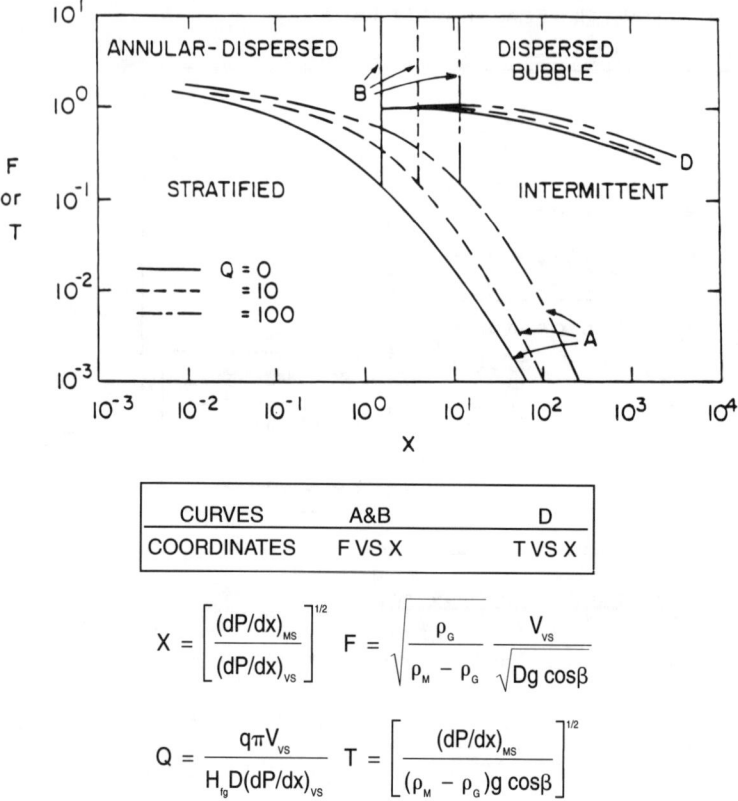

Figure 3.15 Generalized flow pattern map boiling in horizontal tubes. (From Dukler and Taitel, 1991b. Copyright © 1991 by University of Houston, Houston, TX. Reprinted with permission.)

The value of X is modified by substituting $(dP/dz)_{MS}$ for $(dP/dz)_L$, and $(dP/dz)_{VS}$ for $(dP/dz)_G$. Figure 3.16 shows the maps (in conventional coordinates of V_{LS}–V_{GS}) for boiling of pure water in a 2.5-cm (1-in.)-diameter horizontal pipe at near-atmospheric pressure with a heat flux of 100 W/cm^2 (3.17 × 10^5 Btu/hr ft^2). Dukler and Taitel (1991b) claimed such modifications as tentative ones and stated that they should be considered as first steps in the process of generalization.

In boiling flow, the void fraction of an element of fluid increases as it proceeds along the channel; the flow patterns, consequently, vary accordingly and will be different from those of adiabatic or isothermal flow. For example, Figure 3.17 shows the development of flow patterns of a high-velocity boiling flow moving vertically upward in a heated channel. In the initial, local boiling section, a superheated liquid layer exists next to the wall, while the bulk liquid may be subcooled. Adjacent to the superheated liquid layer, in which bubbles nucleate and grow, is a

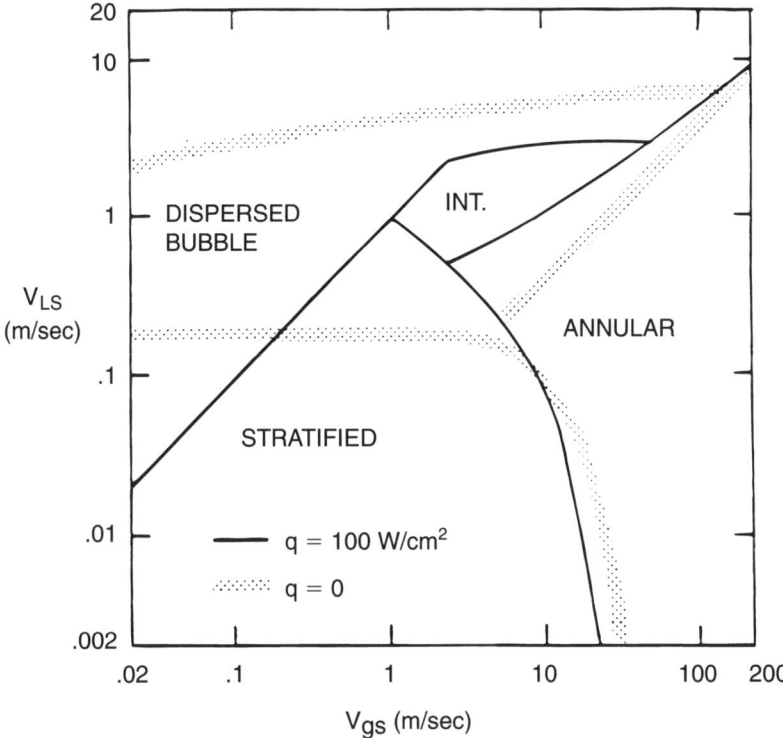

Figure 3.16 Flow pattern map for boiling of water in 2.5-cm-diameter horizontal tube at 1.0 atm pressure. (From Dukler and Taitel, 1991b. Copyright © 1991 by University of Houston, Houston, TX. Reprinted with permission.)

two-phase bubble layer in which bubbles of all ages, either still attached to the wall or carried along with the stream, are in the process of collapsing by recondensation. Gunther (1951) reported visual observations of bubble segregation on or near the wall at low pressure and high subcooling, with the detached bubbles traveling at approximately 80% of flow stream velocity. Jiji and Clark (1964) also observed and empirically correlated the "bubble boundary layer" development in subcooled flow boiling. A typical photograph taken by Jiji is shown in Figure 3.18. Figure 3.19 shows the configuration of the bubble layer as affected by flow rates at high subcooling (Tong et al., 1966b).

In the bulk boiling section, which appears when the bubble boundary layers have developed so as to fill the core and all of the liquid has reached saturation, there are two important flow patterns: bubbly flow and annular flow.* In bubbly flow, the bubbles are "evenly" distributed in the saturated liquid. The superheated

---

* Slug flow can develop only in a relatively low-heat-flux channel having a moderate flow velocity and slightly subcooled inlet temperature.

144 BOILING HEAT TRANSFER AND TWO-PHASE FLOW

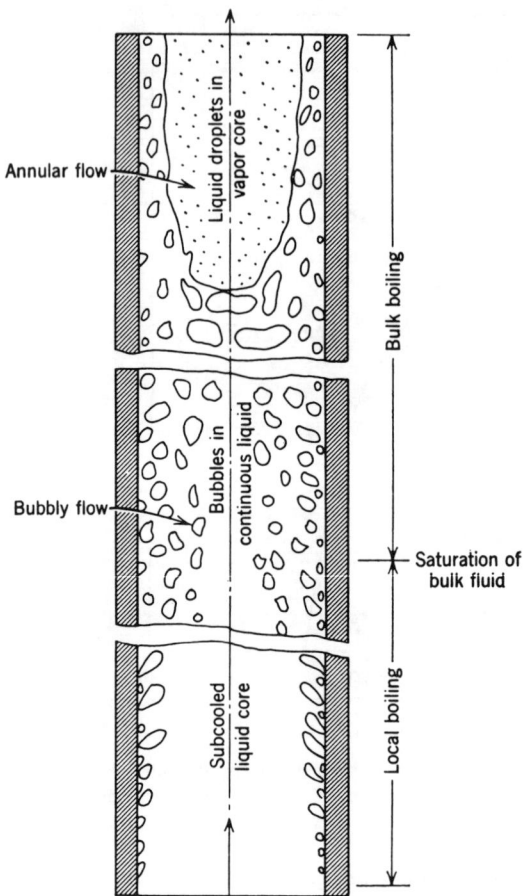

Figure 3.17 Flow patterns in a vertical heated channel.

T_i = 500 psia
V_i = 1 ft/sec
$T_{sat} - T_i$ = 200°F
q/A = 0.804 × 10⁶ Btu/hr ft²

—Leading edge

Figure 3.18 A typical photograph defining the bubble boundary layer (p_i = 500 psia, V_i = 1 ft/sec, $T_{sat} - T_i$ = 200°F, q/A = 0.804 × 10⁶ Btu/hr ft²). (From Jiji and Clark, 1964. Copyright © 1964 by American Society of Mechanical Engineers, New York. Reprinted with permission.)

HYDRODYNAMICS OF TWO-PHASE FLOW **145**

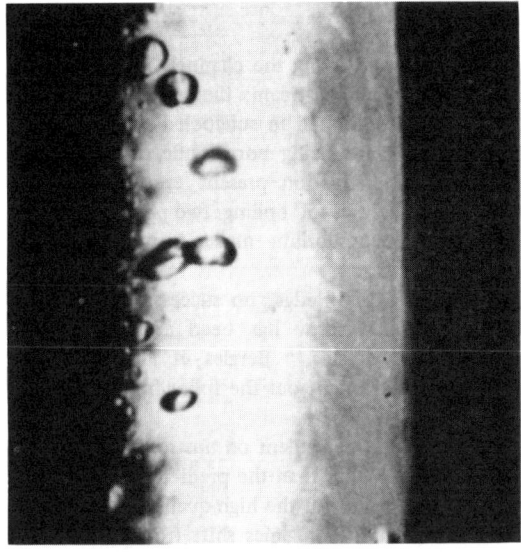

(a)

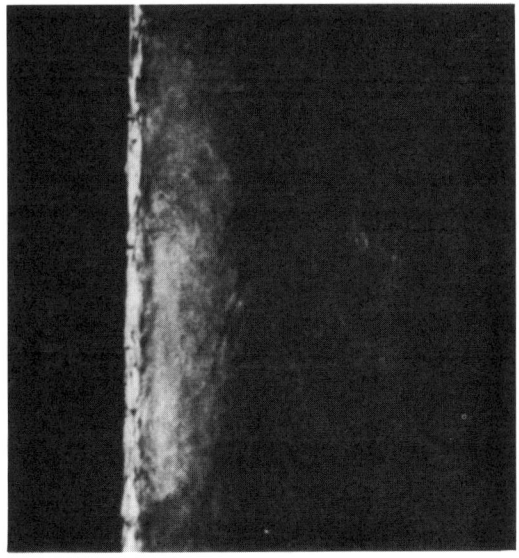

(b)

Figure 3.19 Configuration of bubble layer as affected by flow rate at high subcooling (Freon-118): (a) low-velocity boiling flow; (b) high-velocity boiling flow. (From Tong et al., 1966b. Copyright © 1966 by American Society of Mechanical Engineers, New York. Reprinted with permission.)

146 BOILING HEAT TRANSFER AND TWO-PHASE FLOW

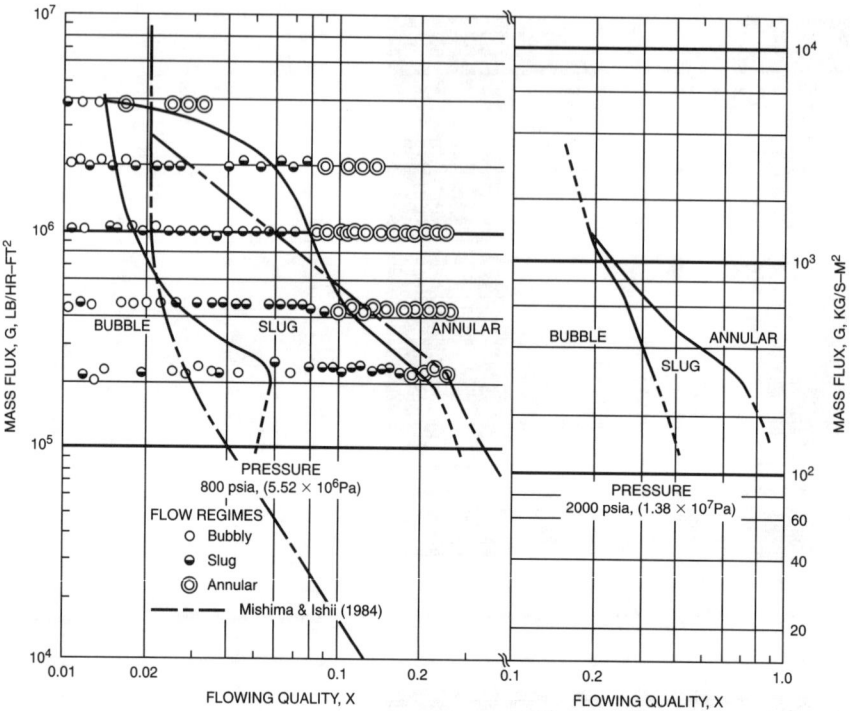

Figure 3.20 Flow patterns for rectangular channels at two different pressures. (From Hosler, 1968. Copyright © 1968 by American Institute of Chemical Engineers, New York. Reprinted with permission.)

liquid layer still exists near the heating surface, feeding the flow with vapor bubbles that no longer collapse. As the void fraction increases, the bubbles gradually concentrate in the central, high-velocity region of the flow by the *Bernoulli effect,* and the annular flow pattern forms. This Bernoulli effect is a speeding up of the less dense phase relative to the more dense phase in an accelerating flow. Exposed to the same pressure gradient, the less dense phase attains a higher kinetic energy, and hence a higher velocity, than the more dense phase. In annular flow, most of the liquid remains in the annulus, next to the wall, while the rest is dispersed as droplets in the continuous vapor core. Since the liquid annulus in vertical flow is attached to the wall as a climbing film, annular flow is also often called *climbing film flow.* Flow patterns for high-pressure water flows in a vertical rectangular channel, 0.34 cm (0.134 in.) × 2.54 cm (1 in.) × 61 cm (24 in.) long with one side heated, were reported by Hosler (1968), as shown in Figures 3.20, 3.21, and 3.22. Figure 3.21 is a plot of bubble-to-slug transition lines for various test pressures; Figure 3.22 is a plot of slug-to-annular transition lines for different pressures. Thus the boundaries between flow regimes shift from adiabatic lines depending on the

HYDRODYNAMICS OF TWO-PHASE FLOW 147

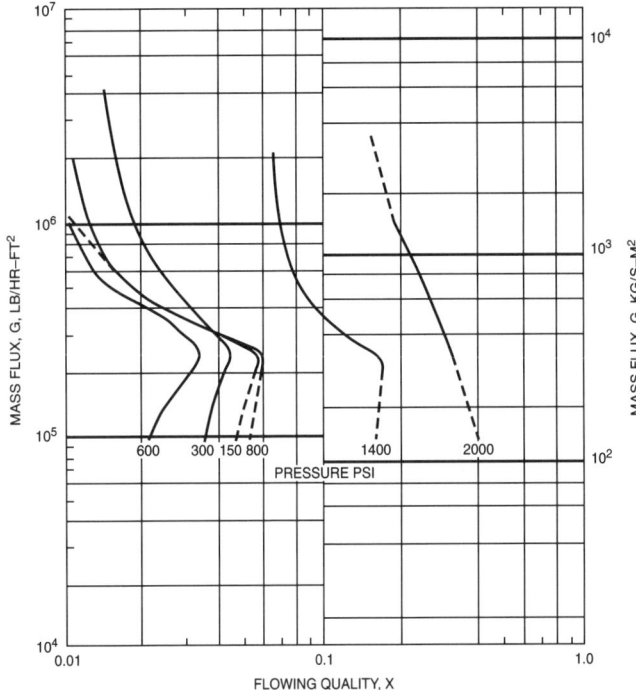

Figure 3.21 Bubble-to-slug transition lines for various test pressures. (From Hosler, 1968. Copyright © 1968 by American Institute of Chemical Engineers, New York. Reprinted with permission.)

system pressure, subcooling, flow rate, and heat flux. The differences between a boiling two-phase flow and its adiabatic counterparts are due mainly to the transient nature of the boiling case. The production of bubbles and their distribution in the flow create a nonhomogeneous flow pattern, and change its composition as it proceeds downstream, which causes the flow not to be fully developed (Hsu and Graham, 1976). It has also been pointed out that the real quality can be higher than the predicted equilibrium quality in the low-quality region and the reverse for the high-quality region, as shown in Figure 3.23 (Hsu and Graham, 1976). Suffice it to say that boiling two-phase flow is more complicated than the adiabatic flow case. The development of flow patterns for such flow will be discussed further in Chapter 5.

3.3 VOID FRACTION AND SLIP RATIO IN DIABATIC FLOW

In addition to knowledge of the flow pattern in gas–liquid two-phase flow, it is important to establish quantitatively the relative amount of each phase, or the void

148 BOILING HEAT TRANSFER AND TWO-PHASE FLOW

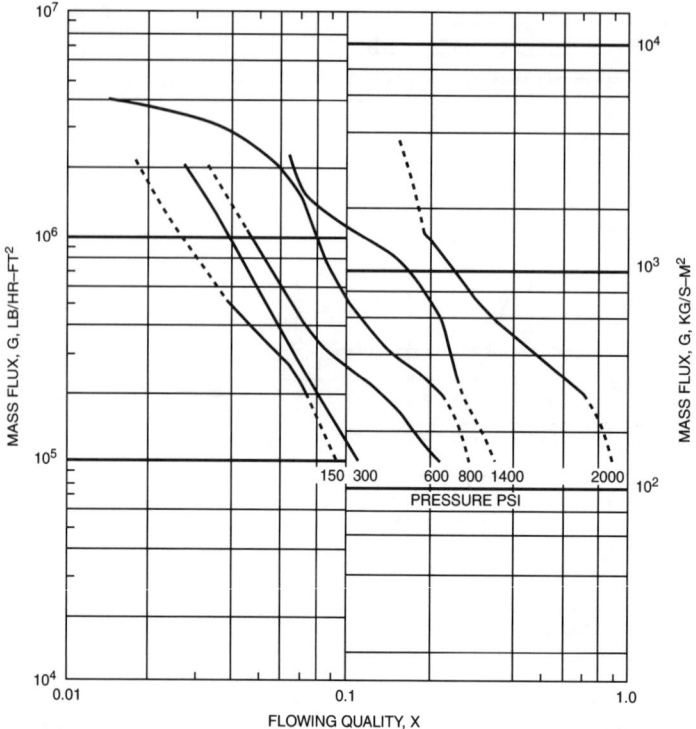

Figure 3.22 Slug-to-annular transition lines for different pressures. (From Hosler, 1968. Copyright © 1968 by American Institute of Chemical Engineers, New York. Reprinted with permission.)

fraction at different locations in the flow. This becomes particularly important when two-phase flow is used as a coolant in a nuclear reactor, where the moderating and absorbing characteristic of the two phases differ considerably. The void distributions of two-component (air–water), two-phase flow, measured by Gill et al. (1963), is shown in Figure 3.24. When the channel-average void fraction is high, the void concentrates at the center and forms a vapor core of annular flow; if the channel-average void fraction is low, the void concentrates away from the center and forms a bubble layer near the wall. This latter trend exists in both the adiabatic (isothermal) low-void, two-component case and the one-component, subcooled boiling flow case. If one-dimensional flow is assumed, such quantities as void fraction and others described below are the averages over the cross section of the flow channel. The channel-average void fraction, $\langle \alpha \rangle$, is defined as the ratio of local vapor volume to the total flow volume at a certain cross section; i.e.,

$$\langle \alpha \rangle = \frac{A_G}{A_G + A_L} \qquad (3\text{-}30)$$

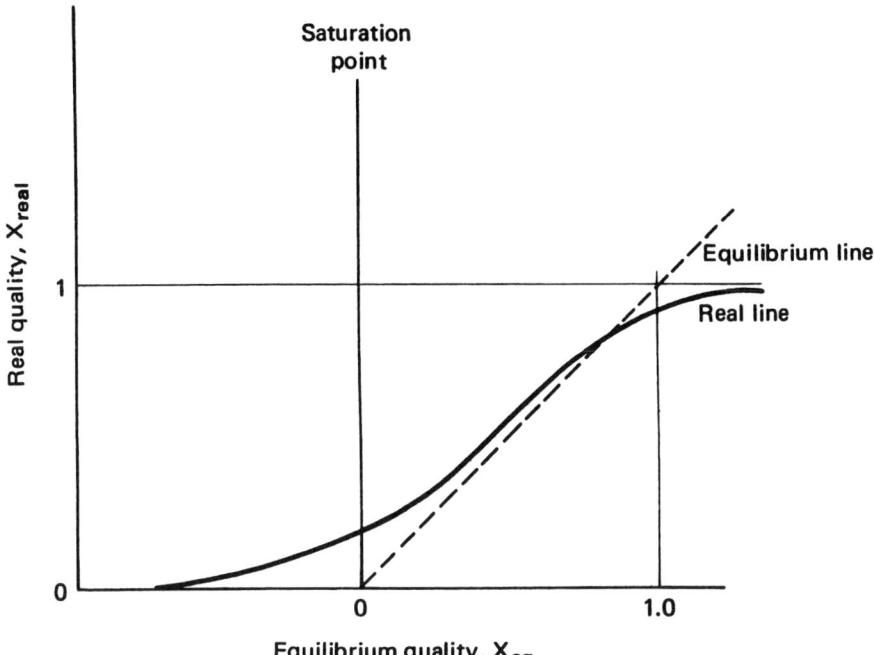

Figure 3.23 Nonequilibrium boiling two-phase flow. (From Hsu and Graham, 1976. Copyright © 1976 by Hemisphere Publishing Corp., New York. Reprinted with permission.)

As a corollary, this expression gives the definition of local average density:

$$\overline{\rho}_{\text{loc.}} = \langle \alpha \rangle \rho_G + (1 - \langle \alpha \rangle) \rho_L \tag{3-31}$$

The vapor volumetric flow ratio, β, or *volumetric vapor quality,* is defined as the ratio of vapor volumetric flow rate to the total volumetric flow rate; that is,

$$\langle \beta \rangle = \frac{Q_G}{Q_G + Q_L}$$

The relationship between a local (or static) quantity, $\langle \alpha \rangle$, and a flow (or dynamic) quantity, $\langle \beta \rangle$, is

$$C = \frac{\langle \alpha \rangle}{\langle \beta \rangle} \tag{3-32}$$

or

150 BOILING HEAT TRANSFER AND TWO-PHASE FLOW

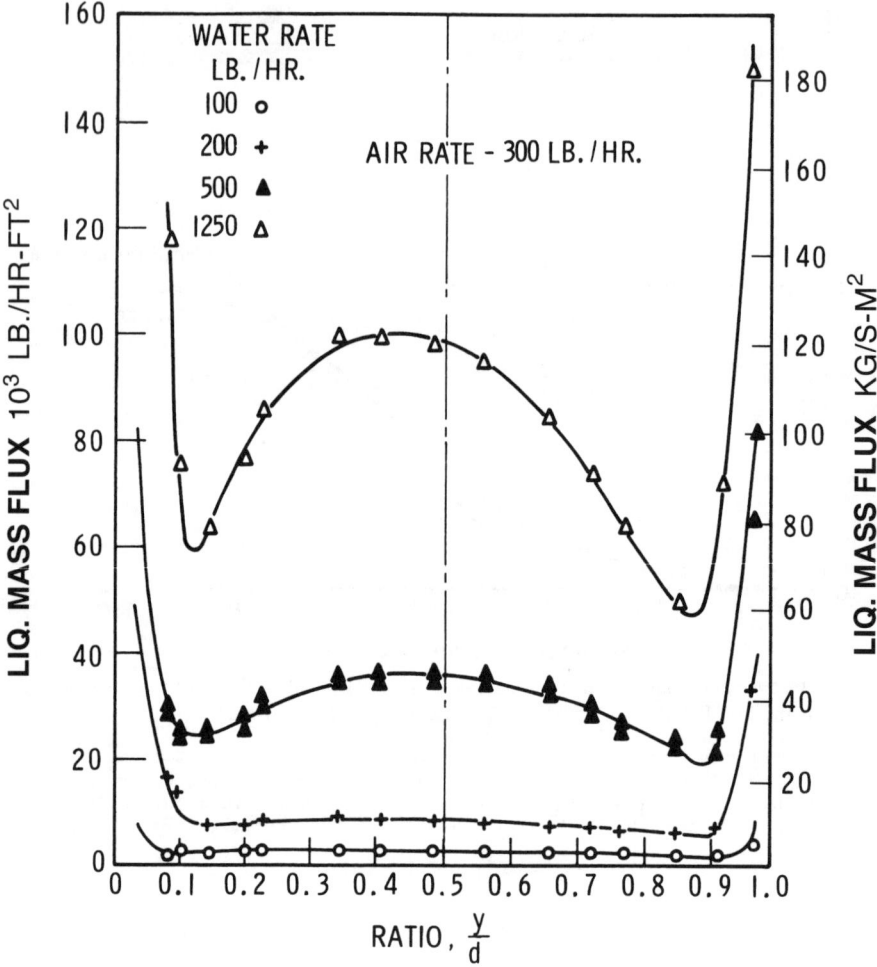

Figure 3.24 Measured void distribution in low and high void flows. (From Gill et al., 1963. Reprinted with permission.)

$$C = \langle \alpha \rangle + \frac{1 - \langle \alpha \rangle}{S} \qquad (3\text{-}33)$$

where the slip ratio, S, is defined as the ratio of the vapor velocity to the liquid velocity,

$$S = \frac{V_G}{V_L} \qquad (3\text{-}34)$$

and

$$V_G = \frac{Q_G}{A_G} \quad V_L = \frac{Q_L}{A_L} \tag{3-35}$$

Three effects cause the vapor to slip:

1. The buoyancy effect—Being lighter than liquid, the bubble slips through the adjacent liquid that flows downstream.
2. The velocity profile effect—For a convex flow velocity profile, for instance, in a steady bubbly flow, the centerline velocity is higher than the average velocity. With bubbles usually concentrating at the center, they attain a higher velocity than the liquid.
3. The Bernoulli effect—In a rapidly expanding flow, the two phases accelerate differently. For low initial velocities, the ratio of final velocities at the end of expansion, V_G/V_L, can be approximated by $(\rho_L/\rho_G)^{1/2}$.

The flow (dynamic) quality, X, is defined as the ratio of vapor flow weight to total flow weight in unit time; that is,

$$X = \frac{A_G G_G}{A_G G_G + A_L G_L} \tag{3-36}$$

where vapor mass flux $G_G = V_G \rho_G$
liquid mass flux $G_L = V_L \rho_L$
total flow weight $\dot{w} = A_G G_G + A_L G_L$

Similarly, flow average density can be defined as

$$\bar{\rho}_{\text{flow, nonslip}} = [XV_G + (1-X)V_L]^{-1}$$

From these definitions, it follows that

$$S = \left(\frac{X}{1-X}\right)\left(\frac{1-\langle\alpha\rangle}{\langle\alpha\rangle}\right)\left(\frac{\rho_L}{\rho_G}\right) \tag{3-37}$$

The actual static quality, X', is defined as the ratio of local void weight to the total local mixture weight; that is,

$$X' = \frac{H_{\text{in}} + (4Lq''/D_e G) - kH_L}{H_G - kH_L} \tag{3-38}$$

where the parameter k is (Thom et al., 1966)

$$k = 1 - \frac{0.15q''}{G} \tag{3-39}$$

Thus, the local (static) quantities represent the stationary properties affecting the static pressure and the local reactivity in a nuclear reactor. The flow (dynamic) quantities represent the transport properties affecting the energy, momentum, and mass balances of a flow.

3.3.1 Void Fraction in Subcooled Boiling Flow

The void in subcooled local boiling can be described in two regions, as shown in Figure 3.25. In region I, the degree of subcooling is high and the void is thin along the wall as bubbles are attached to it; in region II, the subcooling is reduced while the void is increased, and bubbles detach from the heated surface to be swept downstream in a process of slow recondensation. The void fraction in the first region depends on surface flux condition and may be called wall voidage, and that in the second region depends mainly on the bulk flow characteristic and thus is called detached voidage (Bowring, 1962). The wall voiding of region I has been studied experimentally by Griffith et al. (1958). It was observed that the bubbles appear on the surface in strands and with a width approximately equal to the height. Since the thermal boundary-layer thickness is proportional to (k_L/h) and the velocity boundary-layer thickness is approximately proportional to $(k_L/h)\,\mathrm{Pr}_L$, they assumed the width and the height of bubble strands are related to the velocity boundary-layer thickness. From the data obtained at 3.4, 6.9, and 10.3 MPa (500, 1,000, and 1,500 psia), they developed a semiempirical correlation of the wall void volume per unit area:

$$a = \frac{q''_b k_L \,\mathrm{Pr}_L}{1.07 h^2 (T_{\mathrm{sat}} - T_b)} \tag{3-40}$$

where a is in feet when other properties are expressed in English units and 1.07 is an empirical constant given by Maurer (1960). The wall voidage region is defined as $a < 0.004$ in. (or 0.10 mm), which corresponds to the cross-sectional average void fraction,

$$\langle \alpha \rangle < \frac{0.004 \times 4}{12 D_e} \tag{3-41}$$

To predict the detached voidage in region II, the experimental work of Thom et al. (1966), as modified by Tong (1967a), is recommended. Thom et al. measured the local void fraction in subcooled boiling water flows in a 1.52-m (5-ft)-long tube

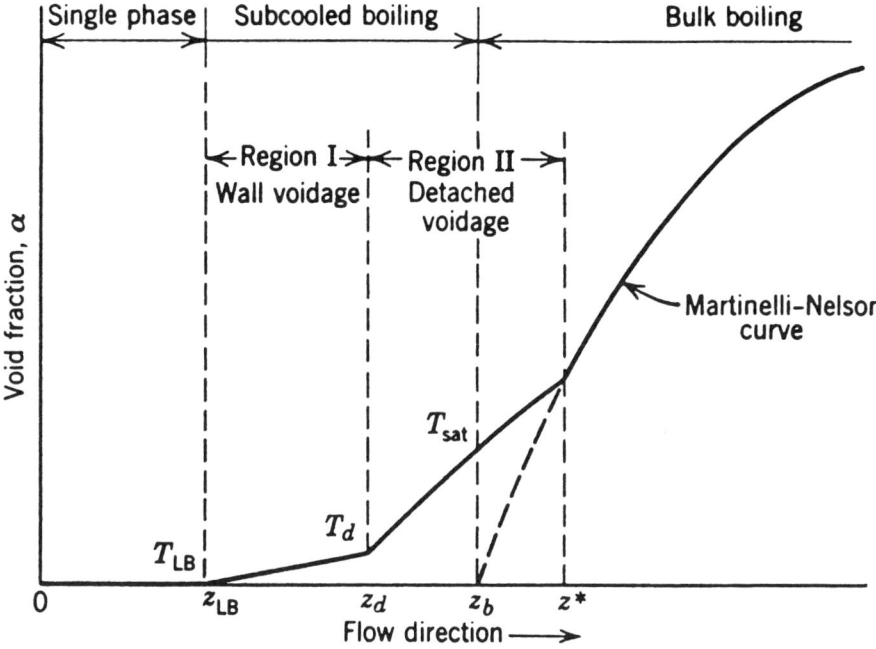

Figure 3.25 Void fraction in various (subcooled) boiling regions.

of 0.975-cm (0.384-in.) I.D. under pressures of 5.2 and 6.9 MPa (750 and 1,000 psia), at mass fluxes of 635–1,550 kg/m² h (470,000–1,150,000 lb/ft² hr), and at uniform heat fluxes of 48,000–158,670 kcal/m² h (130,000–430,000 Btu/ft² hr). They developed a simpler correlation of the *apparent density*, ρ_a, to correlate their data within ±15% up to 6.9 MPa (1,000 psia). Later, Tong (1967a) developed a function of pressure for γ and extended Thom's correlation to a range from atmospheric pressure up to 13.8 MPa (2,000 psia). The modified correlation is

$$\rho_a = \rho_L - \frac{\gamma X'(\rho_L - \rho_G)}{1 + X'(\gamma - 1)} \tag{3-42}$$

and

$$\gamma = \exp[50.959 - 8.369y + 0.25475y^2]^{1/2} \tag{3-43}$$

where $y = \ln(p)$ and p = pressure in psia; X' and k are defined in Eqs. (3-38) and (3-39).

Note that the local boiling void calculated this way is independent of channel length. The prediction of the point of bubble departure (void detachment) is, however, important in predicting the subcooled boiling void. Rouhani (1967) assumed that the heat is removed by vapor generation, the detached bubbles, and single-phase convection, and developed a correlation to predict the subcooled water boiling void based on data obtained at 1.0–4.9 MPa (140–700 psia). Koumoutsos et al. (1967), in their study of bubble departure in subcooled water flow at atmospheric pressure, found that a "neck" is formed at the bottom of a spherical bubble before its departure. Staub (1967) and Levy (1967) determined the point of bubble departure caused by the unbalance of forces acting on the bubble. Staub took into account surface tension, buoyancy, and flow shear, while Levy considered only surface tension and shear effects. Thus both approaches agree at high flow rates. Levy substituted the value of the true quality, X', for the nonequilibrium condition into Zuber and Findlay's equation (Zuber and Findlay, 1965), giving

$$\langle \alpha \rangle = \left(\frac{X'}{\rho_G} \right) \left\{ 1.13 \left[\frac{X'}{\rho_G} + \frac{(1-X')}{\rho_L} \right] + \left(\frac{1.18}{G} \right) \left[\frac{\sigma g g_c (\rho_L - \rho_G)}{\rho_L^2} \right]^{1/4} \right\}^{-1} \quad (3\text{-}44)$$

Larsen and Tong (1969), using the bubble boundary-layer concept, as shown in Figure 3.26, developed an analytical model for predicting the cross-sectional average void in a subcooled boiling flow. They provided a valid framework in which various physical parameters, such as vapor condensation rate, liquid Prandtl number, and local density of the bubbly layer, can be individually adjusted based on further experimental evidence. By using this model, Hancox and Nicoll (1971) developed a similar model for subcooled void predictions with better accuracy. A comparison of percentage of deviations between the data sources and predictions are shown in Table 3.2 (Farrest et al., 1971).

More models have been developed since 1971 for accurate estimation of the void fraction in predicting the thermal hydraulic behavior of light water reactors (LWRs), during either normal operation or an accident. Chexal et al. presented an assessment of eight void fraction models for vertical configurations using steady-state steam–water test data (Chexal et al., 1987). These models are mostly some kind of drift flux model (see Sec. 3.4), namely:

The Chexal-Lellouche model (a flow regime-independent drift flux model)
The Liao, Parlos, and Griffith model (a flow regime-based drift flux formulation including bubbly, churn-turbulent, and annular flows)
The Yeh-Hochreiter model (an algebraic slip model based on FLECHT experiments)
The Wilson bubble rise model (a void fraction of steam bubbling through stagnant water)
The Ohkawa-Lakey model (a drift flux model with empirically derived coefficients)
The Dix model (a drift flux model devised for analyzing boiling water reactors (BWRs) at operating conditions)

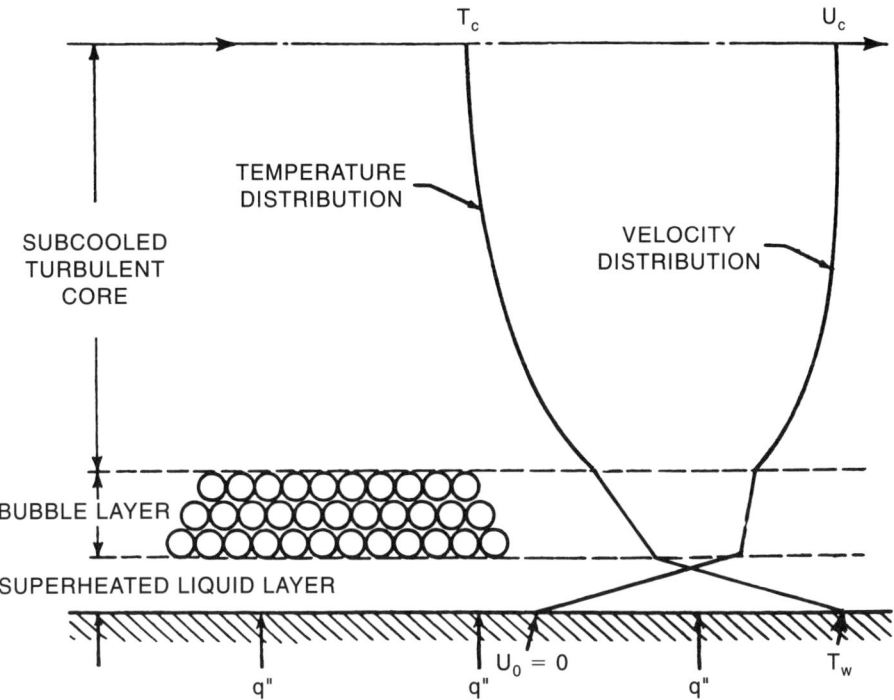

Figure 3.26 Velocity and temperature distribution in a subcooled boiling flow (bubble boundary-layer concept). (From Larson and Tong, 1969. Copyright © 1969 by American Society of Mechanical Engineers, New York. Reprinted with permission.)

The GE ramp model (an empirical drift flux model)
The Ishii model (a flow regime-based drift flux model)

Such models are currently used in the nuclear industry. The details of the assessment are available in the reference.

3.3.2 Void Fraction in Saturated Boiling Flow

A general expression for average void fraction in an adiabatic two-phase flow or in saturated boiling was suggested by Zuber and Findlay (1965). Defining the local superficial velocities, $V_{SG} (= Q_G/A)$ and $V_{SL} (= Q_L/A)$, as volumetric flux densities, they assumed the velocity and void distributions to be

$$\frac{V}{V_o} = 1 - \left(\frac{r}{R}\right)^n \tag{3-45}$$

Table 3.2 Comparison of subcooled models with data (percent deviation)

	Data Sources					
Prediction	Maurer (1960)	Marchaterre and Hoglund (1962)	Egen et al. (1957)	St. Pierre and Bankoff (1967)	Christensen (Farrest et al., 1971)	Foglia (Farrest et al., 1971)	Aver. percent deviation for all sets of data
Bowring (1962)	6.7	5.9	6.9	3.4	5.0	4.5	5.4
Levy (1967)	6.7	10.9	7.1	5.4	4.6	7.1	7.0
Rouhani (1967)	9.7	15.8	7.0	6.0	4.6	10.9	9.0
Larson and Tong (1969)	12.1	11.7	6.9	11.4	9.0	9.6	10.1
Ahmad (Hancox and Nicoll, 1971)	6.7	5.9	6.4	3.8	3.4	4.8	5.2
Hancox and Nicoll (1971)	5.3	6.3	6.1	4.0	3.0	5.1	5.0
Aver. percent deviation from all predictions	7.9	9.4	6.7	5.7	4.9	7.0	

Source: Farrest et al. (1971). Copyright © 1971 by Elsevier Science Ltd., Kidlington, UK. Reprinted with permission.

$$\frac{\alpha - \alpha_w}{\alpha_o - \alpha_w} = 1 - \left(\frac{r}{R}\right)^n \tag{3-46}$$

where subscript o refers to quantities at the center and subscript w refers to quantities at the wall. They also defined a distribution parameter C_o [which is equal to $1/C$ as defined in Eq. (3-32)],

$$C_o = \left[\frac{\left(\frac{1}{A}\right)\int_A \alpha V \, dA}{\left(\frac{1}{A}\right)\left(\int_A \alpha \, dA\right)\left(\frac{1}{A}\right)\left(\int_A V \, dA\right)}\right] \tag{3-47}$$

$$= 1 + \left(\frac{1}{n+1}\right)\left(1 - \frac{\alpha_w}{\langle\alpha\rangle}\right)$$

For an adiabatic flow, that is, $\alpha_w = 0$,

$$C_o = \frac{\langle\beta\rangle}{\langle\alpha\rangle} = \frac{\langle V_{sG}\rangle}{\langle\alpha\rangle\langle V\rangle} = \left(\frac{n+2}{n+1}\right) \tag{3-48}$$

For various flow regimes, they used the terminal bubble rise velocity as the local bubble slip velocity.

In the slug flow region, Griffith and Wallis (1961) suggested the terminal velocity,

$$V_\infty = 0.35\left(\frac{g \, \Delta\rho D}{\rho_L}\right)^{1/2} \tag{3-49}$$

In the turbulent bubbly flow region, Harmathy (1960) suggested

$$V_\infty = 1.53\left(\frac{\sigma g g_c \, \Delta\rho}{\rho_L^2}\right)^{1/4} \tag{3-50}$$

and the profile-weighted mean velocity of the vapor, $\overline{V}_G$, is defined as

$$\overline{V}_G = \frac{\langle V_{sG}\rangle}{\langle\alpha\rangle} = C_o\langle V\rangle + V_\infty \tag{3-51}$$

It can be seen that, for small terminal bubble rise velocities, the distribution parameter C_o can be obtained by

$$C_o = \frac{\overline{V_G}}{\langle V \rangle} \tag{3-52}$$

The approximate correlation for all saturated flow regimes suggested by Zuber and Findlay (1965) is

$$\overline{V_G} = \frac{\langle v_{sG} \rangle}{\langle \alpha \rangle} = 1.13 \langle v \rangle + 1.41 \left(\frac{\sigma g g_c \, \Delta \rho}{\rho_L^2} \right)^{1/4} \tag{3-53}$$

By using the above values of C_o and V_∞, the cross-sectional average void fraction can be calculated as

$$\langle \alpha \rangle = \frac{X'}{(C_o \, \Delta \rho X'/\rho_L) + (C_o + V_\infty \rho_L/G)(\rho_G/\rho_L)} \tag{3-54}$$

where X' is a local static quality. The value of C_o is greater than unity when $\alpha_o > \alpha_w$ and less than unity when $\alpha_o < \alpha_w$.

For a single-phase turbulent flow the ratio of the maximum to the average flow velocity is approximately 1.2, and the value of C_o may also be close to 1.2 for a bubbly flow. Zuber and Findlay (1965) pointed out that, as the mixture velocity increases, the value of the exponent increases and flatter profiles result.

Lockhart and Martinelli (1949) suggested an empirical void fraction correlation for annular flow based mostly on horizontal, adiabatic, two-component flow data at low pressures, Martinelli and Nelson (1948) extended the empirical correlation to steam–water mixtures at various pressures as shown in Figure 3.27. The details of the correlation technique are given in Chapter 4. Hewitt et al. (1962) derived the following expression to fit the Lockhart-Martinelli curve:

$$\ln(1 - \alpha) = -1.482 + 4.915(\ln r) - 5.955(\ln r)^2 + 2.765(\ln r)^3$$
$$+ 6.399(\ln r)^4 - 8.768(\ln r)^5 \tag{3-55}$$

where $r = (\Delta p_{LP})/(\Delta P_{GP})$, the ratio of the single-phase pressure drop of the liquid component to that of the gas component at their respective superficial mass fluxes, $G(1 - X)$ and GX.

Smith (1970) developed a saturated void fraction correlation by equating the dynamic head of a liquid annulus to that of the droplets' mixed vapor core. Smith's correlation (Fig. 3.28) is applicable to both boiling and nonboiling flows in a wide range of pressures. Comments on the range of application for various saturated void models are shown in Table 3.3.

HYDRODYNAMICS OF TWO-PHASE FLOW **159**

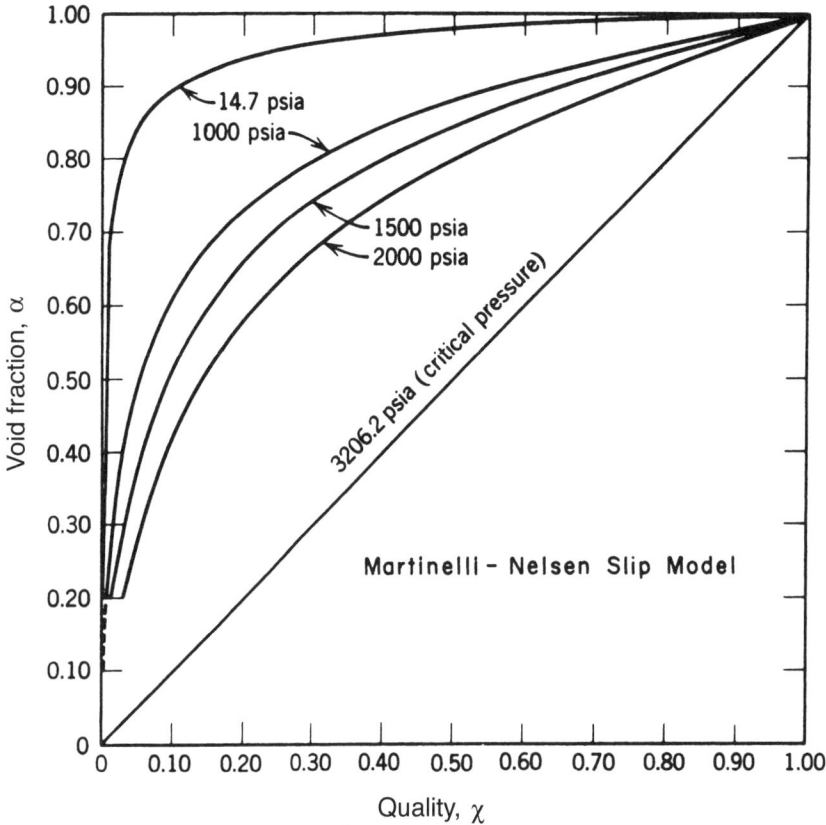

Figure 3.27 Void fraction in saturated steam–water mixture. (From Martinelli and Nelson, 1948. Copyright © 1948 by American Society of Mechanical Engineers, New York. Reprinted with permission.)

3.3.3 Diabatic Liquid Metal–Gas Two-Phase Flow

For liquid metal boiling at atmospheric pressure (normal liquid-metal reactor conditions), the slug and annular flow regimes appear to prevail rather than the bubbly flow regime (Fauske, 1971). Since there is generally good wetting of the heater surface in the case of liquid metals, and the incipient boiling superheats are generally large, a single large volume of vapor frequently is produced, rather than a series of discrete, more nearly normal-size bubbles. In fact, in long narrow vessels or channels, the sudden vapor generation rate in the heated region may be great enough to temporarily expel the liquid metal from the vessel or channel, which causes a slug flow pattern (Schlechtendahl, 1969; Holtz and Singer, 1967). The wall

160 BOILING HEAT TRANSFER AND TWO-PHASE FLOW

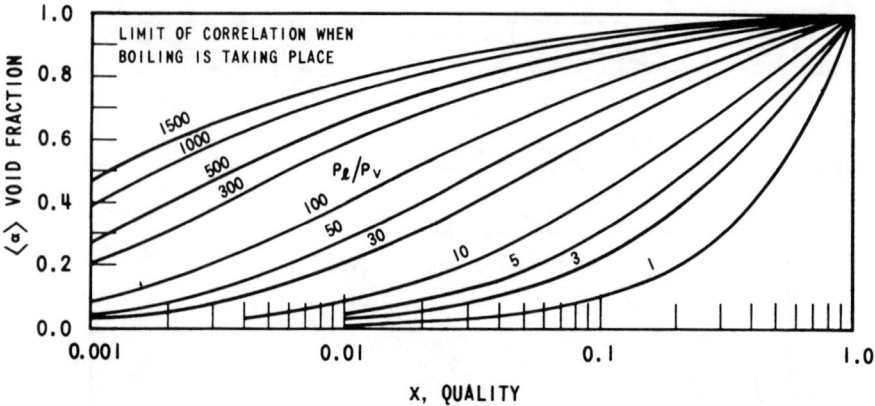

Figure 3.28 Void fraction in two-phase flow. (From Smith, 1970. Copyright © 1970 by Mechanical Engineering Publishing Ltd., Suffolk, UK. Reprinted with permission.)

Table 3.3 Comparison of saturated void models

Reference	Principles	Comment on range of application
Martinelli and Nelson (1948)	No radial pressure gradient over the cross section, i.e., $\Delta p_L = \Delta p_G$	Too high at 60–2,000 psia; too low at atmospheric pressure
Levy (1960)	Equal frictional and elevation head losses	Inaccurate at low pressures
Thom et al. (1966)	Slip ratio is a function of pressure only	Not accurate at quality < 3%
Zuber and Findlay (1965)	$\left(\dfrac{u}{u_o}\right) = 1 - \left(\dfrac{r}{R}\right)^n$ $\dfrac{\alpha - \alpha_w}{\alpha_0 - \alpha_w} = 1 - \left(\dfrac{r}{R}\right)^n$	Not accurate at low void fraction
Smith (1970)	Equal dynamic head of liquid annuli and two-phase mixture core in an annular flow at thermal equilibrium, i.e., $\rho_L V_L^2 = \rho_m V_m^2$	Not for subcooled boiling; excellent at low pressures

superheat, $t_w - t_{sat}$, is usually higher at the instant of boiling inception than at any time during stable boiling. Whereas stable nucleate boiling wall superheats for liquid metals are generally less than those for ordinary fluids, incipient boiling wall superheats are often much greater (Chen, 1968, 1970; Holland and Winterton, 1973; Dwyer, 1976). The flow pattern resulting from the vaporization of superheated liquid alkali metal (e.g., sodium) in a vertical channel was measured indirectly and shown to be that of a single bubble if the incipient boiling bulk liquid superheat is greater than about 5–10°C (9–18°F) (Holtz and Singer, 1967). However, it appears that an asymmetric radial temperature profile may result in the vapor bubble filling only a portion of the cross section of the channel (as little as 30–50%), as opposed to the symmetric case, where essentially complete filling has been observed. As the incipient superheat is increased, this asymmetry effect diminishes. Because of the extremely small vapor-to-liquid density ratios for metals at the temperatures of interest, such two-phase flow is expected to have a large void fraction at low quality. For example, at a quality of 5%, the void fraction for mercury calculated by employing the momentum exchange model (Levy, 1960) for a temperature range of 538–649°C (1,000–1,200°F) varies from 70% to 60%, respectively (Hsia, 1970). Thus, in the boiling process the low void fraction–characterized slug flow regime would be rather short, and annular flow could conceivably exist in most of the boiling region—with some of the liquid in the form of small droplets entrained in the vapor core.

3.3.4 Instrumentation

3.3.4.1 Void distribution measurement. The local void fraction is an important parameter in the reduction and analysis of hydraulic test data in two-phase flow. In addition to photographic visualization and fiber optic video probes for observation (Donaldson and Pulfrey, 1979), the following other methods are currently available, each with its own unique set of attributes and drawbacks (Delhaye et al., 1973; Delhaye, 1986; Andreychek et al., 1989).

Radiation attenuation techniques In this approach a gamma or X ray is used that gives the average fluid density along the path of the beam. The source and detector can be located external to the flow channel to measure an average reading across the width of the channel. A traversing mechanism can also be used to vary the relative location of the source, the detector, and the channel, thereby measuring an average over a selected chordal distance (Schrock, 1969; Hsu and Graham, 1976).

Capacitance or conductance measurement This method is applied where the working fluid acts as a capacitive or conductive element in a circuit (Jones et al., 1981). Use of fiber optics sensors has been developed recently (Moujaes and Dougall, 1987, 1990). These methods are used to measure film thickness in annular flow. Further discussion appears in Section 3.3.4.4. For other regimes, the use of the electrical impedience imaging method has also been introduced (Lin et al., 1991).

Measurement based on heat flux effects This approach uses local probing devices such as hot-wire anemometers and microthermocouples. The hot-wire anemometer can be either a constant-temperature system or a constant-heat-flux system. Because of the difference in heat transfer between the exposed fluid (liquid or gas)

162 BOILING HEAT TRANSFER AND TWO-PHASE FLOW

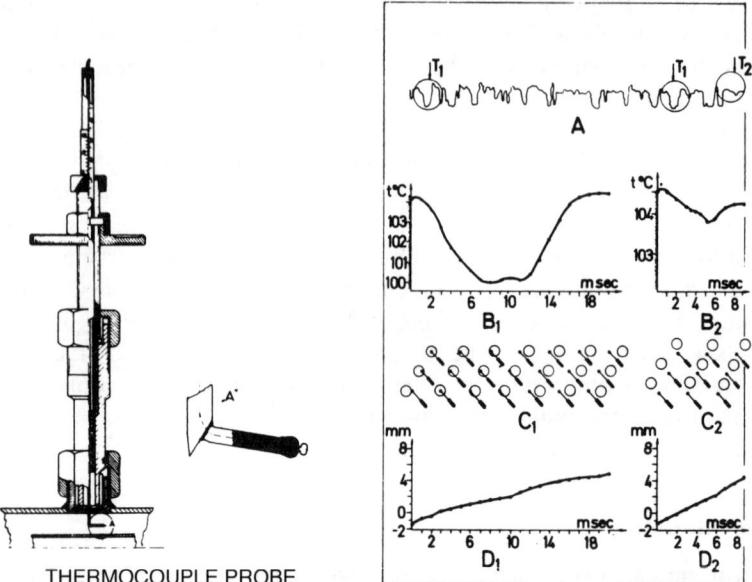

Figure 3.29 Microthermocouple probe and responses to the passing of bubbles. (From Stefanovic et al., 1970. Copyright © 1970 by Hemisphere Publishing Corp., New York. Reprinted with permission.)

and the wire, the variation of heat flux or temperature will settle on two distinctly different levels. In this manner, both the void fraction and the local velocity of each phase may be measured simultaneously. This technique provides a very clear signal for slug flow and for bubbly flow with large bubbles (Hsu et al., 1963).

Microthermocouples, on the other hand, are useful in measuring the superheat of the liquid layer that coexists with saturated bubbles (Stefanovic et al., 1970). The micro-size wires (e.g., 40-μm-diameter wires in an isolated tube of 1-mm or 0.04-in. outside diameter) provide a time response of less than 1 msec. The thermocouples pick up the temperature variations when a bubble is passing by, registering the temperature oscillations produced at the departure of the couple from the bubble. As shown in Figure 3.29, curve A is part of the thermocouple signal response when the thermocouple is at a distance of 2.5 mm (0.1 in.) from the heated tube, passing through a bubble with a diameter of 3.5 mm (0.14 in.). B_1 and B_2 show signals on the oscilloscope for T_1 and T_2, respectively, as a function of time in microseconds. These temperature signals were synchronized with photographed relative positions of the thermocouples with respect to the junction as shown in C_1 and C_2. D_1 and D_2 are the distances between the bubble center and the thermocouples.

Measurement based on electromagnetic effect Experimental investigation has been done to examine the use of an electromagnetic flow meter to measure the continuous phase of a two-phase mixture, combined with theoretical corrections for the presence of gas voids (Bernier and Brennen, 1983; Knoll, 1991). Although no flow meter currently exists that is adequate to conduct such a measurement, air–water experiments verified that such measurements would be accurate at low void fractions (<30%). Obviously, this technique could be applied to liquid-metal two-phase systems (Heineman et al., 1963).

Measurement of holdup through quick-closing valves This method is especially useful for measuring void fractions in transient flow. In this case the measurement time must be short compared with the period of the transient to be investigated. For instance, in nuclear power plants, the period of interest could be the first few seconds of the transient. The experimental technique is required to have a measurement time of the order of 0.01 sec. It uses two quick-closing valves operated simultaneously and located at the inlet and outlet of the channel (Biagioli and Premoli, 1967). This technique was first applied for measuring the mean density of boiling channel in steady-state conditions.

3.3.4.2 Interfacial area measurement.
Knowledge of the interfacial area is indispensable in modeling two-phase flow (DeJesus and Kawaji, 1990), which determines the interphase transfer of mass, momentum, and energy in steady and transient flow. Ultrasonic techniques are used for such measurements. Since there is no direct relationship between the measurement of ultrasonic transmission and the volumetric interfacial area in bubbly flow, some estimate of the average bubble size is necessary to permit access to the volumetric interfacial area (Delhaye, 1986). In bubbly flows with bubbles several millimeters in diameter and with high void fractions, Stravs and von Stocker (1985) were apparently the first, in 1981, to propose the use of pulsed, 1- to 10-MHz ultrasound for measuring interfacial area. Independently, Amblard et al. (1983) used the same technique but at frequencies lower than 1 MHz. The volumetric interfacial area, Γ, is defined by (Delhaye, 1986)

$$\Gamma \triangleq 4n \int_0^\infty \pi a^2 \, f(a) \, da \tag{3-56}$$

where n is the bubble number density, a is the bubble radius, and $f(a)$ is the probability density function of the radius. Delhaye (1986) showed that

$$\Gamma = 4\alpha_{atn} \left[\int_0^\infty \frac{a^2 \, f(a) \, da}{\int_0^\infty S(ka) \, a^2 \, f(a) \, da} \right] \tag{3-57}$$

where $S(ka)$ is the scattering coefficient and α_{atn} is the attenuation coefficient. Depending on the value of ka (k is the wave number of the ultrasound), the volumetric interfacial area, Γ, can be calculated without the knowledge of $f(a)$ in one case (when $ka \gg 100$), or can be calculated with the known value of the Sauter mean diameter, d_{SM}, or the void fraction, α, in other cases (when $ka < 100$). Stravs and von Stocker (1985) proposed measuring Γ, d_{SM}, and α by two different ultrasonic wavelengths. A study was made by Jones et al. (1986) to determine the conditions under which reliable information concerning bubbly two-phase flow could be obtained with this technique in an acoustic field generated by a piston-type transducer.

3.3.4.3 Measurement of the velocity of a large particle. The investigation of the turbulence characteristics in the liquid phase of a bubbly flow has generated detailed studies on the use of thermal anemometry and optical anemometry in gas–liquid two-phase flows. These techniques have been proved to be accurate and reliable for the measurement of the instantaneous liquid velocity in bubble flow. However, the velocity of the gas bubbles—or, more precisely, the speed of displacement of the gas–liquid interfaces—is still an active research area. Three techniques that have been proposed to achieve such measurement were reviewed by Delhaye (1986), as discussed in the following paragraphs.

Measurement by laser doppler anemometry When a large, moving, solid or fluid particle is illuminated by two intersecting laser beams, the reflected or refracted beams interfere and produce a beating frequency at any point in the medium surrounding the particle. The beating frequency results from different Doppler shifts due to the variation of the speed of displacement of the particle surface all over a given particle (bubble or drop). Measurements in rod bundles using this technique were reviewed by Kried et al. (1979).

Measurement with a grating anemometer In this technique, a set of fringes is created in space using a laser beam passing through Ronchi gratings instead of by interference of two light beams as in laser Doppler anemometry. Such a grid is common optical equipment made of a number of alternating transparent and opaque gratings set on glass. The simultaneous measurement of the size and velocity of bubbles or drops has been described using the principles of geometric optics by Semiat and Dukler (1981). Figure 3.30 shows the configuration for velocity and size measurements of a dispersed phase by means of two photodetectors. Bubble velocities are measured by means of principles of geometric optics and simultaneously the time of passage of the bubbles through the beam is measured. The diameter of the bubble can be calculated from the combined measured values. A view of the laser beam in the X–Z plane is shown in Figure 3.30. Photodetector #1 focuses on the optical sample space and detects the velocity by measuring the frequency of fringe crossing. A narrow slot (75 μm wide) and the full width of the laser beam is located in the Z–Y plane as shown facing into the beam (sketch #1). The projection of this slot along the axis of the laser beam is designated as the slot beam. This slot is fitted with a ribbon of optical fibers that conduct light to photodetector #2. As a bubble (or drop) passes across the slot beam, the light is refracted

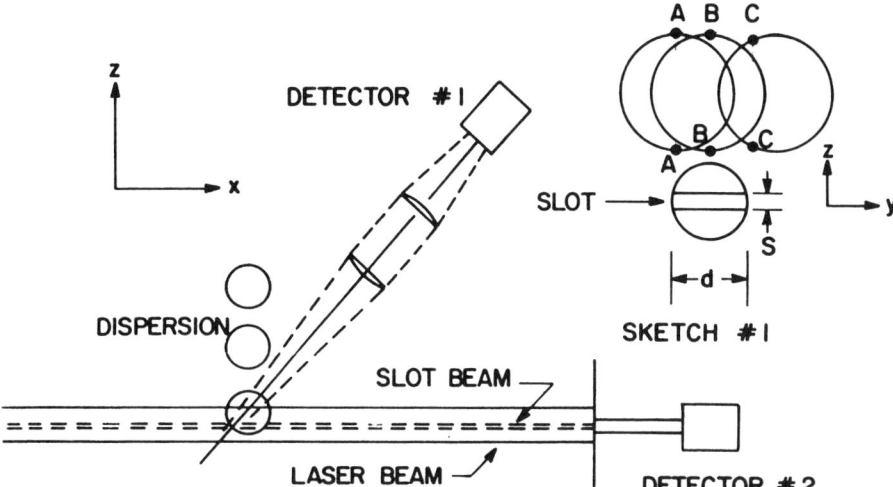

Figure 3.30 Configuration for velocity and size measurement of a dispersed phase. (From Semiat & Dukler, 1981. Copyright © 1981 by American Institute of Chemical Engineers, New York. Reprinted with permission.)

from the optical fibers and the output signal from detector #2 drops. Thus, the signal from detector #2 can be processed to calculate the passing time τ. It can be shown that, within known error, only those bubbles whose centers pass through the optical sample space will produce a simultaneous velocity signal. As shown by sketch #1 in Figure 3.30, three relative positions are shown of the laser beam of diameter d, the slot beam of width s, and a spherical bubble with diameter D, which is moving downward in the Z direction. For bubbles located at A or B, the time that the slot beam interrupts is related to the distance A–A or B–B, both being the diameter of the bubble. At location C the interrupt time is determined by the chord length C–C. However, no velocity signal for this location will be observed at detector #1, because the curvature of the bubble in the Z–Y plane causes refraction sufficient to miss the pinhole in front of the detector.

Measurement with an optical probe Another technique for measuring interfacial velocities was used by Sekoguchi et al. (1985). A single fiber optic probe (e.g., 125 μm in diameter) is illuminated by a He-Ne laser beam through a beam splitter as shown in Figure 3.31. The incoming laser beam is focused on the flat-ended tip of the optical fiber by a lens. Two reflected light beams, one from the flat-ended tip of the optical fiber and the other from the approaching gas–liquid interface, return through the optical fiber, the lens, and the beam splitter to the photodetector. The arrival of the interface produces a Doppler signal that stops when the interface hits the tip of the probe and triggers the void signal. These two properly filtered signals give the interfacial velocities and the phase density function. Sekoguchi et

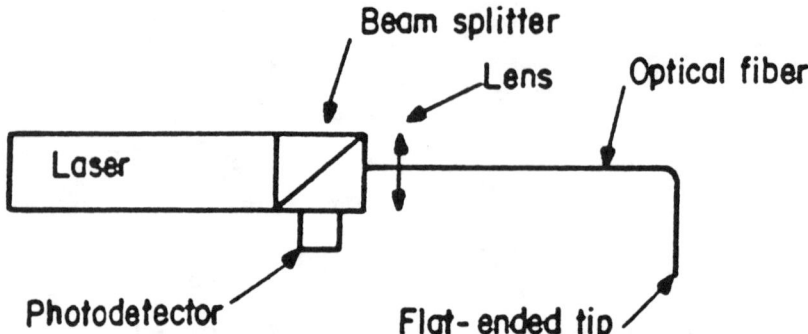

Figure 3.31 Single fiber optic probe used for particle velocity measurement. (From Sekoguci et al., 1985. Copyright © 1985 by American Society of Mechanical Engineers, New York. Reprinted with permission.)

al. (1985) used a rack of six single-fiber optical probes covering a distance of 2.25 to 4.01 mm (0.09 to 0.16 in.) from the wall of a tube, 12 mm ($\frac{1}{2}$ in.) in diameter, where an air–water mixture was flowing.

3.3.4.4 Measurement of liquid film thickness. A variety of techniques have been used to measure the time variation of local film thickness and the data on wave structure that can be deduced therefrom (Dukler and Taitel, 1991a):

Needle contact probes These are probably the simplest and least expensive devices. A needle is mounted on a micrometer and insulated from ground, except for the tip, by a nonconducting varnish. The needle is moved into the wavy liquid film flows along a conducting plate, which is grounded. As the needle is moved, the fraction of time during which contact with the liquid top takes place is noted, and is related to the probability that the film thickness is greater than some value. This technique can provide information on the minimum, maximum, and mean thickness with reasonable reliability.

Electrical conductance probes These can be either flush probes or wire probes. Flush probes are imbedded in a nonconducting wall, with one electrode connected to a voltage source and the second through a precision resistor to ground (Telles and Dukler, 1970; Chu and Dukler, 1974). Wire probes use closely spaced, nearly parallel conducting wires of small diameter, which are positioned normal to the flow (Brown et al., 1978).

Capacitance probes A capacitometer, used as a transducer, measures the gap capacitance between a plate whose area is small compared to the wave and a grounded conducting surface across which the liquid film flows (Nakanishi et al., 1979).

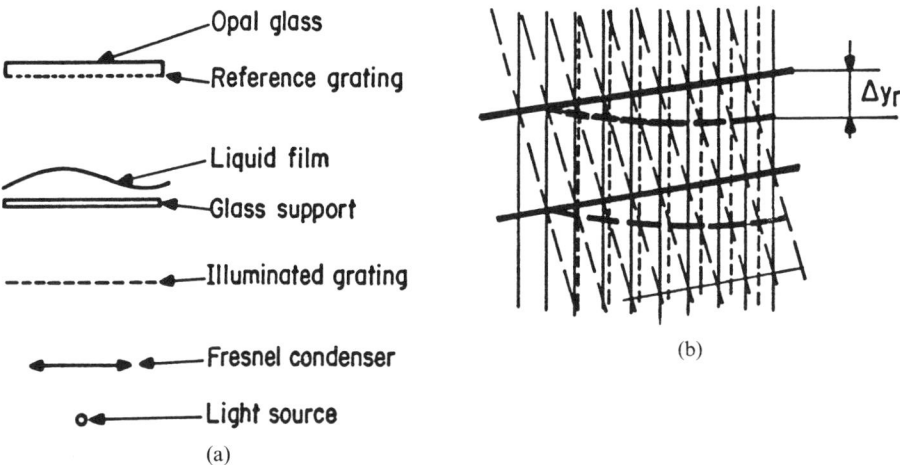

Figure 3.32 Measurement of liquid film thickness by Moiré fringes: (*a*) experimental setup; and (*b*) Moiré pattern with liquid films of varying thicknesses. (From Kheshgi and Scriven, 1983. Copyright © 1983 by Elsevier Science Ltd., Kidlington, UK. Reprinted with permission.)

Fluorescence method This method for nonconducting liquids uses fluorescein dye. The blue light activates fluorescence, and the green light, which has intensity directly proportional to the film thickness (Hewitt, 1969–1970), is emitted.

X-ray absorption method This method is similar to void fraction measurement by radiation attenuation.

Two other acoustical and optical techniques were described by Delhaye (1986):

Measurement by moiré fringes The Moiré fringes are formed by two superposed, parallel Ronchi gratings. Keshgi and Scriven (1983) measured the thickness of liquid films by using Moiré fringes obtained as shown in Figure 3.32*a*. If the free surface of the liquid film is no longer parallel to the gratings, the Moiré fringes are then distorted as shown in Figure 3.32*b*, where y_r is the fringe displacement on the reference grating plane and can be correlated with the film thickness.

Measurement by means of a wall optical sensor The thickness of a film and the slope of its free surface can also be measured by means of a wall optical sensor, as proposed by Ohba et al. (1984). This sensor consists of a cluster of seven optical fibers mounted flush with the wall (Fig. 3.33). A laser beam passed through the central fiber is reflected by the free surface onto the other fiber tips, which collect the light and transmit it to two photodiodes. The light intensities received by these two detectors enable the film thickness and the inclination angle to be determined.

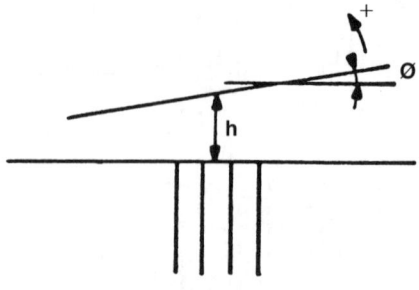

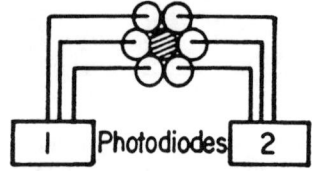

Figure 3.33 Measurement of liquid film thickness by means of a wall optical sensor as used by Ohba et al. (1984). (From Delhaye, 1986. Copyright © 1986 by Hemisphere Publishing Corp., New York. Reprinted with permission.)

3.4 MODELING OF TWO-PHASE FLOW

By a combination of rigorous model development, advanced computational techniques, and a number of small- and large-scale supporting experiments, considerable progress has been made in understanding and predicting two-phase phenomena. Some of the results were reviewed and summarized in symposium proceedings, e.g., by Jones & Bankoff (1977); ANS/ASME/NRC (1980); Wallis (1980); and Ishii and Kocamustafaogullari (1983). Current approaches toward developing multiphase models depend on some form of averaging (e.g., Besnard and Harlow, 1988; Ahmadi and Ma, 1990). Two general models have been used in the development of the basic equations for two-phase flow:

1. Homogeneous model/drift flux model
2. Separate-phase model (two-fluid model)

The first model is more suited to dealing with mixed flow such as bubble flow and slug flow, while the second is more suited to cases where flow is separated, as in stratified and annular flow.

3.4.1 Homogeneous Model/Drift Flux Model

The homogeneous model treats the mixture as a whole, and consequently the physical properties are represented by the average value of the mixture. This treatment assumes that the gas and liquid phases possess the same velocity (or the slip velocity is neglected). This model was used extensively in the past, because of its simplicity. The six unknowns to be determined, as in single-phase flow, are the velocity vectors (in three directions), pressure, temperature, and density. The number of equations (six) is thus equivalent to the number of unknowns:

Conservation of mass equation
Conservation of momentum equations (three equations)
Conservation of energy equation
Equation of state

If this model is further simplified by considering unidirectional flow, the number of equations is reduced to four (Wallis, 1969). Another example is Bankoff's variable-density, single-fluid model for two-phase flow (Bankoff, 1960). Since it is based on an intimate mixture, both mechanical equilibrium (i.e., same velocity) and thermal equilibrium (same temperature) between the two phases must logically be assumed (Bouré, 1975).

In reality, the slip velocity may not be neglected (except perhaps in a microgravity environment). A drift flux model has therefore been introduced (Zuber and Findlay, 1965) which is an improvement of the homogeneous model. In the drift flux model for one-dimensional two-phase flow, equations of continuity, momentum, and energy are written for the mixture (in three equations). In addition, another continuity equation for one phase is also written, usually for the gas phase. To allow a slip velocity to take place between the two phases, a drift velocity, u_{GJ}, or a diffusion velocity, u_{GM} (gas velocity relative to the velocity of center of mass), is defined as

$$u_{GJ} = u_G - J \tag{3-58}$$

where J is the mixture average superficial velocity, or the total volumetric flux, which is constant in one-dimensional, steady flow:

$$J = u_G \alpha_G + u_L \alpha_L \tag{3-59}$$

and

$$u_{GM} = u_G - \frac{G}{\rho_M} \tag{3-60}$$

where the velocity of the center of mass, or the mixture velocity, is G/ρ_M, the mixture mass flux is

$$G = u_G \alpha_G \rho_G + u_L \alpha_L \rho_L \tag{3-61}$$

and the mixture density ρ_M is

$$\rho_M = \alpha_G \rho_G + \alpha_L \rho_L \tag{3-62}$$

Using the above definitions and integrating over the channel cross section, with some manipulations, Zuber's kinematic equation results (Hsu and Graham, 1976):

$$\frac{\langle u_{GS} \rangle}{\langle \alpha_G \rangle} = C_o \langle J \rangle + \frac{\langle u_{GJ} \alpha_G \rangle}{\langle \alpha_G \rangle} \quad (3\text{-}63)$$

where C_o is the concentration parameter,

$$C_o = \frac{\langle \alpha_G J \rangle}{\langle \alpha_G \rangle \langle J \rangle}$$

In this way, nonequilibrium phenomena in steady or quasi-steady state can be taken into account.

The model is considered a generalized homogeneous model (Bouré, 1976), and it seems to be well suited to system codes describing complex systems, such as nuclear reactors, which cannot take into account a detailed separate-phase flow model because of their complexity.

3.4.2 Separate-Phase Model (Two-Fluid Model)

As both phases occupy the full flow field concurrently, two sets of conservation equations correspond to these two phases and must be complemented by the set of interfacial jump conditions (discontinuities). A further topological law, relating the void fraction, α, to the phase variables, was needed to compensate for the loss of information due to model simplification (Bouré, 1976). One assumption that is often used is the equality of the mean pressures of the two phases,*

$$P_G = P_L$$

The intrinsic constitutive laws (equations of state) are those of each phase. The external constitutive laws are four transfer laws at the walls (friction and mass transfer for each phase) and three interfacial transfer laws (mass, momentum, energy). The set of six conservation equations in the complete model can be written in equivalent form:

Mixture conservation of mass equation
Mixture conservation of momentum equation
Mixture conservation of energy equation
Slip equation (concerning the difference in velocity)

---

* It should be noted that this assumption is open to criticism, in particular with respect to propagation phenomena.

Two thermal nonequilibrium equations (the difference between the enthalpy of a phase and the saturation enthalpy)

Being handicapped by the present state of knowledge on the transfer laws, Bouré (1975) suggested reducing the number of transfer laws to be specified by adopting a number of restrictions.

Delhaye proposed the following general two-phase equations (Vernier and Delhaye, 1968; Delhaye, 1969a).

Conservation of mass:

$$\frac{\partial \rho_i}{\partial t} + \text{div } \rho_i \mathbf{u}_i = 0 \qquad i = G \text{ or } L \tag{3-64}$$

Conservation of momentum:

$$\frac{\partial \rho_i \mathbf{u}_i}{\partial t} + \text{div } \rho_i \mathbf{u}_i \mathbf{u}_i = \text{div } \bar{\bar{\tau}}_i + \rho_i \mathbf{F} \tag{3-65}$$

Conservation of energy:

$$\frac{\partial}{\partial t}\left[\rho_i\left(\frac{\mathbf{u}_i^2}{2} + E_i\right)\right] + \text{div } \rho_i\left(\frac{\mathbf{u}_i^2}{2} + E_i\right)\mathbf{u}_i$$
$$= \text{div}(\bar{\bar{\tau}}_i \cdot \mathbf{u}_i - \mathbf{q}_i) + \rho_i \mathbf{F} \cdot \mathbf{u}_i \tag{3-66}$$

where $\mathbf{u}_i$ = velocity vector
$\mathbf{F}$ = body force
$\bar{\bar{\tau}}_i$ = stress tensor
$\mathbf{q}_i$ = heat flux

Interfacial conditions:

$$(\text{mass}) \qquad \sum_{i=L,G} \phi_i = 0 \tag{3-67}$$

$$(\text{momentum}) \qquad \sum_{i=L,G} \phi_i - \text{div } \sigma \bar{\bar{U}} - \sigma \mathbf{n}_G\left(\frac{2}{R}\right) = 0 \tag{3-68}$$

$$(\text{energy}) \qquad \sum_{i=L,G} \phi_i - \text{div } \sigma \mathbf{u}_i \bar{\bar{U}} - \sigma \mathbf{u}_i\left(\frac{2}{R}\right)\mathbf{n}_G = 0 \tag{3-69}$$

where ϕ_i = mass flux across the interface
σ = surface tension
$\bar{\bar{U}}$ = metric tensor of the space
$\mathbf{n}_G$ = normal vector in gas-phase direction

For other discussions of two-phase models and numerical solutions, the reader is referred to the following references: thermofluid dynamic theory of two-phase flow (Ishii, 1975); formulation of the one-dimensional, six-equation, two-phase flow models (Le Coq et al., 1978); lumped-parameter modeling of one-dimensional, two-phase flow (Wulff, 1978); two-fluid models for two-phase flow and their numerical solutions (Agee et al., 1978); and numerical methods for solving two-phase flow equations (Latrobe, 1978; Agee, 1978; Patanakar, 1980).

3.4.3 Models for Flow Pattern Transition

In the framework of two-phase mixtures flow mechanisms, for a circular duct geometry, the most important parameters are the flow directions and the relative directions of the two phases; the angle of slope with respect to the horizontal; the average velocity of each phase; pressure, temperature, density, and viscosity; surface tension; and the dimensions (diameter and length) of the duct. In complex geometries, instead of the last two parameters, other geometric parameters typical of the configuration have to be considered. Section 3.2.2 described criteria for flow pattern transitions. Models for predicting flow pattern transitions in steady gas–liquid flow in pipes were summarized by Barnea (1987). An attempt was made to represent the true physics that was observed in experiments. Reasonable success was achieved for horizontal and slightly inclined tubes, vertical upward and downward flows, and inclined upward and downward flows (Taitel and Dukler, 1976b; Taitel et al., 1978, 1980; Taitel and Barnea, 1990). Unified models were presented with a whole range of pipe inclinations for the transition from annular to intermittent (slug) flow and from dispersed bubble flow, the stratified-nonstratified transition (Barnea et al., 1985; Barnea, 1987), and subregions within the intermittent flow (i.e., slug and elongated bubble flow) (Taitel and Barnea, 1990). Different parameters were used for boundaries of different flow patterns, as shown in Figures 3.3 and 3.6. Barnea et al. (1983) also reported some data on flow pattern transition for gas–liquid, horizontal and vertical upward flow on small-diameter (4–12-mm, or 0.16–0.5-in.) tubes, which offered tests against the flow pattern models of Taital and Dukler (1976b) for horizontal two-phase flow and of that of Taitel et al. (1980) for vertical upward flow. An improvement to these models resulted by taking into account the special effects caused by surface tension in small-diameter tubes.

Another determination theory for flow patterns was suggested by Beattie (1983), based on the extrapolation to two-phase flows of concepts developed in single-phase flow analyses, such as the mixing length, Reynolds number analogy, and laminar sublayers. The kinematic flow characteristics in the laminar sublayer were selected as flow regime classification parameters. These flow characteristics, obtained in terms of velocity profile, apparent viscosity, and apparent density, were used to characterize the flow regimes and were then employed directly to explain the trends of two-phase mixture friction factor, heat transfer coefficient, and void fraction (Cumo and Naviglio, 1988). Thus it was considered a boundary-layer technique for determining the flow regions. Different relationships were proposed for the evaluation of mixture properties with correspondent flow regimes, e.g., film thickness-dependent flows (very thin film sublayers), surface tension-dependent flows (with attached wall bubbles sublayer), and viscosity-dependent flows (dry wall, or wavy gas–liquid interface, or rigid/nonrigid surface bubbles). Although such techniques have not been well developed for two-phase flow, others have in the past attempted to use similar concepts (e.g., Figure 3.26).

3.4.4 Models for Bubbly Flow

A steady homogeneous model is often used for bubbly flow. As mentioned previously, the two phases are assumed to have the same velocity and a homogeneous mixture to possess average properties. The basic equations for a steady one-dimensional flow are as follows.

Continuity equation:

$$m = \rho_M J A = \text{constant} \tag{3-70}$$

where $J = \dfrac{Q_L + Q_G}{A}$

$\rho_M = \alpha \rho_G + (1 - \alpha)\rho_L$

$\alpha = \dfrac{Q_G}{Q_G + Q_L}$

Momentum equation:

$$m\left(\frac{dJ}{dz}\right) = -A\left(\frac{dP}{dz}\right) - S\tau_w - A\rho_M g \sin\theta \tag{3-71}$$

where θ = pipe inclination upward from the horizontal
τ_w = average wall shear stress = $(f/2)\rho_M J^2$
S = periphery on which the stress acts
f = Fanning friction factor

174 BOILING HEAT TRANSFER AND TWO-PHASE FLOW

Energy equation:

$$\frac{dq}{dz} - \frac{dw}{dz} = m\frac{d}{dz}\left(h_M + \frac{J^2}{2} + gz\sin\theta\right) \quad (3\text{-}72)$$

where $\dfrac{dq}{dz}$ = heat transfer per unit length of pipe

$\dfrac{dw}{dz}$ = shaft work output, which is usually zero

h_M = mixture specific enthalpy

However, bubble nonhomogeneous distribution exists in two-phase shear flow. As yet, the following general trends in void fraction radial profiles are being identified for bubbly upward flow (Zun, 1990): concave profiles (Serizawa et al., 1975); convex profiles (Sekoguchi et al., 1981), and intermediate profiles (Sekoguchi et al., 1981; Zun, 1988). Two theories are currently dominant:

1. The theory of bulk liquid turbulent structure control in combination with the transverse lift force (Drew et al., 1978; Lahey, 1988)
2. The bubble deposition theory, which postulates bubble transverse migration due to transverse lift in combination with bubble lateral dispersion (Zun, 1985, 1988)

As mentioned earlier, the drift-flux model is used in slip flows. Equation (3-63) can be written in the form

$$\langle u_{GS}\rangle = C_o\langle\alpha\rangle\langle J\rangle + \langle\alpha\rangle V_{GJ} \quad (3\text{-}73)$$

where V_{GJ} is the effective drift velocity. It can be shown that

$$\frac{\langle\alpha\rangle}{\langle\beta\rangle} = \frac{1}{C_o + (V_{GJ}/\langle J\rangle)} \quad (3\text{-}74)$$

Here the C_o term represents the global effect due to nonuniform voids and velocity profiles. The $V_{GJ}/\langle J\rangle$ term represents the local relative velocity effect (Todreas and Kazimi, 1990).

3.4.5 Models for Slug Flow (Taitel and Barnea, 1990)

Slug flow can be subdivided into two main parts (Fig. 3.34): a liquid slug zone of length ℓ_s, and a film zone of length ℓ_f. The liquid slug zone can be aerated by dispersed bubbles, with a liquid holdup, R_s. As the gas starts to penetrate through

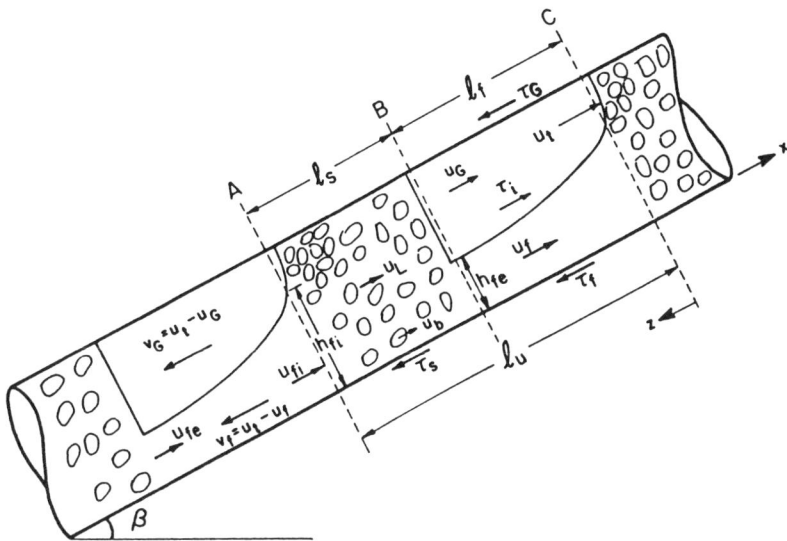

Figure 3.34 Slug-flow geometry. (From Taitel and Barnea, 1990. Copyright © 1990 by Academic Press, Orlando, FL. Reprinted with permission.)

the slug zone, the slugs become wavy annular, which starts the transition to annular flow (Barnea et al., 1980). The velocities shown in Figure 3.34 are

u_L = average liquid velocity in the slug
u_b = average velocity of the dispersed bubble in the slug
u_t = translational velocity of the elongated bubble
u_f = liquid velocity in the film zone (varies along the pipe)
u_G = gas velocity in the film zone
t_u = time for a slug unit (of length ℓ_u) to pass through a fixed point = ℓ_u/u_t
t_f = time for a film zone (of length ℓ_f) to pass through a fixed point = ℓ_f/u_t
t_s = time for a slug zone (of length ℓ_s) to pass through a fixed point = ℓ_s/u_t

The following equations were derived (Taitel and Barnea, 1990).

1. Liquid mass balance over a slug unit:

$$M_L = \frac{1}{t_u}\left(u_L A R_s \rho_L t_s + \int_0^{t_f} u_f A R_f \rho_L \, dt\right) \qquad (3\text{-}75)$$

The amount of liquid in the film zone that moves upstream (backward) relative to the gas–liquid interface is given by

$$(u_t - u_L)\rho_L A R_s = (u_t - u_f)\rho_L A R_f \qquad (3\text{-}76)$$

176 BOILING HEAT TRANSFER AND TWO-PHASE FLOW

Thus,
$$u_{LS} = u_L R_s + u_t(1-R_s)\left(\frac{\ell_f}{\ell_u}\right) - \left(\frac{u_t}{\ell_u}\right)\int_0^{\ell_f} \alpha_f \, dx \qquad (3\text{-}77)$$

2. Mass balance on the mixture, at a cross section in the slug zone (both liquid and gas are assumed incompressible):

$$u_s = u_{LS} + u_{GS} = u_L R_s + u_b \alpha_s \qquad (3\text{-}78)$$

3. Average void fraction of a slug unit:

$$\alpha_u = \left(\alpha_s \ell_s + \int_0^{\ell_f} \frac{\alpha_f \, dx}{\ell_u}\right) \qquad (3\text{-}79)$$

By combining Eqs. (3.77) and (3.79),

$$\alpha_u = \frac{-u_{LS} + u_L R_s + u_t \alpha_s}{u_t} \qquad (3\text{-}80)$$

or

$$\alpha_u = \frac{u_{GS} - u_b \alpha_s + u_t \alpha_s}{u_t} \qquad (3\text{-}81)$$

Note that the average void fraction of a slug unit depends only on the liquid and gas flow rates, the dispersed velocity u_b, the translational velocity u_t, and the void fraction within the liquid slug, α_s, and it is independent of the bubble shape or bubble length, the liquid slug length, as well as the film thickness in the film zone (Barnea, 1990).

4. Momentum equations for the liquid film and the gas above it (relative to a coordinate system moving with a velocity u_t):

$$\rho_L v_f \left(\frac{\partial v_f}{\partial z}\right) = \left(\frac{-\partial P}{\partial z}\right) + \frac{\tau_f S_f}{A_f} - \frac{\tau_i S_i}{A_f} + \rho_L g \sin\beta - \rho_L g \cos\beta\left(\frac{\partial h_f}{\partial z}\right) \qquad (3\text{-}82)$$

$$\rho_G v_G \left(\frac{\partial v_G}{\partial z}\right) = \left(\frac{-\partial P}{\partial z}\right) + \left(\frac{\tau_G S_G}{A_G}\right) + \frac{\tau_i S_i}{A_G} + \rho_G g \sin\beta - \rho_G g \cos\beta\left(\frac{\partial h_f}{\partial z}\right) \qquad (3\text{-}83)$$

where

$$v_f = u_t - u_f \qquad v_G = u_t - u_G$$

$$\tau_f = f_f\left(\frac{\rho_L |u_f| u_f}{2}\right) \qquad \tau_G = f_G\left(\frac{\rho_G |u_G| u_G}{2}\right)$$

and

$$\tau_i = f_i \left[\frac{\rho_G |u_G - u_f|(u_G - u_f)}{2} \right] \tag{3-84}$$

Here f_f, f_G, and f_i are friction factors between the liquid and the wall, the gas and the wall, and at the gas–liquid interface, respectively. Of these factors, the last one, f_i, is hard to determine. Some crude correlations and assumptions have to be used, as suggested by Taitel and Barnea (1990). From Eqs. (3-80) and (3-81), one can derive an equation for h_f (or δ) as a function of z:

$$\frac{dh_f}{dz} = \frac{(\tau_f S_f/A_f) - (\tau_G S_G/A_G) - \tau_i S_i(1/A_f + 1/A_G) + (\rho_L - \rho_G)g \sin\beta}{(\rho_L - \rho_G)g \cos\beta - \rho_L v_f(u_t - u_L)(R_s/R_f^2)(dR_f/dh_f) - \rho_G v_G[(u_t - u_b)(1 - R_s)/(1 - R_f^2)]} \tag{3-85}$$

where, for the case of stratified film flow,

$$\frac{dR_f}{dh_f} = \left(\frac{4}{\pi D}\right)\sqrt{1 - \left(\frac{2h_f}{D} - 1\right)^2} \tag{3-86}$$

For large z, the limiting value of h_{fe} is the equilibrium liquid level h_E, which is obtained when $dh_f/dz = 0$, that is, the numerator of Eq. (3-85) equals zero. The liquid holdup in the front of the liquid film, R_{fi}, equals the value of R_s, and u_{fi} equals u_L; h_s is the liquid level corresponding to R_s. Thus the integration of Eq. (3-85) starts normally with $h_f = h_{fi} = h_s$ at $z = 0$, and h_f decreases from h_s toward the limit of h_E. If, however, the critical liquid level h_c is less than h_s, where h_c is the level that equates the denominator to zero, then (dh_f/dz) becomes positive. In this case, the liquid level reduces instantaneously to the critical level, and the integration of h_f starts with $h_{fi} = h_c$ at $z = 0$.* Further, in the event that h_c or h_s is less than the equilibrium level h_E, then h_E is reached immediately. Several simplifications have been used with Eq. (3-85), which are summarized in the cited reference. To proceed in obtaining a solution of the formulations, Taitel and Barnea (1990) gave auxiliary relations for additional variables: u_t, u_b, R_s and ℓ_s, or the slug frequency, v_s. For the procedure of obtaining the solutions, the reader is referred to their work.

3.4.6 Models for Annular Flow

3.4.6.1 Falling film flow.
The extent of basic modeling of two-phase annular flow is still very limited, because annular flow is the pattern that is least well understood

---

* For the vertical case, the denominator is never zero, and a critical film thickness does not exist.

(Dukler and Taitel, 1991b). The simplest configuration of annular flow is a vertical falling film with concurrent downward flow. Given information on interfacial shear and considering the film to be smooth, the film thickness and heat transfer coefficient between the wall and the liquid film can be predicted from the following basic equations (Dukler, 1960):

Shear stress:
$$\tau = \frac{1}{g_c}(\mu_L + \varepsilon \rho_L)\left(\frac{du}{dy}\right) \quad (3\text{-}87)$$

Local liquid film Reynolds number:
$$\text{Re}_{LX} = \frac{4\rho_L}{\mu_L}\int_0^\delta u\, dy \quad (3\text{-}88)$$

Heat flux at any position y of film:
$$q = -(k_L + \varepsilon_H c_{PL} \rho_L)\left(\frac{dT}{dy}\right) \quad (3\text{-}89)$$

where c_{PL}, μ_L, ρ_L = specific heat, viscosity, and density of liquid, respectively
δ = film thickness measured in the y direction
$\varepsilon, \varepsilon_H$ = eddy viscosity and eddy thermal conductivity, respectively

As developed by Dukler, Deissler's expression for ε was used for the region near the wall, and, von Karman's relationship was used for highly developed turbulent flow.

Numerical solutions agreed with experimental data then available. When waves appear, however, everything changes, as friction increases dramatically, entrainment can take place, and mass and heat transfer are enhanced. Information on the measurement of wave structure is provided by Dukler (1977). There exists a highly irregular wavy interface (even in the absence of gas flow), and the surface is covered by a complex array of large and small waves moving over a substrate that is less than the mean film thickness. Because of this, reliable experimental measurements of velocity distribution are exceedingly difficult in such a small film height and short passage time.

Wasden and Dukler (1989) performed a series of numerical experiments for velocity profiles. The results were in reasonably good agreement with wave shapes, wall shear stress profiles, and wave velocities measured for a falling liquid at Re = 880. These numerical experiments pointed to the shortcomings of the many methods used to model large waves on falling films that are based on parabolic velocity profiles. An example of comparison among various polynomial representations of the streamwise profile is given in Figure 3.35. The streamline map for an evolving wave is presented in Figure 3.35a; the locations A, B, C of acceleration and velocity fit are shown in Figure 3.35b, and a comparison of velocity versus distance from the wall, y, are presented in Figure 3.35c. Significantly deviations are shown of the parabolic fits predicted by Kapitza (1964) from the computed velocities with a least-squares cubic fit at locations A (near the front) and B (beneath the peak).

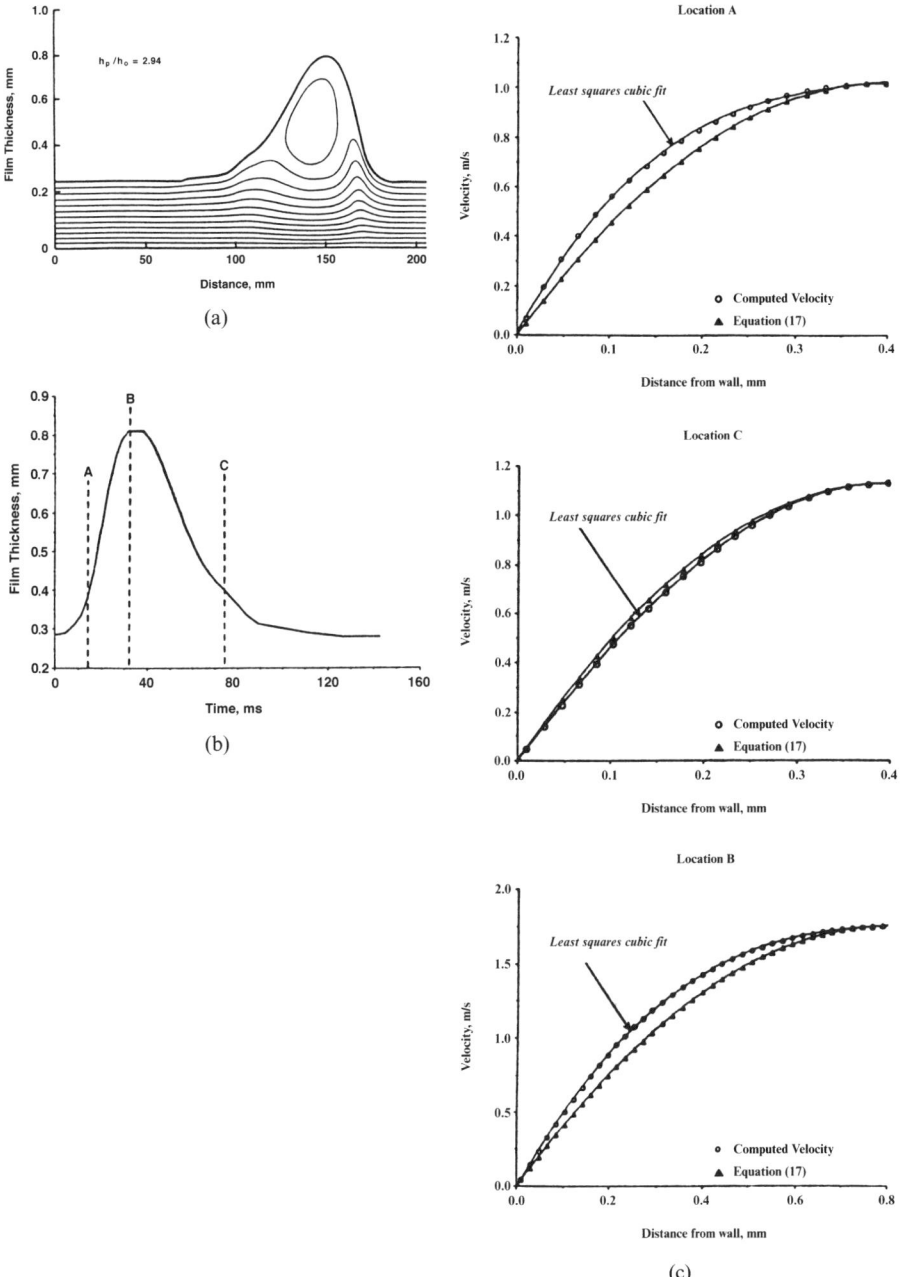

Figure 3.35 Comparison of velocity fit examples of falling liquid film with peak/substrate ≃ 3: (a) streamline map for evolving wave; (b) location of acceleration and velocity fit examples; (c) curve fits of velocity profiles. (From Wasden and Dukler, 1989. Copyright © 1989 by American Institute of Chemical Engineers, New York. Reprinted with permission.)

Due to lack of understanding of the wave structure and motions, modeling of the interfacial shear remains empirical.

3.4.6.2 Countercurrent two-phase annular flow. Countercurrent two-phase annular flow is of interest in reactor analysis. When the gas is blown upward through the center of a vertical tube in which there is a falling film, a shear stress that retards the film is set up at the interface. As long as the film remains fairly smooth and stable, this shear stress is usually small and the film thickness (and consequently the void fraction) is also virtually unchanged from the value that is obtained with no gas flow (Wallis, 1969). However, for a given liquid rate there is a certain gas flow at which very large waves appear on the interface, the whole flow becomes chaotic, the gas pressure drop increases markedly, and liquid is expelled from the top of the tube. This condition is known as *flooding*. Wallis (1969) suggested a general form of empirical flooding correlation in vertical tube as

$$(J_G^*)^{1/2} + m(J_L^*)^{1/2} = C \tag{3-90}$$

where J_G^*, J_L^* are dimensionless groups that relate momentum fluxes to the hydrostatic forces,

$$(J_G^*) = j_G \rho_G^{1/2} [gD(\rho_L - \rho_G)]^{-1/2} \tag{3-91}$$

$$(J_L^*) = j_L \rho_L^{1/2} [gD(\rho_L - \rho_G)]^{-1/2} \tag{3-92}$$

D is a characteristic length; j_G, j_L are the volume fluxes of the gas and liquid, respectively; and m, C are constants. For turbulent flow, m is equal to unity. The value of C is found to depend on the design of the ends of the tubes and the way in which the liquid and gas are added and extracted. It may have values ranging from 0.725 to 1. For viscous flow in a liquid, m and C are functions of the dimensionless inverse viscosity, N_F, where

$$N_f = \frac{g^{1/2} D^{3/2} \rho_L}{\mu_L} \tag{3-93}$$

3.4.6.3 Inverted annular and dispersed flow. Inverted annular flow occurs in a subcooled boiling flow when the liquid enters a heated section that has a high heat flux and/or wall temperature. Vapor is generated very rapidly near the wall, creating an annulus of vapor phase around a liquid core (Fig. 3.36). The vapor accelerates faster than the liquid, causing instabilities in the liquid core and resulting in breakup of the core into droplets (Varone and Rohsenow, 1990). The following basic model equations were presented by Varone and Rohsenow.

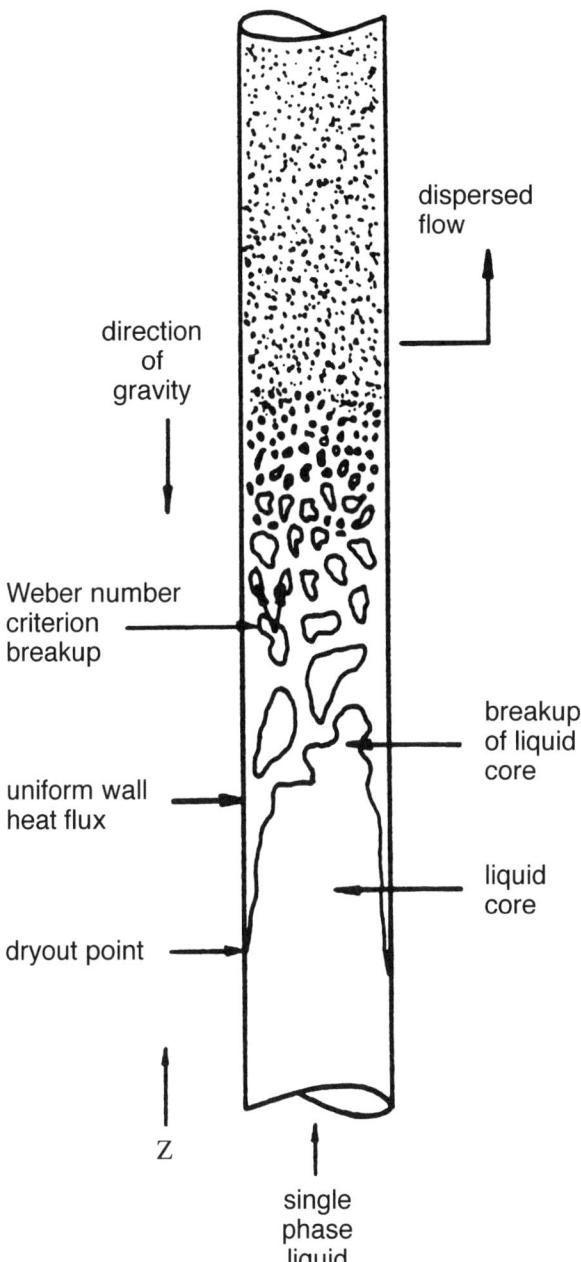

Figure 3.36 Droplet formation in inverted annular dryout. (From Varone and Rohsenow, 1990. Copyright © 1990 by Massachusetts Institute of Technology, Cambridge, MA. Reprinted with permission.)

1. Liquid velocity gradient:

$$\frac{du_L}{dz} = -\frac{g}{u_L}\left(1 - \frac{\rho_G}{\rho_L}\right) + \frac{3}{4}\left(\frac{C_D}{d}\right)\left(\frac{\rho_G}{\rho_L}\right)u_L(S-1)^{1/2} \qquad (3\text{-}94)$$

In the inverted annular flow region, it is assumed that the vapor generation near the wall occurs fast enough that the velocities of the two phases are about equal, or $S = (u_G/u_L) \approx 1$.

2. The wall energy balance:

$$T_W - T_G = \frac{q_w''}{h_{w-G}\alpha} - \frac{E(1-\alpha)H_{fg}\,v_{d-w}\,\rho_L\beta_1}{2h_{w-G}\alpha\beta_2} \qquad (3\text{-}95)$$

where v_{d-w} is droplet deposition velocity; h_{w-G} is wall–vapor (forced-convection) heat transfer coefficient; β_1, β_2 are parameters in the wall-drop effectiveness calculation; and E is the wall-drop heat transfer effectiveness.

The second term on the right-hand side of Eq. (3-95) represents the effect of the wall-drop heat transfer. The heat transfer analysis of dispersed-flow film boiling is discussed in Section 4.4.3.

3.4.7 Models for Stratified Flow (Horizontal Pipes)

A separated flow model for stratified flow was presented by Taitel and Dukler (1976a). They indicated analytically that the liquid holdup, R, and the dimensionless pressure drop, ϕ_G, can be calculated as unique functions of the Lockhart-Martinelli parameter, X (Lockhart and Martinelli, 1949).* Considering equilibrium stratified flow (Fig. 3.37), the momentum balance equations for each phase are

$$-A_L\left(\frac{dP}{dx}\right) - \tau_{wL} S_L + \tau_i S_i = 0 \qquad (3\text{-}96)$$

$$-A_G\left(\frac{dP}{dx}\right) - \tau_{wG} S_G - \tau_i S_i = 0 \qquad (3\text{-}97)$$

and the shear stresses are evaluated as

---

* The assumed dependence of pressure drop on X was first explained by Johannessen (1972) in evolving a theoretical model for stratified flow including some unnecessary simplifications.

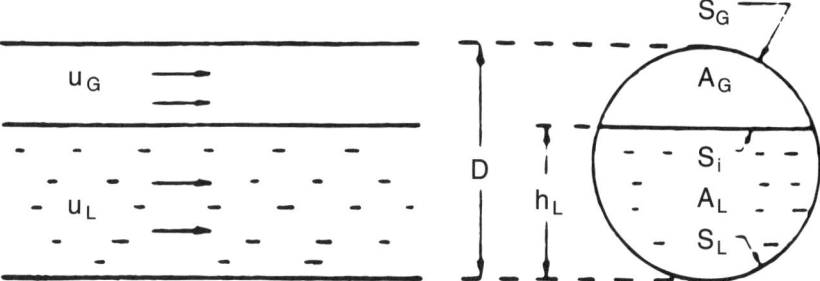

Figure 3.37 Equilibrium stratified flow. (From Taitel and Dukler, 1976a. Copyright © 1976 by Elsevier Science Ltd., Kidlington, UK. Reprinted with permission.)

$$\tau_{wL} = \frac{f_L \rho_L u_L^2}{2} \quad \tau_{wG} = \frac{f_G \rho_G u_G^2}{2}$$

$$\tau_i = \frac{f_i \rho_G (u_G - u_L)^2}{2}$$

Normally, $u_G \gg u_L$.

The Taitel and Dukler model used two basic approximations:

1. The shear stress at the interface is equal to the shear of the gas at the wall.
2. The wall shear stress can be calculated on the basis of the fully developed pipe flow correlation (the hydraulic diameter concept).

Cheremisinoff and Davis (1979) relaxed these two assumptions by using a correlation developed by Cohen and Hanratty (1968) for the interfacial shear stress, using von Karman's and Deissler's eddy viscosity expressions for solving the liquid-phase momentum equations while still using the hydraulic diameter concept for the gas phase. They assumed, however, that the velocity profile is a function only of the radius, r, or the normal distance from the wall, y, and that the shear stress is constant, $\tau = \tau_w$.

3.4.8 Models for Transient Two-Phase Flow

The analysis of transient flows is necessary for safety analysis of nuclear reactors. Such efforts usually result in the development of large computer codes (e.g., RELAP-5, RETRAN, COBRA, TRAC). Rather than going into the details of such codes, this section gives the principles and basic models involved in the analysis.

The two basic models, as described previously (Section 3.4), have also been

184 BOILING HEAT TRANSFER AND TWO-PHASE FLOW

used for transient analysis. Since the drift flux/homogeneous model is more suited for dispersed bubble flow and slug flow, and the two-fluid model for stratified and annular flow, occasionally the model may be changed according to the flow pattern at hand. However, a single model is often preferred, owing to the extreme complexity that otherwise results. The two-fluid model was shown to possess a serious limitation concerning its well posedness and stability, which hinders obtaining theoretical convergent solutions in the general case (Dukler and Taitel, 1991b). Numerical schemes, in practice, have been used by converting to a difference equation and using a finite grid that introduces an artificial stability and limits the size of the wavelength handles. The following basic equations of the drift flux model for transient flow were given by Dukler and Taitel (1991b).

Continuity equation for the gas:

$$\frac{\partial}{\partial t}(\rho_G A \alpha_G) + \frac{\partial}{\partial z}(\rho_G A \alpha_G u_G) = \frac{\partial}{\partial t} m_G A \tag{3-98}$$

where $(\partial m_G/\partial t)$ is the gas generation rate per pipe volume.

Continuity equation for the mixture:

$$\frac{\partial}{\partial t}(\rho_M A) + \frac{\partial}{\partial z}(AG) = 0 \tag{3-99}$$

Note that

$$\frac{\partial}{\partial t}(m_G A) = -\frac{\partial}{\partial t}(m_L A)$$

By using Eqs. (3-58)–(3-62) and the relation of gas concentration,

$$c \equiv c_G = \frac{\alpha_G \rho_G}{\rho_M}$$

$$\frac{u_{GJ}}{u_{GM}} = \frac{\rho_M}{\rho_L}$$

Eq. (3-98) becomes

$$\frac{\partial c}{\partial t} + \frac{G}{\rho_M}\left(\frac{\partial c}{\partial z}\right) + \left(\frac{1}{\rho_M A}\right)\frac{\partial}{\partial z}(\rho_M A c u_{GM}) = \frac{\partial}{\partial t}\left(\frac{m_G}{\rho_M}\right) \tag{3-100}$$

The momentum and energy equations for the mixture can also be derived as

$$\frac{\partial}{\partial t}(AG) + \frac{\partial}{\partial z}\left(\frac{AG^2}{\rho_M}\right) + \frac{\partial}{\partial z}\left[A(u_{GM})^2 \rho_M\left(\frac{c}{1-c}\right)\right]$$

$$= -f_M\left(\frac{4A}{D}\right)\frac{G|G|}{2\rho_M} - \rho_M Ag \sin\beta - A\frac{\partial P}{\partial z} \quad (3\text{-}101)$$

where f_M is the mixture friction factor, defined as

$$\tau_S = f_M\left(\frac{4A}{D}\right)\frac{G|G|}{2\rho_M}$$

and τ, S are the wall shear stress and perimeter, respectively.

$$\frac{\partial H}{\partial t} + \frac{G}{\rho}\frac{H}{\partial z} + \frac{1}{A\rho_M}\left[Au_{GM}\rho_M c(H_G - H_L)\right] - \frac{1}{\rho_M A}\frac{\partial}{\partial t}(AP)$$

$$- \left[\frac{G}{\rho_M^2} + \frac{c(\rho_L - \rho_G)}{\rho_L \rho_G}u_{GM}\right]\frac{\partial P}{\partial z} - \frac{qS}{\rho_M A} - \frac{f_M G^2|G|S}{2A(\rho_M)^3}$$

$$+ \frac{1}{2\rho_M}\frac{\partial m_L}{\partial t}\left[(\hat{u} - u_G)^2 - \left(\frac{G}{P_M} - u_L\right)^2\right] - \frac{\tau_i S_i}{\rho_M A}(u_G - u_L) = 0 \quad (3\text{-}102)$$

where H is enthalpy. In the above equation, the last three terms represent mechanical dissipation, transfer of kinetic energy, and work interaction between the two fluids, all of which usually are neglected compared to the other terms. Three examples follow.

3.4.8.1 Transient Two-Phase Flow in Horizontal Pipes. For transient flow, the pattern transitions involve the equilibrium liquid level, which depends not only the flow rates, fluid properties, and pipe diameter, but also on time and position from the entry. Thus, depending on the nature of the transient, flow pattern transition under conditions of transient flow can take place at liquid and gas rates different from those for equilibrium conditions. Furthermore, flow patterns can appear that would not exist if the flow rate changes along the same path were carried out slowly (Taital et al., 1978). Figure 3.38 illustrates such a case of fast gas transient from A to B in 1.0 sec. (The liquid level remains essentially unchanged over the time it takes for the gas rate to increase enough to cause a transition.) Using the equilibrium flow pattern map [air–water in a 3.8-cm (1.5-in.)-diameter pipe], if the gas rate is increased very slowly (from A to B), the system will pass through a series of quasi-equilibrium states along A–B, and transitions will be observed at flow rates where A–B crosses curves two–two and one–one. For a fast gas transient, the initial

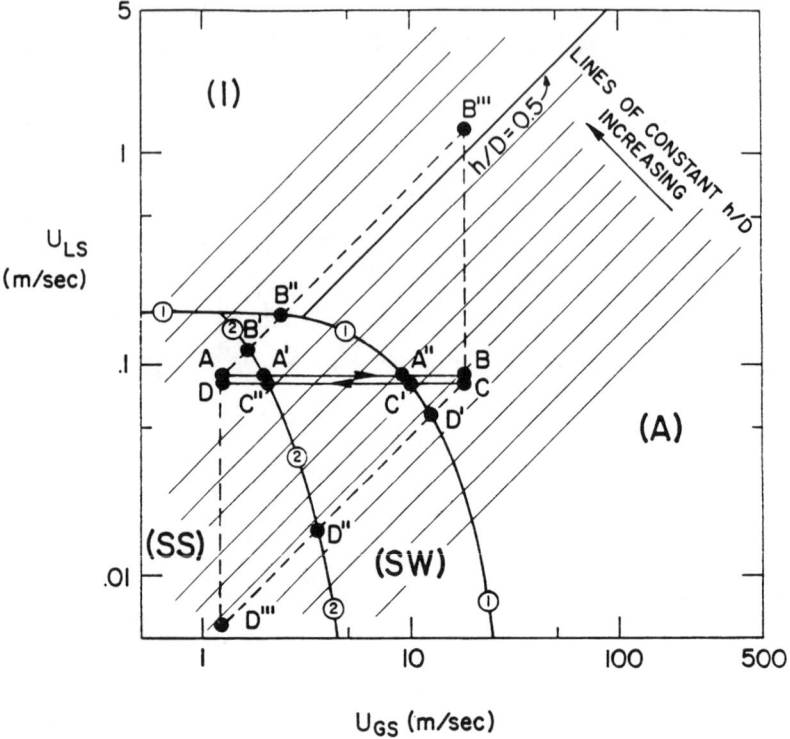

Figure 3.38 Fast gas transient in horizontal flow. (From Taitel et al., 1978. Copyright © 1978 by American Institute of Chemical Engineers, New York. Reprinted with permission.)

liquid level remains unchanged until the gas rate has exceeded that necessary for transition. Thus, instead of $AA'A''B$ in the figure, the process moves along $AB'B''B'''$, along which h/D remains constant until the final gas rate is reached, and then the liquid level relaxes to its value at B along path B'''–B (Taitel et al., 1978).

3.4.8.2 Transient slug flow. Severe slugging is a transient slug flow pattern. Instead of regular slugs of relatively short length ($<100D$), slugs propagating in the pipe, separated by evenly spaced elongated bubbles, the transient slugging is of a different nature, and appears as a very long slug that is being pushed by the gas behind it. When a slug exits the pipe, an increase in the pressure drop occurs, which is a cause for a sufficient increase of the gas flow rate to trigger another new slug in the stratified flow zone, and so on (Taitel and Barnea, 1990).

3.4.8.3 Transient two-phase flow in rod bundles. In analyzing transient two-phase flows in rod bundles, such as the case resulting from a postulated loss-of-coolant or flow accident in a nuclear reactor, Ishii and Chawla (1978) developed a multichannel drift flux model using the drift velocity in both the axial and transverse directions. The constitutive equations for the transverse drift velocity were derived by taking into account the void and flux profiles, interfacial geometry, shear stresses, and interfacial momentum transfer, since these macroscopic effects govern the two-phase diffusions. Basically, the model consists of a set of three mixture conservation equations of mass, momentum, and energy, with one additional continuity equation for the vapor phase for each flow channel. For the determination of the crossflow, a transverse momentum equation for a mixture at each interchannel boundary is also used. The inclusion of the transverse relative motion in this model was intended to be an important improvement (Ishii and Chawla, 1978) over other earlier models for rod bundles, such as those of Rowe (1970, 1973) or Bowring (1967a).

3.5 PRESSURE DROP IN TWO-PHASE FLOW

3.5.1 Local Pressure Drop

Since the pressure drop in two-phase flow is closely related to the flow pattern, most investigations have been concerned with local pressure drop in well-characterized two-phase flow patterns. In reality, the desired pressure drop prediction is usually over the entire flow channel length and covers various flow patterns when diabatic condition exist. Thus, a summation of local Δp values is necessary, assuming the phases are in thermodynamic equilibrium. The addition of heat in the case of single-component flow causes a phase change along the channel; consequently, the vapor void increases and the phase (also velocity) distribution as well as the momentum of the flow vary accordingly.

The pressure drop of a two-phase flow generally consists of three components: frictional loss, momentum change, and elevation pressure drop arising from the effect of the gravitational force field. The local Δp therefore is normally written as

$$\left(\frac{dp}{dz}\right)_{tot} = \left(\frac{dp}{dz}\right)_{fric} + \left(\frac{dp}{dz}\right)_{mom} + \left(\frac{dp}{dz}\right)_{elev}$$

Using a homogeneous model proposed by Owens (1961) for low void fractions ($\alpha < 0.30$) and high mass flux, as is usually encountered in a water-cooled reactor, the momentum change (or acceleration) pressure gradient term is obtained from

$$\left(\frac{dp}{dz}\right)_{mom} = \frac{G^2}{g_c}\frac{d\bar{v}}{dz}$$

where G is the mass flux and $\bar{v}$ is the average specific volume of the mixture at the location considered, given by

$$\bar{v} = \frac{W_G v_G + W_L v_L}{W_G + W_L} = v_L\left[1 + \left(\frac{X}{v_L}\right)(v_G - v_L)\right]$$

and the elevation pressure gradient is given by

$$\left(\frac{dp}{dz}\right)_{elev} = \frac{1}{\bar{v}}\left(\frac{g}{g_c}\right)$$

The frictional pressure gradient is obtained by different correlations described in following sections. In a horizontal flow, $(dp/dz)_{elev} = 0$, it is an ideal case to perform experiments excluding the term of elevation pressure drop. Because of nonhomogeneity of the slug flow, the acceleration pressure gradient term is different from that shown above; it is given in Section 3.5.2.2.

The pressure drop for a given channel length, z, becomes

$$p = \int_0^L \left(\frac{dp}{dz}\right)_{tot} dz$$

3.5.2 Analytical Models for Pressure Drop Prediction

3.5.2.1 Bubbly flow. In bubbly flow, the holdup is generally known and/or near homogeneous flow condition exists, and the frictional pressure drop can be correlated through similarity analysis (Dukler et al., 1964). The development shows that the frictional loss is expressed by a Fanning-type equation,

$$\left(\frac{dp}{dx}\right)_{TPF} = \left(\frac{2}{D}\right) C_{f,TP}\, \rho_{TP}(U_{GS} + U_{LS})^2 \tag{3-103}$$

where

$$\rho_{TP} = \rho_G\left[\frac{(1-\lambda)^2}{\alpha}\right] + \rho_L\left(\frac{\lambda^2}{1-\alpha}\right) \tag{3-104}$$

and

$$\lambda = \frac{U_{LS}}{U_{LS} + U_{GS}}$$

It was shown that a normalized version of the two-phase friction factor, $C_{f,TP}/C_{fo}$, is uniquely related to λ. The normalizing friction factor, C_{fo}, is calculated from single-phase friction factor correlations using a Reynolds number calculated as if both phases flow as liquid,

$$\text{Re}_o = \frac{D\rho_{TP}(U_{LS} + U_{GS})}{\mu_L} \tag{3-105}$$

where the actual in situ average velocities of gas and liquid, which are not readily calculable, can be related to the volumetric rates of gas and liquid flow, Q_G and Q_L, respectively:

$$Q_G = U_G A_G = U_G A\alpha \tag{3-106}$$

where α is the space average void friction. The superficial gas and liquid velocities in Eq. (3-103) are defined by

$$U_{GS} A = U_G A\alpha = Q_G \tag{3-107}$$

$$U_{LS} A = U_L A(1-\alpha) = Q_L \tag{3-108}$$

In a horizontal flow with a homogeneous model, $U_G = U_L$, thus

$$\frac{U_{GS}}{\alpha} = \frac{U_{LS}}{1-\alpha} \tag{3-109}$$

In a vertical upward gas–liquid flow, a continuous swarm of bubbles flows upward with the liquid stream due to a buoyancy effect, and the gas slips past the liquid with a relative velocity U_o (rise velocity):

$$U_G = U_L + U_o \tag{3-110}$$

Consequently, Eq. (3-109) becomes

$$\frac{U_{GS}}{\alpha} = \frac{U_{LS}}{1-\alpha} + U_o \tag{3-111}$$

The rise velocity, U_o, in general depends on the bubble size, or the bubble Reynolds number; but as bubble size increases, as in two-phase upflow, U_o approaches an asymptotic value that is independent of Reynolds number. The following expressions have been accepted for a single bubble rising in an infinite medium, and for one rising in a swarm of surrounding bubbles, respectively (Duckler and Taitel, 1991b):

$$U_{o,\infty} = 1.53 \left[\frac{\sigma g(\rho_L - \rho_G)}{\rho_L^2} \right]^{1/4} \tag{3-112}$$

$$U_o = (1-\alpha)^{1/2}(1.53)\left[\frac{\sigma g(\rho_L - \rho_G)}{\rho_L^2} \right]^{1/4} \tag{3-113}$$

where σ is the interfacial tension and g is gravitational acceleration. Substituting in Eq. (3-111), after some rearrangement, gives

$$\frac{U_{GS}}{U_{LS}} = \frac{\alpha}{1-\alpha} + \frac{\alpha(1-\alpha)^{1/2}(1.53)[\sigma g(\rho_L - \rho_G)]^{1/4}}{U_{LS}\rho_L^{1/2}} \qquad (3\text{-}114)$$

Given the gas and liquid flow rates and the pipe diameter, it is now possible to calculate U_{LS} and U_{GS} from Eqs. (3-107) and (3-108), and the space average void fraction, α, from Eq. (3-114).

3.5.2.2 Slug flow. As a very complex, unsteady, turbulent two-phase flow, slug flow typically has high acceleration and decelerations of liquid, which causes steep pressure drops. Dukler and Hubbard (1975) considered the basic hydrodynamic structure of a slug unit as consisting of several regions (Fig. 3.39):

The slug "mixing zone"—This is a short zone at the slug "nose," of length l_m.
The main slug body—The length of this zone is $(l_s - l_m)$, where l_s represents the slug length. The slug moves as a fully developed, homogeneous mixture.
The film zone—Liquid is shed from the back side of the slug and forms a film of length l_f below a gas zone.
The gas zone—A gas "bubble" is trapped between consecutive slugs, riding on top of the liquid film.

The liquid in the film alongside the Taylor bubble flows in the opposite direction, with negligible interfacial shear from the gas on the bubble. The average gradient due to friction and acceleration across a slug unit is

$$\left(\frac{dp}{dz}\right)_{TP} = \left(\frac{dp}{dz}\right)_{ls}\left(\frac{l_s}{l}\right) \qquad (3\text{-}115)$$

The acceleration loss, Δp_a, results from the force needed to accelerate the liquid in the film around the Taylor bubble from its velocity, U_{LTB}, to that of the liquid slug, U_{LLS}.

$$\Delta p_a = \rho_L U_{LTB}(1-\alpha_{TB})(U_{LLS} + U_{LTB}) \qquad (3\text{-}116)$$

The frictional pressure gradient in the liquid slug can be calculated using Eqs. (3-103) and (3-104), when the flow rates of liquid and gas are written in terms of the flow rate on the slug. Thus,

$$\frac{dp}{dz} = \frac{2}{D}C_{f,TP}\rho_{TP}(U_{LLS})^2 \qquad (3\text{-}117)$$

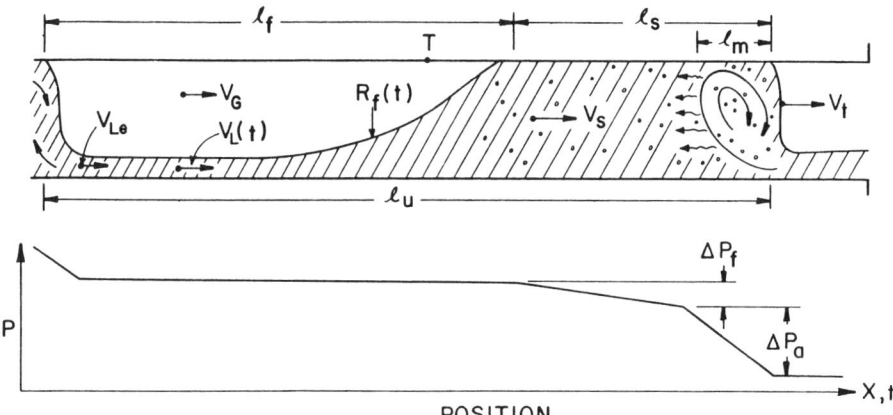

Figure 3.39 Physical model for slug flow. (From Dukler and Hubbard, 1975. Copyright © 1975 by American Chemical Society, Washington, DC. Reprinted with permission.)

where

$$\rho_{TP} = \frac{\rho_G (1-\lambda)^2}{\alpha_{LS}} + \frac{\rho_L \lambda^2}{1-\alpha_{LS}} \quad (3\text{-}118)$$

and

$$\lambda = \frac{U_{LLS}(1-\alpha_{LS})}{U_{LS} + U_{GS}} \quad (3\text{-}119)$$

U_{LLS}, U_{LTB}, α_{LS}, and α_{TB} can be found from the hydrodynamic model, and $C_{f,TP}$ is calculated in the same manner as discussed previously for bubbly flow.

3.5.2.3 Annular flow. Modeling the interfacial shear is central to the problem of modeling hydrodynamics and transport during annular flow. The mechanisms are not clear, and the extent of basic modeling that has appeared is still very limited (Dukler and Taitel, 1991b). Only empirical treatments are currently available (see Sec. 3.5.3.3).

3.5.2.4 Stratified flow. A separated flow model for stratified flow was presented by Taitel and Dukler (1976a) in which the holdup and the dimensionless pressure drop, $\phi_{GS}^2 = (dp/dz)_{TP}/(dp/dz)_{GS}$ is calculated as a function of the Lockhart-Martinelli parameter only. (The results, however, differ from those of Martinelli and compare better with experimental data.) This model uses two basic approximations:

1. That the shear stress at the interface is equal to the shear of the gas at the wall
2. That the wall shear stress can be calculated on the basis of fully developed pipe flow correlation, provided the correct hydraulic diameter is used

Cheremisinoff and Davis (1979) relaxed these two assumptions and used the following expressions for force balances on the gas and liquid phases, assuming no acceleration (Fig. 3.40):

$$A_G\left(\frac{\Delta P}{L}\right) = \tau_{wG}\tilde{p}_G + \tau_i w_i \quad (3\text{-}120)$$

and

$$A_L\left(\frac{\Delta P}{L}\right) = \tau_{wL}\tilde{p}_L - \tau_i w_i \quad (3\text{-}121)$$

where $\dfrac{\Delta P}{\Delta L}$ = pressure gradient
$\tilde{p}_G, \tilde{p}_L$ = perimeters for the gas and liquid phases, respectively
A_G, A_L = flow cross section of the gas and liquid phases, respectively
τ_i, w_i = interfacial stress and interfacial area width, respectively

The gas-phase wall stress can be written in terms of the usual friction factor, C_{fG}, to give

$$\tau_{wG} = C_{fG}\left(\frac{\overline{U}_G^2 \rho_G}{2}\right) \quad (3\text{-}122)$$

where C_{fG} is a function of the gas phase Reynolds number,

$$\text{Re}_G = \frac{D_{eG}\overline{U}_G}{\nu_G}$$

$\overline{U}_G$ is based on the area occupied by the gas phase, and D_{eG} is the equivalent diameter for the gas phase. Since a smooth tube was employed, the applicable conventional friction factor, C_{fG}, is

$$C_{fG} = 0.046\,\text{Re}_G^{-0.2} \quad (3\text{-}123)$$

For the liquid phase, Cheremisinoff and Davis (1979) solved the momentum equation using von Karman's and Deissler's eddy viscosity expressions.

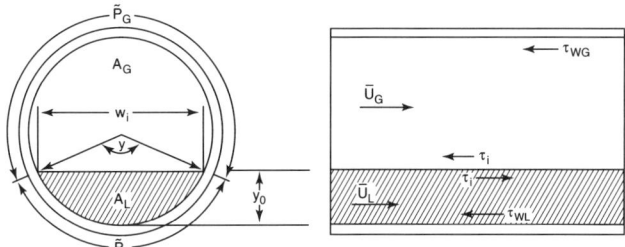

Figure 3.40 The stratified flow system under consideration. (From Cheremisinoff and Davis, 1979. Copyright © 1979 by American Institute of Chemical Engineers, New York. Reprinted with permission.)

The dimensionless liquid flow rate, W^+, can be expressed as

$$W^+ = \frac{W_L}{\mu_L R}$$

$$= 2\int_0^{1/2}\int_0^{y^+} u^+(y^+)\left(\frac{1-y^+}{R^+}\right) dy^+\, d\theta \qquad (3\text{-}124)$$

where dimensionless radius   × $R^+ = \dfrac{u^* R}{\nu_L}$

friction velocity   × $u^* = \left(\dfrac{\tau_{wL}}{\rho_L}\right)^{1/2}$

position of the interface in nondimensional form × $y^+ = \dfrac{yu^*}{\nu}$

Equation (3-124) has been integrated numerically, and results can be presented. Figure 3.41 is a plot of liquid holdup E_L versus W^+ for various values of the parameter R^+.

For the interfacial shear stress with roll waves, the following expression was used:

$$\tau_i = \frac{C_{f,i}\,\rho_G(\overline{U_G})^2}{2} \qquad (3\text{-}125)$$

where, according to Cohen and Hanratty (1968), the interfacial friction factor can be linearly correlated with the liquid Reynolds number,

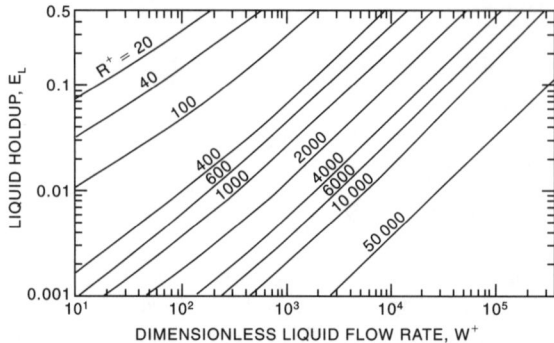

Figure 3.41 The in-situ liquid volume fraction as a function of dimensionless liquid flow rate W^+ and dimensionless tube size parameter R^+. (From Cheremisinoff and Davis, 1979. Copyright © 1979 by American Institute of Chemical Engineers, New York. Reprinted with permission.)

$$C_{f,i} = 0.0080 + 2.00 \times 10^{-5} \, \text{Re}_L \qquad (3\text{-}126)$$

for $100 \leq \text{Re}_L \leq 1{,}700$, where $\text{Re}_L = \Gamma_L/\nu_L$ and $\Gamma_L =$ volumetric flow per unit width of parallel-plate channel.

An iterative procedure to calculate the pressure drop was suggested by Cheremisinoff and Davis (1979) for turbulent/turbulent stratified flow:

1. Choose a trial value of the holdup E_L and calculate the geometric parameters D_{eG}, D_{eL}, $\tilde{p}_G$, $\tilde{p}_L$, and w_i for the tube diameter under consideration.
2. Calculate $W^+ = W_L/\mu_L R$ from the liquid mass flow rate W_L and determine R^+ by interpolation from Figure 3.41.
3. Calculate the friction velocity and then τ_{wL} from R^+ and compute τ_i and τ_{wG} from Eqs. (3-125) and (3-122).
4. Solve Eqs. (3-120) and (3-121) separately for $(\Delta P/L)$. If the two calculated values do not agree to within some specified accuracy, a new value of E_L is assumed, and the procedure is repeated.

The model is a significant improvement over the Lockhart and Martinelli correlations for pressure drop and holdup (discussed in Sec. 3.5.3). A severe limitation of the model, however, is the dependency on the empirical expression for $C_{f,i}$ [Eq. 3-126]. This expression is based on air–water data only, and has not been shown to apply to other systems.

3.5.3 Empirical Correlations

So far the pressure drop in two-phase flow in pipes and rod bundles has often been predicted by empirical correlations, despite the development of analytical models as described in the previous sections. Thus, in the highly subcooled boiling region, where only attached wall voidage is present, a homogeneous model was assumed in the pressure drop calculations. With high-void-fraction flow, the empirical correction for calculating the two-phase frictional ΔP, $(\Delta P)_{TPF}$, was developed by Baroczy (1966). His work is considered an extension of that of Lockhart and Martinelli (1949) and Martinelli and Nelson (1948), all of whom defined the two-phase pressure drop in terms of a single-phase ΔP:

$$\phi_{LO}^2 = \frac{(\Delta P/\Delta L)_{TPF}}{(\Delta P/\Delta L)_{LO}} \tag{3-127}$$

where ϕ_{LO}^2 = two-phase frictional pressure drop multiplier

$\left(\dfrac{\Delta P}{\Delta L}\right)_{TPF}$ = two-phase frictional pressure drop per unit length

$\left(\dfrac{\Delta P}{\Delta L}\right)_{LO}$ = single-phase ΔP obtained at the same mass flux when the fluid is entirely liquid

By defining a property index $[(\mu_L/\mu_G)^{0.2}(\rho_L/\rho_G)]$, Baroczy obtained a correlation for ϕ_{LO}^2 that was independent of pressure. He also observed that his correlation could be used with the gas-phase pressure drop, $(\Delta P/\Delta L)_G$, by noting that

$$\frac{(\Delta P/\Delta L)_L}{(\Delta P/\Delta L)_G} = \frac{(\mu_L/\mu_G)^{0.2}}{\rho_L/\rho_G} \tag{3-128}$$

Baroczy's correlation is given in two sets of curves:

1. A plot of the two-phase multiplier ratio, ϕ_{LO}^2, as a function of property index at one mass flux (Fig. 3.42)
2. Plots of a two-phase multiplier as a function of property index, quality, and mass flux (Fig. 3.43).

It is noted that additional scales for the property index are shown in Figure 3.42 to correspond to liquid metals as well as water and Freon-22 at different temperatures, indicating the applicability of the correlation to other fluids.

For a diabatic flow case, as in the high heat flux, boiling water system typical of reactor cores, Tarasova et al. (1966) proposed the following correlation for the effect of wall heat flux on friction factors by a correction factor:

$$\frac{(\phi_{LO}^2)_{diabatic}}{(\phi_{LO}^2)_{adiabatic}} = 1 + 0.99\left(\frac{q''}{G}\right)^{0.7} \tag{3-129}$$

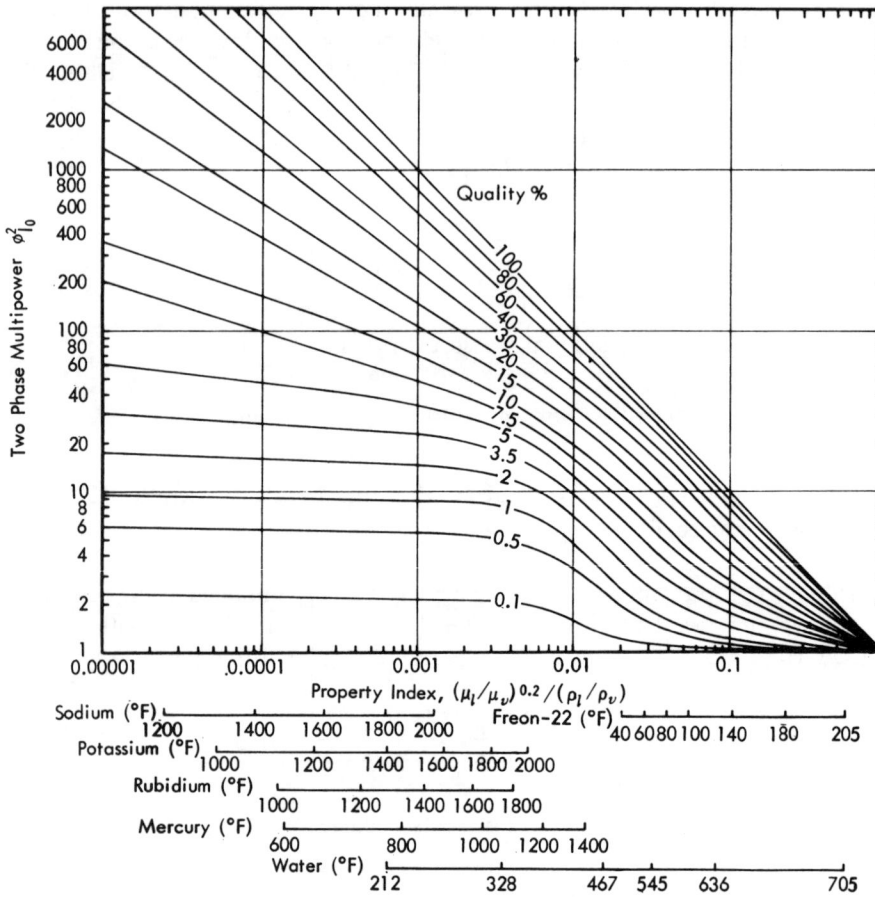

Figure 3.42 Two-phase friction pressure drop correlation for $G = 1 \times 10^6$ lb/hr ft^2. (From Baroczy, 1966. Copyright © 1966 by Rockwell International, Canoga Park, CA. Reprinted with permission.)

where the heat flux, q'', is given in Btu/hr ft^2, and the mass flux, G, is expressed in lb/hr ft^2, for a range of conditions:

$$710 < P < 2,840 \text{ psia} \qquad 0.37 \times 10^6 < G < 1.9 \times 10^6 \text{ lb/hr ft}^2$$

$$0.032 \times 10^6 < q'' < 0.53 \times 10^6 \text{ Btu/hr ft}^2$$

Determinations of ΔP for two-phase adiabatic flow in various flow patterns are given in following sections.

3.5.3.1 Bubbly flow in horizontal pipes. High-velocity flow in horizontal pipes presents a minimum effect of the gravitational field and reduces one potential pa-

HYDRODYNAMICS OF TWO-PHASE FLOW 197

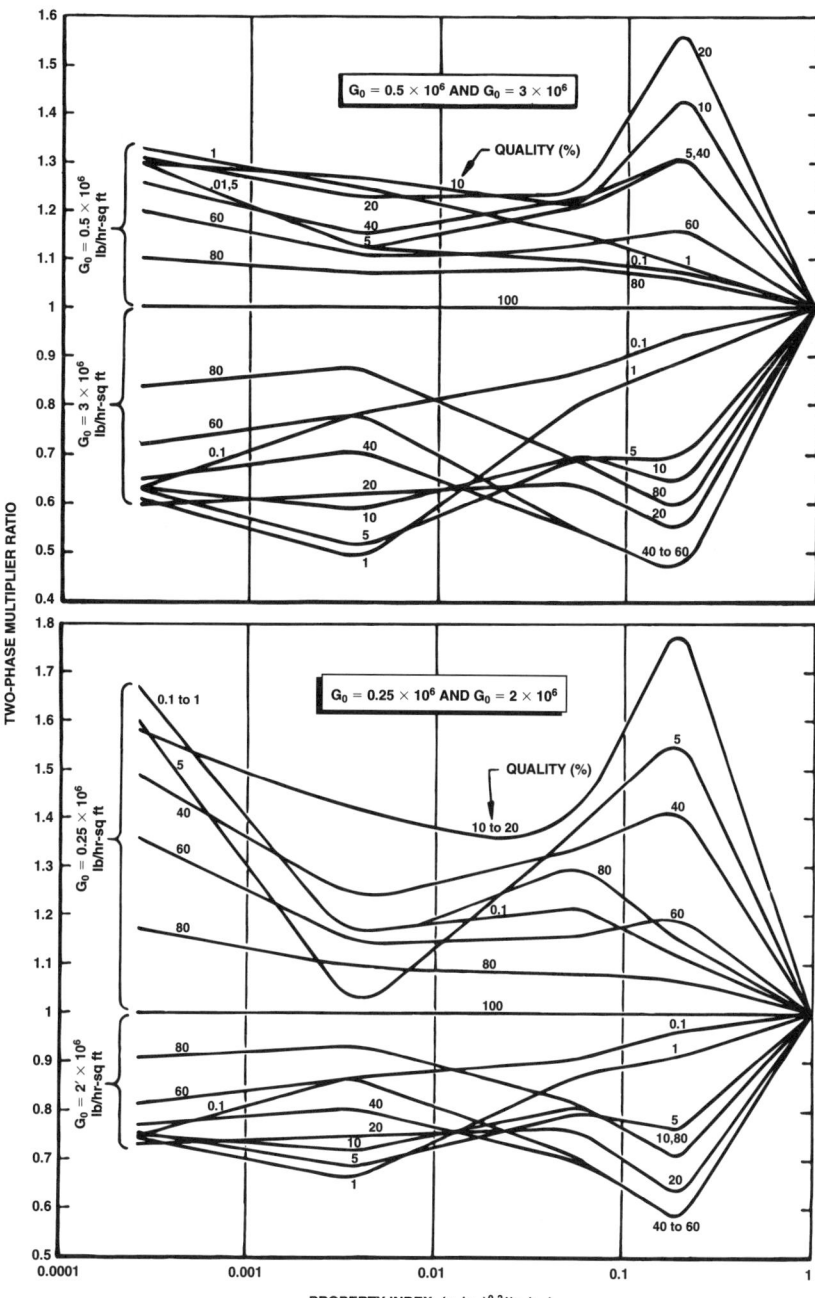

Figure 3.43 Two-phase friction Δp versus property index with mass flux correction. (From Baroczy, 1966. Copyright © 1966 by Rockwell International, Canoga Park, CA. Reprinted with permission.)

rameter in the flow study. The validity of the homogeneous model (see Sec. 3.4.4) for bubbly flow has been substantiated by Kopalinsky and Bryant's (1976) experimental study:

$$\frac{dp}{dx}\left(1 - \frac{\alpha\rho u^2}{p}\right) + \frac{4C_f}{D}\frac{(\rho u^2)}{2} = 0 \qquad (3\text{-}130)$$

where $(\alpha\rho u^2/p)$ is the acceleration term; the second term is attributed to the effects of shear stress at the boundaries and is identified with an average friction coefficient defined by

$$\overline{C}_f = \frac{1}{x - x_1} \int_{x_1}^{x} C_f \, dx$$

It should be noted that the acceleration component is dominant in the last part of the pipe, where, because of the rapid pressure drop and the low absolute pressure, the specific volume of the gas increases sharply. This effect is more pronounced at high mass flow rates with large values of mass flow ratio, β ($= m_g/m_l$). As shown in Figures 3.44a and 3.44b, the average friction coefficient is affected by the mixture mass flow rate $\dot{m}$, the mass flow ratio β, and the diameter of the pipe D. The Re is defined as

$$\text{Re} = \frac{\rho D u}{\mu_L} = (1 + \beta)\,\text{Re}_L$$

The experimental equipment is shown in Figure 3.45, where the approximate pressure tap locations are also illustrated. The range of variables studied was as follows (Kopalinsky and Bryant, 1976):

	Diameter of pipe	
	25.4 mm	50.8 mm
β	0.006–0.0033	0.0006–0.0033
α at inlet	0.09–0.42	0.17–0.48
α at the exit	0.33–0.73	0.33–0.73
G_L, kg/cm²	4,880–10,700	4,880–7,800
Re	1.4×10^5–3.0×10^5	2.4×10^5–4.3×10^5

Two empirical equations, combining data for both pipe sizes, are as follows.

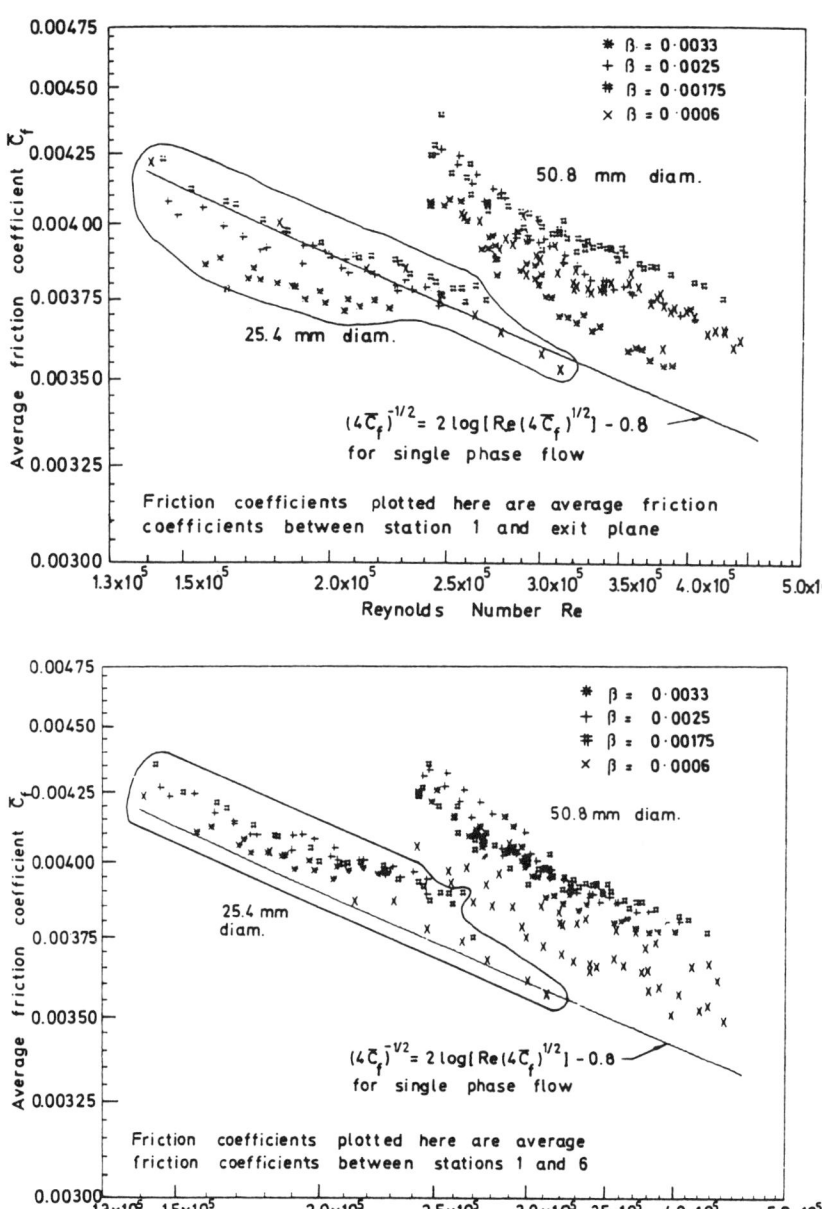

Figure 3.44 Average friction coefficients: (*a*) between station 1 and exit plane; (*b*) between stations 1 and 6. (From Kopalinsky and Bryant, 1976. Copyright © 1976 by American Institute of Chemical Engineers, New York. Reprinted with permission.)

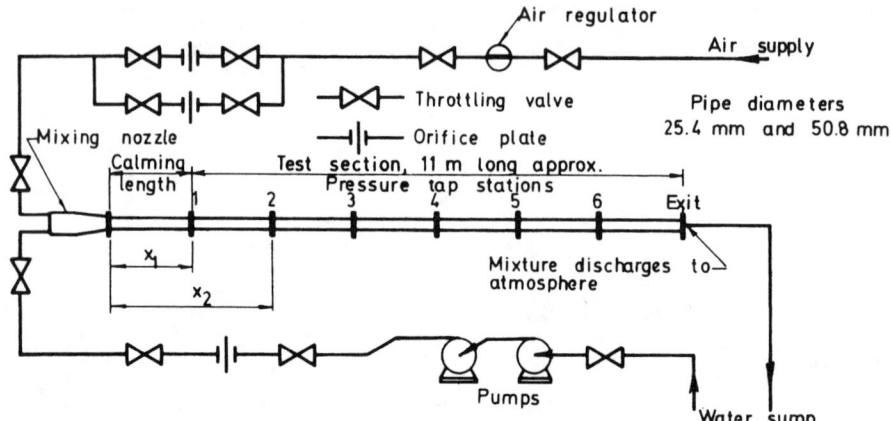

Figure 3.45 Schematic flow diagram. (From Kopalinsky and Bryant, 1976. Copyright © 1976 by American Institute of Chemical Engineers, New York. Reprinted with permission.)

Including the exit plane (i.e., a region at atmospheric pressure), $\Delta x \cong 11$ m,

$$\overline{C}_f = 0.00516 + 0.278X - 81.894X^2$$
$$- 0.000869 \ln(\mathrm{Re} \times 10^{-5}) - 0.1977 \times 10^{-5}\left(\frac{\Delta x}{D}\right) \quad (3\text{-}131)$$

where X = quality of the mixture.
Excluding the exit plane, $\Delta x \cong 9.2$ m,

$$\overline{C}_f = 0.00504 + 0.374X - 84.819X^2$$
$$- 0.000894 \ln(\mathrm{Re} \times 10^{-5}) - 0.2094 \times 10^{-5}\left(\frac{\Delta x}{D}\right) \quad (3\text{-}132)$$

Because of the rapid changes in the flow variables adjacent to the exit plane, Eq. (3-131), which includes high-Mach-number data, should be used in preference to Eq. (3-132) when $M > 0.7$. With this provision, the above equations can be applied to flows in differing pipe lengths, provided the flow conditions (i.e., the pressure ranges) are similar to those obtaining in the experiments.

3.5.3.2 Slug flow. The momentum equations for slug flow are given in Eqs. (3-82) and (3-83) (Sec. 3.4.5). Since the slug is not a homogeneous structure, the local axial Δp is not constant. For practical purposes, we need the average Δp over a slug unit, $\Delta p_u/l_u$.

$$\Delta p_u = \rho_s g \sin\beta\, l_s + \left(\frac{\tau_s \pi D}{A}\right) l_s + \Delta P_{\mathrm{mix}} \quad (3\text{-}133)$$

where ΔP_{mix} is the pressure-loss near-wake region behind the long bubble,

$$\Delta P_{mix} = \Delta P_{acc} = \rho_L R_S A(u_t - u_L)(u_L - u_{fe}) \tag{3-134}$$

or

$$\Delta P_u = \rho_s g \sin\beta \, l_s + \left(\frac{\tau_s \pi D}{A}\right) l_s + \rho_L g \sin\beta \, l_f$$

$$+ \left(\frac{\tau_L S_L}{A}\right) l_f + \left(\frac{\tau_G S_G}{A}\right) l_f \tag{3-135}$$

3.5.3.3 Annular flow. In annular flow, as mentioned in Section 3.4.6.1, modeling of the interfacial shear remains empirical. For adiabatic two-phase flow, Asali et al. (1985) suggested that the friction factor, f_i/f_s, is dependent on a dimensionless group for the film thickness, δ_g+, as defined in Eq. (3-136), and the gas Reynolds number, Re_G:

$$\delta_g^+ = \frac{\delta(\tau_i/\rho_G)^{1/2}}{v_G} \tag{3-136}$$

The final design equations recommended by Asali et al. are as follows.
For low liquid flow rates, $Re_{LF} < 300$, and high gas rates, $U_G > 25$ m/s (8.2 ft/s),

Concurrent two-phase upflow: $\quad \dfrac{f_i}{f_s} - 1 = C_1 \, (\delta_g^+ - 4) \tag{3-137}$

where

$$\delta_g^+ = 0.34(Re_{LF})^{0.6} \left(\frac{v_L}{v_G}\right) \left(\frac{\rho_L \tau_i}{\rho_G \tau_C}\right)^{1/2} \tag{3-138}$$

and

$$\tau_C = \frac{2}{3}\tau_W + \frac{1}{3}\tau_i$$

f_s = friction factor for single-phase flow

Concurrent two-phase downflow: $\quad \dfrac{f_i}{f_s} - 1 = C_2\left(\delta_g^+ - 5.9\right) \tag{3-139}$

where

$$C_1 = C_2 = 0.045$$

For high liquid flow rates, $\text{Re}_{LF} > 300$, and at all gas rates, roll waves appear on the film, accompanied by an atomization of liquid from the wave crests.

Concurrent two-phase upflow : $$\frac{f_i}{f_s} - 1 = 0.45(\text{Re}_G)^{-0.2}(\delta_\theta^+ - 4) \qquad (3\text{-}140)$$

Concurrent downflow : $$\frac{f_i}{f_s} - 1 = 0.45(\text{Re}_G)^{-0.2}(\delta_g^+ - 5.9) \qquad (3\text{-}141)$$

and

$$\delta_g^+ = 0.19(\text{Re}_{LF})^{0.7}\left(\frac{\nu_L}{\nu_G}\right)\left(\frac{\rho_L \tau_i}{\rho_G \tau_C}\right)^{1/2} \qquad (3\text{-}142)$$

It should be recognized that these are highly empirical correlations. Such correlations are limited but will have to do until a better understanding is available of the mechanism by which τ_i is increased (Dukler and Taitel, 1991b).

For diabatic flow, that is, one-component flow with subcooled and saturated nucleate boiling, bubbles may exist at the wall of the tube and in the liquid boundary layer. In an investigation of steam–water flow characteristics at high pressures, Kirillov et al. (1978) showed the effects of mass flux and heat flux on the dependence of wave crest amplitude, δ_c, on the steam quality, X (Fig. 3.46). The effects of mass and heat fluxes on the relative frictional pressure losses are shown in Figure 3.47. These experimental data agree quite satisfactorily with Tarasova's recommendation (Sec. 3.5.3).

3.5.3.4 Correlations for liquid metal and other fluid systems. For liquid metal-vapor two-phase flows, data were reported for potassium (Tippets et al., 1965; Wichner and Hoffman, 1965; Baroczy, 1968; Alad'yev et al., 1969; Chen and Kalish, 1970) and for sodium (Lurie, 1965; Lewis and Groesbeck, 1969; Fauske and Grolmes, 1970). It was concluded that both the potassium and the sodium data agreed with the Lockhart-Martinelli frictional multiplier correlation as shown in Figure 3.48. Also shown in this figure is a simple correlation based on a simplified annular flow model (Lottes and Flinn, 1956). Only two-phase mercury flow has been shown not to obey the Lockhard-Martinelli correlation (possibly because of its wetting ability), and in this case a fog model was recommended (Koestel et al., 1963).

Two-phase flows containing other types of fluids of interest are those of helium and refrigerants. The former fluid is used for cooling different superconductivity devices, while the latter are used in the refrigeration industry. The pressure drop in a two-phase flow of helium in a tube of 1.6 mm (0.06 in.) I.D. under adiabatic conditions and with heat supply were reported by Deev et al. (1978). They indicated that although the actual measured ΔP (at $P = 1.0$ to 1.8×10^5 N/m²) differed

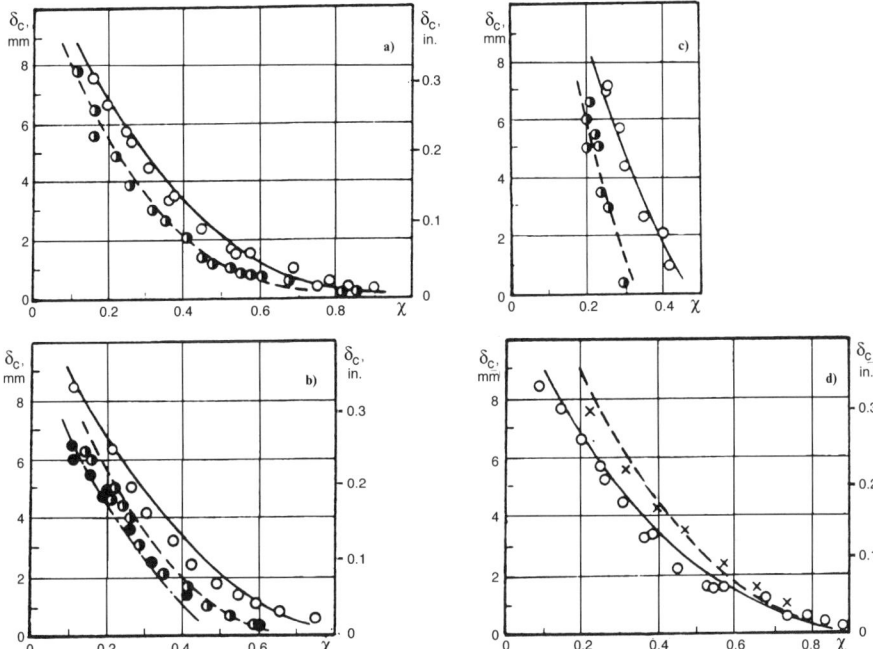

Figure 3.46 Experimental and calculated values of wave crest amplitude: (a) $p = 6.86$ MN/m²; (b) $p = 9.8$ MN/m²; (c) $p = 13.7$ MN/m²; (d) $p = 6.86$ MN/m². At $q = 0$: ○, 500; ◐, 750; ●, 1,000 kg/m² s; at $q = 0.23$ MW/m²: ×, 500 kg/m² s. All lines are calculated from an empirical correlation. (From Kirillov et al., 1978. Copyright © 1978 by National Research Council of Canada, Ottawa, Ont. Reprinted with permission.)

significantly from that of a homogeneous model, the experimental data agreed adequately with that of the Martinelli-Nelson correlation for water under high pressures ($P = 200 \times 10^5$ N/m²) (Martinelli and Nelson, 1948) (Fig. 3.49). Similar pressure drop data for flow-boiling pure and mixed refrigerants inside a tube of 0.9 cm (0.35 in.) diameter were reported by Ross et al. (1987). Their results in the range of the following parameters show good agreement with calculated ΔP based on the Martinelli-Nelson method as modified by Chisholm (1967):

$P = 1.7–8.0$ bar $(1.7–8.0 \times 10^5$ N/m²)
$G = 150–1,200$ kg/m² s $(1.11–8.88 \times 10^5$ lb/hr ft²)
$q'' = 10–95$ kW/m² $(3,180–30,270$ Btu/hr ft²)
Re $= 3,000–50,000$
Pr $= 3–4$

Figure 3.50a displays the comparison between measurement and prediction for the pure fluids, and Figure 3.50b shows a similar comparison for the mixtures. While some increased scatter occurs, the method predicts within its intended accuracy.

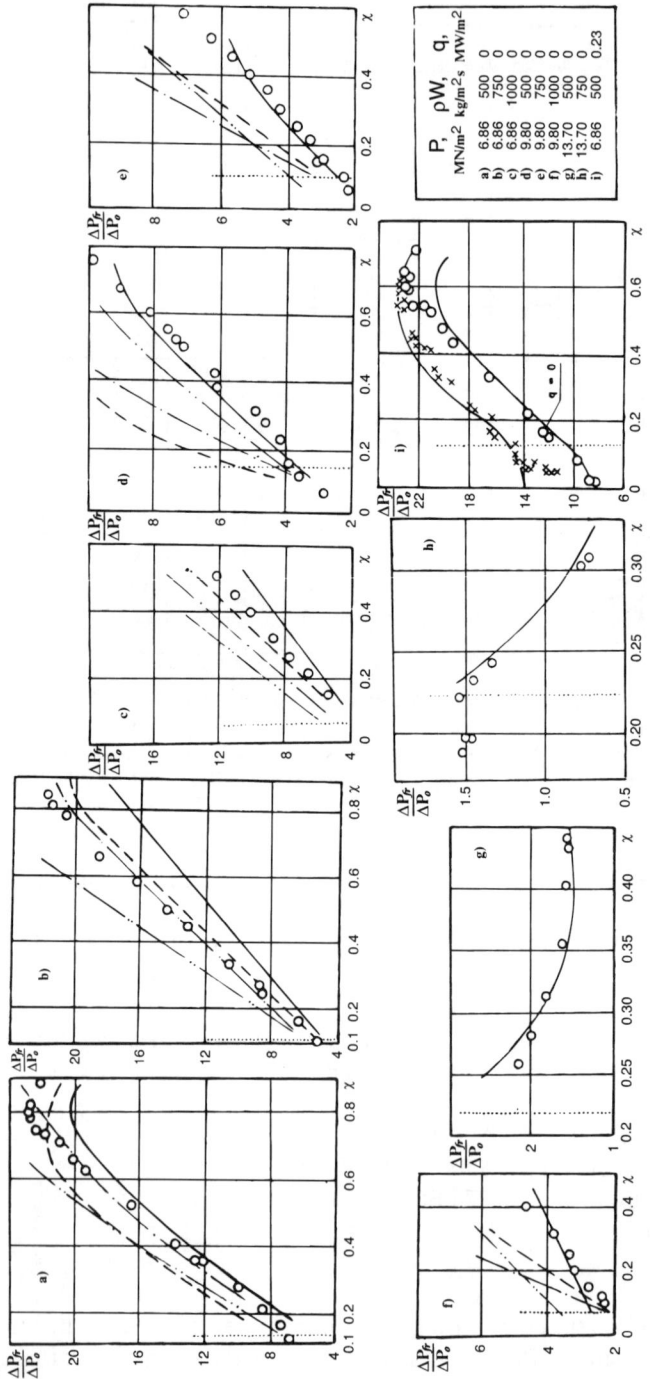

Figure 3.47 Experimental and calculated values of the relative frictional pressure losses: ———, calculation from CISE (1963); ———, calculation from Gill et al. (1963); ———, calculation from Schraub (1968); ———, calculation from Kirillov et al. (1978). (From Kirillov et al., 1978. Copyright © 1978 by National Research Council of Canada, Ottawa, Ont. Reprinted with permission.)

HYDRODYNAMICS OF TWO-PHASE FLOW **205**

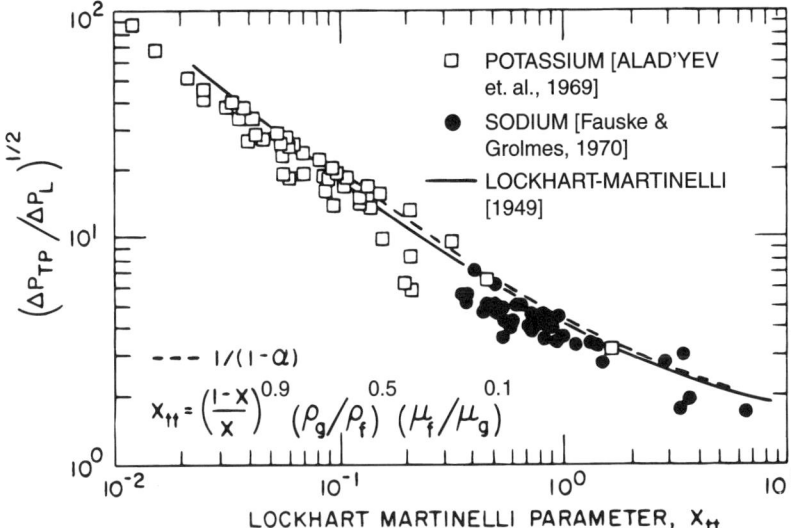

Figure 3.48 Comparison of potassium and sodium two-phase friction pressure drop data with Lockhart-Martinelli correlation, and with a simple correlation [1/(1 − α)]. (From Fauske and Grolmes, 1970. Copyright © 1970 by American Society of Mechanical Engineers. Reprinted with permission.)

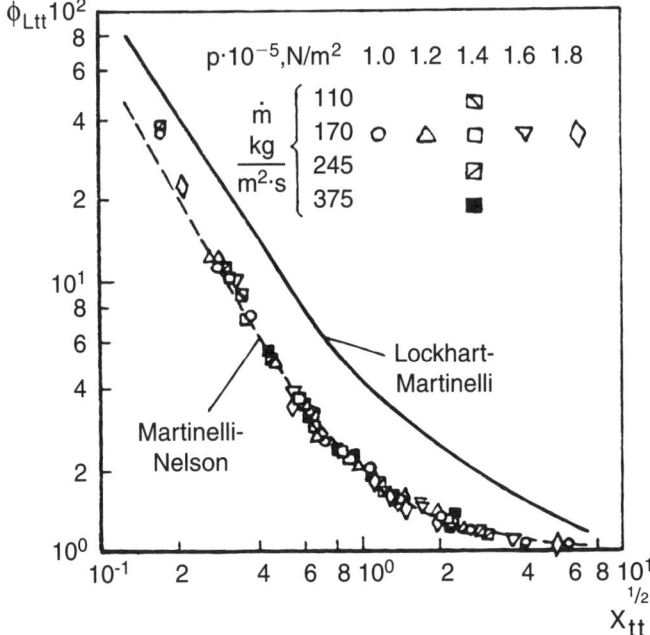

Figure 3.49 Comparison of experimental and calculated data. (From Deev et al., 1978. Copyright © 1978 by National Research Council of Canada, Ottawa, Ont. Reprinted with permission.)

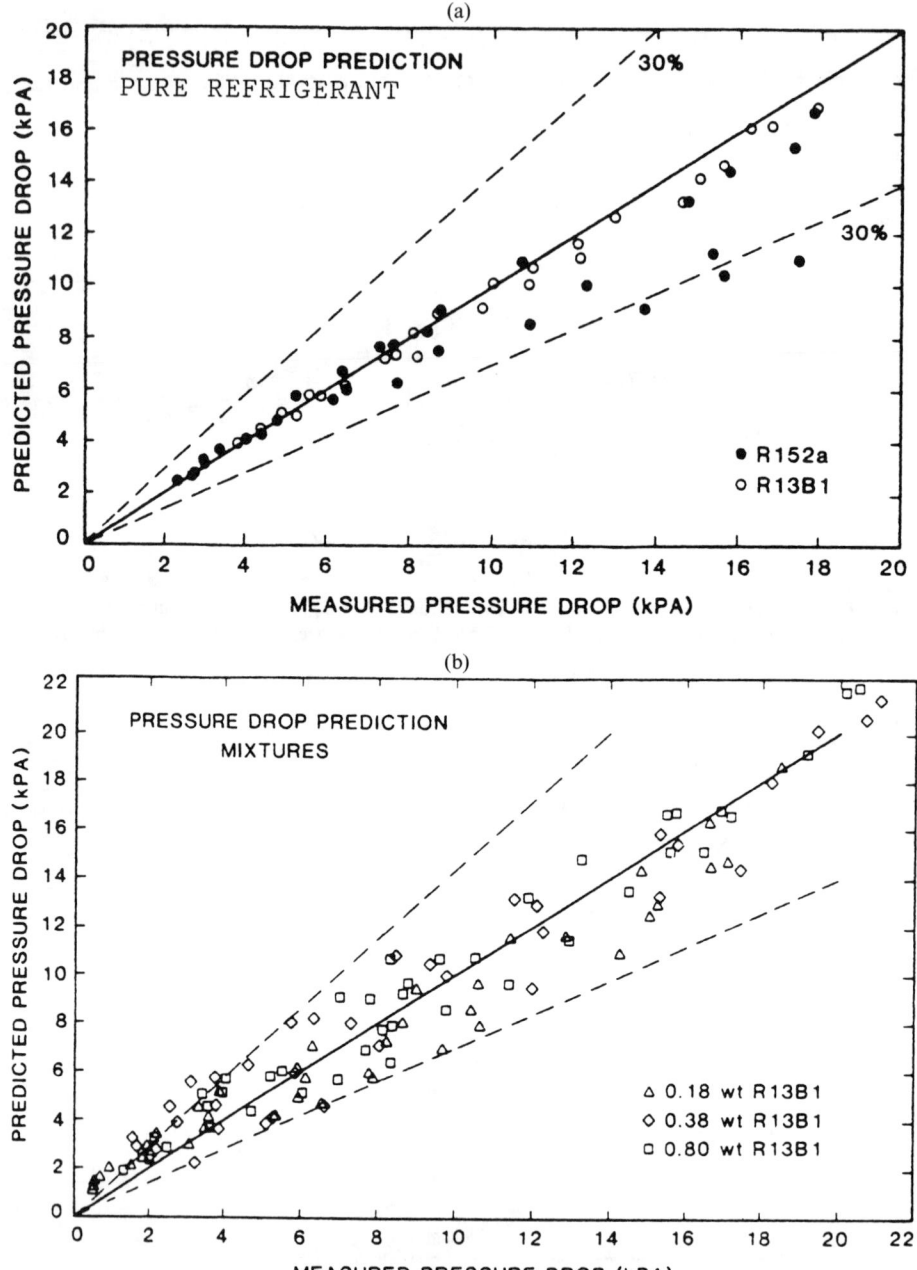

Figure 3.50 Comparison of measured pressure drop to that predicted by the Martinelli-Nelson-Chisholm method: (A) pure refrigerant; (B) refrigerant mixtures. (From Ross et al., 1987. Copyright © 1987 by Elsevier Science Ltd., Kidlington, UK. Reprinted with permission.)

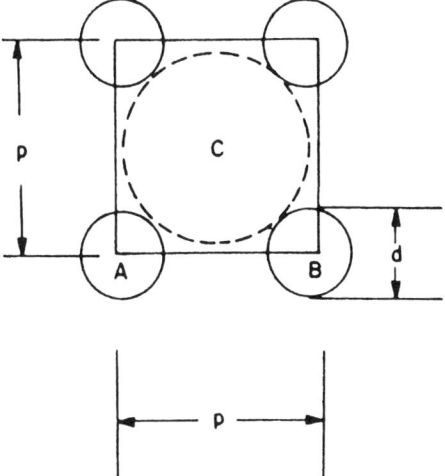

Figure 3.51 A four-rod cell.

The authors thus considered it to be reasonable to apply the conventional ΔP prediction method to boiling mixtures of refrigerants.

3.5.4 Pressure Drop in Rod Bundles

3.5.4.1 Steady two-phase flow. In rod (or tube) bundles, such as one usually encounters in reactor cores or heat exchangers, the pressure drop calculations use the correlations for flow in tubes by applying the equivalent diameter concept. Thus, in a square-pitched four-rod cell (Fig. 3.51), the equivalent diameter is given by

$$D_e = \frac{4(p^2 - \pi d^2/4)}{\pi d} = \frac{4d}{\pi}\left[\left(\frac{p}{d}\right)^2 - \left(\frac{\pi}{4}\right)\right]$$

The pressure drop for steady-state vertical upflow is given by

$$\Delta P_{T(1-2)} = \int_1^2 \left(\frac{g}{g_c \bar{v}}\right) dz + \left(\frac{G^2}{g_c}\right)(\bar{v}_2 - \bar{v}_1)$$
$$+ \left(\frac{G^2}{2g_c D_e}\right)\int_1^2 f_o\left(\frac{f_{TP}}{f_o}\right)\bar{v}\, dz \quad (3\text{-}143)$$

Venkateswararao et al. (1982), in evaluating the flow pattern transition for two-phase flow in a vertical rod bundle, suggested the calculation of pressure gradient for annular flow by

$$\frac{dp}{dz} + [\alpha\rho_G + (1-\alpha)\rho_L]g + \frac{P_L \tau_w}{A_L + A_G} = 0 \qquad (3\text{-}144)$$

where P_L is the wetted perimeter, $(A_L + A_G)$ is the cross-sectional area for the combined gas and liquid flow, and τ_w is shear along the rod surface. For flow in the four-rod cell as shown in Figure 3.51, these values are $P_L = \pi d$, $(A_L + A_G) = p^2 - (\pi/4)d^2$, and $\tau_w = \tfrac{1}{2}f_w \rho_L u_L^2 = f_w \rho_L u_{LS}^2/2(1-\alpha)^2$.

By using Wallis's (1969) assumption of $f_w = 0.005$, Eq. (3-144) becomes

$$\frac{dp}{dz} + [\alpha\rho_G + (1-\alpha)\rho_L]g + \left[\frac{0.0025}{(1-\alpha)^2}\right]\frac{(\rho_L u_{LS}^2)}{(d/\pi)[(p/d)^2 - (\pi/4)]} \qquad (3\text{-}145)$$

In open rod bundles, transverse flow between subchannels is detectable by variations in hydraulic conditions, such as the difference in equivalent diameter in rod and shroud areas (Green et al., 1962; Chelemer et al., 1972; Rouhani, 1973). The quality of the crossflow may be somewhat higher than that of the main stream (Madden, 1968). However, in view of the small size of the crossflow under most circumstances, such variation generally will not lead to major error in enthalpy calculations. The homogeneous flow approximation almost universally used in subchannel calculations appears to be reasonable (Weisman, 1973). The flow redistribution has a negligible effect on the axial pressure drop.

To measure all the parameters pertinent to simulating reactor conditions, Nylund and co-workers (1968, 1969) presented data from tests carried out on a simulated full-scale, 36-rod bundle in the 8-MW loop FRIGG at ASEA, Vasteras, Sweden (Malnes and Boen, 1970). Their experimental results indicate that the two-phase friction multiplier in flow through bundles can be correlated by using Becker's correlation (Becker et al., 1962),

$$\phi^2 = 1 + A_F \left(\frac{x}{p}\right)^{0.96} \qquad (3\text{-}146)$$

where x = steam quality
p = system pressure in bars
$A_F = 2{,}234 - 0.348G \pm 640$
G = total mass flux in kg/s m^2

The equation for the mass flux effect, A_F, has been obtained by correlating the measured friction multiplier values by means of regression analyses (Fig. 3.52). It is assumed that the two-phase friction loss in the channel is essentially unchanged by the presence of spacers. However, the increase in total pressure drop is determined by its presence in rod bundles (Janssen, 1962).

HYDRODYNAMICS OF TWO-PHASE FLOW **209**

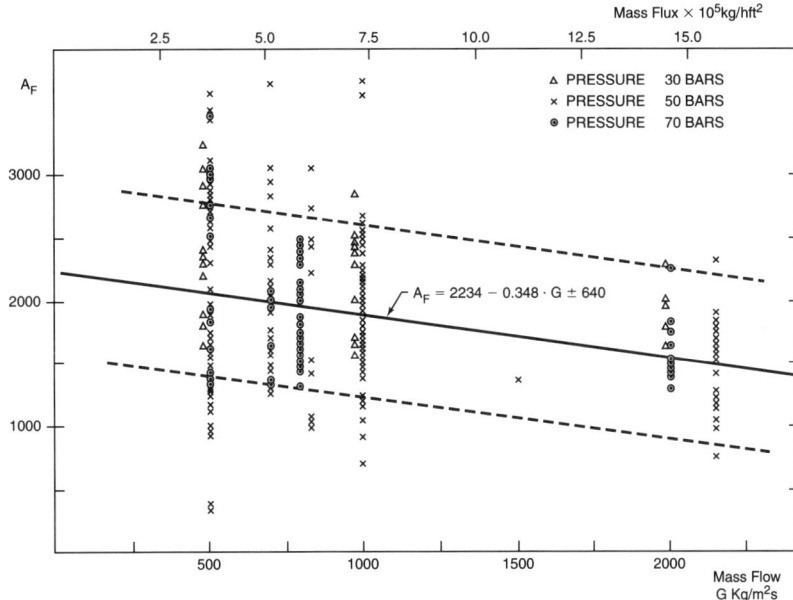

Figure 3.52 Mass flow modified coefficient in the Becker two-phase friction multiplier. (From Malnes and Boen, 1970. Copyright © 1970 by Office for Official Publication of the European Community, Luxembourg. Reprinted with permission.)

3.5.4.2 Pressure drop in transient flow. There are two types of transient hydraulic models: conventional analog-type representations based on transfer functions, and digital finite-method solutions of three basic partial differential equations in space and time for continuity, momentum, and energy (see Sec. 3.4.8). HYKAMO code (Schoneberg, 1968) is an example of the former type, while ROMONA (Bakstad and Solberg, 1967, 1968), HYDRO (Hasson, 1965), and FLICA (Fajean, 1969) are examples of the latter type. Malnes and Boen (1970) carried out a comparison of full-scale, 36-rod-bundle experimental results with the transient model RAMONA with reasonable success. They also called out the importance of using correct correlations in the model to predict transient behavior of two-phase flow. In addition to the necessary correlations, the model comprises the following basic assumptions:

1. Choking does not exist, or the velocity of sound (pressure propagation velocity) is assumed to be infinite.
2. The correlations obtained in steady state for one- and two-phase flow are valid during a transient.

Critical flow and unsteady two-phase flow will be discussed in Section 3.6.

3.5.5 Pressure Drop in Flow Restriction

3.5.5.1 Steady-state, two-phase-flow pressure drop.
The accurate prediction of two-phase-flow pressure drop at restrictions, such as orifices and area changes, is necessary in the design of a two-phase flow system, which, for instance, particularly affects the design performance of a natural-circulation nuclear reactor. At present, the integrated momentum and kinetic energy before and after the restriction cannot be calculated analytically, because the velocity distributions in various flow patterns and the extent of equilibrium between the vapor and liquid phases are generally not known. Semiempirical or empirical approaches are therefore followed in correlating data; the common assumptions are one-dimensional flow and no phase change taking place over the restriction.

Flow through abrupt expansion Using the one-dimensional flow assumption for a single-phase incompressible fluid, the energy equation becomes

$$p_1 + \frac{\rho u_1^2}{2g_c} = p_2 + \frac{\rho u_2^2}{2g_c} + K\left(\frac{\rho u_1^2}{2g_c}\right) \tag{3-147}$$

where K is a loss coefficient (or Borda-Carnot coefficient) that can be obtained by momentum balance as

$$K = \left(1 - \frac{A_1}{A_2}\right)^2$$

Subscripts 1 and 2 refer to the positions before and after the restriction, respectively. Combining the continuity equation and Eq. (3-147) yields the total pressure change,

$$p_2 - p_1 = \frac{+2A_1}{A_2}\left(1 - \frac{A_1}{A_2}\right)\left(\frac{\rho u_1^2}{2g_c}\right) \tag{3-148}$$

where the plus sign indicates pressure recovery after the expansion. For frictionless flow, the pressure rise is due to the momentum or velocity change only, which can be obtained from

$$(\Delta p)_{\text{mom}} = \left[1 - \left(\frac{A_1}{A_2}\right)^2\right]\left(\frac{\rho u_1^2}{2g_c}\right) \tag{3-149}$$

Hence the net expansion loss is*

---

*For cases of nonuniform velocity distribution, Kays and London (1958) suggested using momentum correction and energy correction factors in the above equations. However, these factors are very difficult to evaluate, so the homogeneous model is used here.

$$(\Delta p)_{\text{loss}} = (\Delta p)_{\text{mom}} - (p_2 - p_1)$$

$$= \left[1 - 2\left(\frac{A_1}{A_2}\right) + \left(\frac{A_1}{A_2}\right)^2\right]\left(\frac{\rho u_1^2}{2g_c}\right) \quad (3\text{-}150)$$

For two-phase flow, additional assumptions are made that thermodynamic phase equilibrium exists before and after the restriction (or expansion), and that no phase change occurs over the restriction. Romie (Lottes, 1961) wrote the equation for the momentum change across an abrupt expansion as

$$p_1 A_1 + \frac{w_{L1} u_{L1}}{g_c} + \frac{w_{G1} u_{G1}}{g_c} = p_2 A_2 + \frac{w_{L2} u_{L2}}{g_c} + \frac{w_{G2} u_{G2}}{g_c} \quad (3\text{-}151)$$

and the equations of continuity for the liquid and gas flows as

$$w_{L2} = w_{L1} = w(1 - X) = \bar{\rho} u A_1 (1 - X)$$
$$w_{G2} = w_{G1} = wX = \bar{\rho} u A_1 X$$

where w is the total flow rate and u is the upstream velocity, assuming the total mass flow rate to be liquid. Additional continuity considerations and the definition of void fraction give

$$u_{L1} = \frac{u(1 - X)}{1 - \alpha_1}$$

$$u_{L2} = \frac{(A_1/A_2)u(1 - X)}{1 - \alpha_2}$$

$$u_{G1} = \frac{uX\rho_L}{\alpha_1 \rho_G}$$

$$u_{G2} = \frac{(A_1/A_2)(uX)\rho_L}{\alpha_2 \rho_G}$$

When these equations are combined, the static pressure change is found to be

$$p_2 - p_1 = \left(\frac{A_1}{A_2}\right)\frac{\rho_L u^2}{g_c}$$

$$\times \left\{X^2 \left(\frac{\rho_L}{\rho_G}\right)\left(\frac{1}{\alpha_1} - \frac{A_1}{A_2 \alpha_2}\right) + (1 - X)^2 \left[\left(\frac{1}{1 - \alpha_1}\right) - \frac{(A_1)}{A_2(1 - \alpha_2)}\right]\right\} \quad (3\text{-}152)$$

If it is further assumed that the void fractions upstream and downstream from the expansion are the same, Eq. (3-152) becomes

$$p_2 - p_1 = \frac{\rho_L u^2}{g_c} \frac{A_1}{A_2} \left(1 - \frac{A_1}{A_2}\right) \left[\frac{X^2 \rho_L}{\alpha \rho_G} + \frac{(1-X)^2}{1-\alpha}\right] \quad (3\text{-}153)$$

Lottes (1961) found that the predictions based on Romie's analysis shown above agreed with ANL data within ±4% at 600 psia (4.1 MPa) and within ±6% at 1,200 psia (8.2 MPa), for a natural-circulation boiling system. In addition, Lottes also found that Hoopes's data for flow of steam–water mixtures through orifices appeared to verify Romie's analysis for $(A_1/A_2) \simeq 0$.

On the basis of extensive analysis of available data, Weisman et al. (1978) concluded that, for abrupt expansions, α_1 and α_2 should be evaluated by assuming slip flow. They recommended Hughmark's (1962) relationship for obtaining α from x,

$$\frac{1}{X} = 1 - \frac{\rho_L}{\rho_G(1 - c/\alpha)}$$

where c is the flow factor, defined by $c = \alpha + (1 - \alpha)/S$. Hughmark (1962) concluded that c is a function of parameter Z, where

$$Z = \left[\frac{D_e G}{\mu_L(1-\alpha) + \mu_G \alpha}\right]^{1/6} \left[\frac{G^2}{(\rho_{\text{hom}})^2 g D_e}\right]^{1/8} \left[\frac{(1-X)\rho_G + X\rho_L}{(1-X)\rho_G}\right]^{1/4} \quad (3\text{-}154)$$

ρ_{hom} is the mixture density obtained by assuming no slip, and μ_L and μ_G are the liquid and vapor viscosities, respectively.

Flow through abrupt contraction The two-phase pressure drop at an abrupt contraction usually can be predicted by using a homogeneous flow model and the single-phase pressure coefficient given by Kays and London (1958). Owing to the strong mixing action along the jet formed by the contraction, the mixture of the two phases is finely homogenized. Data measured by Geiger (1964) for the two-phase-flow pressure drop at sudden contraction in water at 200–600 psia (1.4–4.1 MPa) can be correlated by a homogeneous model:

$$\Delta p_c = \left(1 + \frac{X v_{fg}}{v_L}\right)^{-1} \rho_L u^2 \frac{[1 - (A_1/A_2)^2 + K_c]}{2g_c} \quad (3\text{-}155)$$

where $K_c = [(1/C_c) - 1]^2$, C_c is a contraction coefficient, and v_{fg} and v_L are the specific volume change during evaporation and that of the liquid phase, respectively.

Later, Weisman et al. (1978) also found that assuming homogeneous flow everywhere provided nearly as good a correlation of the data as the slip flow model. The total pressure drop across a contraction can be approximated by

$$\Delta p_c = \frac{A_2^2 G^2}{2g_c A_1^2} \left[\left(\frac{1}{A_{VC}} - 1 \right)^2 + \left(1 - \frac{A_1}{A_2} \right)^2 \right] \left[\frac{X}{\rho_G} - \frac{(1-X)}{\rho_L} \right] \quad (3\text{-}156)$$

where A_{VC} = vena contracta area ratio. This equation is considered a useful tool for design work (Tong and Weisman, 1979).

Flow through orifices For the liquid or vapor phase flow alone passing through the orifice, expressions similar to a single-phase incompressible flow case can be written for the mass flow rates of both phases:

$$w_L = A_c' (2g_c \rho_L \, \Delta p_{LPF})^{1/2} \quad (3\text{-}157)$$

$$w_G = A_c' (2g_c \rho_G \, \Delta p_{GPF})^{1/2} \quad (3\text{-}158)$$

where $A_c' = CA_{or}/[1 - (A_{or}/A_o)^2]^{1/2}$, A_{or} is the orifice area, A_o is the pipe cross-sectional area, and C is a discharge coefficient. The mass flow rates in a two-phase flow can then be similarly expressed in terms of the two-phase pressure drop,

$$w_L = A_{cL}' (2g_c \rho_L \, \Delta p_{TPF})^{1/2} \quad (3\text{-}159)$$

$$w_G = A_{cG}' (2g_c \rho_G \, \Delta p_{TPF})^{1/2} \quad (3\text{-}160)$$

Combining Eqs. (3-157)–(3-160) yields

$$A_c' = A_{cL}' + A_{cG}' = \frac{A_c' (2g_c \rho_L \, \Delta p_{LPF})^{1/2}}{(2g_c \rho_L \, \Delta p_{TPF})^{1/2}} + \frac{A_c' (2g_c \rho_G \, \Delta p_{GPF})^{1/2}}{(2g_c \rho_G \, \Delta p_{TPF})^{1/2}}$$

which can be reduced to

$$\left(\frac{\Delta p_{TPF}}{\Delta p_{GPF}} \right)^{1/2} = \left(\frac{\Delta p_{LPF}}{\Delta p_{GPF}} \right)^{1/2} + 1 \quad (3\text{-}161)$$

When the single-phase pressure drops have been predicted and the quality of the flow is known, Eq. (3-161) gives the two-phase orifice pressure drop.

Murdock (1962) has tested the two-phase flows of steam–water, air–water, natural gas–water, natural gas–salt water, and natural gas–distillate combinations in 2.5-, 3-, and 4-in. pipes with orifice-to-pipe diameter ratios ranging from 0.25 to

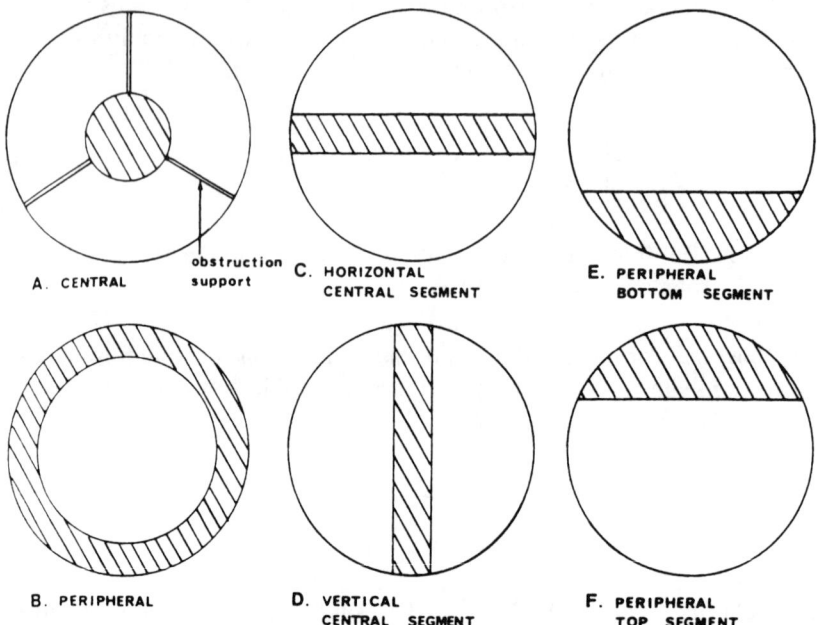

Figure 3.53 Shape and location of the obstruction. (From Salcudean et al., 1983b. Copyright © 1983 by Elsevier Science Ltd., Kidlington, UK. Reprinted with permission.)

0.50. Pressures ranged from atmospheric to 920 psia (6.3 MPa), differential pressures from 10 to 500 in. (25 to 1270 cm) of water, and liquid mass fractions from 2% to 89%. Fluid temperatures ranged from 50 to 500°F (10° to 260°C), and Reynolds numbers were from 50 to 50,000 for the liquid and from 15,000 to 1,000,000 for the gas. The following empirical form of Eq. (3-161) was obtained:

$$\left(\frac{\Delta p_{TPF}}{\Delta p_{GPF}}\right)^{1/2} = 1.26 \left(\frac{\Delta p_{LPF}}{\Delta p_{GPF}}\right)^{1/2} + 1 \qquad (3\text{-}162)$$

which correlates the data to a standard deviation of 0.75%.

Horizontal two-phase flow with obstructions Earlier experimental results suggested that the pressure drop during two-phase flow through fittings may be presented by one equation (Chisholm and Sutherland, 1969; Chisholm, 1971) and that the effect of the obstruction on the changes in phase and velocity distribution depends on the obstructed area, shape, flow regimes, etc. By examining the influence of the degree of flow blockage and the shape of the flow obstructions on pressure drops, Salcudean et al. (1983) computed the two-phase multipliers for the different obstructions. Figure 3.53 shows the shape and location of the obstruction in channels

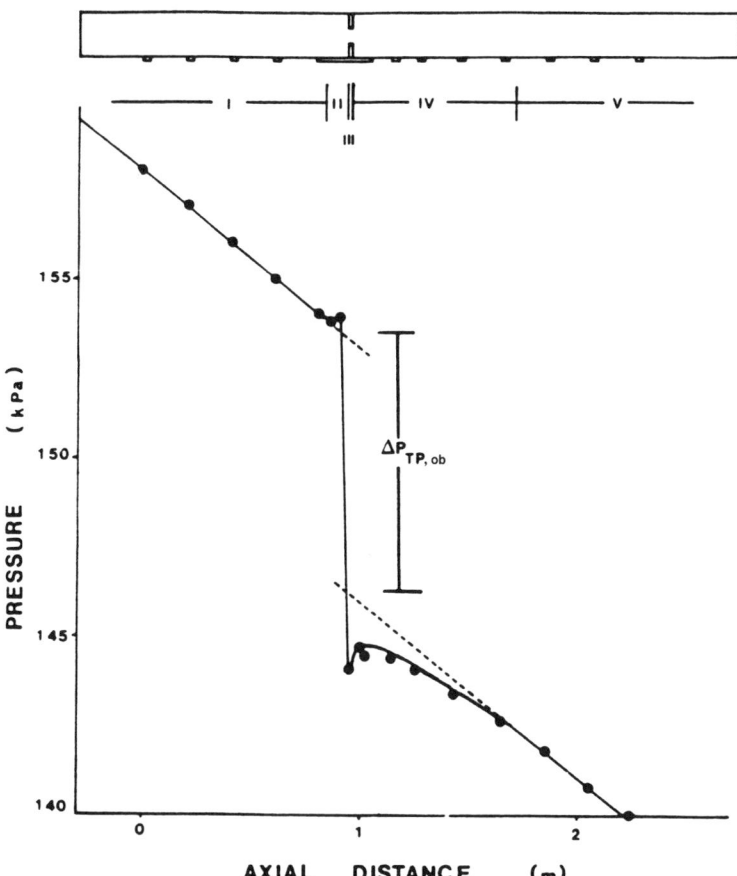

Figure 3.54 Pressure profile for two-phase flow. (From Salcudean et al., 1983b. Copyright © 1983 by Elsevier Science Ltd., Kidlington, UK. Reprinted with permission.)

that were studied, and results are illustrated in Figure 3.54 by a dimensionless pressure drop, $(\Delta p_{TP,ob})^+$, defined as

$$(\Delta p_{TP,ob})^+ = \left(\frac{(\Delta p_{TP,ob})}{(\rho_L U_{LS}^2/2)} \right)$$

where U_{LS} is the superficial liquid velocity, and $\Delta p_{TP,ob}$ is the obstruction pressure drop as shown in the figure, representing the increase in pressure drop due to the

216 BOILING HEAT TRANSFER AND TWO-PHASE FLOW

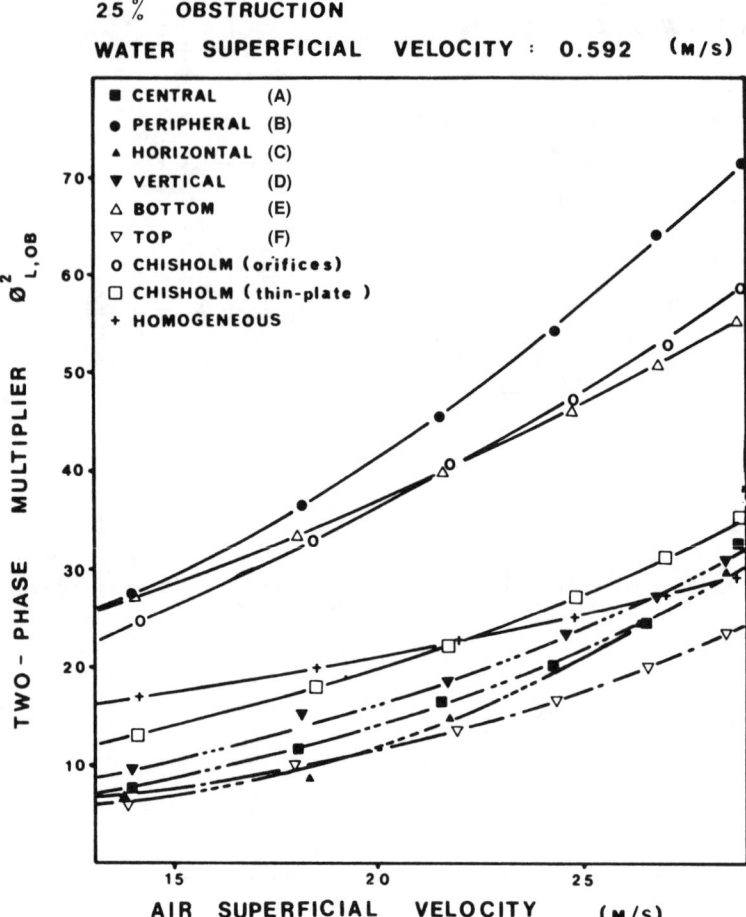

Figure 3.55 Two-phase obstruction pressure drop multiplier for 25% obstructions (for shape designations, see Fig. 3.53). (From Salcudean et al., 1983b. Copyright © 1983 by Elsevier Science Ltd., Kidlington, UK. Reprinted with permission.)

presence of a flow obstruction. The two-phase multiplier for flow through obstructions, $(\phi_{L,ob})^2$, can be found to relate to $(\Delta p_{TP,ob})^+$ as

$$(\phi_{L,ob})^2 = \left(\frac{\Delta p_{TP,ob}}{K_{ob}(\rho_L U_{LS}^2/2)} \right) \quad (3\text{-}163)$$

where K_{ob} is an overall pressure loss coefficient and is evaluated by the single-phase pressure drop,

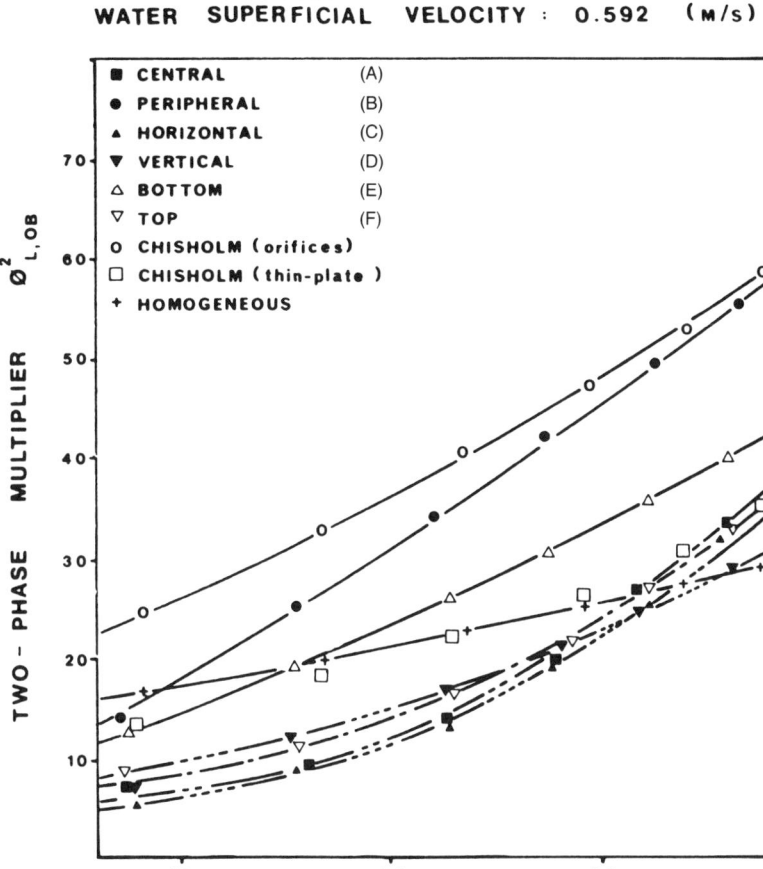

Figure 3.56 Two-phase obstruction pressure drop multiplier for 40% obstructions (for shape designations, see Fig. 3.53). (From Salcudean et al., 1983b. Copyright © 1983 by Elsevier Science Ltd., Kidlington, UK. Reprinted with permission.)

$$\Delta p_{ob} = K_{ob}\left(\frac{\rho_L U_{LS}^2}{2}\right)$$

Figures 3.55 and 3.56 show $(\phi_{L,ob})^2$ for 25% and 40% obstructions, respectively, in different shapes of obstructions (defined in Fig. 3.53) as a function of U_{GS} with a fixed $U_{LS} = 0.592$ m/s (1.94 ft/sec)

3.5.5.2 Transient two-phase-flow pressure drop. Calculation of transient behavior in a complex flow network containing a compressible fluid in two-phase states was demonstrated by Turner and Trimble (1976). Using a NAIAD computer code, these authors show an outline of the finite-difference and numerical methods. By assuming that the pressure, p, is constant over any cross section of a flow path, A, perpendicular to the flow, or z direction, the one-dimensional conservation equations were written in terms of averages $\langle \ \rangle$ of local density, ρ, velocity, u, with z component u_z, internal energy U, and enthalpy H over such cross sections. The conservation equations are

$$\frac{\partial \rho}{\partial t} + \frac{1}{A}\frac{\partial(A\langle\rho u_z\rangle)}{\partial z} = 0$$

$$\frac{\partial\langle\rho u_z\rangle}{\partial t} + \frac{\partial p}{\partial z} + \frac{1}{A}\left(\frac{\partial\langle A\rho u_z^2\rangle}{\partial z}\right) = -F - \langle\rho\rangle g \sin\theta$$

$$\frac{\partial\langle\rho(U + u^2/2)\rangle}{\partial t} + \frac{\partial\langle A\rho u_z(H + u^2/2)\rangle}{A\,\partial z} = q - \langle\rho u_z\rangle g \sin\theta$$

where $\langle\rho u_z\rangle$ = average mass flux

$\langle A\rho u_z^2\rangle$ = average momentum flux

$\left\langle\rho\left(\dfrac{U + u^2}{2}\right)\right\rangle$ = average energy density

$\left\langle\rho u_z\left(\dfrac{H + u^2}{2g_c J}\right)\right\rangle$ = average energy flux

$\left(\dfrac{H + u^2}{2g_c J}\right) = H_o$, stagnation enthalpy

q = power input per unit volume
F = frictional pressure gradient
θ = angle of elevation of z direction
J = thermal-to-mechanical energy conversion factor

In addition, Turner and Trimble defined a slip equation of state combination as the specification of mass flux, momentum flux, energy density, and energy flux as single-valued functions of the geometric parameters (area, equivalent diameter, roughness, etc.) at any z location, and of mass flux, pressure, and enthalpy,

$$\langle H\rangle = \langle\rho u H\rangle/G$$

An example of a slip equation of state combination is for thermodynamic equilibrium with a slip ratio of

$$K = \frac{u_G}{u_L} = \left(\frac{\rho_L}{\rho_G}\right)^e$$

where all vapor moves at a velocity u_G, all liquid at a velocity u_L, and e is a constant. Fausk's (1962), Moody's (1965), and the homogeneous models would correspond to values of e of $\frac{1}{2}$, $\frac{1}{3}$, and 0, respectively, in the above equation. As differentiated from the study by Malnes and Boen (1970), Trimble and Turner (1976) did take the choking condition into consideration in their model. An implicit, locally stable, finite-difference scheme was formulated by using the matrix form of the conservation equations. A calculation was performed (Turner and Trimble, 1976) to simulate a blowdown experiment by Edwards and co-workers (Edwards and Mather, 1973; Borgartz et al., 1969). In this experiment, a glass disk at the end of a closed horizontal steel pipe 4.096 m (13.4 ft) long by 206 mm (8.1 in.) in diameter was broken to allow the enclosed, pressurized water to escape. The initial temperatures were in the range 234–258°C (453–496°F), and the pressure was 5.8 MPa (850 psia). Pressure and temperature were measured at eight axial locations. Adiabatic homogeneous flow was assumed, with the transient initiated by reducing the receiver pressure to atmospheric and using initial conditions of 242°C (468°F) and 5.8 MPa (850 psia), as shown in Figure 3.57. The calculated and experimental pressure transients for different locations in the first 20 ms are shown in Figure 3.58. The agreement of pressures at the midtube measuring stations is reasonably good, while that at the close end and at near the break is less good. The assumption of thermodynamic equilibrium in the calculation is undoubtedly the principal reason for these differences.

3.6 CRITICAL FLOW AND UNSTEADY FLOW

The phenomenon of critical flow is well known for the case of single-phase compressible flow through nozzles or orifices. When the differential pressure over the restriction is increased beyond a certain critical value, the mass flow rate ceases to increase. At that point it has reached its maximum possible value, called the *critical flow rate,* and the flow is characterized by the attainment of the critical state of the fluid at the "throat" of the restriction. This state is readily calculable for an isentropic expansion from gas dynamics. Since a two-phase gas–liquid mixture is a compressible fluid, a similar phenomenon may be expected to occur for such flows. In fact, two-phase critical flows have been observed, but they are more complicated than single-phase flows because of the liquid flashing as the pressure decreases along the flow path. The phase change may cause the flow pattern transition, and departure from phase equilibrium can be anticipated when the expansion is rapid. Interest in critical two-phase flow arises from the importance of predicting dis-

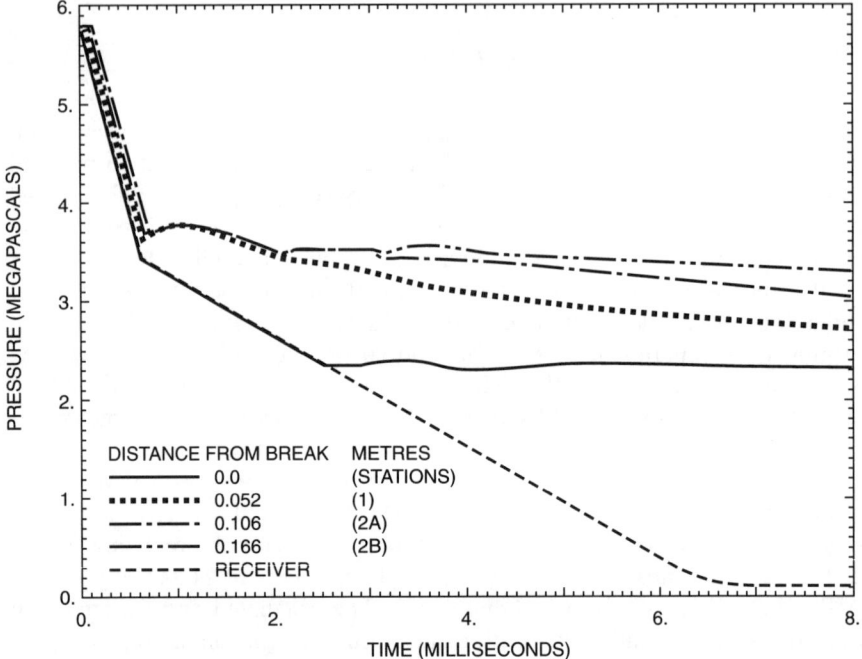

Figure 3.57 Pressure variation at different locations for the blowdown test. (From Turner and Trimble, 1976. Copyright © 1976 by OECD Publishing & Information Center, Washington, DC. Reprinted with permission.)

charge rates of highly pressurized steam–water mixtures through breaks in vessels or pipes. The high storage energy in a nuclear core could melt the core down in a loss-of-coolant accident (such as a pipe break) if emergency coolant injection fails. Therefore the knowledge of the discharge rate of a two-phase flow is important for the design of an emergency cooling system and thus for safety analysis of the extent of damage in accidents.

It has become useful to distinguish between two cases of two-phase critical flow, that is, flow through long pipes versus flow through short pipes and orifices. The mechanism of such flows through long pipes is different from that through short pipes or orifices. In the former case, the flow usually approaches the line of an equilibrium expansion; in the latter case, the liquid does not have enough time for expansion, causing it to be in metastable states. These were well recorded in the review given by Fauske (1962).

3.6.1 Critical Flow in Long Pipes

Since critical flow is determined by conditions behind the wave front, some phase change must be considered. In a long pipe line, where there is adequate time for

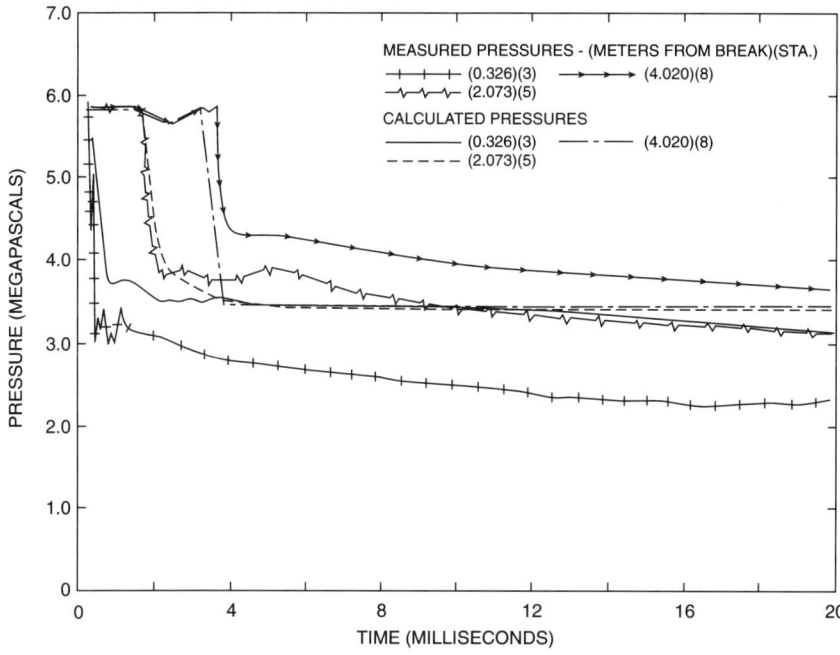

Figure 3.58 Pressures at stations 3, 5, and 8 for initial 20 msec. (From Turner and Trimble, 1976. Copyright © 1976 by OECD Publishing & Information Center, Washington, DC. Reprinted with permission.)

bubble nucleation and growth, thermodynamic equilibrium can be assumed. The homogeneous equilibrium model is the simplest analytical model that can be postulated, assuming:

1. Both phases move at the same velocity.
2. The fluid is in thermal equilibrium.
3. Flow is isentropic and steady.

Applying the continuity and energy conservation equations, the mass flux can be found by

$$G = [(2g_c J)(H_o - H)\rho^2]^{1/2} \tag{3-164}$$

and

$$(G_{cr})^2 = (-1/X)\left[\frac{dp}{d(1/\rho_G)}\right]$$

where, as before,

$$H_o = \text{stagnation enthalpy} = H + \frac{u^2}{2g_c J}$$

$$H = \text{local fluid enthalpy} = H_f - X H_{fg}$$

For fixed stagnation conditions, the critical mass flux is obtained by finding the downstream pressure for which G exhibits a maximum. The results of such calculations for steam–water systems are shown in Figure 3.59. Considering the available data for blowdown in long pipes up to that time, and examining the nonequilibrium behavior during depressurization, Moody (1975) concluded that if pipe length is more than 5 in., equilibrium states can be expected. Figure 3.59 can predict the critical mass flux pretty well. This is in agreement with Sozzi and Sutherland's (1975) conclusions; Caraher and DeYoung's (1975) evaluation of semi-scale blowdown data; as well as Edwards's (1968) observations (Tong and Weisman, 1979). Although homogeneous theory using stagnation (upstream reservoir) conditions provided a good prediction of critical flow rate, Moody also pointed out that it provided a poor prediction of pressure at the exit of the blowdown pipe (Moody, 1975). He concluded that mass fluxes are limited by homogeneous choking near the pipe entrance, but that a transition to slip flow occurred before reaching the exit and a second choked condition was produced near the exit. If critical mass fluxes are to be evaluated on the basis of local conditions near the pipe exit, a slip flow model must be used. In other words, in the region adjacent to the exit, thermodynamic equilibrium may not yet be established.

Fauske (1962) developed a "phases in equilibrium but separated flow" model for a long pipe, which could be used even at exit conditions, by assuming:

1. Average velocities of different magnitude exist for each phase (i.e., a slip flow exists).
2. The vapor and the liquid are in phase equilibrium throughout the flow path.
3. Critical flow is attained when the flow rate is no longer increased with decreasing downstream static pressure, $(\partial G/\partial p)_{H_o} = 0$.
4. The pressure gradient attains a maximum value at that time for a given flow rate and quality.

In the absence of friction, the momentum equation for an isentropic annular flow can be written as

$$\frac{G^2}{g_c}\left(\frac{dv}{dz}\right) + \frac{dp}{dz} = 0 \qquad (3\text{-}165)$$

or

$$\frac{G^2}{g_c} = -\left(\frac{dp}{dv}\right)\bigg|_s$$

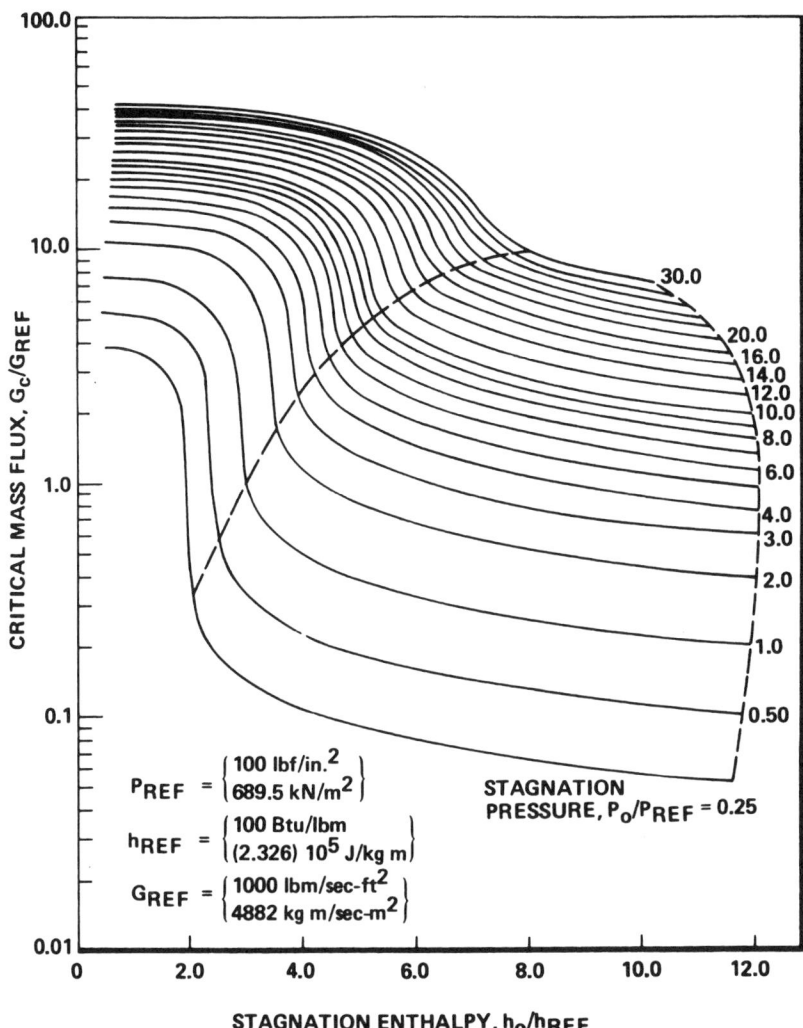

Figure 3.59 Critical mass flux–homogeneous, equilibrium steam–water systems. (From Moody, 1975. Copyright © 1975 by American Society of Mechanical Engineers, New York. Reprinted with permission.)

where the mean specific volume, v, is given by

$$v = \frac{X^2 v_G}{\alpha} + \frac{(1-X)^2 v_L}{1-\alpha} \qquad (3\text{-}166)$$

and v_G and v_L are the specific volumes of the vapor and liquid, respectively. By introducing slip ratio S, defined as

$$S = \frac{u_G}{u_L} = \left(\frac{X}{1-X}\right)\left(\frac{1-\alpha}{\alpha}\right)\left(\frac{v_G}{v_L}\right)$$

Eq. (3-166) becomes

$$v = \left[(1-X)v_L + Xv_G\right]\left[\frac{1+X(S-1)}{S}\right] \quad (3\text{-}167)$$

The maximization of the pressure gradient as stated by assumption 4 above is achieved by varying the slip ratio, keeping other quantities constant:

$$\frac{\partial v}{\partial S} = (X - X^2)\left(v_L - \frac{v_G}{S^2}\right) = 0$$

Hence, at critical flow, the slip ratio becomes

$$S = \left(\frac{v_G}{v_L}\right)^{1/2} = \left(\frac{\rho_L}{\rho_G}\right)^{1/2} \quad (3\text{-}168)$$

The mass flux is obtained by substituting Eq. (3-167) into Eq. (3-165),

$$\frac{g_c}{G^2} = -\frac{d}{dp}\left\{\frac{[(1-X)v_L S + Xv_G][1 + X(S-1)]}{S}\right\}$$

and, by neglecting the insignificant term dv_L/dp,

$$G^2 = \frac{-g_c}{[(1-X+SX)X](dv_G/dp) + [v_G(1+2SX-2X) + v_L(2XS - 2S - 2XS^2 + S^2)](dX/dp)} \quad (3\text{-}169)$$

Fauske (1966) achieved reasonable agreement with low-pressure experimental data using Armand's slip ratio (Armand, 1959), where

$$\alpha = \frac{(0.833 + 0.167X)Xv_G}{(1-X)v_L + Xv_G} \quad (3\text{-}170)$$

A computer code has been programmed by Fauske (1962) for evaluating liquid metal critical flow with this procedure.

Instead of using just energy conservation, Moody (1975) derived a revised model that takes into account all the conservation laws. He found that critical flow rate is given by a determinantal equation that gives G as a function of p, X, and S.

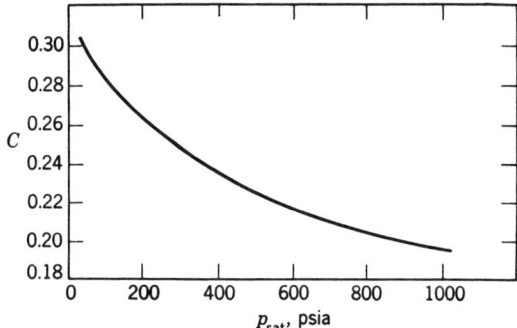

Figure 3.60 Values of constant C in Eq. (3-172) as a function of pressure.

Moody concluded that critical G occurs when S satisfies another determinantal equation. Note that the critical slip ratio S is not constant for a given pressure, as it would be if it were a function of density ratio only, but varies with quality. In view of the uncertainty in the value of S to be used under break conditions, as well as possible nonequilibrium effects, Tong and Weisman (1979) suggested that computations of saturated blowdown rates through long pipes be based on reservoir conditions and the homogeneous model. For subcooled, low-quality blowdown at low pressures, the empirical correction factor, N, suggested by Henry et al. (1968) for the homogeneous model, can be used:

$$G_{cr}^2 = \frac{-g_c}{N[X(\partial v_G/\partial p)_S + v_G(\partial X/\partial p)_S]} \tag{3-171}$$

where $N = 20X$ for $p < 350$ psia (2.4 MPa) and $X < 0.02$.

3.6.2 Critical Flow in Short Pipes, Nozzles, and Orifices

In 1947, Burnell recognized the existence of a metastable state in the flow of flashing water through nozzles. Hypothesizing that the water surface tension retards the formation of vapor bubbles and thus causes the water to be superheated, he developed a semiempirical method for predicting the flow of flashing water through square-edged orifices and correlated the discharge mass flux as

$$G_{cr} = \{2g_c \rho_L [p_{upst} - (1-C)p_{sat}]\}^{1/2} \tag{3-172}$$

where C is an empirical constant, which is a function of p_{sat} as given in Figure 3.60. When a reservoir discharges through a short pipe, orifice, or nozzle, the fluid can be subcooled just upstream of the break. Zaloudek (1963) reported the flow rate-versus-pressure drop characteristics of water critical flow in 14 different short pipes

226 BOILING HEAT TRANSFER AND TWO-PHASE FLOW

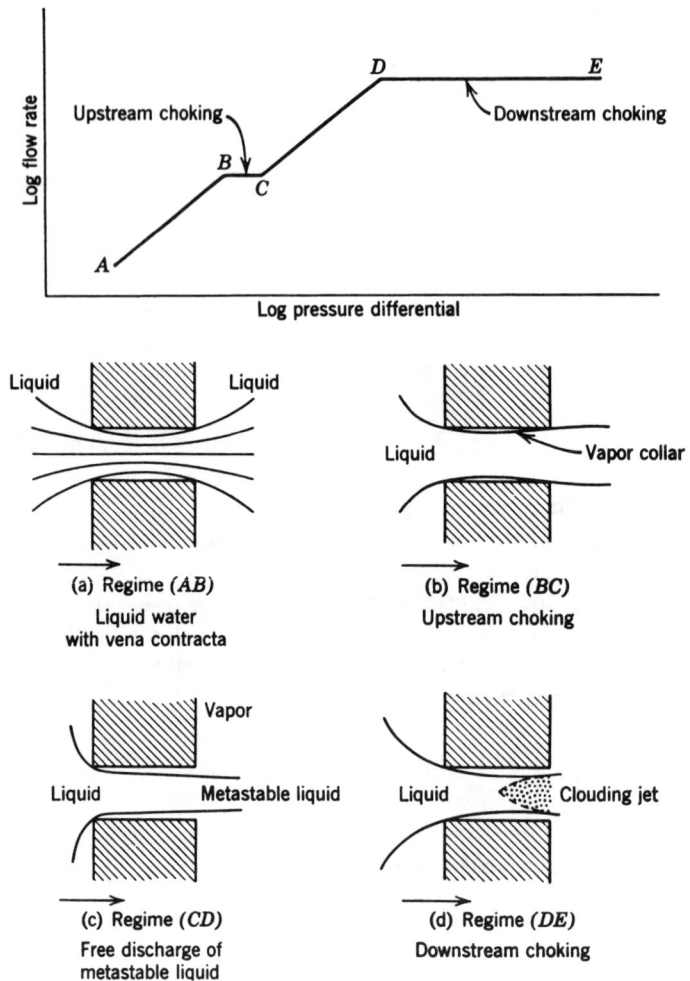

Figure 3.61 Critical flow patterns in a short pipe. (From Zaloudek, 1963. Reprinted with permission of U.S. Department of Energy, Washington, DC, subject to the disclaimer of liability for inaccuracy and lack of usefulness printed in cited reference.)

with $L/D < 6$. Two types of choking mechanism were observed visually, as shown in Figure 3.61. The first critical flow rate occurs at point B, where the pressure at the vena contracta near the entrance of the short tube approaches the saturation pressure of the water. Visual observations showed that at this point, local flashing occurs in the region between the vena contracta and the solid wall. This *upstream choking* is illustrated in Figure 3.61*b*. When a greater differential pressure, corresponding to point C, is reached, the flow pattern changes drastically to a free discharge of metastable liquid as shown in Figure 3.61*c*. In this flow pattern, the pressure throughout the pipe nearly equals the downstream pressure and the flow rate is again pressure differential dependent. At point D, liquid flashes violently at the downstream end of the pipe and causes a flow choking (*downstream choking* or *second choking*), as shown in Figure 3.61d. The flow rate at which it occurs will not increase upon further increase of the pressure difference. Second choking is recognized as occurring in compressible single-phase gas flow. Although the preceding description is in terms of decreasing downstream pressure, it was found (Zaloudek, 1963) that the same characteristic points can be produced equally well in reverse order by increasing the downstream pressure to the desired value. Thus the inception of the various flow patterns is not path dependent. Zaloudek also found that when choking occurs at the vena contracta, critical mass flux is given by

$$G_{cr,1} = C_1[2g_c\rho_L(p_{ups} - p_{sat})]^{1/2} \quad (3\text{-}173)$$

where C_1 is an empirically determined constant, which is found to range from 0.61 to 0.64. When downstream choking occurs, the critical mass flux agrees with Burnell's correlation, Eq. (3-172), where the surface tension-dependent constant, C, can be given by (Burnell, 1947)

$$C = 0.294\,\frac{\sigma(\text{for }p_{sat}\text{ upstream})}{\sigma(\text{for }p_{sat}\text{ at 175 psia})}$$

Sozzi and Sutherland (1975) also observed nonequilibrium effects in blowdown through short nozzles. They found a strong dependence on nozzle length, with the shorter nozzles giving higher flow rates, and that fluid passing through a very short length will not have sufficient time to nucleate completely before leaving the tube. Agreement between observations and homogeneous predictions was obtained for nozzle lengths of about 5 in. Sozzi and Sutherland also found that critical mass flux decreases with increasing throat diameter.

Henry and Fauske (1971) developed a model for critical flow in nozzles and short tubes, which allows for nonequilibrium effects and considers a two-phase mixture upstream of the break by using an empirical correlation to relate actual dX/dp to the value (dX_e/dp) under equilibrium conditions. For a dispersed flow, they assumed that

$$dX/dp = N\left(\frac{dX_e}{dp}\right) \quad (3\text{-}174)$$

where the experimental parameter N is given by

$$N = \frac{X_e}{0.14} \quad \text{for } X_e < 0.14$$

$$N = 1.0 \quad \text{for } X_e \geq 0.14$$

They concluded that the critical mass flux is given by

$$G_{cr}^2 = \frac{X_o v_G}{np} - (v_G - v_{Lo}) \left[\frac{(1-X_o)N}{(S_{Ge}-S_{Le})} \frac{dX_{Le}}{dp} - \frac{X_o c_p[(1/n)-(1/\gamma')]}{p(S_{Go}-S_{Lo})} \right] \quad (3\text{-}175)$$

where X_o = quality at stagnation conditions
n = polytropic exponent for vapor compression
γ' = isentropic exponent for vapor compression = C_p/C_v
C_p, C_v = specific heats for vapor at constant pressure and volume, respectively
S_{Ge}, S_{Le} = entropy under equilibrium conditions of vapor and liquid, respectively
S_{Go}, S_{Lo} = entropy at stagnation conditions of vapor and liquid, respectively

Other models have also been proposed, such as a relaxation model by Bauer et al. (1978) and a mechanistic model by Re'ocreaux (1977). The readers are referred to these references for details.

3.6.3 Blowdown Experiments

3.6.3.1 Experiments with tubes.
The blowdown experiments of Edwards and co-workers (Borgartz et al., 1969; Edwards and O'Brien, 1970; Edwards and Mather, 1973; Flanagan and Edwards, 1978) have been described in Section 3.5.4.2, in the discussion of transient two-phase-flow pressure drop. Other similar hot water decompression experiments were reported by Zaker and Wiedermann (1966). They found that if nonequilibrium states occur, these last between 0.5 and 1.0 msec. Similar experiments with hot water up to 2,000 psia (13.8 MPa) and initial temperatures up to saturation have also been reported by Gallagher (1970), and that author found that if metastable states do occur, they persist for only ~1.0 msec and are independent of geometry. Therefore, it is concluded that if a fluid particle is expelled in a time of ~1.0 msec or less, we should expect metastability and must accommodate nonequilibrium states in the flow rate prediction (Lahey and Moody, 1977). Blowdown rate data (Fauske, 1965; Uchida and Nariai, 1966) on saturated water from large reservoirs and straight tubes of varying length and diameter, and reservoir pressures, are shown in Figure 3.62 (Lahey and Moody, 1977). It was noted that the blowdown rate decreases rapidly with increasing tube length in the interval $0 < L < 2.0$ in. ($0 < L < 5.1$ cm). Beyond a tube length of 2.0 in. (5.1 cm), blowdown rate decreases at a much slower rate. Values at zero tube length in the figure correspond to ideal liquid, sharp-edged orifice flow with a flow coefficient of $C_c = 0.61$. Also shown is an initial blowdown rate corresponding to saturated water blowdown from 1,600 psia (11 MPa) through a 29.0-in. (73.7-cm)-long, 6.8-in. (17.3-cm)-diameter pipe (Allemann et al., 1970). Numbers shown at the extreme right of Figure 3.62 give ideal critical flow rates for saturated water under the homogeneous equilibrium model and the equilibrium slip model of Moody (1975).

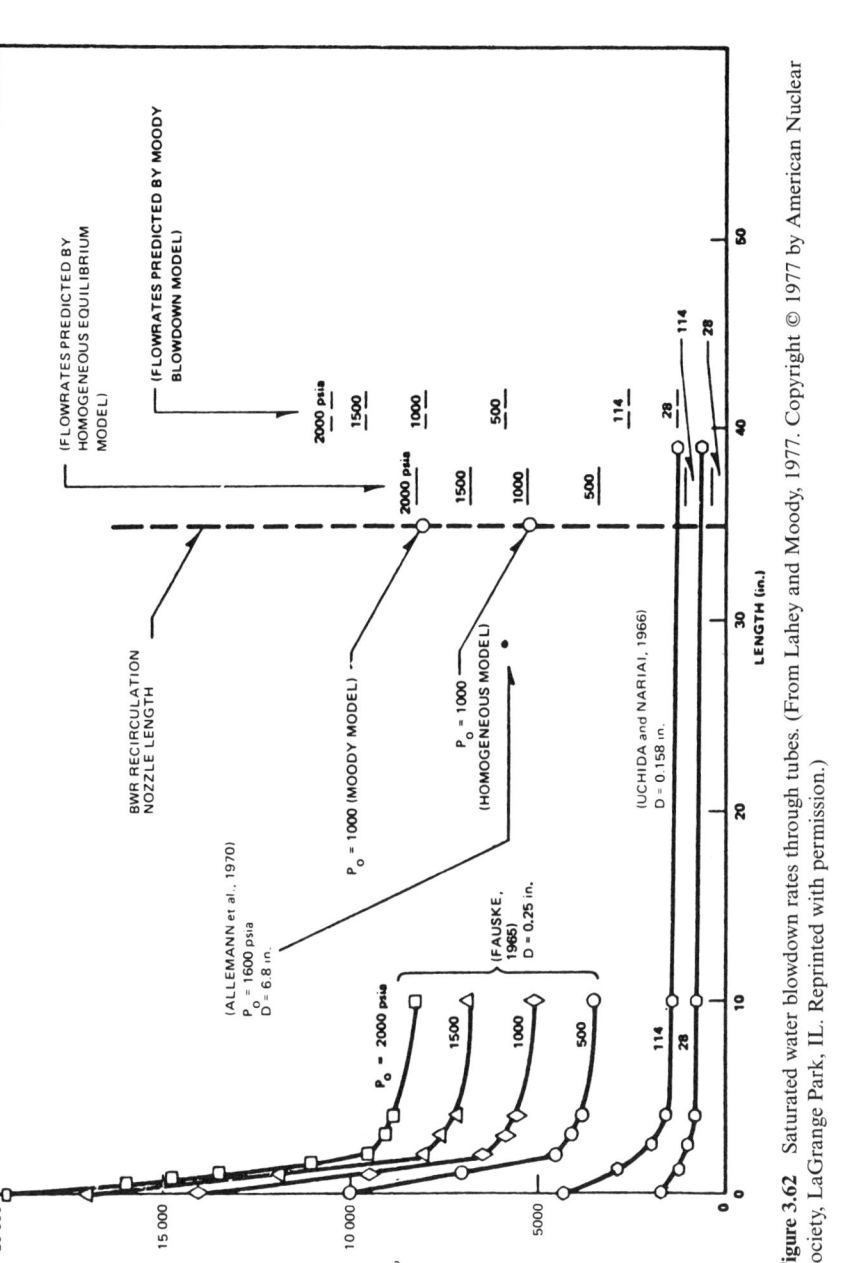

Figure 3.62 Saturated water blowdown rates through tubes. (From Lahey and Moody, 1977. Copyright © 1977 by American Nuclear Society, LaGrange Park, IL. Reprinted with permission.)

3.6.3.2 Vessel blowdown.
The previously mentioned relationships for the critical flow rate of a steam–water mixture can be employed with the conservation of mass and energy for a vessel of fixed volume to determine its time-dependent blowdown properties. The range of problems associated with coolant decompression in water-cooled reactors is quite broad. The types of hypothetical (some are even incredible) reactor accidents may be

1. Rupture of a large-size pipe, capable of causing complete loss of coolant and depressurization of the primary system
2. Sudden development of a large split in the sidewall of the vessel
3. Blowoff of the vessel head

Some of these phenomena are highly transient and involve propagation of relatively strong pressure waves throughout the system, while others are quasi-steady, with pressure waves of only negligible strength. The relative importance of the transient and quasi-steady portions of decompression, under a variety of initial blowdown conditions, had to be determined experimentally (Gallagher, 1970). The system of interest consists of two regions, initially at vastly different pressures, held in equilibrium by a diaphragm (or vessel wall). When the diaphragm (or vessel wall) is ruptured, a highly transient process ensues, after which the two regions eventually come into equilibrium. In the transient phase, two complex processes of closely related flow phenomena exist: (1) the formation of compressional (or shock) waves in the air, and (2) decompressional wave motion in the water (or vapor). A small-scale experiment simulating blowdown of nuclear reactors was conducted at IIT Research Institute (Gallagher, 1970) using a water-to-air shock tube and its auxiliary components. The shock tube consists primarily of a driver section and a driven section, and orifice plates and diaphragm that separate these two sections. Flow obstructions placed in the driver section of the shock tube simulate the geometry of fuel assemblies within a reactor vessel. Some of Gallagher's major conclusions resulting from this study are the following.

1. Initial air shock pressures for one-dimensional flow depend primarily on the enthalpy of the driver fluid.
2. For three-dimensional expansion in the air-filled section with the initial air volume equal to at least 200 times that of the initial water volume and with no partitions, baffles, or other types of compartments present, the air shock pressure is smaller than the final quasi-steady equilibrium pressure buildup. It can therefore be rationalized that the design of containment for a reactor is governed by equilibrium pressure only.
3. Water decompression may be assumed to be an isentropic equilibrium process for unheated blowdown or limited heating of fluid in the core region in order to calculate the transient pressure–time relationships with reasonably good accuracy for the full duration of the blowdown.

4. The presence of metastable decompression paths having transient pressures temporarily much smaller than the temperature equilibrium values could not be absolutely confirmed or refuted. If present, they persist for only ~1 msec and do not seriously affect the decompression time of the driver tube, and the duration is independent of vessel size.
5. The pressure–time variation in the driver section can be approximated by a quasi-steady model consisting of a single lumped-mass system with choked outflow at the minimum throat area when the ratio of the driver volume to the throat area is equal to ~20 ft^3/ft^2 (6.1 m^3/m^2) or more. For smaller values of the ratio, the representation by a quasi-steady model becomes less accurate, and the numerical wave analysis approach is more suitable.

3.6.4 Propagation of Pressure Pulses and Waves

As two-phase flow engineers, we are interested in the effect of the propagation of pressure pulses and waves on critical discharge and on the onset of flow instability. Just as the critical flow rates for two-phase mixtures should be treated differently from those for a single phase, because of the various interfacial transport processes involved, the propagation velocities for single-phase mixtures should not be applied to two-phase mixtures. This is because in the ideal single-phase problem, the perturbations are assumed small and equilibrium is assumed to be maintained all the time. Furthermore, in two-phase flow, pressure pulse propagation velocity should not be confused with sonic velocity. The basic differences between these two propagation rates lie in the different forms of perturbation and the different time scale as compared with the relaxation time of the fluid to reach an equilibrium state. For the propagation of a single pressure pulse, the wave front is usually steep. Fluid subjected to such a moving front does not have time to equilibrate to the rapid change in state, and thus it usually can be considered to be close to the frozen state. In this case, one worries only about the state of change while the pulse is passing and is not concerned with the states after the pulse has passed. On the other hand, sound is propagated through continuous waves of small amplitude, and the fluid is subject to a periodic change of pressure to which it has to respond. Depending on the period of the wave, an equilibrium state can be approached if the frequency is low or the amplitude is infinitesimal. Conversely, for a wave of high frequency or high amplitude, fluid response will lag.

3.6.4.1 Pressure pulse propagation. Following the approach of Shapiro (1953), Hsu (1972) derived a general form of the propagation equation by considering a control volume around the wave front traveling at velocity a in a channel of constant cross section A, for a two-phase mixture, and with the observer traveling with the wave front. The momentum and continuity equations in this case are

$$a[\alpha\rho_G(du_G) + (1 - \alpha)\rho_L(du_L)] = dp \qquad (3\text{-}176)$$

and

$$ad[\alpha\rho_G + (1-\alpha)\rho_L] = \alpha\rho_G(du_G) + (1-\alpha)\rho_L(du_L) \qquad (3\text{-}177)$$

Combining these equations results in

$$a^2 = \frac{dp}{d[\alpha\rho_G + (1-\alpha)\rho_L]}$$

Since α is function of quality X and the slip ratio S, the above equation can be expanded to read:

$$a^2 = \frac{[(1-X)\rho_G + X\rho_L]^2}{X\rho_L^2\left(\dfrac{\partial \rho_G}{\partial p}\right) + (1-X)\rho_G^2\left(\dfrac{\partial \rho_L}{\partial p}\right) - (\rho_L - \rho_G)\left(\dfrac{\partial X}{\partial p}\right) + X(1-X)(\rho_L - \rho_G)\rho_G\rho_L\left(\dfrac{\partial S}{\partial p}\right)} \qquad (3\text{-}178)$$

The terms $\partial\rho_G/\partial p$, $\partial\rho_L/\partial p$, $\partial X/\partial p$, and $\partial S/\partial p$ are determined by interfacial heat transfer, mass transfer, and momentum transfer, respectively.

The propagation characteristics of compression and rarefaction pressure pulses were studied experimentally by Grolmes and Fauske (1969), and by Barclay et al. (1969), the compression waves being found to travel faster than the rarefaction wave. Figure 3.63 shows the wave shapes from the work by Grolmes and Fauske (1969). From these wave profiles, the authors concluded that the mass transfer, or $\partial X/\partial p$ term of Eq. (3-178), can be neglected; i.e., the frozen state can be assumed. It is important to note that the pressure wave propagation in a flow system is different from the propagation in a stagnant two-phase medium (Barclay et al., 1969). This is because the term $\partial S/\partial p$ is very much a function of flow pattern, as different expressions for $\partial S/\partial p$ can be derived depending on whether the flow is stratified ($\partial S/\partial p \neq 0$) or dispersed ($\partial S/\partial p \rightarrow 0$) (Fig. 3.64) (Henry et al., 1969). Later, Henry (1970, 1971) proposed a series of refined models for propagation velocity for various flow patterns to take into account the virtual mass effect on $\partial S/\partial p$ as well as the heat transfer effect on $\partial v_G/\partial p$. When momentum transfer is considered, Henry also included the virtual mass, which is the inertial effect of the accelerating gas and its surrounding liquid. A summary of the models proposed by Henry et al. was published in a report (Henry et al., 1971).

Bubbly flow Instead of assuming that $\partial S/\partial p = 0$ as in his earlier model, Henry proposed that the virtual mass term of the gaseous volumes should be included:

$$-\left(\frac{\partial p}{\partial z}\right) = \rho_G u_G\left(\frac{\partial u_G}{\partial z}\right) + \frac{\rho_L}{C_M}\left[u_G\left(\frac{\partial u_G}{\partial z}\right) - u_L\left(\frac{\partial u_L}{\partial z}\right)\right] \qquad (3\text{-}179)$$

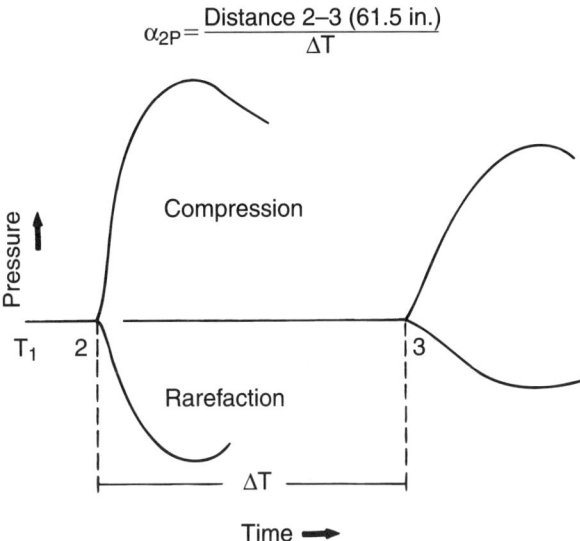

Figure 3.63 Superposition of representataive oscilloscope pressure traces at locations 2 and 3 for compression and rarefaction pressure pulses in low-void-fraction steam–water mixtures. (From Grolmes and Fauske, 1969. Copyright © 1969 by Elsevier Science SA, Lausanne, Switzerland. Reprinted with permission.)

where the last term is the virtual mass term and can be related to the change of slip ratio S. The constant C_M varies depending on the gas phase, and thus was assumed to be dependent on the void fraction. The gas compressibility term is also assumed to be a function of void fraction. Using data from the air–water system with $p = 25$ psia (0.17 MPa) and void fractions up to 0.5, Henry empirically correlated a correction factor, which includes the effects of interfacial momentum transfer C_M and interfacial heat transfer $\partial v_G/\partial P$:

$$\frac{a}{a_{HT}} = 1.035 + 1.671(\alpha) \qquad (3\text{-}180)$$

where a_{HT} is the propagation velocity for homogeneous, isothermal conditions and can be determined by

$$a_{HT}^2 = \left\{ \alpha^2 + \alpha(1-\alpha)\frac{\rho_L}{\rho_G} + \left[(1-\alpha)^2 + \alpha(1-\alpha)\frac{\rho_G}{\rho_L}\right]\frac{np}{\rho_G a_L^2} \right\}^{-1} \frac{np}{\rho_G}$$

with

$$n = \frac{(1-X)C_L + XC_{pG}}{(1-X)C_L + XC_{vG}}$$

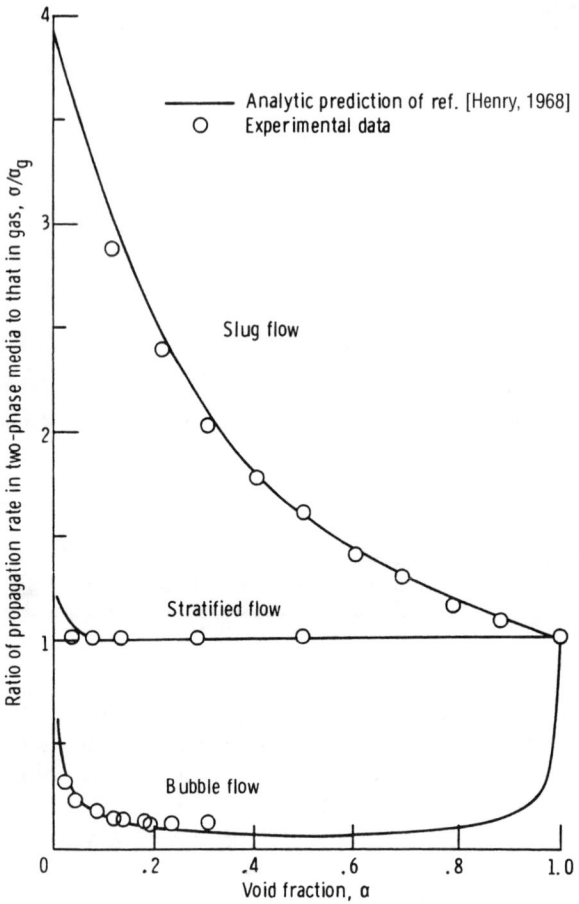

Figure 3.64 Measured and calculated pressure wave propagation velocity in air–water mixtures at 25 psia. (From Henry et al., 1969. Copyright © 1969 by Plenum Publishing, New York. Reprinted with permission.)

This empirical equation was successfully checked against data for the 10- to 285-psia (0.07- to 1.9-MPa) pressure range.

Annular flow, smooth interface **(Henry et al., 1969)** Since the interface is relatively small compared to dispersed flow and assumed to be smooth, there is no significant momentum transfer or mass transfer between phases. Under such conditions, the change of slip ratio with pressure is

$$\frac{\partial S}{\partial p} = -\left(\frac{1}{a^2}\right)\left(\frac{1}{\rho_G} - \frac{1}{\rho_L}\right)$$

The resulting propagation equation is

$$a^2 = \frac{[(1-X)\rho_G + \rho_L X]^2 + X(1-X)(\rho_L - \rho_G)^2}{(X\rho_L^2/a_G^2) + (1-X)\rho_G^2/a_L^2} \quad (3\text{-}181)$$

or

$$\frac{a}{a_G} = \left[1 + \left(\frac{1-X}{X}\right)\left(\frac{\rho_G}{\rho_L}\right)^2\right]^{1/2} \quad (3\text{-}181a)$$

Annular flow, wavy interface (Henry, 1971) Under this condition the virtual mass of gas flowing over a wavy surface can be approximated by flow over a surface made of continuous rows of half-cylinders:

$$\frac{dp}{dz} = \rho_L u_L \left(\frac{du_L}{dz}\right) - \rho_G \left[u_G\left(\frac{du_G}{dz}\right) - u_L\left(\frac{du_L}{dz}\right)\right] \quad (3\text{-}182)$$

where the last term on the right side is the virtual mass of the liquid filament in the accelerating gas or vapor stream. The resulting propagation equation is

$$\frac{a}{a_G} = \alpha^{1/2} \quad (3\text{-}183)$$

Mist flow, two components The momentum equation is

$$\frac{dp}{dz} = \rho_L u_L \left(\frac{du_L}{dz}\right) - \left(\frac{\rho_G}{2}\right)\left[u_G\left(\frac{du_G}{dz}\right) - u_L\left(\frac{du_L}{dz}\right)\right] \quad (3\text{-}184)$$

assuming spherical drops. The resulting equation is

$$\frac{a}{a_G} = \frac{[1 + 2\alpha^2(1-\alpha)\rho_L]^{1/2}}{[\alpha^2 + \alpha(1-\alpha)\rho_L/\rho_G]^{1/2}} \approx \left(\frac{2\alpha}{1+\alpha}\right)^{1/2} \quad (3\text{-}185)$$

for $\alpha > 0.5$.

Mist flow, one component In a one-component system with finely dispersed drops in the mist flow, the mass transfer between phases over a large interfacial area has to be considered. For the compression wave the frozen state can be assumed to be subcooled liquid, superheated vapor conditions generated by the wave are fairly stable, and the expressions for the two-component system are valid (Henry, 1971):

$$X = S\left(\frac{\alpha}{1-\alpha}\right)\left(\frac{v_L}{v_G}\right)$$

However, for a rarefaction wave, the vapor becomes subcooled and the liquid becomes superheated. When the wave front passes, the liquid phase is assumed to adjust from the metastable state at an equilibrium rate. If isentropic processes are assumed, the mass transfer rate can be shown to be

$$\frac{\partial X}{\partial p} = -\frac{(1-X)}{S_{fg}}\left(\frac{\partial S_L}{\partial p}\right) \tag{3-186}$$

Inclusion of this mass transfer term results in the propagation equation

$$a^2 = \frac{1 + [2\alpha^2(1-\alpha)\rho_L/(1+\alpha)\rho_G]}{[\alpha^2 + \alpha(1-\alpha)(\rho_L/\rho_G)]/a_G^2 + \rho_L[\alpha(1-\alpha)/XS_{fg}](\partial S_L/\partial p)} \tag{3-187}$$

The need for different expressions for compression and rarefaction waves is consistent with the experimental observation of Barclay et al. (1969) that the compression wave travels faster than the rarefaction wave.

Slug flow For slug flow, the time required for a pressure pulse to sweep across the length of a slug is

$$t = \frac{L_L}{a_L} + \frac{L_G}{a_G}$$

and the void fraction is

$$\alpha = \frac{L_G}{L_G + L_L}$$

Thus, the propagation velocity is

$$\frac{a}{a_G} = \frac{a_L}{\alpha a_L + (1-\alpha)a_G} \tag{3-188}$$

3.6.4.2 Sonic wave propagation.
Sonic waves differ from pressure pulses in two respects:

1. The pressure fluctuation is small, only large enough to avoid being completely attenuated within short distances.
2. The waves are continuous.

Most research has been performed in a two-phase medium with no flow conditions; very little has been done in two-phase-flow systems. Although the flow condition may bring in new variables, such as slip ratio, it is reasonable to assume that the basic phenomena observed in the nonflow condition also occur in the flow condition.

Various important parameters according to basic two-phase, one- or two-component configurations are examined as follows (Hsu, 1972).

Bubbly mixture (gas–liquid, two-component system) Most of the earlier work on acoustic velocity, such as that of Hsieh and Plesset (1961), was based on the following assumptions.

1. There is homogeneous distribution of small bubbles.
2. The velocity is primarily a function of the void fraction.
3. The acoustic velocities are the single-phase values of the liquid and vapor phases, c_L and c_G.

Hsieh and Plesset assumed that the two-phase homogeneous mixture can be represented as a uniform medium with physical properties synthesized from the constituent phases and weighted according to void fraction, α, and quality, X. Using such a model, they were able to show that the gas compression is essentially isothermal and the acoustic velocity can be approximated as

$$\frac{1}{c^2} = \frac{1}{[\rho_G(1-X)/\rho_L][1 + X\rho_L/(1-X)\rho_G]^2}$$
$$\times \left[\left(\frac{X}{1-X} \frac{1-\alpha}{\alpha} \frac{1}{c_L^2} \right) + \frac{\alpha}{1-\alpha} \frac{1}{c_G^2} \right] \quad (3\text{-}189)$$

where $c_G^2 = \partial p/\partial \rho_G = p/\rho_G$ (isothermal). Equation (3-189) is limited to small bubble sizes such that the bubble radius is much smaller than the wavelength and the frequency is well below the bubble resonance. Thus the acoustic velocity is independent of wave frequency in this case.

In a similar analysis (McWilliam and Duggins, 1969), the sonic velocity was shown to decrease with increasing bubble size and decreasing pressure. Van Wijngaarden (1966) derived equations to show that there is a dispersion of the acoustic wave; that is,

$$w = \frac{K}{(1+K^2)^{1/2}}$$

where w and K are frequency and wave number, respectively. He later (Van Wijngaarden, 1968) showed that the wave propagation in a two-phase bubble mixture can be treated by analogy to water waves. For high gas content ($R_o n_b^{1/3} \simeq 1$), the dispersion is not small, where $R_o n_b^{1/3}$ is the ratio between bubble radius and bubble distance (R_o being the radius of an unperturbed bubble and n_b, bubbles per unit volume). Thus, he concluded that the acoustic velocity is a function not only of void but also of frequency and bubble size, a conclusion that was borne out by Karplus's (1958) experiments.

Droplet suspensions (gas–liquid, two-component system) Since the inertia of a liquid suspended in the gas phase is higher than the inertia of the gas, the time for the displacement of liquid under the pressure waves should be considered. Temkin (1966) proposed a model to account for the response of suspension with pressure and temperature changes by considering the suspensions to move with the pressure waves according to the Stokes's law. The oscillatory state equation is thereby approximated by a steady-state equation with the oscillatory terms neglected, which is valid if the ratio of the relaxation time to the wave period is small, or

$$0 < \tau_D w \simeq 1 \qquad \tau_t w = 0$$

where the drag relaxation time $\tau_D = (2R^2 \rho_L / 9\mu_G)$, and the thermal relaxation time $\tau_t = \frac{3}{2}(c_{PL}/c_{PG})(\text{Pr})_G \tau_D$.

In Temkin's analysis, mass transfer is assumed to be absent in the two-component system. He found that an attenuation coefficient, A, is function of the relaxation times τ_D and τ_t and the mass fration of suspension, x_s:

$$A = \frac{2c_o K_2}{x_s w}$$

$$= \frac{w\tau_D}{1 + w^2 \tau_D^2} + (\gamma - 1)\left(\frac{c_{pL}}{c_{pG}}\right)\left(\frac{w\tau_t}{1 + w^2 \tau_t^2}\right) \quad (3\text{-}190)$$

where K_2 is the imaginary part of the wave number, $K = K_1 + iK_2$, and the dispersion coefficient B is

$$B = \frac{c_o/c - 1}{X_s}$$

$$= \frac{1}{1 + w^2 \tau_D^2} + \frac{(\gamma - 1)(c_{pL}/c_{pG}) w\tau_t}{1 + w^2 \tau_t^2} \quad (3\text{-}190)$$

This expression shows the difference of sonic velocity c in a particle suspension from that in the air, c_o as a function of the mass fraction of suspension, X_s, the relaxation times, τ_D and τ_t, and the frequency, w. These equations show that the acoustic velocity in a droplet suspension is a strong function of frequency and drop size. Experimental data for A and B for a suspension of alumina particles in air were shown to compare well with the prediction (Temkin, 1966).

Vapor–liquid, one-component system In contrast to two-component, gas–liquid mixture systems, in one-component, boiling systems, where liquid and vapor coexist, the change of pressure causes condensation or evaporation. Thus the mass transfer must be considered. Karplus (1961) suggested that the sound velocity in mixtures of water and steam be determined from $\partial p/\partial \rho$ along constant entropy lines, assuming thermodynamic equilibrium. The calculation showed that the velocity is lower than in single-phase water or steam, and there is a discontinuity in velocity at both the saturated liquid and saturated vapor lines. It may be noted that the experimental values of sonic velocity, c, are higher than the analytical prediction based on the equilibrium assumption. Clinch and Karplus (1964) analyzed the propagation of pressure waves in a hydrogen liquid–vapor system of droplet suspension, and found that for the mass transfer the relaxation time ratios wR_d^2/v_t and wR_d^2/v should be considered, where R_d represents the radius of the droplets; v and v_t are kinematic viscosity and thermal diffusivity, respectively. Thus, for very low frequency, equilibrium can be assumed and the equation for gas–liquid systems can be used. For very high frequency, where there is no time for mass and heat transfer, the acoustic velocity approaches that of the dominating phase. But for intermediate frequency, the acoustic velocity is dependent on X, quality, the relaxation time ratio, and the spacing between them (Fig. 3.65) (Clinch and Karplus, 1964). It was also found that a trace of noncondensable gas in the vapor phase can significantly reduce the compressibility. Thus, for a vapor–liquid system, insufficient time for reaching equilibrium (i.e., high frequency) and a trace of gas can both cause an increase in the acoustic velocity. On the other hand, for a two-phase mixture in flowing condition with slip, the sonic velocity will be lower than that in a stationary two-phase medium (Hsu, 1972).

3.6.4.3 Relationship among critical discharge rate, pressure propagation rate, and sonic velocity. The three propagation or discharge rates, G_{cr}, c, and a, in a single phase are closely related to the isentropic equilibrium compressibility of the fluid. In two-phase flow or a two-phase mixture, these three terms are still determined by the compressibility of the mixture, which is no longer under equilibrium and homogeneous conditions. The nonhomogeneity is determined from knowledge of the flow pattern, and the extent of nonequilibrium is determined by knowing the momentum, energy, and mass transport processes. The critical discharge rate and the pressure propagation rate should be related if one considers the existence of a stationary wave front in a discharging flow. At the wave front, the pressure propagation rate moving upstream is balanced by the outflow of the discharging fluid. Because of the interfacial transport that affects slip ratio, quality, etc., the two phases should not be considered separately. The interfacial transports tend to adjust the system, at least partially, to accommodate the change in pressures. The

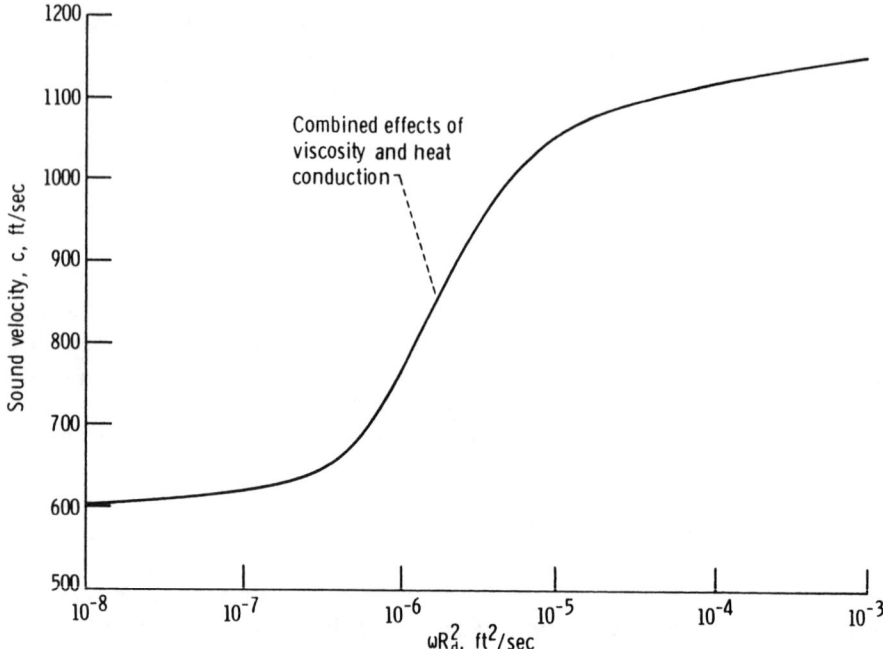

Figure 3.65 Sound velocity in hydrogen mixture against wR_d^2, at atmospheric pressure. Critical quality $X_C = 0.375$. (From Clinch and Karplus, 1964. Reprinted with permission of NASA Science and Technical Information, Linthicum Heights, MD.)

time available for this accommodation should be associated with the characteristic time L/u. Thus a longer channel should be closer to equilibrium situation than a short channel (Fig. 3.66) (Fauske, 1965). It is difficult to assign a correct set of local properties and parameters, such as local quality, local slip ratio, etc., at the wave front to make the critical discharge flow rate equal to the local pressure propagation rate but of opposite direction. Furthermore, for a two-phase flow, the choke condition may not always be met at the throat.

As to the relation between the sonic velocity and the propagation velocity of the single pressure pulse, the imposed pressure disturbances are different—one being a continuous wave and another being a single impulse. The difference is analogous to the steady periodic heating of a block by a cyclical change in surface temperature as compared with momentarily changing the surface temperature to a new level. The responses to these two kinds of disturbance are different, both in depth of penetration and in lag time. As mentioned before, two-phase flow involves the time required for the interfacial transport to respond to the changing boundary condition. It is therefore easier to define the relaxation time for a sonic wave propagation, τ_D and τ_t. When the single pressure pulse is imposed, the shape of the pulse

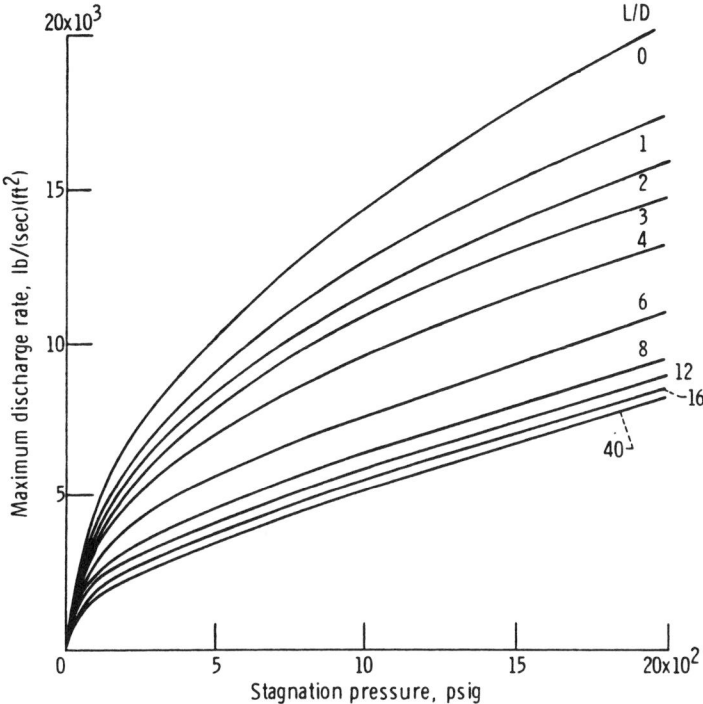

Figure 3.66 Maximum discharge rates of saturated water for 0.25-in.-I.D. tube. (From Fauske, 1965. Copyright © 1965 by American Institute of Chemical Engineers, New York. Reprinted with permission.)

should really be included in considering the extent of nonequilibrium involved. Ruggles et al. (1989), based on data on the propagation of pressure perturbations in bubbly air–water flows from their work at Rensselaer Polytechnic Institute (Cheng et al., 1983, 1985; Ruggles, 1987; Ruggles et al., 1988) established such a relationship by a two-fluid model and through the use of the Fourier decomposition techniques. The relationship between the celerity deduced from the linear dispersion relation and the critical flow velocity was demonstrated and shown to be unique when a simple quasi-static interfacial pressure model was used, such that

$$\delta p_G = \delta p_L \, (w < 0.5 w_r, R_b > 0.5 \text{mm})$$

where the expression in parentheses indicates the conditions for which the perturbations of gas pressure and liquid pressure are approximately equal, w_r being the bubble resonance frequency, and R_b, the equilibrium bubble radius.

3.7 ADDITIONAL REFERENCES FOR FURTHER STUDY

Additional references are given here on recent research work related to the subject of this chapter, which are recommended for further study.

For phenomena in horizontal two-phase flow, the current state of knowledge was summarized in detail by Hewitt in an invited lecture at the Tsukaba International Conference (Hewitt, 1991). Spedding and Spence (1993) determined flow patterns for co-current air–water horizontal flow in two different pipes of different diameters. They indicated that several of the existing flow regime maps did not predict correctly the flow regimes for the two diameters (9.35 and 4.54 cm, or 3.7 and 1.8-in.), and some theoretical and empirical models were deficient in handling changes in physical properties and geometry. Thus, a need was shown to develop a more satisfactory method of predicting phase transition. Jepson and Taylor (1993) compiled a flow regime map for the air–water two-phase flow in a 30-cm (11.8-in.)-diameter pipeline, for which the transitions differ substantially from those for small-diameter pipes and are not predicted accurately by any theoretical model so far. Several changes in the distribution of the phases in large-diameter pipes are reported and are especially prominent in slug and annular flow. Crowley et al. (1992) suggested a one-dimensional wave model for the stratified-to-slug or -annular flow regime transition. The authors provided a complete set of equations and a solution methodology. They subsequently presented their one-dimensional wave theory and results in dimensionless form (Crowley et al., 1993). A typical design map was illustrated of the transition in dimensionless form for a fixed pipe inclination ($\theta = 0$ in this case), constant wall–liquid and wall–gas friction factors, with both phases in turbulent flow. The dimensionless variables derived in this analysis were liquid-phase Froude number, gas-phase Froude number, liquid–gas density ratio, and interfacial friction factor ratio. Binder and Hanratty (1992) also developed an analytical framework for defining the dimensionless groups that control the degree of stratification of particles and rate of deposition. Because of the approximations made in the analysis, exact quantitative agreement with measurements cannot be made. Further calculation should explore the effects of ignoring the influences of flow nonhomogeneities and errors associated with turbulence effects.

For vertical two-phase flow, Govan et al. (1991) studied flooding and churn flow. Jayanti and Hewitt (1992) proposed an improved model for flooding that is in good agreement with experiments at both low and high liquid flow rates. Comparison was made to the flooding model of McQuillan and Whalley (1985), which gave satisfactory results at low liquid flow rates, and to the bubble entrainment model of Barnea and Brauner (1985), which yielded satisfactory results at high liquid flow rates. Additional work was reported by Lacy and Dukler (1994) on a flooding study in vertical tubes.

The annular flow model is useful for diabatic flow beyond critical heat flux (CHF). Hewitt and Govan (1990) introduced a model for the CHF state that is applicable in low quality and in the subcooled boiling regions, where vapor blanketing governs, as well as in the annular (film dryout) region. With reference to the CHF for boiling in a bottom-closed vertical tube, Katto (1994a) presented an analytical study on the limit conditions of steady-state countercurrent annular flow in a vertical tube. Serizawa et al. (1992) summarized the current understanding of dispersed flow. In a practical application, Fisher and Pearce (1993) used an annular flow model to predict the liquid carryover into the superheater of a steam generator. This model represented the steam/water flows in sufficient detail that the final disappearance of liquid film at the wall and the position of complete dryout can be located. It not only included the evaporation from the liquid films and entrained droplets but also took into account the thermal nonequilibrium caused by the presence of dry surfaces.

A new correlation by Hahne et al. (1993) for the pressure drop in subcooled flow boiling of refrigerants employed nondimensional parameters, e.g., density ratio, (ρ_L/ρ_G), and the ratio of heated and wetted perimeter, (D_h/D_w). A measurement strategy was used to eliminate the hydrostatic component by subtracting the two values of ΔP for upflow and downflow from each other. The resultant value, called reduced pressure drop, is the sum of accelerational, $(\Delta P)_{acc}$, and frictional $(\Delta P)_f$, components. Using the usual Lockhart and Martinelli definition (1949), the nondimensional pressure drop, ϕ^2, containing the reduced pressure drop $(\Delta P)_r$, and the single-phase ΔP for liquid flowing with the same mass flux, are in the form

$$\phi^2 = \frac{\left(\frac{\Delta P}{\Delta z}\right)_{r,TP}}{(\Delta P/\Delta z)_{Lo}} \tag{3-191}$$

$$= 1 + C(Bo)^m(Ja)^n \left(\frac{\rho_L}{\rho_G}\right)\left(\frac{D_h}{D_w}\right)$$

For water the constants were found to give

$$\phi^2 = 1 + 80(Bo)^{1.6}(Ja)^{-1.2}\left(\frac{\rho_L}{\rho_G}\right)\left(\frac{D_h}{D_w}\right)$$

For R_{12} and R_{134} the expression becomes

$$\phi^2 = 1 + 500(Bo)^{1.6}(Ja)^{-1.2}\left(\frac{\rho_L}{\rho_G}\right)\left(\frac{D_h}{D_w}\right)$$

Data for extended ranges of parameters should be used to increase the range of validity of this form of correlation.

The understanding of heat and mass transfer phenomena occurring during critical flow of two-phase mixtures is important in the safety analysis of PWRs, BWRs, and LMRs (liquid metal-cooled reactors). Flow limiting phenomena were discussed by Yadigaroglu and Andreani (1989) in LWR safety analysis. Ellias and Lellouche (1994) reviewed two-phase critical flow from the viewpoint of the needs of thermal-hydraulic systems codes and conducted a systematic evaluation of the existing data and theoretical models to quantify the validity of several of the more widely used critical flow models. This will enhance the understanding of the predictive capabilities and limitations of the critical flow models currently used in the power industry.

CHAPTER
FOUR
FLOW BOILING

4.1 INTRODUCTION

Flow boiling is distinguished from pool boiling by the presence of fluid flow caused by natural circulation in a loop or forced by an external pump. In both systems, when operating at steady state, the flow appears to be forced; no distinction will be made between them, since only the flow pattern and the heat transfer are of interest in this section.

To aid in visualizing the various regimes of heat transfer in flow boiling, let us consider the upward flow of a liquid in a vertical channel with heated walls. When the heat flux from the heating surfaces is increased above a certain value, the convective heat transfer is not strong enough to prevent the wall temperature from rising above the saturation temperature of the coolant. The elevated wall temperature superheats the liquid in contact with the wall and activates the nucleation sites, generating bubbles to produce incipience of boiling. At first, nucleation occurs only in patches along the heated surfaces, while forced convection persists in between. This regime is termed *partial nucleate boiling*. As the heat flux is increased, more nucleation sites are activated and the number of boiling surfaces increases until *fully developed nucleate boiling,* when all surfaces are in the nucleate boiling stage. Any further increase in heat flux activates more nucleation sites until the critical flux is reached, as was discussed under pool boiling. A typical relationship between heat flux and bubble population (the product of frequency and nucleation sites) in flow boiling is shown in Figure 4.1. Beyond critical heat flux, an unstable region of heat transfer, termed *partial film boiling* or *transition boiling,* occurs. This is gradually converted to *stable film boiling* as the surface temperature increases above the Leidenfrost point. The mode of heat transfer and the flow pattern are

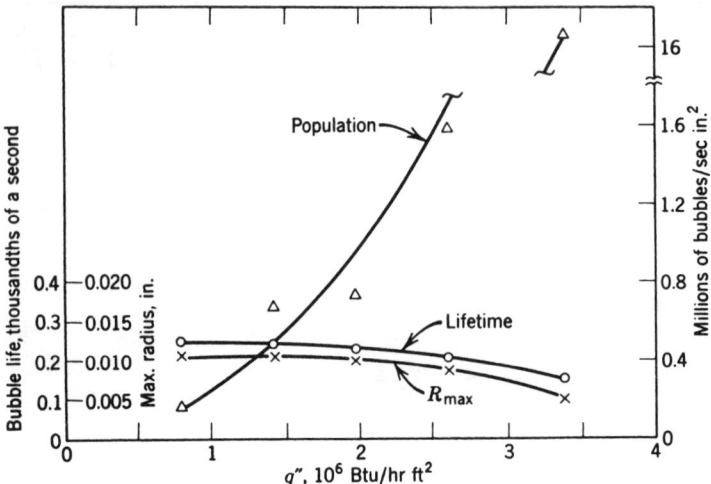

Figure 4.1 Bubble histories for forced-convection subcooled boiling. (From Gunther, 1951. Copyright © 1951 by American Society of Mechanical Engineers, New York. Reprinted with permission.)

intimately related, so a change in one leads to a corresponding change in the other. Figure 4.2 shows the various flow patterns encountered over the length of the vertical tube, together with the corresponding heat transfer regions (Collier, 1981). Region *B* signifies the initiation of vapor formation in the presence of subcooled liquid; the heat transfer mechanism is called *subcooled nucleate boiling*. In this region, the wall temperature remains essentially constant a few degrees above the saturation temperature, while the mean bulk fluid temperature is increasing to the saturation temperature. The amount by which the wall temperature exceeds the saturation temperature is called the *degree of superheat*, ΔT_{sat}, and the difference between the saturation and local bulk fluid temperature is the *degree of subcooling*, ΔT_{sub}. The transition between regions *B* and *C*, from subcooled nucleate boiling to saturated nucleate boiling, is clearly defined from a thermodynamic viewpoint, where the liquid reaches the saturation temperature ($x = 0$). However, as shown on the left side of the figure, before the liquid mixed mean (liquid core) temperature reaches the saturation temperature, vapor is seen to form as a result of the radial temperature profile in the liquid. In this case, subcooled liquid can persist in the liquid core even in the region defined as *saturated nucleate boiling*. In other conditions, vapor formation may not occur at the wall until after the mean liquid temperature has exceeded the saturation temperature (as in the case of liquid metals). Vapor bubbles growing from wall sites detach to form a bubbly flow. With the production of more vapor, the bubble population increases with length, and coalescence takes place to form slug flow and then gives way to annular flow farther along the channel (regions *D* and *E*). Close to this point the formation of vapor at wall sites may cease, and further vapor formation will be a result of evaporation

FLOW BOILING 247

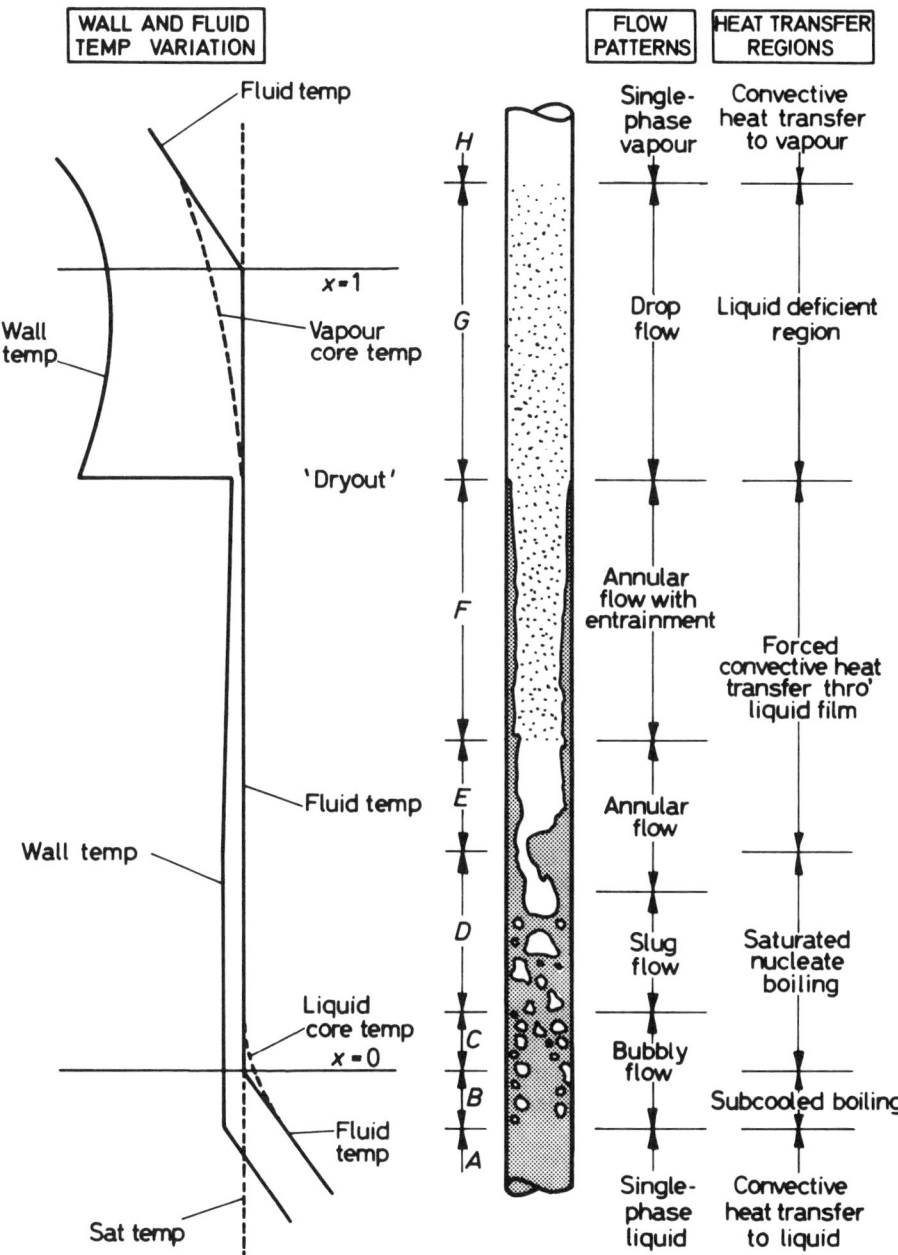

Figure 4.2 Regions of heat transfer in convective boiling. (From Collier and Thome, 1994. Copyright © 1994 by Oxford University Press, New York. Reprinted with permission.)

at the liquid film–vapor core interface. Increasing velocities in the vapor core will cause entrainment of liquid in the form of droplets (region F). Since nucleation is completely suppressed, the heat transfer process becomes that of two-phase forced convection and evaporation. The depletion of the liquid from the film by entrainment and by evaporation finally causes the film to dry out completely (dryout point). Droplets continue to exist in region G (liquid-deficient region), and the corresponding flow pattern is called *drop flow*. Drops in this region, which is shown as region H, are slowly evaporated until only single-phase vapor is present.

4.2 NUCLEATE BOILING IN FLOW

As in nucleate pool boiling, heat is transferred from the heated surface to the liquid by several mechanisms:

1. Heat transport by the latent heat of bubbles, q''_{b1}
2. Heat transport by continuous evaporation at the root of the bubble and condensation at the top of the bubble, while the bubble is still attached to the wall, q''_{b2} (microlayer evaporation)
3. Heat transfer by liquid–vapor exchange caused by bubble agitation of the boundary layer, $q''_{b.l.}$ (microconvection)
4. Heat transfer by single-phase convection between patches of bubbles $q''_{f.c.}$

In the study and analysis of the flow boiling process, the problem is to identify the contribution of each mechanism in the various regimes of nucleate flow boiling (Sec. 4.1).

4.2.1 Subcooled Nucleate Flow Boiling

4.2.1.1 Partial nucleate flow boiling.
The transition from forced convection to nucleate boiling, constituting the regime of partial nucleate boiling, is shown in Figure 4.3. The heat flux at the incipience of boiling, which is on the forced-convection line, is defined as $q''_{f.c.}$ (or q''_{conv} as shown in the figure). Several methods of determining the fully developed nucleate boiling, q''_{FDB}, have been suggested using saturated pool boiling data (McAdams et al., 1949; Kutateladze, 1961). Figure 4.3 illustrates the method suggested by Forster and Grief (1959),

$$q''_{FDB} = 1.4 q''_o$$

where q''_o is located at the intersection of the forced-convection and pool boiling curves. The boiling curve in the transition region then becomes a straight line connecting q''_{conv} at the incipient boiling point and q''_{FDB}. By testing the heat transfer of flow boiling and pool boiling on stainless steel tubes cut from the same stock to assure similar surface conditions, Bergles and Rohsenow (1964) found that the

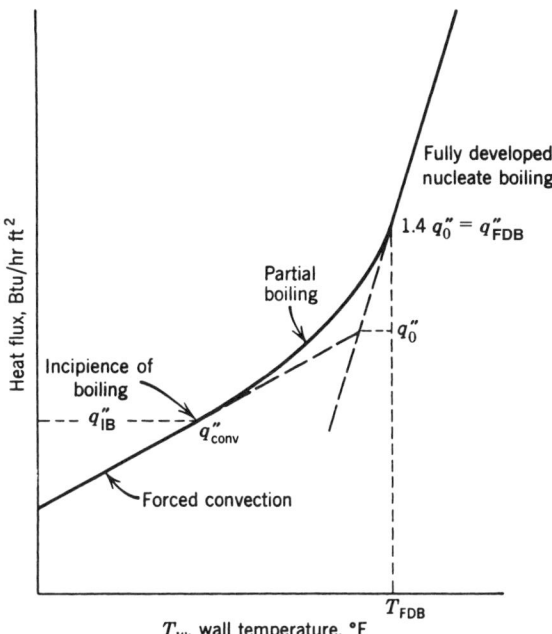

Figure 4.3 Boiling curve for partial nucleate boiling.

fluid mechanics of flow boiling is different from that of saturated pool boiling, because the degree of subcooling influences pool boiling strongly. This observation led to the conclusion that the curves for flow boiling should be based on actual flow boiling data. Bergles and Rohsenow's data for flow boiling and pool boiling are shown in Figure 4.4. They suggested the following simple interpolation formula for the boiling curve in the transition region:

$$\frac{q''}{q''_{f.c.}} = \left\{ 1 + \left[\left(\frac{q''_B}{q''_{f.c.}} \right) \left(1 - \frac{q''_{Bi}}{q''_B} \right) \right]^2 \right\}^{1/2} \tag{4-1}$$

where q''_B can be calculated from fully developed boiling correlations at various wall temperatures, and q''_{Bi} is the fully developed boiling heat flux at T_{LB} of incipient local boiling,* as shown in the figure.

Partial nucleate flow boiling of ordinary liquids Bergles and Rohsenow (1964), using data obtained from several commercially finished surfaces, have developed a criterion for the incipience of subcooled nucleate boiling by solving graphically the

---
* Subcooled nucleate boiling is frequently called *local boiling* or *surface boiling*.

250 BOILING HEAT TRANSFER AND TWO-PHASE FLOW

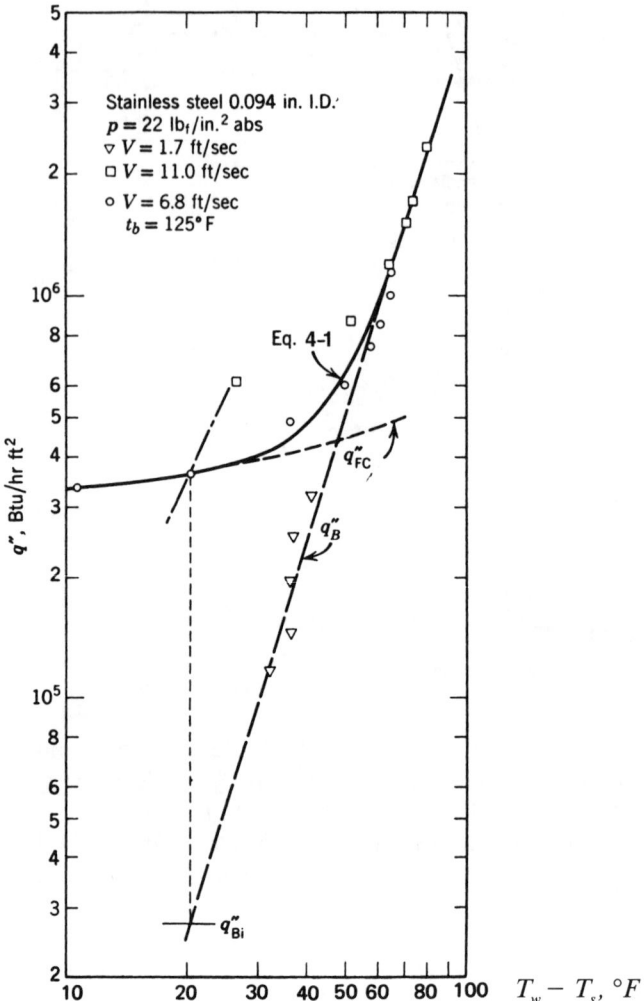

Figure 4.4 Procedure for construction of curve for partial nucleate boiling. (From Bergles and Rohsenow, 1964. Copyright © 1964 by American Society of Mechanical Engineers, New York. Reprinted with permission.)

bubble growth equation and the liquid temperature profile as postulated by Hsu (1962), who stated that bubble nuclei on cavities in the heated wall will grow only if the lowest temperature on the bubble surface is greater than the required wall superheat from the equilibrium equation

$$q''_{IB} = 15.60(p)^{1.156}(T_w - T_{sat})^{2.3/(p)^{0.0234}} \qquad (4\text{-}2)$$

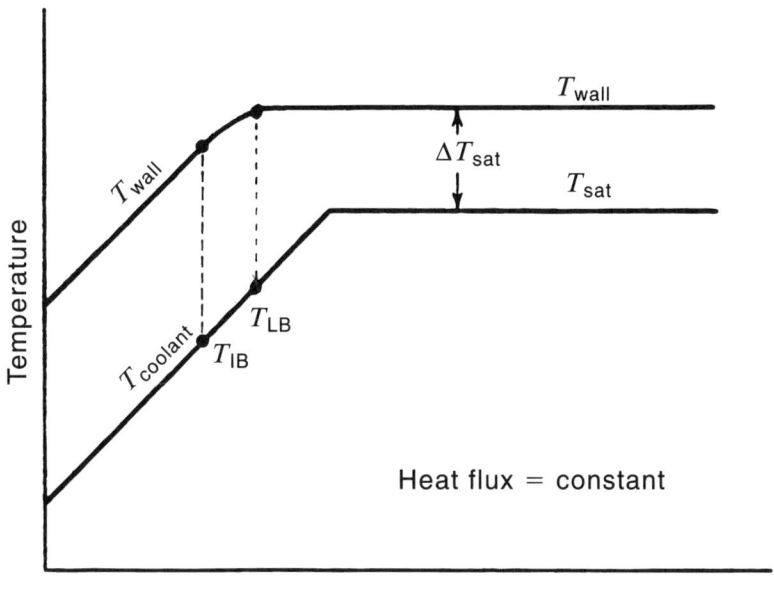

Figure 4.5 Wall and coolant temperatures in flow boiling.

where q''_{IB} is the incipient local boiling heat flux, in Btu/hr ft², p is in psia, and T is in °F. Davis and Anderson (1966) carried out an analytical solution resulting in an expression for q''_{IB}:

$$q''_{IB} = \left(\frac{k_L H_{fg} \rho_G}{8\sigma T_{sat}}\right)(T_w - T_{sat})^2 \qquad (4\text{-}3)$$

Equations (4-2) and (4-3) are in good agreement and adequately predict the onset of nucleation in Bergles and Rohsenow's experiments (1964). Equation (4-3) assumes that a sufficiently wide range of active cavity sizes is available. Otherwise, an estimate of the largest active cavity size available on the heating surface must be made. Bergles and Rohsenow found reasonable agreement with data for water and benzene using a maximum active cavity size of 1 μm radius.

The relationship between the wall temperature and the coolant temperature can be seen in Figure 4.5. The wall temperature starts to bend at the incipience of subcooled boiling, where the coolant temperature is defined as T_{IB}. The wall temperature follows a curve of partial boiling and then reaches an approximately constant value at a fully developed nucleate boiling where the coolant temperature

252 BOILING HEAT TRANSFER AND TWO-PHASE FLOW

is defined as T_{LB}. Treshchev (Borishansky and Paleev, 1964) suggested a correlation for the prediction of incipient, fully developed nucleate boiling water, q''_{LB} or q''_{FDB},

$$q''_{LB} = (1.04 \times 10^3)(\Delta T_{sub})V^{0.8} \text{ Btu/hr ft}^2 \tag{4-4}$$

where ΔT_{sub} is in °F and V is in ft/sec. Based on the comparison of the predicted values using the above correlations and several others in the literature with the experimentally measured values for water in a range of system pressures of 715–2,145 psia (4.9–14.6 MPa) and mass fluxes of 0.94–1.44 × 10⁶ lb/hr ft² (4.57–7.0 × 10⁶ kg/hr m²), Eq. (4-3) is recommended for predicting the incipience of partial nucleate boiling, and Eq. (4-4) for predicting fully developed nucleate boiling.

Partial nucleate flow boiling of liquid metals For liquid metals, the situation becomes complicated because of their higher thermal conductivities and consequently much less steep temperature profiles compared to ordinary liquids at a given heat flux. For instance, for potassium at 1 atm and $q'' = 100,000$ Btu/hr ft² (314 kW/m²), the minimum cavity radius, $r_{c,m}$, is $> 4 \times 10^{-3}$ in. (0.1 mm), with a corresponding wall superheat, ΔT_{sat}, of 4°F (2.2°C). These larger cavities are much more likely to be flooded, especially with well-wetting fluids such as the alkali metals. It becomes questionable whether any nonflooded cavities of radius $r_{c,m}$ would exist to initiate boiling at the predicted wall superheat (Chen, 1968). This lack of potential sites was borne out by experimental evidence that indicated incipient superheats of as much as 100°F (55°C) (Chen, 1970), which are higher than predicted by Hsu's criterion. Extending the "equivalent cavity" model of Holtz (1966), Chen (1968) suggested a deactivation model by which the incipient boiling superheat can be estimated as a function of deactivation conditions (P' and T') and the boiling pressure, p. In addition, Chen reported the parametric effects of the flow rate or velocity, V, while maintaining the other variables (i.e., heat flux q'', oxide content in the liquid metal, gas content in the liquid as well as in nucleation cavities, and nucleation cavity sizes and geometry) reasonably constant. According to the concept of cavity deactivation, the superheat, ΔT_{sat}, that is required on a surface having a multitude of cavities with a wide range of cavity sizes and geometries, at a pressure P, is dependent on the most stringent deactivation conditions (P' and T') to which the liquid-filled system has previously been subjected. The analysis is based on idealized cavities of conical shape as shown in Figure 4.6, where conditions are illustrated at initial filling, during preboiling operations when cavities are partially flooded (deactivated), and at incipient vaporization.

$$\Delta P_{sat} = P_v - P = \left(\frac{\sigma}{\sigma'}\right)\left(P' - P'_v\right) - \left(\frac{G_o}{r_d^3}\right)\left(T + \frac{T'\sigma}{\sigma'}\right) \tag{4-5}$$

$$r_d^3 - r_d^2 \left(\frac{2\sigma'}{P' - P'_v} - \frac{G_o T'}{P' - P'_v} \right) = 0 \qquad (4\text{-}6)$$

$$\Delta T_{sat} = T - T_{sat} \qquad (4\text{-}7)$$

where P is system pressure at boiling and P_v is the vapor pressure in the bubble corresponding to the superheat temperature, ΔT_{sat}; G_o is an empirical parameter that accounts for gas partial pressure in the nucleating cavity. These equations suppose that flow effects are absent ($V = 0$) and are applied to those cavities with minimum θ ($\theta \to 0$).

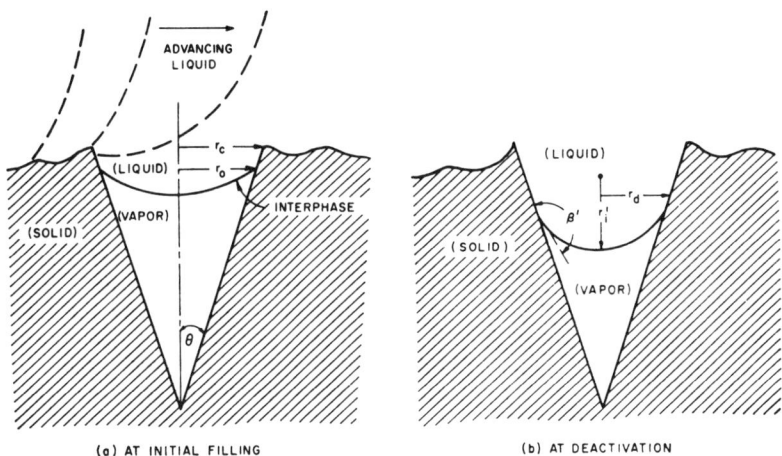

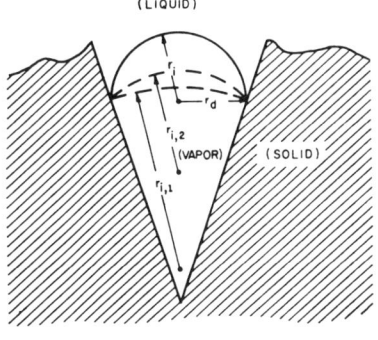

Figure 4.6 Deactivation model for idealized cavity. (From Chen, 1968. Copyright © 1968 by American Society of Mechanical Engineers, New York. Reprinted with permission.)

254 BOILING HEAT TRANSFER AND TWO-PHASE FLOW

The effects on ΔT_{sat} of deactivation pressure P' (at 100°F or 55°C subcooling), of the boiling pressure P [at P' 24.7 psia (0.17 MPa) and $T' = 1377°F$ (747°C)], and of the flow velocity V are shown in Figures 4.7, 4.8, and 4.9, respectively. These experimental data were obtained in a 0.622-in. (1.58-cm)-I.D. stainless steel pipe, electrically heated via direct Joule heating with a 10-ft (3.0-m)-long preheater (Chen, 1970). Additional experiments (Logan et al., 1970; Dwyer et al., 1973a, 1973b; France et al., 1974) included such important parametric effects on incipient boiling superheat as the P–T history, the dissolved inert gas, and the location of boiling inception. Of importance is the last reference to the application of liquid metal-cooled reactors, because the nature of coolant (sodium) boiling subsequent to an abnormal reactor incident is of concern in safety analysis, and the formation of voids in the nuclear reactor core and the physical phenomenon is most heavily influenced by the magnitude of the liquid superheat in the system. A brief description of experiments by France et al. (1974) is given here.

Sodium superheat experiments were performed in a forced-convection facility employing system parameters in the range of interest for application to loop- and pot-type liquid metal-cooled fast breeder reactors (LMFBRs). The test section was

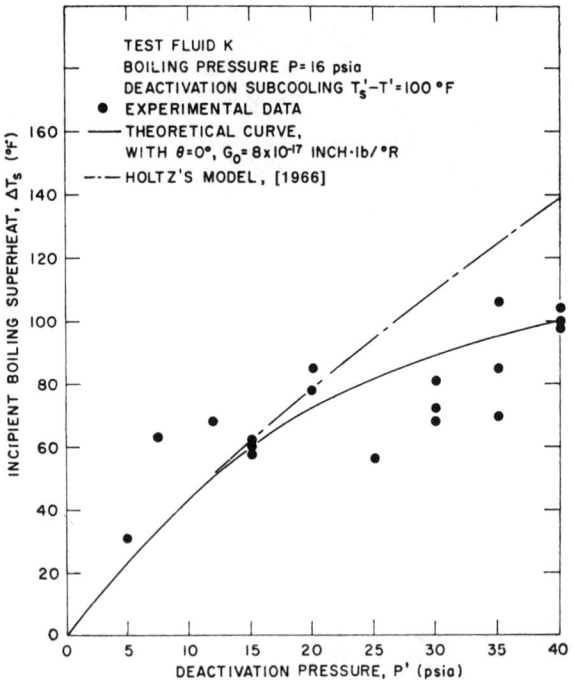

Figure 4.7 Incipient boiling superheats after deactivations at 100°F subcooling. (From Chen, 1968. Copyright © 1968 by American Society of Mechanical Engineers, New York. Reprinted with permission.)

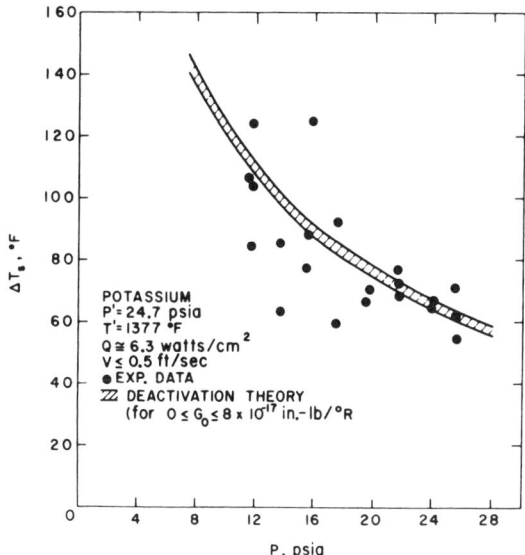

Figure 4.8 Variation of incipient boiling superheat with increasing boiling pressure. (From Chen, 1970. Copyright © 1970 by American Society of Mechanical Engineers, New York. Reprinted with permission.)

representative of a single reactor fuel element, with sodium flowing vertically upward in an annulus, heated indirectly from the inside wall (0.23 in. or 0.58 cm in diameter) only. Steady-state operating parameters prior to a flow coast down approach to boiling included: velocity of 17 ft/sec (6.7 m/s), heat flux of 7×10^5 Btu/hr ft² (2.2 kW/m²), test section inlet temperature of 600°F (315°C), outlet to plenum 900°F (482°C) or 700°F (371°C), and plenum gas pressure of 15 psia (0.1 MPa). Another system variable controlled was sodium inert gas content, by the length of duration in steady-state operation to reestablish the system pressure–temperature history.

Five series of tests were carried out, with the first three representing cases of real reactor operation. Series A simulated a LMFBR loop-type system operating at steady state for a period of time long enough to establish mass equilibrium of inert gas, and boiling inception following a system transient (flow coast down). Test series B was performed with the primary objective of obtaining a high test section superheat at boiling inception. This series was conducted with all parameters identical to those of test series A except that tests were run with only 0.8 hr or 1.5 hr of steady-state loop operation subsequent to sodium fill, in contrast to 30.5 hr or over 100 hr prior to test in series A. This difference in procedure affected the inert gas content of the sodium, and series B was performed with far less argon dissolved in the sodium. Series C was performed utilizing a low plenum temperature (700°F or 371°C), which represented pot-type reactor operational conditions.

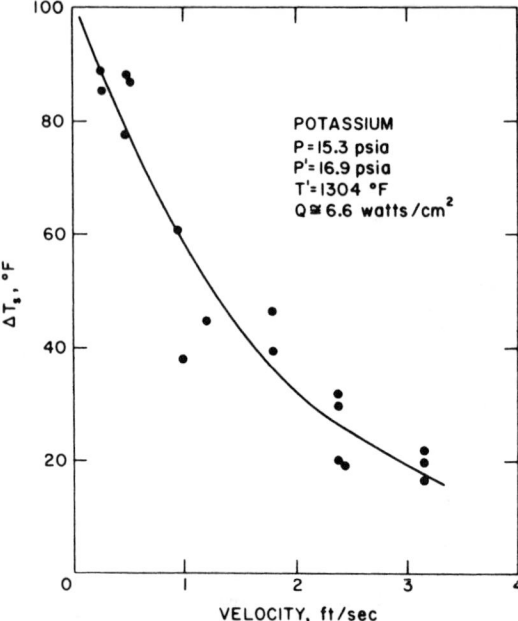

Figure 4.9 Variation of incipient boiling superheat with flow velocity, at near-atmospheric pressure. (From Chen, 1970. Copyright © 1970 by American Society of Mechanical Engineers, New York. Reprinted with permission.)

The last two series of tests, D and E, repeated series A and C, respectively, in all conditions. After series D tests were run, the loop was dumped and refilled after several days before test series E was initiated. The test results demonstrated that in series A, a typical LMFBR loop-type system operating at steady state for a period of time, the measured superheat was zero, which was consistent with previous predictions (Holtz et al., 1971) and attributable to inert gas in the system. Although extension of the experiments to include variations in system parameters but still within the range of interest to LMFBRs were not demonstrated, such experiments would be expected to yield the same zero superheat result due to the inert gas (France et al., 1974). Because far less argon gas was dissolved in the sodium than in series A, no inert gas bubbles were observed in series B. As mentioned before, nucleation in test series B was believed to have occurred from test section wall cavities with a low inert gas pressure. The maximum test section superheats of 120°F (67°C) and 150°F (83°C) were found in the two tests of series B, respectively, with corresponding incipient boiling superheats of 0°F (0°C) and 60°F (33°C). The superheat results of tests in series C (simulating a pot-type LMFBR having a sodium-argon interface several hundred degrees Fahrenheit lower than the maximum system temperature) show moderate incipient boiling superheat (20°F or 11°C) in one test and zero incipient boiling superheat (but 100°F, or 56°C, maximum test section superheat at boiling inception) in the other test. The test result of moderate incipient superheat was confirmed by the similar test in series E. These data also compared well with theoretical predictions including the effect of inert gas on the wall cavity nucleation sites (France et al., 1974).

4.2.1.2 Fully developed nucleate flow boiling. In fully developed nucleate flow boiling, the heat flux is affected by pressure and wall temperature but not by flow velocity. The difference between the wall temperature and the coolant temperature has been found to be constant for a given p and q'' by several investigators. For a relatively clean surface, this difference is (Thom et al., 1966)

$$\Delta T_{sat} = T_w - T_{sat} = 0.072 \exp\left(\frac{-p}{1,260}\right) q''^{0.5} \; °F \tag{4-8}$$

where the system pressure, p, is in psia and q'' is in Btu/hr ft².

From Figure 4.5, the value of T_{LB} can be calculated as

$$T_{LB} = T_{sat} + \Delta T_{sat} - \left(\frac{q''}{h_{conv}}\right) \tag{4-9}$$

where

$$h_{conv} = 0.023\left(\frac{k_L}{D_e}\right)(Re_L)^{0.8}(Pr_L)^{0.4} \tag{4-10}$$

Brown (1967) noted that a vapor bubble in a temperature gradient is subjected to a variation of surface tension which tends to move the interfacial liquid film. This motion, in turn, drags with it adjacent warm liquid so as to produce a net flow around the bubble from the hot to the cold region, which is released as a jet in the wake of the bubble (Fig. 4.10). Brown suggested that this mechanism, called thermocapillarity, can transfer a considerable fraction of the heat flux, and it appears to explain a number of observations about the bubble boundary layer, including the fact that the mean temperature in the boundary layer is lower than saturation (Jiji and Clark, 1964).

For low-heat-flux boiling, such as on the shell side of a steam generator as used in PWRs, Elrod et al. (1967) reported heat transfer data at pressures of 535–1,550 psia (3.6–10.5 MPa). It is feasible to apply their data directly to the design of steam generators, as their results were tested at parameters common to this design.

For boiling in rod bundles, the heat transfer coefficient in the subcooled flow boiling of water was measured from 7- and 19-rod bundles by Kor'kev and Barulin [1966] in the following parameter ranges: $p = 1,400$ psia (9.5 MPa); $V = 1.3$–10 ft/sec (0.5–3.9 m/s); $X = -0.4$ to 0.0. Their data yielded the following empirical correlation:

258 BOILING HEAT TRANSFER AND TWO-PHASE FLOW

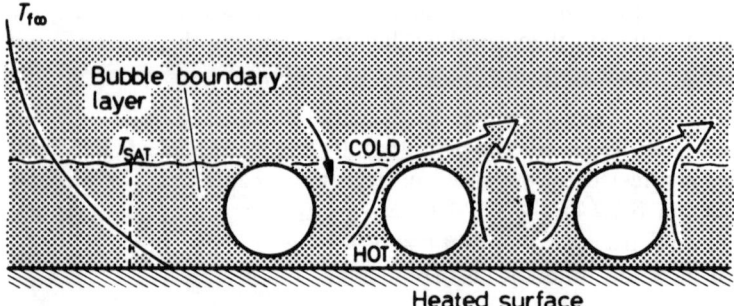

Figure 4.10 Thermocapillarity mechanism of subcooled boiling. (From Brown, 1967. Reprinted with permission of Massachusetts Institute of Technology, Cambridge, MA.)

$$h = (60 - 0.085T_{sat})^{-1} \times 10^6 \left(\frac{q''}{10^6}\right)^{0.7} \text{ Btu/hr ft}^2 \text{ °F} \qquad (4\text{-}11)$$

where h is in Btu/hr ft² °F, q'' is in Btu/hr ft², and T_{sat} is in °F.

4.2.2 Saturated Nucleate Flow Boiling

As shown in Figure 4.2 (Sec. 4.1), saturated nucleate flow boiling covers regions C and D, where nucleate boiling is occurring at the wall and where the flow pattern is typically bubbly, slug, or low-vapor-velocity annular flow. For most practical purposes, the assumption of thermodynamic equilibrium between phases is often used except in cases of small values of reduced pressure, P_r, and cases of boiling liquid metals. Figure 4.11 shows the various regions of two-phase heat transfer in forced-convection boiling on a plot of heat flux versus mass quality, X, with negative values of X representing subcooled liquid. Lines marked (i), (ii), etc., represent constant heat flux lines of increasing value. Thus, as the flow system is heated through a constant flux of (ii), it will follow, in the direction of flow, the path from single-phase forced-convective region A to subcooled boiling region B; as the liquid temperature reaches the saturation temperature of $X = 0$, it enters the saturated nucleate boiling regions C and D. With further increase of the quality, X, the system enters the two-phase forced-convective heat transfer regions E and F, corresponding to the same regions as shown in Figure 4.2. Also shown in Figure 4.11 are lines of DNB (departure from nucleate boiling) and dryout, where the critical heat flux (CHF) has been exceeded. As defined earlier in this book, if the initial condition before reaching CHF is one of nucleation in the subcooled or low-mass-quality region, the transition is called departure from nucleate boiling. This can occur in either the subcooled region or the saturated nucleate boiling region, and the resulting mechanism is one of film boiling as shown in the figure. If the initial condi-

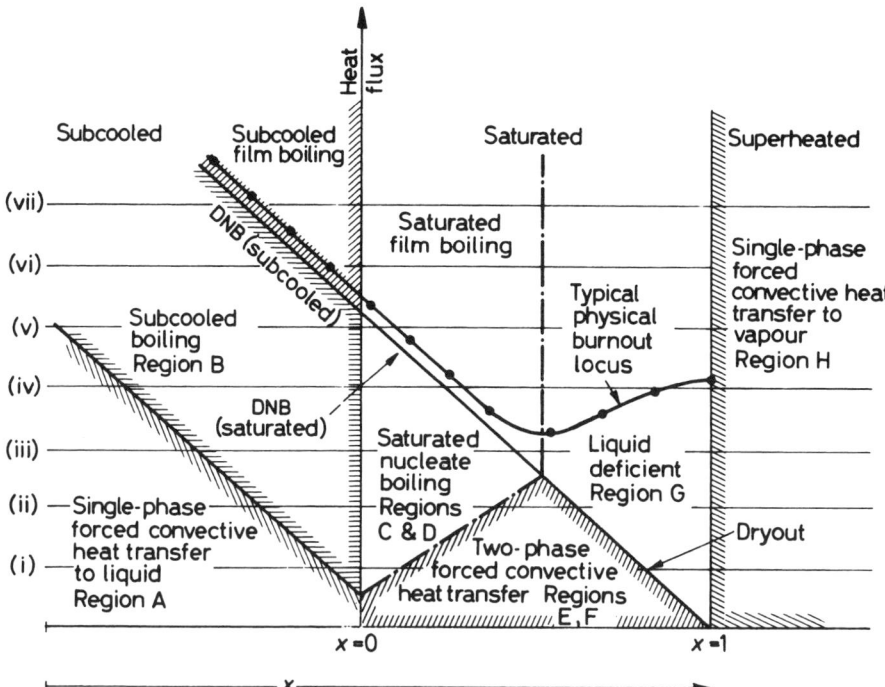

Figure 4.11 Regions of two-phase forced-convective heat transfer as a function of quality with increasing heat flux as ordinate. (From Collier and Thome, 1994. Copyright © 1994 by Oxford University Press, New York. Reprinted with permission.)

tion is one of evaporation at the liquid film–vapor core interface, then in higher-mass-quality areas the liquid-deficient region is entered and the transition is called dryout. Region H, on the right of the line $X = 1$, represents the presence of a single, vapor phase, and the mechanism becomes forced-convective heat transfer to vapor. The thermodynamic boundary line $X = 0$ marks the onset of saturated boiling through which the coolant temperature, heating wall temperature, and heat transfer coefficient variations are smooth and continuous. The equations used to correlate experimental data in the subcooled regions remain valid in this region provided the coolant temperature equals T_{sat}. Just as the heat transfer mechanism in the subcooled region is independent of the degree of subcooling and essentially of the mass flux, the heat transfer process in this region is also independent of the mass quality, X, and the mass flux, G.

4.2.2.1 Saturated nucleate flow boiling of ordinary liquids. To maintain nucleate boiling on the surface, it is necessary that the wall temperature exceed a critical value for a specified heat flux. The stability of nucleate boiling in the presence of a temperature gradient, as discussed in Section 4.2.1.1, is also valid for the suppression of nucleate boiling with decreasing wall superheat in the two-phase region. In other words, if the wall superheat is less than that given in Eq. (4-2) or (4-3) for the imposed surface heat flux, then nucleation does not take place. The value of ΔT_{sat} in these equations can be calculated from the ratio (q''_{IB}/h_{tp}) where h_{tp} is the two-phase heat transfer coefficient in the absence of nucleation. By assuming that all the temperature drop occurred across the boundary sublayer and using the Dengler and Addoms (1956) equation for h_{tp}, Eq. (4-3) becomes

$$q''_{IB} = \frac{2\sigma T_{sat}}{h_{fg} \rho_G} \left(\frac{49(h_{Lo})^2}{k_L X_{tt}} \right) \qquad (4\text{-}12)$$

based on a complete range of active cavities on the heating surface. In this equation, h_{Lo} is the heat transfer coefficient for the total flow, assumed to be all liquid phase, and X_{tt} is the Martinelli parameter, defined in Chapter 3 as

$$X_{tt} = \frac{(dp/dz)_L}{(dp/dz)_G} = \left(\frac{1-X}{X} \right)^{0.9} \left(\frac{\rho_G}{\rho_L} \right)^{0.5} \left(\frac{\mu_L}{\mu_G} \right)^{0.1} \qquad (4\text{-}13)$$

When the active cavity size spectrum is limited, the treatment of Davis and Anderson (1966) (Sec. 4.2.1.1) should be employed here as well.

A number of relationships for h_{tp} and h_{Lo} have been proposed and in some cases extended to cover the saturated nucleate boiling region. Chen (1963) has carried out a comparison of then-available correlations along with those of Dengler and Addoms (1956) using a representative selection of 594 experimental data points from various investigators. The result was that none of the examined correlations could be considered satisfactory. Chen therefore proposed a new correlation, which proved very successful in correlating all the above-mentioned data for water and organic systems. His correlation covers both the saturated nucleate boiling region and the two-phase forced-convection region, assuming that both of these mechanisms occur to some degree over the entire range of the correlation and that such contributions are additive. The assumption of superposition is similar to that used by Rohsenow (1953) in the partial boiling region for subcooled liquids,

$$h_{tp} = h_{NB} + h_{f.c.}$$

where h_{NB} and $h_{f.c.}$ are contributions due to nucleate boiling and forced convection, respectively. For the convection component $h_{f.c.}$, Chen suggested a Dittus-Boelter type of equation,

$$h_{f.c.} = 0.023 (\text{Re}_{tp})^{0.8} (\text{Pr}_{tp})^{0.4} \left(\frac{k_{tp}}{D} \right)$$

where the thermal conductivity (k_{tp}) and the Reynolds (Re_{tp}) and Prandtl (Pr_{tp}) numbers are effective values associated with the two-phase fluid. Since heat is effectively transferred through a liquid film in annular and dispersed flow, Chen argued that the liquid thermal conductivity, k_L, can be used for k_{tp}. The values of the Prandtl number for liquid and vapor are normally of the same magnitude, so it may be expected that the value of Pr_{tp} will also be close to Pr_L. A parameter F is defined such that

$$F = \left(\frac{Re_{tp}}{Re_L}\right)^{0.8} = \left[\frac{Re_{tp}\,\mu_L}{G(1-X)D}\right]^{0.8} \tag{4-14}$$

and the equation for $h_{f.c.}$ becomes

$$h_{f.c.} = 0.023 F \left[\frac{G(1-X)D}{\mu_L}\right]^{0.8} \left(\frac{\mu c_p}{k}\right)_L^{0.4} \left(\frac{k_L}{D}\right) \tag{4-15}$$

where F is the only unknown and may be expected to be a function of the Martinelli factor, X_{tt}, as indicated in Eq. (4-14). For the evaluation of the nucleate boiling component, h_{NB}, Forster and Zuber's (1955) analysis with pool boiling was used. However, the actual liquid superheat across the boundary layer is not constant but falls. The mean superheat of the fluid, ΔT_o, in which the bubble grows, is lower than the wall superheat ΔT_{sat}. The difference between these two values, which is small in the case of pool boiling and was neglected by Forster and Zuber, cannot be neglected in the forced-convection boiling case. Thus,

$$h_{NB} = 0.00122 \left[\frac{(k_L)^{0.79}(c_{pL})^{0.45}(\rho_L)^{0.49}}{(\sigma)^{0.5}(\mu_L)^{0.29}(H_{fg})^{0.24}(\rho_G)^{0.24}}\right](\Delta T_o)^{0.24}(\Delta P_o)^{0.75} \tag{4-16}$$

The ratio of the mean superheat, ΔT_o, to the wall superheat, ΔT_{sat}, is defined by Chen as a suppression factor, S. Thus,

$$S = \left(\frac{\Delta T_o}{\Delta T_{sat}}\right)^{0.99} = \left(\frac{\Delta T_o}{\Delta T_{sat}}\right)^{0.24}\left(\frac{\Delta P_o}{\Delta P_{sat}}\right)^{0.75} \tag{4-17}$$

and Eq. (4-16) becomes

$$h_{NB} = 0.00122 S \left[\frac{(k_L)^{0.79}(c_{pL})^{0.45}(\rho_L)^{0.49}}{(\sigma)^{0.5}(\mu_L)^{0.29}(H_{fg})^{0.24}(\rho_G)^{0.24}}\right](\Delta T_{sat})^{0.24}(\Delta P_{sat})^{0.75} \tag{4-18}$$

262 BOILING HEAT TRANSFER AND TWO-PHASE FLOW

The value of S ranges from zero to unity, as the flow varies from low velocity ($S \to 1$) to high velocity ($S \to 0$). Chen suggested that S can be represented as a function of the local two-phase Reynolds number, Re_{tp}, and the functions F and S were determined empirically from experimental data as shown in Figures 4.12 and 4.13. The final correlation is a combination of Eqs. (4-15) and (4-18), which agrees with the above-mentioned data within ±15%. This form of correlation is, at present, still the best available for the saturated forced-convective boiling regions and is recommended for use with most single-component, nonmetallic fluids (Cumo and Naviglio, 1988) with the exception of refrigerants, for which modifications can be used as will be discussed later. Although Chen's original derivation was for saturated boiling in the annular flow region, applications have been extended as the experimental data grew and modifications have been developed to cover subcooled boiling and nonaqueous fluids.

Several analytical studies have sought to extend the application of the basic method of Chen. For fluids of Prandtl number different from unity, Bennett and Chen (1980) extended the analysis by a modified Chilton-Colburn analogy to give

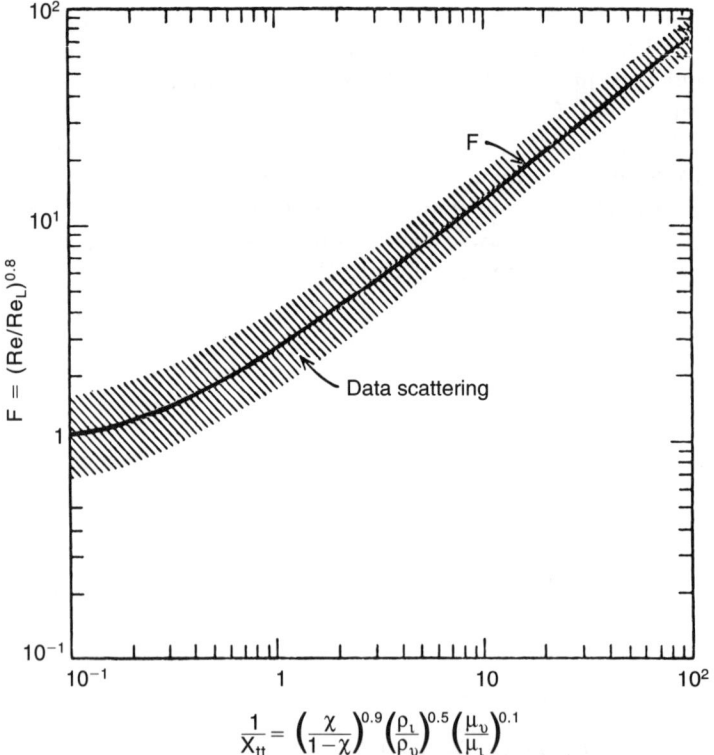

Figure 4.12 Forced-convection factor F [Eq. (4-14)]. (From Chen, 1966. Copyright © 1966 by American Chemical Society, Washington, DC. Reprinted with permission.)

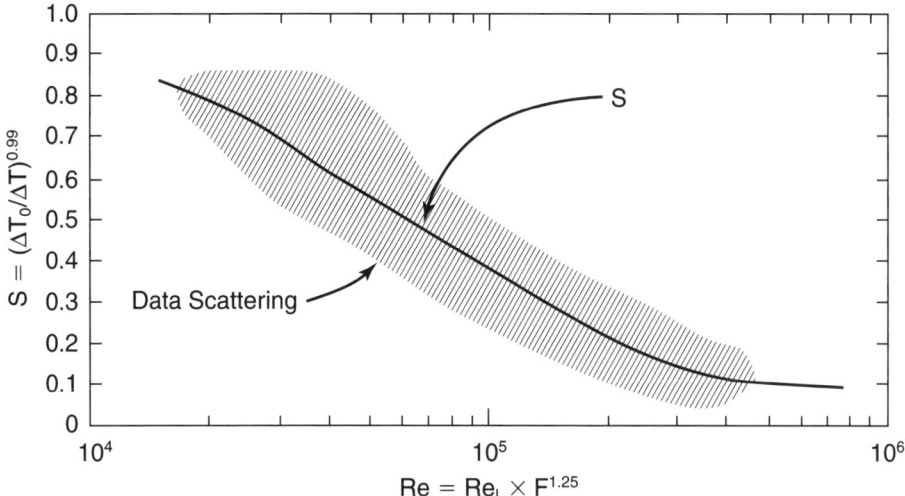

Figure 4.13 Boiling suppression factor S [Eq. (4-17)]. (From Chen, 1966. Copyright © 1966 by American Chemical Society, Washington, DC. Reprinted with permission.)

$$F = (\Pr)_L^{0.296}\left[\frac{(dp/dz)}{(dp/dz)_L}\right]^{0.444} \quad (4\text{-}19)$$

Another modification of the F factor was proposed by Sekoguchi et al. (1982) for both the subcooled and saturated boiling regions. From measured data over a wide range of parameters, they extracted values of q''_{NB} by subtracting $q''_{f.c.}$ from the total heat flux, q'', corresponding to Eq. (4-12). Here $q''_{f.c.}$ was obtained by using the Colburn correlation for h_{Lo} and the following relationship for F:

$$F = \frac{1-X}{1-\varepsilon}$$

where X is mass quality and ε is obtained from the Sekoguchi et al. (1980) derived correlation for the void fraction. The S factor on the nucleate boiling contribution was also closely examined by Bennett et al. (1980). They derived the factor analytically by postulating an exponential temperature profile over the heating surface and a bubble growth region whose thickness, δ, appeared to be independent of shear stress:

$$S = \left[\frac{k_L}{(h_{Lo})F\delta}\right]\left[1 - \exp\left(\frac{-h_{Lo}F\delta}{k_L}\right)\right] \quad (4\text{-}20)$$

where the bubble growth region (thermal boundary layer) thickness, δ, is expressed as

$$\delta = (0.041)\left[\frac{\sigma}{g(\rho_L - \rho_G)}\right]^{0.5} \qquad (4\text{-}21)$$

Modifications for boiling fluorocarbon refrigerant systems were suggested by Herd et al. (1983), whereby an improvement over the Chen's original correlation for refrigerants was obtained. Kandlikar (1983) also presented a correlation based on experimental data for water, R-11, R-12, R-113, R-114, cyclohexane, neon, and nitrogen in vertical as well as horizontal tubes. The author included the effect of Froude number on both convective and boiling terms in the case of horizontal flow and a fluid-dependent parameter, Fr_L, in the nucleate boiling contribution term for both horizontal and vertical flows, where

$$\text{Fr}_L = \frac{G^2}{(\rho_L)^2 gD}$$

Instead of the Martinelli parameter, X_{tt}, a convection number, Co, was used for the F factor neglecting the vapor viscosity effects:

$$\text{Co} = \left(\frac{1-X}{X}\right)^{0.8}\left(\frac{\rho_G}{\rho_L}\right)^{0.5} \qquad (4\text{-}22)$$

The boiling number, Bo, Eq. (2-79) was used to represent the nucleate boiling term. Thus Kandlikar's correlation for vertical flow is expressed as

$$h_{tp} = C_1(\text{Co})^{C_2}(h_{Lo}) + C_3(\text{Bo})^{C_4}\,\text{Fr}_L(h_{Lo}) \qquad (4\text{-}23)$$

and for horizontal flow as

$$h_{tp} = C_1(\text{Co})^{C_2}(25\,\text{Fr}_L)^{C_5}(h_{Lo}) + C_3(\text{Bo})^{C_4}(25\,\text{Fr}_L)^{C_6}(h_{Lo})\,\text{Fr}_L \qquad (4\text{-}24)$$

The constants C_1–C_6 and Fr_L for each fluid are evaluated from the experimental data. The correlation thus obtained compared reasonably well with experimental data (Wright, 1961; Jallouk, 1974; Shah, 1976, 1982; Steiner and Schlunder, 1977; Gungor and Winterton, 1986). In a later article, Kandlikar (1989) simplified the above equations into one for both vertical and horizontal tubes:

$$h_{tp} = h_{Lo}[C_1(\text{Co})^{C_2}(25\,\text{Fr}_L)^{C_5} + C_3(\text{Bo})^{C_4}\,F_k] \qquad (4\text{-}25)$$

The constants C_1–C_5 are given in Table 4.1 relative to the value of Co. The factor F_k is a fluid-dependent parameter for which values are listed in Table 4.2 for various fluids (Kandlikare, 1989). Unfortunately, the large number of empirical constants

Table 4.1 Constants in Eq. (4-25)

Constants	Co < 0.65 (convective region)	Co > 0.65 (nucleate boiling region)
C_1	1.1360	0.6683
C_2	−0.9	−0.2
C_3	667.2	1,058.0
C_4	0.7	0.7
C_5	0.3	0.3
(C_5 with $Fr_L > 0.04$)	0	0

Source: Kandlikar (1989). Copyright © 1989 by American Institute of Mechanical Engineers, New York. Reprinted with permission.

Table 4.2 Factor F_k in Eq. (4-25) for various fluids

Fluid	H_2O	R-11	R-12	R-13B1	R-22	R-113	R-114	R-152a	N_2	Ne
F_k	1.00	1.30	1.50	1.31	2.20	1.30	1.24	1.10	4.70	3.50

Source: Kandlikar (1989). Copyright © 1989 by American Society of Mechanical Engineers, New York. Reprinted with permission.

may discourage users. The equation is recommended for use only for flow boiling with refrigerants for which constants have been determined as shown.

4.2.2.2 Saturated nucleate flow boiling of liquid metals. Modifications to Chen's correlation are also required for the boiling of liquid metals. Chen (1963) suggested using a modified form of the Lyon-Martinelli equation for convective single-phase liquid metal heat transfer, while the Forster-Zuber equation is assumed to remain valid for the nucleate boiling component, h_{NB}. Thus the macroconvective contribution, $h_{f.c.}$, to the total heat transfer coefficient can be evaluated by

$$h_{f.c.} = [\delta + 0.024(F \operatorname{Re}_L^{0.8})(\beta \operatorname{Pr}_L)^\alpha] \left(\frac{\gamma k_L}{D_h} \right) \quad (4\text{-}26)$$

where the following values are recommended for the constants:

$$\delta = 7 \quad \beta = 1 \quad \gamma = 1 \quad \alpha = 0.8$$

4.3 FORCED-CONVECTION VAPORIZATION

The increasing void fraction and acceleration of the flow also produce changes in the flow regime with downstream location. As shown in Figure 4.2, for vertical upward flow, bubbly flow at the onset location subsequently changes to slug, churn, and then annular flow. When there is a large difference in the liquid and vapor densities (such as with alkali metals or water at low pressures), the transition from bubbly flow to the annular configuration associated with churn or annular flow can occur over a very short portion of the tube length. The core vapor velocity can be so high and the turbulence at the vapor–liquid interface so strong that the heat transfer mechanism changes character—a mechanism of forced convective heat transfer through liquid film. Heat is then conducted through the thin liquid layer, and evaporation takes place at the interface of the liquid layer and the vapor core. For lower values of X_{tt}, the heat transfer coefficient is strongly flow dependent, a characteristic of nonboiling heat transfer. Hence, as discussed in the previous section, nucleate boiling is suppressed in that region. For higher values of X_{tt}, the thickness of the liquid layer is increased and nucleate boiling is no longer suppressed by the convection. The character of the heat transfer mechanism becomes that of fully developed nucleate boiling, where the effect of flow rate on h_{tp} is negligible.

4.3.1 Correlations for Forced-Convection Vaporization

A correlation for the heat transfer coefficient in the region of forced-convection vaporization can be expressed in the form

$$\frac{h_{tp}}{h_{Lo}} = A\left(\frac{1}{X_{tt}}\right)^n \qquad (4\text{-}27)$$

where various values of the constants A and n have been determined by a number of investigators: $A = 2.50$, $n = 0.75$ by Schrock and Grossman (1959); $A = 2.72$, $n = 0.58$ by Wright (1961); and $A = 3.5$, $n = 0.50$ by Dengler and Addoms (1956). Obviously, these correlations are empirical in nature, and some of the data were obtained with reference to different ranges of parameters. Thus, in applications, use of the original references is suggested.

A specific reference to the design of steam generators is given here. The following correlations are recommended based on 762 experimental data points at (Campolunghi et al., 1977a) $p = 70$ bar (7 MPa), and G in the range of 300–4,500 kg/s m² (2.22–33.3 × 10⁵ lb/hr ft²):

$$h_{sat.b.} = 150(\Pr_G)\left(\frac{k_L}{D_e}\right)\left[\frac{q''\sigma/g(\rho_L - \rho_G)}{H_{fg}\mu_L}\right]^{0.65}\left(1 + \frac{H_{fg}G \times 10^{-4}}{q''}\right)\left(1 + \frac{1}{X_{tt}}\right)^{0.13}$$

$$(4\text{-}28)$$

where $h_{sat.b.}$ is the saturated boiling (bulk boiling) heat transfer coefficient. The subcooled boiling heat transfer coefficient, $h_{sub.b.}$, is expressed as

$$h_{\text{sub.b.}} = \frac{q''}{\left\{\Delta T^* + (\Delta T_{\text{f.c.}} - \Delta T^*)\left[\left(\frac{H_{fg} - H(z)}{H_{fg} - H_{\text{IP}}}\right)\right]^{0.5}\right\}} \quad (4\text{-}29)$$

where

$$\Delta T^* = \frac{q''}{h_{\text{sat.b.}}} \qquad \Delta T_{\text{f.c.}} = \frac{q''}{h_{\text{f.c.}}}$$

and H_{IP} is the liquid enthalpy at the onset of nucleate boiling.

For very long, helically coiled steam generator tubes, and for conditions typical of liquid-metal fast breeder reactors (LMFBRs), where steam is generated on the tube side, an overall heat transfer correlation for the whole boiling length (from $X = 0$ to $X = 1.0$) has been deduced experimentally (Campolunghi et al., 1977b):

$$\bar{h} = 11.226(q'')^{0.6} \exp(0.0132\bar{p})\,\text{W/m}^2\,\text{K} \quad (4\text{-}30)$$

with $1{,}000 < G < 2{,}500$ kg/s m² ($7.4\text{--}18.5 \times 10^5$ lb/hr ft²), $10 < q'' < 300$ kW/m² ($3{,}180\text{--}95{,}400$ Btu/hr ft²), and $80 < p < 170$ bar ($8\text{--}17$ MPa). For parallel flow in a square lattice and for typical conditions of PWR steam generators, where steam is generated on the shell side, an overall heat transfer correlation for the whole boiling length (from $X = 0$ to $X = 1.0$) has been proposed (Caira et al., 1985):

$$h_{tp} = F_2 h_{\text{f.c.}} - S_2 h_{\text{N.B.}} \quad (4\text{-}31)$$

where $F_2 = \left[1 + \frac{a}{X_{tt}} + \left(\frac{1}{X_{tt}}\right)^2\right]^{0.4}$

$S_2 = \dfrac{1}{(F_2)^{0.5}}$

$h_{\text{N.B.}} = 44.405(q'')^{0.5} \exp(0.0115p)$

$h_{\text{f.c.}} = (0.0333E + 0.0127)\left(\dfrac{k_L}{D_e}\right)\left(\dfrac{GD_e}{\mu_L}\right)^{0.8} (\text{Pr})^{0.4}$

with $180 < G < 1{,}800$ kg/s m² ($1.33\text{--}13.3 \times 10^5$ lb/hr ft²), $30 < q'' < 300$ kW/m² ($9{,}540\text{--}95{,}400$ Btu/hr ft²), and $35 < p < 80$ bar ($3.5\text{--}8$ MPa), where $a = 1$ for $G > 1,500$ kg/s m² (11×10^5 lb/hr ft²)

$$a = \frac{1,500}{G} \text{ for } G < 1,500 \text{ kg/s m}^2 (11 \times 10^5 \text{ lb/hr ft}^2)$$

$$E = \frac{\text{free flow area in infinite array}}{\text{bundle global area}}$$

4.3.2 Effect of Fouling Boiling Surface

The porous deposit on the fouling surface increases the thermal resistance in forced convection due to the semistagnant layer of water in the porous deposit. In a reactor core such a surface deposit does exist and is usually called crud. The effective thermal conductivity of the layer of water and crud is taken as 0.5 Btu/hr ft °F (0.87 W/m °C). In nucleate boiling, however, the crud behaves quite differently. First, it increases the number of nucleation sites, which enhances the boiling heat transfer. Second, it acts as a heat pipe on the heating surface, as the cold water is sucked onto the wall by the capillary force of crud pores, and the vapor is blown away through the gap between crud particles. Indeed, at a low pressure of 30 psia (0.2 MPa), a large number of nucleation sites releasing very small bubbles were observed with boiling water on the fouled surface. A clean surface, on the other hand, nucleated at a relatively small number of sites, producing considerably larger bubbles, especially at low pressures. This is why crud can improve boiling heat transfer better at low pressures than at high pressures. A similar situation occurs with artificial porous metallic surfaces. To calculate the wall superheat of a crudded surface, the outer surface temperature of the crud is maintained at saturation:

$$\Delta T_{sat} = (T_w - T_{sat})_{crud} = \frac{q''}{k_{c,B}/s} \tag{4-32}$$

where $k_{c,B}$ = effective thermal conductivity of crud in nucleate boiling
s = thickness of crud.

The values of thermal conductivity of crud in nucleate boiling are listed in Table 4.3.

4.3.3 Correlations for Liquid Metals

No and Kazimi (1982) derived the wall heat transfer coefficient for the forced-convective two-phase flow of sodium by using the momentum-heat transfer analogy and a logarithmic velocity distribution in the liquid film. The final form of their correlation is expressed in terms of the Nusselt number based on the bulk liquid temperature, Nu_b:

Table 4.3 Values of thermal conductivity of crud in nucleate boiling

Type of data	Pressure, psia	$(T_w - T_{sat})$ Thom, clean surface, °F	Flow velocity, ft/sec	Typical values of crud test				Avg. k of crud in boiling, Eq. (2), Btu/hr ft °F
				$(T_w - T_{sat})$, °F	Crud thickness, mil	Crud solidity	Heat flux, Btu/hr ft²	
Loop[a]	2,000	10.5	12	15	1.0	0.30	500,000	2.5
Loop[b]	1,000	25.7	12	65	3.7	0.40	630,000	3.0
Loop[c]	30	30.0	—	~10	1.0	—	175,000	1.5

[a] Westinghouse Electric Corp. (1969)
[b] R. V. Macbeth, personal communication, 1970.
[c] Owens and Schrock (1960).

$$Nu_b = 0.152\left(\frac{F_1}{F_2}\right)(Pr_L)(Re_L)^{0.9}\phi_L \qquad (4\text{-}33)$$

where $Nu_b = \left(\dfrac{q_w''}{T_w - T_m}\right)\left(\dfrac{D_h}{k_L}\right)$

$ = Nu\, F_1$

and values of F_1 and F_2 are (Re_L)-dependent:

For $Re_L < 50$,
$$F_1 = 1.5$$
$$F_2 = 0.7071(Re_L)^{0.5}(Pr_L)$$

For $50 < (Re_L) < 1{,}125$,
$$F_1 = 1.563$$
$$F_2 = 5\,Pr_L + \left(\frac{5}{E}\right)ln\left[1 + E\,Pr_L\left(\frac{\delta^+}{5} - 1\right)\right]$$

where $E = 0.00375(Pe_L)[1 - \exp(-0.00375\,Pe_L)]$

$\delta^+ = 0.4818(Re_L)^{0.585}$

For $Re_L > 1{,}125$,
$$F_1 = 1.818$$
$$F_2 = 5(Pr_L) + \left(\frac{5}{E}\right)[1 + 5E(Pr_L)] + \left(\frac{6}{E\gamma}\right)ln\left[\frac{2M + \gamma - 1}{1 + \gamma - 2M} \times \frac{1 + \gamma - \beta}{\beta + \gamma - 1}\right]$$

where $\beta = 60\left(\dfrac{M}{\delta^+}\right)$

$\gamma = \left[1 + \left(\dfrac{10M}{E\delta^+\,Pr_L}\right)\right]^{0.5}$

$\delta^+ = 0.133(Re_L)^{0.7614}$

$M = 1 - (\alpha)^{0.5}$

and the parameter ϕ_L is taken from the empirical relation between Zeigarnick and Litvinov's work (1980) and the Lockhart and Martinelli correlation as

$$\phi_L = (\phi_L)_{tt}^{0.88} \tag{4-34}$$

where $(\phi_L)_{tt} = \left[1 + \dfrac{20}{X_{tt}} + \left(\dfrac{1}{X_{tt}}\right)^2\right]^{1/2}$

$X_{tt} = \left(\dfrac{1-X}{X}\right)^{0.9}\left(\dfrac{\rho_G}{\rho_L}\right)^{0.5}\left(\dfrac{\mu_L}{\mu_G}\right)^{0.1}$

The above correlation, along with Chen's correlation and that of NATOF Code (Granziera and Kazimi, 1980), which used a modified Chen correlation, are compared with the Zeigarnick and Litvinov data in Figure 4.14. Their experiments were made in the following range of parameters: heat flux at the wall up to 3.5×10^5 Btu/hr ft² (1.1 MW/m²), mass flux 1.1×10^5 to 2.95×10^5 lb/hr ft² (150 to 400 kg/s m²), quality up to 0.45, and operating pressure 1–2 atm (0.1–0.2 MPa). In these experiments, flow boiling stabilization over a sufficiently long period was achieved either by drilling artificial, double-reentrant, angle-type cavities at the surface or by injection of a small amount of inert gas at the test tube entrance. A

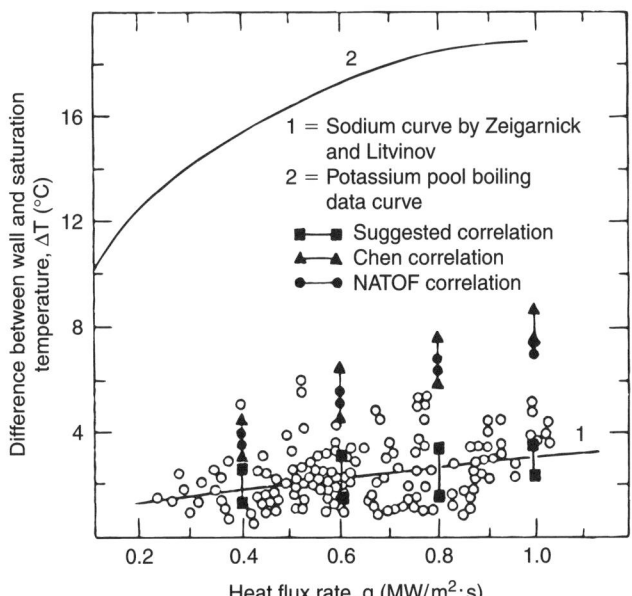

Figure 4.14 Comparison between sodium data and correlations. (From No and Kazimi, 1982. Copyright © 1982 by American Nuclear Society, LaGrange Park, IL. Reprinted with permission.)

special feature of the experiments was the direct measurement of saturated pressure and thus a more accurate determination of the saturation temperature, instead of measuring the latter by means of thermocouples, which the authors observed are inherently associated with uncertainties caused by significant local pressure drops around the thermocouple location in alkali metal, two-phase flow. It was also found that the phase change in sodium occurs by evaporation from the vapor–liquid interface without bubble generation at the wall. This suggested that the macroscopic contribution in sodium is highly dominant over nucleate boiling, and the proposed correlation represents the macroscopic heat transfer coefficient (No and Kazimi, 1982). As shown in the figure, the suggested correlation is in excellent agreement with the data over the whole range, while the Chen and NATOF code correlations predict a lower heat transfer coefficient. Comparison was also made with data of Longo (1963), but the suggested correlation predicts well only in the high heat transfer coefficient region. No and Kazimi argued that the data in the low heat transfer coefficient region were affected by unstable flow conditions and the uncertainties in the saturation-temperature measurements. No and Kazimi's correlation is therefore recommended for the calculation of sodium boiling heat transfer coefficient.

Forced-convective annular flow boiling with liquid mercury under wetted conditions was studied by Hsia (1970) because of interest in the design of a mercury boiler for space power conversion systems. Previous experiments exhibited some nonreproducibility of thermal performance, which was thought to be due to partially or nonwetted conditions. Hsia's data were therefore taken in a single horizontal tantalum tube (0.67-in. or 1.7-cm I.D.), which was shown to have perfect wetting at elevated temperature (>1,000°F or 538°C) with the following range of parameters:

Mass flux = 81.5–192 lb/sec ft² (396–934 kg/s m²)
T_{sat} = 975–1,120°F (524–604°C)
Re = 0.875–2.18 × 10^4
q'' = 0.25–2.08 × 10^5 Btu/hr ft² (0.79–6.5 × 10^5 W/m²)

The local heat flux data are shown in Figure 4.15 as a function of the wall superheat ($T_w - T_{sat}$). The effects from mass flux and boiling pressure (or T_{sat}) are also indicated. For each curve, the slope at low heat fluxes is nearly that associated with liquid-phase forced-convection mercury heat transfer. Like the boiling curves of other wetting liquids (e.g., water or potassium), the change of heat transfer mechanism from the forced-convection-dominated region to the boiling-dominated region is marked by a break to a steeper slope of the curve. Annular flow boiling is shown to occur over a wide range of heat fluxes (starting from a very low quality of less than 10%) up to the onset of the critical heat flux point. Due to the small vapor-to-liquid density ratios for mercury at temperatures of interest as in these tests, the flow was expected to have a large void fraction at low quality. At a quality of 5%, for instance, the void fraction calculated by the momentum exchange model (Levy, 1960) for a temperature range of 1,000–1,200°F (538–649°C) varies from 70% to 60%, respectively. Consequently, in these tests the low

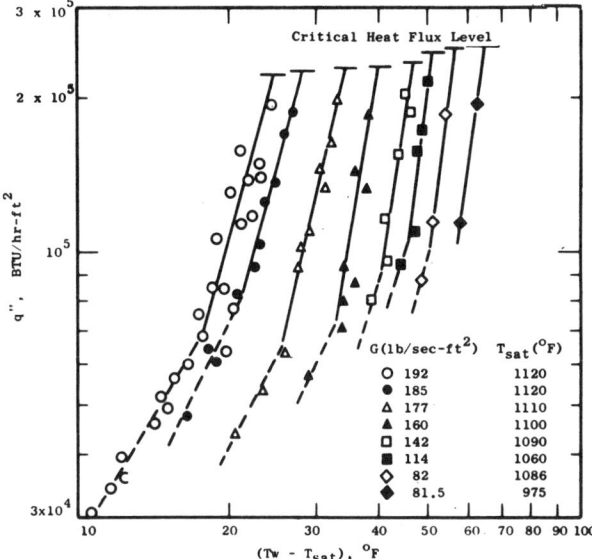

Figure 4.15 Mercury boiling heat transfer at wetted conditions inside a tantalum tube with helical insert. (From Hsia, 1970. Copyright © 1970 by American Society of Mechanical Engineers, New York. Reprinted with permission.)

void fraction-characterized slug or bubble flow regime was expected to be very short or even absent. By looking at two possible vaporization mechanisms as mentioned before, the bubble nucleation (near the wall) model or the film evaporation (at the interface) model, the values of boiling heat transfer coefficient so predicted, h_{pred}, were compared with measured coefficients, h_{meas}. Hsia (1970) concluded that bubble nucleation was the most likely heat transfer mechanism to occur in the mercury annular flow boiling region. An empirical correlation for the local heat transfer coefficient of this data with a scatter band of ±20% is

$$h_B = 3.09 \times 10^{-4}(G)^{0.05}(P_{sat})^{0.54}(q_b'')^{0.85} \quad (4\text{-}35)$$

where h_B is in Btu/hr ft² °F, G is in lb/sec ft², and q'' is in Btu/hr ft². This equation is recommended for use within the range of parameters from which the data were obtained.

Binary liquid metal systems were used in liquid-metal magnetohydrodynamic generators and liquid-metal fuel cell systems for which boiling heat transfer characteristics were required. Mori et al. (1970) studied a binary liquid metal of mercury and the eutectic alloy of bismuth and lead flowing through a vertical, alloy steel tube of 2.54-cm (1-in) O.D., which was heated by radiation in an electric furnace. In their experiments, both axial and radial temperature distributions were measured, and the liquid temperature continued to increase when boiling occurred. A radial temperature gradient also existed even away from the thin layer next to the wall. These characteristics are peculiar to the two-component liquid and are different from the one-component system due to the presence of the phase diagram. The relations between the boiling heat flux and the temperature difference between the wall and the liquid at the center ($\Delta T = T_w - T_c$) were obtained where boiling occurs for various heat fluxes, flow rates, and pressure levels. Their boiling heat transfer data are shown in Figure 4.16 along with data for other two-component, potassium amalgam systems by Tang et al. (1964), a one-component, mercury system by Kutateladze et al. (1958), and potassium systems by Hoffman (1964) and by General Electric Co. (GE Report, 1962). Note that for the data for the other systems shown in this figure, $\Delta T = T_w - T_{sat}$ was used, or T_c was assumed to equal T_{sat}. The binary system data indicate that the heat flux q'' is proportional to the 1.3 power of the temperature difference, which appears to be in good agreement among the different amalgams shown in the figure. Although no general correlation incorporating such variables as mass flux, locations, or local quality is established, it is shown that a larger temperature difference is required in a binary system than in pure potassium forced-convection boiling at a given heat flux.

4.4 FILM BOILING AND HEAT TRANSFER IN LIQUID-DEFICIENT REGIONS

The heat transfer mechanism of a vapor–liquid mixture in which the critical heat flux has been exceeded can be classified as partial or stable film boiling. The differ-

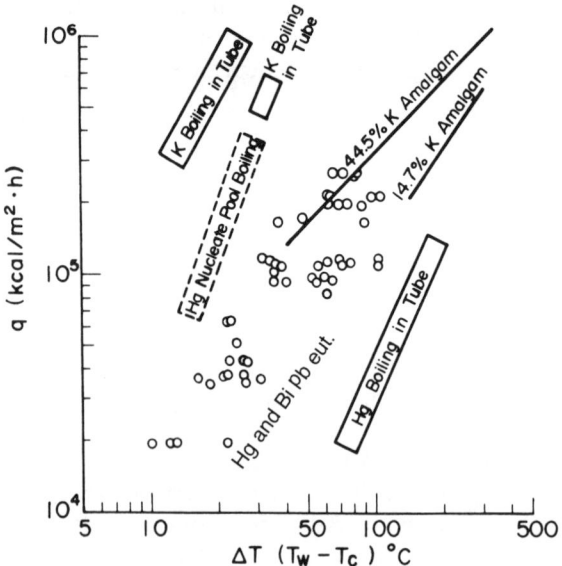

Figure 4.16 Boiling heat transfer data for mercury and amalgams. (From Mori et al., 1970. Copyright © 1970 by American Society of Mechanical Engineers, New York. Reprinted with permission.)

ence between the two lies in the magnitude of the surface temperature, which, in turn, depends on flow rate and quality. For a relatively low surface temperature, the liquid droplets in the flow are still able to "wet" the heating surface when striking it and thus can be evaporated by direct contact with the surface. It has been reported that the heat transfer coefficient of a steam–water mixture at 30 psia (0.2 MPa) and at a wall temperature less than 50°F (28°C) above saturation temperature is between three and six times the value expected for dry steam flowing under the same conditions, which is due to *partial film boiling*. On the other hand, with a relatively high surface temperature, the liquid droplets can no longer "wet" the heating surface. Thus, at a wall superheat greater than 50°F (28°C), the measured heat transfer coefficient is almost identical with that for dry steam, even though a considerable amount of liquid remains in droplet form. This is called the *spheroidal state* or *Leidenfrost point* (Leidenfrost, 1756), where the surface is so hot that the momentum of the rapidly evaporating vapor between the liquid droplet and the hot surface forms a steam cushion to support the droplet and prevent the liquid from wetting the surface. This is a state of *stable film boiling*.

It should be realized that the Leidenfrost superheat, $\Delta T_{LDF} = (T_{LDF} - T_{sat})$, is a function not only of pressure but also of droplet size, flow conditions, and force fields. Furthermore, experimental results obtained by Berger (Drew & Mueller, 1937) for stagnant ether droplets falling on a horizontal, heated surface indicated a possible effect of surface material and roughness, as the minimum surface temperature necessary for the spheroidal state changes from 226°F (108°C) on a smooth surface of zinc to 240°F (116°C) on that of a rough surface, and from 260°F (127°C) on a smooth surface of iron to 284°F (140°C) on that of a rough surface.

4.4.1 Partial Film Boiling (Transition Boiling)

Partial film (transition) boiling is a microscopically unstable mode of heat transfer in which both nucleate and film boiling exist. The local wall temperature fluctuates as these two different boiling mechanisms occur intermittently. The amplitude of fluctuations is controlled by the thermal diffusivity of the heated wall, and it can be greatly reduced by disturbing the boundary sublayer through a surface roughness of 0.002–0.003-in. (0.05–0.075-mm) height (Quinn, 1963), indicating that the wall temperature fluctuation is caused by the local shifts between two different heat transfer mechanisms. The average heat transfer coefficient of transition boiling from the wall to the bulk liquid can be obtained only by measuring a microscopically averaged (both temporal and spatial) wall temperature. For water at 2,000 psia (13.6 MPa), a conservative correlation of transition boiling heat transfer coefficient, h_{TB}, was reported by Tong (1967b) to be used in water-cooled nuclear reactors. For $T_w < 800°F$ (427°C),

$$h_{TB} = h_{FB} + 16,860 \exp[-0.01(T_w - T_{sat})] \text{ Btu/hr ft}^2 \qquad (4\text{-}36)$$

where h_{FB} is the stable film boiling heat transfer coefficient, which in this case is assumed to be 890 Btu/hr ft². For $T_w > 800°F$ (427°C),

$$h_{TB} = h_{FB} = 0.0193 \left(\frac{k}{D_e}\right)_f (\text{Re})_f^{0.8} (\text{Pr})_f^{1.23} \left(\frac{\rho_G}{\rho_{bk}}\right)^{0.68} \left(\frac{\rho_G}{\rho_L}\right)^{0.068} \quad (4\text{-}37)$$

where f refers to film temperature, which equals $(T_w + T_{bk})/2$. With a constant heat source and a heating wall of small heat capacitance, T_w increases rapidly and the time interval of the existence of transition boiling is very short. In the case of a nuclear fuel rod, however, the heat flows from a UO_2 pellet to a metallic clad through a (high-resistance) gap between the pellet and the clad, causing a fairly long transition boiling period. In the above equations, the wall temperature of 800°F (427°C) is conservatively estimated as the incipient point of stable film boiling (Leidenfrost point) for heating-up conditions. Cooling-down conditions, which are also important in the safety analysis of water-cooled nuclear reactors, will be discussed in Section 4.4.4.

4.4.2 Stable Film Boiling

As was shown before, the Leidenfrost temperature is the second transformation of heat transfer mechanisms. Empirical correlations have been established by film boiling data obtained from water at high pressure levels. For a wide range of steam–water mixture velocities, the correlation for h_{FB} reported by Bishop et al. (1965), as shown in Eq. (4-37), is recommended for use in design.

$$\left(\frac{h_{FB}D}{k}\right)_f = 0.0193(\text{Re})_f^{0.8}(\text{Pr})_f^{1.23}\left(\frac{\rho_G}{\rho_{bulk}}\right)^{0.68}\left(\frac{\rho_G}{\rho_L}\right)^{0.068}$$

The ranges of parameters used in developing this correlation are

$q'' = 0.11 \times 10^6\text{--}0.61 \times 10^6$ Btu/hr ft² (346–1,918 kW/m²)
$G = 0.88 \times 10^6\text{--}2.5 \times 10^6$ lb/hr ft² (1,190–3,380 kg/s m²)
$p = 580\text{--}3,190$ psia (3.95–21.7 MPa)
I.D. $= 0.10\text{--}0.32$-in. (0.25–0.81 cm)
T(coolant) $= 483\text{--}705°F$ (250–374°C)
$T_w = 658\text{--}1,109°F$ (348–598°C)

For larger-diameter tubes of 0.92 cm (0.36 in.) and 1.28 cm (0.50 in.), at pressures of 2,000–2,600 psia (13.6–17.7 MPa), Lee (1970) tested once-through steam generation and found that the above equation can be used to predict the heat transfer coefficient if $q''/G > 0.2$ Btu/lb. For $q''/G < 0.2$ Btu/lb, a new correlation was suggested by Lee (1970):

$$T_w - T_{sat} = 191.5 \left\{ \frac{10q''}{G[X - (1-X)/4.15]} \right\}^2 \; °C \qquad (4\text{-}38)$$

4.4.2.1 Film boiling in rod bundles. Film boiling data for flow normal to a heated rod were obtained by Bromley et al. (1953) with horizontal carbon tubes of different diameters: 0.387 in. (1.0 cm), 0.496 in. (1.26 cm), and 0.637 in. (1.62 cm). They used four liquids, benzene, carbon tetrachloride, ethanol, and N-heptane, in a velocity range of 0–14 ft/sec (4.27 m/s). Their data can be represented fairly well by two groups. For $V/(gD)^{1/2} < 2$, the equation for film boiling in a pool applies (Sec. 2.4.4.1). For $V/(gD)^{1/2} > 2$,

$$h_{FB} = 2.7 \left[\frac{V k_G \rho_G H_{fg}}{D(T_w - T_{sat})} \right]^{1/2} \qquad (4\text{-}39)$$

4.4.3 Mist Heat Transfer in Dispersed Flow

Dispersed flow heat transfer consists of heat transfer to a continuous vapor phase containing a dispersion of fine liquid droplets. The latter have diameters of the order of 50–1,000 μm and occupy only about 0–10% of the total volume of the mixture. Despite such low liquid volumetric concentrations, the liquid mass fraction can be as high as 90% of the total mass flow rate. As defined in the previous section, film boiling occurs when the channel wall temperature is high enough that the liquid no longer wets the heating surface, called Leidenfrost or dryout point. In addition to annular flow dryout, another type of dryout occurs in inverted annular flow depending on mass flux and heat flux or wall temperature. Annular flow dryout occurs when mass flux is high enough to maintain wall temperatures low enough that nucleate boiling occurs at the heating surface. Figure 4.17a shows the development of dispersed flow for such a case, and usually with several flow regimes preceding dryout as shown earlier in Figure 4.2. Inverted annular dryout occurs when liquid enters a heated section that has a high heat flux and/or wall temperature. Vapor is generated rapidly near the wall, creating an annulus of vapor around a liquid core, as shown in Figure 4.17b. Because the vapor accelerates faster than the liquid, it causes instabilities in the liquid core resulting in breakup of the core into droplets. Such a flow regime is characterized by low vapor mass and volume fractions.

The primary mode of heat transfer at the wall is forced convection of the vapor phase. As the liquid does not wet the heating surface during film boiling, heat transfer due to drop–wall collisions is relatively small, resulting in low wall–drop heat transfer (only a few percent of the total heat input). Most of the droplet evaporation occurs because of vapor–drop heat transfer. Just after dryout, the

278 BOILING HEAT TRANSFER AND TWO-PHASE FLOW

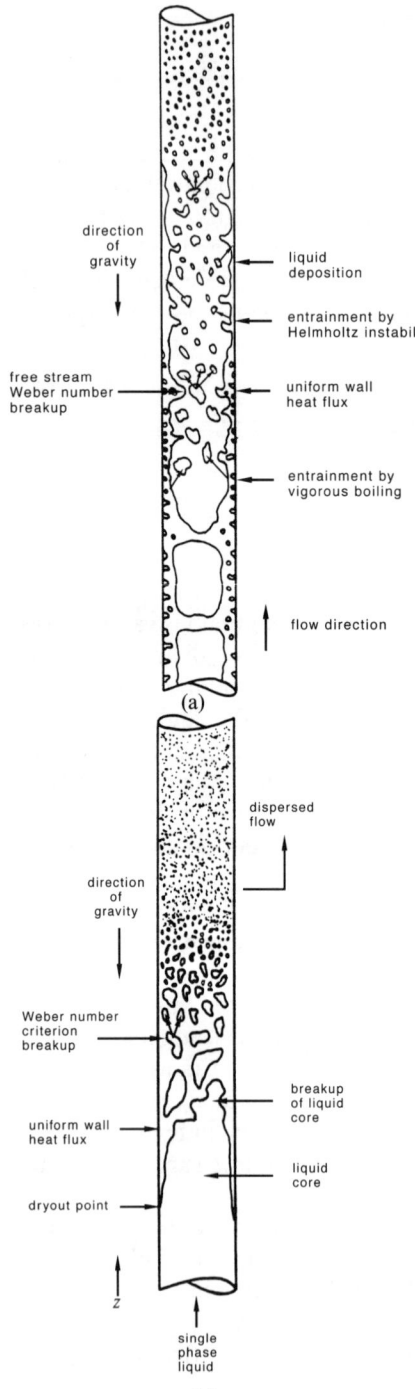

Figure 4.17 Droplet formation in dispersed flow: (*a*) in annular flow dryout; (*b*) in inverted annular dryout. (From Varone and Rohsenow, 1990. Reprinted with permission of Massachusetts Institute of Technology, Cambridge, MA.)

wall–vapor convective heat transfer rate is greater than the vapor–drop heat transfer, the latter being nearly zero since both phases are close to the saturation temperature. As a result, the vapor phase becomes superheated, and only then can heat transfer to the droplets occur. This is why flows in dispersed flow film boiling are generally in thermal nonequilibrium, and actual vapor flowing mass fraction, or quality, is less than the quality when thermal equilibrium is assumed. Flows with low dryout qualities and high mass fluxes have temperature behavior close to that of complete thermal equilibrium. Flows with high dryout quality and low mass flux have temperature behavior which approaches that of complete nonequilibrium (Varone and Rohsenow, 1990). The vapor and wall temperatures are dependent on the heat transfer from the vapor to the droplets, which in turn are dependent on the relative velocity between the vapor and drops, or the slip velocity, and droplet size distribution. The last two parameters are dependent on how dryout occurs. Generally, when inverted annular flow dryout (Fig. 4.17b) occurs, the drops, or globs of fluid, tend to be few and fairly large and break up into the frothy mixture as the vapor accelerates, which finally evolves into dispersed flow. When annular flow precedes dryout, dispersed flow occurs immediately after the dryout point. Initial drop sizes are relatively small, resulting in better interfacial heat transfer and less vapor superheating.

Although drop–wall heat transfer is usually small for high wall temperatures, this temperature may decrease sufficiently to permit drops to wet the wall if the vapor convective heat transfer becomes high enough. In this case, the wall temperature is at the rewet temperature, which is dependent on mass flux, quality, and the surface conditions of the wall (the degree of roughness or oxidation). The dispersed flow film boiling is important in the safety analysis of water reactors because during the reflood stage of a loss-of-coolant accident, the highest reactor cladding temperatures occur downstream of the quench front, resulting in a flow regime likely to be dispersed flow film boiling. This regime also occurs in such heat transfer equipment as once-through steam generators and cryogenic systems. The literature pertaining to dispersed flow film boiling heat transfer is therefore quite large. Most of the data for basic dispersed flow research are for vertical upflow in a circular tube with constant heat flux. Under the direction of Professor Rohsenow, the M.I.T. group has contributed a significant part of the research in the span of some 20 years: Dougall and Rohsenow (1963), Laverty and Rohsenow (1964), Forslund and Rohsenow (1966), Hynek (1969), Plummer et al. (1974), Iloege et al. (1974), Kendall and Rohsenow (1978), Yoder and Rohsenow (1980), Hull and Rohsenow (1982), Hill and Rohsenow (1982), and Varone and Rohsenow (1986). Other contributors include Bennett et al. (1967b), Cumo et al. (1971), Groeneveld (1972), Groeneveld et al. (1976), Chen et al. (1979), and Kumamaru et al. (1987) (using water flowing in a section of rod bundles).

4.4.3.1 Dispersed flow model. To calculate the actual quality, vapor temperature, and wall temperature, or heat flux, as functions of axial position beyond dryout

280 BOILING HEAT TRANSFER AND TWO-PHASE FLOW

with known dryout conditions. Yoder and Rohsenow (1980) found a model for an analysis based on mass, momentum, and energy balance using empirical correlations where necessary, restricted to steady-state vertical upflow in a circular tube. The following heat transfer mechanisms were considered important:

1. Heat transfer directly from the tube wall to the vapor
2. Heat transfer directly from the tube wall to the drops during drop–wall collisions
3. Heat transfer from the vapor to the entrained liquid droplets

Yoder showed that radiation heat transfer and axial conduction heat transfer in the tube wall have a negligible effect on predicting wall temperatures. The following equations were used by Yoder and Rohsenow (1980) as well as previous investigators such as Bennett et al. (1967b), Hynek (1969), and Groeneveld (1972).

Liquid velocity gradient:

$$\frac{du_L}{dz} = -\frac{g}{u_L}\left(1 - \frac{\rho_G}{\rho_L}\right) + \frac{3}{4}C_D\left(\frac{\rho_G}{\rho_L}\right)\left(\frac{u_L}{d}\right)(S-1)^2 \qquad (4\text{-}40)$$

where u_L is liquid velocity and S is slip velocity.

Droplet diameter gradient:

$$\frac{d(d)}{dz} = -2\left[\frac{h_{vd}(T_G - T_{sat})}{u_L \rho_L H_{fg}} + \frac{dv_L}{3D_t u_L}E\right] \qquad (4\text{-}41)$$

where v_L is droplet deposition velocity, h_{vd} is vapor–drop heat transfer coefficient, D_t is tube diameter, and E is wall–drop heat transfer effectiveness, where

$$E = \frac{q''_{wd}}{\dot{n}_p(\pi/6)D^3\rho_L H_{fg}}$$

and where $\dot{n}_p$ is drop flux impinging on the wall.

Quality gradient:

$$\frac{d(W_L)}{dz} = \frac{\pi}{2}\dot{n}\rho_L d^2\left[\frac{d(d)}{dz}\right] \qquad (4\text{-}42)$$

where

$$W_L = \dot{n}\rho_L\left(\frac{\pi d^3}{6}\right) \qquad (4\text{-}43)$$

and $\dot{n}$ = droplet flux, which is assumed to be constant at a given axial position. By definition, $(1 - X) = (W_L/W)$, Eq. (4-42) can be simplified to yield an expression for the actual quality gradient:

$$\frac{dX}{dz} = -3\left[\frac{(1-X)}{d}\right]\frac{d(d)}{dz} \qquad (4\text{-}44)$$

Vapor temperature gradient:

$$\frac{dT_G}{dz} = \frac{4q''_w}{GXD_t c_{pG}} - \left[\frac{H_{fg}}{c_{pG}} + (T_G - T_{sat})\right]\left(\frac{dX}{X\,dz}\right) \qquad (4\text{-}45)$$

where G is the mass flux and c_{pG} is the specific heat of the vapor.

Wall energy balance:

$$T_w - T_G = \frac{q''_w}{H_{wv}\alpha} - \left[\frac{(1-\alpha)H_{fg}\,v_L\,\rho_L\beta_1}{2h_{wv}\alpha\beta_2}E\right] \qquad (4\text{-}46)$$

where β_1, β_2 are parameters in wall–drop effectiveness calculation, and vapor properties are evaluated at the bulk vapor temperature, T_G.

4.4.3.2 Dryout droplet diameter calculation. Based on the work of Tatterson et al. (1977), Cumo and Naviglio (1988), and others showing that most of the liquid mass is contained in a small percentage of large drops, the model assumes that the droplet diameter distribution can be represented by one average drop size. Varone and Rohsenow (1990) suggested that the drop diameter be evaluated separately for annular flow dryout and invested annular dryout cases. For annular flow dryout, Figure 4.17a illustrates how the formation of drop sizes begins immediately upon the formation of annular flow. Four processes affecting the average drop diameter at dryout can be identified.

1. Boiling in the liquid film can throw large chunks of liquid into the vapor core, characterized by the Weber number based on the local relative velocity between the vapor core and the liquid film.
2. Helmholtz instabilities may cause drops to be formed by roll waves erupting from the film into the vapor core.
3. Drops may break up after being entrained in the vapor core. Vapor is generated continuously as a result of heat addition, leading to acceleration of the mixture and increased slip velocity. The sizes of drops formed in this way are characterized by the Weber number based on the slip velocity S.
4. Drop deposition onto the liquid film downstream reduces the number of liquid drops formed at each axial position that remain entrained until dryout.

For the expressions used to calculate the mass average diameter and wall–drop effectiveness, readers are referred to Varone and Rohsenow (1990).

In inverted annular flow dryout, liquid mass flux is low enough and wall heat fluxes are high enough to cause vapor to be generated rapidly near the wall, forming a vapor annulus surrounding a liquid core (Fig. 4.17b). The vapor generation near the wall occurs so quickly that the velocities of the two phases are about equal, or $S = 1$, so the expression for the void fraction at dryout, α_{do}, can be calculated from the known dryout quality, X_{do}:

$$\alpha_{do} = \left[\left(\frac{\rho_G}{\rho_L}\right)\left(\frac{1-X_{do}}{X_{do}} + 1\right)\right]^{-1} \quad (4\text{-}47)$$

When the core breaks up, the size of the drops should be of the order of the pipe diameter, D_t, thus it can be shown that the dryout drop diameter for inverted annular dryout becomes

$$d_{do} = [1 - \alpha_{do}]^{1/2} D_t \quad (4\text{-}48)$$

As the mixture accelerates, the globs of liquid break up downstream and are exposed to an accelerating vapor flow, with increasing relative velocity between the two phases. This relative velocity further breaks the drop into smaller ones depending on the Weber number,

$$\text{We} = \frac{d\rho_G(U_G - U_L)^2}{\sigma}$$

There exists a critical We above which the drop will break, and in this case, a value of 6.5 was used by Varone and Rohsenow (1986).

In comparing the model-predicted and actual wall temperatures, Varone and Rohsenow discovered a need for modification of the wall–vapor heat transfer coefficient because of the presence of liquid droplets near the wall, which may affect the basic turbulent structure of the flow. A ratio of Nusselt number of dispersed flow to that of single-phase flow is therefore used, ranging from about 0.7 to 2.0, depending on the bulk-to-wall viscosity ratio and quality (Fig. 4.18). Further modification of the model included droplet breakup and determination of the critical Weber number, We_c (Varone and Rohsenow, 1990). The modified model predicts wall temperatures with much greater accuracy than the previously mentioned Yoder model (1980) does. Figures 4.19, 4.20, and 4.21 are typical comparisons with data of Bennett et al. (1967b), Era et al. (1966), and Cumo et al. (1971), respectively, as reproduced from Varone and Rohsenow (1990). This model is recommended for the analysis of dispersed flow heat transfer.

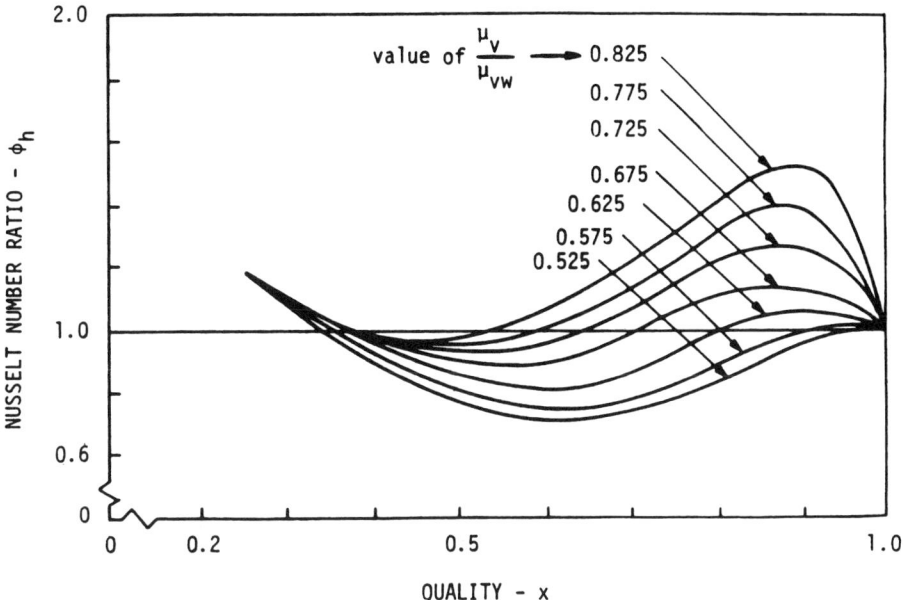

Figure 4.18 Family of curves of Nusselt number ratio versus quality for constant bulk wall vapor viscosity ratios. (From Varone and Rohsenow, 1990. Reprinted with permission of Massachusetts Institute of Technology, Cambridge, MA.)

4.4.4 Transient Cooling

This section describes some of the boiling phenomena that occur in water reactors with respect to safety analyses that require thermal hydraulic considerations.

4.4.4.1 Blowdown heat transfer. Blowdown occurs during a loss-of-coolant accident in a water-cooled nuclear reactor, as the fuel rods in the reactor core are first cooled by the blowdown of the coolant (water). The reactor being shut down, most of the heat stored in fuel is supposed to be transferred away by various boiling mechanisms. At the end of blowdown, the fuel is uncovered by water, being heated by the decay heat of UO_2 and cooled by the steam flow and thermal radiation with a low heat transfer rate. The fuel temperature thus rises again until emergency core cooling becomes effective.

Three boiling heat transfer mechanisms exist during blowdown: nucleate boiling, transition boiling, and stable film boiling. The order of magnitude of the heat transfer coefficients of these mechanisms are 50,000, 5,000, and 200 Btu/hr ft² (1.0×10^5, 1.6×10^4, and 629 W/m²), respectively. As indicated before, the interface between nucleate boiling and transition boiling is determined by the critical heat flux (CHF), while the interface between transition boiling and stable film boiling is specified by a wall temperature that has been empirically determined to be about 800°F (427°C) (Tong, 1967b) for the case of water-cooled reactor condi-

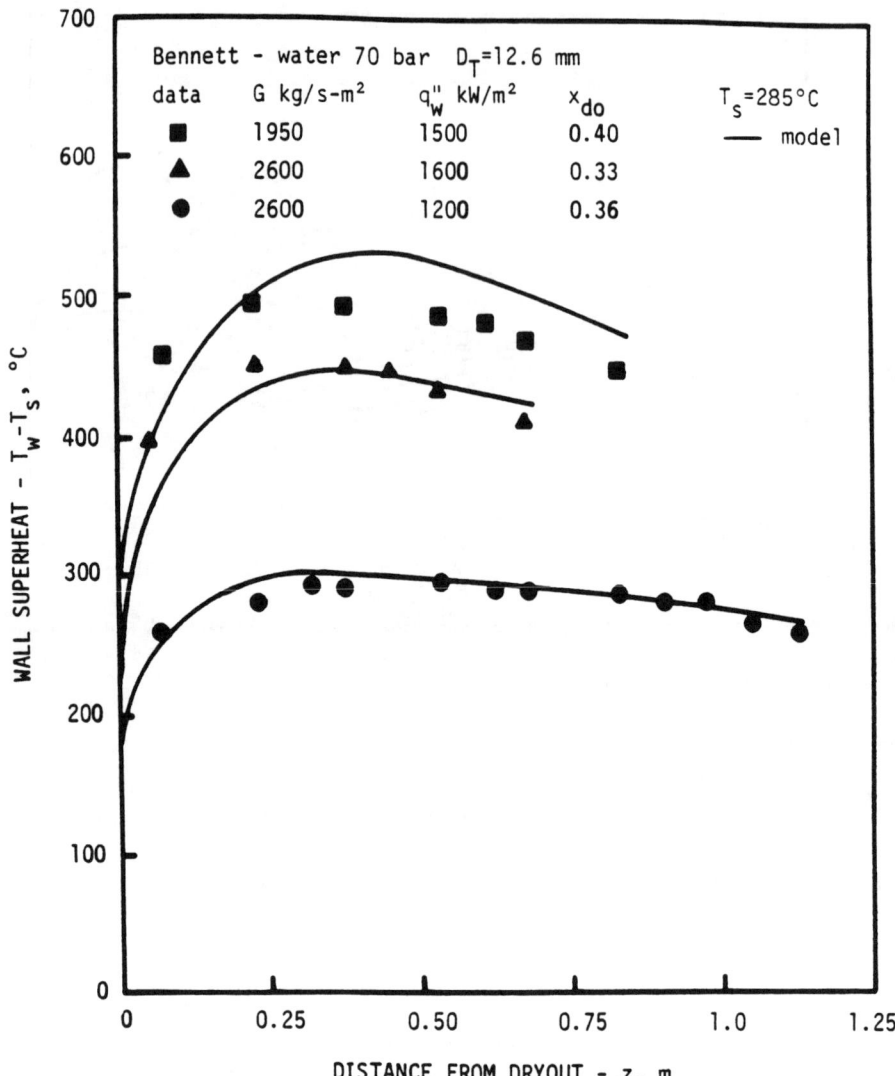

Figure 4.19 Comparison of predicted tube wall temperatures and the data of Bennett et al. (1967b). (From Varone and Rohsenow, 1990. Reprinted with permission of Massachusetts Institute of Technology, Cambridge, MA.)

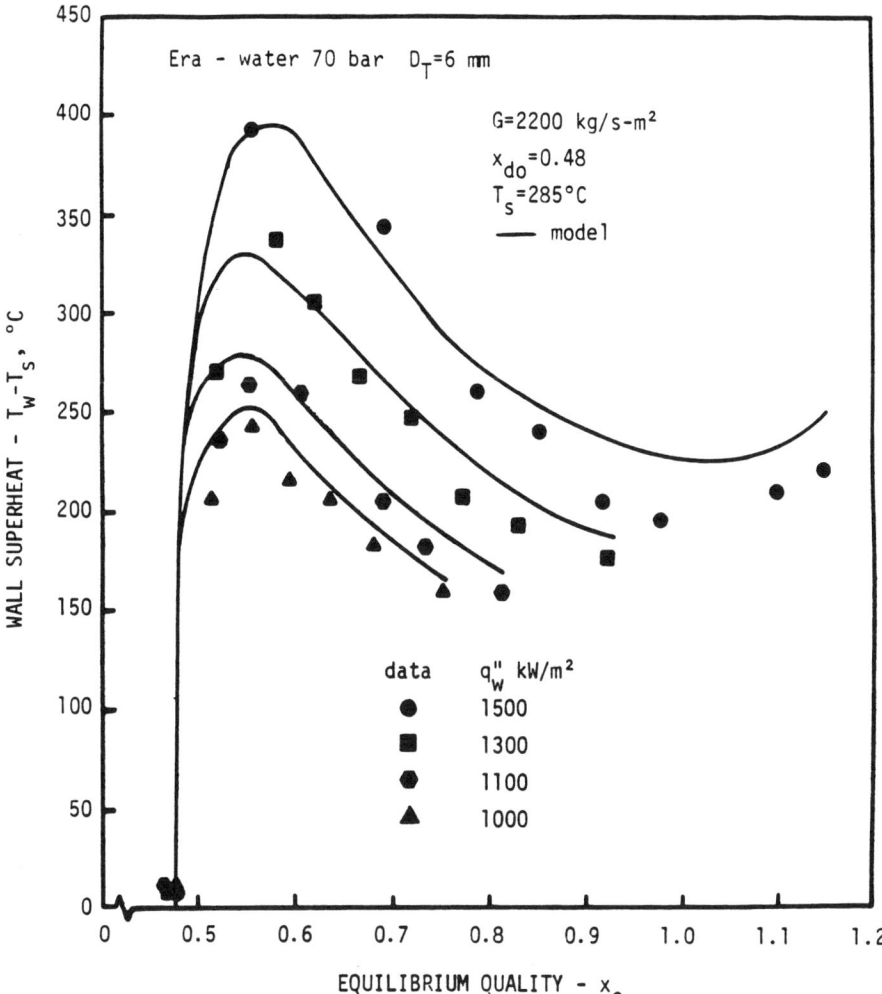

Figure 4.20 Comparison of predicted tube wall temperatures and the data of Era et al. (1966). (From Varone and Rohsenow, 1990. Reprinted with permission of Massachusetts Institute of Technology, Cambridge, MA.)

tions. The effect of these two interfaces on the maximum clad temperature of a typical PWR plant that is reached in a loss-of-coolant accident (LOCA) can be illustrated by the following different scenarios.

1. Dryout at 0 sec after a double-ended cold leg break, $h_{FB} = 200$ Btu/hr ft^2 (629 W/m^2) from 0 sec to fuel uncovering, and $h_f = 25$ Btu/hr ft^2 (78 W/m^2) after hot spot recovery. The maximum clad temperature would be 2,400°F (1,316°C).

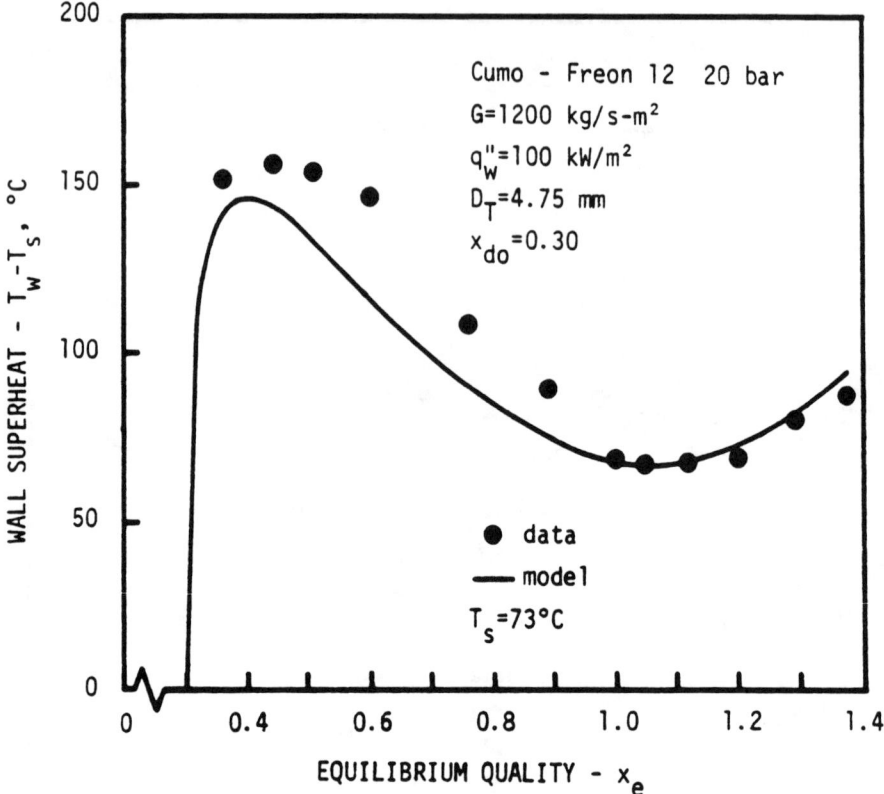

Figure 4.21 Comparison of predicted tube wall temperatures and the data of Cumo and Naviglio (1988). (From Varone and Rohsenow, 1990. Reprinted with permission of Massachusetts Institute of Technology, Cambridge, MA.)

2. Dryout at 0.5 sec, h_{NB} = 50,000 Btu/hr ft² (1.6 × 10⁵ W/m²) until dryout, h_{FB} = 200 Btu/hr ft² (629 W/m²) from dryout until fuel uncovering, and h_f = 25 Btu/hr ft² (78 W/m²) after hot spot recovery. The maximum clad temperature is lowered by 70°F (39°C).
3. Dryout at 0.5 sec, h_{NB} = 50,000 Btu/hr ft² (1.6 × 10⁵ W/m²) until dryout, h_{TB} = 5,000 Btu/hr ft² (1.6 × 10⁴ W/m²) until fuel uncovering, and h_f = 25 Btu/hr ft² (78 W/m²) after hot spot recovery. The maximum clad temperature is lowered by 500°F (278°C).

It is thus clearly demonstrated that the accurate predictions of CHF (or dryout) delay and the existence of transition boiling are very important in the evaluation of a maximum clad temperature in this type of accident. The test results of Tong et al. (1965, 1967a) and of Cermak et al. (1970) indicate the validity of using the steady-state CHF correlation to predict the CHF during a PWR transient cooling or blowdown.

4.4.4.2 Heat transfer in emergency core cooling systems. Of two types of emergency core cooling systems used in water-cooled reactors, top spray and bottom flooding systems, top spray cooling is less effective in heat transfer at high fuel temperatures because the chimney effect of the evaporating steam may hinder the downward cooling liquid flow. Bottom flooding systems, on the other hand, usually cannot effectively fill up a large lower plenum volume of a boiling water reactor, with its bottom entry control rods. Thus bottom flooding is compatible with a pressurized water reactor, and top spray with a BWR system.

The heat transfer behavior in cooling down a channel wall with a vertically upward two-phase flow being heated up is shown in Figure 4.22, which forms a "thermal hysteresis." The originally cold fuel clad temperature is effectively cooled via nucleate boiling heat transfer until it reaches a point of departure from nucleate boiling in a high-quality region during blowdown. Steep rising of temperature results. During core cooling, the fuel clad is initially very hot and cannot be quenched, and thus can be considered as the departure from film boiling (DFB), or second transition point. Quenching will occur only when the hot surface cools sufficiently to become wettable (at a rewet temperature). A comparison of saturated pool boiling curves obtained from heating up and cooling down was made by Bergles and Thompson (1970), who found that quenching of a dirty surface would occur at higher wall temperature and transfer a higher heat flux than quenching of a clean surface. Kutateladze and Borishansky [1966] reported flow DFB data for water and isopropyl alcohol, which shows that the DFB heat flux increases with liquid velocity. They suggested a saturated pool quench correlation of

$$q''_{DFB} = 300 H_{fg} (\rho_G)^{0.5} [\sigma(\rho_L - \rho_G)]^{0.25} \quad (4\text{-}49)$$

The effect of subcooling on DFB was reported by Witte et al. (1969) from a heated silver sphere inserted in a stagnant water-filled tube. For a low-wall-temperature sphere, their correlation is

$$(T_w - T_{sat})_{DFB} = 3.6(\Delta T)_{sub} + 20 \text{ °C} \quad (4\text{-}50)$$

At higher wall temperature (i.e., 1,600°F or 871°C) in a pool of water at 1 atm, Bradfield (1967) suggested

$$(T_w - T_{sat})_{DFB} = 6.15(\Delta T)_{sub} + 355 \text{ °F} \quad (4\text{-}51)$$

These two correlations indicate that the magnitude of wall superheat at quenching is a strong function of the initial wall temperature. Later, Stevens et al. (1970), using a moving copper sphere of 3/4-in. (1.9-cm) diameter with a velocity of 10–20 ft/s (3–6.1 m/s) quenched in subcooled water, found that the flow DFB behavior in a moving liquid is quite different from that in a stagnant liquid. The higher

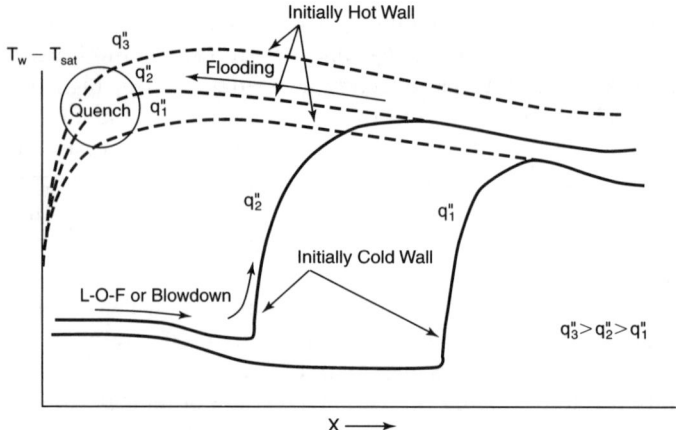

Figure 4.22 Hysteresis effect of wall temperature at various qualities during blowdown and core cooling.

liquid (or initial wall) temperature has lower surface tension, which enhances DFB, while the velocity of moving liquid does not seem to affect the quench temperature.

Bottom flooding systems The heat transfer coefficients and mechanisms during bottom flooding are shown in Figure 4.23. It can be see that there are four types of cooling mechanisms before quenching, steam cooling, liquid-droplet dispersed flow, liquid-chunk dispersed flow, and stable film boiling, which occur in this sequence with reducing elevation of hot rod facing a coolant flow with an increasing liquid content. The corresponding void fraction and flow pattern map is given in Figure 4.24 (Cermak et al., 1970). The criterion for the occurrence of each individual mechanism depends on the local liquid content and steam velocity. Liquid-chunk dispersed flow usually occurs at a liquid content of 10–70%.

Top spray systems During top-spray cooling of an overheated core, the wall temperature is usually higher than the Leidenfrost temperature, which causes water to be sputtered away from the wall by violent vapor formation and then pushed upward by the chimney effect of the steam flow generated at lower elevations (as shown in Fig. 4.25). A spray-cooling heat transfer test with BWR bundles was reported by Riedle et al. (1976). They found the dryout heat flux to be a function of spray rate and system pressure. The collapsed level required to keep the bundle at saturation for various pressures compared reasonably well with that in the literature (Duncan and Leonard, 1971; Ogasawara et al., 1973).

4.4.4.3 Loss-of-coolant accident (LOCA) analysis.

Analyses of the effects of breaking instrument tubes at the PWR vessel lower plenum (Fletcher and Bolander, 1986) One safety concern for nuclear reactors is related to

FLOW BOILING **289**

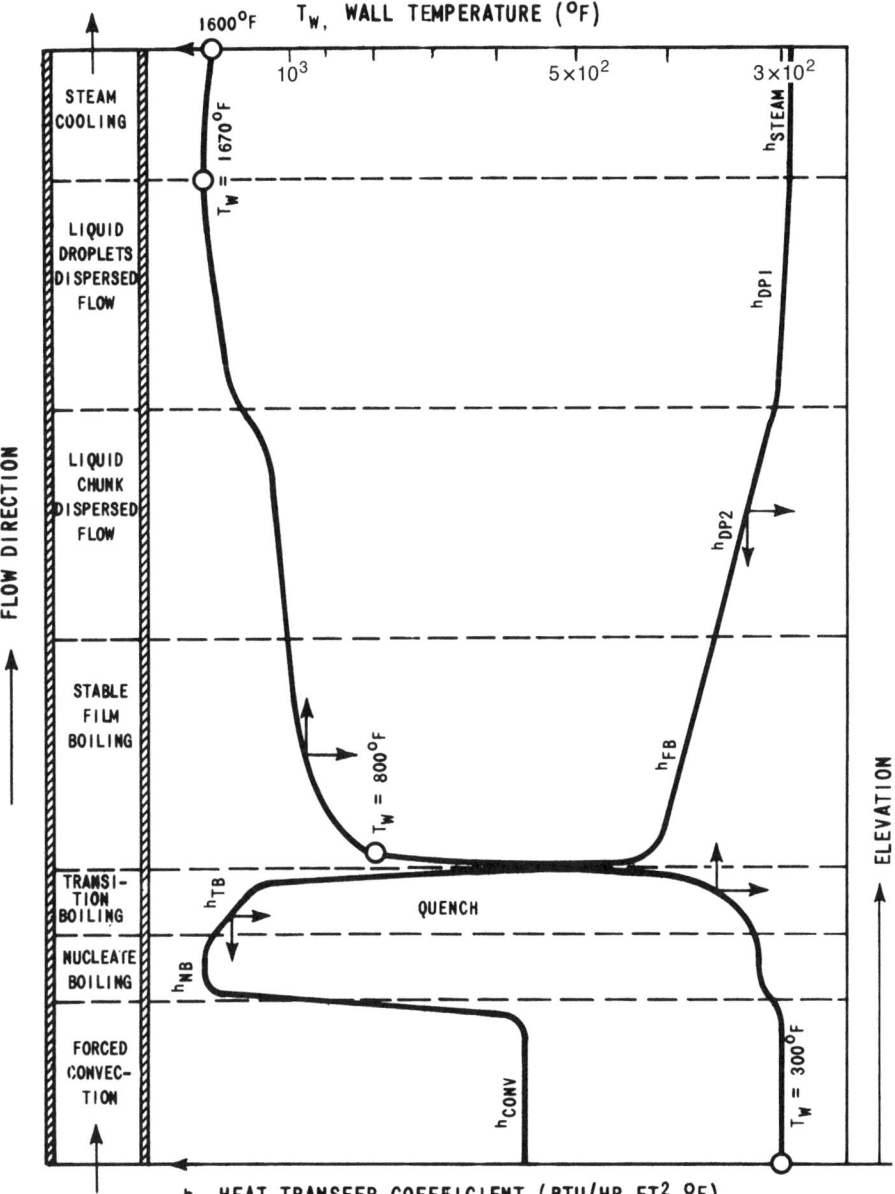

Figure 4.23 Heat transfer coefficient and mechanisms during bottom flooding.

290 BOILING HEAT TRANSFER AND TWO-PHASE FLOW

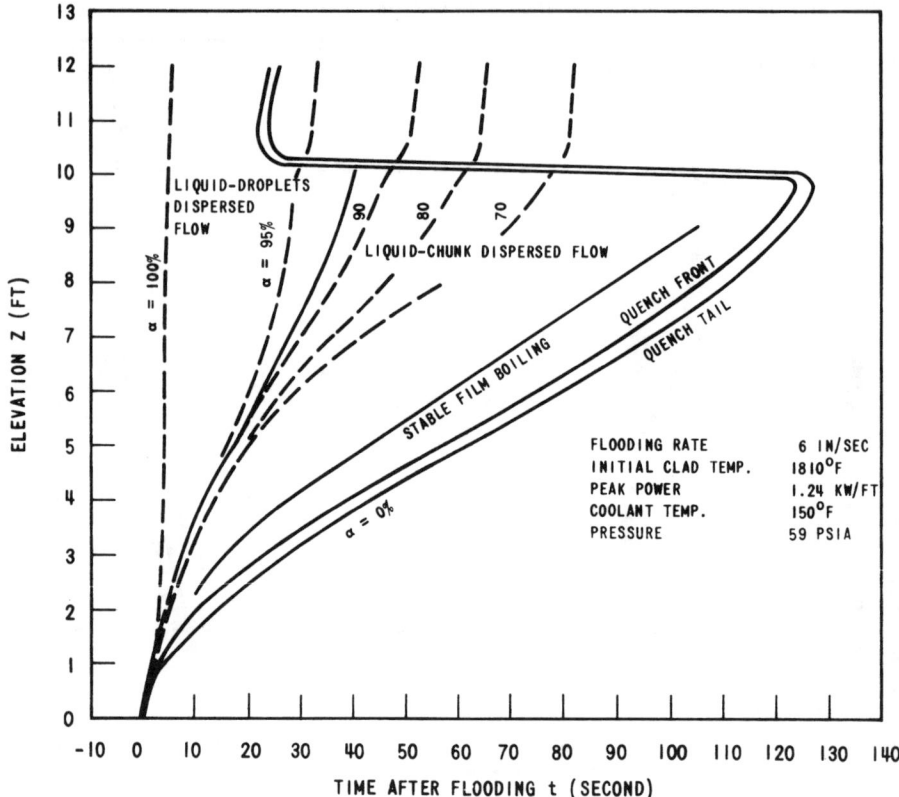

Figure 4.24 Void fraction map. (From Cermak et al., 1970. Copyright © 1970 by American Society of Mechanical Engineers, New York. Reprinted with permission.)

a possible seismic event. Such an event might cause instrument tubes to be broken at the flux mapping seal table of a pressurized water reactor (Fig. 4.26), which could result in uncovering and heatup of the reactor core. In 1980, the U.S. Nuclear Regulatory Commission (USNRC) performed an analysis of instrument tube breaks in Westinghouse four-loop PWRs, using the RELAP4/MOD7 computer code for many combinations of break sizes and conservative emergency core cooling (ECC) availabilities (Fletcher and Bolander, 1986). The computer model was based on the Zion-1 PWR geometry, but with boundary and initial plant conditions adjusted for the worst-case conditions expected from within the group of all Westinghouse four-loop PWRs. All instrument tube breaks were assumed to occur at the reactor vessel, with instrument lines expelled, and thus did not account for line losses between the reactor vessel and the seal table. The 1980 analysis also used three assumptions of ECC flow characteristics, all of which are conservative compared to best-estimate (i.e., no failure) injection flow, assuming one of the four

FLOW BOILING **291**

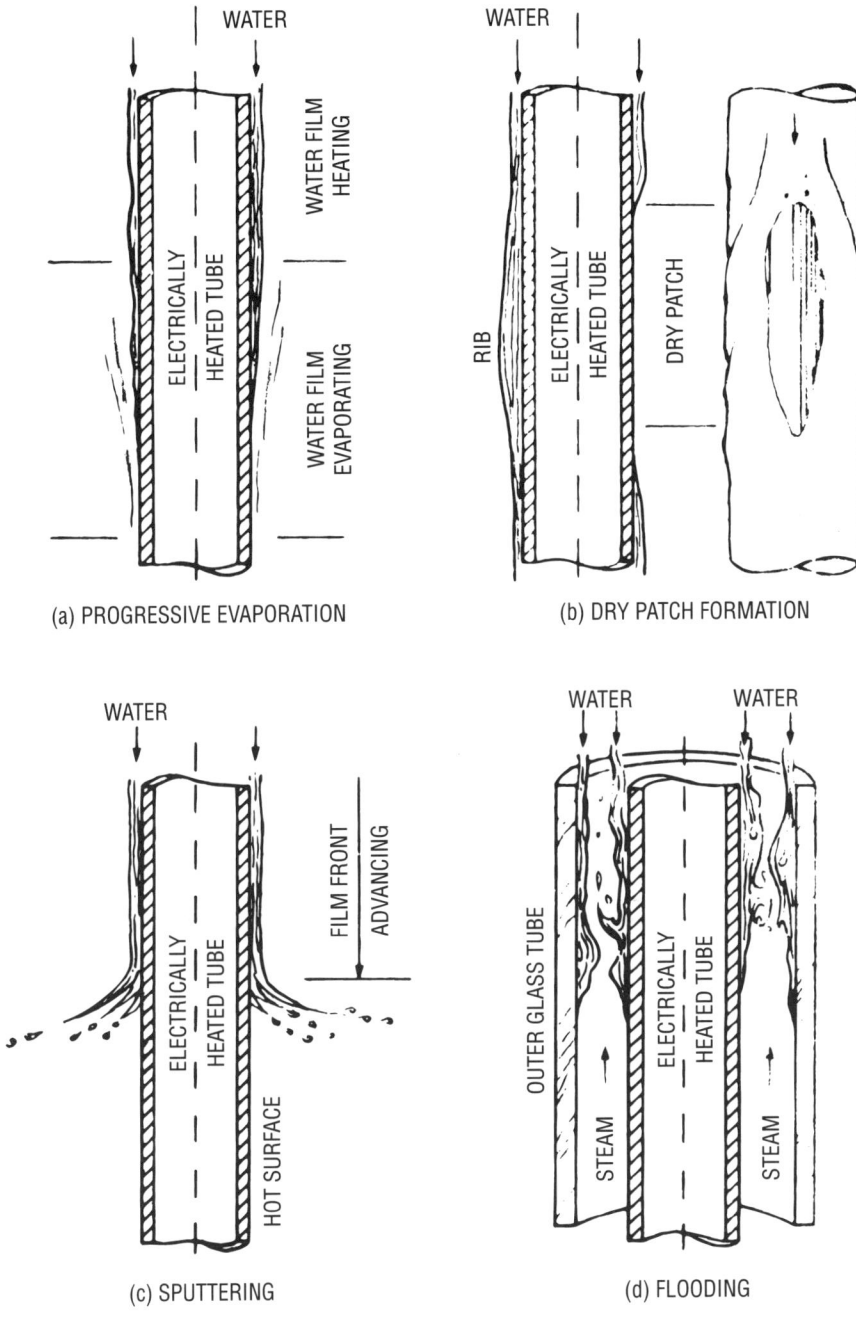

Figure 4.25 Heat transfer mechanism in top spray.

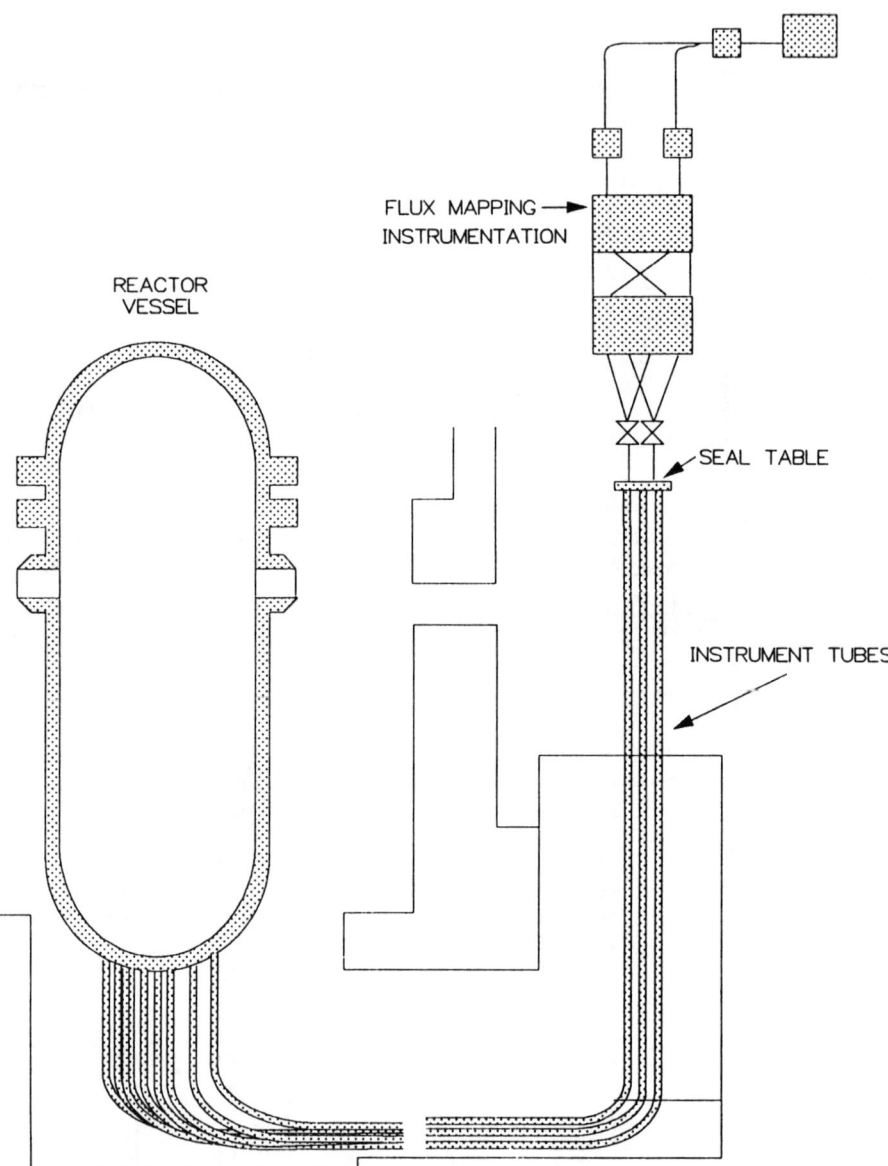

Figure 4.26 Representative instrument tube routing in a PWR. (From Fletcher and Bolander, 1986. Reprinted with permission of U.S. Nuclear Regulatory Commission, subject to the disclaimer of liability for inaccuracy and lack of usefulness printed in the cited reference.)

cold-leg injection lines spilled to containment, as was common at that time for large-break LOC accidents. The "full ECC" flow rate was based on delivery from two high-pressure injection (HPI) and two charging pumps. Results of the 1980 analysis indicated that, assuming "full ECC" availability, no uncovering of the core would occur following the breaking of seven small or three large instrument tubes at the reactor vessel. Assuming that only half of the "full ECC" is available (the limiting case), no uncovering of the core was found to occur following the breaking of two small or one large instrument tube at the reactor vessel. Assuming "minimum safeguards ECC" (based on delivery of only one HPI and one charging pump), no uncovering of the core was found to occur following the rupture of five small or two large instrument tubes at the reactor vessel.

Later, in 1985, a comprehensive evaluation program was performed by EG&G Idaho, Inc. (Fletcher and Bolander, 1986), using RELAP5/MOD2 simulation of instrument tube break sequences and the Semiscale experimental facility in assessing computer code results. An example of such system effect calculations is given in Table 4.4, which lists the sequence of events of "Transient 3." The break size was 0.001963 ft^2 (0.000182 m^2). The purposes of this calculation were (1) to confirm that the primary coolant inventory depletion rate would disappear prior to the onset of reflux cooling-loop natural circulation, and (2) to provide a basis for selecting separate effects calculation parameters (lower plenum pressure and subcooling) to support adjustment of the 1980 analysis at the seal table. The calculated primary system pressure response is shown in Figure 4.27. As expected, with this small break size, the depressurization rate is small; reactor trip is not encountered until 4,685 sec. The pressure oscillations from about 1,000 to 2,000 sec and the following repressurization are caused by subcooled nucleate boiling and its

Table 4.4 Sequence of events, RELAP 5 calculation of Transient 3

Event	Time, sec
Break opens; complete shear of one large instrument tube at the reactor vessel; instrument line expelled	0
Charging flow initiated	102
Reactor trip signal; reactor coolant pumps tripped	4,685
Main steam throttle valves fully closed	4,686
Power decay begins; control rods bottomed	4,688
Main feedwater control valves fully closed	4,695
Safety injection system available	4,726
Auxiliary feedwater flow available	4,745
Reactor coolant pump coastdown completed	4,792
To maintain steam generator secondary level, throttling of auxiliary feedwater began	5,200
Pressurizer empty	6,808
Calculation terminated; ECC flow exceeded break flow	12,598

Source: Fletcher and Bolander (1986). Reprinted with permission of U.S. Department of Energy, Washington, DC, subject to the disclaimer of liability for inaccuracy and lack of usefulness printed in cited reference.

294 BOILING HEAT TRANSFER AND TWO-PHASE FLOW

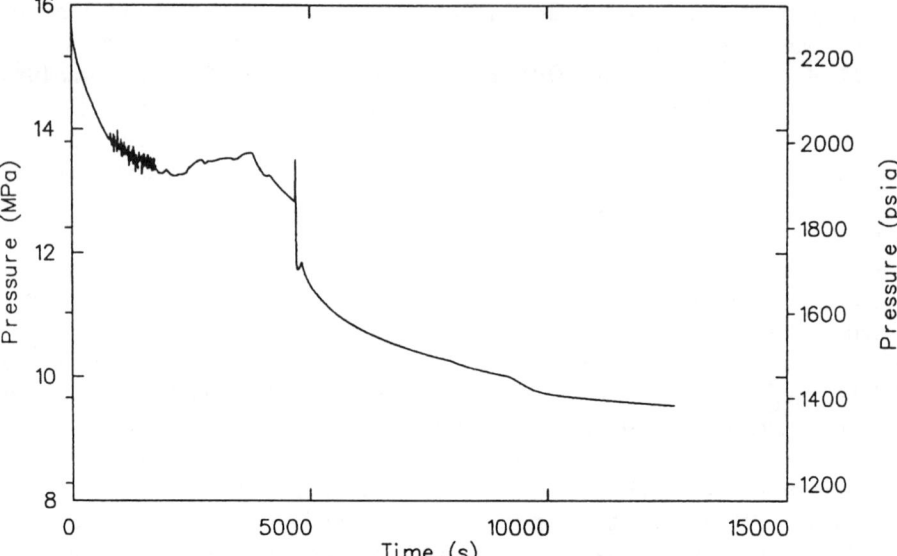

Figure 4.27 RELAP-5 modeling of primary system pressure variations during Transient 3 (Table 4.4). (From Fletch and Bolander, 1986. Reprinted with permission of U.S. Nuclear Regulatory Commission, subject to the disclaimer of liability for inaccuracy and lack of usefulness printed in the cited reference.)

cessation. Some of the authors' conclusions from the evaluation can be summarized as follows.

1. Uncovering and heatup of the core requires a much larger number of tubes to be broken at the seal table than are postulated to break as a result of a seismic event.
2. System thermal-hydraulic phenomena associated with instrument tube breaks at the reactor vessel and seal table are comparable.
3. The key findings of the 1980 USNRC analysis of instrument tube breaks at the reactor vessel may be adjusted for breaks at the seal table by applying a multiplier of 3.55. That is, 3.55 tubes broken at the seal table are the equivalent to 1 tube broken at the reactor vessel.

ORNL small-break LOCA tests Experimental investigation of heat transfer and reflood analysis was made under conditions similar to those expected in a small-break LOCA. These tests were performed in a large, high-pressure, electrically heated test loop of the ORNL Thermal Hydraulic Test Facility. The analysis utilized a heat transfer model that accounts for forced convection and thermal radiation to steam. The results consist of a high-pressure, high-temperature database of experimental heat transfer coefficients and local fluid conditions.

Rod bundle heat transfer analysis (Anklam, 1981a) A 64-rod bundle was used with an axially and radially uniform power profile. Bundle dimensions are typical of a 17 × 17 fuel assembly in a PWR. Experiments were carried out in a steady-state mode with the inlet flow equal to the steaming rate. Generally, about 20–30% of the heated bundle was uncovered. Data were taken during periods of time when the two-phase mixture level was stationary and with parameters in the following ranges:

2.6 MPa (375 psia) ≤ pressure, p ≤ 7.1 MPa (1,030 psia) Maximum rod surface temperature, T_s > 1,000 K (1,340°F) 3,500 < Reynolds numbers, Re ≤ 10,000 0.8 kW/m (0.24 kW/ft) ≤ linear power (uniform) ≤ 1.4 kW/m (0.43 kW/ft)

The observed total (convective and radiative) heat transfer coefficients were between 0.01 and 0.019 W/cm² K (17.6 and 33.0 Btu/hr ft² °F) at an estimated accuracy of within ±15%.

High-pressure reflood analysis (Anklam, 1981b) A series of six high-pressure reflood tests under conditions similar to those expected in a small-break LOCA was also used to produce a database for high-pressure reflood cases. Primary parametric variations were in pressures, ranging from 2.09 MPa (303 psia) to 6.94 MPa (1,006 psia), and in flooding rates, ranging from 2.9 cm/s (1.1 in./sec) to 16.5 cm/s (6.5 in./sec). This database was intended to be of particular use in the evaluation of thermal-hydraulic computer codes that attempt to model high-pressure reflood [such as the EPRI rewetting model (Chambré and Elias, 1977)]. Before reflood, the makeup water supplied to the test section was sufficient to offset what was being boiled off. Thus, the two-phase mixture level was stationary, the test loop was in a quasi-steady state, and reflood was initiated by increasing the makeup flow. Test results showed that in most cases the quench front velocity was 40–50% of the flooding rate. In high-flooding-rate tests (>5.0 cm/s or 2.0 in./sec), results indicated the presence of significant quantities of entrained liquid in the bundle steam flow. In a low-flooding-rate test (2.9 cm/s or 1.1 in./sec), however, little or no liquid entrainment was indicated. None of the six tests showed evidence of liquid carryover, probably due to deentrainment of liquid in the test-section upper plenum. In tests where the flooding rate exceeded 13.0 cm/s (5.1 in./sec) and initial surface temperature exceeded 800 K (980°F), the collapsed liquid level was observed to exceed the quench level, which suggested that inverted annular film boiling may have existed. The fuel rod simulator quench temperatures varied between 718 and 788 K (833 and 959°F). Because of the difference between the experimental heater rods and actual fuel rods, Anklam (1981b) cautioned not to use the data by extrapolating to the case of a nuclear reactor, but to use the data only for benchmarking predictive thermal-hydraulic computer codes where the experimental heater rods could be incorporated correctly into the code's heat transfer model.

296 BOILING HEAT TRANSFER AND TWO-PHASE FLOW

Simulated steam generator tube ruptures during LOCA experiments (Cozznol et al., 1978) The potential effects of steam generator tube ruptures during large-break LOCA were investigated in the Semiscale Mod-1 system, which is a small-scale nonnuclear experimental facility with components that represent the principal physical features of a commercial PWR system. The core (composed of an array of electrically heated rods) is contained in a pressure vessel that also includes a downcomer, a lower plenum, and an upper plenum. The system is arranged in a $1\frac{1}{2}$-loop configuration with the intact loop containing an active pressurizer, steam generator, and pump, and the broken loop containing passive simulators for the steam generator and pump (Fig. 4.28). For the steam generator tube rupture tests, secondary-to-primary flow was simulated by injecting liquid into the intact-loop hot leg between the steam generator inlet plenum and the pressurizer, using a constant-pressure water source at a temperature typical of a PWR steam generator secondary fluid.

For small-tube-rupture flows that simulated the flow from the single-ended rupture of up to 16 steam generator tubes in a PWR system initiated at the start of vessel refill, the core thermal response was strongly dependent on the magnitude

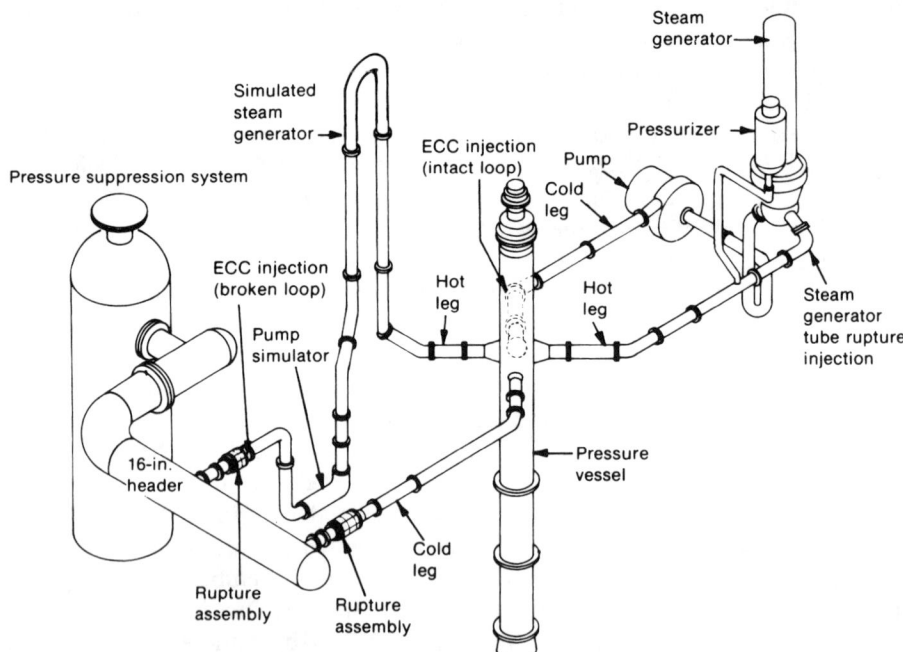

Figure 4.28 Semiscale Mod-1 system cold-leg break configuration—isometric diagram. (From Cozzuol et al., 1978. Reprinted with permission of U.S. Nuclear Regulatory Commission, subject to the disclaimer of liability for inaccuracy and lack of usefulness printed in the cited reference.)

of the secondary-to-primary flow rate. The peak cladding temperatures observed during the tube rupture injection period for the small-tube-rupture flow rate cases increased to a maximum of about 1,258 K for the case of simulating 16 steam generator tubes rupture. The principal reason for the higher peak cladding temperature in this case was that reflooding of the core was considerably retarded due to the increased steam binding in the intact-loop hot leg resulting from the secondary-to-primary flow. For relatively large-tube-rupture flows that simulated the flow from the single-ended rupture of 20 or more steam generator tubes in a PWR system initiated at the start of vessel refill, the core thermal response was characterized by an early top-downward quenching of the upper part of the core due to steam generator secondary liquid that penetrated the core from the intact-loop hot leg, and a delayed bottom-upward quenching of the lower part of the core resulting from bottom reflooding. The peak cladding temperatures for the relatively large-tube-rupture flow rate cases decreased as the tube rupture flow rates were increased from 20 to 60 tubes. A peak cladding temperature of about 1,208 K was observed for the 20-tube-rupture case. Thus a narrow band of tube rupture flows (flow from between about 12 and 20 tube ruptures) resulted in significantly higher peak cladding temperatures than were observed for the other rupture flow cases. The rupture of 12 or 20 tubes corresponds to only about 0.08% of the total number of tubes in three of four steam generators in a four-loop PWR. Even though relatively high peak cladding temperatures were observed for tests simulating tube rupture flow rates within this band, the maximum peak cladding temperature observed experimentally was considerably below the temperature necessary to impair the structural integrity of PWR fuel rod cladding. This kind of information is what the safety analysis is looking for. However, the test results in the Semiscale Mod-1 system cannot be related directly to a PWR system because of the large differences in physical size and the scaling compromises in the Mod-1 system. The results have been used to identify phenomena that control or strongly influence core thermal behavior during the period of secondary-to-primary mass flow in a LOCA with steam generator tube ruptures, and are used in evaluating the ability of computer codes to predict the thermal-hydraulic phenomena that occur as a result of such tube ruptures.

4.4.5 Liquid-Metal Channel Voiding and Expulsion Models

In the safety analysis of LMFBRs, as opposed to light water cooled reactors (LWRs), the mechanics of sodium ejection in the event of channel blockage, pump failure, or power transients are of primary importance. A reduction of sodium density can result in either a positive or negative (nuclear) reactivity, depending on the location and extent of the vapor void. Consequently, an accurate description of the voiding process with respect to space and time is necessary. Cronenberg et al. (1971) presented a single-bubble model for such sodium expulsion which was based on a slug-type expulsion with a liquid wall film remaining on the wall during expulsion, as indicated by Grolmes and Fauske (1970) for the ejection of Freon-11, and by Spiller et al. (1967) in their experiments with liquid potassium. This "slug" model is similar to the ejection model of Schlechtendahl (1967), but takes into account heat transfer in both upper and lower liquid slugs, as well as the wall film. It calculates the convective heat transfer from the cladding of fuel element to the coolant, until at some point in space and time the coolant reaches a specified superheat when the inception of boiling occurs. The stages of spherical bubble growth are approximated by a thin bubble assumed to occupy the entire coolant channel area except for the liquid film remaining at the cladding wall. The coupled solution to the energy and hydrodynamic equations of the coolant, and the heat transfer equations of the fuel element, are solved continuously during the voiding process (Cronenberg et al., 1971). This is done by first determining the two-phase interface position from the momentum equation, and then solving the energy equation for the vapor, which is generated at a rate proportional to the heat transferred by conduction across the slug interfaces and the wall film, neglecting the sensible heat of the vapor and the frictionless pressure work. The authors indicated some physical features of the sodium expulsion process by these calculations: the growth of the vapor slug is dominated by heat conduction from the liquid slugs in the early growth period, i.e., the length of wall film (Z) $<<$ the equivalent channel diameter, D_e, and by heat conduction from the exposed walls, in the late growth period, $Z/D_e >> 1$, provided that in the latter case the exposed wall is covered by an adhering film of residual liquid. The nature of the expulsion and reentry will therefore depend strongly on whether or not liquid film is present. Chugging will most likely occur for an accident condition of a sudden loss of flow if there exists a long blanket or plenum region, since the cladding temperature in such an unheated region will then be below the saturation temperature. Reentry may occur if film vaporization in the core region ceases due to film breakup or dryout. Superheat is also an important parameter, which affects both the time to initiate boiling and the rate of voiding. The superheat effect is most pronounced for low heat fluxes and at the beginning of the voiding process.

Ford et al. (1971a) used the same model in the study of the sudden depressurization of Freon-113 in a 0.72-cm (0.28-in.)-diameter glass tube. Figure 4.29 shows the experimental interface position as a function of time for a superheat of 109°F (61°C), which compared quite well with the theoretical solution computed in the manner described above. Cronenberg et al. noted such agreement even during the early stages of ejection, despite the assumption of a disk of vapor filling the tube cross section. They believed such assumption of a slug geometry, even during the early stages of vapor growth, approximated the surface-volume relationship of the vapor region satisfactorily, at all times. Alternatively, Schlechtendahl (1969) and Peppler et al. (1970) employed a spherical bubble geometry in the early growth stages. The bubble initiating at the tube wall quickly assumed an intermediate shape between a disk and a sphere.

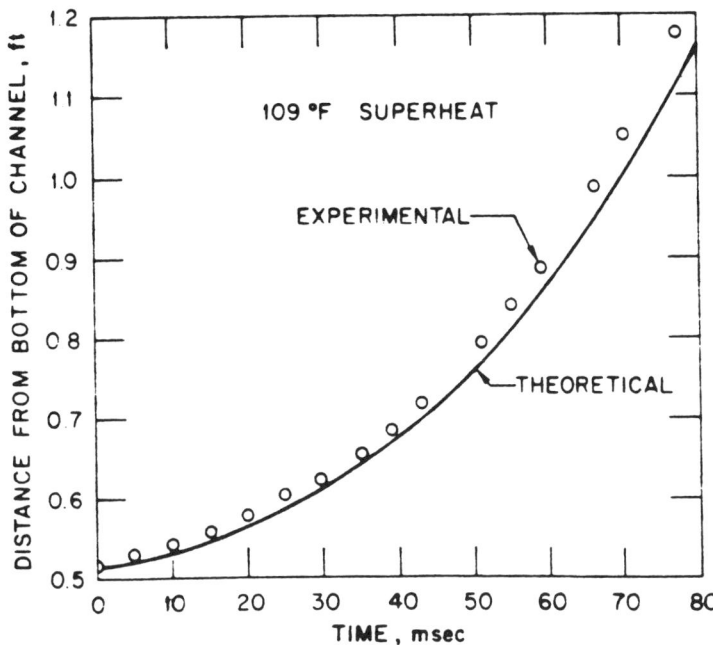

Figure 4.29 Voided channel height versus time for slug expulsion of Freon-113. (From Cronenberg et al., 1971. Copyright © 1971 by Elsevier Science SA, Lausanne, Switzerland. Reprinted with permission.)

4.5 ADDITIONAL REFERENCES FOR FURTHER STUDY

Additional references are given here for recent research work on the subject related to this chapter that are recommended for researchers' outside study:

A general correlation for subcooled and saturated boiling in tubes and annuli, based on the nucleate pool boiling equation by Cooper (1984), was proposed by Liu and Winterton (1991). A total of 991 data points for water, R-12, R-11, R-113, and ethanol were used to compare with their correlation, and some improvement was found over Chen's original correlation (1963a) and that by Gungor and Winterton (1986). Caution should be given to the fact that the behavior of the nucleation site density in flow boiling is significantly different from that in pool boiling. Zeng and Klausner (1993) have shown that the mean vapor velocity, heat flux, and system pressure appear to exert a strong parametric influence. The vapor bubble departure diameter in forced-convection boiling was also analyzed by Klausner et al. (1993), who demonstrated that results for pool boiling are not applicable to flow boiling. In overviews of the onset of nucleation of pure fluids under forced convection and pool boiling, Braver and Mayinger (1992) and Bar-Cohen (1992) both included recent advances toward a better understanding of the individual phenomena influencing boiling incipience. As opposed to the thermal equilibrium model in boiling incipience, such as those of Hsu (1962, Hsu & Graham, 1976), and of Han and Griffith (1965a) (Sec. 2.4.1.2), mechanical models have been advanced which are based on force balance at the liquid–vapor interface within the cavity (Mizukami et al., 1992). The contact angle hysteresis was taken into account, and reentrant cavities were introduced.

A much-needed bubble ebullition cycle study in forced-convective subcooled nucleate boiling conditions was reported recently by Bibean and Salcudean (1994) using high-speed photography. Experiments were performed using a vertical circular annulus at atmospheric pressure for mean flow velocities of 0.08–1.2 m/s (0.3–3.9 ft/sec) and subcoolings of 10–60°C (18–108°F), and filmed conditions were relative to the onset of nucleate boiling and the onset of significant void. Bibean and Salcudean observed the following.

1. Bubble growth occurs rapidly and is followed by a period when the bubbles radius remains relatively constant. Bubbles do not grow and collapse on the wall, but start to slide away from their nucleation sites almost immediately after nucleation.
2. Bubbles later eject into the flow for subcooling below 60°C (108°F). Bubbles become elongated as they slide on the wall and condense while sliding along the wall. These bubbles are shaped like inverted pears, with the steam touching the wall just prior to ejection.
3. The bubble diameter at ejection is smaller than the maximum diameter, which varies between 0.08 and 3 cm (0.03 and 1.2-in.).
4. The bubble behavior is mapped into two regions for increasing heat flux with constant subcooling and flow rate. The first region occurs near the onset of nucleate boiling, where bubbles slide along the wall for more than 0.8 cm (0.3 in.) and up to a distance of 5 cm (2 in.) and oscillate in size before being ejected. The second region occurs well after the onset of nucleate boiling, when the average maximum axial distance is 0.14 cm (0.06 in.). In this region the average axial distance traversed by the condensing bubbles after ejection is 0.06 cm (0.02 in.). The variation in the bubble size and lifetime with respect to subcooling has not been previously documented for low- and medium-subcooled conditions. However, nonmonotonic changes in a bubble's characteristics as a function of subcooling were previously reported for pool boiling conditions (Judd, 1989).

The wide-ranging precautions taken to protect this planet's atmosphere have accelerated the worldwide search for replacements for fully halogenated chlorofluorocarbons. While correlations are available for heat flux and wall superheat at boiling incipience of water and well-wetting fluids, further work should be directed toward testing the correlations with experimental data on new refrigerants and possibly refrigerant mixtures (Spindler, 1994). Kandlikar (1989b) also developed a flow boiling map for subcooled and saturated flow boiling inside circular tubes. It depicted the relationship among the heat transfer coefficient, quality, heat flux, and mass flux for different fluids in the subcooled and saturated flow boiling regions. The particular areas were also indicated where further investigation is needed to validate the trends. Chisholm (1991), after reviewing the forms of correlations for convective boiling in tubes, presented a dimensionless group,

$$\gamma = \frac{h_c}{h_L}\left(\frac{1}{F_t}\right) = \frac{(\psi_L)^2}{(\psi_{Lt})^2} \tag{4-52}$$

where h_c = convective boiling heat transfer coefficient
h_L = convective heat transfer coefficient if liquid component flows alone
F_t = Chen's multiplier (Chen, 1963a) at the thermodynamic critical point (or where the properties of the different phases are identical)
ψ_L = two-phase multiplier with respect to the heat transfer coefficient if the liquid component flows alone
ψ_{Lt} = (ψ_L) at the thermodynamic critical point

Thus, an alternative form of correlation can be given as

$$(\psi_L)^2 = [1 + X^{-2/(2-n)}]^{0.8} \tag{4-53}$$

Here γ was shown to be essentially independent of the Lockhart-Martinelli parameter, X, for values of $(1/X)$ greater than unity. Further study, however, is necessary to develop a generalized equation for the coefficient γ.

Variations of the Chen correlation (Chen, 1963a) have been developed for several widely different channel geometries, e.g., offset strip fins (Mandrusiak and Carely, 1989) and perforated plate fins (Robertson, 1983). The use of other special types of configurations include spray cooling (Yang et al., 1993). Devices that can augment heat transfer are finding challenging applications in a variety of situations. One approach is to insert a twist tape inside the channel. Notably, one of the formidable engineering problems raised by fusion technology is the heat removal from fusion reactor components such as divertors, plasma limiters, ion dumps, and first-wall armor. Using subcooled boiling in tubes for this purpose has been the subject of investigation by Akoski et al. (1991). As the next step, enhancement of water subcooled flow boiling heat transfer in tubes was studied by Weisman et al. (1994).

Controversial issues of the effect of dispersed droplets on the wall-to-fluid heat transfer were discussed by Andreani and Yadigaroglu (1992). They reviewed the dispersed flow film boiling (DFFB) phenomena and their modeling, stating that although available models might be able to account for history effects within the DFFB region, these models were not capable of accounting for any additional (upstream) history effects. Inverted-annular film boiling regime may be described as consisting of a long liquid column or "liquid core," which may contain vapor bubbles, above the quench front, separated from the wall by a vapor film. Modeling of IAFB depends critically on the interfacial heat transfer law between the superheated vapor and the (usually) subcooled liquid core. The net interfacial heat transfer determines the rate of vapor generation and therefore also the film thickness (Yadigaroglu, 1993). Rapid steam generation accelerates the low-viscosity, low-density vapor more easily than the denser core and produces a high steam velocity. When the velocity in the vapor annulus reaches a certain critical value, the liquid core becomes unstable and breaks up into large segments. Following a transition zone, dispersed droplet flow is established. Two-fluid models are well suited for describing correctly the IAFB situation to satisfy the need for a more fundamental approach. Models of Analytis and Yadigaroglu (1987), and of Kawaji and Benerjee (1987) are examples of such models using the two-fluid model. The former predicted the experimental trends correctly, but enhancements of the heat transfer downstream of the quench front were still necessary to match the data. The latter assumed laminar flow in the liquid core and computed the heat transfer coefficient from the interface to the bulk of the liquid from an available analytical solution of transient conduction of heat in a circular cylinder. In general, there are too many adjustable parameters and assumptions influencing the results, as well as considerable difficulties in measuring flow parameters in addition to the value of the heat transfer coefficient to verify these results (Yadigaroglu, 1993). Nelson and Ünal (1992) presented an improved model to predict the breakdown of IAFB considering the various hydrodynamic regimes encountered in IAFB. The main difficulty in all the models mentioned lies in the determination of the superheat of the vapor, a quantity that is extremely difficult to measure. Recent investigation of the dispersed-drop flow model and post-CHF temperature conditions for boiling in a rod bundle are reported by Ünal et al. (1991a, 1991b). George and France (1991) found that thermal nonequilibrium effects are again important even at low wall superheats (25–100°C or 45–180°F).

CHAPTER FIVE

FLOW BOILING CRISIS

5.1 INTRODUCTION

Flow nucleate boiling has an extremely high heat transfer coefficient. It is used in various kinds of compact heat exchangers, most notably in nuclear water reactors. This high heat transfer flux, however, is limited by a maximum value. Above this maximum heat flux, benign nucleate boiling is transformed to a film boiling of poor heat transfer. As mentioned before, this transition of boiling mechanism, characterized by a sudden rise of surface temperature due to the drop of heat transfer coefficient, is called the *boiling crisis* (or *boiling transition*). The maximum heat flux just before boiling crisis is called *critical heat flux* (CHF) and can occur in various flow patterns. Boiling crisis occurring in a bubbly flow is sometimes called *departure from nucleate boiling* (DNB); and boiling crisis occurring in an annular flow is sometimes called *dryout*.

Water reactor cores are heated internally by fission energy with a constant high heating rate. The deteriorated heat transfer mechanism in a boiling crisis would overheat and damage the core. Thus a reactor core designer must fully understand the nature of boiling crisis and its trend in various flow patterns, and then carefully apply this information in design to ensure the safety of nuclear power reactors.

In this chapter, the subject matter is handled in three phases.

Phase 1: To understand the mechanisms of flow boiling crisis by means of

Visual study of boiling crisis in various flow patterns
Microscopic analysis of boiling crisis in each known flow pattern

Phase 2: To evaluate the gross operating parameter effects on CHF in simple channels for

Local p, G, X effects and channel size effects
Coupled p-G-X effects
Boiling length effects

Phase 3: To apply the CHF correlations to rod bundles in reactor design through

Subchannel analysis for PWR cores
CHF predictions in BWR fuel channels
Recommendation of approaches for evaluating CHF margin in reactor design

5.2 PHYSICAL MECHANISMS OF FLOW BOILING CRISIS IN VISUAL OBSERVATIONS

To understand the physical mechanisms of flow boiling crisis, simulated tests have been conducted to observe the hydraulic behavior of the coolant and to measure the thermal response of the heating surface. To do this, the simulation approaches of the entire CHF testing program are considered as follows.

1. Prototype geometries are used in the tests for developing design correlations for the equipment, while simple geometries are used in the tests for understanding the basic mechanisms of CHF.
2. Operating parameters of the prototype equipment are used for running the prototype tests. These parameters are also used in running the basic study tests, but in a wider range to prove the validity of the basic correlations under broader conditions.
3. The prototype test results are correlated by using the operating parameters directly, while the basic study test results (including photographic records and experimental measurements) are used in microscopic analysis based on the local flow characteristics and their constitutive correlations.

5.2.1 Photographs of Flow Boiling Crisis

Boiling crisis is caused essentially by the lack of cooling liquid near a heated surface. In subcooled bubbly flow, the heating surface of a nucleate boiling is usually covered by a bubble layer. An increase in bubble-layer thickness can cause the boiling surface to overheat and precipitate a boiling crisis. Thickening of the bubble layer, therefore, indicates the approach of departure from nucleate boiling.

In saturated annular flow boiling, on the other hand, a liquid film annulus normally covers the heating surface and acts as a cooling medium. Thinning of the liquid film annulus, therefore, indicates approaching dryout. The behavior of bubble layers and liquid annuli are of interest to visual observers.

Tippets (1962) took high-speed motion pictures (4,300 pictures per second) of boiling water flow patterns in conditions of forced flow at 1,000 psia (6.9 MPa) pressure in a vertical, heated rectangular channel. Pictures were taken over the range of mass fluxes from 50 to 400 lb/sec ft^2 (244 to 1,950 kg/m^{2}s), of fluid states from bulk subcooled liquid flow to bulk boiling flow at 0.66 steam quality, and of heat fluxes up to and including the CHF level. In very low-quality and high-pressure flows, Tippets observed from the edges of heater ribbons a vapor stream where surface temperatures are fluctuating. This phenomenon indicates that DNB occurs under the vapor stream. In high-quality flows, Tippets also noticed the profile of a wavy liquid film along the heater surface where surface temperatures are fluctuating. This indicates that dryout occurs under the unstable liquid film. Six frames of Tippet's motion pictures are reproduced in Figure 5.1, and the test opera-parameters are described in Table 5.1. In his article, Tippets did recommend that "close up high speed photography normal (and parallel) to the heated surface should be done," to explore in close detail the nature of the liquid film on the heated surface immediately prior to and through inception of the critical heat flux condition and into transition boiling.

A visual study of the bubble layer in a subcooled flow boiling of water was carried out by Kirby et al. (1965). Their experiments were performed at mass fluxes of 0.5, 1.0, and 1.5 × 10^6 lb/hr ft^2 (0.678, 1.356, and 2.034 × 10^3 kg/m^2 s), subcooling of 4–38°F (2.2–21.2°C), at inlet pressures of 25–185 psia (0.17–1.28 MPa). They found that a frothy steam–water mixture could be produced only under special conditions when the subcooling was less than 9°F (5°C) and the heat flux about one-tenth of the critical heat flux. At higher subcoolings, the bubbles condensed close to the wall; and at higher heat fluxes, the bubbles coalesced on and slid along the heater surface with a liquid layer in between. Even with a careful inspection of the fast movie film, no change in flow pattern was evident during the boiling crisis. The interval between the time when the heater temperature started rising and the time when the heater started melting is about 50 msec. Kirby et al. noted that several bubbles passed over the boiling crisis location during this interval.

A visual study was made by Hosler (1963) with water flowing over a heating surface in a rectangular channel. A front view of the heating surface is shown in Figure 5.2. This figure shows flow patterns at various local enthalpies at various surface heat fluxes. Pictures were taken at 600 psia (4.14 MPa) and a mass flux of 0.25 × 10^6 lb/hr ft^2 (339 kg/m^2 s). For higher pressures and higher mass fluxes of PWR operation range, e.g., 2,000 psia (17.9 MPa) and 2.5 × 10^6 lb/hr ft^2 (3,390 kg/m^2 s), the bubble sizes are expected to be much smaller than in these pictures. The bubbles of a high mass flux under high pressures at higher surface heat flux would appear like (or even behave similar to) those under lower pressures at a lower surface heat flux as in Figure 5.2. Therefore, this figure could simulate the flow pattern immediately before boiling crisis at a water flow of high pressure and high mass flux (say, 2,000 psia and 2.5 × 10^6 lb/hr ft^2 or 17.9 MPa and 3,390 kg/m^2 s).

306 BOILING HEAT TRANSFER AND TWO-PHASE FLOW

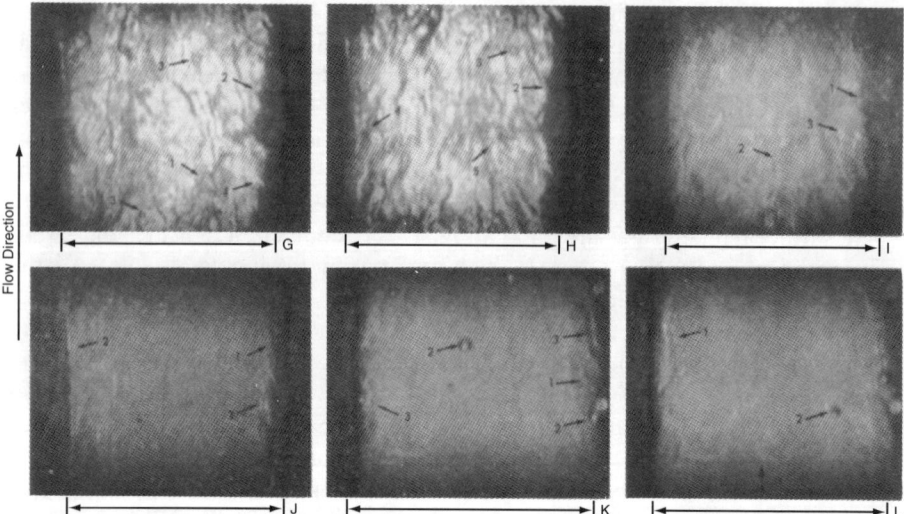

Figure 5.1 Typical boiling water flow patterns observed at 1,000 psia (operating conditions listed in Table 5.1). (From Tippets, 1962. Copyright © 1962 by American Society of Mechanical Engineers, New York. Reprinted with permission.)

Gaertner (1965) studied nucleate pool boiling on a horizontal surface in a water pool under atmospheric pressure. He increased the surface heat flux gradually. The vapor structures on the surface progressed from discrete bubbles to vapor columns and vapor mushrooms, and finally to vapor patches (dryout). The observed pictures of vapor mushroom and vapor patch are also sketched in Figure 5.3.

Because of the damagingly high temperature of the heater surface at DNB in a water flow, most studies of bubble behavior near the boiling crisis have been conducted on a Freon flow, where the surface temperature is much lower than in a water flow. The validity of the simulations of boiling crisis has been established in many studies, such as those of Stevens & Kirby (1964), Cumo et al. (1969), Tong et al. (1970), Mayinger (1981), and Celata et al. (1985).

Tong et al. (1966b) conducted a photographic study of boiling flow of Freon-113 in a vertical rectangular channel. While investigating the microscopic mechanism of the boiling crisis, they did find vapor mushrooms at DNB in a low-mass-flux flow and at a low pressure as shown in Figure 5.4. These vapor mushroom-vapor patch mechanisms are similar to those found by Gaertner (1965) in water pool boiling. Tong et al. also found the mechanism of DNB under a thin dense bubble layer in a highly subcool and high-velocity Freon flow as shown in Figure 5.5.

In a subcooled Freon flow, Tong (1972) caught a front view of a heating surface with nucleate boiling and film boiling existing simultaneously on the same

Table 5.1 Operating conditions for motion-picture frames in Figure 5.1

	G, $\frac{\text{lb}}{\text{sec ft}^2}$	X	q'' $\frac{10^6 \text{ Btu}}{\text{hr ft}^2}$	$\frac{q}{q_c}$	Notes (arrow designations in parentheses)
G					Frames exposed 0.07 sec apart. Wave structure on liquid film on window (1). Profile of wavy liquid film against heater surface (2). Tiny spherical bubbles in liquid film on window (3). Vapor streamers from edge of heater ribbons (4). Shadow of focusing target visible at right side (5). Fluctuating surface temperature.
H	50	0.656	0.588	1.0	
I	100	0.465	0.857	1.0	Profile of wavy liquid film against heater surface (1). Finely divided waves on liquid film on window (2). Vapor streamers from edges of heater ribbons (3). Fluctuating surface temperature.
J	200	0.160	0.957	1.0	Profile of wavy liquid film against right-hand heater surface (1). Edge of heater ribbon visible at left (2). Vapor streamers from edges of heater ribbons (3). Fluctuating surface temperature.
K	400	0.074	0.987	1.0	Frame exposed 0.01 sec before power "trip." Profile of wavy irregular liquid film against heater surface (1). Spherical bubbles in foreground are outside channel (2). Vapor streamers from edges of heater ribbons (3). Fluctuating surface temperature.
L	400	0.037	1.141	1.0	Frame exposed 0.01 sec before power "trip." Vapor streamers forming in liquid film at edge of heater ribbon (1). Spherical bubbles in foreground are outside channel (2). Fluctuating surface temperature.

surface during a slow reduction of the power input as shown in Figures 5.6 and 5.7. The thermocouple on the heating surface showed a temperature rise at the connection of the nucleate and film boiling regions. This temperature rise confirmed the occurrence of DNB at the end of the observed nucleate boiling region.

Two visual studies of Freon boiling crisis were conducted at the University of Pittsburgh (Lippert, 1971) and at Michigan University (Mattson et al., 1973). Both programs succeeded in identifying the DNB under the saw-shaped bubble layer of subcooled Freon flows as shown in Figures 5.8 and 5.9, respectively

The bubble behavior near the boiling crisis is three-dimensional. It is hard to show a three-dimensional view in side-view photography, because the camera is focused only on a lamination of the bubbly flow. Any bubbles behind this lamination will be fussy or even invisible on the photograph, but they can be seen by the naked eye and recorded in sketches as shown in Section 5.2.3. For further visual studies, the details inside bubble layers (such as the bubble layer in the vicinity of the CHF) would be required. Therefore, close-up photography normal and parallel to the heated surface is highly recommended.

308 BOILING HEAT TRANSFER AND TWO-PHASE FLOW

Pressure: 600 PSIA (4.14 × 10⁶ Pa)
Mass velocity: 0.25 × 10⁶ lb/hr ft² (339 kg/m²s)
Inlet temperature: 400°F (204°C)
Geometry: 0.134 in. × 1.00 in. × 24 in. long rectangular channel
(0.34 cm × 2.54 cm. × 61 cm
Location: 21.5 in. from inlet
(54.6 cm)

$$\text{A} \qquad \text{B} \qquad \text{C}$$
$\dfrac{\phi}{10^6} = 0.129 \dfrac{\text{BTU}}{\text{Hr FT}^2} (0.407 \dfrac{\text{W}}{\text{m}^2})$ $\qquad$ $\dfrac{\phi}{10^6} = 0.145 \dfrac{\text{BTU}}{\text{Hr FT}^2} (0.457 \dfrac{\text{W}}{\text{m}^2})$ $\qquad$ $\dfrac{\phi}{10^6} = 0.174 \dfrac{\text{BTU}}{\text{Hr FT}^2} (0.548 \dfrac{\text{W}}{\text{m}^2})$
$\Delta T_S = 20°\text{F} (11\ \text{C})$ $\qquad\qquad\qquad$ $\Delta T_S = 12°\text{F} (6.7\ \text{C})$ $\qquad\qquad\qquad$ $X = 0.004$

$$\text{D} \qquad \text{E} \qquad \text{F}$$
$\dfrac{\phi}{10^6} = 0.212 \dfrac{\text{BTU}}{\text{Hr FT}^2} (0.668 \dfrac{\text{W}}{\text{m}^2})$ $\qquad$ $\dfrac{\phi}{10^6} = 0.239 \dfrac{\text{BTU}}{\text{Hr FT}^2} (0.753 \dfrac{\text{W}}{\text{m}^2})$ $\qquad$ $\dfrac{\phi}{10^6} = 0.287 \dfrac{\text{BTU}}{\text{Hr FT}^2} (0.905 \dfrac{\text{W}}{\text{m}^2})$
$X = 0.041$ $\qquad\qquad\qquad\qquad$ $X = 0.063$ $\qquad\qquad\qquad\qquad$ $X = 0.102$

Figure 5.2 Photographs of boiling water flow patterns. (From Hosler, 1965. Copyright © 1965 by American Institute of Chemical Engineers, New York. Reprinted with permission.)

(a) Vapor Mushroom with Stems   (b) Vapor Patch at CHF (without Stems)

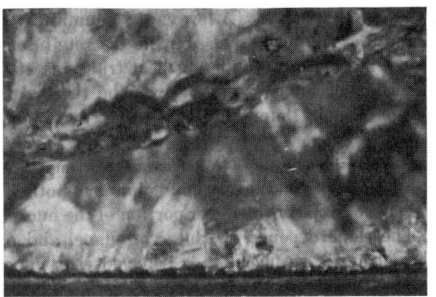 

q/A: 9.22 × 10⁵W/m² 293,200 Btu/hrft²
ΔT : 21.2°C (38.2°F)
Surface : 4/0 copper
Field of view: 2.66 × 1.80cm (1″ × 0.71″)

q/A: 12.0 × 10⁵W/m² 381,000 Btu/hrft²
ΔT : 36.6°C (65.9°F)
Surface : 4/0 copper
Field of view: 4.30 × 3.0cm (1.7″ × 1.2″)

Authors' Interpretation

Figure 5.3 Visual observation of boiling crisis in water pool. (From Gaetner, 1963. Copyright © 1963 by General Electric Co., San Jose, CA. Reprinted with permission.)

5.2.2 Evidence of Surface Dryout in Annular Flow

Hewitt (1970) studied the behavior of a water film on the heated central rod in an annular test section carrying flowing steam at low and high pressures. The water was introduced through a porous wall. The critical heat flux was measured, as was the residual film flow on the heated rod. A set of typical measurements is shown in Figure 5.10. It can be seen that at film breakdown (when dry patches appeared), the measured residual liquid film flow rate was very small and the power was very close to the burnout power. Visual and photographic studies revealed a relatively stable condition at the breakup of the climbing film, with rivulets around dried-out patches. This test clearly indicates that boiling crisis in an annular flow pattern is the result of progressive loss of water from the film by evaporation, and reentrainment and local value of critical heat flux are of secondary importance.

5.2.3 Summary of Observed Results

Based on visual observations and measurements in various basic tests, observed impressions are recorded in sketches as a summary. These sketches can be justified

310 BOILING HEAT TRANSFER AND TWO-PHASE FLOW

G = 0.36 × 10⁶ lb/hrft² (486 kg/m²s);
ΔT = 4°F Subcooling; (2.2 C);
P = 10 psig (p/p$_c$ = 0.05) (0.17MPa);
q/A = 64000 Btu/hr/ft² (2.0 × 10⁵ W/m²s).

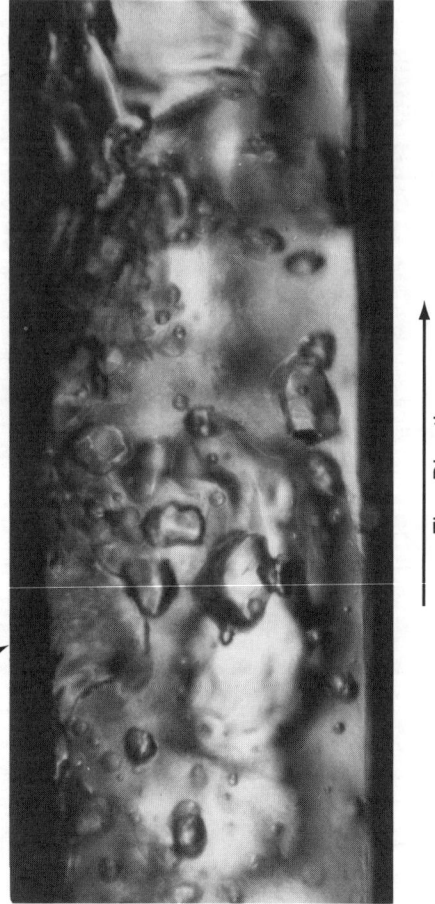

Figure 5.4 Enlarged view of DNB at low-pressure Freon flow. (From Tong et al., 1966b. Copyright © 1966 by American Society of Mechanical Engineers, New York. Reprinted with permission.)

by the agreement of the model analysis of the sketches and the measured data. Such justifications are given in a later section as microscopic analysis.

Flow boiling crisis are closely related to the type of flow pattern present at its occurrence. The local void distributions of various flow patterns can be generally divided into two categories as sketched in Figure 5.11. Accordingly, the flow boiling crisis are roughly classified into two categories.

Category 1: Boiling crisis in a subcooled or low-quality region This category occurs only at a relatively high heat flux. The high heat flux causes intensive boiling, so that the bubbles are crowded in a layer near the heated surface as shown in Figure 5.5. The local voidage impairs surface cooling by reducing the amount of incoming

FLOW BOILING CRISIS **311**

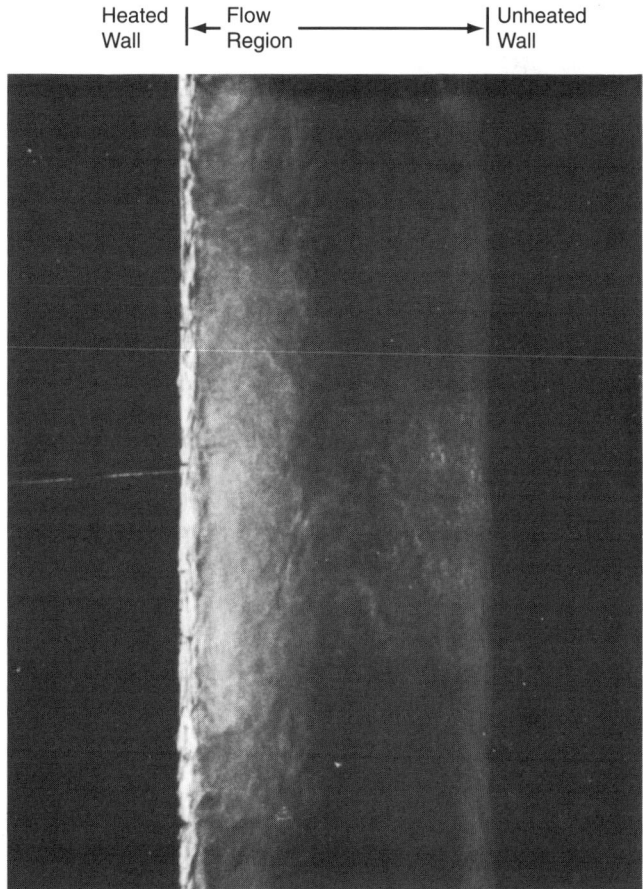

Figure 5.5 Vertical upward, high-velocity boiling flow of Freon-113: mass flux 1.06×10^6 lb_m/hr ft^2 (1,432 kg/m^2 s); bulk subcooling 66°F (37°C); pressure 41 psig (0.38 MPa); heat flux 14,300 Btu/hr ft^2 (45,000 W/m^2). (From Tong, 1965. Reprinted with permission.)

liquid. At boiling crisis, the local void spreads as a vapor blanket on the heating surface and the nucleate boiling disappears, replaced by a film boiling. Thus was the name, departure from nucleate boiling (DNB), established. This type of boiling crisis usually occurs at a high flow rate, and the flow pattern is called inverted annular flow. The observed boiling crisis mechanisms in the subcooled or low-quality, bubbly flows are sketched in Figure 5.12, listed in order of decreasing subcooling:

(A) The surface overheating at CHF in a highly subcooled flow is caused by poor heat transfer capability of the liquid core.

312 BOILING HEAT TRANSFER AND TWO-PHASE FLOW

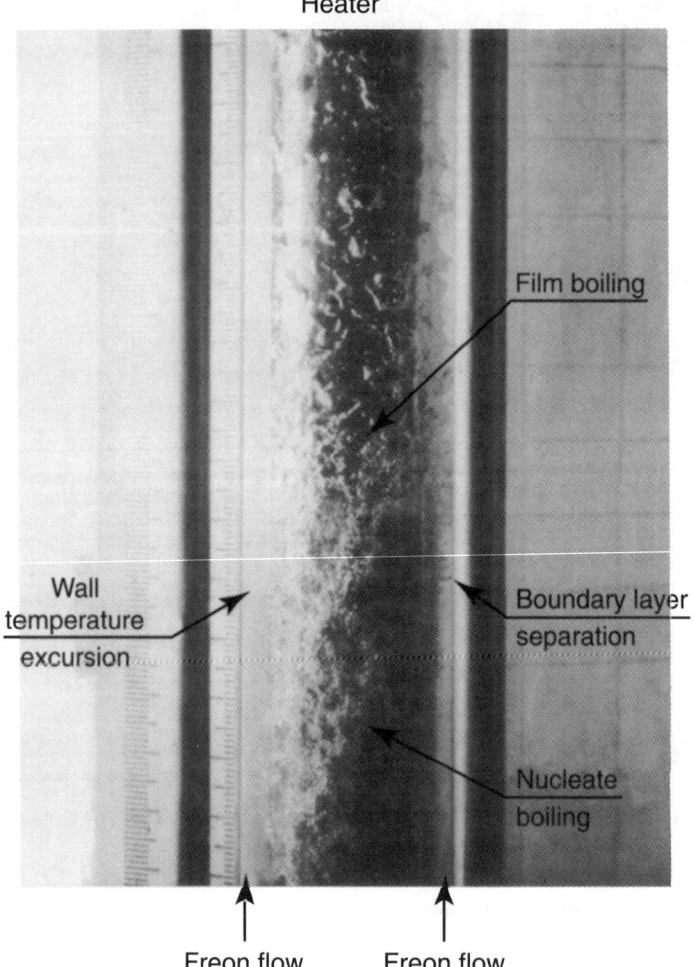

Figure 5.6 Boiling crisis in subcooled Freon flow: mass flux 1.5×10^6 lb_m hr ft^2 (2,030 kg/m^2 s), pressure 40 psia (0.27 MPa), $\Delta T_{sc} = 30°F$ (16.7°C) at boiling crisis. (From Tong 1972. Reprinted with permission of U.S. Department of Energy, subject to the disclaimer of liability for inaccuracy and lack of usefulness printed in the cited reference.)

(B) The DNB in a medium- or low-subcooling bubbly flow is caused by near-wall bubble crowding and vapor blanketing.

(C) The DNB in a low-quality froth flow is caused by a bubble burst under the bubbly liquid layer.

The two-phase mixture in the boundary layer near the heating surface can hardly be in thermodynamic equilibrium with the bulk stream. Thus, the magni-

FLOW BOILING CRISIS 313

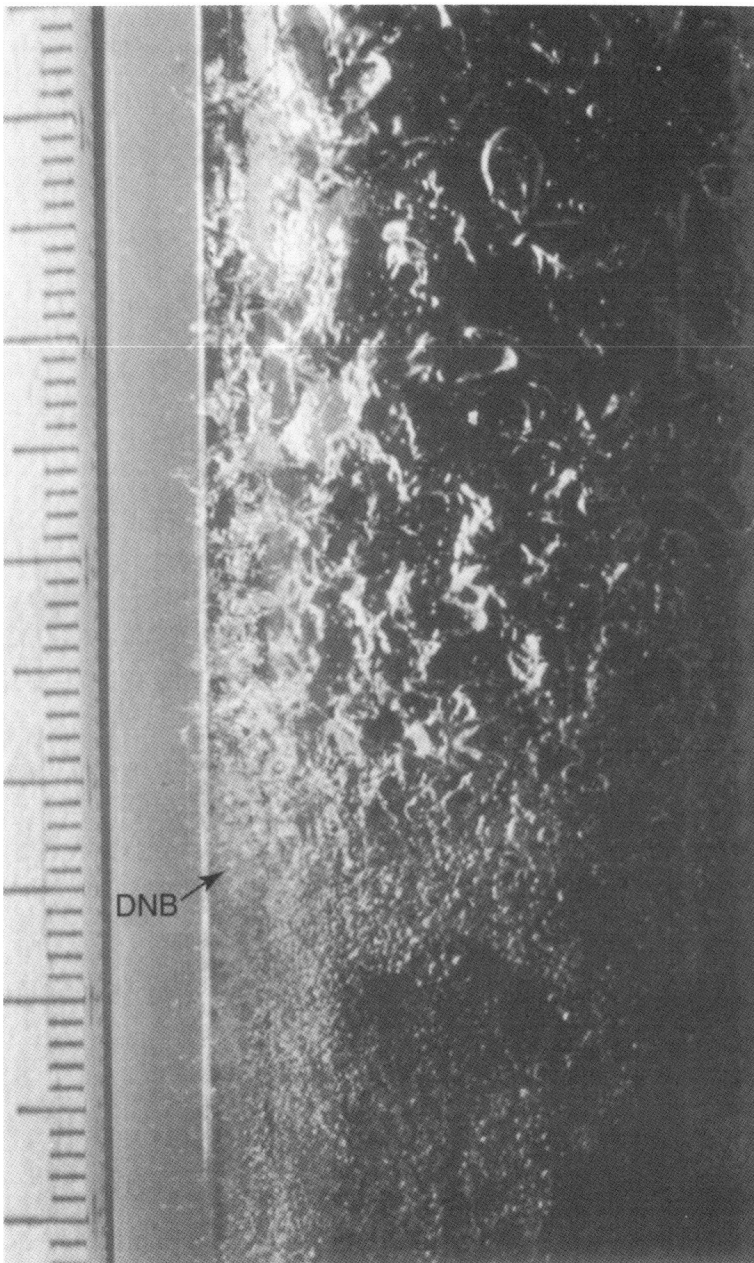

Figure 5.7 Enlarged view of boiling crisis between nucleate and film boiling. (From Tong, 1972. Reprinted with permission of U.S. Department of Energy, subject to the disclaimer of liability for inaccuracy and lack of usefulness printed in the cited reference.)

314 BOILING HEAT TRANSFER AND TWO-PHASE FLOW

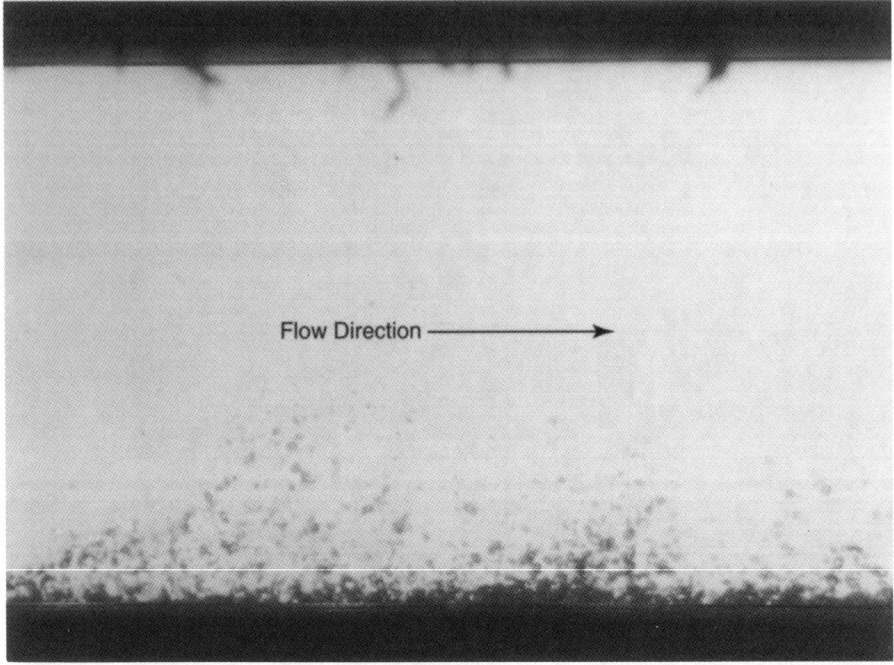

Figure 5.8 Saw-shaped bubble layer at DNB in a high-pressure, subcooled, vertical Freon flow: $p = 190$ psig (1.4 MPa); $G = 3.2 \times 10^6$ lb$_m$/hr ft^2 (4,320 kg/m^2 s); $\Delta T_{sub} = 29°F$ (16°C) subcooled; $(q/A)_c = 1.2 \times 10^5$ Btu/hr ft^2 (3.8 × 10^5 W/m^2). (From Dougall and Lippert, 1973. Reprinted with permission of NASA Scientific & Technical Information, Linthicum Heights, MD.)

tude of the CHF depends on the surface proximity parameters as well as the local flow patterns. The surface proximity parameters include the surface heat flux, local voidage, boundary-layer behavior, and its upstream effects. When this type of boiling crisis occurs, the internal heating source causes the surface temperature to rise rapidly to a very high value. The high temperature may physically burn out the surface. In this case, it is also described as fast burnout.

Category 2: Boiling crisis in a high-quality region This category of boiling crisis occurs at a heat flux lower than the one described previously. The total mass flow rate can be small, but the vapor velocity may still be high owing to the high void fraction. The flow pattern is generally annular. In each of these configurations a liquid annulus normally covers the heated surface as a liquid layer and acts as a cooling medium. If there is excessive evaporation due to boiling, this liquid layer breaks down and the surface becomes dry. Thus this boiling crisis is also called dryout. The magnitude of the CHF in such a high-void-fraction region depends strongly on the flow pattern parameters, which include the flow quality, average

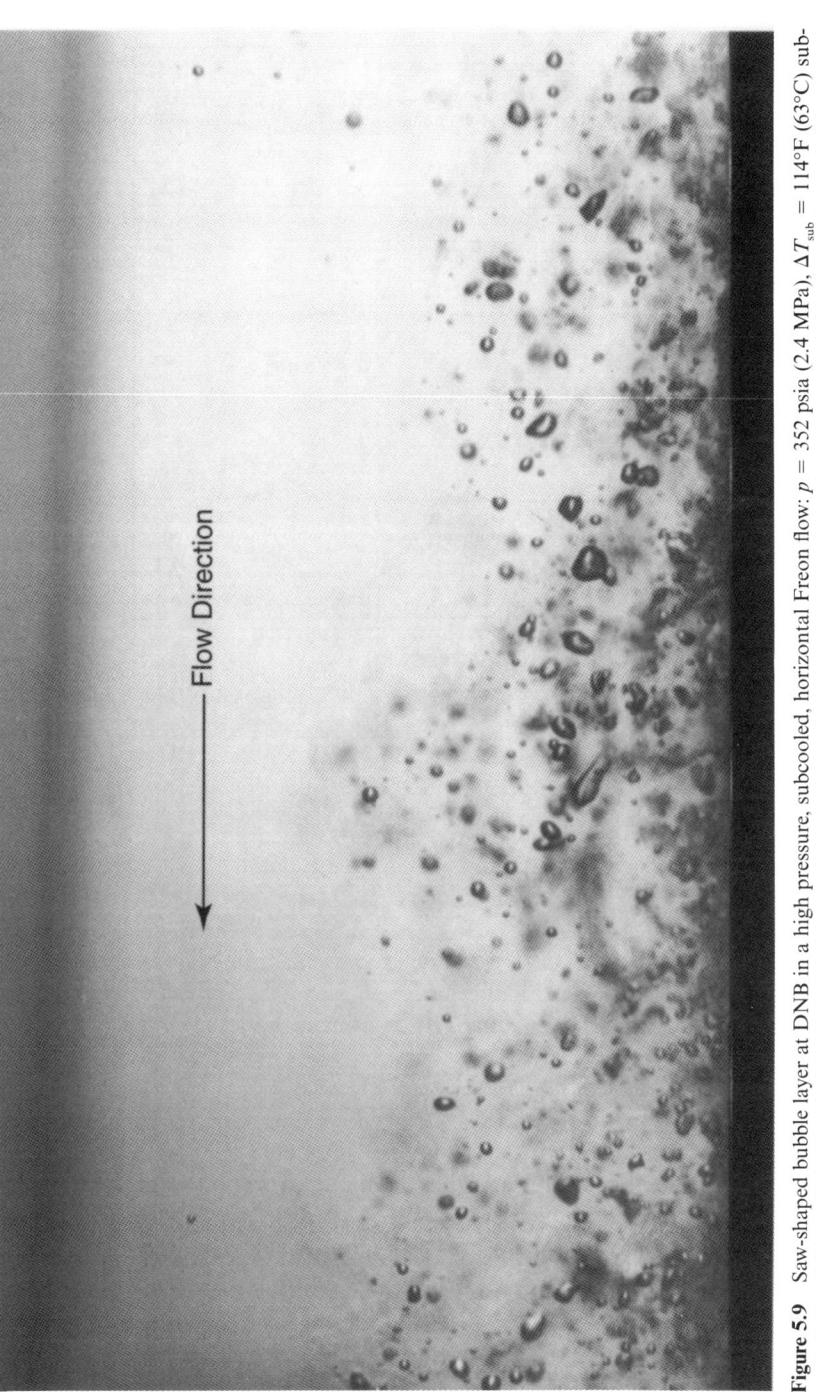

Figure 5.9 Saw-shaped bubble layer at DNB in a high pressure, subcooled, horizontal Freon flow: $p = 352$ psia (2.4 MPa), $\Delta T_{sub} = 114°F$ (63°C) subcooled; $G = 2.96 \times 10^6$ lb$_m$/hr ft^2 (4,000 kg/m^2 s); $(q/A)_c = 2.12 \times 10^5$ Btu/hr ft^2 (666.7 W/m^2). (From Mattson et al., 1973. Copyright © 1973 by American Society of Mechanical Engineers, New York. Reprinted with permission.)

316 BOILING HEAT TRANSFER AND TWO-PHASE FLOW

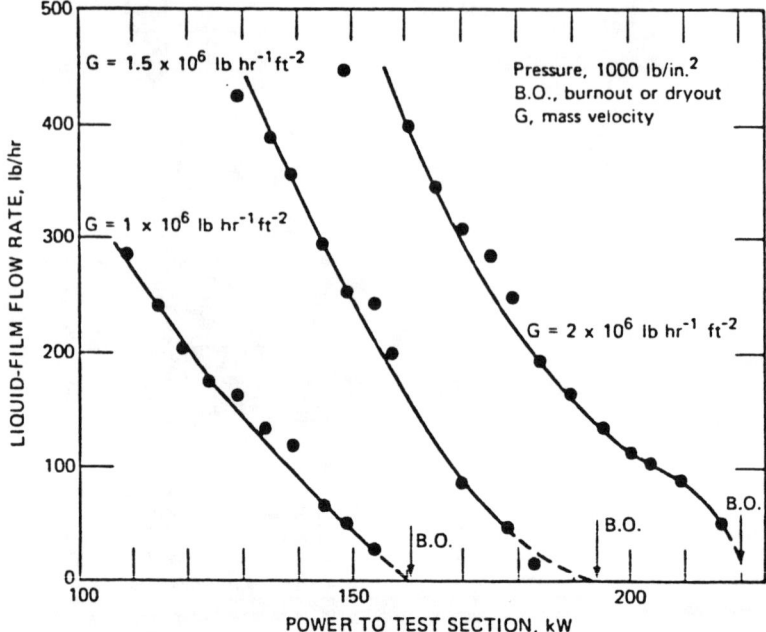

Figure 5.10 Dryout of liquid-film thickness. (From Hewitt, 1970. Copyright © 1970 by Hemisphere Publishing Corp., New York. Reprinted with permission.)

void fraction, slip ratio, vapor velocity, liquid-layer thickness, boiling length, etc. It may have only a weak dependence on the surface parameters. The observed boiling crisis mechanisms in medium- or high-quality annular flows are sketched in Figure 5.13:

(A) The dryout in a medium-quality annular flow is caused by liquid-layer disruption due to surface wave instability.
(B) The dryout in a high-quality annular flow is caused by the dryup of a liquid layer on the heating wall.

When the boiling crisis occurs, the surface temperature rises. Because of the fairly good transfer coefficient of a fast-moving vapor core in an annular flow, the wall temperature rise after a dryout in the high-quality region is usually smaller than that in a subcooled boiling crisis. It is even possible to establish steady-state conditions at moderate wall temperatures, so that physical burnout may not occur immediately. Thus dryout is also described as slow burnout.

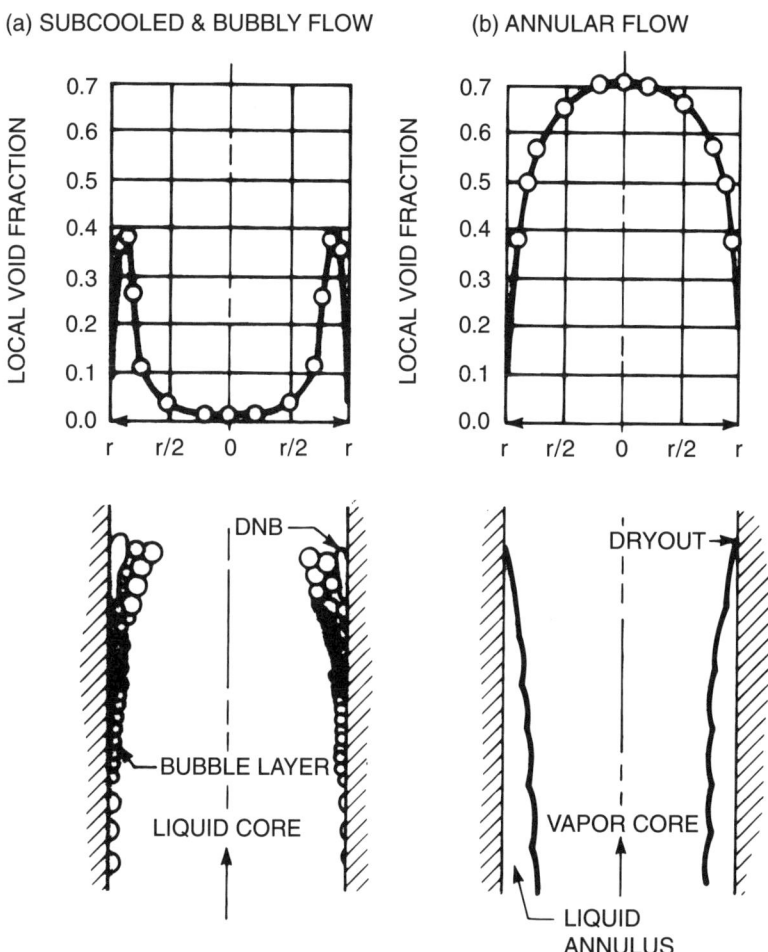

Figure 5.11 Comparison of boiling-crisis mechanisms in various flow patterns. (From Tong and Hewitt, 1972. Copyright © 1972 by American Society of Mechanical Engineers, New York. Reprinted with permission.)

5.3 MICROSCOPIC ANALYSIS OF CHF MECHANISMS

In the microscopic analysis of CHF, researchers have applied classical analysis of the thermal hydraulic models to the CHF condition. These models are perceived on the basis of physical measurements and visual observations of simulated tests. The physical properties of coolant used in the analysis are also deduced from the operating parameters of the test. Thus the insight into CHF mechanisms revealed in microscopic analysis can be used later to explain the gross effects of the operating parameters on the CHF.

318 BOILING HEAT TRANSFER AND TWO-PHASE FLOW

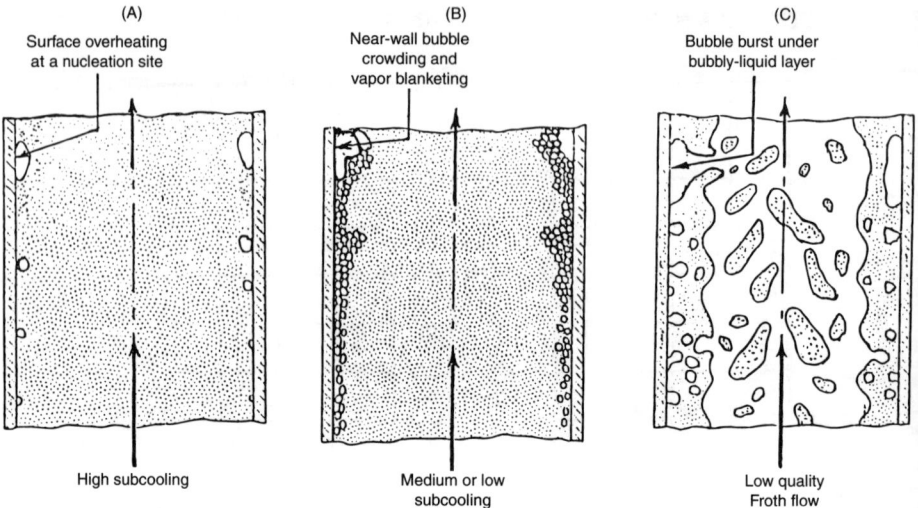

Figure 5.12 Mechanisms of DNB in bubbly flows.

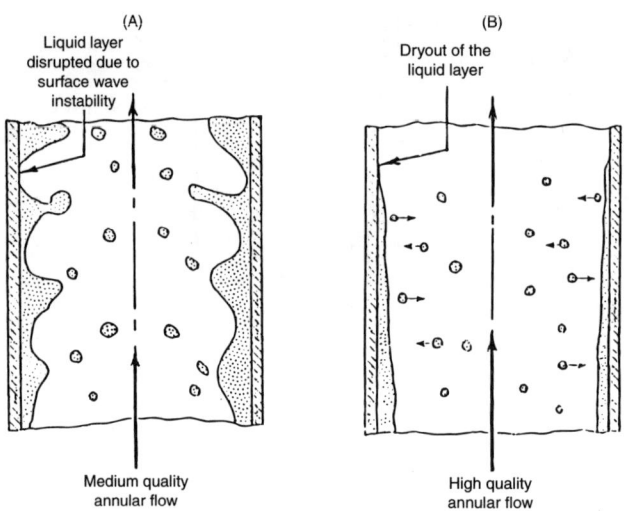

Figure 5.13 Mechanisms of dryout in annular flows.

5.3.1 Liquid Core Convection and Boundary-Layer Effects

A highly subcooled flow channel, even at CHF, is occupied almost entirely by a liquid core as shown in Figure 5.12. It can be postulated that the convective heat transfer of the liquid core is the limiting factor for occurrence of CHF at the wall. The following ways of analyzing the liquid core convective capability have been suggested:

1. The liquid core temperature and velocity distribution analysis was suggested by Bankoff (1961).
2. The boundary-layer separation and Reynolds flux analyses were suggested by Kutateladze and Leont'ev (1964, 1966), Tong (1968b), and Wallis (1969, 1970).
3. The subcooled core bubble-layer liquid exchange and bubble condensation analysis was suggested by Tong (1975).

5.3.1.1 Liquid core temperature and velocity distribution analysis. Bankoff (1961) analyzed the convective heat transfer capability of a subcooled liquid core in local boiling by using the turbulent liquid flow equations. He found that boiling crisis occurs when the core is unable to remove the heat as fast as it can be transmitted by the wall. The temperature and velocity distributions were analyzed in the single-phase turbulent core of a boiling annular flow in a circular pipe of radius r. For fully developed steady flow, the momentum equation is given as

$$\tau_o\left(1 - \frac{y}{r}\right) = -\rho_L \overline{u'v'} + \mu\left(\frac{du}{dy}\right) \tag{5-1}$$

where the Reynolds stress $\overline{u'v'}$ is the time average of the product of the velocity deviations in the axial and radial directions, respectively. Similarly, the energy equation is

$$q''\left(1 - \frac{y}{r}\right) = -\rho_L c_L \overline{T'v'} + k_L\left(\frac{dT}{dy}\right) \tag{5-2}$$

where $\overline{T'v'}$ is the time average of the product of the local temperature deviation and velocity deviation in the radial direction. By assuming the validity of the Reynolds analogy for momentum and energy transfer,

$$\frac{\overline{T'v'}}{(dT/dy)} = \frac{1}{\Pr_t}\left[\frac{\overline{u'v'}}{(du/dy)}\right] \tag{5-3}$$

From these three equations, the dimensionless velocity and temperature distributions can be established (Deissler, 1955; Townsend, 1956):

$$u^+ - u_1^+ = 2.5 \ln\left(\frac{y^+}{y_1^+}\right) \quad y^+ < y_1^+ \tag{5-4}$$

$$u^+ - u_1^+ = T^+ - T_1^+ \quad y^+ > y_1^+ \tag{5-5}$$

where

$$u^+ = \frac{u}{(g_c \tau_o/\rho_L)^{1/2}} \qquad y^+ = \frac{y}{\nu}\left(\frac{g_c \tau_o}{\rho_L}\right)^{1/2}$$

$$T^+ = (T_w - T)\left(\frac{\rho_L c_L}{q''}\right)\left(\frac{g_c \tau_o}{\rho_L}\right)^{1/2}$$

and y_1^+ is the dimensionless thickness of the bubble layer.

Sabersky and Mulligan's (1955) experimental results indicate that Eqs. (5-4) and (5-5) hold all the way to the wall in a subcooled flow boiling system. Since $u_w^+ = T_w^+ = 0$, it follows that $u_c^+ = T_c^+$ inside the core. Combining the preceding equations gives

$$\frac{T^+ - T_1^+}{T^+} = \frac{T_1 - T_c}{T_w - T_c} = \frac{2.5}{u_c^+}\ln\left(\frac{y_c^+}{y_1^+}\right) \tag{5-6}$$

where $y_c = r$

T_c = center temperature of the turbulent core

$$u_c^+ = (u_c^+ T_c^+)^{1/2}$$

$$= \left[\frac{(T_w - T_c)\rho_c c_c u_L}{q''}\right]^{1/2}$$

Hence,

$$\frac{T_1 - T_c}{T_w - T_c} = \frac{2.5(q'')^{1/2}}{[(T_w - T_c)\rho c u]_c^{1/2}}\ln\left(\frac{y_c^+}{y_1^+}\right) \tag{5-7}$$

where $T_w \approx T_{sat}$ and $T_{sat} = T_s$ (at the edge of the superheated layer).

Bankoff calculated T_1 by using Gunter's experimental data and obtained the interesting result that, in each series of runs, T_1 rises steeply toward the saturation temperature as burnout is approached. This gives a fairly thick bubble layer, which increases the degree of superheat near the wall. Bankoff concluded that "burnout occurs when the core is unable to remove the heat as fast as it can be transmitted by the wall layer."

5.3.1.2 Boundary-layer separation and Reynolds flux. Kutateladze and Leont'ev (1964, 1966) suggested that the flow boiling crisis can be analyzed using the concept of boundary-layer separation (blowoff) from a permeable flat plate with gas injection (without condensation), as shown in Figure 5.14. Kutateladze and Leont'ev (1966) also give the critical condition of boundary layer separation from a flat plate with isothermal injection of the same fluid as

FLOW BOILING CRISIS **321**

Figure 5.14 Kutateladze's concept of boundary-layer separation over a flat plate.

$$(\rho_{inj} v_{inj})_{crit} = 2 f_o \rho_o U_o \tag{5-8}$$

where f_o = friction factor for no injection
v = velocity normal to the main stream
U = velocity parallel to the main stream
subscript inj = injected gas
subscript o = main stream fluid

The physical interpretation of flow boiling crisis can be considered primarily a hydrodynamic phenomenon of radial bubble injection to the axial flow with a momentum exchange near the wall. The process leading to crisis begins with the evaporating steam blowing off the liquid layer near the wall in an annular flow having a liquid annulus, which is similar to the boundary-layer separation in a single-phase flow. The rate of radial bubble blowoff can be simply written as

$$\rho_v v_b = \frac{q''}{H_{fg}} \tag{5-9}$$

The flow friction of a channel having subcooled local boiling behaves in a manner similar to the flow friction of a channel having a rough surface. From air–water data obtained at low void fractions (Malnes, 1966), the two-phase flow friction factor without bubble departure was found (Tong, 1968b; Thorgerson, 1969) to be

$$f_{TP} \propto Re^{-0.6} \tag{5-10}$$

where

$$\mathrm{Re} = \frac{\rho_L U_o D_e}{\mu_{sat}}$$

When Eq. (5-10) is combined with Eqs. (5-8) and (5-9), we get

$$q''_{crit} = \frac{C_1 H_{fg} \rho_L U_o}{(\mathrm{Re})^{0.6}} \tag{5-11}$$

By comparing Eq. (5-11) with existing flow boiling crisis data obtained in water at 1,000–2,000 psia (6.9–13.8 MPa) inside a single tube test section with uniform heat flux, Tong (1968) reported that C_1 is a function of the bulk quality and Eq. (5-11) becomes

$$q''_{crit} = \frac{(1.76 - 7.43X + 12.22X^2) H_{fg} (\rho_L U_o)^{0.4} \mu_L^{0.6}}{(D_e)^{0.6}} \tag{5-12}$$

where X is the bulk quality and is negative in subcooled local boiling, and G is in the range 1×10^6 to 3×10^6 lb/hr ft² (1,351 to 4,054 kg/m² s). This equation agrees with the data within ±25% for pressures higher than 1,000 psia (7.0 MPa). While Eq. (5-8) is derived from the boundary-layer separation concept, a similar equation can be derived based on the Reynolds flux concept by using the analysis of Wallis (1969, 1970) and the mixing data of Styrikovich et al. (1970). The similarity of equations indicates that these two concepts are complementary.

Recently, Celata et al. (1994c) modified Eq. (5-12) on the parameter C_1 together with a slight modification of the Reynolds number power, to give a more accurate prediction in the range of pressures below 5.0 MPa (725 psia). The modified equation is

$$q''_{crit} = \frac{C_1 H_{fg} \rho_L U_o}{(\mathrm{Re})^{0.5}} \tag{5-12a}$$

where $C_1 = (0.216 + 4.74 \times 10^{-2} p)\psi$ (p is pressure in MPa)

$\psi = 0.825 + 0.986 X_o$ if the exit quality $X_o > -0.1$

$\psi = 1$ $X_o < -0.1$

$\psi = \dfrac{1}{2 + 30 X_o}$ $X_o > 0$

The modified correlation was developed using a total of 1,865 data points in the following operating ranges:

$0.1 < p < 8.4$ MPa	$14.7 < p < 1,220$ psia
$0.3 < D < 25.4$ mm	$0.012 < D < 1.0$ in.
$0.1 < L < 0.61$ m	$0.33 < L < 2.0$ ft
$2 < G < 90.0$ Mg/m^2 s	$1.48 < G < 66.6 \times 10^6$ lb/hr ft^2
$90 K < \Delta T_{sub,in} < 230 K$	$162°F < \Delta T_{sub,in} < 414°F$

The experimental data were reported by Celata (1991), and Celata and Mariani (1993) together with more than 20 other references by Celata et al. (1994a, 1994c). The modified correlation, Eq. (5-12a), gives a root-mean-square error of 21.2% in the CHF prediction of all 1,865 data points. Celata et al. (1994c) concluded in their assessment of correlations and models for the prediction of CHF in water subcooled flow boiling that the modified correlation is the best correlation compared with three other existing correlations. It does give better confidence in the boundary-layer separation theory for describing CHF phenomena in subcooled flow boiling.

5.3.1.3 Subcooled core liquid exchange and interface condensation. A core heat transfer limiting CHF analysis was developed by Tong (1975) in combining physical effects of the liquid exchange between the subcooled liquid core and the two-phase boundary layer and the interface condensation. The subcooled CHF is compared with the known CHF at zero equilibrium quality. The comparison ratio can be correlated in nondimensional functions: A Jacob number is used to define the effectiveness of liquid–vapor exchange; a Reynolds number indicates the intensities of turbulent mixing; and the reduced pressure determines the bubble size. Thus, the comparison ratio becomes

$$\frac{q''_{sub} - q''_{X=0}}{q''_{X=0}} = C(p/p_c)^{\ell} (\text{Re})^n (\text{Ja})^m \tag{5-13}$$

The values of C, ℓ, n, and m were determined by plotting the CHF data of high-pressure water flowing in circular tubes with uniform heat flux as shown in Figure 5.15, and in one-side-heating annuli as shown in Figure 5.16. Based on the above plots, Eq. (5-13) becomes

$$\frac{q''_{cr.\,sub}}{q''_{cr.\,X=0}} = 1 + 0.00216 \left(\frac{p}{p_c}\right)^{1.8} (\text{Re})^{0.5} (\text{Ja}) \tag{5-14}$$

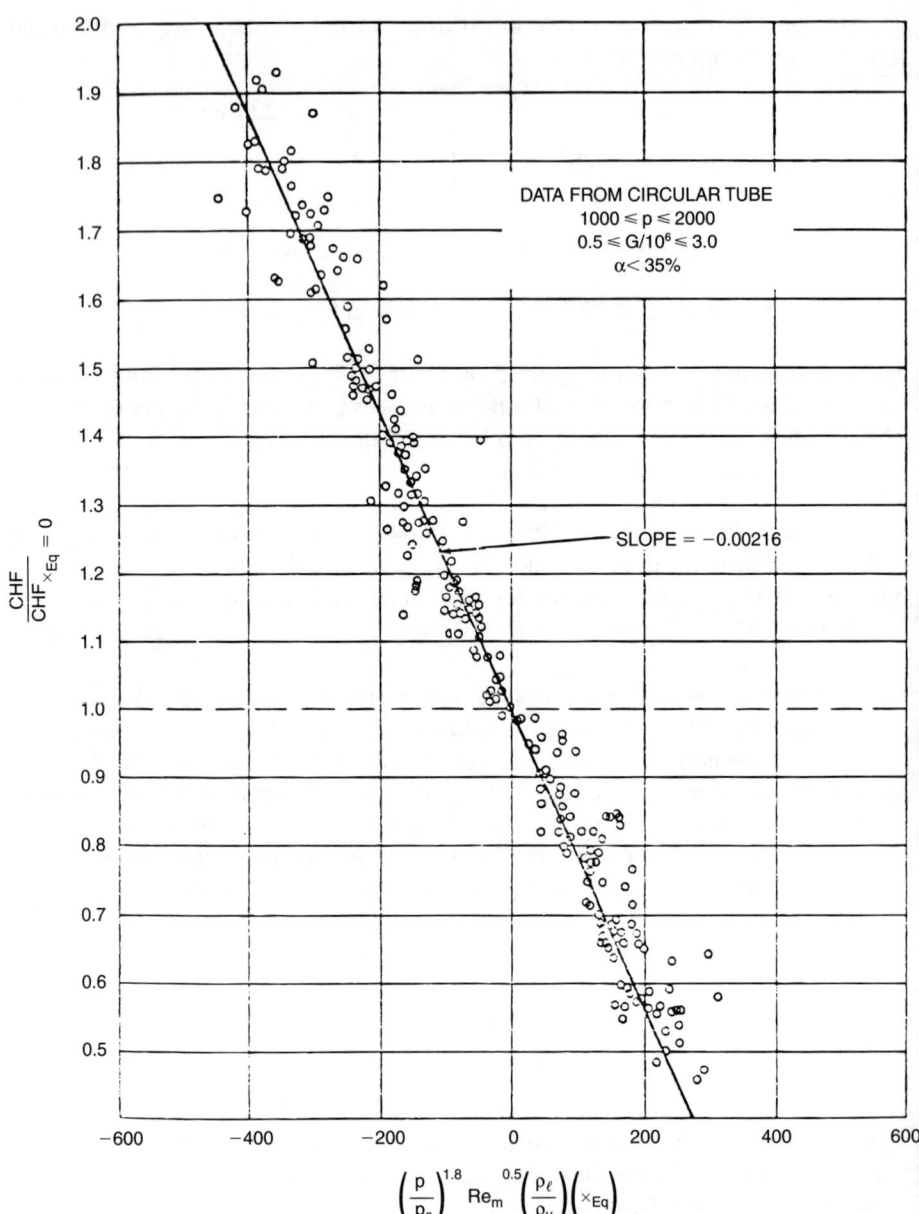

Figure 5.15 Liquid–vapor exchange effect on CHF (circular tubes). (From Tong, 1975. Copyright © 1975 by American Society of Mechanical Engineers, New York. Reprinted with permission.)

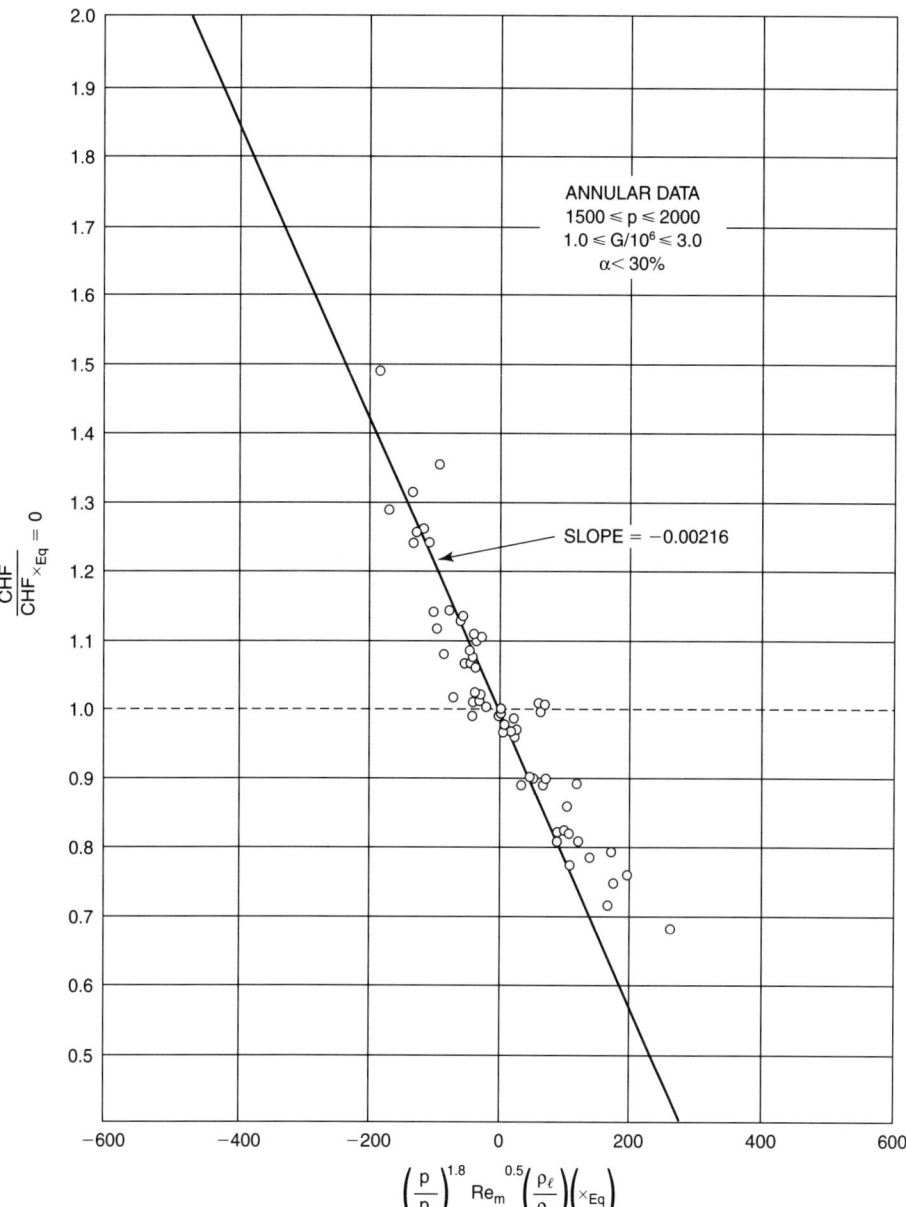

Figure 5.16 Liquid–vapor exchange effect on CHF (annular data). (From Tong, 1975. Copyright © 1975 by American Society of Mechanical Engineers, New York. Reprinted with permission.)

where $\mathrm{Re} = \dfrac{GD_e}{\mu(1-\alpha)}$

$$\mathrm{Ja} = \frac{\rho_L c_L \Delta T_\mathrm{sub}}{\rho_G H_{fg}} = -X_\mathrm{sub}\left(\frac{\rho_L}{\rho_G}\right)$$

$$\Delta T_\mathrm{sub} = T_\mathrm{sat} - T_\mathrm{core}$$

α is evaluated by using Thom's correlation (Thom et al., 1966), and the subcooled quality is defined as

$$X_\mathrm{sub} = -\frac{c_p \Delta T_\mathrm{sub\,bulk}}{H_{fg}} \qquad (5\text{-}15)$$

Equation (5-14) is valid as long as a subcooled core exists. All existing data indicate that a subcooled core seems to exist at least up to a local void fraction of 35% for a circular tube and 30% for an annulus.

We further consider that the boiling crisis occurs when the bubble-layer shielding effect reaches a maximum at bubble stagnation. A criterion for bubble-layer stagnation is suggested by Tong (1968b) to be

$$\rho_G V_b = c_o f_o G \qquad (5\text{-}16)$$

where c_o = empirical constant including subcooling effect
f_o = Fanning friction factor in a two-phase flow without bubble generation
ρ_G = vapor density
G = mass flux
V_b = bubble velocity normal to the wall

Since $(\rho_G V_b)$ is the mass flux of the bubbles generating from the surface, the CHF in a saturated flow becomes

$$q''_{X=0} = C_1 H_{fg}\, \rho_G V_b \qquad (5\text{-}17)$$

where C_1 is an empirical constant excluding subcooling effect. For a subcooled flow, Eqs. (5-15), (5-16), and (5-17) are combined as

$$\frac{q''_\mathrm{cr}}{GH_{fg}} = \left[1 + 0.00216\left(\frac{p}{p_c}\right)^{1.8}(\mathrm{Re})_m^{0.5}\,\mathrm{Ja}\right] C_o C_1 f_o \qquad (5\text{-}18)$$

where the evaluation of the two-phase friction factor f_o is given by

$$f_o = 8.0(\mathrm{Re}_m)^{-0.6}\left(\frac{D_e}{D_o}\right)^{0.32} \qquad (5\text{-}19)$$

where D_o is 0.5 in. (1.3 cm) to serve as a datum, and

$$\text{Re}_m = \frac{GD}{\mu_L(1-\alpha)}$$

A phenomenological correlation of CHF can be developed by combining the subcooling effect and the flow friction effect on the CHF. This is done by substituting Eq. (5-19) into Eq. (5-18), yielding

$$\frac{q''_{cr}}{GH_{fg}} = 8C_oC_1\left[1 + 0.00216\left(\frac{p}{p_c}\right)^{1.8}(\text{Re}_m)^{0.5}(\text{Ja})\right]\left[(G_mD_e/\nu)^{-0.6}\left(\frac{D_e}{D_o}\right)^{0.32}\right] \quad (5\text{-}20)$$

The critical pressure, p_c, for water is 3,206 psia (22 MPa). The product of constants C_oC_1 is 0.23, which was evaluated from existing water DNB data for circular tubes. As Eq. (5-20) was developed from a uniform heat flux distribution, a shape factor F_c (Tong et al., 1966a) should be applied to the correlation in a case with nonuniform heat flux distribution.

The spacer grid in a rod bundle is also a turbulence promoter that enhances liquid–vapor exchange and bubble condensation. The local intensity of such turbulence is a function of the grid pressure loss coefficient, K, and the distance from the grid, ℓ/D. Thus an empirical spacer factor, F_s, can be defined as

$$F_s = 1 + \phi K e^{-\psi(\ell/D)}(\text{Re})^{0.6}\left(\frac{D_e}{0.5}\right)^{-0.32} \quad (5\text{-}21)$$

where the values of θ and ψ are determined empirically as 0.00005 and 0.0128, respectively. By substituting Eq. (5-21) into the CHF equation (5-20), a generalized CHF correlation is found as

$$\frac{q''_{crit}}{GH_{fg}} = 1.85[1 + 0.00005Ke^{-0.0128(\ell/D)}(\text{Re}_m)^{0.6}(2D_h)^{-0.32}]$$

$$\times\left[1 + 0.00216(p_r)^{1.8}(\text{Re}_m)^{0.5}\left(\frac{\rho_L}{\rho_G}\right)X\right]\left(\frac{G_mD_h}{\nu}\right)^{-0.6}(2D_h)^{0.32} \quad (5\text{-}22)$$

where K is the pressure drop coefficient of a spacer grid in the rod bundles ($K = 0$ for a circular tube, $K = 1.4$ for a mixing vane grid); ℓ is the distance downstream from a spacer grid; and X is the quality, defined according to the following:

For circular tubes,

When $\alpha \leq 0.35$, $X = X_{eq}$

When $\alpha > 0.35$, $X = X_{(\alpha=0.35)} + \frac{1}{1.7}[X_{eq} - X_{(\alpha=0.35)}]$

Table 5.2 Comparison of predictions with CHF data

Channel geometry	Heat flux distribution	Parameter ranges	Standard deviation	Data points	Sources
Circular tube	Uniform	$1{,}000 \leq p \leq 2{,}000$ psia $0.5 \leq G/10^6 \leq 4.4$ $\alpha = 0.35$	0.108	469	WAPD-188 (DeBortoli et al., 1958) ANL-6675 (Weatherhead, 1962) EUR-2490.e (Biancone et al., 1965) AEEW-R-356,479 (Lee, 1966b) NYO-187-7 (Thompson and Macbeth, 1964)
Annulus	Uniform	$1{,}500 \leq p \leq 2{,}000$ psia $1.0 \leq G/10^6 \leq 3.0$ $\alpha = 0.30$	0.163	317	ASME 67-WA/HT-29 (Tong, 1967b) EUR-2490.e (Biancone et al., 1965) GEAP-3899 (Janssen, 1963b)
Rod bundle	Nonuniform	$900 \leq p \leq 2{,}480$ psia $0.5 \leq G/10^6 \leq 4.3$ $\alpha = 0.61$	0.063	201	WCAP-5727 (Cermak et al., 1971)

For annuli,

When $\alpha \leq 0.30$, $X = X_{eq}$

When $\alpha > 0.30$, $X = X_{(\alpha=0.30)} + \dfrac{1}{1.7}[X_{eq} - X_{(\alpha=0.30)}]$

Comparisons of the predictions from Eq. (5-22) and rod bundle and circular tube CHF data are given in Table 5.2.

5.3.2 Bubble-Layer Thermal Shielding Analysis

In a subcooled or low-quality flow boiling with high flow rates and at high pressures (conditions similar to those in water-cooled nuclear reactors), a bubble layer is often observed as shown in Figure 5.5. The bubble layer is a highly viscous layer of crowded, tiny bubbles flowing parallel to the heating surface and thus serves as a thermal shield impairing the incoming liquid for cooling the heated surface. The boiling crisis occurs at an overheating of the heated surface under the bubble layer.

Several approaches to analyzing the thermal shielding effect of a bubble layer on CHF are presented here chronologically.

1. The analysis of critical enthalpy in a bubble layer was suggested by Tong et al. (1966a).

2. The analysis of turbulent mixing at the core–bubble layer interface was suggested by Weisman and Pei (1983).
3. The analysis of mass and energy balance on the bubble layer was suggested by Chang and Lee (1989).

These different approaches are complementary to each other in basic concept. However, these analyses have not provided clear insight information of the bubble layer at the CHF about the bubble shape (spherical or flat elliptical), bubble population and its effect on turbulent mixing, and bubble behavior. The bubble behavior in a bubble layer could involve bubble rotation caused by flow shear, normal bubble velocity fluctuation, and bubble condensation in the bubble layer caused by the subcooled water coming from the core. Further visual study and measurements in this area may be desired.

5.3.2.1 Critical enthalpy in the bubble layer (Tong et al., 1966a). The cooling liquid comes from the main stream through the bubble layer to cool the heating wall as shown in Figure 5.17. The subcooling of the incoming liquid is reduced by mixing with the saturated liquid, and at the same time it quenches the bubbles to make the space for the bubbles generated from the heated wall in maintaining a volume balance. As a result, the average enthalpy of the bubble layer increases along the wall in the flow direction. As the incoming liquid gradually becomes saturated near the wall, it can no longer quench the bubbles but flashes to void. Then the bubble layer swells and the wall temperature increases suddenly, whereby DNB occurs. Therefore, the average enthalpy obtained from an energy balance of the bubble layer near the wall is a measure of the closeness to the onset of DNB.

From the simplified model of Figure 5.17, the energy equation for the bubble layer is written in terms of average conditions across the liquid layer:

$$\frac{d(\rho V P_h s H)}{dz} + h P_h (T - T_b) = q'' P_h$$

or

$$\frac{d(\rho V P_h s H)}{dz} + \frac{h P_h}{c_p}(H - H_b) = q'' P_h \qquad (5\text{-}23)$$

where ρ = average density of bubble layer
V = average velocity of bubble layer
P_h = perimeter of heater
s = thickness of bubble layer
H = average enthalpy of bubble layer
H_b = enthalpy of bulk flow
h = heat transfer coefficient from wall to bubble layer
c_p = average specific heat of bubble layer

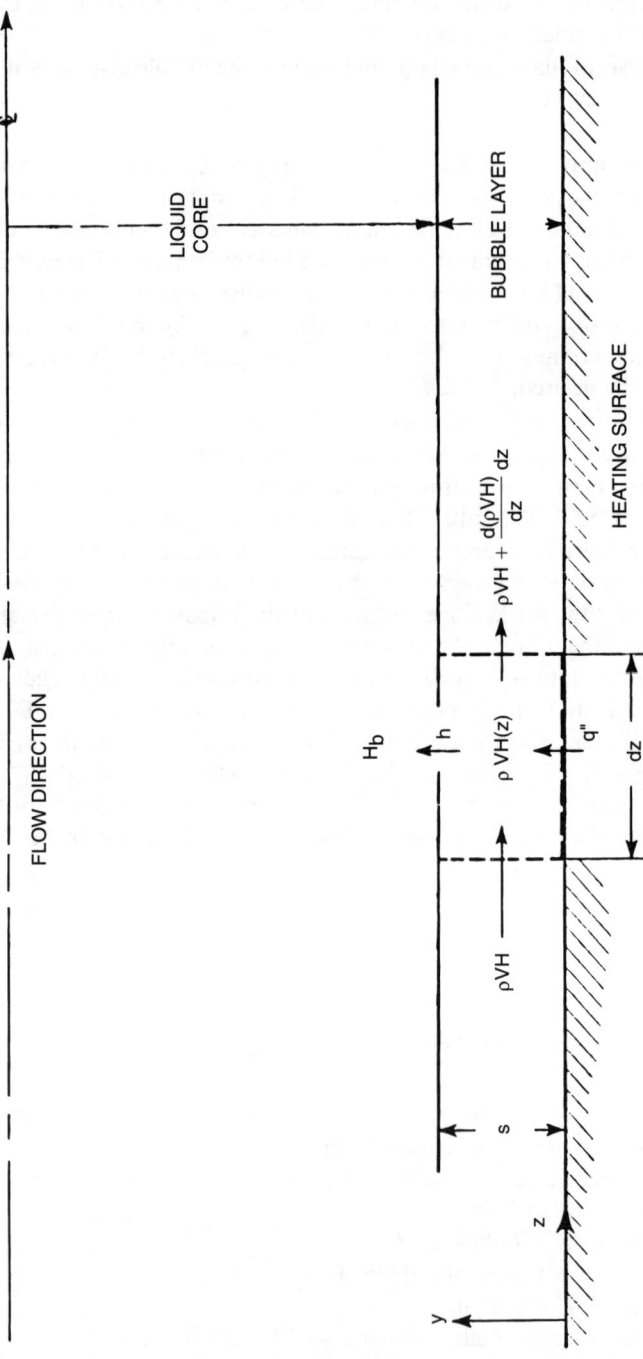

Figure 5.17 Physical model of heat balance near wall. (From Tong, 1972. Reprinted with permission of U.S. Department of Energy, subject to the disclaimer of liability for inaccuracy and lack of usefulness printed in the cited reference.)

T = average temperature of bubble layer
T_b = temperature of bulk flow
q'' = surface heat flux
z = distance from inception of local boiling, measured in the direction of flow

In order to evaluate Eq. (5-23) in the neighborhood of the boiling crisis for given local bulk conditions (i.e., fluid pressure, bulk mass flow rate, quality, and equivalent diameter), the following simplifications are made.

1. The bubble layer is assumed to have constant void fraction along the length before DNB, with a balanced rate of bubble detachment and bubble condensation in the layer. Hence, the average properties ρ, μ, and c of the bubble layer are assumed to be independent of position.
2. The thickness, s, and average velocity, V, of the bubble layer are approximately constant along the flow direction before DNB. The layer thickness includes the thin layer of superheated liquid in contact with the wall and is considered a homogenized inside layer.
3. The average temperature of liquid in the bubble layer is approximately constant before DNB.
4. The heat transfer coefficient h from the bubble layer to the bulk flow is constant before DNB.

Equation (5-23) can then be simplified to the form

$$\frac{d(H - H_b)}{dz} + C(H - H_b) = C \frac{c_p q''}{h} \tag{5-24}$$

where the constant $C = h/\rho V s c_p$ is independent of position. However, this factor is expected to be different for different flow regimes, and C may therefore be a function of mass flux and local quality at the DNB point, both of which are representative measures for the flow regime.

For the case of a uniform heat flux, the solution of Eq. (5-24), subject to the initial condition $[H(0) - H_b] = 0$ at the beginning of local boiling, is

$$H(z) - H_b = \frac{c_p q''}{h}[1 - \exp(-Cz)] \tag{5-25}$$

This solution is readily extended to the case of a nonuniform heat flux distribution for Eq. (5-24):

$$H(z) - H_b = c\left(\frac{c_p}{h}\right)\int_0^z q''(z')e^{-c(z-z')}\,dz' \tag{5-26}$$

Postulating that the inception of DNB is determined by a limiting value of the enthalpy of the bubble layer, we can write

$$[H(z) - H_b]_{DNB,U} = [H(z) - H_b]_{DNB,non} \quad (5\text{-}27)$$

The equivalent uniform DNB heat flux for the case of a nonuniform flux distribution then becomes

$$q''_{cr\,EU} = \frac{C}{1 - \exp(-C\ell_{DNB})} \int_0^{\ell_{DNB}} q''(z) \exp[-C(\ell_{DNB} - z)]\,dz \quad (5\text{-}28)$$

where $z = \ell_{DNB}$, which is the "subcooled boiling length" measured from the inception of local boiling to the location of the DNB. Consequently, a shape factor F_c can be defined as the ratio of the CHF for the uniform flux case to the CHF for the nonuniform flux case:

$$F_c = \frac{q'_{cr\,EU}}{q''_{cr\,non}} \quad (5\text{-}29)$$

or

$$F_c = \frac{C}{q''_{non}[1 - \exp(-C\ell_{DNB,EU})]} \int_0^{\ell_{DNB,non}} q''(z) \exp[-C(\ell_{DNB,non} - z)]\,dz \quad (5\text{-}30)$$

where $\ell_{DNB,EU}$ = axial location at which DNB occurs for uniform heat flux, starting from inception of LB (in.)
$\ell_{DNB,non}$ = axial location at which DNB occurs for nonuniform heat flux (in.)
$\ell_{DNB,EU} \simeq \ell_{DNB,non}$
$C = \dfrac{h}{\rho V s c_p}$

Because the components of the analytical expression for C are not sufficiently known to permit an analytical evaluation, C is determined empirically as a function of the local quality at the point of DNB, X_{DNB}, (under nonuniform heat flux conditions) and the bulk mass flux, G. The empirically determined expression for C is

$$C = 0.15 \frac{(1 - X_{DNB})^{4.31}}{(G/10^6)^{0.478}} \quad (1/\text{in.}) \quad (5\text{-}31)$$

For convenience of application, a simplification of subcooled boiling length ℓ_{DNB} is suggested to be measured from the inlet rather than from the inception of local boiling. This simplification has two justifications. First, the empirical expres-

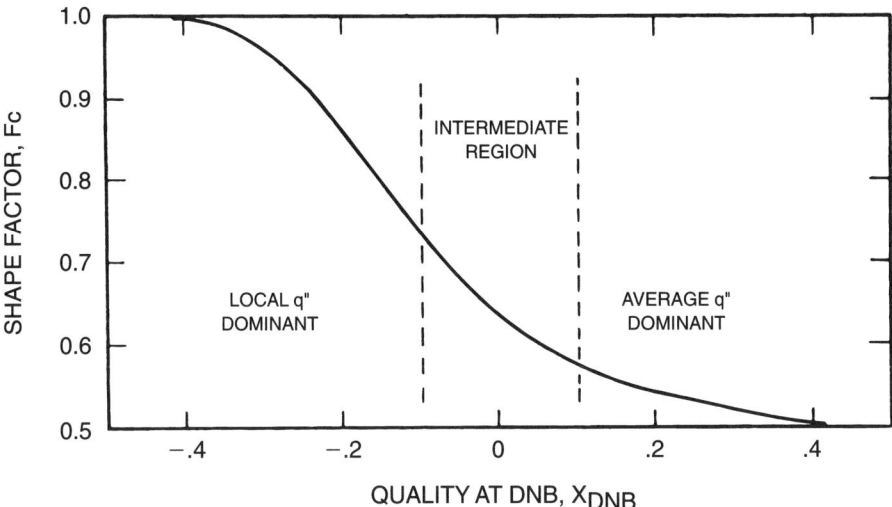

Figure 5.18 Shape factor F_c as a function of quality at DNB point for hot-patch heat flux distribution. (From DeBortoli et al., 1958. Reprinted with permission of U.S. Department of Energy, subject to the disclaimer of liability for inaccuracy and lack of usefulness printed in the cited reference.)

sion for C is determined on the basis of an ℓ_{DNB} measured from inlet, so the use of the same simplification in the prediction will be consistent. Second, the rapid decay with upstream distance of the memory effect makes the conditions near the inlet have little effect on F_c. Axial conduction in the heater is ignored in determining the nonuniform DNB test section flux shape.

Since the boiling crisis originates at the interface between the fluid and the heating surface, the bulk stream conditions alone are not sufficient for determining DNB for the case of a nonuniform heat flux distribution. The flow in the boundary bubble-layer region in particular, coming from the upstream surface, carries accumulated liquid enthalpy and bubbles to the downstream surface. In this way, the effect of the upstream heat flux distribution is conveyed to the downstream boundary layer where DNB occurs. The effect acts as a memory of past history and is called the *upstream memory effect* on boiling crisis. Achievement of the shape factor enables reactor designers to utilize existing uniform heat flux CHF data and their correlations for the design of nuclear reactors where different nonuniform heat flux distributions prevail.

The general nature of the relationship between the shape factor F_c and quality for a given flux shape is illustrated in Figure 5.18. Note that this figure was calculated from only one set of hot-patch data obtained from a 27-in. (0.7-m)-long rectangular test section with a $\frac{1}{2}$-in. (1.3-cm) or $1\frac{1}{2}$-in. (3.8-cm) hot patch located $\frac{3}{8}$ in. (1 cm) from the exit. The peak-to-average flux ratio was 2:1 at the hot patch (DeBortoli et al., 1958). In the subcooled region or at low steam qualities, C is large and thus the product of $C(\ell_{DNB} - z)$ is large for a given value of $(\ell_{DNB} - z)$. This reduces the magnitude of the weighting function of the memory effect, and hence the local heat flux q'' primarily determines boiling crisis. Such a correlation may be called a "local condition hypothesis." For a hot patch of higher peak-to-average flux ratio, the CHF is reduced more for a longer hot patch and in a lower-quality region (Styrikovich et al., 1963).

At high qualities, C is small, and the memory effect becomes strong for a greater upstream length from the point of boiling crisis. In this region, the *average* $\bar{q}'$ (or ΔH) primarily determines the crisis. This finding is in good agreement with findings reported by others (Cook, 1960; Alessandrini et al., 1963; Janssen and Kervinen, 1963; Kirby, 1966). Upstream-effect approach in correlation CHF data is said to be on the basis of an "integral concept." Such a concept can take many analytic forms, the most common being the bulk boiling length effect. The bulk boiling length is the CHF location measured from the inception of bulk boiling. It is, by convention, called simply the *boiling length*, L_B, which is smaller than the aforementioned subcooled boiling length, ℓ_{DNB}, by a subcooling length, ℓ_{sc}; that is,

$$L_B = \ell_{DNB} - \ell_{sc}$$

CHF correlations based on boiling length effect in the high-quality region are described in the following paragraphs.

Lahey and Gonzalez-Santalo (1977) extended the memory-effect analysis into the quality region by giving an analytical justification for the CISE concept of boiling length (Bertoletti et al., 1965) by using L'Hôpital's rule for the limiting form of Eq. (5-30) in a quality region:

$$\lim_{C \to 0} F_c = \frac{q''_{cr,EU}}{q''_{cr,non}} = (q''_{cr,non} L_B)^{-1} \int_{\ell_{sc}}^{L_B + \ell_{sc}} q''(z)\, dz \tag{5-32}$$

with the energy balance in the boiling length, L_B,

$$\int_{\ell_{sc}}^{L_B + \ell_{sc}} P_h q''(z)\, dz = (GA_c H_{fg} X_{cr}) \tag{5-33}$$

where P_h again is the heated perimeter of the boiling channel. Equation (5-33) shows that X_{cr} is a function of L_B and $q''(z)$. Let

$$\hat{W}_B = (GA_c H_{fg} X_{cr}) \tag{5-34}$$

where $\hat{W}_B$ is the critical power over the boiling length, and it, too, is a function of L_B and $q''(z)$. The above formulation justifies that the critical power, $\hat{W}_B$ (over the boiling length) is a function of the boiling length at a fixed G, H_{fg}, $q''(z)$, and channel geometry, as shown in Figure 5.19.

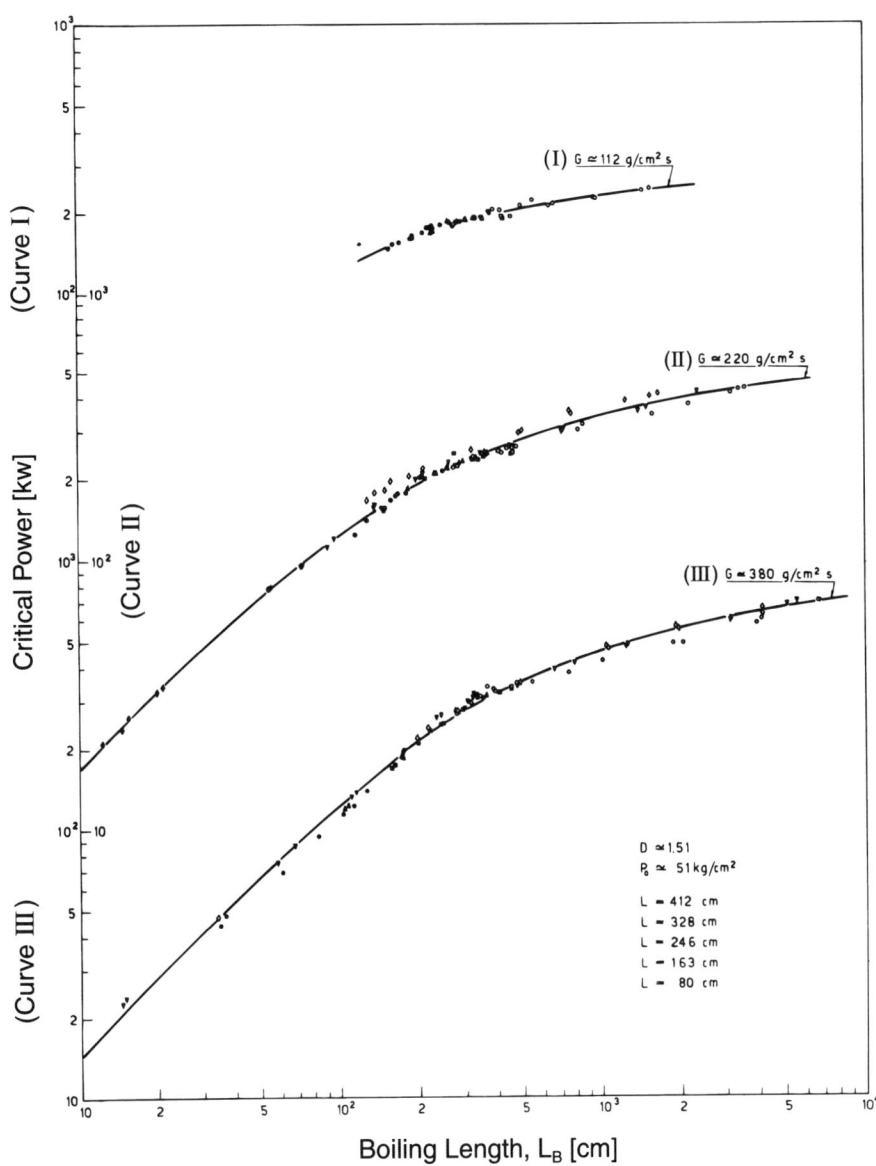

Figure 5.19 Boiling length effect on critical power. (From Bertoletti et al., 1964b. Reprinted with permission of CISE, Milan, Italy.)

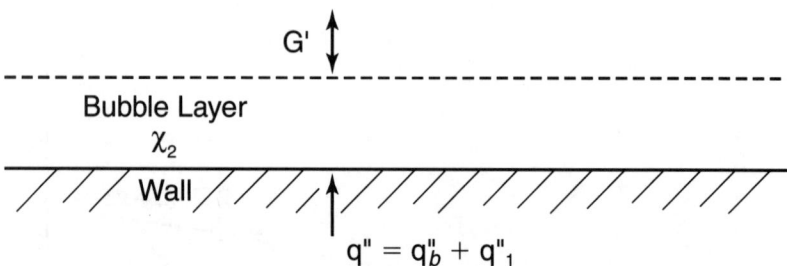

Figure 5.20 Radial flow interchange in bubble layer.

5.3.2.2 Interface Mixing

Weisman and Pei Model Weisman and Pei (1983) assumed that CHF occurs when the steam void fraction in the bubble layer just exceeds the critical void fraction (they estimated it to be 0.82) at which an array of ellipsodal bubbles can be maintained without significant contact between the bubbles. The steam void fraction in the bubble layer is determined by a balance between the outward flow of vapor and the inward flow of liquid at the bubble layer–core interface.

By assuming no bubble layer mass gradient along the flow direction, z, a simplified radial mass balance equation for the bubble layer can be established according to Figure 5.20 as

$$G'(X_2 - X_1) = \frac{q''_b}{H_{fg}} \qquad (5\text{-}35)$$

where G' is the effective mixing mass flux in and out of the bubble layer; X_1 and X_2 are the actual qualities in the core and the bubble layer, respectively; and q'_b is the heat flux for evaporation to produce the outward vapor flow.

The value of q''_b is related to q'' by

$$q''_b = \frac{q''(H_\ell - H_{\ell d})}{H_f - H_{\ell d}} \qquad (5\text{-}36)$$

where $H_{\ell d}$ = enthalpy of mixture at bubble detachment point [model of Levy (1966)]
H_f = enthalpy of saturated liquid
H_ℓ = local enthalpy of liquid at given axial location

This allows Eq. (5-35) to be written in dimensionless form as

$$\frac{q''_{DNB}}{H_{fg} G'} = (X_2 - X_1) \left(\frac{H_f - H_{\ell d}}{H_\ell - H_{\ell d}} \right) \qquad (5\text{-}37)$$

where X_1 and X_2 are the qualities at DNB in the core and bubble layer, respectively, and X_2 is the value corresponding to $\alpha_2 = 0.82$.

The quantity G' of the effective mixing mass flux is determined by the turbulent velocity fluctuations at the bubble-layer edge. The distance of the edge of the bubble layer from the wall is taken as the distance at which the size of the turbulent eddies is k times the average bubble diameter. Weisman and Pei have determined empirically that k equals 2.28. Only a fraction of the turbulent velocity fluctuations produced are assumed to be effective in reaching the wall. The effective velocity fluctuations are those in which the velocity exceeds the average velocity away from the wall produced by evaporation heat flux q''_b. At the bubble layer–core interface, the effective mass flux to the wall is computed as

$$G' = \psi i_b G \qquad (5\text{-}38)$$

The perimeter i_b, representing the turbulent intensity at the bubble layer–core interface, is calculated as the product of the single-phase turbulent intensity at the bubble-layer edge and a two-phase enhancement factor. The resulting expression is

$$i_b = 0.462(k)^{0.6} (\mathrm{Re})^{-0.1} \left(\frac{D_b}{D} \right)^{0.6} \left[1 + \frac{a(\rho_L - \rho_G)}{\rho_G} \right] \qquad (5\text{-}39)$$

where a = empirical constant (a is given as 0.135)
D_b = bubble diameter
D = tube diameter
Re = Reynolds number
ρ_L, ρ_G = liquid and vapor density, respectively

Through consideration of the velocity fluctuations that are effective in reaching the wall, the parameter ψ was computed by Pei (1981) to be

$$\psi = \frac{1}{2\pi} \exp\left[-\frac{1}{2} \left(\frac{v_b}{\sigma_{v'}} \right)^2 \right] - \frac{1}{2} \left(\frac{v_b}{\sigma_{v'}} \right) \mathrm{erfc}\left(\frac{1}{2} \frac{v_b}{\sigma_{v'}} \right) \qquad (5\text{-}40)$$

where v_b = radial velocity produced by vapor generation
$\sigma_{v'}$ = standard deviation of radial fluctuation velocity
$\phantom{\sigma_{v'}} = \left(\dfrac{G}{\bar{\rho}} \right) i_b$

Weisman and Pei found that their approach could be used over the following ranges:

P = 20–205 bar (2.0–20.5 MPa)
G = 3.5–49 × 10^6 kg/m² h (0.72–10.1 × 10^6 lb/ft² hr)
Tube length = 0.35–360 cm (0.14–142 in.)
Tube diameter = 0.115–3.75 cm (0.04–1.48 in.)
$\langle \alpha \rangle_{CHF} \leq 0.6$

For the approximately 1,500 points in this range of uniform heat flux CHF experiments, the root-mean-square error was ~10%.

Weisman and Pei applied their approach to data from tubes with nonuniform axial heat flux. They found the predictive scheme to hold without the need for any nonuniform heat flux correction factor. The accuracy of their prediction was only slightly less than that of the W-3 correlation, Eq. (5-106), for the data they examined.

Weisman and Pei also applied their approach to DNB tests with fluids other than water. Without any revision of the approach derived for water, good predictions were obtained for DNB data from systems using liquid nitrogen, anhydrous ammonia, Refrigerant-113, and Refrigerant-11 (see Sec. 5.3.4.2).

An improved CHF model for low-quality flow The Weisman-Pei model was later improved by employing a mechanistic CHF model developed by Lee and Mudawwar (1988) based on the Helmholtz instability at the microlayer–vapor interface as a trigger condition for microlayer dryout (Fig. 5.21). The CHF can be expressed by the following equation due to the energy conservation of the microlayer (Lin et al. 1989):

$$q''_{crit} = G_m \delta_m \left[\frac{H_{fg} + (H_{L,sat} - H_m)}{L_m} \right] \quad (5\text{-}41)$$

where G_m = liquid mass flux flowing into the microlayer
δ_m = thickness of microlayer
L_m = length of the microlayer
H_m = liquid enthalpy flowing into the microlayer

If the local liquid enthalpy flowing into the microlayer is assumed to be independent of bulk subcooling, it can be approximated by the saturated liquid enthalpy, $H_{L,sat}$, near the dryout point:

$$H_{L,sat} = H_m$$

Equation (5-41) reduces to the form

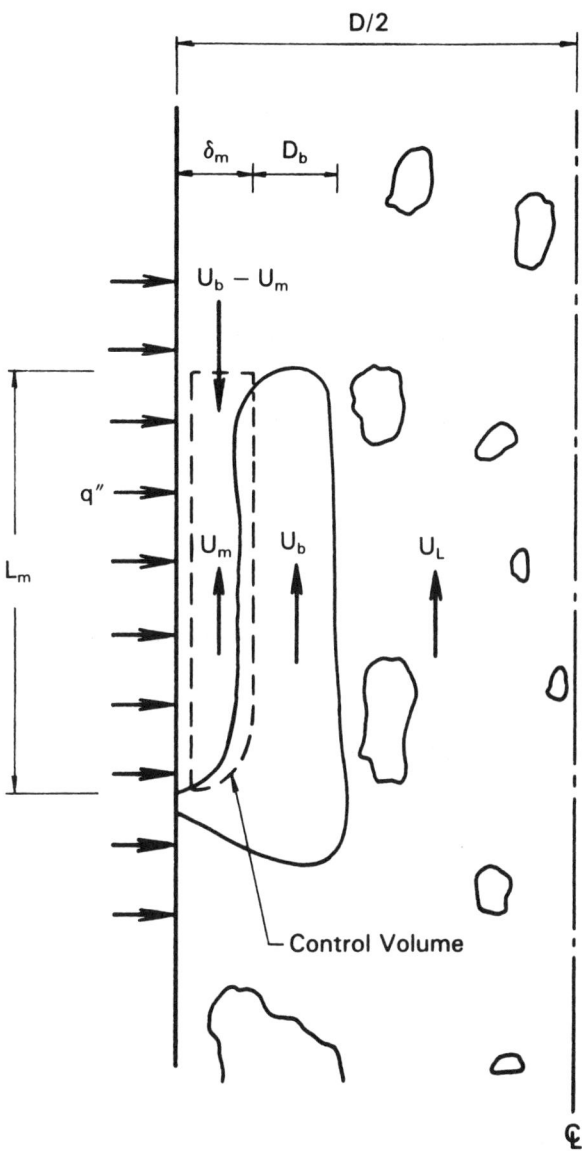

Figure 5.21 Schematic of the onset of microlayer dryout. (From Lee and Mudawwar, 1988. Copyright © 1988 by Elsevier Science Ltd., Kidlington, UK. Reprinted with permission.)

$$q''_{crit} = \frac{G_m \delta_m H_{fg}}{L_m} \tag{5-42}$$

Assuming L_m to be equal to the Helmholtz critical wavelength, we have, in a form similar to Eq. (2-72),

$$L_m = \frac{2\pi\sigma(\rho_L + \rho_G)}{\rho_L \rho_G (U_b - U_m)^2}$$

where U_b and U_m are vapor blanket velocity and liquid velocity in the microlayer, respectively. Since U_b is always much larger than U_m, the expression for L_m can be reduced to

$$L_m \simeq \frac{2\pi\sigma(\rho_L + \rho_G)}{\rho_L \rho_G U_b^2} \tag{5-43}$$

and

$$G_m = \rho_L (U_b - U_m) \simeq \rho_L U_b \tag{5-44}$$

The velocity of the vapor blanket, U_b, in the turbulent stream can be determined by the force balance (Lin et al., 1989),

$$U_b = S_1 + S_2 + \frac{U_{bL}}{3} \tag{5-45}$$

where U_{bL} = liquid velocity at $y = \delta_m + \left(\dfrac{D_b}{2}\right)$ (Fig. 5.21)

$$S_1 = \left\{ \frac{S_3}{2} + \left(\frac{U_{bL}}{3}\right)^3 + \left[\left(\frac{S_3}{2}\right)^2 + S_3 \left(\frac{U_{bL}}{3}\right)^3\right]^{1/2} \right\}^{1/3}$$

$$S_2 = \left\{ \frac{S_3}{2} + \left(\frac{U_{bL}}{3}\right)^3 - \left[\left(\frac{S_3}{2}\right)^2 + S_3 \left(\frac{U_{bL}}{3}\right)^3\right]^{1/2} \right\}^{1/3}$$

$$S_3 = \frac{\pi\sigma g D_b (\rho_L^2 - \rho_G^2)}{12 \rho_L \rho_G \mu_L}$$

Hence, U_b is a function of U_{bL}, D_b, and fluid properties. The equivalent diameter of the vapor blanket, D_b, can be obtained from the correlation for the bubble departure diameter (Cole and Rohsenow, 1969). To calculate the liquid velocity, U_{bL},

Lin et al. (1989) proposed using a homogeneous two-phase flow model which is an improvement over the original Lee and Mudawwar approach:

$$U_{bL} = U_t \left\{ 5 \ln \left[\frac{\rho_{t.p.}(\delta_m + D_b/2)U_t}{\mu_{t.p.}} \right] - 3.05 \right\} \tag{5-46}$$

where U_t = friction velocity = $\left(\dfrac{\tau_w}{\rho_{t.p.}} \right)^{1/2}$

$$\tau_w = \frac{(0.046 \, \text{Re}_{t.p.}^{-0.2})G^2}{2\rho_{t.p.}}$$

$\text{Re}_{t.p.}$ = effective Reynolds number for two-phase flow

$$= \frac{GD_w}{\mu_{t.p.}}$$

$\rho_{t.p.}$ = homogeneous fluid density

$$= \rho_L(1 - \alpha) + \rho_G \alpha$$

$\mu_{t.p.}$ = mean two-phase viscosity

$$= \rho_{t.p.} \left[\frac{X_t \mu_G}{\rho_G} + \frac{(1 - X_t)\mu_L}{\rho_L} \right]$$

X_t, the true quality, and α, the void fraction, have to be predicted under the conditions of subcooled or low-quality flow boiling. Again, with the homogeneous two-phase flow model,

$$\alpha = \frac{X_t}{X_t + (1 - X_t)(\rho_G/\rho_L)}$$

To determine the thickness of the microlayer, δ_m, a force balance on the vapor blanket was used in the direction normal to the wall. The inertial force, F_I, due to vapor generation was used to balance the force due to vapor blanket circulations, F_R:

$$F_I = \left(\frac{q''_{crit}}{H_{fg}} \right)^2 \left(\frac{D_b L_m}{\rho_G} \right)$$

$$F_R = -C\rho_G (U_b - U_{bL}) \left(\frac{\partial U_L}{\partial y} \right)(\pi L_m) \left(\frac{D_b}{2} \right)^2$$

and

$$\frac{\partial U_L}{\partial y} = (5U_t/2)\left[\frac{1}{(\delta_m)} + \frac{1}{(D_b + \delta_m)}\right]$$

$$C = a_1(Y_r)^{a_2} \operatorname{Re}^{[a_3 - a_4 \alpha/(1-X_t)]}$$

where a_1, a_2, a_3, and a_4 are empirical constants, and Y_r is a parameter accounting for the wall effects on vapor blanket circulation,

$$Y_r = \frac{\delta_m + D_b}{\delta_m}$$

By combining the above equations, critical heat flux, q''_{crit}, can be predicted by iterative calculations if the values of constants a_1, a_2, a_3, and a_4 are determined. After regression analysis of a database of 812 water CHF experimental data points from vertical round tubes, these constants were determined to be 7,000, -0.4, -0.8, and 0.5, respectively (Lin et al., 1989).

5.3.2.3 Mass and energy balance in the bubble layer.

In analyzing the bubble-layer thermal shielding effect on CHF, Chang and Lee (1989) wrote coupled mass and energy balance equations for the bubble layer including the most important mixing mass flux equation. The mixing mass flux equation was derived from momentum balance equations, leading to the relative motion equation. The mass balance equation on the bubble layer is of the form

$$\frac{dG_b}{dz} = (G_{cb} - G_{bc})(P_i/A_c) \tag{5-47}$$

where G_{cb} and G_{bc} are the inward and outward mass fluxes at the interface of the core and the bubble layer; P_i is the perimeter at the interface; A_c is the cross-sectional area of the core; G_b is the axial mass flow rate of the bubble layer; and (dG_b/dz) is arbitrarily assumed to be zero at CHF condition. Hence

$$G_{cb} = G_{bc} G^* \tag{5-48}$$

where G^* is the limited mixing mass flux, which is equivalent to G' in Weisman and Pei's analysis. Thus Eq. (5-35) becomes

$$G^*(X_b - X_c) = (q''_b/H_{fg}) \tag{5-49}$$

where q''_b is the boiling heat flux, which is determined by the "relative motion" equation obtained from the momentum balance equations on two regions and the energy balance on the bubble layer, respectively. Consequently, Chang and Lee established a CHF correlation for water in a round tube with axially uniform heat flux as

$$q''_{CHF} = \left[(\rho_c - \rho_b)g + \frac{\tau_{wi} P_i}{A\alpha_c(1-\alpha_c)} - \left(\frac{G}{\rho_c \alpha_c}\right)^2 \left(\frac{d\rho_c}{dz}\right) \right]$$

$$\times \left[\frac{\rho_c \alpha_c^2 (1-\alpha_c) A(X_b - X_c) H_{fg}}{GP_w} \right] \quad (5\text{-}50)$$

where the subscripts c and b refer to the core and the bubble layer, respectively. α_c is the void fraction of the liquid core, τ_{wi} is the shear stress at the interface, and the authors suggested that it could be evaluated by the models of Larsen and Tong (1969), Beattie (1975), Beattie and Whalley (1982), or Levy (1966).

This correlation compared well with uniform flux round tube data of Thompson and Macbeth (1964), Zenkevich et al., (1969), and Becker et al. (1971). The average ratio of predicted to measured CHF values was 1.06 with a standard deviation of 0.12. The authors noted that the profile-fit method (Levy, 1966) for evaluating the enthalpy ($H_{\ell d}$) at bubble detachment would be inadequate for the case where the condensation effect of detached bubbles is dominant in a bubble layer. Thus a mechanistic model employing bubble condensation is required.

5.3.3 Liquid Droplet Entrainment and Deposition in High-Quality Flow

A phenomenological analysis for predicting the dryout heat flux in an annular flow was suggested by Isbin and co-workers (1961). Their method was to balance the rates of liquid entrainment and evaporation against the rate of droplet deposition. The result of this approach depends on the accuracy of the analytical evaluation of the liquid entrainment and deposition mass fluxes (G_E and G_D). Others in England, such as Bennett et al. (1967b), Hewitt and Hall Taylor (1970); Whalley et al. (1973), and Whalley (1976) all demonstrated that CHF in annular flow is a dryout of the liquid film on the wall. This has been shown in Figure 5.13 as an integral phenomenon of liquid entrainment, evaporation, and deposition.

In the analytical approaches to boiling crisis in annular flow, there appear to be two important aspects: the thickness of the liquid film and the exchange of liquid droplets between the core of the flow and the liquid film. Vanderwater (1956) suggested that the thickness of the liquid film depends on the balance of the liquid reentrainment rate, the liquid evaporation rate, and the liquid droplet deposition rate. The liquid balance equations for the core and the liquid film in a pipe length dz (Fig. 5.22) are given as follows:

$$\frac{d(A_c V_d R_{Lc} \rho_L)}{dz} = P_c R_{Lc} \rho_L (V_{te} - V_{td}) \quad (5\text{-}51)$$

$$\frac{d(A_f V_f \rho_L)}{dz} = P_c R_{Lc} \rho_L (V_{td} - V_{te}) - \frac{P_f q''}{H_{fg}} \quad (5\text{-}52)$$

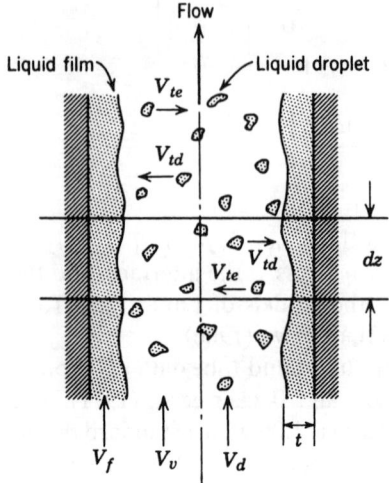

Figure 5.22 Physical model of annular flow. (From Vanderwater, 1956. Reprinted with permission of the copyright holder.)

where R_{Lc} is the liquid fraction in the core; A_c and A_f are the cross-sectional areas of the core and the film, respectively; P_c and P_f are the core perimeter and the outer perimeter of the film, respectively; and V_{td} and V_{te} are the liquid cross-flow velocities at the interface due to the droplet deposition and the liquid reentrainment, respectively. Since it is reasonable to assume that the boiling crisis occurs when the thickness of the liquid film becomes zero, the critical heat flux can be predicted by solving Eqs. (5-51) and (5-52). However, the characteristics of the liquid cross-flow velocities V_{td} and V_{te} at the interface due to the droplet deposition and the liquid reentrainment must be known.

In order to determine the liquid exchange mass flux at the interface due to the droplet deposition and the liquid reentrainment, Quandt (1962) measured the dye concentration in an isothermal annular flow. His steady-state model is similar to Vanderwater's as shown in Figure 5.22, except for his assumption that $V_{td} = V_{te} = V_t$. Hence, the concentration balance of dye can be expressed as

$$\frac{d(A_c V_d R_{Lc} \rho_L C_d)}{dz} = P_c \rho_L V_t R_{Lc}(C_f - C_d) \tag{5-53}$$

$$\frac{d(A_f V_f \rho_L C_f)}{dz} = P_c \rho_L V_t R_{Lc}(C_d - C_f) \tag{5-54}$$

where C_d is the dye concentration of the droplets and C_f is the dye concentration of the liquid film. Equations (5-53) and (5-54) can be rewritten in terms of the film thickness, $t = A_f/P_c$, as

$$\left(\frac{A_c V_d}{P_c V_t}\right)\left(\frac{dC_d}{dz}\right) = C_f - C_d \tag{5-55}$$

$$\left(\frac{tV_f}{R_{Lc}V_t}\right)\left(\frac{dC_f}{dz}\right) = C_d - C_f \tag{5-56}$$

These equations can be solved simultaneously, using the boundary conditions that, at $z = 0$, $C_f = C_o$ and $C_d = 0$, where C_o is the initial dye concentration introduced into the liquid. The solutions are

$$\left(\frac{C_f}{C_o}\right) = 1 + R_{Lt}(e^{-kz} - 1) \tag{5-57}$$

$$\left(\frac{C_d}{C_o}\right) = (1 + R_{Lt})(1 - e^{-kz}) \tag{5-58}$$

where k = ratio of transverse and axial liquid flow rates per unit length

$$= \frac{V_t R_{Lc} \rho_L}{V_f t R_{Lt} \rho_L}$$

R_{Lt} = fraction of the total liquid entrained in the core

$$= \frac{A_c R_{Lc}}{A_c R_{Lc} + A_f}$$

As $z \to \infty$, $C_d = C_f = C_m$. From the definitions of these quantities, it can be demonstrated that

$$\frac{C_m}{C_o} = (1 - R_{Lt})$$

Hence, Eq. (5-57) becomes

$$\left(\frac{C_f}{C_m} - 1\right) = \left[\frac{R_{Lt}}{(1 - R_{Lt})}\right] e^{-kz} \tag{5-59}$$

Quandt (1962) measured the values of $(C_f/C_m - 1)$ at various axial positions of an air–water mixture flow in a 0.25-in. × 3-in. channel and converted the raw data to the exchange mass flux, $\rho_L V_t$, as shown in Figure 5-23. He also measured the film velocity V_f by injecting a pulse of dye into the liquid film and recording its transport time between two photocells. Such measured data are shown in Figure 5-24. By using the measured values for V_f, the liquid film thickness t may be calculated as

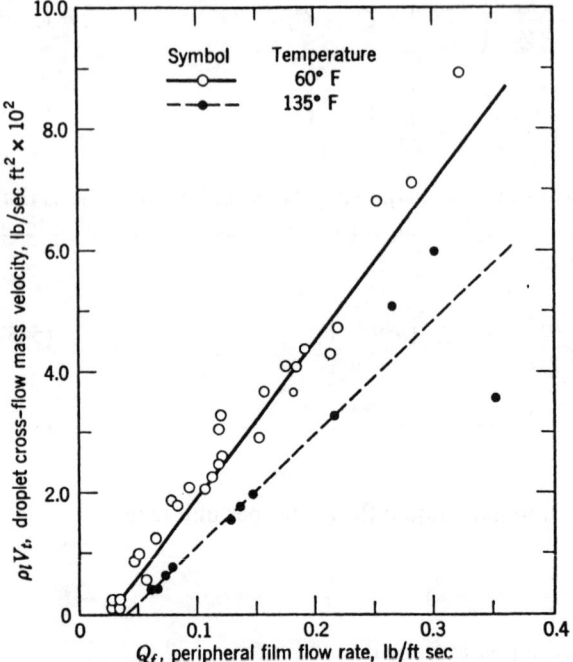

Figure 5.23 Droplet cross-flow mass flux versus peripheral film flow rate. (From Quandt, 1965. Copyright © 1965 by American Institute of Chemical Engineers, New York. Reprinted with permission.)

$$t = \frac{W_L(1 - R_{Lt})}{P_c V_f \rho_L} \quad (5\text{-}60)$$

where W_L is the total liquid flow rate in the channel.

The condition for boiling crisis when t is zero can now be readily predicted. However, it must be noted that the value of R_{Lt} in a boiling flow is not a constant but is a function of location. Furthermore, both R_{Lt} and k may be geometry-dependent. The values measured by Quandt (1962) are valid only for the geometry he tested, and no general expressions have been developed. The dryout heat flux resulting from a balance among liquid entrainment, evaporation, and deposition can be given in a generalized form as

$$\frac{q''_{crit}}{H_{fg}} = G_F + G_D - G_E \quad (5\text{-}61)$$

where G_F = liquid mass flux due to the liquid film coming down the upstream of CHF per unit heating area

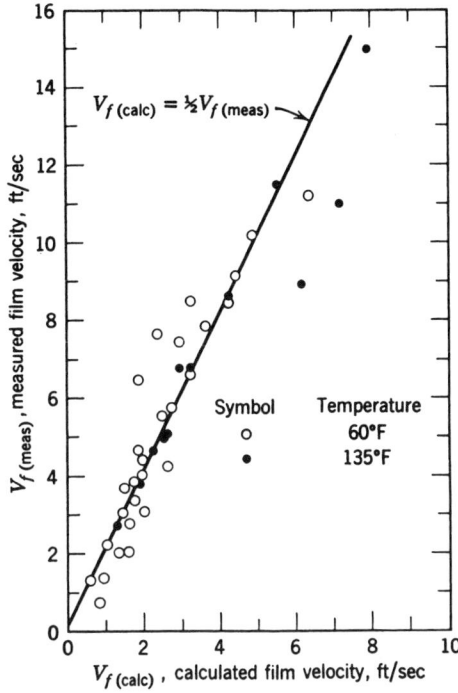

Figure 5.24 Measured versus calculated film velocity. (From Quandt, 1965. Copyright © 1965 by American Institute of Chemical Engineers, New York. Reprinted with permission.)

G_D = liquid mass flux due to deposition per unit heating area
G_E = liquid mass flux due to entrainment per unit heating area

Values of G_F, G_D, and G_E in Eq. (5-61) are local quantities. Bennett et al. (1967a) defined a hydrodynamic equilibrium curve for the entrained liquid mass flux $G_E(\Delta E = G_E \Delta Z)$ on the basis of the hypothesis that, for a given total mass flux and steam quality in a long channel, there exists an equilibrium distribution of water rate between the vapor core and the liquid film, as shown in Figure 5.25. In a boiling annular flow, the entrained liquid mass flux increases with increasing local quality until the equilibrium curve is reached, and then begins to decrease as net deposition occurs. Burnout (or dryout) occurs at X_{BO}, when the entrained liquid mass flux equals the total liquid mass flux of the tube. By using these curves, Bennett et al. (1967a) demonstrated that, when a cold patch is located at an upstream point, the entrained liquid mass flux, G_E, is increased, and the dryout quality is reduced by $(\Delta X_{BO})_1$. They also concluded that, when a cold patch is located at a downstream point, the entrained liquid mass flux is decreased, and the dryout quality is increased by $(\Delta X_{BO})_2$. Bennett et al. found that when the cold patch was near the end of the tube, the power to cause a boiling crisis could exceed that for a uniformly heated tube of the same total length. However, in some tests with inlet

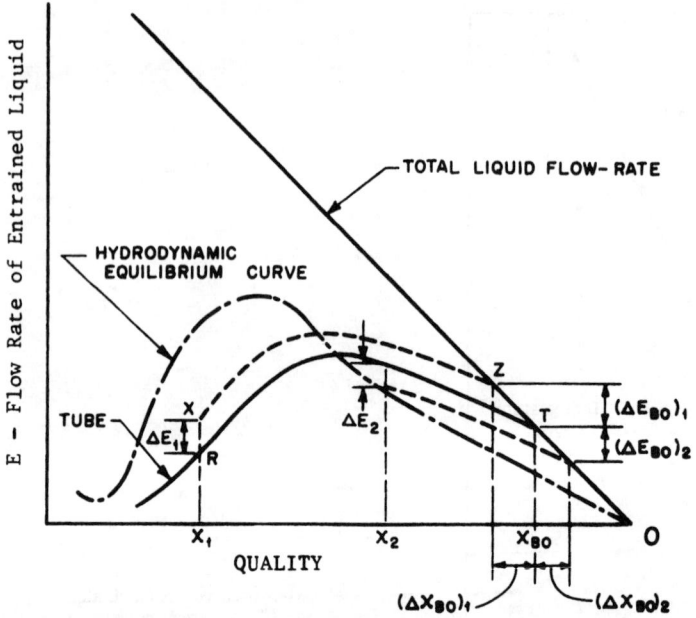

Figure 5.25 Liquid entrainment and cold-patch effect. (From Bennett et al., 1967a. Reprinted with permission of Institute of Chemical Engineers, Rugby, Warmickshire, UK.)

peaking, a midchannel boiling crisis was observed, in which case no significant improvement of critical power in the quality region could be achieved by a cold-patch arrangement.

At very high qualities the liquid film is thin and the rate of entrainment is low. The entrained liquid mass flux curve is almost parallel with the total liquid mass flux in Figure 5.26; i.e., the liquid evaporation rate is supported solely by the liquid deposition rate. If the boiling heat flux $q'' < q''_D$, where $q''_D = G_D H_{fg}$, the boiling crisis can be averted by a deposition liquid mass flux, G_D, as shown in Figure 5.26, and therefore is called deposition-controlled CHF.

Since the liquid film thickness is zero at dryout, Bennett et al. (1967a) suggested that the mass balance on the wall for a small length increment, ΔZ, upstream of the dryout, be in the form

$$-\Delta Z\left(\frac{dE}{dz}\right) + G_D \, \Delta Z = q'' \frac{\Delta Z}{H_{fg}} \tag{5-62}$$

or

$$\frac{q''_{crit}}{H_{fg}} = G_D - \left(\frac{dE}{dz}\right) \tag{5-63}$$

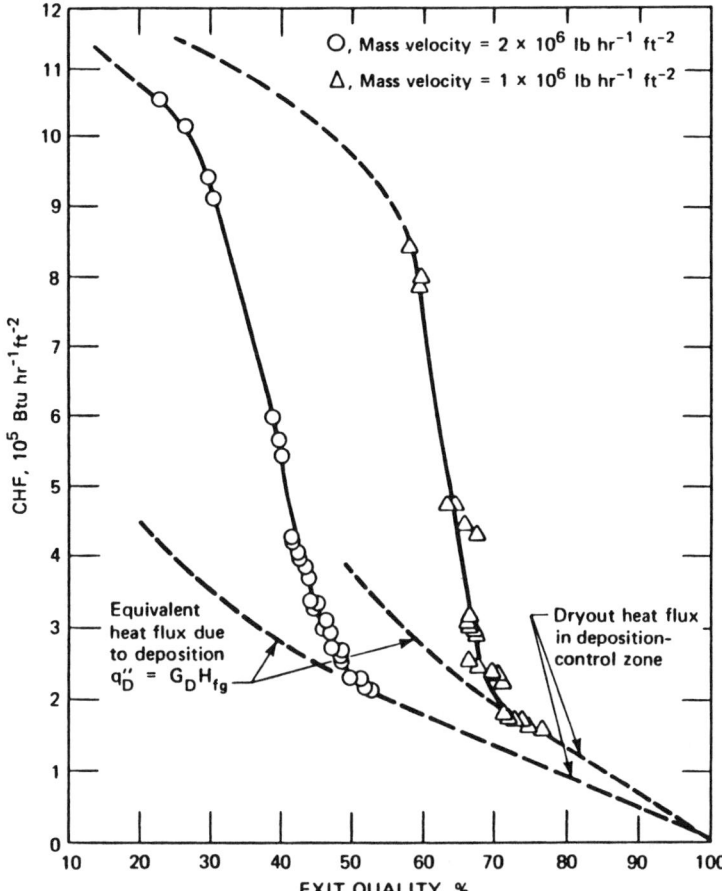

Figure 5.26 Deposition-controlled CHF: pressure (P) = 1,000 psia; diameter (D) = 0.243 in. (From Hewitt, 1970. Copyright © 1970 by Hemisphere Publishing Corp., New York. Reprinted with permission.)

and

$$\frac{1}{H_{fg}}\left(\frac{dq''_{crit}}{dz}\right) = \frac{dG_D}{dz} - \frac{d^2E}{dz^2} \quad (5\text{-}64)$$

For a midchannel dryout in a nonuniform heat flux distribution, $G_D > 0$ in both upstream and downstream regions of the dryout, and at the dryout $G_D = 0$ and $(dG_D/dz) = 0$. The wall in the downstream region of the dryout is wetted; it follows, then, that $d^2E/dz^2 > 0$. Equation (5-64) becomes

$$\frac{dG_D}{dz} > \frac{1}{H_{fg}}\left(\frac{dq''_{crit}}{dz}\right)$$

Since the deposition rate decreases with increasing quality or length, dG_D/dz is negative. Consequently, dq''_{crit}/dz) must also be negative; i.e., a midchannel dryout can occur only at a downward slope of the heat flux distribution curve. Bennett et al. (1967a) also hypothesized that the midchannel dryout occurs along the deposition-controlled CHF curve, which is usually lower than the uniformly distributed heat flux dryout curve, as shown in Figure 5.27. On the basis of the above analysis, the local values of G_D and G_E appear to be complicated functions of location and flow parameters. Evaluation of the local values of G_D and G_E is given in the later work of Whalley et al. (1973, 1974):

$$G_D = kC_E \tag{5-65}$$

where C_E is the concentration of entrained droplets in the vapor core, which can be evaluated as

$$C_E = \frac{G(1-X) - G_F}{(GX/\rho_G) + [G(1-X) - G_F]/\rho_L} \tag{5-66}$$

and k is a mass transfer coefficient (Cousins et al., 1965), (Cousins and Hewitt, 1968), which is also a function of the surface tension as shown in Figure 5.28. This figure was determined empirically as

$$G_E = kC_{E,eq} \tag{5-67}$$

where $C_{E,eq}$ is the equilibrium concentration of droplets entrained in an annular flow (Hutchinson and Whalley, 1973), and

$$C_{E,eq} = f\left(\frac{\tau_i t}{\sigma}\right) \tag{5-68}$$

where τ_i is the interfacial shear stress and t is the average liquid film thickness. According to Wallis (1970), the interfacial friction factor can be expressed as $(1 + 360t/d)$ times the friction factor for the gas core flowing in the absence of the liquid film, where d is the diameter of the tube. Turner and Wallis (1965) suggested that the value of t can be evaluated from the pressure gradient as

$$\frac{4t}{d} = \left[\frac{(dp/dz)_{liq}}{(dp/dz)_{TP}}\right]^{1/2} \tag{5-69}$$

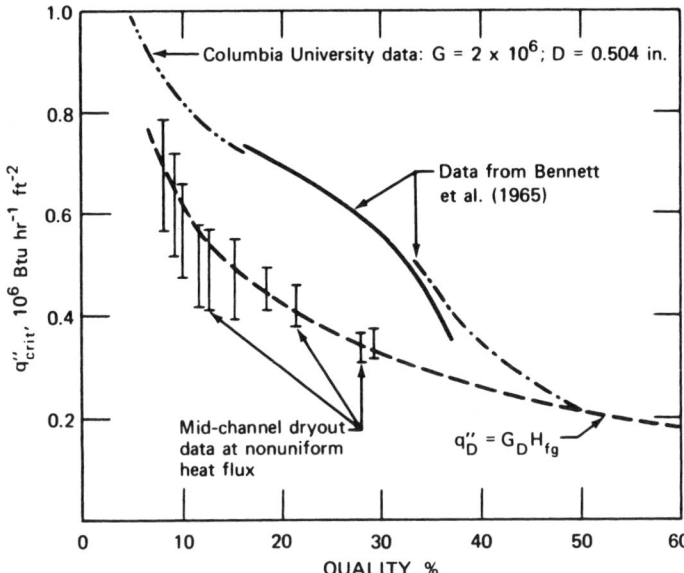

Figure 5.27 Midchannel dryout on deposition-controlled curve. Curves for uniform flux data are shown for comparison. (From Bennett et al., 1967a. Reprinted with permission of Institute of Chemical Engineers, Rugby, Warwickshire, UK.)

Equation (5-68) is also determined empirically as shown in Figure 5.29. It can be seen that the mechanism of liquid droplet entrainment and deposition at CHF in annular flow is qualitatively validated by the data trend plotted in Figure 5.29.

5.3.4 CHF Scaling Criteria and Correlations for Various Fluids

5.3.4.1 Scaling Criteria. The benefit of scaling is to simplify CHF testing facilities. For instances, if Freon is used to simulate water, the system pressure can be reduced by a factor of 6, and the heating power can be reduced by a factor of 10. These reductions make a simulated boiling crisis test much cheaper, faster, and easier to conduct. In developing scaling criteria, various hypothetical mechanisms of boiling crisis models can be validated or disproved. Thus the understanding of boiling crisis models can be improved. Since the boiling crisis is related to heat transfer, phase conversion, and momentum exchange, its simulation requires geometric and dynamic similitudes. A geometric similitude implies a similar flow field, channel geometry, and bubble size, where the bubble size is a function of fluid properties and system pressure.

A dynamic similitude embodies both hydrodynamic and thermodynamic similitudes. A hydrodynamic similitude requires a similar flow pattern and similar veloc-

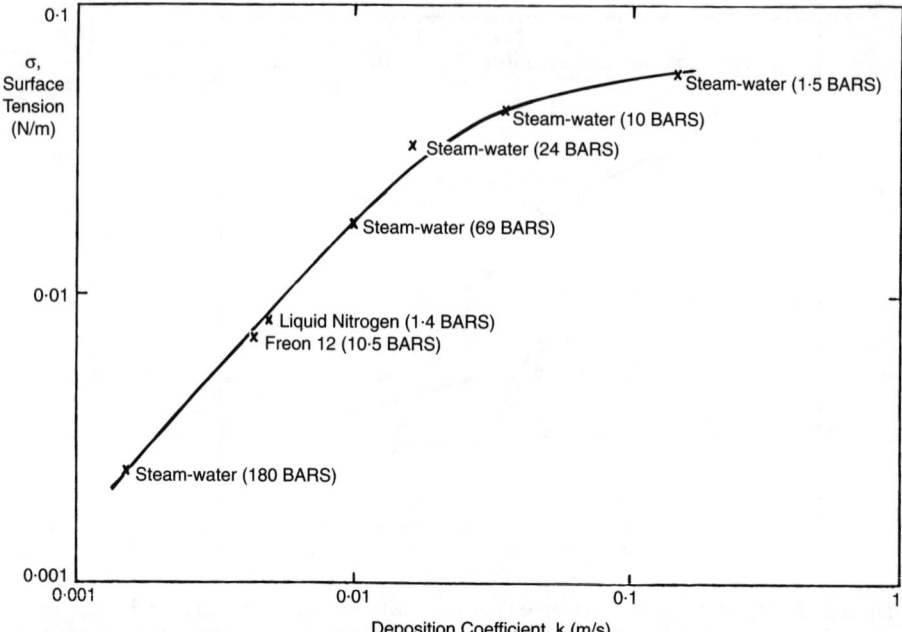

Figure 5.28 Variation of deposition coefficient with surface tension. (From Whalley et al., 1973. Reprinted with permission.)

ity and shear distributions in the pipe. The flow pattern is a function of the Froude number, V^2/gD; the Weber-Reynolds number, $\mu_L^2/\sigma D\rho_L$; the volumetric flow rate ratio, $Q_G/(Q_L + Q_G)$; and the ratio of the liquid density to the vapor density, ρ_L/ρ_G. The velocity and shear distributions in the pipe are functions of the Reynolds number, GD/μ. A thermodynamic similitude requires the same heat transfer mechanism and equivalent thermodynamic properties of the coolant, such as system pressure, p; coolant temperature, T; and specific volume, v. These are mostly functions of the reduced pressure, $p_R = p/p_c$.

For a flow boiling crisis, an empirical scaling factor was suggested by Stevens and Kirby (1964). It is a graphical correlation developed from uniform-flux round-tube data. They suggested an empirical function:

$$E = GD^{1/4}\left(\frac{D}{L}\right)^{0.59} \times 10^{-4} \text{ lb(in.)}^{1/4}(\text{hr})^{-1}(\text{ft})^{-2} \quad (5\text{-}70)$$

$$X_{cr} = \left(\frac{4L}{DG}\right)\left(\frac{q''_{crit}}{H_{fg}}\right) - (\Delta H_{in}/H_{fg}) \quad (5\text{-}71)$$

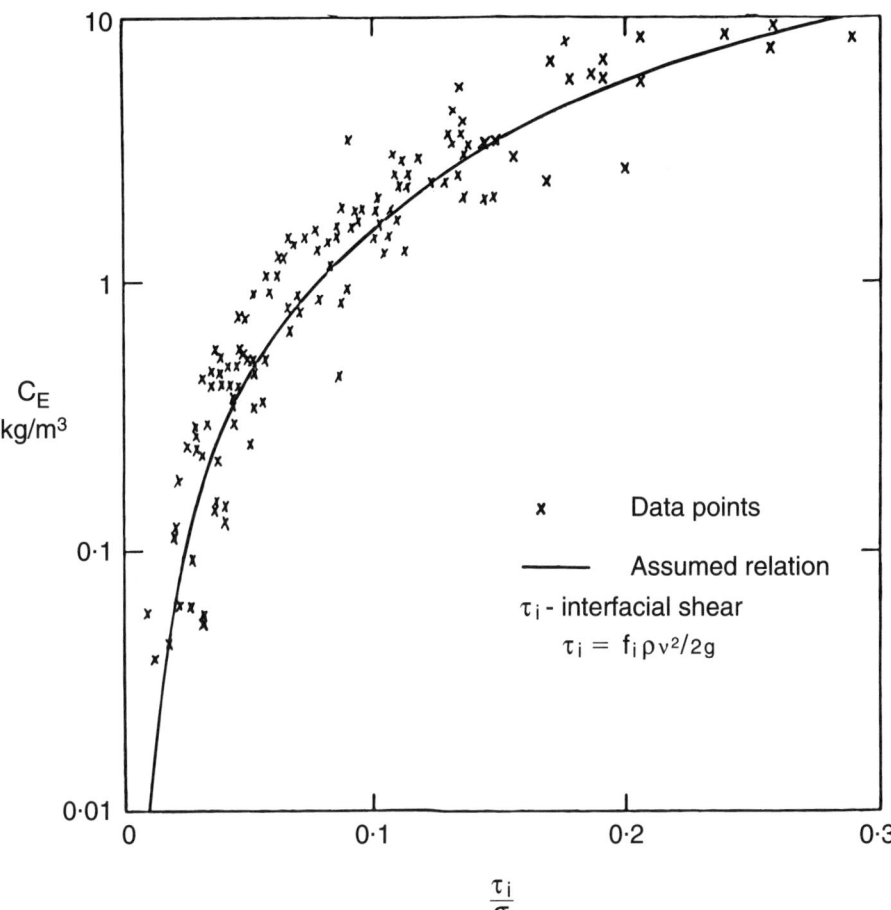

Figure 5.29 Entrainment correlation. (From Whalley et al., 1973. Reprinted with permission.)

They then found that all Freon-12 data can be plotted as X_{cr} versus E (Fig. 5.30), and that all water data can be superimposed on the same curve by changing the abscissa from E to KE, where

$$K = \frac{E_{(F\text{-}12)}}{E_{(H_2O)}} = \frac{G_{(F\text{-}12)}}{G_{(H_2O)}} = 0.658$$

The graphical correlation shown in Figure 5.30 was originally developed for a steam–water mixture of high quality. Coffield et al. (1967) extended its validity into the subcooled region by comparing their subcooled Freon-113 DNB data with

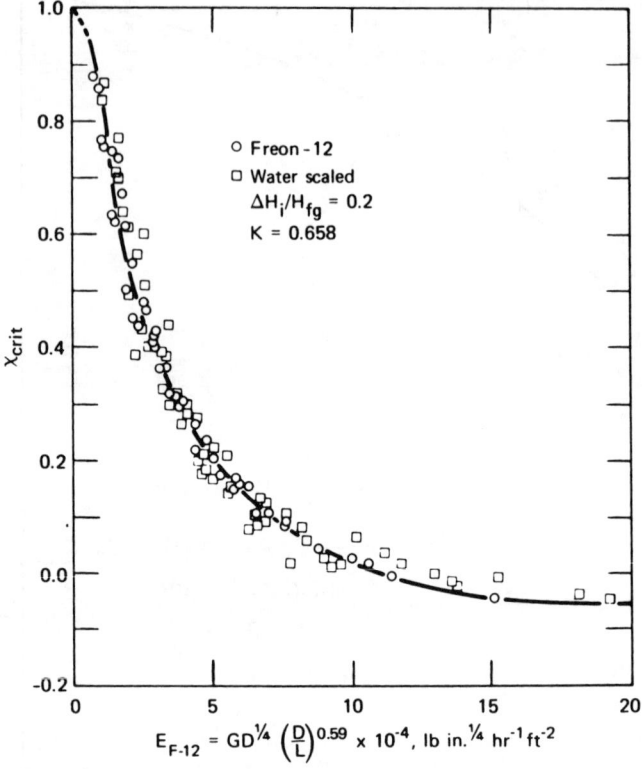

Figure 5.30 Empirical scaling factor. (From Steven and Kirby, 1964. Reprinted with permission of UK AEA Technology, Didcot, Oxfordshire, UK.)

existing water results. The graphical scaling factor K was found to be a function of the equivalent water pressure, or ρ_L/ρ_G as listed in Table 5.3. Tong et al. (1970) found that the critical quality, X_{cr}, is also a function of inlet subcooling. On the basis of the subcooled Freon-113 data, the inlet subcooling effect can be expressed by the following equation:

$$X_{cr} - (X_{cr})\big|_{\text{at } \Delta H_i/H_{fg}=0.2} = -0.58\left(\frac{\Delta H_i}{H_{fg}}\right) + 0.116 \tag{5-72}$$

where

$$\frac{\Delta H_i}{H_{fg}} = \frac{H_{\text{sat}} - H_i}{H_{fg}}$$

Table 5.3 Variation of graphical scaling factor K with equivalent water pressure

Graphical factor, K	Equivalent water pressure		ρ_L/ρ_G
	psia	MPa	
0.684	1,000	6.8	20.6
0.690	1,500	10.2	11.8
0.715	2,000	13.6	7.2

Source: Tong et al. (1970). Copyright © 1970 by Hemisphere Publishing Corp., New York. Reprinted with permission.

The DNB heat fluxes in rod bundles of nonuniform axial and radial heat fluxes measured in the flows of water and Freon-11 at an equivalent water pressure range of 1,500–2,100 psia (10.2–14.3 MPa) were used to develop the graphical scaling factor (Tong et al., 1970). The scaling factor K was obtained by the graphical method of Stevens and Kirby, where $K = E_{(F-11)}/E_{(H_2O)}$. It was found that K is a function of the system pressure as $K = 1.10$ at 1,500 psia (10.2 MPa) and $K = 1.25$ at 2,100 psia (14.3 MPa). Comparisons of the predictions and the measured data show excellent agreement. It should be noted that in the axially nonuniform heat-flux test section, the DNB locations in the water tests agree closely with those in the Freon tests. Staub (1969) found that the CISE dryout correlation (Bertoletti et al., 1965) for water is valid for Freon-12 when a constant of 319 replaces the constant of 168 in the following water correlation:

$$X_{cr} = \frac{1-p_R}{1.1(G/10^6)^{1/3}} \frac{L_B}{\{L_B + 168[(1/p_R)-1]^{0.4} D^{1.4}(G/10^6)\}} \quad (5\text{-}73)$$

where all units are in the British system. This replacement indicates that scaling factors can be developed from the constants of a CHF correlation where these constants have physical significance. A pressure correction factor to the above equation,

$$\frac{1-(1/4-p/p_c)}{(9/4-p/p_c)^2}$$

was suggested by M. Cumo (1972).

In summary, we agree with the conclusions of Mayinger (1981) in his scaling and modeling criteria in two-phase flow and boiling heat transfer:

> Scaling by use of dimensionless numbers only is limited in two-phase flow to simple and isolated problems, where the physical phenomenon is a unique function of a few parameters. If there is a reaction between two or more physical occurrences, dimensionless scaling numbers can mainly serve for selecting the hydrodynamic and thermodynamic conditions of the modelling tests. In

356 BOILING HEAT TRANSFER AND TWO-PHASE FLOW

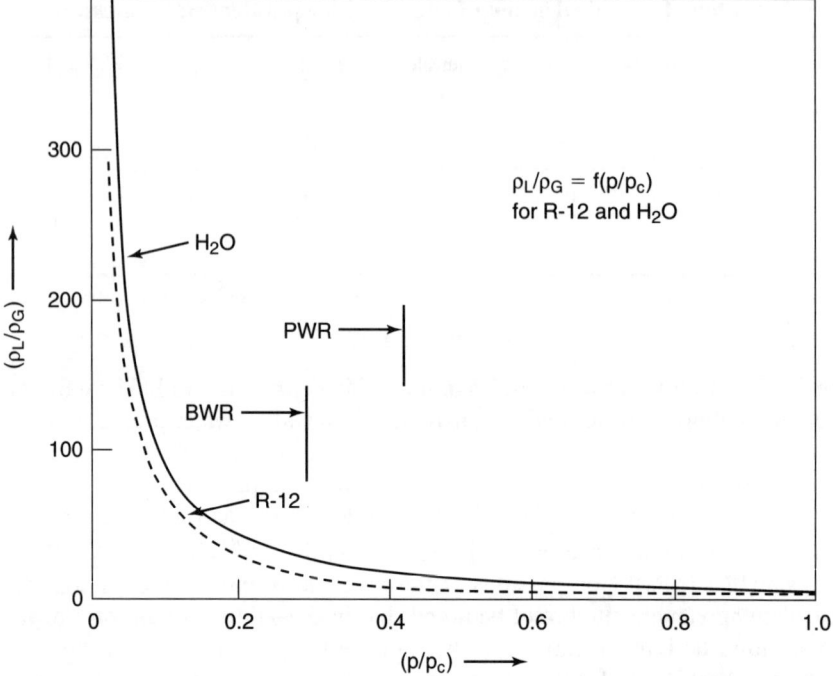

Figure 5.31 Density ratio, ρ_L/ρ_G, of water and Freon-12 as a function of reduced pressure. (From Mayinger, 1981. Copyright © 1981 by Hemisphere Publishing Corp., New York. Reprinted with permission.)

this case, the primary influencing parameters have to be separated from the secondary ones as the former determine the system behavior more dominantly than the latter.

To develop the scaling criteria for complicated thermohydraulic behavior of a two-phase flow, such as boiling crisis of a rod bundle, we would have to scale via a computer code that analyzes and describes the thermohydraulic behavior clearly and in detail. Testing separately in complicated prototype geometry and simple geometry could identify which dimensionless numbers are primary ones describing the behavior of the testing fluid and which ones describing the behavior are related to geometry. For example, Mayinger (1981) plotted scaling factors of the physical properties of water and Freon-12 as shown in Figures 5.31, 5.32, and 5.33. For scaling the bubble size in Freon-114 flow, Cumo (1969) suggested

$$D = 0.42\left(\frac{p}{p_{cr}}\right) + 0.39 \text{mm}$$

Mayinger (1981) also quoted Boure's suggestion of dimensionless numbers for scaling critical heat flux in a simple geometry as listed below.

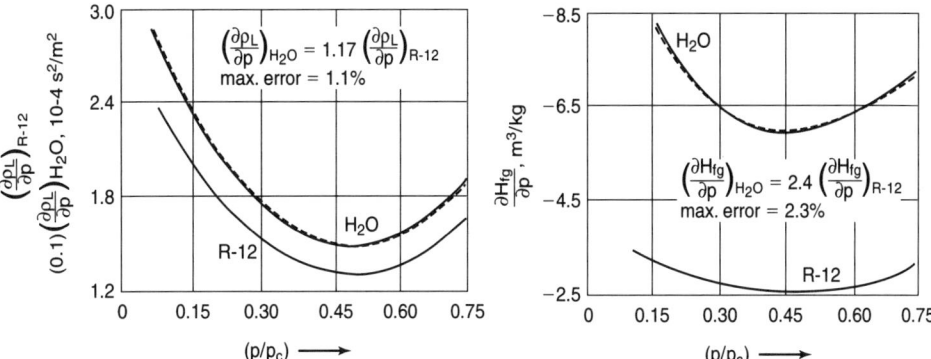

Figure 5.32 Comparison of derived thermodynamic properties (water and Freon-12). (From Mayinger, 1981. Copyright © 1981 by Hemisphere Publishing Corp., New York. Reprinted with permission.)

$$N_1 = K_1 \frac{\rho_L}{\rho_G}$$

$$N_2 = K_2 \frac{\Delta H_{in,l}}{H_{fg}}$$

$$N_3 = K_3 \frac{q''}{GH_{fg}}$$ (5-74)

$$N_4 = K_4 \frac{G}{\rho_L (g\ell)^{1/2}}$$

The scaling factors K_1–K_4 depending on the geometry of the channel as well as the pressure. Kocamustafaogullari and Ishii (1987) studied scaling of two-phase flow transients using a reduced pressure system and simulant fluid. They concluded that power scaling is more practical than time scaling for modeling a high-pressure water system by a low-pressure water system. It was shown that real-time scaling is not possible for an axially scaled-down model. The modeling of a typical LWR system in loss-of-coolant accident by a low-pressure water system and a Freon system is possible for simulating pressure transients. However, simulation of phase-change transitions is not possible with a low-pressure water system without distortions in power or time.

5.3.4.2 CHF correlations for organic coolants and refrigerants. The idea of using organic liquids as reactor moderators and/or coolants was studied in the 1960s. While the spectrum of organic coolant advantages was wide and included many factors of different importance, their tendency to decompose under radiation or at

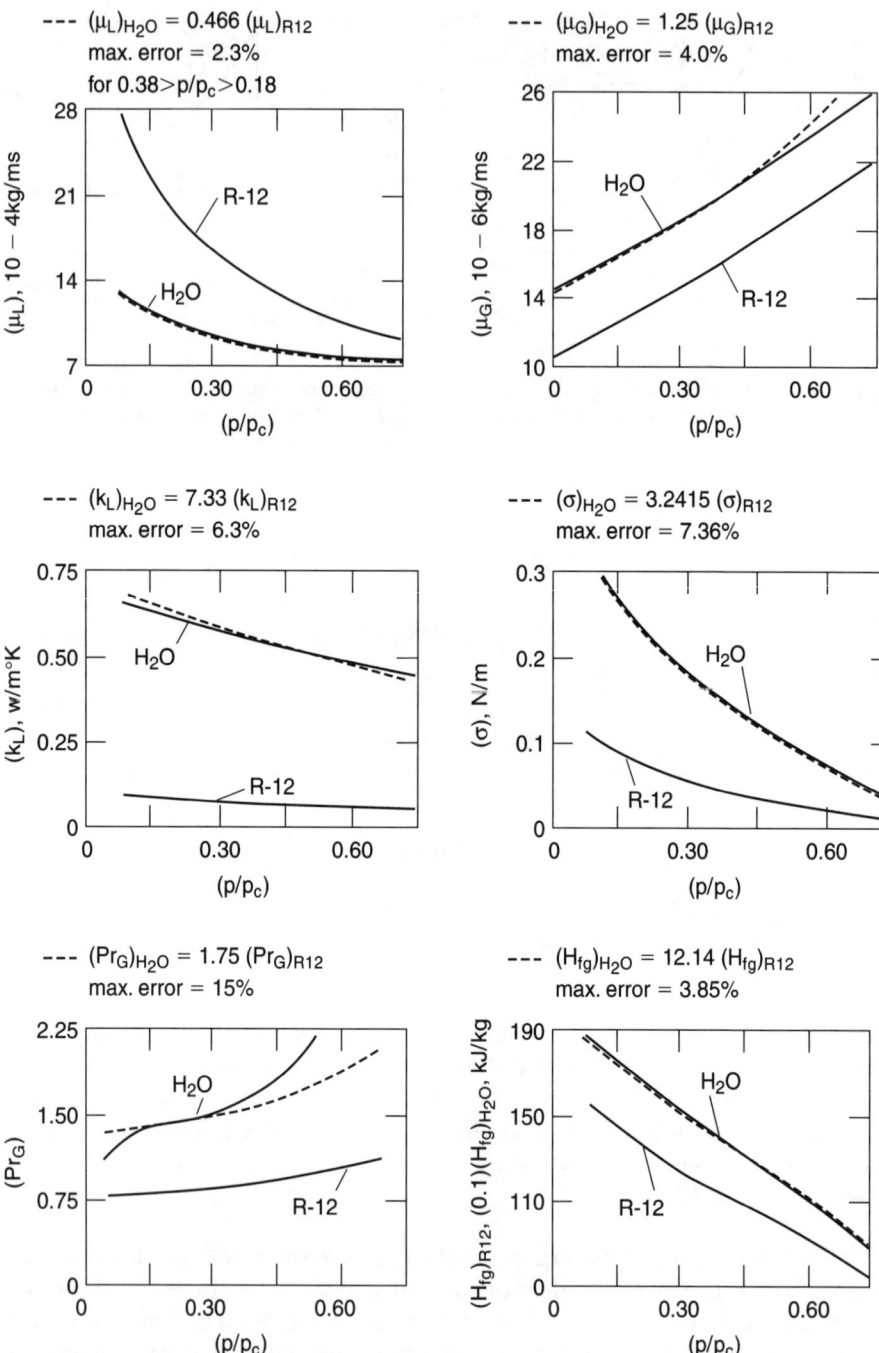

Figure 5.33 Comparison of thermodynamic and transport properties of water and Freon-12. (From Mayinger, 1981. Copyright © 1981 by Hemisphere Publishing Corp., New York. Reprinted with permission.)

working temperatures, and their high melting points and low thermal conductivities, were unfavorable characteristics that could not compete with liquid metals. The organic liquids used during that period belong mainly to the aromatic hydrocarbon polymer group (biphenyl is a typical example). Commercial products were provided by different companies under a variety of names; e.g., Santowax-R was a mixture of *ortho-, meta-,* and *para*-terphenyls produced by Monsanto.

Because Freons, or fluorochloro-compounds, have been used in refrigeration equipment, their evaporation and condensation characteristics have been studied, as summarized recently by Carey (1992). They have also been shown to be good fluids to simulate water in boiling crisis tests (Sec. 5.3.4.1). Correlations for CHF of these compounds are therefore of interest.

Correlations for CHF in polyphenyl Two empirical correlations were developed by Core and Sato (1958) for predicting the CHF of polyphenyl flow in an annulus. The correlations are as follows.

1. For diphenyl with the range of parameters
   $p = 23–406$ psia (0.16–2.76 MPa)
   $V = 1–17$ ft/sec (0.3–5.18 m/s)
   $\Delta T_{sub} = 0–328°F$ (0–182°C)   (5-75)
   $q''_{crit} = 454\, \Delta_{sub} V^{0.65} + 116{,}000$ Btu/hr ft²

2. For Santowax-R with the range of parameters
   $p = 100$ psia (0.68 MPa)
   $V = 5–15$ ft/sec (1.5–4.6 m/s)   (5-76)
   $q''_{crit} = 552\, \Delta T_{sub} V^{2/3} + 152{,}000$ Btu/hr ft²

Robinson and Lurie (1962) reported several other empirical correlations for CHF in organic coolants. These correlations are

1. For Santowax-R with the range of parameters
   $p = 30–150$ psia (0.2–1.0 MPa)
   $G = 0.8 \times 10^6–3.3 \times 10^6$ lb/hr ft² (1.1 × 10³–4.5 × 10³ kg/m² s)   (5-77)
   $q''_{crit} = 14.5\, \Delta T_{sub} G^{0.8} + 198{,}000$ Btu/hr ft²

2. For Santowax-R + 27wt% radialytic heptyl benzene with the range of parameters
   $p = 24–79$ psia (0.16–0.54 MPa)
   $G = 10^6–2.9 \times 10^6$ lb/hr ft² (1.35 × 10³–3.92 × 10³ kg/m² s)   (5-78)
   $q''_{crit} = 16.9\, \Delta T_{sub} G^{0.8} + 353{,}000$ Btu/hr ft²

3. For the above mixture + 2.9 to 7.4wt% diphenyl with the range of parameters
   $p = 22–124$ psia (0.15–0.84 MPa)
   $G = 1.1 \times 10^6–5.3 \times 10^6$ lb/hr ft² (1.5 × 10³–7.2 × 10³ kg/m² s)   (5-79)
   $q''_{crit} = 23.0\, \Delta T_{sub} G^{0.8} + 282{,}000$ Btu/hr ft²

Correlations for CHF in refrigerants Weisman and Pei (1983) suggested a theoretically based predictive procedure for CHF at high-velocity water flow in both uniformly and nonuniformly heated tubes, which was found to yield equally good results with experimental data for four other fluids: R-11, R-113, liquid nitrogen, and anhydrous ammonia.

$$\frac{q''_{crit}}{G \, \Delta H_{fg}} \left(\frac{h_L - h_{LD}}{h_f - h_{LD}} \right) = (x_2 - x_1) \psi i_b \tag{5-80}$$

where h_L, h_{LD} = enthalpy of the liquid and at the point of bubble detachment, respectively
h_f = saturated liquid enthalpy
i_b = turbulent intensity at bubbly layer – core interface

$$= (0.462)(Re)^{-0.1}(K)^{0.6} \left(\frac{D_p}{D} \right)^{0.6} \left[\frac{1 + a(\rho_L - \rho_G)}{\rho_G} \right] \tag{5-81}$$

K, a = constants

$$\psi = \frac{1}{\sqrt{2\pi}} \exp\left[-\left(\frac{V_r}{2\sigma_{v'}}\right)^2\right] - \left(\frac{V_r}{2\sigma_{v'}}\right) \text{erfc}\left(\frac{V_r}{\sqrt{2}\sigma_{v'}}\right)$$

V_r = radial velocity created by vapor generation
σ'_v = standard deviation of v' (radial fluctuating velocity)
X_1 = average quality in core region
X_2 = average quality in bubbly layer, corresponding to the critical void fraction α_{CHF} = 0.82

The above method was developed using an assumption that the two-phase mixture could be treated as a homogeneous fluid, which was acceptable only at high mass flux, above 7.2×10^5 lb/hr ft^2 (3.5×10^6 kg/m^2 s).

5.3.4.3 CHF correlations for liquid metals.

Flow boiling of sodium Noyes and Lurie (1966) attempted to correlate experimental data of flowing sodium by the method of superposition,

$$(q'')_{cr, sub, f.c.} = (q'')_{cr.sat, pool} + (q'')_{\substack{\text{subcooling} \\ \text{effect on} \\ \text{pool boiling}}} + (q'')_{\substack{\text{nonboiling} \\ \text{f.c.}}} \tag{5-82}$$

where the effects of the interactions among various contributions are thus neglected, and the terms can all be approximated using established correlations. This

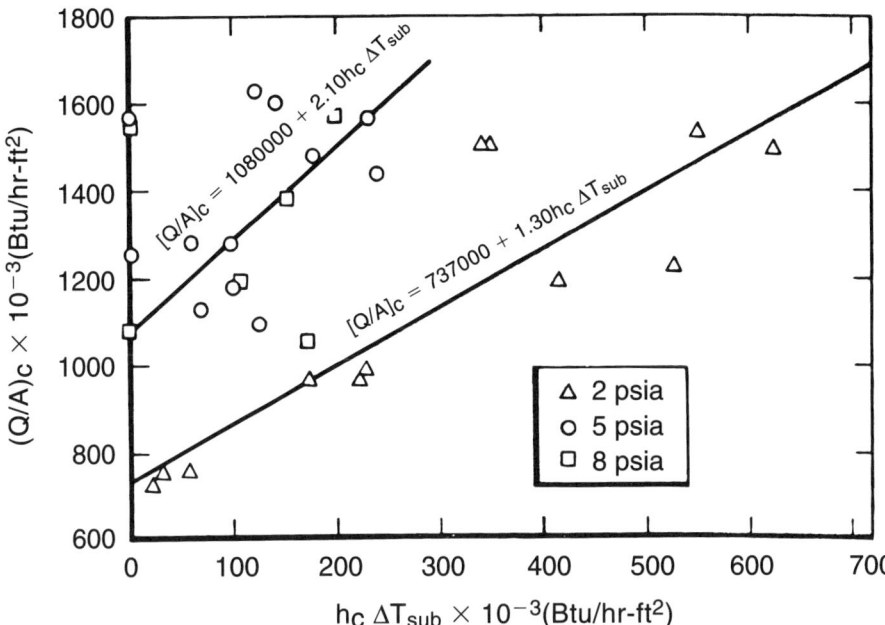

Figure 5.34 Critical heat flux of boiling sodium under subcooled forced convection versus nonboiling convection heat flux. (From Lurie, 1966. Copyright © 1966 by Rockwell International, Canoga Park, CA. Reprinted with permission.)

hypothesis suggested plotting subcooled critical heat flux versus the nonboiling convection heat flux ($h_c \Delta T_{sub}$) (Fig. 5.34). As shown in the figure, rather large scatter occurred in the data. The CHF measured during bulk vapor boiling was correlated by Noyes and Lurie for sodium at the exit quality and system pressure shown in Figure 5.35. The CHF decreased sharply with an increase in exit quality and a decrease in system pressure. These data, along with the subcooled forced-convection boiling CHF data, are summarized in Figure 5.36, as a function of the dimensional group in $\dot{m}(h - h_{sat})$, which are believed to be two of the more important parameters.

Flow boiling of potassium The critical heat flux in flow boiling of potassium has been reported by Hoffman (1964), as shown in Figure 5.37. Because these data are for high exit qualities, they are related to the dryout type of boiling crisis. It was found that the CHF of potassium flow agrees well with a correlation developed for water by Lowdermilk et al. (1958).This correlation is dimensional, and the curves representing the correlation are also shown in the figure relative to the specific geometries tested. The boiling crisis is evidenced by a sharp rise in wall temperature with an equally sudden reduction in overall pressure drop; a wall temperature increase of 50°F is taken as the power cutoff criterion. No wall temperature oscilla-

362 BOILING HEAT TRANSFER AND TWO-PHASE FLOW

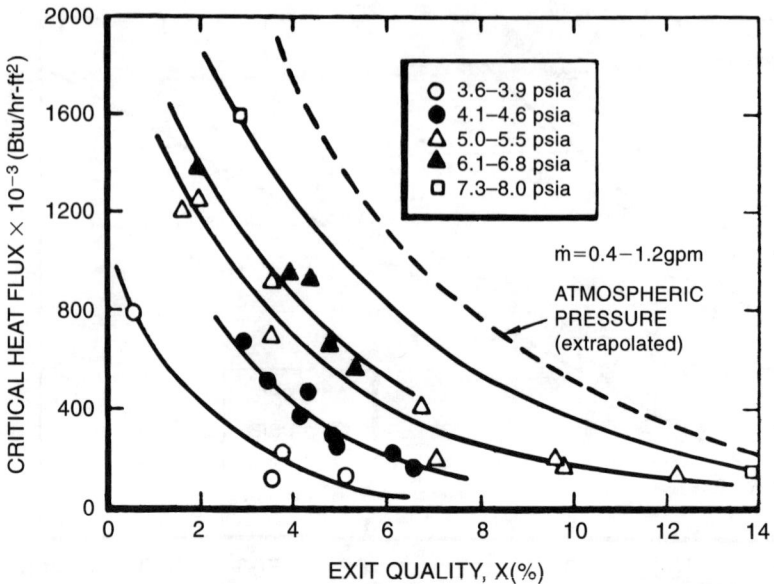

Figure 5.35 Critical heat flux for two-phase sodium flow. (From Lurie, 1966. Copyright © 1966 by Rockwell International, Canoga Park, CA. Reprinted with permission.)

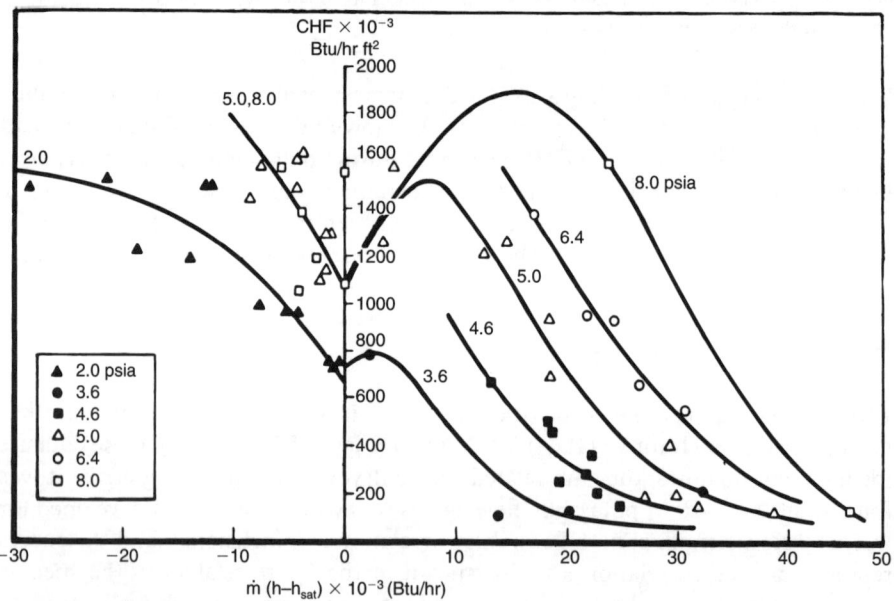

Figure 5.36 Critical heat flux of sodium under forced convection. (From Lurie, 1966. Copyright © 1966 by Rockwell International, Canoga Park, CA. Reprinted with permission.)

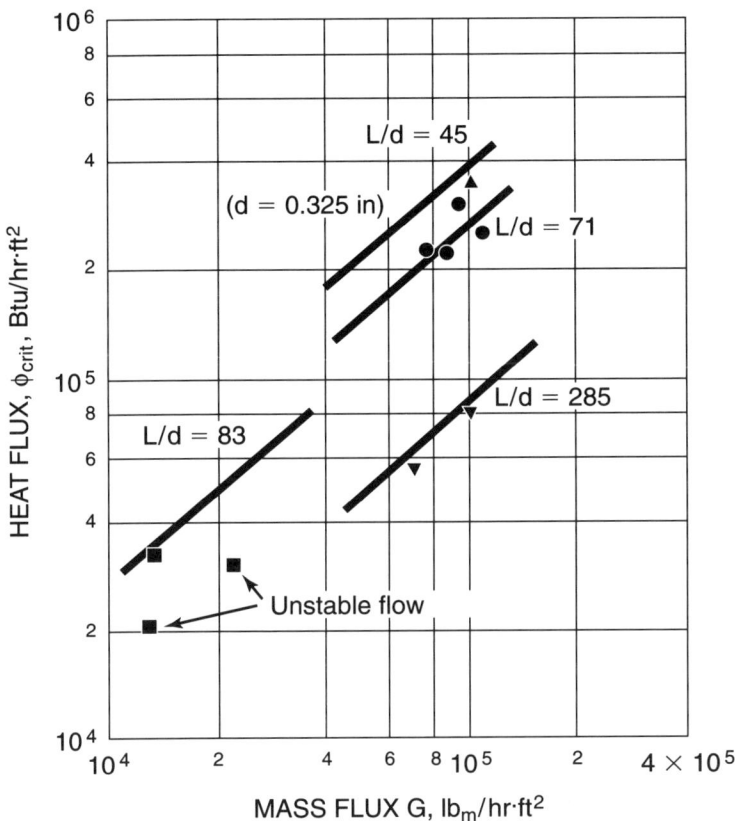

Figure 5.37 Critical heat flux with boiling potassium. (From Hoffman and Keyes, 1965. Reprinted with permission of Oak Ridge National Laboratory, Oak Ridge, TN.)

tion is observed before the boiling crisis. The similarity between the thermophysical and transport properties of these two fluids can be invoked to support a modeling postulate. This is also shown in the CHF data for boiling water in a rod bundle as a prelude to boiling potassium (Jones and Hoffman, 1970). Figure 5.38 shows the effect of mass flux on CHF of potassium and water in a seven-rod bundle. The water data of Jones (1969), obtained at 24.3 and 9.3 psia (0.17 and 0.06 MPa) are shown by the continuous and dashed lines, respectively, in the figure; while the potassium data (points) are from Huntley (1969) and Smith (1969) at various pressures, as indicated at the top of the figure. The discrepancy between these two fluids decreases with increasing G. Figure 5.39 shows the effect of exit quality on the CHF of two fluids in the same test configuration.

It should be mentioned that boiling within a liquid metal-cooled reactor (such as a sodium-cooled reactor) is an accident condition and may give rise to rapid fuel failure. In designing a reactor core, on the other hand, sodium boiling should

364 BOILING HEAT TRANSFER AND TWO-PHASE FLOW

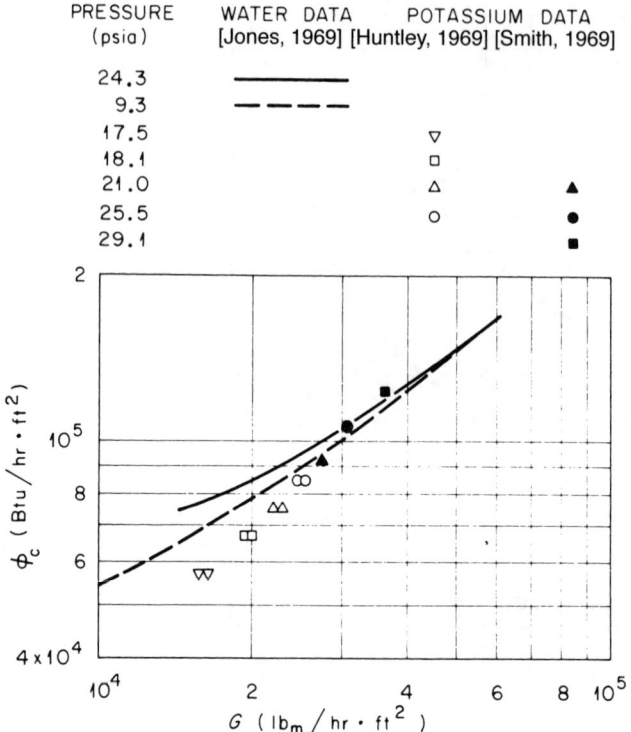

Figure 5.38 Effect of mass flux on the CHF of potassium and water in a seven-rod bundle. (From Jones and Hoffman, 1970. Copyright © 1970 by American Society of Mechanical Engineers, New York. Reprinted with permission.)

not be included, and the design should almost certainly include considerable margin before boiling might be initiated (Graham, 1971).

Flow boiling of other alkali metals CHF data for other alkali metals were reported by Fisher et al. (1964, 1965), who tested rubidium and cesium in axial and swirl flow and potassium in swirl flow. The data were correlated by postulating a mist or fog flow model for the hydrodynamic situation in the heated section in which CHF occurs. These investigations were motivated by the potential use of alkali metals as Rankine cycle working media in space applications and have not been pursued further, because there is no longer interest in such concepts.

The properties of liquid metals can cause flow instability (oscillation) because of vapor pressure—temperature relationship. Most liquid metals, especially alkali metals, show a greater change in saturation temperature, corresponding to a given change of pressure, than does water. In a vertical system under gravitational force, the change of static pressure could appreciably alter the saturation temperature such that "explosion"-type flow oscillation would occur that would result in liquid

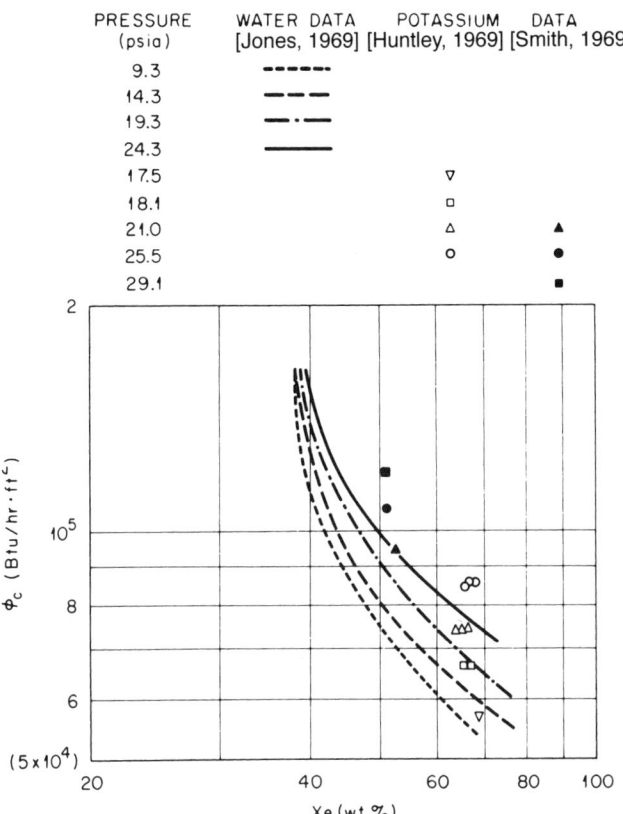

Figure 5.39 Effect of exit quality on the CHF of potassium and water in a seven-rod bundle. (From Jones and Hoffman, 1970. Copyright © 1970 by American Society of Mechanical Engineers, New York. Reprinted with permission.)

expulsion. This could be considered another form of departure from nucleate boiling. Studies of the expulsion dynamics of sodium were made easier by the fact that information obtained from nonmetallic fluid tests can be shown to be applicable to a liquid metal system (Singer and Holtz, 1970; Grolmes and Fauske, 1970). To prevent this type of flow instability in boiling liquid metals, the following recommendations are worth mentioning.

1. Provision of finite inlet quality condition, which can be accomplished by allowing flashing across the flow restriction at the boiler inlet.
2. Addition of appropriate nucleation sites on the heated surface or a localized hot spot.
3. Introduction of a vortex or swirl flow, which effectively reduces the pressure at the center and thus reduces the superheat of the boiling liquid metal near the wall.

5.4 PARAMETER EFFECTS ON CHF IN EXPERIMENTS

Reactor operating parameters are used in predicting the boiling crisis in order to maintain a proper safety margin during the reactor operations. There are seven core thermal hydraulic parameters for operating a nuclear reactor: surface heat flux (q''), system pressure p, mass flux G, local enthalpy H_{loc}, inlet enthalpy H_{in}, channel diameter D_e, and channel length L. They are not all independent, but are mutually related through the energy balance of the heating channel. For a uniform heat flux distribution, the energy balance equation of a test section can be written as

$$(H_{ex}) - (H_{in}) = \frac{A_h \bar{q}''}{A_c G} = \left(\frac{4L}{D_e}\right)\left(\frac{\bar{q}''}{G}\right) \tag{5-83}$$

This equation brings an additional complexity into the analysis of boiling crisis, because it converts an independent parameter into a dependent parameter. Whenever one of the operating parameters is changed, another parameter must change according to the energy balance. The influences of these two simultaneously changing parameters cannot be differentiated. This is often called a *parametric distortion* and must be clarified in the presentation of uniform-flux boiling crisis data by noting the accompanying variable along with the main variable at each data point.

For a nonuniform heat flux distribution, the energy equation of the test section becomes

$$(H_{ex}) - (H_{in}) = \left(\frac{4}{GD_e}\right) \int_0^L q'' \, dz \tag{5-84}$$

In the quality flow region, where the CHF occurs at the exit of the channel, experimental results indicate that the critical power of a nonuniformly heated test section is a function of H_{in}, L, D_e, G, and p, and suggest that an approximate relationship for boiling crisis in an axially nonuniformly heated test section can be written as

$$q''_{crit} = F_1(H_{in}, L, D_e, G, p) \tag{5-85}$$

In the subcooled or low-quality region, the boiling crisis of a nonuniformly heated section can occur upstream of a boiling channel. The equation for predicting the local CHF can be written as

$$q''_{crit} = F_2(H_{loc}, \ell_{cr}, D_e, G, p, F_c) \tag{5-86}$$

where F_c is a shape factor for nonuniform flux distribution, as given previously (Tong et al., 1966a). For a given heat flux distribution, the local heat flux in Eq. (5-86) is connected directly to the inlet enthalpy through an energy balance that can be stated as

$$H_{\text{loc}} - H_{\text{in}} = \frac{4}{GD_e} \int_0^{\ell_{\text{cr}}} q'' \, dz \tag{5-87}$$

where ℓ_{cr} is the location of the boiling crisis. The axial heat flux distribution q'' is usually given in a reactor design as a function of location z.

The gross effects of the operating parameters on CHF are correlated empirically in the range of parameters tested. The selection of which operating parameters to use in a specific correlation should be made under the guidance of insight into the CHF mechanism revealed by microscopic analyses as well as by visual observations obtained in simulated tests as described previously.

It should be noted that the units of operating parameters in the correlations must be maintained in the same system (i.e., either in English or in SI units), consistent with those used by the originator(s) of the correlation. Any parameter quantity, before being used in the correlation for CHF predictions, must be checked for the right unit system. Since the empirical constants used in the correlation may involve a nonlinear function of operating parameters, the original units have to be kept to maintain their proper effects in the correlation. Examples of such empirical constants are C in the energy equation for the bubble layer [Eq. (5-31)], and a and b in the CISE-1 correlation [Eq. (5-141)].

5.4.1 Pressure Effects

In a uniform heat flux test section, the CHF cannot vary by one variable without affecting another accompanying variable. Figure 5.40 is reproduced from an article by Aladyev et al. (1961). This figure actually indicates the combined effects of pressure and inlet subcooling at a constant exit quality. The CHF occurs at the exit, and the exit enthalpy is kept at saturation. Because the critical flux varies with pressure, the inlet temperature must also vary. Hence the high CHF at low pressure is achieved by means of a low inlet temperature; and the favorable physical properties of water and steam under low pressures also help the heat transfer at the core–bubble layer interface.

Another curve showing the effect of pressure on the CHF is demonstrated by Figure 5.41, which is reproduced from an article by Macbeth (1963a). The CHF varies with pressure and the accompanying variable of exit enthalpy. The trend in this figure is completely different from that of Figure 5.40. For the design of a boiling system with a predetermined inlet temperature, the curve in Figure 5.41 could provide useful information about the individual pressure effect on the CHF of the system, where the effect of increasing local quality, X_{cr}, at lower pressures seems to be compensated by the more favorable heat transfer properties of the water and steam at the saturation of lower pressures. This information clearly suggests that a coupled effect of the system pressure and the local quality at CHF should be used in the CHF correlation. Judging from the shape of the curve in Figure 5.41, the form of the coupling term appears to be nonlinear.

368 BOILING HEAT TRANSFER AND TWO-PHASE FLOW

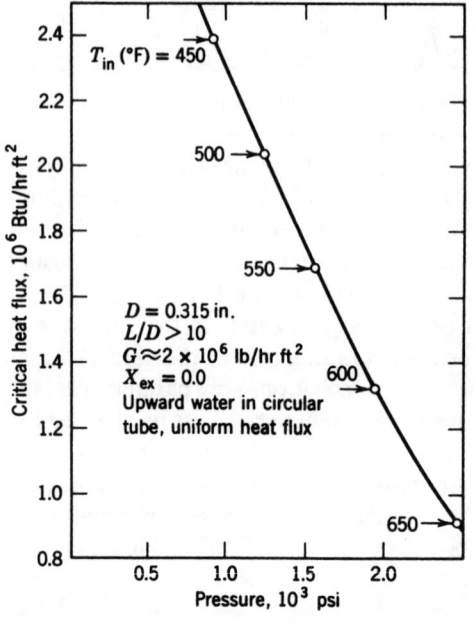

Figure 5.40 Effect of pressure and inlet subcooling on CHF. (From Aladyev et al., 1961. Copyright © 1961 by American Society of Mechanical Engineers, New York. Reprinted with permission.)

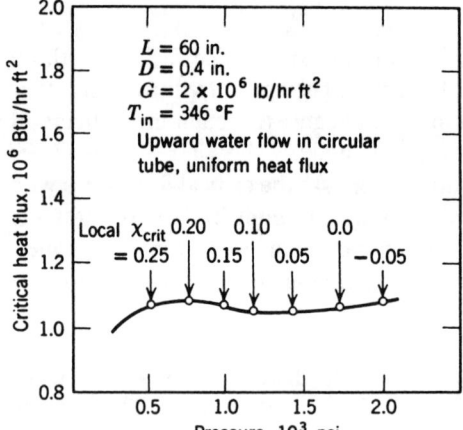

Figure 5.41 Pressure and local enthalpy effects on CHF. (From Macbeth, 1963a. Reprinted with permission of UK AEA Technology, Didcot, Oxfordshire, UK.)

The water wall superheat at CHF in flow boiling was measured by Bernath (1960) at various pressures, and the data were reduced into the form:

$$\text{Wall superheat} = \left(T_{w,\text{BO}} - T_{\text{sat}} + \frac{V}{4}\right) {}^\circ\text{C}$$

where V is expressed in feet per second. The values of wall superheat are plotted against the reduced saturation temperature in Figure 5.42.

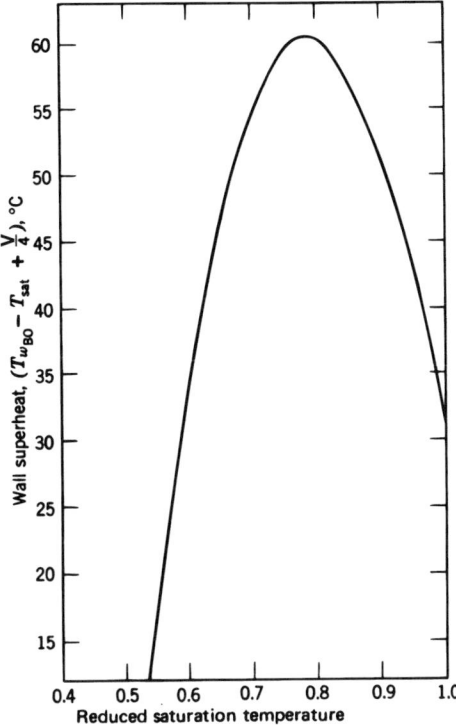

Figure 5.42 Generalized wall superheat at burnout in flow boiling. Note: V in the expression for wall superheat is in feet per second. (From Bernath, 1960. Copyright © 1960 by American Institute of Chemical Engineers, New York. Reprinted with permission.)

Cumo et al. (1969) reported that the pressure effect on the bubble diameter is linear in a Freon-114 flow, as shown in Figure 5.43. They tested the two-phase Freon-114 flow in a vertical rectangular test section at a mass flux of 100 g/cm² s (0.737×10^6 lb/ft² hr). The average bubble diameters at various system pressures were obtained from high-speed photographic recordings. The effect of reduced pressure, p/p_{cr}, on the average diameter of Freon bubbles is correlated as

$$D = -0.42 \left(\frac{p}{p_{cr}} \right) + 0.39 \text{ mm} \tag{5-88}$$

The bubble sizes at various system pressures affect the flow pattern, which in turn affects CHF as a coupled effect.

5.4.2 Mass Flux Effects

5.4.2.1 Inverse mass flux effects. Critical heat fluxes at three different mass fluxes obtained on uniformly heated test sections at Argonne National Laboratory (Weatherhead, 1962) are plotted in Figures 5.44 and 5.45. The crossing over of the curves in Figure 5.44 is generally referred to as the *inverse mass flux effect*, in a

370 BOILING HEAT TRANSFER AND TWO-PHASE FLOW

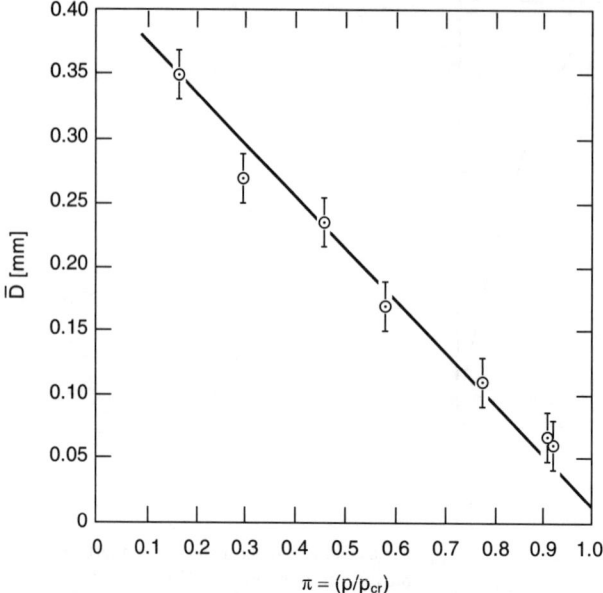

Figure 5.43 Bubble mean diameter versus reduced pressure for Freon-114; the straight line represents $D = -0.42\pi + 0.39$ mm. (From Cumo et al., 1969. Copyright © 1969 by American Society of Mechanical Engineers, New York. Reprinted with permission.)

presentation based on the local condition concept, where the local quality effect is shown along with the coupled flow pattern effect. However, that the same three sets of data do not cross over each other, up to inlet enthalpy at saturation, in Figure 5.45, shows that there is no inverse mass flux effect in a presentation based on the system parameter concept, where local quality effect is built into the mass effect.

A further study of the inverse mass flux effect was made by Griffel and Bonilla (1965) using a systematic test of flow boiling crisis with water in circular tubes having uniform heat flux distributions. A typical plot of their data (Fig. 5.46) shows that the effects of local enthalpy and mass flux become coupled at a given pressure. The inverse mass flux effect occurs at a high-quality region in which a high steam velocity promotes liquid droplet entrainment. This finding indicates that the mass flux effect on CHF is different in various flow patterns, and that an experimentally determined coupled effect of the mass flux and local quality should be used in correlating CHF data. The inverse mass flux effect becomes even stronger at a very high steam velocity, where the rapid fall of heat flux at 1,000 psia (6.8 MPa) was found to occur at a constant value of the steam velocity of 50–60 ft/sec (15–18 m/s). Mozharov (1959) defined a critical steam velocity V_G^* at which the entrainment of water droplets from the liquid film on the pipe wall increases significantly. This critical velocity is given as

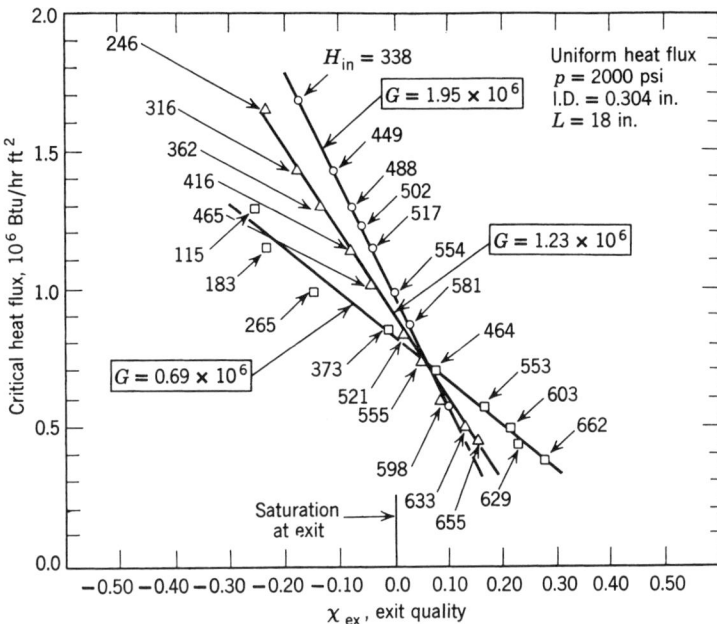

Figure 5.44 Mass flux effect on critical heat flux (local condition concept). (From Weatherhead, 1962. Reprinted with permission of U.S. Department of Energy, subject to the disclaimer of liability for inaccuracy and lack of usefulness printed in the cited reference.)

$$V_G^* = 115 \left(\frac{\sigma}{\rho_G}\right)^{1/2} \left[\frac{X}{D(1-X)}\right]^{1/4} \text{ m/s} \qquad (5\text{-}89)$$

where σ = surface tension, in kg/m
ρ_G = density of steam, in kg/m^3
X = steam mass quality
D = pipe inside diameter, in m

Increases in local quality and in steam velocity reduce the liquid film thickness of an annular flow, and thereby decrease the CHF (dryout heat flux). Bennett et al. (1963) studied the inverse mass flux effect in the quality region and also found that the CHF drops suddenly when a "critical steam velocity" is reached. Critical steam velocity is a function of pressure, as shown in Table 5.4. Note that the values given were obtained from a single test series showing the trend; they may not be valid in general, because the critical steam velocity is also a function of tube size and length, inlet enthalpy, and heat flux. The values in the table agree roughly with the steam velocity calculated from mid channel dryout data reported by Waters et al. (1965).

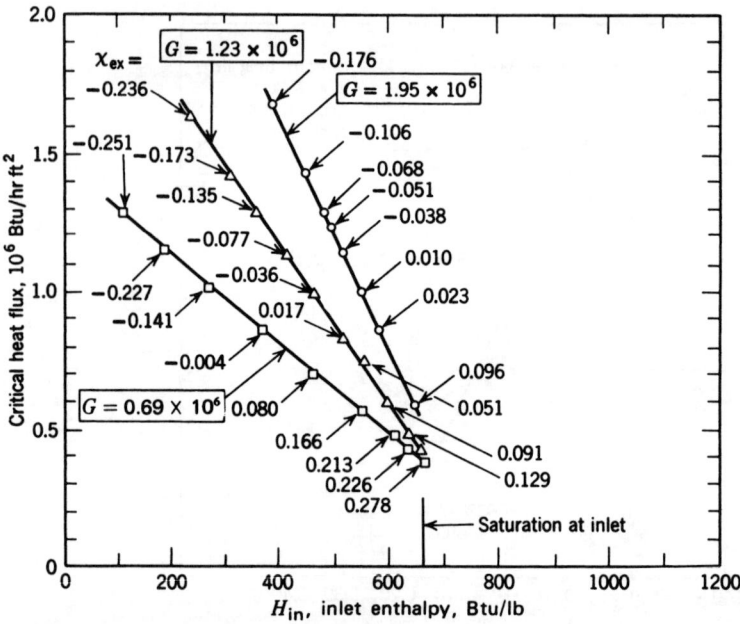

Figure 5.45 Mass flux effect on critical heat flux (system parameter concept) (based on same data as Fig. 5.44). (From Weatherhead, 1962. Reprinted with permission of U.S. Department of Energy, subject to disclaimer of liability for inaccuracy and lack of usefulness printed in the cited reference.)

Table 5.4 Critical steam velocity in flow boiling

Pressure		Critical steam velocity	
psia	10^6 MPa	ft/sec	m/s
200	1.38	311	95.0
500	3.45	130	39.6
750	5.18	85	25.9
1,000	6.90	52	15.8
1,250	8.63	38	11.6

The breakdown of a liquid film along a heating surface was studied by Simon and Hsu (1970). Liquid films can be divided into thin and thick regions. In the thin-film region, the product of breakdown heat flux and heating length is a function of fluid properties (including the temperature coefficient of surface tension) and the logarithm of the ratio of the initial film thickness to the zero heating film thickness. In the thick-film region, the motion is of a roll-wave type that causes the heating surface to be intermittently dry and wet. Kirby (1966) studied the dryout of annular flow (or climbing-film flow) and reported the following.

Two types of dryout exist in the high and medium mass fluxes, respectively. The boundary between these two mass flux regimes lies in the range from 0.1×10^6 to 0.5×10^6 lb/hr ft² (136 to 678 kg/m² s), depending on pressure and channel length (Macbeth, 1963a).

The dryout heat flux from a uniform heat flux distribution can be correlated as a function of p, X, and $(D^{1/2}G)$ but is not generally valid for a nonuniform heat flux distribution.

5.4.2.2 Downward flow effects. The CHF correlation of a downflow at high or medium mass flux in an annulus is given by Mirshak and Towell (1961) for $V > 10$ ft/sec (3 m/s) in a steady flow:

$$\frac{Q}{A} = 92,700(1 + 0.145V)(1 + 0.031\,\Delta T_{sub}) \tag{5-90}$$

where Q/A is in Pcu/hr ft² (1 Pcu ≡ 1.8 Btu), V is in ft/sec, and ΔT_{sub} is degree subcooling in °C. Their experimental data show that at high flow velocities, an increase in subcooling increases CHF, similar to the situation in upflows. However, in a downflow at low flow velocities, subcooling does not increase the CHF appreciably, because the buoyancy effect in this case is significant. The dryout flux (CHF) at very low mass fluxes behaves differently than at either medium or high mass fluxes. Low flow CHF occurs along with a flow instability or flooding that lowers the magnitude of the CHF. This effect is more pronounced in a downward flow than in an upward flow. It can be seen in the plots of CHF at the vicinity of zero mass flow rate (i.e., flooding) shown in Figures 5.47 and 5.48. Mishima and Nishihara (1985) suggested a flooding CHF for thin rectangular channels of

$$q''_{cr,F} = \frac{C^2 A_c H_{fg}\sqrt{2\rho_G g\,\Delta\rho W}}{A_h[1 + (\rho_G/\rho_L)^{1/4}]^2} \tag{5-91}$$

where A_h is the heated area and A_c is the flow channel cross-sectional area. The constant C can be determined by comparing Eq. (5-91) with the experimental data:

$C = 0.73$ for a test section heated from one side
$C = 0.63$ for a test section heated from two opposite sides

For an inlet temperature less than 70°C, the weak subcooling effect can be correlated as

$$\frac{q''_{cr}}{q''_{cr,F}} = 1 + 2.9 \times 10^5 \left(\frac{\Delta H_{in}}{H_{fg}}\right)^{6.5} \tag{5-92}$$

Since a downflow in flooding is under the choked condition by the upward-moving void, the void may become stagnant along the wall. Thus the heat transfer

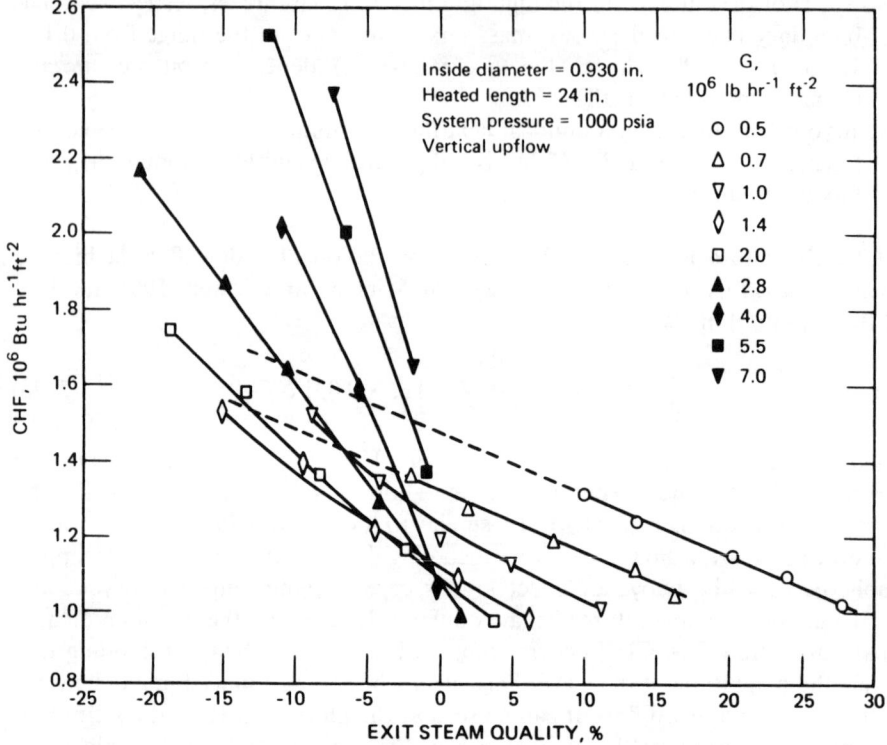

Figure 5.46 Quality and flow rate effects on CHF. (From Griffel and Bonilla, 1965. Copyright © 1965 by Elsevier Science SA, Lausanne, Switzerland. Reprinted with permission.)

mechanism in this case could be worse than in pool boiling, where natural convection exists around the heating surface. Such a comparison can be shown in a plot of nondimensional CHF against nondimensional mass flux (Figs. 5.49 and 5.50). The nondimensional CHF, q^*, and the mass flux, G^*, are defined in Eqs. (5-93) and (5-94), respectively:

$$q^* = \frac{q''}{H_{fg} (\lambda \rho_G g \Delta \rho)^{1/2}} \tag{5-93}$$

$$G^* = \frac{G}{(\lambda \rho_G g \Delta \rho)^{1/2}} \tag{5-94}$$

where the length scale λ of the Taylor instability is given by

$$\lambda = \left(\frac{\sigma}{g \Delta \rho}\right)^{1/2}$$

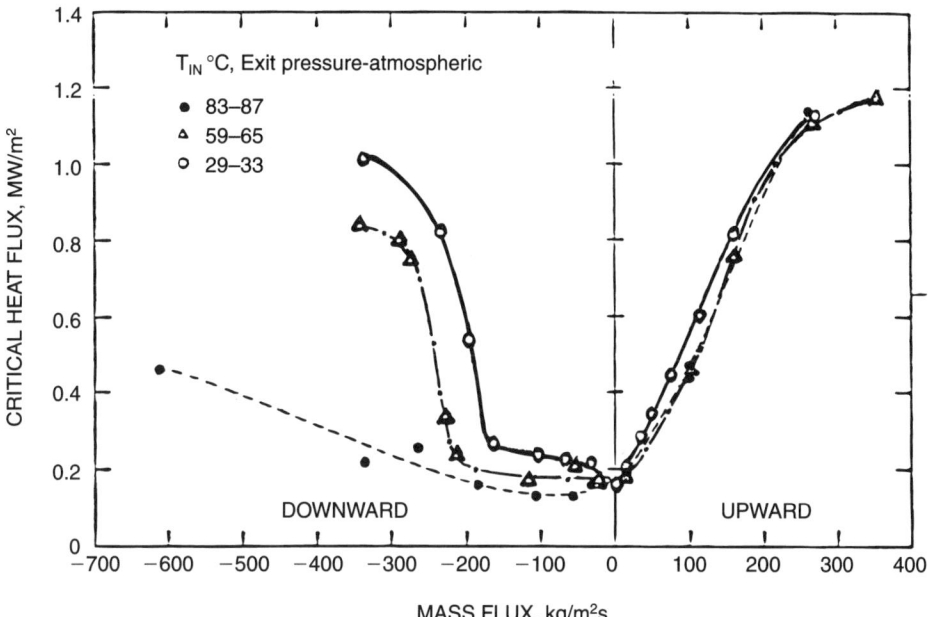

Figure 5.47 Critical heat flux for test section heated from one side. (From Mishima and Nichihara, 1985. Copyright © 1985 by Elsevier Science SA, Lausanne, Switzerland. Reprinted with permission.)

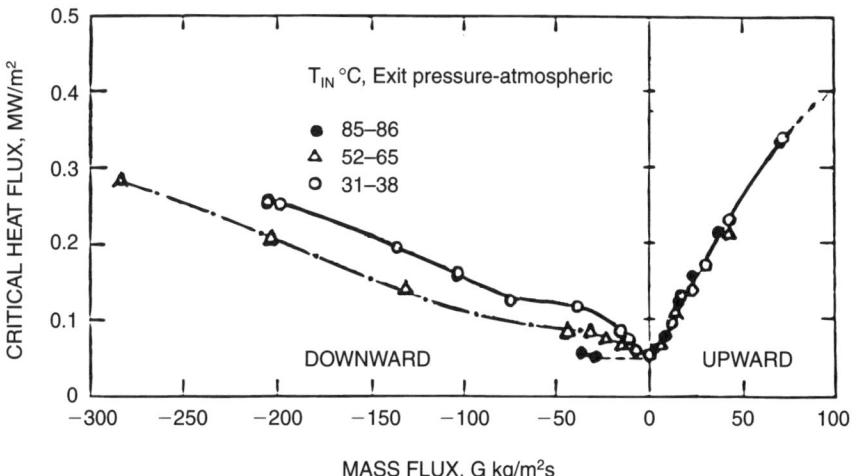

Figure 5.48 Critical heat flux for test section heated from two (opposite) sides. (From Mishima and Nishihara, 1985. Copyright © 1985 by Elsevier Science SA, Lausanne, Switzerland. Reprinted with permission.)

376 BOILING HEAT TRANSFER AND TWO-PHASE FLOW

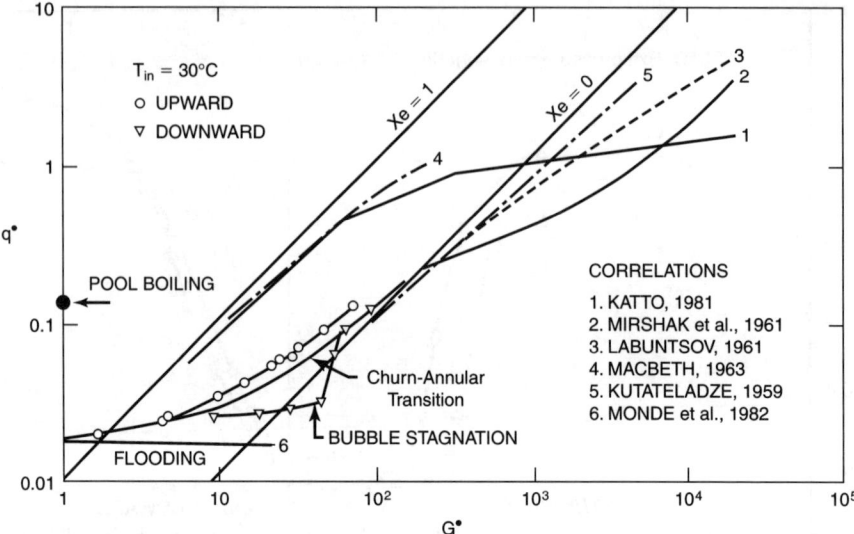

Figure 5.49 Nondimensional plot of the CHF data at an inlet temperature 30°C for test section heated from one side for several available correlations. (From Mishama and Nishihara, 1985. Copyright © 1985 by Elsevier Science SA, Lausanne, Switzerland. Reprinted with permission.)

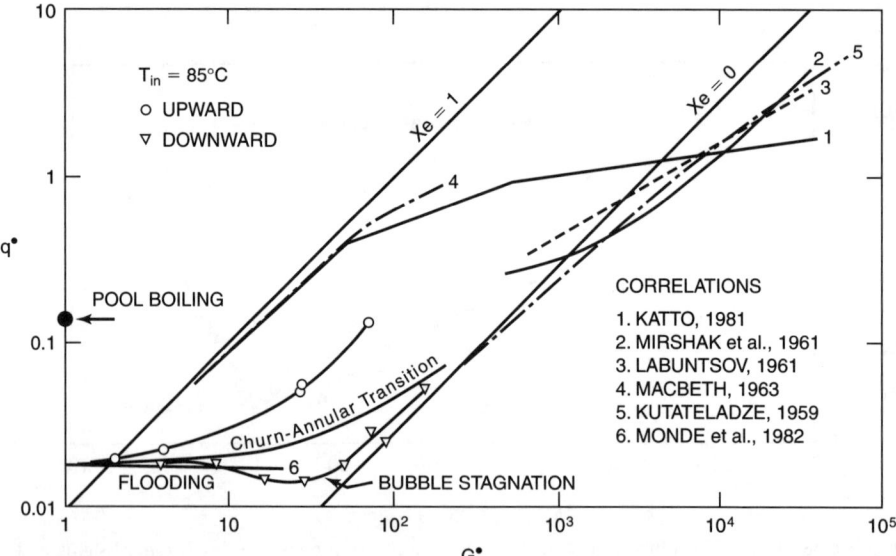

Figure 5.50 Nondimensional plot of CHF data at an inlet temperature 85°C for test section heated from one side for several available correlations. (From Mishima and Nishihara, 1985. Copyright © 1985 by Elsevier Science SA, Lausanne, Switzerland. Reprinted with permission.)

5.4.3 Local Enthalpy Effects

The effect of local enthalpy at CHF is due primarily to the wall voidage, which impairs the critical flux, and secondarily to the bulk voidage, which affects the flow pattern. The coupled effects of local subcooling and flow velocity in a subcooled bubbly flow were first reported by Griffel and Bonilla (1965), neglecting the pressure effect:

$$q''_{crit} = (384,000 + 0.0553G)(8 + \Delta T_{sub})^{0.27} \text{ Btu/hr ft}^2 \qquad (5\text{-}95)$$

which was developed in the following ranges of parameters:

$D = 0.22\text{–}1.48$ in. (0.56–3.76 cm)
$L = 24\text{–}78$ in. (0.61–2.0 m)
$\Delta T_{sub} = 0\text{–}117°F$ (0–61°C)
$G = 0.5 \times 10^6\text{–}7.0 \times 10^6$ lb/hr ft² (676–9,460 kg/m² s)
$p = 500\text{–}1{,}500$ psia (3.4–10.2 MPa)

Other data were obtained from subcooled water flowing through small tubes (0.08–0.12 in. or 0.2 to 0.3 cm in diameter) at pressures ranging from 290 to 2,900 psia (2.0 to 20 MPa) (Povarnin and Semenov, 1960). This correlation, coupling the pressure effect with local subcooling, is in the form

$$q''_{crit} = q''_o (1 + B\,\Delta T_{sub}) \left[1 + \left(\frac{V}{V_o} \right) \right]^{0.8} \qquad (5\text{-}96)$$

where q''_o = CHF at zero subcooling and zero velocity and under a given pressure (Btu/hr ft²)
B = pressure-dependent coefficient (1/°F), given in Table 5.5
V = velocity (ft/sec)
V_o = a pressure-dependent coefficient (ft/sec), also given in Table 5.5

Table 5.5 Values of q''_o and pressure-dependent coefficients B and V_o in Eq. (5-96)

p, atm	q''_o, Btu/hr ft²	B, 1/°F	V_o, ft/sec
20	0.923×10^6	0.0378	19.69
35	1.022×10^6	0.0306	22.97
50	1.070×10^6	0.0306	22.80
70	1.052×10^6	0.0306	22.47
100	0.930×10^6	0.0297	20.67
150	0.627×10^6	0.0279	13.78
200	0.262×10^6	0.0270	4.92

Several empirical uniform-flux correlations were simplified to the form

$$q''_{crit} = C_1(C_2 - X_o)$$

where the effect of X_o on q''_{crit} is very strong. However, this strong X_o effect is limited to a medium-quality region.

5.4.4 CHF Table of p-G-X Effects

To summarize the p-G-X effects, Groenveld et al. (1986) used the Canadian AECL's CHF databank, containing over 15,000 CHF data points for water over a very wide range of testing conditions, to make a reference CHF table. This reference table is derived specifically for upward water flow in a uniformly heated 8-mm (0.3-in.) tube, and shows a matrix of combined p-G-X effects on CHF. The table predicts the general trend of p-G-X effects well, as the values from the table compare favorably with the prediction of Bowring (1972) and Biasi et al.'s (1968) correlations (Groenveld and Snoek, 1986). However, to apply them to reactor design, proper adjustments for detailed geometrical effects of a prototype rod bundle would be needed. Such adjustments are usually expressed in empirical formulas that are as complicated as the existing CHF design correlations. Thus, in this case the table method does not provide much more convenience.

5.4.5 Channel Size and Cold Wall Effects

5.4.5.1 Channel size effect. Since the bubble size does not scale down in accordance with channel size, the fraction of cross-sectional area occupied by wall voidage in a small channel is considerably greater than in a large channel. In a small channel, the size effect on subcooled boiling crisis can be considered as the effect of throttling of a liquid core due to the wall voidage. Likewise, the size effect on a quality-region boiling crisis may be considered as the effect of throttling of a vapor core due to the liquid annulus on the wall. As a result, the channel size effect changes the turbulent mixing intensity at the interface between the core and the wall layer (whether a bubble layer or a liquid annulus), and thus affects the CHF. However, the effect of channel size on CHF in the high-quality region is weaker than that in the subcooled region. Data on channel size effect on CHF in a uniform heat flux distribution by Macbeth (1963b) are plotted in Figures 5.51 and 5.52. It should be noted that in Figure 5.51 the local enthalpy is reduced as the diameter increases at a constant G and L; and in Figure 5.52 the inlet enthalpy is increased as the diameter increases at a constant G and L. Such an effect of changes in the accompanied parameter (H_{loc} or H_{in}) partially distorts the effect of channel diameter, D, on CHF.

It should also be noted that in single-phase flow heat transfer, the effect of channel size is expressed by equivalent diameter. This concept, however, should be

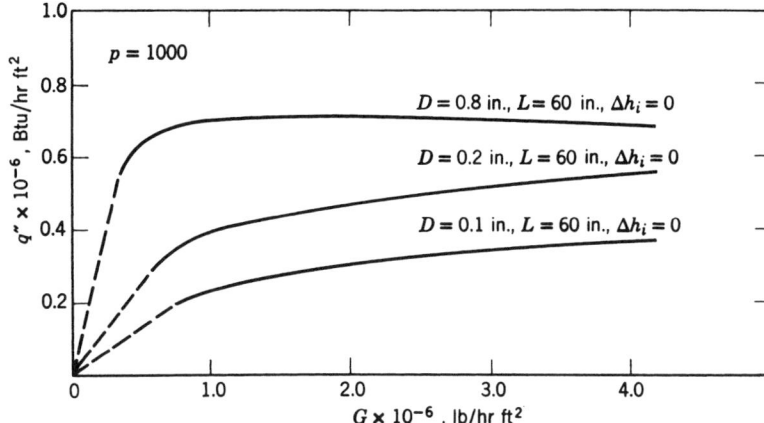

Figure 5.51 Parametric effects (D, L, G) on burnout. (From Macbeth, 1963b. Reprinted with permission of Office of Official Publications of European Communities, Luxembourg.)

examined carefully before being employed to correlate the boiling crisis of a two-phase flow. The conventional concept of equivalent diameter, D_e, has been used to describe the rectangular channel and the rod bundle only for an approximation. A correction factor may be needed to account for the slowdown effect of the flow near the corner of a rectangular channel and in the gap between the rods of a rod bundle. In an annulus, or in other channels that have a large fraction of unheated wall, the equivalent diameter, D_h, may be redefined as four times the flow area divided by the *heated* perimeter.

5.4.5.2 Effect of unheated wall in proximity to the CHF point. Data have been obtained from a tubular test section (Babcock, 1964) and an annular test section (Janssen and Kervinen, 1963) with the inner surface heated and the outer surface unheated. Both sets of data are plotted in Figures 5.53 and 5.54 according to different concepts. In Figure 5.53 (the local condition concept), the apparent unheated wall effect gives the annular channel a lower heat flux. In Figure 5.54 (the system parameter concept), the unheated wall effect gives a higher heat flux. Inspection of the accompanying changing parameters reveals that there exists an additional effect of the changing inlet enthalpy in the former case (also discussed in Sec. 5.4.6) and of the changing exit quality in the latter. When a fuel assembly consists of both types of channels in parallel and with a common inlet plenum in a nuclear reactor, the CHF given in Figure 5.54 should indicate the correct trend in this case.

However, in a channel with an unheated wall, a thick liquid film may build up on the unheated wall where it cannot cool the heated surface, and this portion of the cold liquid flow is "wasted." Hence, in this channel the coolant effectiveness is

380 BOILING HEAT TRANSFER AND TWO-PHASE FLOW

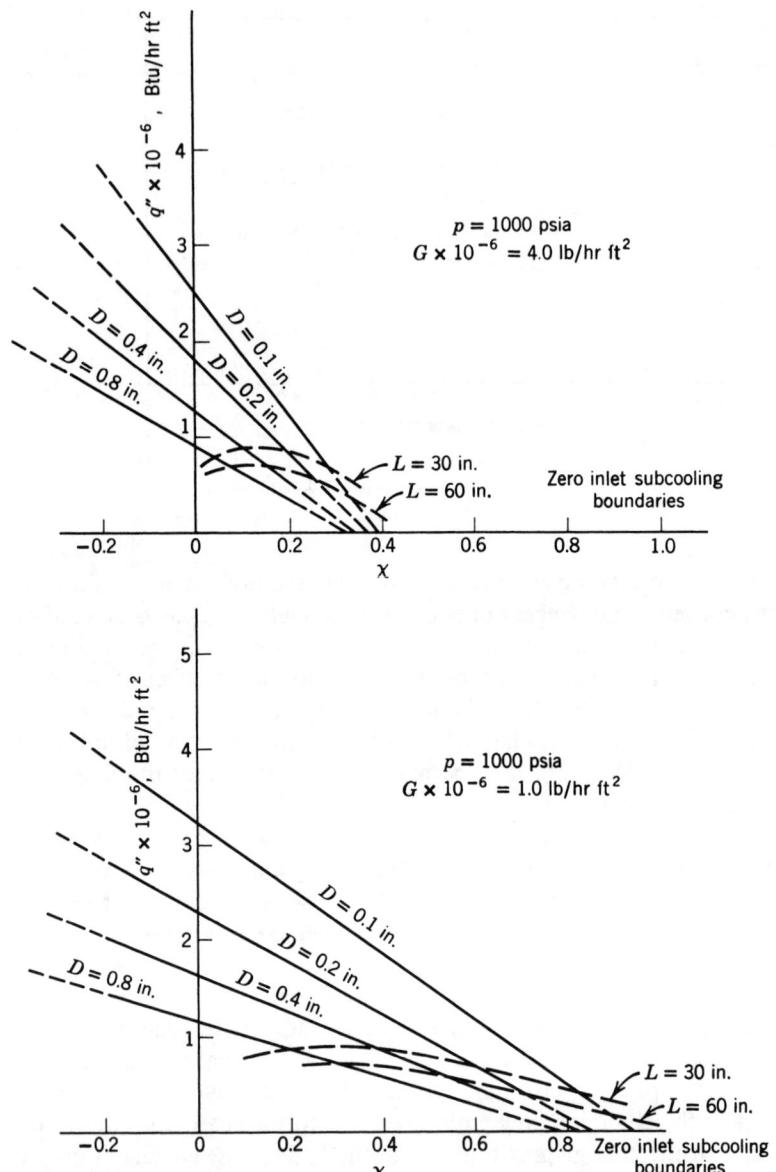

Figure 5.52 Parametric effects (D, L, X) on burnout. (From Macbeth, 1963b. Reprinted with permission of Office of Official Publications of European Communities, Luxembourg.)

FLOW BOILING CRISIS **381**

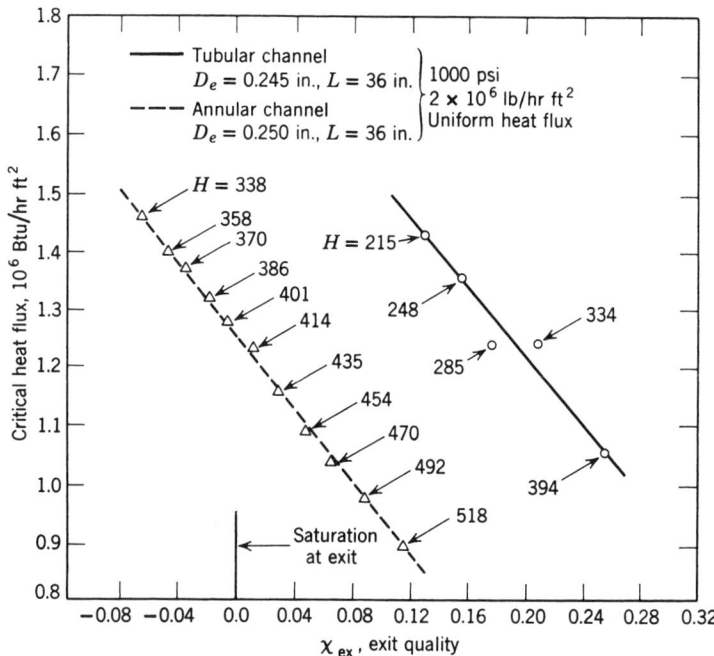

Figure 5.53 Unheated wall effect on critical heat flux (local condition concept). Data from Babcock (1964) and Janssen and Kervinen (1963a).

reduced. Tong (1968a) determined the cold wall effect empirically from annulus data:

$$\frac{(CHF)_{\text{cold wall}}}{(CHF)_{W-3,Dh}} = 1.0$$

$$-Ru\left[13.76 - 1.372e^{1.78X} - 4.732\left(\frac{G}{10^6}\right)^{-0.0535} -0.0619\left(\frac{p}{10^3}\right)^{0.14} - 8.509(D_h)^{0.107}\right] \quad (5\text{-}97)$$

where all parameters are in English units, $R_u = 1 - (D_e/D_h)$, and D_e and D_h are equivalent diameters based on wetted perimeter and heated perimeter, respectively. The ranges of parameters are

$X_{\text{DNB}} \leq 0.10$
$1.0 \leq \dfrac{G}{10^6} \leq 5.0$ lb/hr ft² (1,356 ≤ G ≤ 6,780 kg/m² s)
$1{,}000 \leq p \leq 2{,}300$ psia (6,900 ≤ p ≤ 15,860 kPa)
$L \geq 10$ in. (25.4 cm)
Gap ≥ 0.10 in. (0.25 cm)

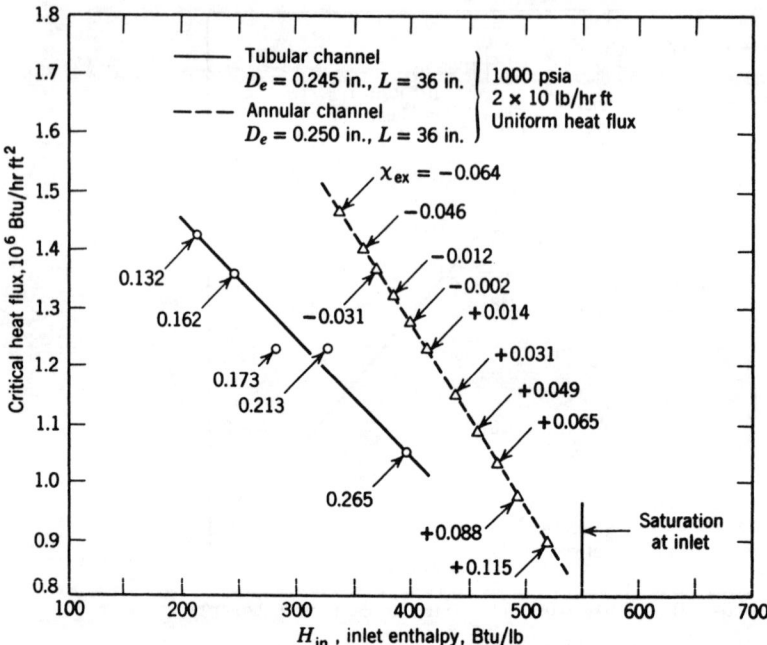

Figure 5.54 Unheated wall effect on critical heat flux (system parameter concept). Data from Babcock (1964) and Janssen and Kervinen (1963a).

This adverse unheated wall effect on the coolant effectiveness can be reduced by installing a "rough liner" or a square ring of 0.080-in. (0.20-cm) height with spacing of 0.9–2.6 in. (2.3–6.6 cm) on the unheated surface of the annulus to trip the "cold" liquid film. The test results obtained at about 1,000 psia by Janssen and Kervinen (1963) are as follows.

At low steam qualities (<0.05), the film trippers show no advantage.
At high qualities, the advantages are as shown in Table 5.6.

These results appear to indicate that flow turbulences in the whole channel can improve the critical heat fluxes.

5.4.5.3 Effect of dissolved gas and volatile additives. The effect of dissolved gas content on flow boiling crisis has been found to be insignificant. This is somewhat different from pool boiling. Experiments at high pressure by Zenkevich and Subbotin (1959) and by Debortoli and Masnovi (1957) with dissolved gas concentrations up to approximately 140 cm³/liter (14% by volume) support this conclusion. Gunther (1951) found no effect of dissolved air content on boiling crisis in his low-pressure studies. The Heat Transfer Laboratory of Columbia University has tested

Table 5.6 Improvement of CHF by film trippers

Quality X	Flow 10⁶ lb/hr ft²	kg/m² s	Percent improvement in q''_{crit}
0.15	0.5	678	—
0.15	1.7	2,305	48
0.25	0.5	678	75
0.25	1.7	2,305	>100

Source: Based on data of Janssen and Kervinen (1963).

Table 5.7 CHF at constant X_o but variable H_{in} (based on L/D)

	Critical heat flux, q''_{crit}					
	$X_o = 0$		$X_o = 0.2$		$X_o = 0.4$	
L/D	10⁶ Btu/hr ft²	kW/m²	10⁶ Btu/hr ft²	kW/m²	10⁶ Btu/hr ft²	kW/m²
188					0.63	1,990
50	1.40	4,420	1.10	3,470	0.85	2,680
20	1.40	4,420	1.20	3,780	1.02	3,220
15	1.48	4,670	1.30	4,100	1.15	3,630
7.5	1.62	5,110	1.60	5,050	1.59	5,020

the effect of dissolved helium on boiling crisis. The result showed that, for an amount of helium as much as four times the saturation quantity, the CHF decreased by only 5%.

5.4.6 Channel Length and Inlet Enthalpy Effects and Orientation Effects

5.4.6.1 Channel length and inlet enthalpy effects. Inlet enthalpy determines the enthalpy level of the flow along the entire channel length. It affects the local enthalpy at CHF directly. Besides, a high inlet enthalpy introduces a "soft inlet," which may lead to a flow instability and thus lower CHF. If the exit quality, X_o, is kept constant, the channel length will vary corresponding to the given inlet enthalpy. Two concepts deal with such effects, the local parameter concept and the inlet parameter concept.

Local parameter concept Styrikovich et al. (1960) studied the length effect on the critical heat flux at a constant exit quality with an 8-mm (0.3-in.) round tube at a pressure of 1,500 psia (10.2 MPa) operating in a stable system. Their measured CHF at various L/D values (with accompanying H_{in} changes) are listed in Table 5.7. This can be considered a demonstration of small length effect and/or H_{in} effect on CHF at a fixed local (exit) quality. The high CHF obtained from a very small L/D test section may be due to the entrance turbulence effect. A large value of L/D indicates a long test section that has a low inlet enthalpy effect as well as the length effect per se.

Gaspari et al. (1970) plotted the rod critical power input over a boiling length against the boiling length, L_B, as shown in Figure 5.55. It can be seen that a 10% reduction in critical power can be observed in every double of channel length. This may be considered a "true length" effect in local parameter CHF correlations.

Inlet parameter concept CHF correlations based on the inlet enthalpy must be accompanied by a length term to predict the CHF in test sections of various lengths, as shown in Figure 5.56, where Δi represents the inlet enthalpy (ΔH_{in}). The predicted burnout conditions shown in the figure are based on Becker's correlation (Becker et al., 1973):

$$\left(\frac{q}{A}\right)_{BO} = \frac{G(450 + \Delta H_{in})}{40(L/D) + 156(G)^{0.45}} \left[1.02 - \left(\frac{p}{p_{cr}} - 0.54\right)^2\right] \quad (5\text{-}98)$$

The equation is recommended for the following parameter range:

$120 < p < 200$ bar ($1{,}740 < p < 2{,}900$ psia)
$2{,}000 < L < 8{,}500$ mm ($79 < L < 335$ in.)
$8 < d < 25$ mm ($0.31 < d < 1$ in.)
$G(p) < G < 7{,}000$ kg/m² s [$G(p) < G < 5.18 \times 10^6$ lb/hr ft²]
$0 < X_{BO} < 0.60$

where ΔH_{in} is in kJ/kg, and the function $G(p)$ is obtained from Figure 5.56. This correlation was compared only with data obtained from uniform-flux, round-tube test sections of diameter 10 mm (0.4 in.). The agreement was excellent, with a root-mean-square error of 5.7% as shown in Figure 5.57. Becker's correlation indicates the inseparable relationship of inlet enthalpy (or inlet subcooling) and channel length. As mentioned before, only in a CHF correlation developed from a short test section can this length effect be neglected, and therefore it is not useful in a long channel.

Aladyev et al. (1961) demonstrated that, with a compressible volume connected at the inlet of a test section, the flow oscillates and hence lowers the CHF. Flow fluctuation in the test section also depends on the compressibility of fluid upstream and on the pressure drop through the test section. Because the compressibility of water is approximately a function of temperature alone, the inlet temperature affects the boiling crisis.

Collins et al. (1971) carried out an experimental program to investigate parallel-channel instability in a full-scale simulated nuclear reactor channel op-

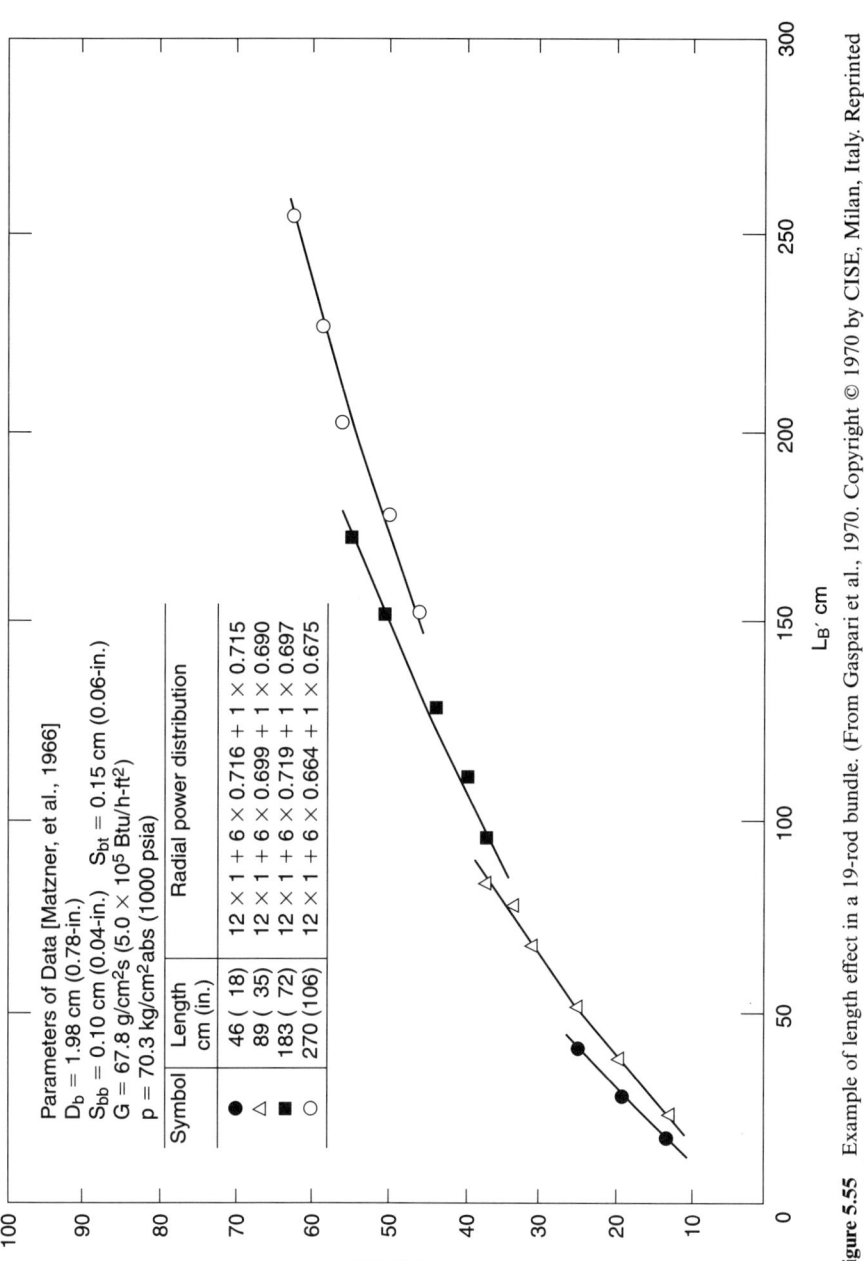

Figure 5.55 Example of length effect in a 19-rod bundle. (From Gaspari et al., 1970. Copyright © 1970 by CISE, Milan, Italy. Reprinted with permission.)

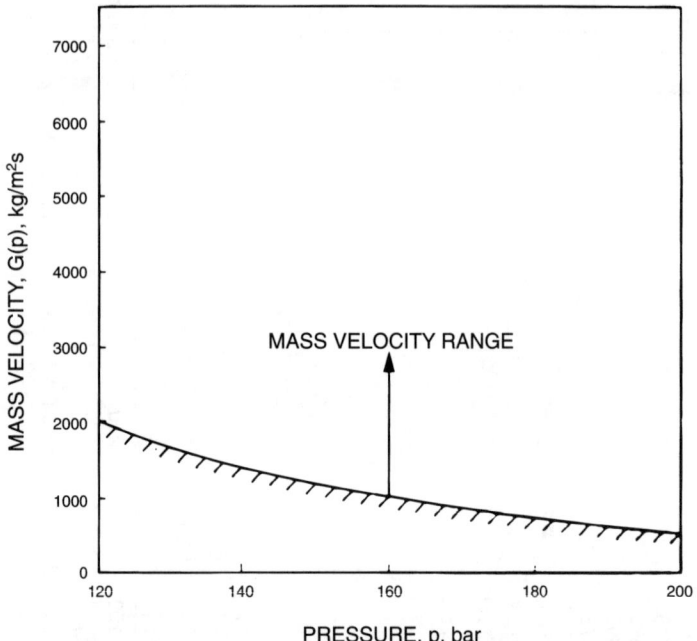

Figure 5.56 Mass flux range for burnout correlation. (From Becker et al., 1973. Copyright © 1973 by Elsevier Science Ltd. Kidlington, UK. Reprinted with permission.)

erating in vertical, high-pressure, boiling-water upflow. A vertically mounted test section, containing two 19-rod, electrically heated bundles with a total heated length of 97.5 in. (2.5 m), was operated in parallel. The effects of pressure, exit and inlet feeder pressure drop ratio, and inlet subcooling on the steady-state critical power and the threshold of periodic dryout (with flow fluctuation) were obtained. The generalized rod bundle stability characteristics are given in Figure 5.58, which shows that the bundle dryout power is lowered at the same exit quality as the ratio of $\Delta p_o/\Delta p_i$ is increased. The periodic dryout heat flux is also reduced as the ratio of $\Delta p_o/\Delta p_i$ is increased, as shown in Figure 5.59. Thus the reactor core thermal design limit should be lowered if the ratio of $\Delta p_o/\Delta p_i$ is 3 or more.

The length effect is stronger in the high-quality region than in the low-quality region. The following findings have been reported for a long boiling channel.

Local flow velocity fluctuations at the exit portion have been observed in a long boiling channel as a result of the large fluid compressibility inside the channel (Proskuryakov, 1965). Exit velocity fluctuation frequency is usually the same as the natural frequency of the channel.

Additional precautions to be included in the thermal design of a very long boiling channel of $(L/D) > 250$ were suggested by Dolgov and Sudnitsyn (1965).

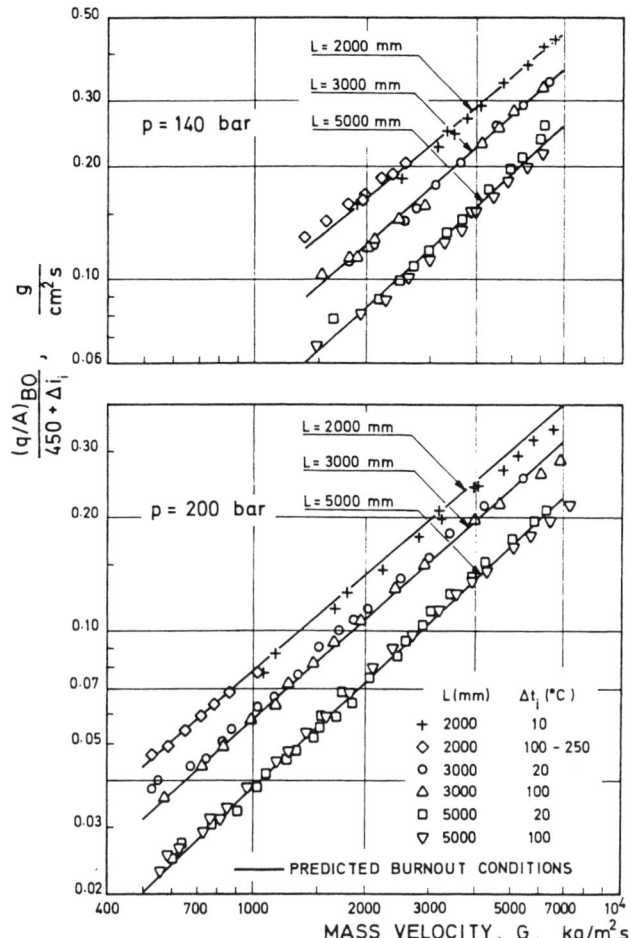

Figure 5.57 Measured and predicted burnout conditions. (From Becker et al., 1973. Copyright © 1973 by Elsevier Science Ltd., Kidlington, UK. Reprinted with permission.)

In a test section operated in a quality region with the same H_{in}, L, D, G, and p, the critical powers will be approximately the same for both the uniform and nonuniform flux distributions, provided the peak-to-average flux ratio is not greater than 1.6 (Janssen and Kervinen, 1963).

5.4.6.2 Critical heat flux in horizontal tubes. Horizontal CHF data are rather meager, so correlations for predicting such cases are less accurate than for vertical flows. Groeneveld et al. (1986) suggested that use be made of a correction factor, K, such that

388 BOILING HEAT TRANSFER AND TWO-PHASE FLOW

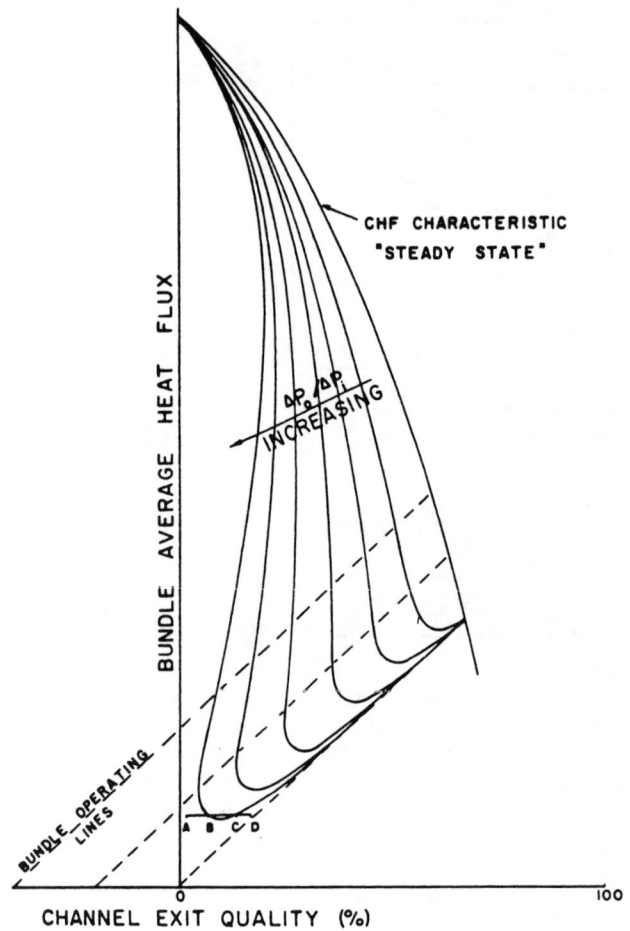

Figure 5.58 Generalized rod bundle stability characteristics. (From Collins et al., 1971. Copyright © 1971 by American Society of Mechanical Engineers, New York. Reprinted with permission.)

$$q''_{\text{crit, hor}} = K q''_{\text{crit, vert}} \qquad (5\text{-}99)$$

where $q''_{\text{crit,hor}}$ and $q''_{\text{crit,vert}}$ are the CHF in horizontal and in vertical flow, respectively. For flow with high mass fluxes, the effect of tube orientation on CHF is negligible; for intermediate and low flows, the CHF for horizontal flow can be considerably lower than that for vertical flow. Figure 3.14 (Sec. 3.2.3) shows the possible dryout occurrences in a horizontal flow. Dryout occurs in low-quality regions where bubbles coalesce and form a continuous vapor cushion along the upper portion of the tube (Becker, 1971). As vaporization continues downstream, the increased vapor velocity causes the formation of high-amplitude waves at the liquid–vapor inter-

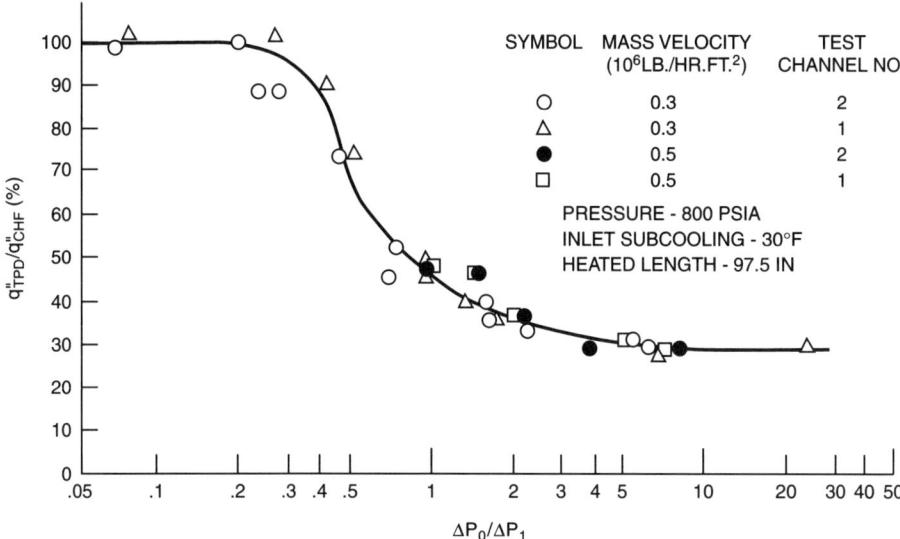

Figure 5.59 Variation of periodic dryout heat flux q''_{TPD}/q''_{CHF} with feeder pressure drop ratio $\Delta p_o/\Delta p_i$ (two parallel heated channels). (From Collins et al., 1971. Copyright © 1971 by American Society of Mechanical Engineers, New York. Reprinted with permission.)

face. The impingement of a fast-moving vapor stream on such waves causes liquid to be entrained in the vapor core region, with some of entrained liquid depositing onto the upper portion of the tube, which again covers the full circumference to establish annular flow (Fig. 3.14).

At low qualities, bubbles at the wall may form ribbons of vapor along the upper surface of the channel under low-flow conditions. These vapor ribbons act as barriers and inhibit the replenishment of liquid lost by draining and evaporation. A premature CHF condition can thus occur (Hetsroni, 1993). The mechanism of CHF at intermediate qualities is different from that at very low qualities in that alternating large splashing waves carry the liquid to the upper surface of the horizontal channel. Since there is no droplet entrainment, no liquid replenishment occurs at the top of the channel, the liquid film at the upper surface is subjected to drainage and evaporation, and therefore to dryout, if sufficient time elapses before the next splashing wave arrives. For high qualities, the flow pattern will most likely be annular (see Fig. 3.14). Because of drainage along the circumferential direction, the annular liquid film at the top of the channel is thinner than the rest. At the bottom of the channel, larger-amplitude waves give rise to a significant droplet entrainment into the vapor core. At CHF, the liquid film at the top becomes completely depleted, resulting in dryout.

The correction factor K is thus strongly dependent on the flow conditions (Wong et al., 1990). For values of mass flux below a limit, G_{min}, the flow is fully stratified and hence $q''_{crit,hor}$ is zero, or $K = 0$. On the other hand, if mass fluxes are high, or $G > G_{max}$, the effect of the tube orientation on CHF, as mentioned before, becomes insignificant and hence $q''_{crit, hor}$ may be assumed to be the same as $q''_{crit,vert}$. Estimates of threshold values of G_{min} and G_{max} can be derived from the flow regime map of Taitel and Dukler (1976b). A simple exponential expression was suggested by Wong et al.:

$$K = 1 - \exp\left[-\left(\frac{T_i}{A}\right)^B\right] \qquad (5\text{-}100)$$

where T_i represents T_1 to T_6,

$$T_1 = C_1(\text{Re}_L)^{-0.2}\left(\frac{1-X_a}{1-\alpha}\right)\left[\frac{G^2}{gD\rho_L(\rho_L - \rho_G)\alpha^{0.5}}\right]$$

turbulent/buoyant force (annular flow)

$$T_2 = C_2\left(\frac{\rho_G}{\rho_L}\right)\left[\frac{G^4}{g\sigma\rho_L^2(\rho_L - \rho_G)}\right]\left[\frac{(\rho_L/\rho_G)(1-\alpha)X_a - \alpha(1-X_a)}{(1-\alpha)}\right]^4$$

drag/buoyant force (bubbly flow)

$$T_3 = C_3(\text{Re}_L)^{-0.2}(\rho_G)^2\frac{u_L^2(u_G - u_L)^2}{g(\rho_L - \rho_G)} \qquad \text{turbulent/buoyant force}$$

(bubbly flow)

$$T_4 = C_4\left(\frac{\rho_G}{\rho_L}\right)\frac{G^4}{g\sigma(\rho_L)^2(\rho_L - \rho_G)}\left[\frac{(\rho_L/\rho_G)(1-\alpha)X_a - \alpha(1-X_a)}{\alpha(1-\alpha)}\right]^4$$

drag/buoyant force (droplet flow)

$$T_5 = C_5\frac{G}{\sqrt{gD\rho_G(\rho_L - \rho_G)}}\left(\frac{X_a}{\alpha}\right)\left(\frac{D}{L_h}\right) \qquad \text{transit time ratio (bubbly flow)}$$

$$T_6 = C_6\frac{G}{\sqrt{gD\rho_L(\rho_L - \rho_G)}}\left(\frac{1-X_a}{1-\alpha}\right)\left(\frac{D}{L_h}\right) \qquad \text{transit time ratio}$$

(droplet flow)

All six parameters were tested in the derivation of the correction factor (Wong et al., 1990). The best correlation appears to be with the use of parameter T_1 and constants $B = 0.5$ and $A = 3.0$ in Eq. (5-100), that is,

$$K = 1 - \exp\left[-\left(\frac{T_1}{3}\right)^{0.5}\right]$$

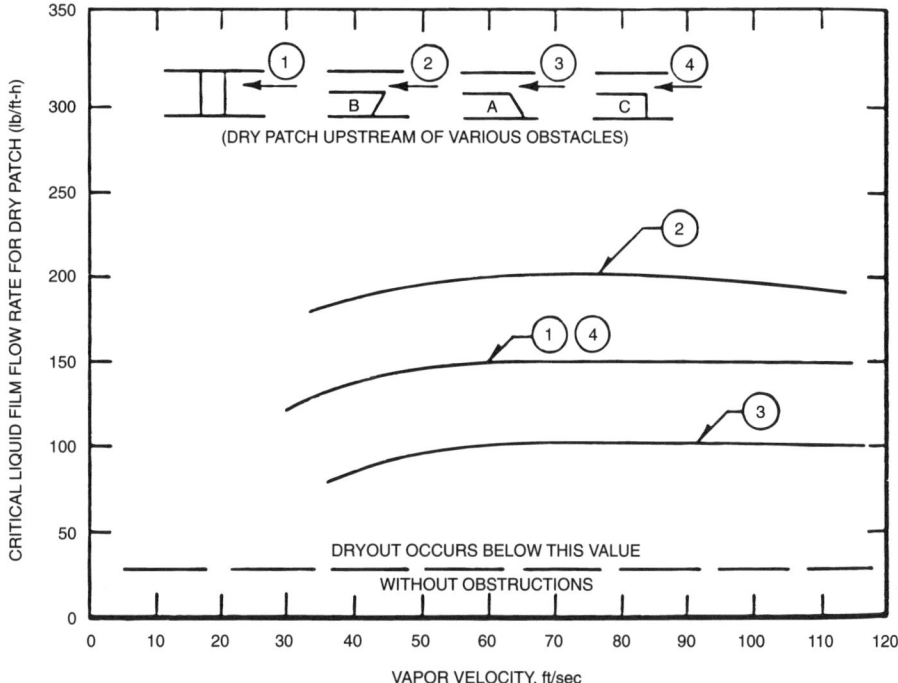

Figure 5.60 Comparison of critical film flow rates for various obstacles. (From Lahey and Moody, 1977. Copyright © 1977 by American Nuclear Society, LaGrange Park, IL. Reprinted with permission.)

5.4.7 Local Flow Obstruction and Surface Property Effects

5.4.7.1 Flow obstruction effects. Obstacles of various shape and size were investigated in an adiabatic air/water experiment (Shiralkar and Lahey, 1973) to determine the critical liquid film flow rate (i.e., the rate below which upstream dryout occurs). It is seen in Figure 5.60 that some shapes are much better than others. In general, the more streamlined the obstacle, e.g., shape 3, the lower the critical liquid film flow rate and, thus the more difficult it is to cause an upstream dryout. The hydrodynamic situation that occurs in a LWR fuel rod bundle is completely analogous. Hence, the components contained in a grid-type spacer can strongly affect the thermal performance of a fuel rod bundle, and the spacer designer is well advised to make these components as streamlined as possible.

5.4.7.2 Effect of surface roughness. CHF for rough surfaces was measured on vertical annular tubes cooled by a downward flow of subcooled water by Durant and Mirshak (1959, 1960). An increase in the apparent critical heat flux of as much as 100% over a smooth surface was obtained at the same coolant velocity, temperature, and pressure. The heated surfaces were 304 SS and Zircaloy-2 tubes about

1 ft (0.3 m) long. The equivalent diameters of channels were 0.25–0.75 in. (6.35–19.0 mm). The empirical correlation describing this effect is

$$q''_{crit} = 480,000(1 + 0.0365V)(1 + 0.005\Delta T_{sub})(1 + 0.0131p)(0.3 + 0.7R_f) \quad (5\text{-}101)$$

where all parameters are in English units in the following ranges:

q''_{crit} = critical heat flux, 1.8×10^6–36×10^6 Btu/hr ft^2
V = flow velocity, 10–32 ft/sec
ΔT_{sub} = subcooling, 36–148°F
p = pressure, 33–70 psia
R_f = ratio of rough pipe friction factor to smooth pipe friction factor, 1.0–2.9

This beneficial effect may be due partly to the increase of agitation near the wall, enhancing the liquid cooling, and partly to an increased number of nucleation sites on the surface.

The effect of a rough surface on boiling crisis in the quality region is reversed. Jansson et al. (1963) performed experiments on a heated rod in an annulus whose surface was blasted with coarse-grit sand to 300-μin. roughness. The CHF was reduced by as much as 35% at low flow and by as much as 50% at the higher flow. Janssen and co-workers offered the explanation that roughening of the heated surface (the disturbance increases the liquid reentrainment in an annular flow pattern) can also be expected to reduce the thickness of the liquid film on this surface and thus bring about a low CHF. On the other hand, others have found that the CHF can be improved 20% by increasing the roughness from 20 to 200 μin. on an Inconel heater surface (Tong, 1972). Unfortunately, the uniformity of surface roughness was not reported. To interpret the behavior of nucleate boiling correctly, information about the size distribution of the nucleation sites is always required. Table 5.8 lists work by various investigators on the surface roughness effect on CHF in a flow boiling of water. The following conclusions can be derived from the results listed in the table.

In subcooled flow boiling, a large roughness increases flow agitation and hence improves CHF. If the roughness is smaller than the thickness of the bubble layer, the agitation effect on CHF is insignificant.

In a high-quality two-phase flow in a straight tube, the surface roughness increases the liquid entrainment and thus reduces the liquid film thickness and brings about a low CHF.

5.4.7.3 Wall thermal capacitance effects. The wall thermal capacitance effect on CHF in a boiling water flow can be observed only at low pressures, where the bubble size is large and the wall temperature fluctuation period is long. These conditions were satisfied in a test in water at 29–87 psia (200–600 kPa) (Fiori and Bergles, 1968). Two test sections of 0.094-in. (2.39-mm) I.D. with wall thicknesses

Table 5.8 Surface roughness effects

Reference	Height of roughness, in. (mm)		Subcooled or saturated	Type of tube	Increase/decrease in CHF
Durant and Mirshak (1959, 1960)	0.0130	(0.330)	Subcooled	Straight tube	Increase by 100%
Tong et al. (1967b)	0.0025	(0.064)	Subcooled	Straight tube	No increase
Janssen et al. (1963)	0.0003	(0.008)	Saturated	Straight tube	Decrease by 35%
Moussez (Tong, 1972)	0.0002	(0.005)	Saturated	Tube with twisted tapes	Increase by 20%
EURAEC (1966)	0.0025	(0.064)	Saturated	Straight tube	Decrease by 20%

Source: Tong (1972). Reprinted with permission of U.S. Department of Energy, subject to the disclaimer of liability for inaccuracy and lack of usefulness printed in the cited reference.

of 0.012 in. (0.30 mm) and 0.078 in. (1.98 mm) were used. The increase of CHF using the thick-walled tube was found to be as much as 58%.

To evaluate the effect of surface material properties on the CHF, it is useful to review the effect on CHF in pool boiling. Carne and Charlesworth (1966) tested the CHF on various vertical surfaces at atmospheric pressure in a saturated pool boiling of *n*-propanol and reported that CHF is definitely a function of the thermal conductivity, k, and the thickness, t, of the heating surface. Ivey and Morris (1965) reported that the oxidized surfaces of an aluminum wire appeared to yield a higher CHF than clean metallic surfaces did. Farber and Scorah (1948) tested CHF in a pool boiling of water around 0.04 in. (1.02 mm) wires of nickel, tungsten, Chromel A, and Chromel C with a hard deposit on the surfaces and found the CHF to be two to three times higher than for a clean surface. It can be seen, therefore, that the surface condition effect on pool boiling CHF is influenced by the size of the heater.

The surface deposit (or scale) in flow boiling has been observed on many occasions, but no adverse effect on CHF has been reported. Knoebel et al. (1973) developed a subcooled DNB correlation from annular geometry with stainless steel and aluminum for $\Delta T_{sub} > 45°F$ (25°C):

$$q''_{crit} = 1,360(We/Re)^{0.573}_{H_2O}(\rho c_p \Delta T_{sub})^{0.759}_{H_2O}(\rho c_p)^{0.621}_{SS/Al}(k)^{0.190}_{SS/Al} \quad (5\text{-}102)$$

where all variables are in English units. They found that the CHF from aluminum is 20% greater than that from stainless steel. Furthermore, an aluminum heater wall thickness of 36 mil (0.9 mm) gives 20% higher CHF than 20-mil (0.5-mm) aluminum.

5.4.7.4 Effects of ribs or spacers. A spacer rib test was conducted by the Savannah River Laboratory (Mirshak and Towell, 1961). The results show that critical heat flux reduction due to contact of a longitudinal spacer rib can be as much as 32%.

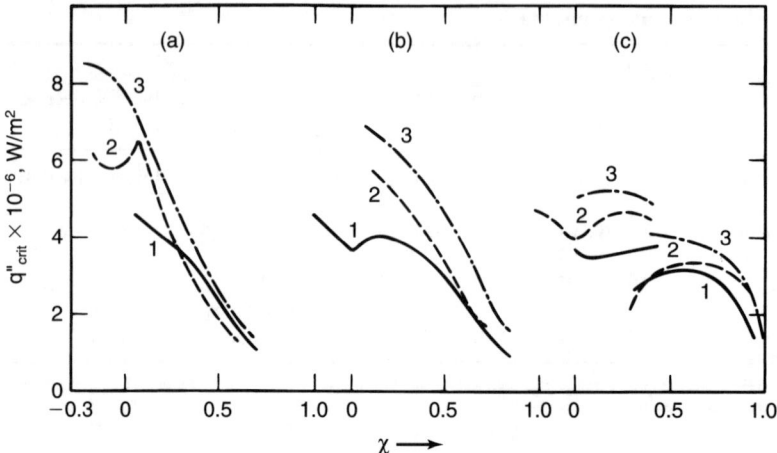

Figure 5.61 Influence of the length of the hot spot L_1 on q''_{crit} when the pulsation development is limited ($P = 100$ atm, $\varepsilon = 2$, $L = 160$ mm, $d = 8$ mm): (a) $G = 2{,}000$ kg/m² s; (b) $G = 850$; (c) $G = 400$; curve 1, $L_1 = 64$ mm; curve 2, $L_1 = 16$ mm; curve 3, $L_1 = 4$ mm. Data from Styrikovich et al. (1963).

The CHF was measured for vertical flat surfaces locally insulated by vertical ribs and cooled on one side by downward-flowing water. The results of 22 tests (at 30–50 psia, 2–3.3 MPa); 17–40 ft/sec, 5.2–12 m/s; and 30–67°C, 54–120°F subcooling) were correlated by the equation

$$\frac{q''_{crit,\ w/o\ rib}}{q''_{crit,\ w/rib}} = 1 + 28\left(\frac{W}{\sqrt{ky}}\right)\exp(-50C) \quad \text{for} \quad \left(\frac{W}{\sqrt{ky}}\right) < 0.02 \quad (5\text{-}103)$$

where C = clearance between rib tip and the heated surface, in.
W = (width of rib tip)/2, ft
k = thermal conductivity of heated surface, Btu/hr ft °F
y = thickness of heated surface, ft

5.4.7.5 Hot-patch length effects. The magnitude of CHF depends on the length of hot patch. Shorter hot patch gives higher CHF (Styrikovich et al., 1963) (Fig. 5.61, peak-to-average heat flux ratio, $\varepsilon = 2$). A power spike of a length less than an inch (25 mm) located at 80% of the test section length does not affect significantly the critical power (Swenson et al., 1962a). This operation can be explained by the memory effect (Tong, 1965). The differences in CHF indicate the different strengths of the memory effect due to the different local qualities and flux distributions. Hill et al. (1974) measured the power spike effect on CHF for a realistic spike (20% power spike in 4 in. at PWR conditions). The measured spike effect was so small that it lies within the repeatable uncertainty of the DNB measurements. Therefore, it is considered to be insignificant in reactor core design.

5.4.7.6 Effects of rod bowing.
Two sets of data are available for rod bowing, one for PWRs and the other for BWRs.

Bowing effect on CHF of a PWR fuel assembly The bowing effect (E_{bow}) on a PWR fuel assembly can be defined as

$$E_{bow} = \frac{(CHF_{meas}/CHF_{pred})_{unbow} - (CHF_{meas}/CHF_{pred})_{bow}}{(CHF_{meas}/CHF_{pred})_{unbow}} \quad (5\text{-}104)$$

The following measurements were made on a PWR fuel assembly at 2,200 psia (15 MPa) (Hill et al., 1975):

$$\frac{q''}{10^6} \leq 0.34 \text{ Btu/hr ft}^2 \text{ (or } 1.1 \times 10^{-3} \text{ kW/m}^2\text{)}$$

$$E_{bow} = 0$$

$$\frac{q''}{10^6} > 0.34 \text{ Btu/hr ft}^2 \text{ (or } 1.1 \times 10^{-3} \text{ kW/m}^2\text{)}$$

$$E_{bow} = 1.12\left[\frac{q''}{10^6} - 0.34\right]$$

Bowing effects on CHF of a BWR fuel assembly Early in 1988, dryout of fuel elements occurred in the Oskarshamn 2 BWR. It was discovered during refueling that one corner element had been damaged in each of four fuel assemblies. The damaged zone covered about 180° of the element periphery facing the cornor subchannel, over a stretch of about 30 cm, with the upper end just below the last downstream spacers. The main cause of the dryout was the reuse of fuel channels for ordinary 64-element fuel assemblies (Hetsroni, 1993). Becker et al. (1990) calculated the flow and power conditions in the damaged fuel assemblies, resulting in the predicted local quality, X, and heat flux as shown in Figure 5.62. For the bundle operating conditions, the heat flux is plotted versus the steam quality along the bundle, as shown in Figure 5.62*b*. The dashed line refers to the highest loaded element (hot rod), for which the axial heat flux distribution is shown in Figure 5.62*a*. The steam quality is the average value over the cross section, neglecting quality and mass flux variations between the subchannels of the assembly. Becker et al. (1990) used these calculations as a basis for dryout prediction, with good results.

5.4.7.7 Effects of rod spacing.
The rod spacing effect on CHF of rod bundles has been found to be insignificant (Towell, 1965) at a water mass flux of about 1×10^6 lb/hr ft² (1,356 kg/m² s) under 1,000 psia (6.9 MPa) and with an exit quality of 15–45% in the star test section, the center seven-rod mockup of a 4-ft (1.22-m)-

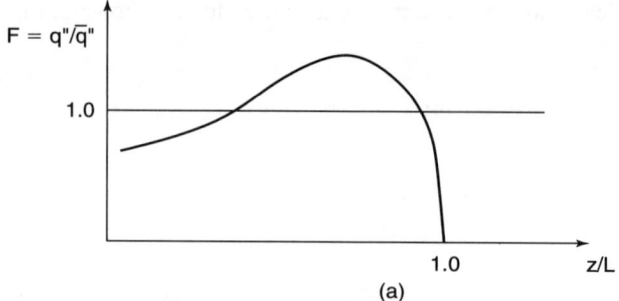

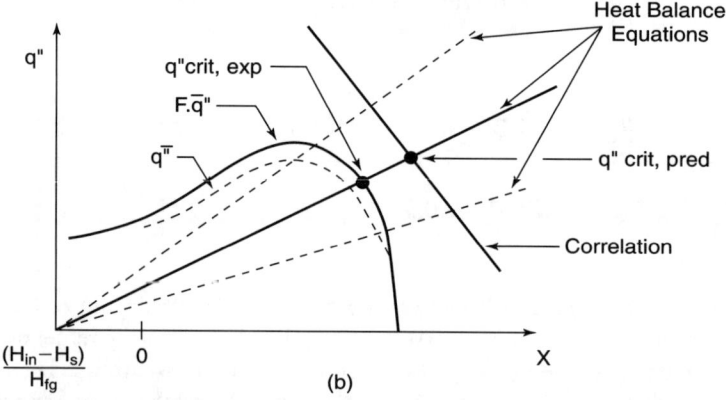

Figure 5.62 Becker's analysis of CHF for damaged fuel assembly in BWR ($q''_{crit,\,pred}$ versus $q''_{crit,\,exp}$). (From Becker et al., 1990. Copyright © 1990 by Elsevier Science Ltd., Kidlington, UK. Reprinted with permission.)

long, 19-rod bundle. Rod spacings of 0.018–0.050 in. (0.45–1.27 mm) were tested, and no effect on the CHF was found at constant exit enthalpy whether the adjoining surfaces were heated or not. This finding was also in agreement with the test results of Lee and Little (1962) in a "dumbbell" section at 960 psia (6.6 MPa) with a 10% mean exit quality and a vertical upflow of water. The gap between the rod surfaces was varied from 0.032 to 0.220 in. (0.81 to 5.6 mm). It was also found to be true by Tong et al. (1967a) in square and triangular rod arrays having the same water-to-fuel ratio. Their tests were conducted in a subcooled or low-quality flow of water at 2,000 psia (13.8 MPa).

5.4.7.8 Coolant property (D_2O and H_2O) effects on CHF. "Fluid property effects" here refer to fluids of heavy water versus light water as used in water-cooled reactors. For other fluids, readers are referred to Section 5.3.4. These effects were taken into account by a phenomenological correction (Tong, 1975) as shown by the equation

$$\frac{q''_{crit}}{GH_{fg}} = (1 + 0.00216 \, Pr^{1.8} \, Re_m^{0.5} \, Ja) C_o C_1 f_o \tag{5-105}$$

where $Re = \dfrac{GD_e}{\mu(1-\alpha)}$

$Ja = \dfrac{\rho_L c_L \Delta T_{sub}}{\rho_G H_{fg}}$

$\Delta T_{sub} = T_{sat} - T_{core}$

$f_o = 8.0 (Re_m)^{-0.6} \left(\dfrac{D_e}{D_o}\right)^{0.32}$

For a common geometry, under the same pressure and at the same flow rate, the property effects on CHF can be evaluated according to this equation:

$$q''_{crit} = K_1 \rho_L H_{fg} (Re_m)^{-0.6} [1 + K_2 (Re_m)^{0.5} Ja] \tag{5-106}$$

where K_1 and K_2 are constants. For property ratios of these two fluids of

$$\frac{\rho_{D_2O}}{\rho_{H_2O}} = 1.11, \qquad \frac{\mu_{D_2O}}{\mu_{H_2O}} = 1.10$$

$$\frac{H_{fg,D_2O}}{H_{fg,H_2O}} = 0.92 \qquad \frac{c_{L,D_2O}}{c_{L,H_2O}} = 0.97$$

the CHF ratio becomes

$$\frac{q''_{crit,D_2O}}{q''_{crit,H_2O}} = (1.11)(0.92)(0.99)^{0.6} \left[\frac{1 + K_2(1.005)(1.17)(Re_m)^{0.5} Ja}{1 + K_2 (Re_m)^{0.5} Ja}\right]$$

Thus, for $[K_2(Re_m)^{0.5} Ja] \to 1$, $q''_{crit,D_2O}/q''_{crit,H_2O} \simeq 1.11$, and for $[K_2(Re_m)^{0.5} Ja] \to 2$, $q''_{crit,D_2O}/q''_{crit,H_2O} \simeq 1.14$.

5.4.7.9 Effects of nuclear heating. Both out-of-pile loop experiments and in-pile reactor operating measurements are available. The rod bundle data obtained in an operating reactor (Farmer and Gilby, 1971) agree with those obtained in an out-of-pile loop, as shown in Figure 5.63.

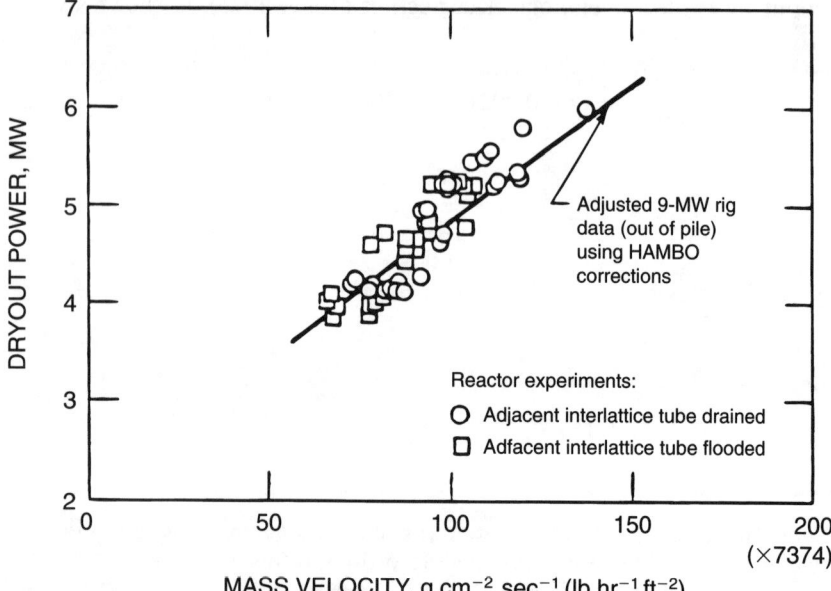

Figure 5.63 Comparison of reactor dryout experiments and out-of-pile data; pressure 900 psia (6.2 MPa), subcooling enthalpy 21 Btu/lb (63 J/g). (From Farmer and Gilby, 1971. Copyright © 1971 by United Nations Pub., New York. Reprinted with permission.)

5.4.8 Flow Instability Effects

During the early days of BWR technology, there was considerable concern about nuclear-coupled instability—that is, interaction between the random boiling process and the void-reactivity feedback modes. Argonne National Laboratory (ANL) conducted an extensive series of experiments which indicated that, while instability is observed at lower pressures, it is not expected to be a problem at the higher pressures typical of modern BWRs (Kramer, 1958). Indeed, this has proved to be the case in many operating BWRs in commercial use today. The absence of instability problems due to void-reactivity feedback mechanisms is because BWR void reactivity coefficients $[\partial k/\partial \langle \alpha \rangle]$ are several orders of magnitude smaller at 1,000 psia (6.9 MPa) than at atmospheric pressure and, thus, only small changes in reactivity are experienced due to void fluctuations. Moreover, modern BWRs use Zircaloy-clad UO_2 fuel pins, which have a thermal time constant of about 10 sec, and consequently the change in voids due to changes in internal heat generation resulting from reactivity tends to be strongly damped (Lahey and Moody, 1977).

In addition to nuclear-coupled instability, the reactor designer must consider a number of flow instabilities. A discussion of various instabilities is given in Chapter 6. Only those instability modes of interest in BWR technology are mentioned here:

Density-wave oscillations
Pressure drop oscillations
Flow regime-induced instability

The phenomenon of density-wave oscillations has received rather thorough experimental and analytical investigation. This instability is due to the feedback and interactions among the various pressure drop components and is caused specifically by the lag introduced through the density head term due to the finite speed of propagation of kinematic density waves. In BWR technology, it is important that the reactor is designed so that it is stable from the standpoint of both parallel channel and system (loop) oscillations. This instability will be examined at length in the next chapter.

Pressure drop oscillations (Maulbetsch and Griffith, 1965) is the name given the instability mode in which Ledinegg-type stability and a compressible volume in the boiling system interact to produce a fairly low-frequency (0.1 Hz) oscillation. Although this instability is normally not a problem in modern BWRs, care frequently must be exercised to avoid its occurrence in natural-circulation loops or in downflow channels.

Another instability mode of interest is due to the flow regime itself. For example, it is well known that the slug flow regime is periodic and that its occurrence in an adiabatic riser can drive a dynamic oscillation (Wallis and Hearsley, 1961). In a BWR system, one must guard against this type of instability in components such as steam separation standpipes. The design of the BWR steam separator complex is normally given a full-scale, out-of-core proof test to demonstrate that both static and dynamic performance are stable.

5.4.9 Reactor Transient Effects

Transient boiling crisis was tested during a power excursion in pool boiling of water. Tachibana et al. (1968) found that the transient CHF increases as the power impulse time decreases, as shown in Figure 5.64. This CHF increase may be due to the increase in the number of nucleation sites being activated simultaneously, since examination of high-speed motion pictures revealed that all bubbles on the heating surface remained in the first-generation phase until the critical condition was reached. These results agree with observation of Hall and Harrison (1966) that in a rapid exponential power impulse with $\Delta t < 1$ msec, film boiling was always preceded by a short burst of nucleate boiling at heat fluxes about 5 to 10 times the steady-state values under the same water conditions. The power excursion effect on CHF decreases as the initial exponential period increases and approaches zero at 14–30 msec, as reported by Rosenthal and Miller (1957) and by Spiegler et al. (1964).

In flow boiling of water, however, Martenson (1962) found that the transient CHF values were slightly higher than the steady-state values predicted from the

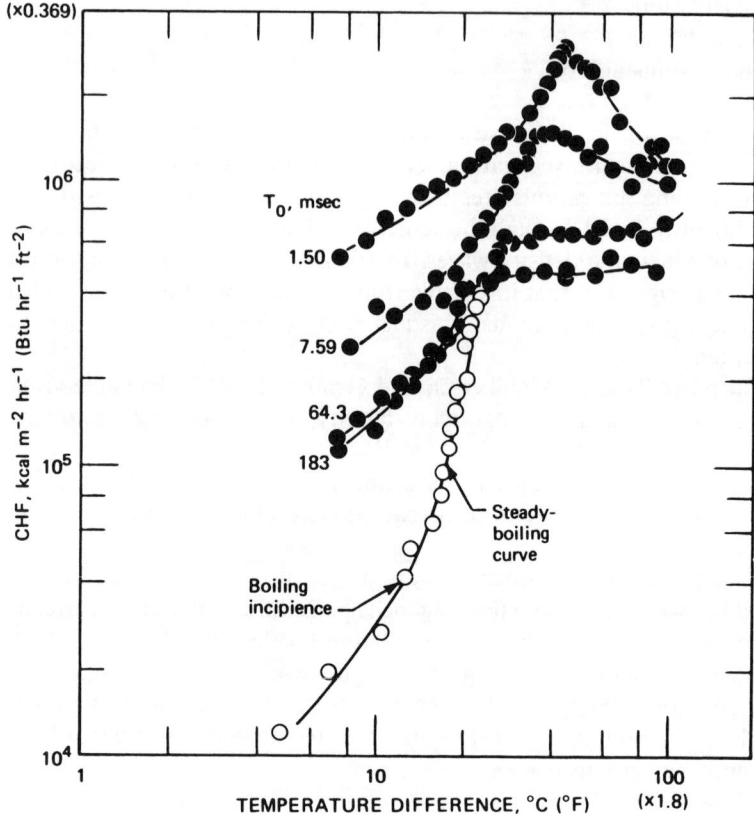

Figure 5.64 Transient CHF. (From Tachibana et al., 1968. Copyright © 1968 by Atomic Energy Society of Japan, Tokyo. Reprinted with permission.)

Bernath (1960) correlation. The transient CHF was also tested by Schrock et al. (1966) in a water velocity of 1 ft/sec (0.3 m/s). They also reported transient CHF values that were higher than those under steady-state conditions. Borishanskiy and Fokin (1969) tested transient CHF in flow boiling of water at atmospheric pressure. They found that the transient CHF in water was approximately the same as the steady-state value. On the basis of Bernath's correlation (Bernath, 1960) and Schrock et al.'s (1966) data, Redfield (1965) suggested a transient CHF correlation as follows:

$$q''_{\text{crit}} = \left[12,300 + \frac{67V}{D_e^{0.6}}\right]\left[102.5\ln(p) - 97\left(\frac{p}{p+15}\right) + 32 - T_{\text{bulk}}\right]\exp\left(\frac{4.25}{\Delta t}\right)$$

(5-107)

where V = coolant velocity, ft/sec
D_e = channel equivalent diameter, ft
Δt = initial exponential period, msec

Transient boiling crisis in rod bundles was tested at high pressures by Tong et al. (1965, 1967a), Moxon and Edwards (1967), and Cermak et al. (1970). It was generally concluded from these tests that transient CHF can be predicted by using the steady-state CHF data from rod bundles having the same geometry and tested under the same local fluid conditions. The detailed conditions of the tests are given in Table 5.9. It should be noted that the CHF in a high-quality, low-mass-velocity flow during blowdown is a deposition-controlled CHF that can be delayed considerably by a high liquid droplet deposition rate at a low surface heat flux. Celata et al. (1985), in their tests of flow coastdown in refrigerant R-12, found that the measured time to DNB is usually 1–2 sec longer than that predicted by steady-state DNB correlations. This finding indicates that using steady-state DNB correlations to predict transient DNB during flow coastdown is conservative.

5.5 OPERATING PARAMETER CORRELATIONS FOR CHF PREDICTIONS IN REACTOR DESIGN

Boiling crisis limits the power capability of water-cooled nuclear power reactors. The thermal margin of a reactor core design is determined by the protection and control settings based on certain thermal limiting events, such as the occurrence of a CHF. The power ceiling of a PWR control system is set by the limiting event of DNB ratio of 1. As shown in Table 5.10, the limiting condition for operation (LCO) in a power control system can be set at a power level lowered from the ceiling level (DNB event power) by following amounts to accommodate (1) the DNB correlation uncertainty required margin and (2) the instrumentation monitoring uncertainty and operational transient. Then the rated power can be selected by also reserving a predetermined margin for operational flexibility and a net power margin on LCO to accommodate possible requirements from other design limits. For example, in design of a Westinghouse PWR, the amount of power margin to accommodate the W-3 DNB correlation uncertainty is to ensure the operating power below the DNB ratio of 1.30. This margin provides assurance of the heat transfer mechanism on the fuel rod surface remaining at the benign nucleate boiling with a 95% probability at a 95% confidence level.

The monitoring uncertainty and operational transient margin is to ensure that the minimum DNB ratio is calculated at the "worst operating condition." The assumed worst operating condition consists of a power surge of 12% in a worst power distribution (power skew at top), accompanied by an inlet coolant temperature elevation of 4°F (2°C) and a pressure swing of 30 psi (0.2 MPa). A set of worst hot channel factors in core life should also be used in evaluation of the worst power distribution. Such an assumed worst operating condition is obviously overly

Table 5.9 Transient CHF tests in rod bundles and long tubes

Reference	Test section Geometry	Test section Flux shape	Type of transient operation	Results
Tong et al. (1965)	19-rod bundle 4.5 ft long	Uniform axial; nonuniform radial	Power ramps in a subcooled flow at 1,500 psia Flow coastdown from an originally subcooled flow at 1,500 psia	Transient CHF can be predicted from the steady-state CHF data of same local flow conditions No CHF observed in a simulated PWR flow coastdown rate with simulated power decay after scram in 2 sec
Moxon and Edwards (1967)	37-rod bundle 12 ft long Single tube 12 ft long	Uniform axial Nonuniform axial	Both test sections were tested at 1,000 psia in a quality flow Power ramp at 150%/sec Flow coastdown by tripping the circulating pump	The measured time of CHF is longer than predicted from the steady-state CHF data
Tong et al. (1967a)	19- and 21-rod bundles 60 in. long	Uniform axial; nonuniform radial	Power ramps in a subcooled flow at 1,500 psia Flow coastdown from an originally subcooled flow at 1,500 psia	Transient CHF can be predicted from the steady-state data of same local flow conditions No CHF observed in a simulated PWR flow coastdown rate with simulated power decay after scram in 2 sec
Cermak et al. (1970)	21-rod bundle 5 ft long	Uniform axial; nonuniform radial	Pressure blowdown of originally subcooled boiling flow from the initial pressure of 1,500 psia	Transient CHF can be predicted from the steady-state CHF data of same local flow conditions

Source: Tong (1972). Reprinted with permission of U.S. Department of Energy, subject to the disclaimer of liability for inaccuracy and lack of usefulness printed in the cited references.

conservative, because the assumption that all these conservative input conditions occur concurrently is hardly realistic. Development efforts for recognizing realistic worst operating conditions have focused on establishing more realistic input conditions that would reduce the unnecessary conservatism and still maintain the required degree of safety (Tong, 1988).

The minimum DNB ratio is evaluated at the hottest fuel rod in the hottest flow channel of the core. The fluid conditions in the hottest flow channel of an open-channel PWR should be realistically evaluated by considering the cross-channel

Table 5.10 Thermal margins for reactor power

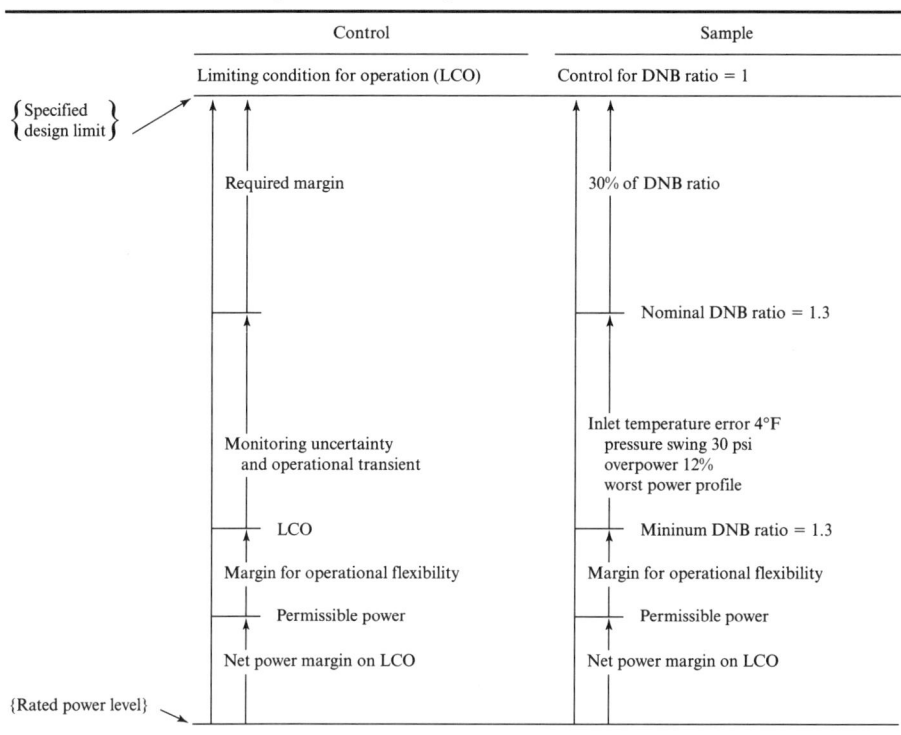

Source: Tong (1988). Copyright ©1988 by Hemisphere Publishing Corp., New York. Reprinted with permission.

fluid mixing between the hot channel and neighboring normal channels. Such an evaluation has to be carried out in a subchannel analysis. A practical example of subchannel analysis result is reproduced in Figure 5.65 by comparing the relative enthalpy rises with and without fluid mixing. The maximum enthalpy-rise hot channel factor of 1.63 in a nonfluid mixing calculation is reduced to 1.54 in a fluid mixing calculation in THINC-II code (Chelemer et al., 1972). Details of subchannel analysis codes are given in the Appendix.

The "worst operating condition" in a common design practice consists of overly conservative assumptions on the hot-channel input. These assumptions must be realistically evaluated in a subchannel analysis by the help of in-core instrumentation measurements. In the early subchannel analysis codes, the core inlet flow conditions and the axial power distribution were preselected off-line, and the most conservative values were used as inputs to the code calculations. In more recent, improved codes, the operating margin is calculated on-line, and the hot-channel power distributions are calculated by using ex-core neutron detector signals for core control. Thus the state parameters (e.g., core power, core inlet temper-

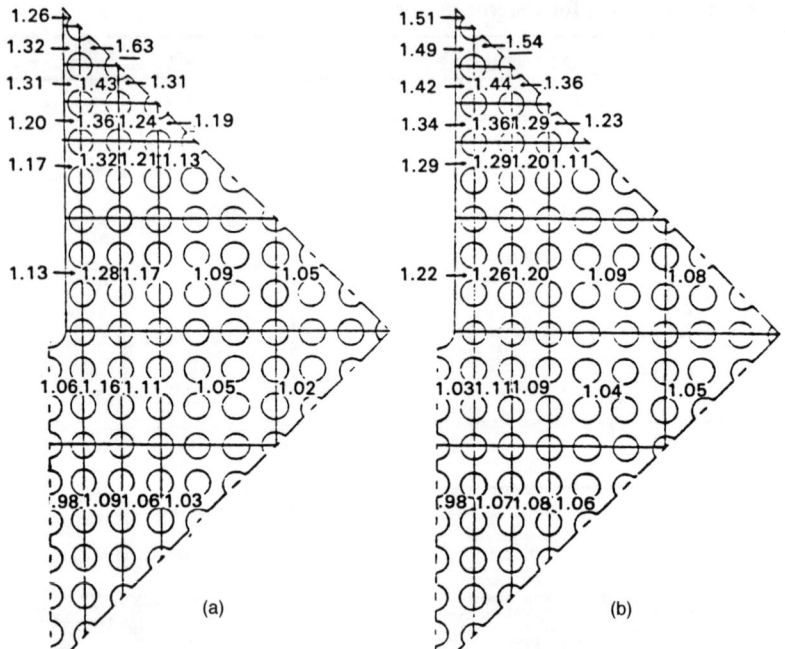

Figure 5.65 Relative enthalpy rises in Yankee core III, hot assembly: (*a*) relative enthalpy rises without fluid mixing; (*b*) relative enthalpy rises computed by THINC-II. (From Chelemer et al., 1972. Copyright © 1972 by Elsevier Science SA, Lausanne, Switzerland. Reprinted with permission.)

ature, system pressure, radial power peaking factor) are continuously monitored or measured on-line; but the system parameters, describing the physical system and boundary conditions, are not monitored on-line during reactor operations. The uncertainties of these input parameters are combined statistically by using the appropriate probability distributions for each uncertainty. Therefore, in defining a thermal design margin, a CHF correlation must be coupled with a subchannel analysis code; and the certainty of the CHF correlation applied to a rod bundle core is essentially a combined uncertainty of the correlation and the code.

The final issue of application is the selection of CHF correlations for reactor design. Of course, the design CHF correlation must accurately predict CHF behavior of the reactor hardware operating in the designed parameter ranges. Most reactor vendors test flow boiling crisis on their typical rod bundle geometries with their proprietary spacer grid design and operate the test section in the same ranges as the reactor operates. From their own test data, each of the vendors then separately develops their CHF correlation in their proprietary subchannel code for application in their product design. While research laboratories and universities have tested the boiling crisis in a wide range of operating conditions and have developed more generalized CHF correlations, they aim at understanding the nature and the trend of boiling crisis. Sometimes a generalized CHF correlation can be used in a publically available subchannel analysis code to evaluate the uncertainties of the thermal margins of various reactor designs for comparison purposes. However, in the absence of exact fuel assembly geometry information and other specific design information, such an evaluation is not suitable for use in the control and protection systems of a reactor. Current publically available CHF correlations for fuel rod bundles are described in the following section. The parameter ranges of application and the assessed prediction uncertainties of each of the CHF correlations are also listed for designers' information.

5.5.1 W-3 CHF Correlation and THINC-II Subchannel Codes

5.5.1.1 W-3 CHF correlation. The insight into CHF mechanism obtained from visual observations and from macroscopic analyses of the individual effect of p, G, and X revealed that the local p-G-X effects are coupled in affecting the flow pattern and thence the CHF. The system pressure determines the saturation temperature and its associated thermal properties. Coupled with local enthalpy, it provides the local subcooling for bubble condensation or the latent heat (H_{fg}) for bubble formation. The saturation properties (viscosity and surface tension) affect the bubble size, bubble buoyancy, and the local void fraction distribution in a flow pattern. The local enthalpy couples with mass flux at a certain pressure determines the void slip ratio and coolant mixing. They, in turn, affect the bubble-layer thickness in a low-enthalpy bubbly flow or the liquid droplet entrainment in a high-enthalpy annular flow.

By fitting the best q''_{crit} data available at that time, Tong (1967a) correlated these coupled relationships, along with other flow pattern-related parameters (D_e and H_{in}), into various correlating functions, $F(X, p)$, $F(X, G)$, $F(D_e)$, and $F(H_{\text{in}})$. Each correlating function was developed by plotting along an independent parameter against the measured CHF, q''_{crit}, with all other parameters held at constant values. The W-3 DNB correlation for uniform heat flux can be expressed as the product of these correlating functions:

$$q''_{\text{crit}} = F(X, p) \times F(X, G) \times F(D_e) \times F(H_{\text{in}}) \qquad (5\text{-}108)$$

where

$$F(X, p) = [(2.022 - 0.0004302 p) + (0.1722 - 0.0000984 p) \\ \times \exp(18.177 - 0.004129 p) X](1.157 - 0.869 X) \qquad (5\text{-}109)$$

$$F(X,G) = (0.1484 - 1.596X + 0.1729X \, |X|)\left(\frac{G}{10^6}\right) + 1.037 \quad (5\text{-}110)$$

$$F(D_e) = [0.2664 + 0.8357 \exp(-3.151 D_e)] \quad (5\text{-}111)$$

$$F(H_{in}) = [0.8258 + 0.000794(H_{sat} - H_{in})] \quad (5\text{-}112)$$

The values of the empirically determined correlating functions are plotted in Figures 5.66, 5.67, 5.68, and 5.69, for use in evaluating the parameter change effects in reactor core design. By substituting the expressions for all the correlating functions into the W-3 DNB correlation, Eq. (5-106) for uniformly heated channels with grid or spacers, we get:

$$\frac{q''_{crit,EU}}{10^6} = \{(2.022 - 0.0004302p) + (0.1722 - 0.0000984p)$$

$$\times \exp[(18.177 - 0.004129p)X]\}[1.157 - 0.869X]$$

$$\times \left[0.1484 - 1.596X + 0.1729X \, |X|)\left(\frac{G}{10^6}\right) + 1.037\right]$$

$$\times [0.2664 + 0.8357 \exp(-3.151 D_e)][0.8258 + 0.000794(H_{sat} - H_{in})]F_s$$

$$\times \text{Btu/hr ft}^2 \quad (5\text{-}113)$$

where the following parameters are used:

F_s = grid or spacer factor, dimensionless
p = 1,000–2,300 psia
G = 1.0 × 10⁶–5.0 × 10⁶ lb/hr ft²
D_e = 0.2–0.7 in.
$X \le 0.15$
$H_{in} \ge 400$ Btu/lb
L = 10–144 in.

and (heated perimeter, D_h)/(wetted perimeter, D_e) = 0.88–to 1.0. For nonuniformly heated channels, a shape factor, F_c, should be used (Sec. 5.3.2), as shown in Eq. (5-114):

$$q''_{crit,non} = \left(\frac{q'_{crit,EU}}{F_c}\right) \quad (5\text{-}114)$$

where $q''_{crit,non}$ is critical heat flux for the nonuniformly heated channel. As shown before, the shape factor can be obtained via Eq. (5-30),

$$F_c = \frac{C}{q''_{loc}[1 - \exp(C \ell_{DNB,EU})]} \int_0^{\ell_{DNB,non}} q''(z) \exp[-C(\ell_{DNB,non} - z)] \, dz$$

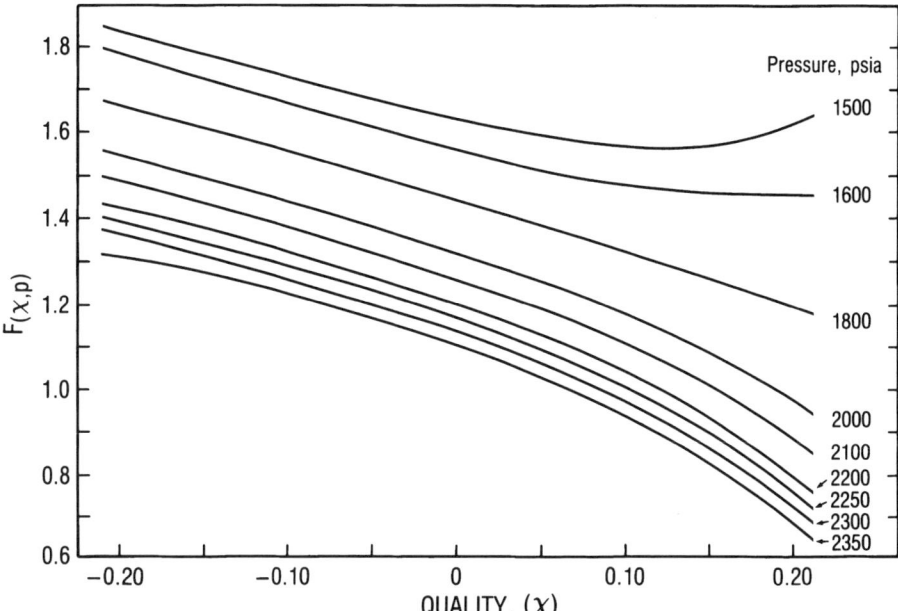

Figure 5.66 Correlating function of coupled quality and pressure effects on CHF, $F(X, p)$, where $F(X, p) = [(2.022 - 0.0004302) + (0.1722 - 0.0000984p) \exp(18.177 - 0.004129p)X] \cdot [1.157 - 0.869X]$. (From Tong, 1968a. Copyright © 1968 by American Nuclear Society, LaGrange Park, IL. Reprinted with permission.)

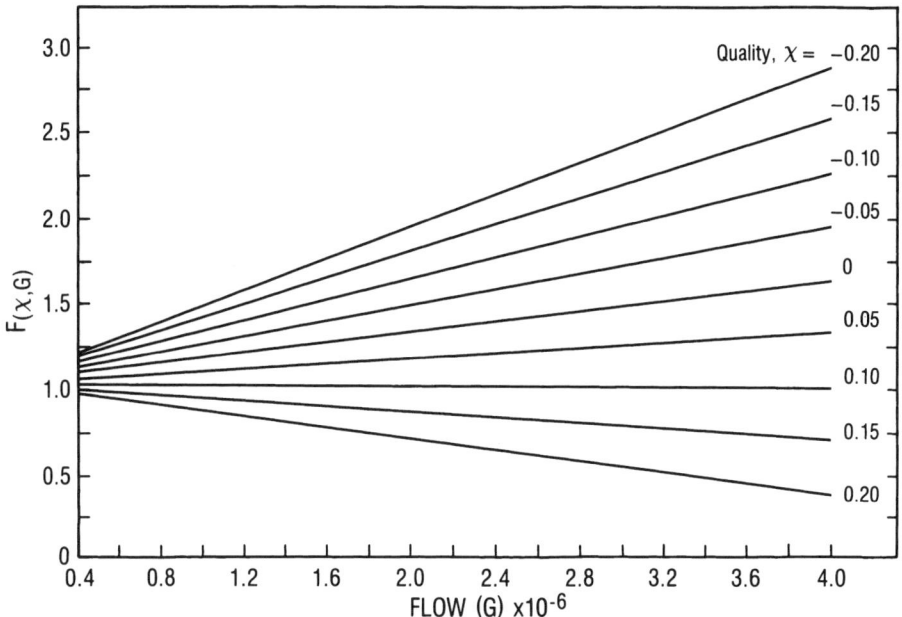

Figure 5.67 Correlating function of coupled flow rate and quality effects on CHF, $F(X, G)$, where $F(X, G) = [(0.1484 - 1.596X + 0.1729X|X|)G/10^6 + 1.037]$. (From Tong, 1968a. Copyright © 1968 by American Nuclear Society, LaGrange Park, IL. Reprinted with permission.)

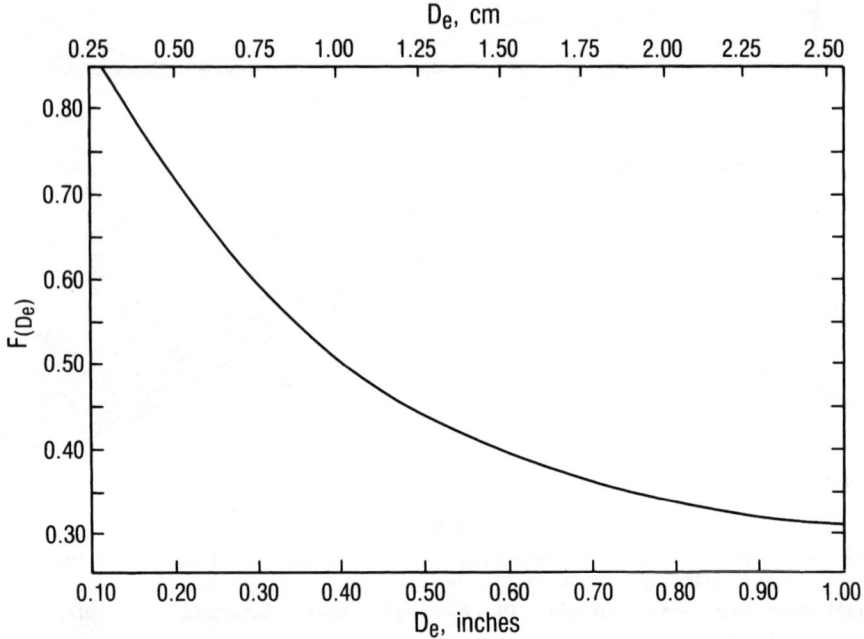

Figure 5.68 Correlating function of hydraulic diameter effect on CHF, $F(D_e)$, where $F(D_e) = [0.2664 + 0.8357 \exp(-3.151 D_e)]$. (From Tong, 1968a. Copyright © 1968 by American Nuclear Society, La-Grange Park, IL. Reprinted with permission.)

where C was also shown to be

$$C = 0.15 \frac{(1 - X_{DNB})^{4.31}}{(G/10^6)^{0.478}} \quad (1/\text{in.}) \tag{5-31}$$

and

$\ell_{DNB,EU}$ = axial location at which DNB occurs for uniform heat flux (in.), starting from inception of local boiling
$\ell_{DNB,non}$ = axial location at which DNB occurs for nonuniform heat flux (in.),
$\ell_{DNB,non} = \ell_{DNB,EU}$
X_{DNB} = quality at DNB location under nonuniform heat flux conditions

To account for the mixing effect of the grid spacer, the mixing vane grid spacer factor, F_s, is used:

$$F_s = 1.0 + 0.03 \left(\frac{G}{10^6}\right)(\alpha'/0.019)^{0.35} \tag{5-115}$$

where consistent units are used as shown:

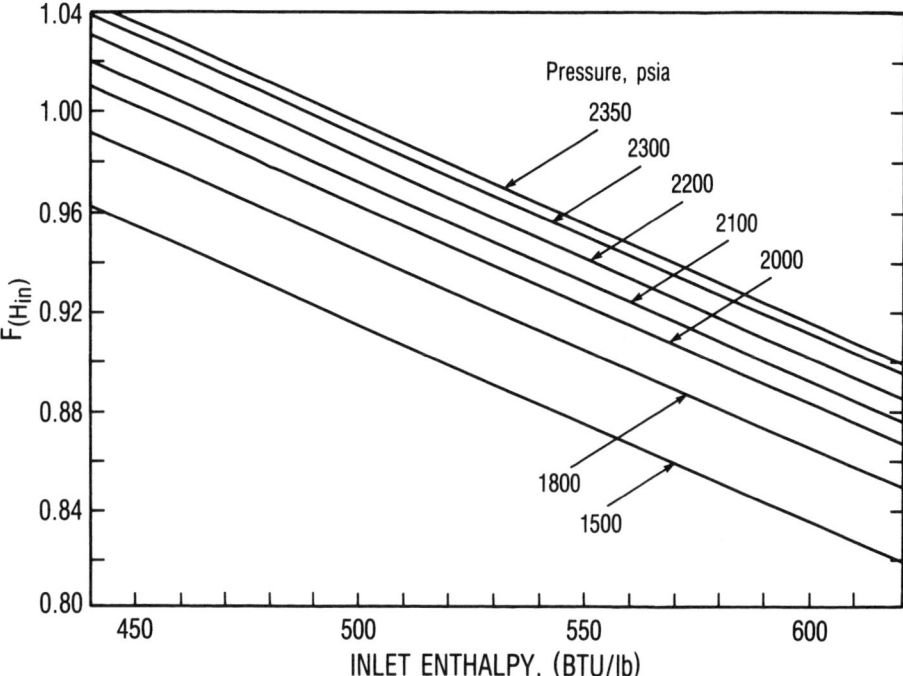

Figure 5.69 Correlating function of inlet enthalpy effect on CHF, $F(H_{in})$, where $F(H_{in}) = [0.8258 + 0.000794(H_{sat} - H_{in})]$. (From Tong, 1968a. Copyright © 1968 by American Nuclear Society, La-Grange Park, IL. Reprinted with permission.)

G = mass flux, lb/hr ft²
α' = nondimensional thermal diffusion coefficient = ε/Vb, = 0.019–0.060 for T_{in} = 500–560° F for a mixing vane grid

In addition, the DNB in a flow channel containing an unheated wall should be corrected by an unheated-wall factor. This factor, as shown in Section 5.4.5.2, is of the form

$$\frac{q''_{\text{DNB, unheated wall}}}{q''_{\text{DNB (W-3), Dh}}} = 1.0 - R_u[13.76 - 1.372e^{1.78X} - 4.732\left(\frac{G}{10^6}\right)^{-0.0575}$$

$$- 0.0619\left(\frac{p}{1000}\right)^{0.14} - 8.509 D_h^{0.107}] \quad (5\text{-}97)$$

and

410 BOILING HEAT TRANSFER AND TWO-PHASE FLOW

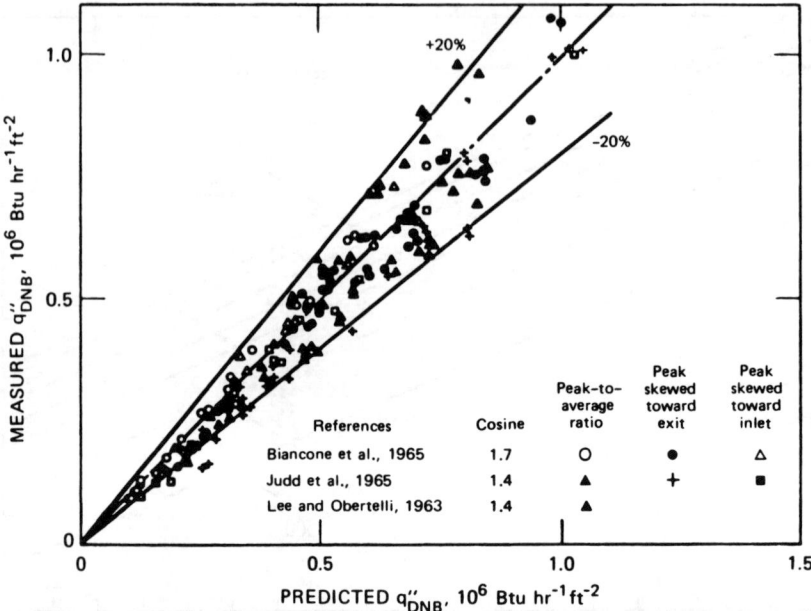

Figure 5.70 Comparison of W-3 prediction and data with nonuniform fluxes. (From Tong, 1967b. Copyright © 1967 by Elsevier Science SA, Lausanne, Switzerland. Reprinted with permission.)

$$R_u = \frac{D_h - D_e}{D_h}$$

Predictions of a nonuniform heat flux, $q''_{DNB,non}$ obtained by using $q''_{DNB(W-3)}$, for uniform heat flux in a single tube and the shape factor, F_c, agree very well with the measured nonuniform flux condition, $q''_{DNB,non}$ of Biancone et al. (1965), Judd et al. (1965), and Lee and Obertelli (1963), as shown in Figure 5.70.

5.5.1.2 THINC II code verification. A systematic experimental verification for both W-3 DNB correlation and THINC-II subchannel analysis in a rod bundle was conducted on twelve 16-rod, full-length fuel element bundles at Westinghouse (Rosal et al., 1974). Effects on DNB due to nonuniform radial and axial heat flux distributions, the geometry of the spacer grids (with and without mixing vane), rod length up to 14 ft (4.3 m), and the axial grid span along the rods were investigated. The experimental ranges tested were pressure, 1,490–2,400 psia (10.3–16.5 MPa); inlet temperature, 431–627°F (221–330°C); and mass flux, 1.02–3.95 × 10⁶ lb/hr ft² (1,383–5,357 kg/m² s). The differences in DNB behavior between a uniformly heated single channel and a nonuniformly heated rod bundle can be attributed to the following effects:

1. Shape of axial and radial heat flux distributions
2. Flow mixing between the neighboring subchannels in a rod bundle
3. Geometry of spacer grid and mixing vane
4. Axial span between grids
5. Boundary of unheated wall

The effect of the shape of axial heat flux distribution on DNB flux in a rod bundle can be evaluated similarly to that in a single channel, Eq. (5-114). The effect of the shape of radial heat flux distribution on DNB flux must be evaluated by a subchannel analysis code, since the local flow conditions depend strongly on flow mixing among the laterally connected flow channels and on the radial power gradient. The spacer grid and/or mixing vane usually act as turbulence promoters and increase the flow mixing. The strength of turbulence decays as the flow goes downstream and away from the grid. Thus, the intensity of flow mixing is also a function of grid span. The effect of flow mixing due to the spacing grids on the DNB heat flux can be evaluated by a spacer factor, F_s [Eq. (5–115)]. The nondimensional thermal diffusion coefficient, α', is defined as

$$\alpha' = \left(\frac{\varepsilon}{Vb}\right)$$

where $b = a \times f(\text{grid span})$ and a is the gap between rods, in feet. Various values of thermal diffusion coefficient have been used in subchannel analyses for rod bundles with specific grid geometries. A coefficient of $\alpha' = 0.019$ was used for a grid without mixing vanes. For a grid with mixing vanes, values of α' varied with grid spans, namely, $\alpha' = 0.108$ for a 10-in. span, 0.061 for a 20-in. span, and 0.051 for a 26-in. span. The unheated wall factor, Eq. (5-97), was used for flow in channels facing the unheated boundary of the test section. The test section configuration of the 16-rod bundle was arranged in a 4 × 4 square array with a heated 8-ft length, with heater rods having either a (cos u) or a $u(\sin u)$ axial heat flux distribution. To achieve such nonuniform axial heat flux, tubes were specially manufactured with the appropriate variable wall thickness while maintaining constant outside diameter. The peak heat flux at the center, cos u, or the peak heat flux skewed toward the top, $u(\sin u)$, of the heated length was obtained by assembling heater rods from two tubes joined at their thin ends by means of a welding insert. The nonuniform radial power distribution was achieved by heating the four center rods at higher power levels than the outer rods. Sheathed thermocouples (0.040 in O.D.) inside the rods were employed as DNB detectors, with readout on Offner oscillograph multichannel recorders.

The objective of this test was to present and analyze suitable experimental results for verifying quantitatively the use of the above-mentioned three corrections with the W-3 correlation for predicting the DNB heat flux in a rod bundle. Uncertainties in the data due to instrument errors and heater rod fabrication tolerances

Table 5.11 Data uncertainties due to instrument errors and fabrication tolerances

Inlet temperature	±1.3°F (0.7°C)
Pressure	±5 psia (34 kPa)
Flow	±1% of measured value
Power	±0.7% of measured value
Heater rod	
Wall thickness	±0.0005 in. (0.013 mm)
Outside diameter	±0.0005 in. (0.013 mm)

Source: Rosal et al. (1974). Copyright © 1974 by Elsevier Science SA, Lausanne, Switzerland. Reprinted with permission.

Table 5.12 Comparison of measured data with predictions for different bundles and types of grids

Test bundle	VI	V	X	IV	IX	VII	VIII
Type of grid	Simple grid	T-H no vane	T-H w/vane	T-H no vane	T-H w/vane	Mixing-vane grid	Mixing-vane grid
Grid span (in.)	10	10	10	20	20	20	26
α' coefficient	0.019	0.019	0.108	0.019	0.061	0.061	0.051
$1 - \frac{\text{(pred. av.)}}{\text{(meas. av.)}}$	−0.1%	6.2%	9.2%	1.8%	7.3%	−1.6%	−5.9%

Source: Rosal et al. (1974). Copyright © 1974 by Elsevier Science SA, Lausanne, Switzerland. Reprinted with permission.

are given in Table 5.11. Combining the instrument errors yields an error on the measured DNB heat flux of ±2%. The error in DNB heat flux arising from the variation in wall thickness was ±3%, while the reproducibility of the DNB data was ±1.6% on average. Thus the maximum error on the DNB heat flux was ±6.6%, and the most probable error was ±4%.

The effect due to mixing promoters (vanes) and grid span was evaluated by testing various thermal diffusion coefficients in a THINC code subchannel analysis and in the equation for spacer factor to find the values that agreed best with the observed results, as shown in Table 5.12. It can be seen from the table that the T-H grid with a mixing vane gives a higher mean discrepancy even with the higher α' values.

To show the power level difference between the predicted DNB heat flux and the measured local DNB heat flux, the latter was taken at the axial location to give the minimum DNB ratio (DNBR) at the actual DNB power instead of that at the actual DNB location. However, the correct ratio of measured to predicted channel powers was maintained equal to the ratio of the heat fluxes, as shown in Figure 5.71. That is,

$$\left(\frac{q''_{\text{meas}}}{q''_{\text{pred}}} \right) = \frac{q''_F}{q''_c} = \frac{q''_E}{q''_{c'}} = \frac{\text{power I}}{\text{power II}} \qquad (5\text{-}116)$$

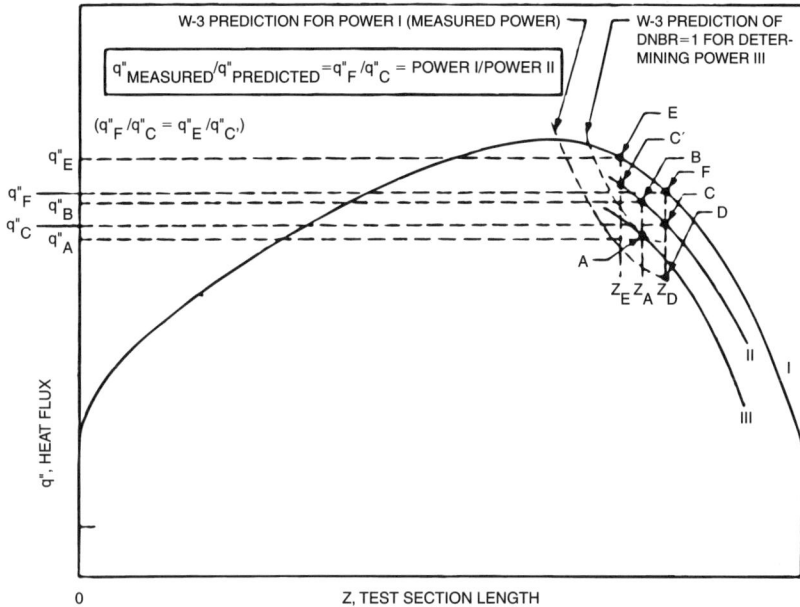

Figure 5.71 Procedure for comparing measured and predicted DNB heat fluxes. Z_A = min. DNBR location at power III (established by DNBR = I with test inlet conditions); Z_D = min. DNBR location at power I (established by test inlet conditions and measured power); Z_E = location of actual measured DNB; $q''F$ = heat flux at Z_D of power I; $q''C$ = heat flux at Z_D of power II, power II = F_s × power III; q''_A = predicted DNB heat flux by W-3 without F_s; and q''_B = predicted DNB heat flux by W-3 with F_s, $q''_B = F_s \times q''_A$. (From Rosal et al., 1974. Copyright © 1974 by Elsevier Science SA, Lausanne, Switzerland. Reprinted with permission.)

where

q''_E is the measured CHF at test power I
$q''_{C'}$ is the heat flux at power II at the measured CHF location Z_E
q''_F is the heat flux at the minimum DNBR location, Z_D, for power I
q''_C is the heat flux at Z_D for power II
Power I is the measured power
Power II is the product of F_S (spacer grid factor) and power III
Power III is analytically established by DNBR = 1 (W-3 prediction with test inlet conditions and without spacer grid).

Also shown in the figure are

q''_A, the predicted DNB heat flux by W-3 without F_S, at the minimum DNBR location, Z_A, for power III (established by DNBR = 1 with test inlet conditions)
q''_B, the product of F_s and q''_A

414 BOILING HEAT TRANSFER AND TWO-PHASE FLOW

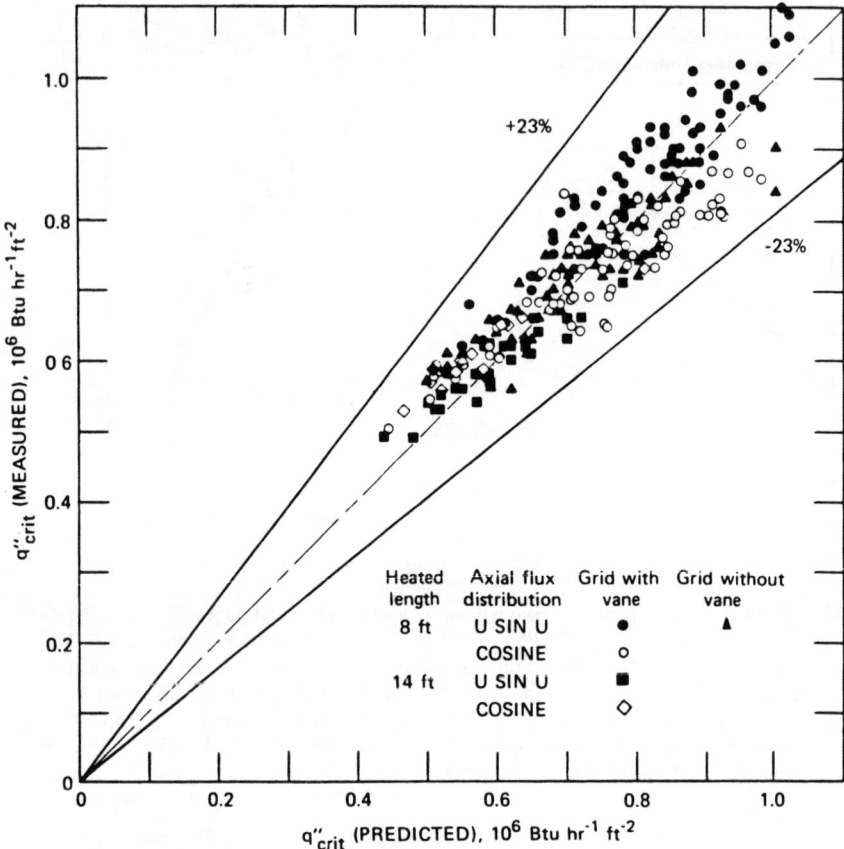

Figure 5.72 Rod-bundle nonuniform axial heat flux CHF data (from twelve 16-rod bundles) with ranges of: mass fluxes, $G = 1.0 \times 10^6$ to 3.8×10^6 lb/hr ft^2; inlet temperatures, 430–620°F; pressure, 1,500–2,400 psia; CHF quality 15%; grid spacing, 10–26 in.; spacer factor used. (From Rosal et al., 1974. Copyright © 1974 by Elsevier Science SA, Lausanne, Switzerland. Reprinted with permission.)

The following conclusions can be drawn from the verification.

1. The nonuniform heat flux prediction of W-3 DNB correlation (developed from single channel data as shown in Fig. 5.70) was verified in axially, nonuniformly heated bundles by comparison of the 284 DNB data points (Fig. 5.72) with the W-3 correlation through a subchannel analysis including spacer factor, showing excellent agreement. The standard deviation was 7.4%.
2. The pressure limit of the W-3 correlation was extended from 2,300 to 2,400 psia (from 15.8 to 16.5 MPa), based on the verification results.
3. The mixing-vane T-H grid gives higher DNB heat flux than the grid without a vane.
4. The shorter grid span gives higher DNB hat fluxes with the same type of grid.

5. The scatter of the rod bundle data, when compared to the W-3 correlation with a spacer factor, is much less than the scatter of single-channel data used to develop the correlation. This is an indication of the improvement in DNB testing techniques.

5.5.2 B &W-2 CHF Correlation (Gellerstedt et al., 1969)

5.5.2.1 Correlation for uniform heat flux. The correlation for uniform heat flux, in specified units, is

$$q''_{crit} = \frac{A'(A - BX)}{C} \times 10^6 \text{ Btu/hr ft}^2 \qquad (5\text{-}117)$$

where $A' = b_1 - b_2 d$
$A = b_3(b_4 G)^{[b_5 + b_6(p-2,000)]}$
$B = b_7 G H_{fg}$
$C = b_8(b_9 G)^{[b_{10} + b_{11}(p-2,000)]}$

and

G is mass flux $\times 10^6$ lb/hr ft^2
p is pressure, psia
X is quality
d is subchannel equivalent diameter, in.

The constants in the above equations, b_1 through b_{11}, have the following values:

b_1	b_2	b_3	b_4	b_5	b_6
1.1551	0.4070	0.3702×10^8	0.5914	0.8304	0.6848×10^{-3}

b_7	b_8	b_9	b_{10}	b_{11}
0.1521	12.7100	3.0545	0.7119	0.2073×10^{-3}

The parameter ranges are

Pressure p, 2,000–2,400 psia (13.8–16.5 MPa)
Local quality X, −0.03 to 0.20
Local mass flux G, 0.75–4.00 × 10^6 lb/hr ft^2 (1,000–5,420 kg/ms^2 s)
Hydraulic diameter d, 0.20–0.50 in. (0.5–1.3 cm)
Heated length L, 72 in. (1.83 m)

Equation (5-117) correlates 207 data points of Gellerstedt et al. (1969) from rod bundles 72 in. (1.83 m) long with 15-in. (0.38-m) grid span with a standard deviation of 7.7%.

5.5.2.2 Correlation for nonuniform heat flux. On the basis of Tong's shape factor formulation (1967a), Wilson et al. (1969) developed another set of constants:

$$F_c = \frac{1.025 C \int_0^{\ell_{CHF}} q''(Z) \exp[-C(\ell_{CHF} - Z)] dZ}{q''_{loc}[1 - \exp(-C \ell_{CHF,EU})]} \quad (5\text{-}118)$$

where

$$C = \frac{0.249(1 - X_{CHF})^{7.82}}{(G/10^6)^{0.457}} \quad (5\text{-}119)$$

The ranges of parameters used were

$p = 2{,}000\text{--}2{,}400$ psia (13.8–16.5 MPa)
$G = 1 \times 10^6\text{--}3.5 \times 10^6$ lb/hr ft² (1,356–4,750 kg/m² s)
$D_e = 0.2\text{--}0.5$ in. (0.5–1.3 cm)
$X_e = 0.02\text{--}0.25$

Combined with Eq. (5-114), they correlated the Wilson et al.'s (1969) data of 81 data points as obtained from rod bundles 72 in. (1.83 m) long having 15-in. (0.38-m) grid span with a standard deviation of 11.5%.

5.5.3 CE-1 CHF Correlation (C-E Report, 1975, 1976)

The CHF correlation developed by C-E Company is in the same form as Eq. (5-117) but with different constants:

$$q''_{crit} = \frac{A'(A - BX)}{C} \times 10^6 \text{ Btu/hr ft}^2 \quad (5\text{-}120)$$

where $A' = b_1 \left(\dfrac{d}{d_m}\right)^{b_2}$

$A = (b_3 + b_4 p) G^{(b_5 + b_6 p)}$
$B = G H_{fg}$
$C = G^{(b_7 p + b_8 G)}$

All parameters use the specified units,

G = mass flux × 10^6 lb/hr ft²
p = pressure, psia
X = quality
d = subchannel equivalent diameter, in.
d_m = matrix channel equivalent diameter, in.

and b_1 through b_8 are constants given by

b_1	b_2	b_3	b_4
2.8922×10^{-3}	-0.50749	405.32	-9.9290×10^{-2}

b_5	b_6	b_7	b_8
-0.67757	-6.8235×10^{-4}	3.1240×10^{-4}	-8.3245×10^{-2}

Parameters were in the following ranges:

p = 1,785–2,415 psia (12.3–16.7 MPa)
X = −0.16 to 0.20
G = 0.87–3.21 × 10^6 lb/hr ft² (1,180–4,350 kg/m² s)
T_{in} = 382–644°F (194–340°C)
D_e = 0.36–0.55 in. (0.9–1.4 cm)
L = 84–150 in. (2.1–3.8 m)

5.5.4 WSC-2 CHF Correlation and HAMBO Code

5.5.4.1 Bowring CHF correlation for uniform heat flux (Bowring, 1972).
For water in round tubes with uniform heat flux, the CHF can be expressed as

$$\frac{q''_{crit}}{10^6} = \frac{DGH_{fg}}{4 \times 10^6 \, C}(A - X_{crit}) \text{ Btu/hr ft}^2 \tag{5-121}$$

where all parameters use the specified units,

G = mass flux, lb/hr ft²
D = equivalent diameter, in.
p = pressure, psia

and

$A = 2.317 F_1/(1 + 3.092 F_2 D^{0.5} G)$
$C = 104.4 F_3 DG/(1 + 0.347 F_4 G^n)$
F_1, F_2, F_3, F_4 are functions of p'_r (= $p/1,000$)

For $p \leq 1{,}000$ psia, on $p'_r \leq 1$,

$$F_1 = \frac{(p'_r)^{18.942} \exp[20.89(1 - p'_r)] + 0.917}{1.917}$$

$$\frac{F_1}{F_2} = \frac{(p'_r)^{1.316} \exp[2.444(1 - p'_r)] + 0.309}{1.309}$$

$$F_3 = \frac{(p'_r)^{17.023} \exp[16.658(1 - p'_r)] + 0.667}{1.667}$$

$$\frac{F_4}{F_3} = (p'_r)^{1.649}$$

(at $p = 1{,}000$ psia, $F_1 = F_2 = F_3 = F_4$).

For $p > 1{,}000$ psia, or $p'_r > 1$,

$$F_1 = (p'_r)^{-0.368} \exp[0.648(1 - p'_r)]$$

$$\frac{F_1}{F_2} = (p'_r)^{-0.448} \exp[0.245(1 - p'_r)]$$

$$F_3 = (p'_r)^{0.219}$$

$$\frac{F_4}{F_3} = (p'_r)^{1.649}$$

5.5.4.2 WSC-2 correlation and HAMBO code verification (Bowring, 1979). The WSC-2 correlation covers the pressure range from 3.4 to 15.9 MPa (500 to 2,300 psia) and is considered to be applicable to pressure tube reactors (PTRs), pressurized water reactors (PWRs), and boiling water reactors (BWRs). It was developed exclusively from subchannel data. All 54 different clusters were analyzed using HAMBO and the correlation optimized for the calculated subchannel conditions.

The basic equation for the correlation is

$$q''_{\text{crit}} = \frac{A + B \, \Delta H_{\text{in}}}{C + ZY'Y} \times 10^6 \text{ Btu/hr ft}^2 \qquad (5\text{-}122)$$

where all parameters use in specified units,

ΔH_{in} = inlet subcooling, Btu/lb
Z = distance from channel inlet, in.
Y = axial heat flux profile parameter
Y' = subchannel imbalance factor

and A, B, C are parameters that depend on subchannel shape, mass fluxes, pressure, etc., and can be calculated by Eqs. (5-123) or (5-124), below. The subchannel shapes are defined by three types:

Type 1: "equilateral-triangular rodded," i.e., a subchannel bounded by three rods forming an equilateral triangle (one or more rods may be unheated)

Type 2: "square rodded," i.e., a subchannel bounded by four rods on a square lattice, or three rods forming a triangle of 90°, 45°, 45° (one or more rods may be unheated)

Type 3: "outer subchannel," i.e., a subchannel bounded by one or more heated or unheated rods and a section of unheated straight or circular surface containing the cluster

For types 1 and 2, the triangular- and square-rodded subchannels,

$$A = \frac{(0.25 G D H_{fg} F_1) Q_1}{1 + Q_2 F_2 G D (Y')^{Q_3}}$$

$$B = 0.25 G D$$

$$C = C'V \left[1 + \left(\frac{Y-1}{1+G} \right) \right] \quad (5\text{-}123)$$

$$C' = \frac{Q_4 F_3 (G D Y')^{1/2}}{D_w}$$

where $D = F_p D_h$
$p'_r = (0.001) p$
$F_1 = (p'_r)^{0.982} \exp[1.170(1 - p'_r)]$
$F_2 = (p'_r)^{0.841} \exp[1.424(1 - p'_r)]$
$F_3 = (p'_r)^{1.851} \exp[1.241(1 - p'_r)]$

and values of Q_1, Q_2, Q_3, Q_4 are given as follows:

Type Shape	(1) Triangular	(2) Square
Q_1	1.329	1.134
Q_2	2.372	1.248
Q_3	−1.0	−2.5
Q_4	12.26	28.76

For type 3, outer subchannels,

$$A = (0.151 H_{fg} F_1) G^{0.5} D^{0.64} [1 - 0.581 \exp(-5.221 F_2 G D_w)]$$

$$B = 0.3077 G D^{1.16}$$

$$C = C'V \left[1 + \left(\frac{Y-1}{1+G} \right) \right] \quad (5\text{-}124)$$

$$C' = 286.8 F_3 G^{0.75} D_w^{*1.264}$$

where $D = F_p D_h$
$$D_w^* = (d^2 + dD)^{1/2} - d$$
$$p_r' = (0.001)p$$
$$F_1 = (p_r')^{1.321} \exp[1.072(1 - p_r')]$$
$$F_2 = (p_r')^{-0.60} \exp[0.497(1 - p_r')]$$
$$F_3 = (p_r')^{3.005} \exp[2.498(1 - p_r')]$$

Axial heat flux parameter Y The parameter Y, which replaces the heat flux shape factor in the CHF correlation, is not only a measure of the nonuniformity of the axial heat flux profile but also a means of converting from the inlet subcooling (ΔH_{in}) to the local quality, X, form of the correlation via the heat balance equation. It is defined as

$$Y = \frac{\text{ave. subchannel or cluster heat flux from entry to } Z}{\text{local subchannel or cluster radial ave. heat flux at } Z}$$

$$= \frac{(1/Z)\int_0^Z q'' \, dZ}{q''} \tag{5-125}$$

An approximate value of Y_1 at $Z = Z_i$ may be calculated by summing over a number of intervals of length. Thus, for a continuous axial heat flux profile, Y_i is given by

$$Y_i = \frac{\sum_{r=2}^{r=i} 0.5(\overline{q_r''} + \overline{q_{(r-1)}''})(Z_r - Z_{(r-1)})}{(q_i'' Z_i)}$$

For a cluster with uniform axial heat flux, $Y = 1$ at all Z. For nonuniform heat flux, Y varies along the length. For example, with a chopped-cosine profile, $Y < 1$ over the first part of the channel, $Y = 1$ at about two-thirds of the length, and $Y > 1$ near the exit of the channel.

Correlation for nonuniform axial heat flux Two methods of correlating nonuniform heat flux have been postulated:

1. Local quality method: Dryout occurs when the local nonuniform heat flux equals the uniform heat flux dryout value at the same local conditions (quality, etc.).
2. Total power method: The dryout power of a nonuniformly heated channel is the same as if the channel were heated uniformly.

CHF data for an axially, uniformly heated round tube have been correlated by an equation of the form

$$q''_{crit} = \frac{A + 0.25GD_h \Delta H_{in}}{C + Z}$$

Using the heat balance, this may be written alternatively as

$$q''_{crit} = \frac{A + 0.25GD_h H_{fg} X}{C}$$

For nonuniform heat flux the first postulate states that the local quality, X, is the same, and the CHF may be written, via the heat balance containing Y, as

$$q''_{crit} = \frac{A + 0.25GD_h \Delta H_{in}}{C + ZY} \quad \text{(local quality method)}$$

It may similarly be shown that the total power postulate is equivalent to

$$q''_{crit} = \frac{A + 0.25GD_h \Delta H_{in}}{CY + ZY} \quad \text{(total power method)}$$

Examination of data for subchannels and other geometries suggests that the local quality postulate tends to be more accurate at high mass fluxes, while the total power postulate is more accurate at low mass fluxes. The above two expressions can be made one if C in the equation is multiplied by a function, $f(Y)$, where

$$f(Y) \to 1.0 \quad \text{as } G \to \infty$$

and

$$f(Y) \to Y \quad \text{as } G \to 0$$

Thus,

$$f(Y) = 1.0 + \frac{Y - 1}{1 + G}$$

was found to give a satisfactory fit to the data for which $B = 0.25GD_h$, as shown in Eq. (5-123). The situation becomes more complicated when $B \neq 0.25GD_h$ and for channels with interchannel mixing. Nevertheless, the method is so simple that it has been incorporated and found to work in the subchannel equation.

Subchannel imbalance factor Y' The parameter Y' was used in the heat balance equation to account for enthalpy transfer between subchannels. It is defined as the fraction of the heat retained in the subchannel and is a measure of this subchannel imbalance relative to that of its neighbors. Thus,

$$Y' = \frac{\text{heat retained in subchannel}}{\text{heat generated in subchannel}}$$

$$= \frac{0.25 G D_h (H - H_{in})}{\int_0^Z q'' \, dZ} \tag{5-126}$$

This parameter was used in conjunction with Z in an earlier form of the correlation with $B = 0.25 G D_h F_p$, since the equation could be written as

$$q''_{crit} = \frac{A - B H_{fg} X}{C}$$

When B was reoptimized, the group ZY in the correlation was retained. For "hot" subchannels losing heat, $Y' < 1$; and for "cold subchannels gaining heat, $Y' > 1$. The use of Y' in the correlation recognizes that dryout in a subchannel is not a purely local phenomenon; for example, a subchannel gaining heat from mixing in a large square-lattice cluster would have different dryout characteristics from an identical subchannel elsewhere in the cluster, operating at identical fluid conditions but losing heat by mixing. A similar effect is inherent in the W-3 DNB correlation (Tong, 1967a), which contains both X and ΔH_{in} as parameters.

In most HAMBO analyses near the dryout power (Bowring, 1979), the value of Y' for a particular subchannel in a cluster varies only very slightly with axial position, total flow rate, inlet subcooling, and pressure. It is thus a characteristic of the subchannel (like equivalent diameter, for example) and is related to its environment in the cluster.

Effect of vaned grids The correlation was originally optimized using experimental data from clusters with relatively nonobstructive grid spacers ($V = 1$). When applied to PWR-type clusters with vane grids, it was found to underpredict the CHF, with the amount increasing with the increasing mass flux. The effect of vaned grids is expressed in the correlation by multiplying the term C by $V (= 0.7)$. In practice, for PWRs this is equivalent to multiplying the dryout heat flux by a factor of approximately $(1.13 + 0.03G)$. Thus the trend is similar to that found by Westinghouse, which may be expressed as a multiplier on heat flux of approximately $(1.0 + 0.05G)$ (Tong, 1969). However, the magnitude of this vane effect over the PWR range is clearly greater than that found by Westinghouse. Since there was some uncertainty in the vane effect due to the paucity of data, a value of 0.85 instead of 0.7 is recommended for V for greater conservatism in assessing PWR vaned-grid assemblies. In summary, $V = 1.0$ for nominal PWR and BWR grids; $V = 0.7$ for "best fit" to vaned-grid data; and $V = 0.85$ for more conservative PWR assessment. Comparison of WSC-2 prediction with data is given in Figure 5.73.

FLOW BOILING CRISIS **423**

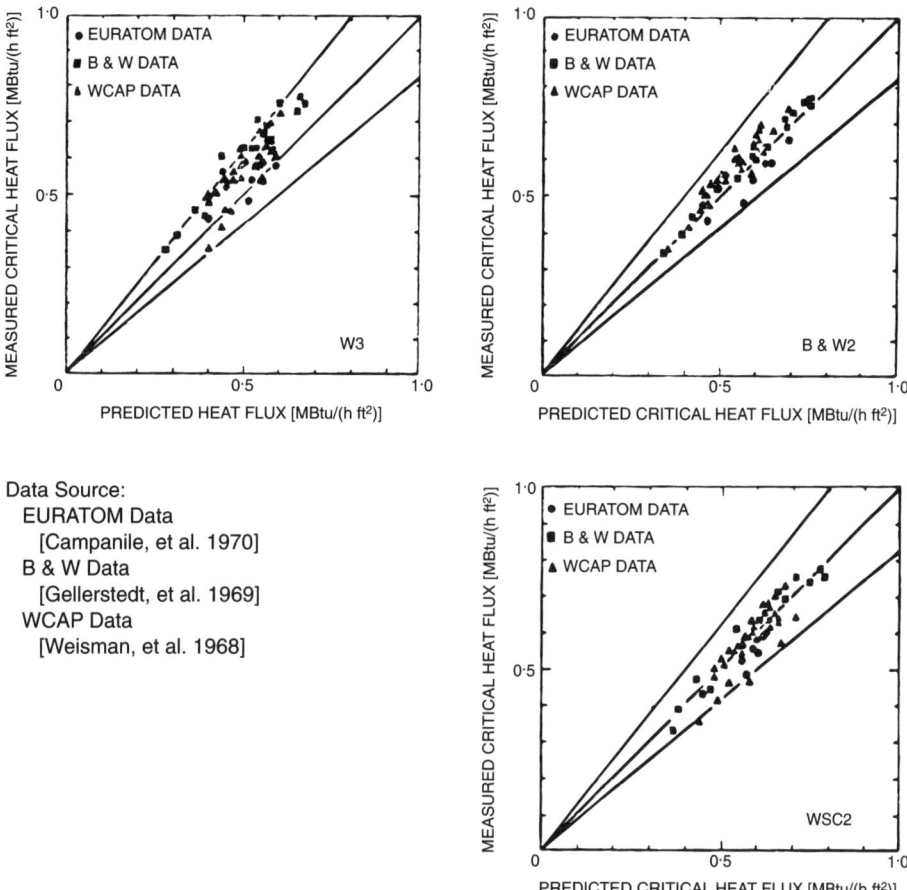

Data Source:
EURATOM Data
[Campanile, et al. 1970]
B & W Data
[Gellerstedt, et al. 1969]
WCAP Data
[Weisman, et al. 1968]

Figure 5.73 Comparison of predicted versus measured dryout heat fluxes for W-3, B&W-2, WSC-2 correlations for PWR-type conditions (upper and lower lines represent ±20% deviation. (From Bowring, 1979. Reproduced with permission of UK AEA Technology, Didcot, Oxfordshire, UK.)

5.5.5 Columbia CHF Correlation and Verification

5.5.5.1 CHF correlation for uniform heat flux. The CHF correlation based on data obtained by Columbia University Heat Transfer Laboratory is (Reddy and Fighetti, 1983)

$$q''_{crit} = \frac{A - X_{in}}{C + [(X - X_{in})/q''_L]} \times 10^6 \text{ Btu/hr ft}^2 \quad (5\text{-}127)$$

where

$$A = P_1(P_R)^{P_2} G^{(P_5 + P_7 P_R)}$$

and

$$C = P_3 P_R^{P_4} G^{(P_6 + P_8 P_R)}$$

All parameters use specified units,

q_L'' is local heat flux $\times\ 10^6$ Btu/hr ft^2
X_{in} and X are inlet and local qualities, respectively
G is mass flux $\times\ 10^6$ Btu/hr ft^2
P_R is reduced pressure (p/p_{crit})
P_1–P_8 are constants as given below:

P_1	P_2	P_3	P_4
0.5328	0.1212	1.6151	1.4066

P_5	P_6	P_7	P_8
−0.3040	0.4843	−0.3285	2.0749

The correlation predicts the source data of 3,607 CHF data points under axially uniform heat flux condition from 65 test sections with an average ratio of 0.995 and RMS deviation of 7.2%.

With cold-wall effect, two cold-wall correction factors, F_A and F_C, are used, and the correlation becomes, (Reddy and Fighetti, 1982)

$$q_{crit}'' = \frac{A F_A - X_{in}}{C F_C + [(X - X_{in})/q_L'']} \quad (5\text{-}128)$$

where

$$F_A = G^{0.1}$$

and

$$F_C = 1.183 G^{0.1}$$

All parameters use English units, i.e., mass flux $G \times 10^6$ lb/hr ft^2, and pressure p in psia. The ranges of parameters are as follows:

Pressure p, 600–1,500 psia (4.14–10.3 MPa)
Local mass flux G, 0.15–1.2 × 10^6 lb/hr ft² (200–1,630 kg/m² s)
Local quality X, 0.0–0.70
Subchannel type: corner channels only

Error statistics are

Number of test sections, 22
Number of data points, 638
Average ratio, 0.997
Root-mean-square error, 6.13%
Standard deviation, 6.13%

5.5.5.2 COBRA IIIC verification (Reddy and Fighetti, 1983). To extend the Columbia CHF correlation to rod bundles with grids, the correlation is written in the following form:

$$q''_{crit} = \frac{A - X_{in}}{CF_g F_C + [(X - X_{in})/q''_L]} \times 10^6 \text{ Btu/hr ft}^2 \quad (5\text{-}129)$$

where F_g is the grid spacer factor (without grid span effect). Optimization of F_g in terms of grid loss coefficient K results in the following grid correction factor:

$$F_g = 1.3 - 0.3K \quad K = 1 \text{ for standard grid}$$

where F_c is the axially nonuniform flux correction factor,

$$F_c = 1 + \frac{y - 1}{1 + G}$$

with

$$y = \frac{1}{L} \int_0^L \left(\frac{q''_Z}{q''_L}\right) dZ$$

and G is mass flux × 10^6 lb/hr ft². This correlation was based on 933 data points.

Comparisons of rod bundle data with the Columbia correlation and other existing correlations were made using COBRA-IIIC code for predictions of all correlations. The DNBR (or CHFR) reported is not the critical power ratio as used by other authors. The DNBR errors reported by Reddy and Fighetti (1983) are based on the following analysis: The measured local heat flux at the experimental location of the first or higher-rank CHF indications is compared with the predicted CHF calculated using local conditions from the subchannel analysis for the

Table 5.13 Comparison of rod bundle data with various correlations

Correlation/reference	Average error, %	RMS error, %	STD deviation, %
WSC-2 (Bowring, 1972, 1979)	7.4	14.6	12.4
CE-1 (C-E Report, 1975, 1976)	2.4	8.3	7.9
B & W (Gellerstedt et al., 1969)	8.3	11.4	7.7
W-3 (Tong, 1967a)	−16.0	21.2	13.6
Columbia (Reddy and Fighetti, 1983)	−0.4	6.2	6.7

Note: Negative average error indicates that the correlation is conservative. The thermal diffusion coefficients of the reactor vendors' proprietary grid geometries are not considered in this comparison.
Source: Reddy and Fighetti (1983). Copyright © 1983 by Electric Power Research Institute, Palo Alto, CA. Reprinted with permission.

measured bundle power and test inlet conditions. The parameter ranges of the 647 data points compared were

Pressure p, 1,975–2,325 psia (13.6–16.0 MPa)
Mass flux G, 0.9–3.2 × 10^6 lb/hr ft² (1,210–4,340 kg/m² s)
Quality X, 0–0.15
Length L, 84 in. (2.13 m)

The results of the comparison are given in Table 5.13.

5.5.5.3 Russian data correlation of Ryzhov and Arkhipow (1985). Ryzhov and Arkhipow (1985) correlated the data from three- and seven-rod bundles with a heated rod length of 0.6 m (1.8 ft) and a grid span of 0.1 m. Rod bundles were tested at p = 4.9–14.7 MPa (710–2,100 psia), G = 300–2,000 kg/m² s (0.22–1.48 × 10^6 lb/ft² hr), T_{in} = 30 to (T_{sat} − 10)°C or 86 to (T_{sat} − 18)°F. They developed a CHF correlation as follows:

$$\frac{\Delta H_{crit}}{H_G}(\text{Fr})^{0.3}\left(1 + \frac{180}{A_h/A_c}\right) = 1.382 - 0.548(\text{Fr})^{0.15}(H_{in}/H_L) \quad (5\text{-}130)$$

where

$$(\text{Fr}) = \left(\frac{G}{\rho_L}\right)\left[\frac{(\rho_L - \rho_G)}{gd}\right]^{1/2}$$

Here

ΔH_{crit} is enthalpy gained along heated length, kJ/kg
H_L and H_G are enthalpy of water and steam, respectively, kJ/kg
ρ_L and ρ_G are density of water and steam, respectively, kg/m³

H_{in} is the inlet enthalpy, kJ/kg
A_h/A_c is the ratio of bundle heated area and flow cross-sectional area
$K(F_{\Delta H})$ is the thermohydraulic imbalance coefficient
$F_{\Delta H} \approx 1$ for testing bundles

Ryzhov and Arkipov reported that their equation correlated their bundle test data with a rms error of 6.4% and also extended to correlate other test data of lengths 0.2–3.7 m (7.90–14.6 in.), number of rods 3–37, thermohydraulic imbalance < 1.15, by a rms error of 7.8%.

5.5.6 Cincinnati CHF Correlation and Modified Model

5.5.6.1 Cincinnati CHF correlation and COBRA IIIC verification.
The Weisman-Pei model (Weisman and Pei, 1983) has been applied to rod bundles and was evaluated by Weisman and Ying (1985) using the publically available COBRA IIIC subchannel analysis code. The Weisman-Pei prediction procedure was evaluated against a series of three tests made with an assembly containing 21 electrically heated rods and simulated control element thimble replacing four normal rods. The support grids were of simple design and had no mixing vanes attached to them. Three axial flux distributions (uniform, flux skewed to inlet, and flux skewed to outlet) were included in the data examined. In order to evaluate the accuracy of the rod bundle data quantitatively, Weisman and Ying used the ratio DNBR, defined as

$$\text{DNBR} = \frac{\text{predicted rod power at DNB}}{\text{observed rod power at DNB}} \quad (5\text{-}131)$$

They found that, for the 155 data points examined, the mean value of DNBR was 0.98 and the standard deviation of DNBR was 0.10.

In the above evaluation, Weisman and Ying (1985) suggested a very useful simplification in the subchannel analysis of CHF. Since the DNB predictions are not identical to the experimental observations, the predictions must be evaluated at several different heat fluxes. To avoid redoing the subchannel analysis at each of these power levels, Weisman and Ying took advantage of the observation that small changes in total power level had little effect on the mass flux in the hot channel. They also noted that for small changes in power level, the ratio of actual enthalpy rise to the enthalpy rise in a closed channel at the same power remained nearly constant. They were able to define a mixing factor, $f_m(z)$, as

$$f_m(z) = \frac{\substack{\text{actual enthalpy rise (from COBRA)/unit length} \\ \text{(in hot channel at experimental conditions)}}}{\substack{\text{enthalpy rise/unit length in a closed channel} \\ \text{(with same heat input as in experiment)}}} \quad (5\text{-}132)$$

The values of $f_m(z)$ were taken as being constant for small changes of power level in a given channel at a given location. The values of $f_m(z)$ were then used to determine the local hot channel conditions for the CHF prediction procedure.

5.5.6.2 An improved CHF model for low-quality flow. As described in Section 5.3.2.2, the Weisman-Pei model was modified by Lin et al. (1989), employing a mechanistic CHF model developed by Lee and Mudawwar (1988):

$$q''_{crit} = \frac{G_m \delta_m H_{fg}}{L_m} \tag{5-42}$$

Using Eqs. (5-42)–(5-46) in Section 5.3.2.2 with iterative calculations, the predicted CHF were compared with Columbia University data (Fighetti and Reddy, 1983). The comparison was made by examining the statistical results of critical power ratios (DNBRs), where

$$\text{DNBR} = \frac{CP_{pred}}{CP_{meas}}$$

The statistical analysis of DNBRs for the data set gives an indication of the accuracy of the correlation. The three statistical parameters, R_{av} (average value of DNBR), RMS (root-mean-square error), and STD (standard deviation), as conventionally defined, were used by Lin et al. (1989). The COBRA IIIC/MIT-1 core subchannel thermal-hydraulic analysis code* (Bowring and Moreno, 1976) was also used to provide correct local subchannel average conditions for the CHF correlations. Two sets of bundle CHF data were used, consisting of experimental data selected from the Heat Transfer Research Facility databank at Columbia University (Fighetti and Reddy, 1983). The first set of data were collected from six simple bundles with no guide tubes and with uniform axial heat fluxes (test sections 13, 14, 21, 48, 201, and 512). The second set of data were collected from test sections 38, 58, and 60, which had no mixing vanes, but did have guide tubes in the bundle. Test section 38 had a uniform axial heat flux distribution, whereas test section 58 had a distribution highly skewed to the outlet and test section 60 had a distribution highly skewed to the inlet. The parametric ranges covered by these two sets of experimental bundle CHF data are listed in Table 5.14. For the calculation of DNBRs in rod bundles, the predicted critical power can be obtained only by trial-and-error iterations of subchannel analysis with different bundle powers. The simplified procedure used by Reddy and Fighetti (1983) was adopted. The statistical results of DNBRs for the two sets of bundle are shown in Tables 5.15 and 5.16 (Lin et al., 1989). It appears that their work brought the theoretical approach a step closer to the prediction of subcooled flow boiling crisis in bundles.

* The COBRA IIIC/MIT-1 code has an improved numerical scheme, and runs faster than the COBRA IIIC code without sacrificing accuracy.

Table 5.14 Parametric ranges for experimental bundle CHF data

Pressure	9.5–17.0 MPa (1,375–2,460 psia)
Local mass flux	1,252–4,992 kg/m² s (0.9–3.7 × 10⁶ lb/ft² hr)
Subchannel hydraulic diameter	0.0120–0.0136 m (0.04–0.045 ft)
Local equilibrium quality	−0.155 to 0.276
Local void fraction	0.0–0.72
Inlet equilibrium quality	−0.9517 to −0.0089

Table 5.15 Statistical results of DNBR for six simple bundles

Correlation	Number of points	R_{av}	RMS	STD
W-3 corr. [Eq. (5-108)] (Tong, 1967a)	271	1.0038	0.0905	0.0905
Columbia [Eq. (5-128)] (Reddy and Fighetti, 1983)	271	0.9917	0.0721	0.0716
Improved model [Eq. (5-42)] (Lin et al., 1989) ($a_1 = 7,000$)	271	1.0351	0.0833	0.0757
Improved model [Eq. (5-42)] (Lin et al., 1989) ($a_1 = 5,000$)	271	0.9946	0.0737	0.0736

Source: Lin et al. (1989). Copyright © 1989 by American Nuclear Society, LaGrange Park, IL. Reprinted with permission.

5.5.7 A.R.S. CHF Correlation

5.5.7.1 CHF correlation with uniform heating. A correlation for uniformly heated round ducts was proposed by A.R.S. (Clerici et al., 1967; Biasi et al., 1967, 1968). The correlation was claimed to combine a very simple analytical form with a wide range of validity and a great prediction accuracy. The correlation consists of two straight lines in the plane q_o'', X_e:

$$q''_{crit} = \frac{1.883 \times 10^3}{D_a G^{1/6}} \left(\frac{y(P)}{G^{1/6}} - X_e \right) \quad \text{for low-quality region}$$

$$q''_{crit} = \frac{3.78 \times 10^3 h(P)}{D^a G^{0.6}} (1 - X_e) \quad \text{for high-quality region} \tag{5-133}$$

where a is a numerical coefficient equal to 0.4 for $D \geq 1$ cm and 0.6 for $D < 1$ cm (0.39 in.); $y(P)$ and $h(P)$ are two functions that depend only on the pressure, defined as

$$y(P) = 0.7249 + 0.099P \exp(-0.032P)$$

$$h(P) = -1.159 + 0.149P \exp(-0.019P) + \left(\frac{8.99P}{10 + P^2} \right)$$

Table 5.16 Statistical results of DNBR for test sections 38, 58, and 60

Test section	No. of pts.	Correlation	R_{av}	RMS	STD
T 38 Uniform axial flux	42	W-3 [Eq. (5-108)] (Tong, 1967a)	0.9944	0.0772	0.0779
		Columbia [Eq. (5-128)] (Reddy and Fighetti, 1983)	1.0388	0.0629	0.0501
		Improved model [Eq. (5-42)] (Lin et al., 1989)	0.9963	0.0552	0.0558
T 58 Axial flux skew to outlet	65	W-3 [Eq. (5-108)] (Tong, 1967a)	0.9662	0.0895	0.0835
		Columbia [Eq. (5-128)] (Reddy and Fighetti, 1983)	1.0913	0.1092	0.0605
		Improved model [Eq. (5-42)] (Lin et al., 1989)	1.0167	0.0639	0.0622
T 60 Axial flux skew to inlet	75	W-3 [Eq. (5-108)] (Tong, 1967a)	0.9330	0.1099	0.0877
		Columbia [Eq. (5-128)] (Reddy and Fighetti, 1983)	1.0474	0.0751	0.0587
		Improved model [Eq. (5-42)] (Lin et al., 1989)	0.9636	0.0686	0.0585
All data	182	W-3 [Eq. (5-108)] (Tong, 1967a)	0.9590	0.0960	0.0871
		Columbia [Eq. (5-128)] (Reddy and Fighetti, 1983)	1.0611	0.0866	0.0615
		Improved model [Eq. (5-42)] (Lin et al., 1989)	0.9901	0.0641	0.0635

Source: Lin et al. (1989). Copyright © 1989 by American Nuclear Society, LaGrange Park, IL. Reprinted with permission.

P is pressure in atm, X_e is exit quality, G is mass flux in g/cm² s, and Eq. (5-133) gives the CHF in SI units (W/cm²). The intersection point of the two straight lines [Eq. (5-133)] does not occur at constant outlet quality, X_e, but changes with pressure and mass flow rate. It is therefore not possible to define the validity range of the two straight lines a priori, and the predicted burnout point is assumed to be the highest value obtained by the intersection of the first equation with the heat balance equation. The range of validity of the correlation is given below:

0.3 cm < D < 3.75 cm (0.12 in. < D < 1.5 in.)
20 cm < L < 600 cm (7.8 in. < L < 236 in.)
1.7 atm < P < 140 atm (0.17 MPa < P < 14 MPa)
10 g/cm² s < G < 600 g/cm² s (0.074 < G < 4.44 × 10⁶ lb/hr ft²

$$X_{in} < 0$$
$$[1 + (\rho_L/\rho_G)]^{-1} < X_e < 1$$

All the restrictions on G and the lower ones on P, L, and D are effective limitations for the correlation, while the upper restrictions on P, L, and D are introduced only because of the lack of comparison data. The restriction $X_{in} < 0$ is introduced in order to avoid the dependence on the length and the inlet conditions.

5.5.7.2 Extension of A.R.S. CHF correlation to nonuniform heating.

To extend the correlation described in the previous section, the saturation boiling length hypothesis is followed. By means of a straightforward transformation, Eq. (5-133) becomes

$$\frac{(W_{crit,sat})_i}{\dot{m} H_{fg}} = \frac{a_i(L_B)_i}{b_i + (L_B)_i} \quad (i = 1, 2) \tag{5-134}$$

where $W_{crit,sat}$ is the critical saturation power, i.e., the burnout power given to the test section from the zero quality point up to the burnout point; L_B is the saturation boiling length, defined as the tube length along which positive quality is developed up to the burnout point; $\dot{m}$ is the total mass flow rate; and a_i and b_i are monomial functions of geometry, mass flux, diameter, and pressure, defined by the following expressions:

$$a_1 = \frac{y(P)}{G^{1/6}} \qquad a_2 = 1$$

$$b_1 = \frac{H_{fg} D^{\alpha+1} G^{7/6}}{7.532 \times 10^3} \qquad b_2 = \frac{D^{\alpha+1} G^{1.6} H_{fg}}{1.512 \times 10^4 h(P)}$$

According to Silverstri (1966), the burnout condition is then reached for all points z where the input saturation power W_{sat} becomes higher than, or at least equal to, the critical saturation power, $W_{crit,sat}$ given by Eq. (5-134). In other words, the burnout condition can be defined by the inequality

$$W_{sat} = \int_{L_o}^{z} q(z') \, dz' \geq W_{crit,sat}(z) \tag{5-135}$$

where $q(z')$ is the heat flux distribution profile and L_o is the subcooled zone length.

Chopped cosine heat flux distribution For a heat flux distribution of a chopped cosine curve, the flux profile and the subcooled zone length are given by

$$q(z) = q_M \sin\left[\frac{\pi(z + z_o)}{M}\right] \tag{5-136}$$

$$L_o = \frac{M}{\pi} \arccos\left[\cos\left(\frac{\pi z_o}{M}\right) + \left(\frac{\pi \dot{m} H_{fg} X_{in}}{M q_M}\right)\right] \quad (5\text{-}137)$$

where z_o is the middle extrapolated length and $M = (L + 2z_o)$.

Tangency between burnout correlation and the calculated integral of the heat flux distribution can be imposed by the following system of equations:

$$q_M\left(\frac{M}{\pi}\right)\left\{\cos\left[\pi \frac{(L_o + z_o)}{M}\right] - \cos\left[\frac{\pi(z + z_o)}{M}\right]\right\} = \frac{a_i \dot{m} H_{fg}(z - L_o)}{b_i + z - L_o} \quad (5\text{-}138)$$

$$q_M \sin\left[\frac{\pi(z + z_o)}{M}\right] = \frac{a_i \dot{m} H_{fg} b_i}{(b_i + z - L_o)^2} \quad (5\text{-}139)$$

By combining Eqs. (5-138) and (5-139) the following second-order equation in q_M is obtained:

$$K_3 (q_M)^2 - 2 K_1 q_M + K_2 = 0 \quad (5\text{-}140)$$

where

$$K_1 = \frac{M \dot{m} H_{fg}}{\pi} \left(\left\{\cos\left(\frac{\pi z_o}{M}\right) - \cos\left[\frac{\pi(z + z_o)}{M}\right]\right\}(a_i - X_{in}) + \frac{\pi a_i b_i}{2M} \sin\frac{\pi(z + z_o)}{M}\right)$$

$$K_2 = [\dot{m} H_{fg}(a_i - X_{in})]^2$$

$$K_3 = \left\{(M/\pi)\left[\cos\left(\frac{\pi z_o}{M}\right) - \cos\frac{\pi(z + z_o)}{M}\right]\right\}^2$$

Between the two roots of Eq. (5-140), one must take into consideration the higher value in order to assure the positivity of $(z - L_o)$. Substituting this value of q_M into Eq. (5-138) yields a transcendental equation; its solution gives the value of z corresponding to the burnout point.

5.5.7.3 Comparison of A.R.S. correlation with experimental data. A validity test of the A.R.S. correlation was performed with 219 experimental burnout data points for nonuniform heating. The data have been arranged in groups with reference to their source in Table 5.17. For every group, the original number of experimental points, n_{ori}, the number of tested runs, n_t, which are within the validity range, and the variation of the most important parameters, G, X_{in}, P, L, and D, have also been reported. The last column of the table gives the error percent over the critical power defined as

Table 5.17 Comparison of ARS correlation with experimental data for nonuniform heating

Source of data	n_{ori}	n_t	G, g/cm² s	X_{in}	P, atm	L, cm	D, cm	ε, %
Lee (1965): AEEW R 355	32	32	200/406	−0.3558/−0.0247	65.5/69	366	0.947	4.89
Lee and Obertelli (1963): AEEW R 309	44	44	100/403	−0.4165/−0.0352	67.3/110	183	0.373	8.45
Lee (1966b): AEEW R 479	66	33	160/403	−0.3208/−0.0052	85/122	101	1.587	6.99
Lee (1966b): AEEW R 479	38	38	161/270	−0.3025/−0.065	85/122	101	1.587	9.10
Bertoletti et al. (1964a): CISE R 90	34	28	109/393	−0.3281/−0.0020	71	65	0.807	7.50
MIT (1964)	44	44	67/270	−0.3295/−0.0603	6.8/15.4	76	0.544	9.43

Source: Biasi et al. (1968). Copyright © 1968 by Elsevier Science Ltd., Kidlington, UK. Reprinted with permission.

$$\varepsilon = 100 \times \left(\frac{W_c - W_m}{W_m}\right)\%$$

where W_c and W_m are predicted and measured critical power, respectively, and the overall rms error is $[\Sigma (\varepsilon_i)^2/n_t]^{1/2}$.

Every group of data from Lee (1965, 1966), Lee and Obertelli (1963), Bertoletti et al. (1964), and MIT (1964) is well described by the correlation, and the overall rms error is 7.95%. These data have been represented in Figure 5.74, where the calculated critical power, W_c, is given as a function of the measured critical power, W_m. Included are three different types of axial flux distribution: a chopped cosine distribution, and linearly increasing and linearly decreasing distributions. It can be seen that nearly all the data examined (81%) lie between the two lines delimiting percent error of ±10%; 45% of the data exhibit an error of less than 5%. The predicted and measured CHF locations for data of Lee (1965) and Lee and Obertelli (1963) are also presented in Tables 5.18 and 5.19. The parametric ranges of the data are given in Table 5.20.

5.5.8 Effects of Boiling Length: CISE-1 and CISE-3 CHF Correlations

5.5.8.1 CISE-1 correlation. CISE (Bertoletti et al., 1964a, 1965) reported that the upstream hydrodynamic effect on boiling crisis in a quality region can be expressed by a boiling length, L_B, defined as the tube length starts from flow saturation to the point of dryout, as shown in Figure 5.19. It can be seen that the boiling length in a uniform flux tube with a subcooled inlet (or a two-phase inlet) is a function of critical power (the power where dryout occurs). This concept was verified analytically by Lahey and Gonzalez Santalo (1977) in Eqs. (5-33) and (5-34). CISE (Ber-

434 BOILING HEAT TRANSFER AND TWO-PHASE FLOW

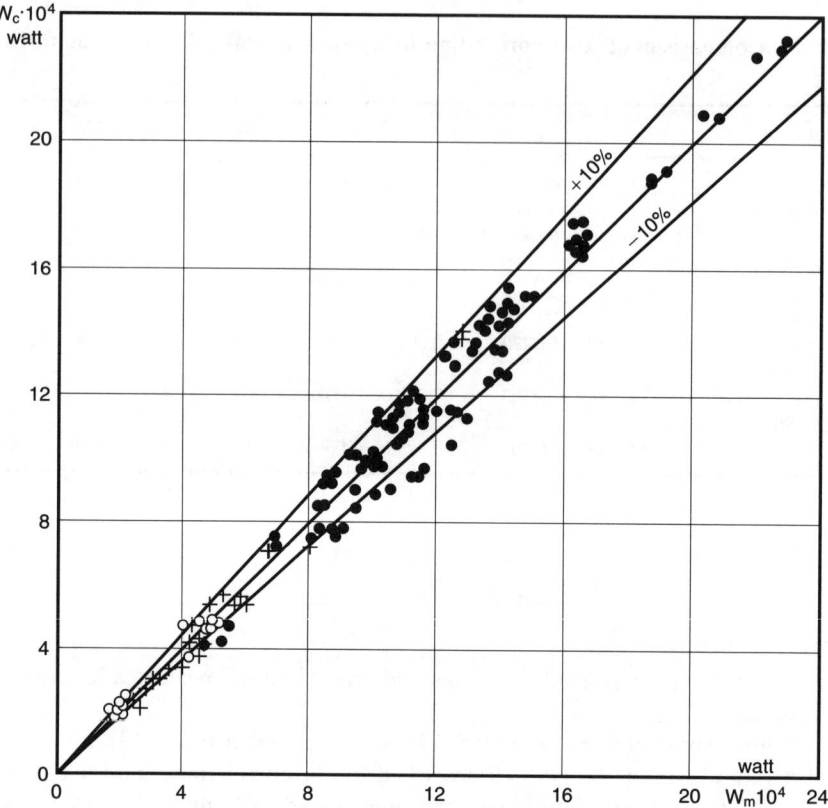

Figure 5.74 Trend of W_c as a function of W_m for data of Lee (1965, 1966b), Lee and Obertelli (1963), Bertoletti et al. (1964), and MIT (1964). (From Biasi et al., 1968. Copyright © 1968 by Elsevier Science Ltd., Kidlington, UK. Reprinted with permission.)

toletti et al., 1965) correlated critical quality, X_{cr}, with boiling length, L_B, in a saturated annular flow as shown in Figure 5.75. Hence the CISE-1 correlation can be written as

$$X_{cr} = \frac{W_B}{GA_c H_{fg}} = \frac{a}{1 + b/L_B} \tag{5-141}$$

where the parameters a and b are, with specified units as shown,

$$a = \frac{(1 - p/p_{cr})}{(G/100)^{1/3}} \quad \text{(CGS units)}$$

$$b = 0.315 \left[\left(\frac{p_{cr}}{p}\right) - 1 \right]^{0.4} G(D)^{1.4} \quad \text{(CGS units)}$$

In the case of uniform axial power distribution, Eq. (5-141) becomes

Table 5.18 AEEW-R-355 burnout locations (cm from the inlet)

Run no.	L_c (calc.)	L_m (meas.)
61	366	315–356
62	366	315–356
64	361	336–356
65	361	336–356
70	359	336–356
71	359	336–356
67	356	336–356
68	355	315–356
56	360	315–356
52	355	315–356
53	356	336–356
49	352	315–356
50	353	315–356
44	352	315–356
46	336	315–336
41	352	315–335
42	352	294–356
39	346	315–356
31	359	315–355
32	340	315–356
33	340	315–356
20	343	295–356
21	343	295–315
17	335	295–315
18	335	295–356
28	326	295–315
29	327	275–315
26	322	295–356
27	322	275–335
23	318	275–335
73	318	235–335
75	318	265–356

Source: Lee (1965). Copyright © 1965 by AEA Technology, Didcot, UK. Reprinted with permission.

$$\frac{W_{crit}}{GA_c H_{fg}} = \frac{a - X_{in}}{1 + b/L} \tag{5-142}$$

or

$$q''_{cr} = \frac{0.794 H_{fg}}{[(p_{cr}/p) - 1]^{0.4} D^{1.4}} \left[\frac{1 - (p/p_{cr})}{(G/100)^{1/3}} - X_o \right] \tag{5-143}$$

Table 5.19 AAEW-R-309 burnout locations (cm from the inlet)

Run no.	L_c (calc.)	L_m (meas.)	Run no.	L_c (calc.)	L_m (meas.)
1	167	152–178	23	150	147–178
2	167	152–168	24	150	147–178
3	167	152–168	25	182	147–178
4	160	152–178	26	182	152–178
5	161	152–178	27	156	147–178
6	161	147–178	28	156	147–157
7	163	152–178	29	156	147–157
8	163	147–178	30	156	147–178
9	163	147–178	31	175	136–168
10	156	147–178	32	175	147–178
11	156	136–178	33	175	147–168
12	156	147–178	34	164	147–157
13	150	136–168	35	182	147–178
14	150	136–178	36	182	147–178
15	150	147–178	37	158	147–168
16	139	136–178	38	158	147–168
17	139	136–178	39	182	147–157
18	139	136–178	40	182	147–178
19	182	136–178	41	182	147–168
20	182	136–178	42	182	147–168
21	131	147–178	43	182	147–168
22	131	147–178	44	182	147–168

Source: Lee and Obertelli (1963). Copyright © 1963 by AEA Technology, Didcot, UK. Reprinted with permission.

This correlation was based on approximately 4,000 experimental data points, and the corresponding inaccuracy was evaluated to be about ±15%. The following nomenclature is used in the CISE correlations with appropriate units:

D = diameter = $4A_c/P$, cm
f_b = average-to-maximum heat flux ratio, dimensionless
G = mass flux, g/cm² s
H = enthalpy, J/g
H_{fg} = latent heat of vaporization, J/g
L = overall heated length, cm
L_B = boiling length, cm
n = number of rods, dimensionless
P = perimeter, cm
p = pressure, kg cm²
W_B = power over the boiling length, kW
W_{crit} = overall power at CHF, kW
X_{cr} = quality (weight percentage) at CHF, dimensionless

Table 5.20 Parametric ranges of the data used

Local mass flux	$0.2\text{--}4.1 \times 10^6$ lb/ft^2 hr (270–5,560 kg/m^2 s)
Pressure	200–2,450 psia (1.38–16.8 MPa)
Local quality	-0.25 to 0.75
Inlet quality	-1.1 to 0
Hydraulic diameter	0.35–0.55 in. (0.89–1.4 cm)
Heated diameter	0.25–0.55 in. (0.64–1.4 cm)
Length	30–168 in. (76–427 cm)
Rod diameter	0.38–0.68 in. (1–1.7 cm)
Number of rods	3×3 to 5×5
Radial profile	Uniform and nonuniform (radial and corner peaking)
Axial profile	Uniform
Subchannel type	Matrix channels only
Rod bundle type	PWR and BWR (with and without unheated rods)

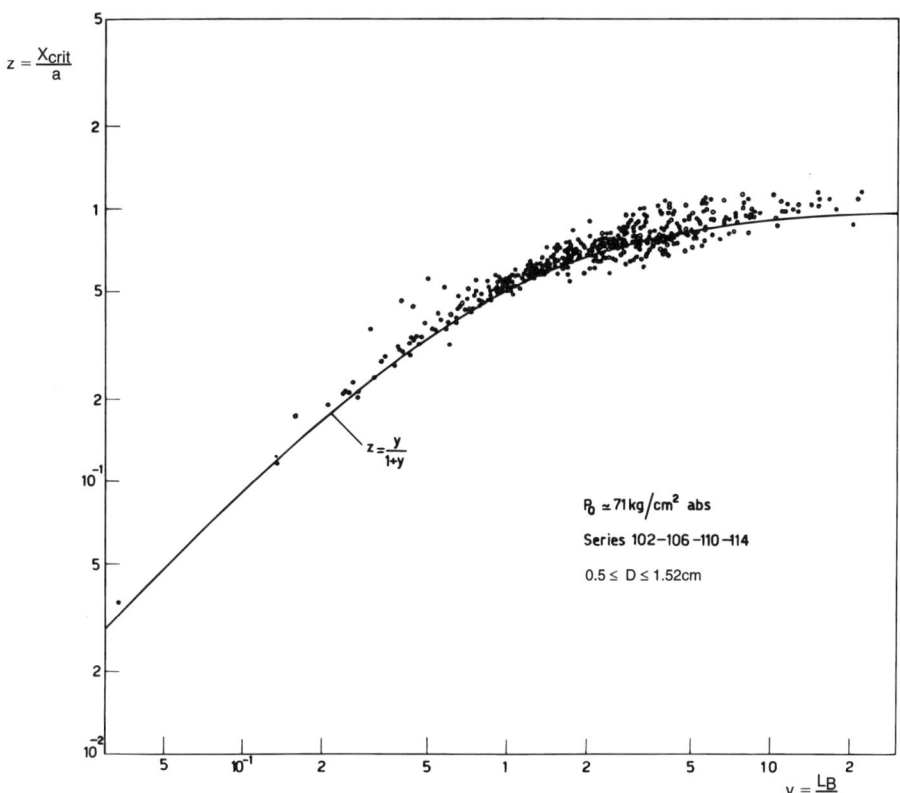

Figure 5.75 CISE experimental data on critical quality, X_{crit}. L_B = boiling length. (From Bertoletti et al., 1965. Copyright © 1965 by CISE, Milan, Italy. Reprinted with permission.)

438 BOILING HEAT TRANSFER AND TWO-PHASE FLOW

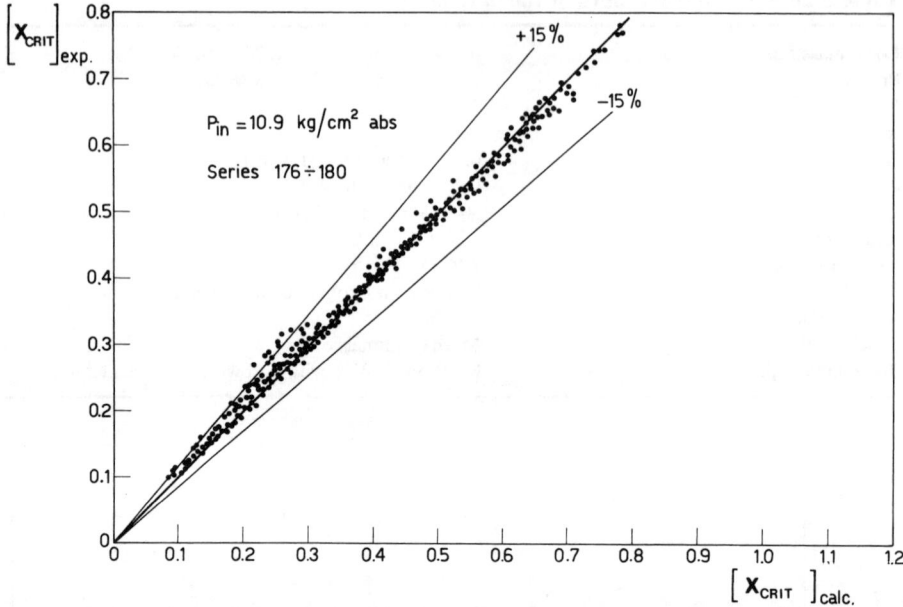

Figure 5.76 AEEW experimental data obtained with Freon-12 checked against the correlation [Eq. (5-141)]. (From Bertoletti et al., 1965. Copyright © 1965 by CISE, Milan, Italy. Reprinted with permission.)

ΔH_{in} = inlet subcooling, J/g
q'' = heat flux, W/cm^2
A_c = cross-sectional flow area, cm^2

Freon-12 data (Stevens and Kirby, 1964) obtained at 10.9 kg cm^2 (p/p_{cr} = 0.27) led to the same expression, where b becomes

$$b_{(Freon)} = (0.600)G\left[\left(\frac{p_{cr}}{p}\right) - 1\right]^{0.4} D^{1.4} \qquad (5\text{-}144)$$

A comparison of measured CHF and predicted heat flux is given in Figure 5.76.

In an annular test section, the flow channel cross session is subdivided into two subchannels surrounding each solid surface. Round tube correlation is applied to each subchannel, i, to obtain W_{Bi}. The correlation for an annulus then becomes

$$\frac{W_{Bi}}{GA_{ci}H_{fg}} = \frac{1 - (p/p_{cr})}{(G/100)^{1/3}}\left[1 + \left(\frac{0.315G}{L_B}\right)(D_i)^{1.4}\left(\frac{p_{cr}}{p} - 1\right)^{0.4}\right]^{-1} \qquad (5\text{-}145)$$

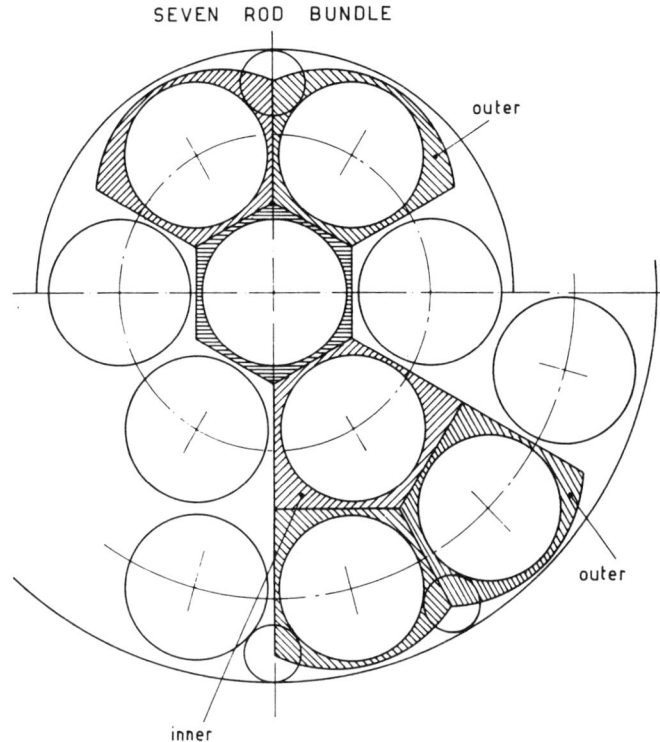

Figure 5.77 Subchannels in 7- and 19-rod bundles.

In the case of a single heated surface with uniform axial power distribution, Eq. (5-145) becomes

$$\frac{W_B}{GA_c H_{fg}} = \frac{1-(p/p_{cr})}{(G/100)^{1/3}} \frac{A_{ci}}{A_{c,\text{tot}}} \left[1 + \left(\frac{b_i}{L_B}\right)\right]^{-1} \qquad (5\text{-}146)$$

or

$$\frac{W}{GA_c H_{fg}} = \left[\frac{1-(p/p_{cr})}{(G/100)^{1/3}} \frac{A_{ci}}{A_{c,\text{tot}}} - X_{\text{in}}\right]\left[1 + \left(\frac{b_i}{L}\right)\right]^{-1} \qquad (5\text{-}147)$$

5.5.8.2 CISE-3 correlation for rod bundles (Bertoletti et al., 1965). If the rod bundle's cross section is divided into subchannels as shown in Figure 5.77, the CISE-3 correlation for rod bundles with nonuniform axial and radial power distributions is

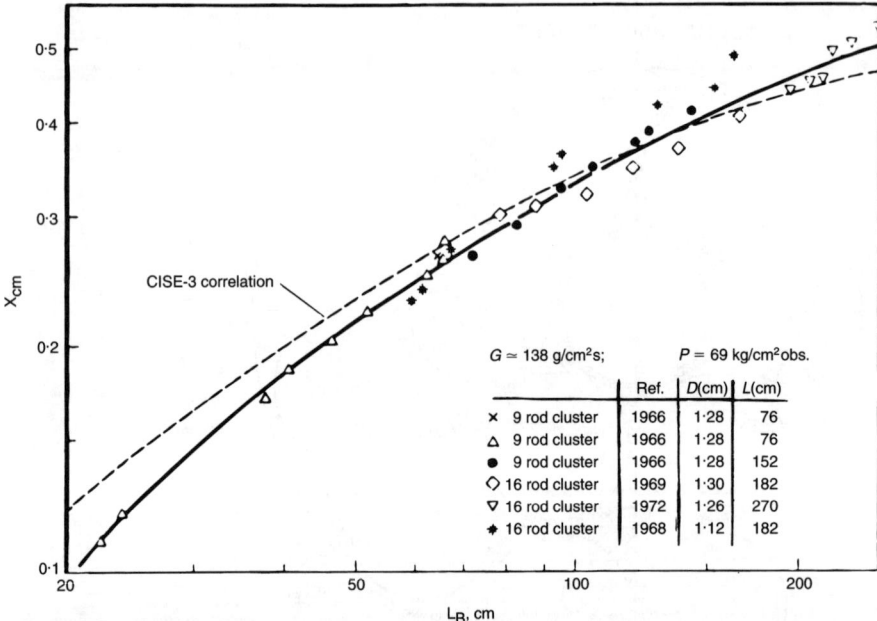

Figure 5.78 Comparison of square lattice rod bundle data from different sources and CISE-3 correlation, Eq. (5-148). Data sources: X, Δ, ●, 9-rod bundle (Hench and Boehm, 1966); ◊, 16-rod bundle (Janssen et al., 1969); ▽, 16-rod bundle (Evangelisti et al., 1972); ✻, 16-rod bundle (Israel et al., 1968). (From Evangelisti et al., 1972. Copyright © 1972 by Elsevier Science Ltd., Kidlington, UK. Reprinted with permission.)

$$\frac{W_B}{GA_c H_{fg}} = \frac{(1 - p/p_{cr})}{(G/100)^{1/3}} \frac{A_{ci} n}{A_{c,tot}} \frac{(f_b/B_i)}{[1 + (0.315 G/L_B)(p_{cr}/p - 1)^{0.4}(D_i)^{1.4}]} \quad (5\text{-}148)$$

where i refers to the rod affected by the crisis and

$$f_{b,i} = \frac{\overline{q}_i''}{q_{max}''}$$

$$B_i = \frac{\overline{q}_i''}{\overline{q}_{bundle}''}$$

n = number of heated rods

In the case of uniform axial heat flux distribution,

$$\frac{W_{cr}}{GA_c H_{fg}} = \left[\frac{(1 - p/p_{cr})}{(G/100)^{1/3}} \frac{A_{ci}}{A_{c,tot}} \frac{nf_b}{B_i} - X_{in}\right] \bigg/ \left[1 + \left(\frac{0.315 G}{L}\right)\left(\frac{p_{cr}}{p} - 1\right)^{0.4} D_i^{1.4}\right]$$

$$(5\text{-}149)$$

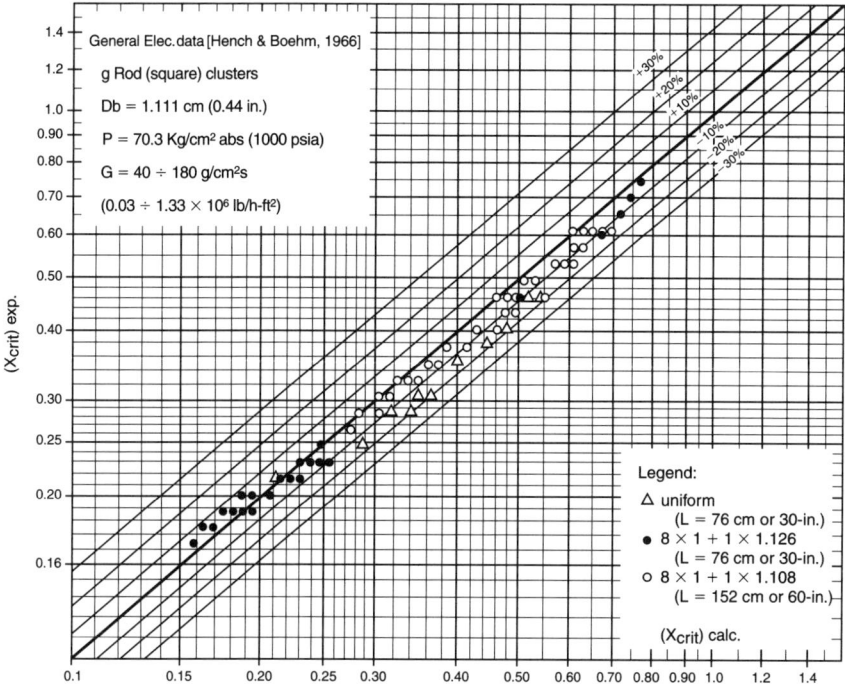

Figure 5.79 CISE-3 correlation predicted X_{cr} compared with experimentally measured X_{cr}. (From Gasperi et al., 1968. Copyright © 1968 by CISE, Milan, Italy. Reprinted with permission.)

The rod bundle CHF data were compared with CISE-3 predictions as shown in Figures 5.78 and 5.79.

5.5.9 GE Lower-Envelope CHF Correlation and CISE-GE Correlation

5.5.9.1 GE lower-envelope CHF correlation. The General Electric Company's design equations for predicting CHF based on the lower envelope of its data (Janssen and Levy, 1962) are as follows. For pressure at 1,000 psia (6.7 MPa):

$$\frac{q''_{cr}}{10^6} = 0.705 + \frac{0.237G}{10^6} \quad \text{(for } X < X_1\text{)} \qquad (5\text{-}150)$$

$$\frac{q''_{cr}}{10^6} = 1.634 - \frac{0.270G}{10^6} - 4.710X \quad \text{(for } X_1 < X < X_2\text{)} \qquad (5\text{-}151)$$

$$\frac{q''_{cr}}{10^6} = 0.605 - \frac{0.164G}{10^6} - 0.653X \quad \text{(for } X_2 < X\text{)} \qquad (5\text{-}152)$$

where

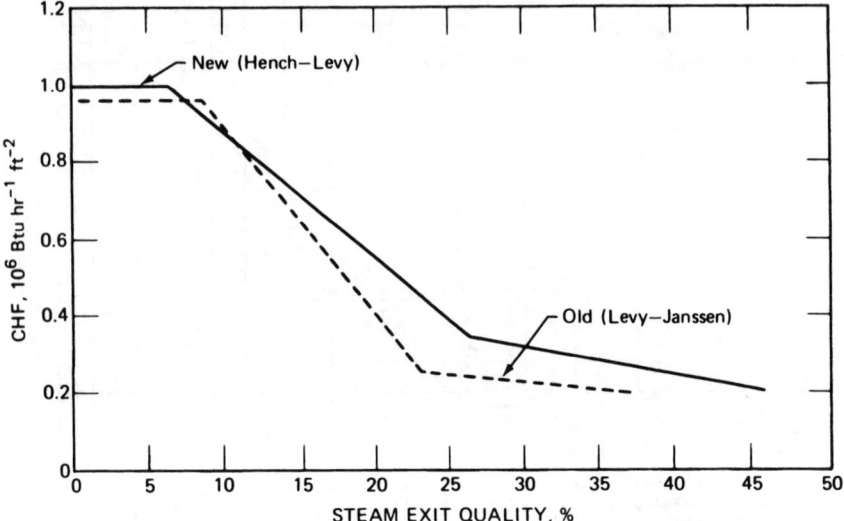

Figure 5.80 General Electric correlations for CHF. Curves are typical hot-channel mass flux of 1.0 × 10⁶ lb/hr ft²). (From Roy, 1966. Copyright © 1966 by Nucleionics Week, Washington, DC. Reprinted with permission.)

$$X_1 = 0.197 - \frac{0.108G}{10^6}$$

$$X_2 = 0.254 - \frac{0.026G}{10^6}$$

For other pressures,

$$q''_{cr}(\text{at } p \text{ psia}) = q''_{cr}(\text{at } 1{,}000 \text{ psia}) + 440(1{,}000 - p) \tag{5-153}$$

The ranges of parameters are

p = 600–1,450 psia (4.1–9.7 MPa)
G = 0.4–6.0 × 10⁶ lb/hr ft² (540–8,108 kg/m² s)
X = negative to 0.45 (negative to 0.45)
D_e = 0.245–1.25 in. (0.62–3.18 cm)
L = 29–108 in. (0.95–3.5 m)

General Electric Company (Roy, 1966) published a part of its new (Hench-Levy) correlation as shown in Figure 5.80.

Slifer and Hench (1971) reported the General Electric Company's lower-envelope correlation for low-mass-flux CHF at pressures less than 1,000 psia (6.7 MPa). For $G < 0.5 \times 10^6$ lb/hr ft² (<675 kg/m² s),

$$\frac{q''_{cr}}{10^6} = 0.84 - X \tag{5-154}$$

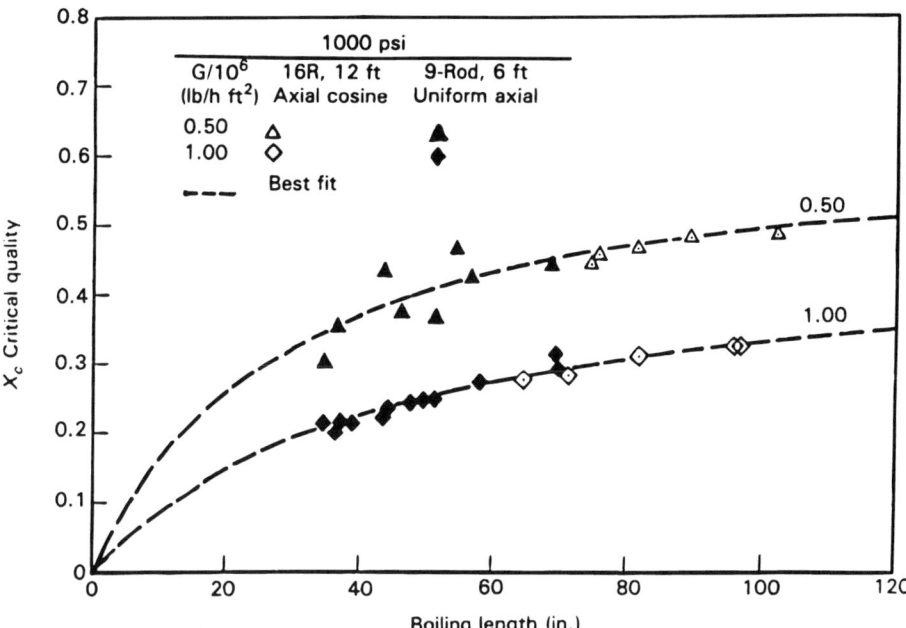

Figure 5.81 Nine- and 16-rod-bundle critical quality versus boiling length at 1,000 psia (6.8 MPa). (From Lahey and Moody, 1977. Copyright © 1977 by American Nuclear Society, LaGrange Park, IL. Reprinted with permission.)

For $0.5 \times 10^6 \leq G \leq 0.75 \times 10^6$ lb/hr ft² (1,012 kg/m² s),

$$\frac{q''_{cr}}{10^6} = 0.80 - X \quad (5\text{-}155)$$

Although these equations were developed from steady-state CHF data, they are also suggested for use in flow depressurization during a loss-of-coolant accident (Slifer and Hench, 1971).

5.5.9.2 GE approximate dryout correlation (GE Report, 1975). The GE rod bundle data (Lahey and Moody, 1977), shown in Figure 5.81, indicate that critical quality is indeed a function of boiling length, as suggested earlier by the CISE-1 correlation, Eq. (5-141). This relationship is also demonstrated by B & W round-tube data as plotted in Figure 5.82. Based on a boiling-length concept, GE in 1973 developed its proprietary GEXL critical power correlation (GE Report, 1973). The GEXL is used for GE core design, replacing previous correlations (Janssen and Levy, 1962; Roy, 1966).

For convenience of outside design reviewers, GE has also developed an approximation of its proprietary core design correlation, called the CISE-GE correlation (GE Report, 1975), which follows the format of the CISE-1 correlation but uses different constants:

444 BOILING HEAT TRANSFER AND TWO-PHASE FLOW

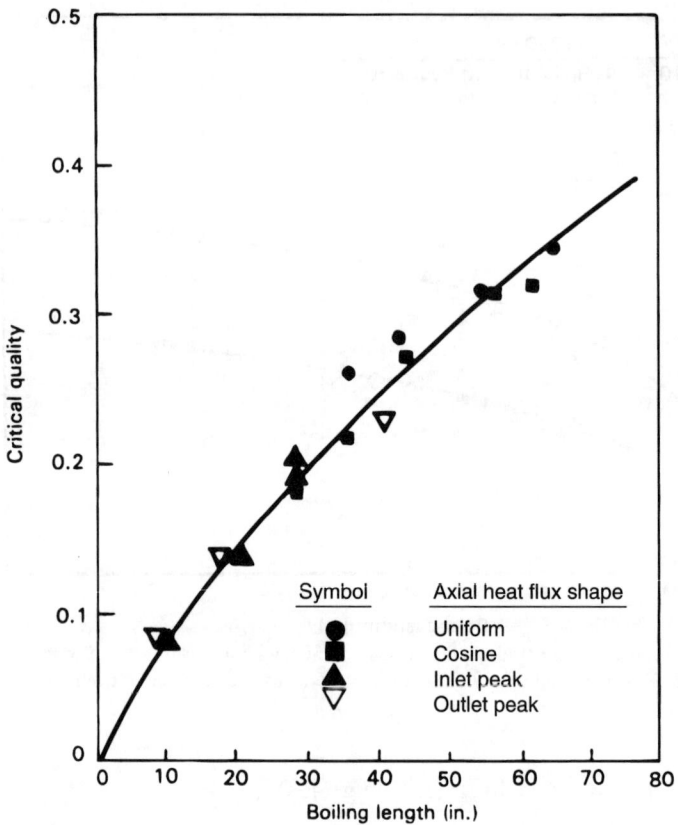

Figure 5.82 Critical quality versus boiling length, data from B&W. (From GE Report, 1973. Copyright © 1973 by General Electric Co., San Jose, CA. Reprinted with permission.)

$$X_{cr} = \frac{AL_B}{B + L_B}\left(\frac{1.24}{R_f}\right)^{1/2} \tag{5-156}$$

where R_f is the radial peaking factor, which should be determined experimentally. Its value is 1.24 in a typical hot assembly. For $100 < G \leq 1{,}400$ kg/m² s ($7.4 \times 10^4 < G \leq 1 \times 10^6$ lb/ft² hr),

$$p \approx 6.9 \times 10^6 \text{ Pa } (1{,}000 \text{ psia})$$

$$A = 1.055 - \left(\frac{0.013}{2.758 \times 10^6}\right)[(p - 4.137 \times 10^6)^2]$$
$$- 1.233(7.37 \times 10^{-4}G) + 0.907(7.37 \times 10^{-4}G)^2$$
$$- 0.205(7.37 \times 10^{-4}G)^3$$

$$B = 0.457 + 2.003(7.37 \times 10^{-4}G) - 0.901(7.37 \times 10^{-4}G)^2$$

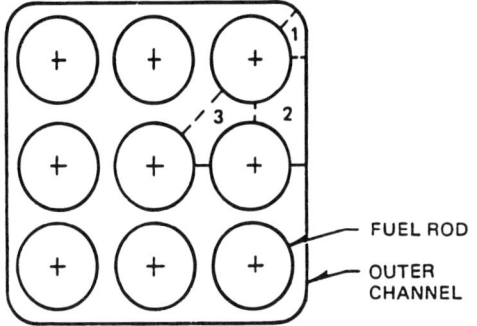

Figure 5.83 Square (BWR-type) lattice subchannels.

In Equation (5-156), it is shown that a bundle-average-type correlation, such as GEXL, is able to predict local burnout point accurately, provided a generalized local peaking parameter, R_f, is properly formulated. Correlations of this type are empirical and, thus, generally are valid only for conditions representative of the data on which they are based. If a designer is asked to investigate the thermal performance of a proposed new lattice with different lattice dimensions and configuration, the empirical bundle-average correlation is normally of little help.

To optimize a given lattice, to investigate a proposed new lattice, or to look at some abnormal lattice configuration, we normally must use subchannel techniques. The essence of subchannel analysis is shown in Figure 5.83. The rod bundle is divided into a number of ventilated flow tubes (subchannels). In the two cases shown in Figure 5.83, it is assumed that the local peaking factor (i.e., power) on each rod is the same, and thus only a small section of the bundle needs to be analyzed; the remainder is known by symmetry. By convention, subchannels 1, 2, and 3 in Figure 5.83 are known as the corner, side, and center subchannels, respectively. In the nine-rod bundle shown, symmetry yields eight corner subchannels, side, and center subchannels.

A review of the available data (Lahey and Schraub, 1969) indicated that there is an observed tendency for the vapor to seek the less obstructed, higher-velocity regions of a BWR fuel rod bundle. This tendency can be seen in the adiabatic data shown in Figure 5.84, where it is noted that the flow quality is much higher in the more open interior (center) subchannels than in the corner and side subchannels. This indicates the presence of a thick liquid film on the channel wall and an apparent affinity of the vapor for the more open subchannels. Later diabatic subchannel data (Lahey et al., 1971, 1972) confirmed this observation. The isokinetic steam–water data presented in Figure 5.84 shows clearly that, even though the power-to-flow ratio of the corner subchannel is the highest of any subchannel in the bundle, the quality in the corner subchannel is the lowest, while that in the center subchannels is the highest. Moreover, the enhanced turbulent two-phase mixing that occurs near the slug–annular transition point [~10% quality at 1,000 psia (6.9 MPa)] is seen clearly in Figures 5.85 and 5.86.

446 BOILING HEAT TRANSFER AND TWO-PHASE FLOW

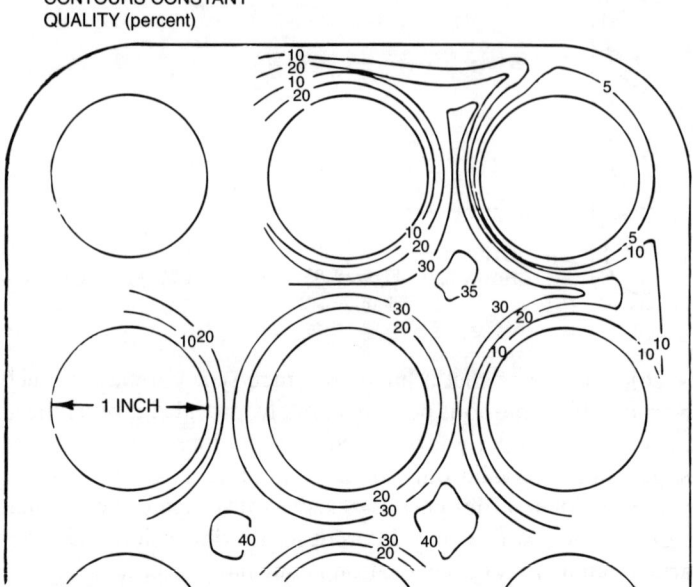

Figure 5.84 Quality contours from isokinetic probe sampling of air–water flow in a 9-rod array. (From Schraub et al., 1969. Copyright © 1969 by General Electric Co., San Jose, CA. Reprinted with permission.)

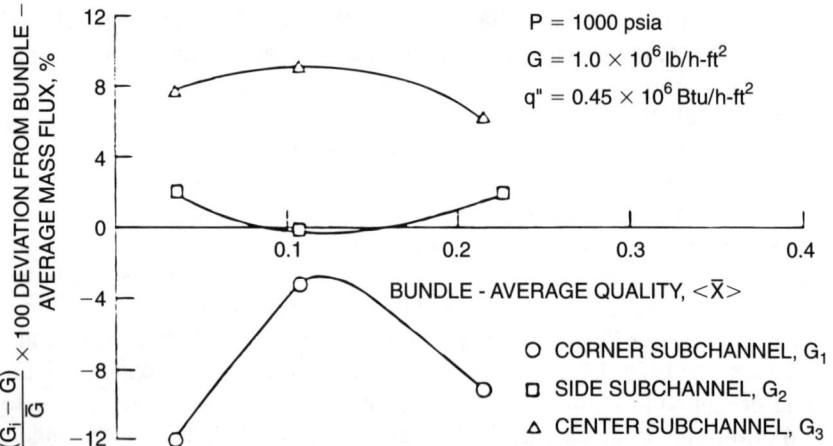

Figure 5.85 Comparison of subchannel flows for three subchannels (corner, side, and center). (From Lahey et al., 1971. Copyright © 1971 by American Society of Mechanical Engineers, New York. Reprinted with permission.)

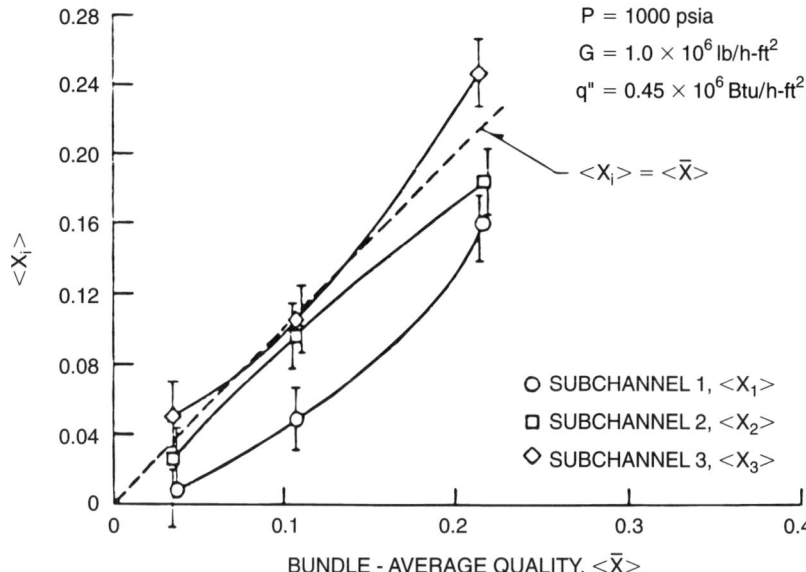

Figure 5.86 Variation of subchannel qualities with average quality for three subchannels (corner, side, and center). (From Lahey et al., 1971. Copyright © 1971 by American Society of Mechanical Engineers, New York. Reprinted with permission.)

In the derivation of a model, the basic approach is always to consider the conservation of mass, momentum, and energy in each subchannel. Figure 5.84 shows that not only axial effects need to be considered, but also the transverse (radial) interchange of mass, momentum, and energy across the imaginary interfaces dividing subchannels. These transverse interchange phenomena are quite complicated and difficult to decompose into more elementary interchange terms. Nevertheless, they normally are decomposed arbitrarily into three components:

1. Flow diversions that occur due to imposed transverse pressure gradients
2. Turbulent (eddy diffusivity) mixing, which occurs due to stochastic pressure and flow fluctuations
3. "Void drift" that, as discussed previously, is apparently due to the strong tendency of the two-phase system to approach equilibrium conditions

A detailed subchannel analysis for BWRs is described by Lehey and Moody (1977).

5.5.10 Whalley Dryout Predictions in a Round Tube (Whalley et al., 1973)

The mass balances on the liquid film and on the entrained liquid in the core of an annular flow can be written as

$$\frac{dG_F}{dz} = 2\pi r\left[G_D - G_E - \left(\frac{q''}{H_{fg}}\right)\right] \tag{5-157}$$

where G_F is the liquid mass flux (in lb/hr ft²) in the liquid film along the wall at z. The integration of G_F should be continued until the liquid film flow rate becomes zero, when it is assumed that dryout has occurred. Whalley (1976) extended the application of his dryout correlation, Eq. (5-157), to rod bundles. The differential equations for the liquid film flow rate and for the entrained liquid film flow rate, W_{LE}, are similar to those for the round tube, except that for the entrained liquid flow rate there are some extra terms to the equation.

First, there is a term to account for turbulent gas-phase mixing between adjacent subchannels. This is accounted for by a term that has the form of a concentration difference between the subchannels multiplied by a mass transfer coefficient and the area available for transfer. This representation was used, as it is similar to the equation used for deposition.

There are also two terms to account for the flow of entrained liquid carried around the bundle by crossflow of gas. Two terms are necessary because if a particular gas crossflow is reversed, then the subchannel from which the flow originates (the donor subchannel) has a different concentration of entrained liquid drops in the gas phase. It is assumed here that the crossflow of gas carries liquid droplets with it, and that the droplets adopt the same crossflow velocities as the gas. It is certain that the droplets do not behave in this way, but the effect of their actual behavior is not known. The equations then are

$$\frac{dW_{LEi}}{dz} = S_{Si}(G_{Ei} - G_{Di})$$

$$+ \sum_{j=1}^{N} k_{cij} S_{Fij}(C_j - C_i) n_{ji} \quad \text{turbulent mixing between subchannels}$$

$$+ \sum_{j=1}^{N} a_{ij} \frac{W_{LEj}}{W_{Gi}} J_{Gji} n_{ji}$$

transfer of entrained liquid between subchannels due to gas crossflow

$$+ \sum_{j=1}^{N} b_{ij} \frac{W_{LEi}}{W_{Gj}} J_{Gji} n_{ji} \tag{5-158}$$

where W_{Lei} = entrained liquid flow rate in each subchannel of type i
S_{Si} = solid surface per unit length in each subchannel of type i
G_{Ei} = entrainment rate from solid surface in a subchannel of type i
G_{Di} = deposition rate from gas core onto solid surface in a subchannel of type i
N = number of subchannel types

k_{cij} = mass transfer coefficient for turbulent interchange of entrained drops between subchannels of type i and type j

S_{Fij} = area of boundary in the fluid per unit length between each subchannel of type i and each subchannel of type j

C_i = concentration of entrained droplets in the gas core of a subchannel of type i

n_{ij} = number of subchannels of type i bordering on each subchannel of type j; n_{ij} is thus zero if subchannel types i and j are not adjacent. Note that n_{ij} is not necessarily equal to $-n_{ji}$.

W_{Gi} = gas flow rate in each subchannel of type i

J_{Gij} = crossflow of gas from each subchannel of type i to each subchannel of type j per unit length of the bundle; J_{Gij} is zero if subchannel types i and j are not adjacent. Note that $J_{Gij} = -J_{Gji}$.

a_{ij}, b_{ij} = constants taking the value of zero or unity:
when $J_{Gij} > 0$, then $a_{ij} = 1, b_{ij} = 0$
when $J_{Gij} < 0$, then $a_{ij} = 0, b_{ij} = 0$
when $J_{Gij} = 0$, then $a_{ij} = 0, b_{ij} = 0$

Then

$$\frac{dW_{LEi}}{dz} = S_{Si}\left[G_{Di} - G_{Ei} - \left(\frac{q_i''}{H_{fg}}\right)\right] \quad (5\text{-}159)$$

where q_i'' is the heat flux on the solid surface of a subchannel of type i. For each type of subchannel, there should be one equation of the form of Eq. (5-158) and one of the form of Eq. (5-159). The subchannels of a 37-rod bundle are rod-centered as shown in Figure 5.87, and their connections are shown in Figure 5.88. The calculated dryout powers for this bundle are shown in Figure 5.89. Comparing with the measured values, the calculated values are shown to be lower. This difference may be due to neglecting the flow mixing effect of the spacer grids.

5.5.11 Levy's Dryout Prediction with Entrainment Parameter

Levy, Healzer, and Abdollahian (1980) predicted the dryout flux in vertical pipes by a semiempirical adiabatic model (Levy and Healzer, 1980) for liquid film flow and entrainment. It starts with a heat balance along the flow direction and a mass balance perpendicular to the flow direction:

$$G(\Delta H_{sub} + XH_{fg}) = \frac{2}{r}\int_0^z q'' \, dz \quad (5\text{-}160)$$

$$\frac{dG_F}{dz} = 2\pi r\left[G_D - G_E - \left(\frac{q''}{H_{fg}}\right)\right] \quad (5\text{-}161)$$

450 BOILING HEAT TRANSFER AND TWO-PHASE FLOW

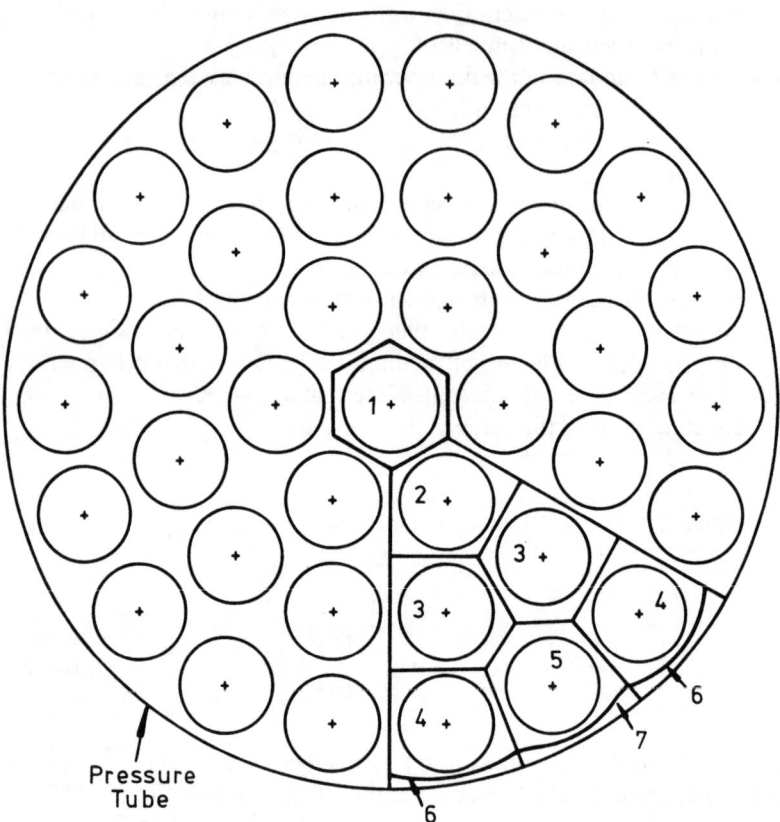

Figure 5.87 Subchannels in a 37-rod bundle. (From Whalley, 1976. Reprinted with permission of UK AERE, Harwell, Kent, UK.)

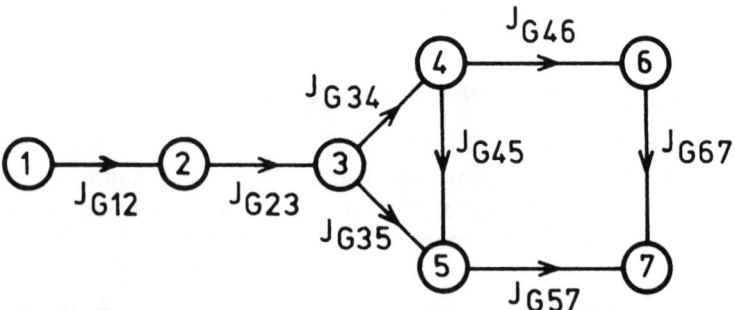

Figure 5.88 Subchannel connection for 37-rod bundle. (From Whalley, 1976. Reprinted with permission of UK AERE, Harwell, Kent, UK.)

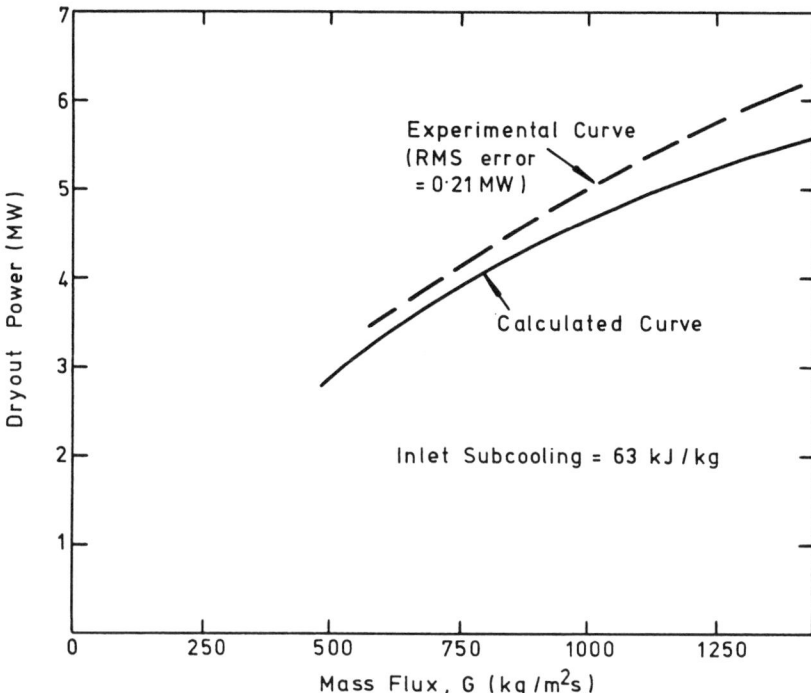

Figure 5.89 Experimental and calculated results for dryout power in a 37-rod bundle. (From Whalley, 1976. Reprinted with permission of UK AERE, Harwell, Kent, UK.)

and

$$G_D = kC_E \tag{5-162}$$

$$G_E = kC_{E,\text{eq}} \tag{5-163}$$

Instead of using a constant value of the mass transfer coefficient k at each pressure given by Harwell (Walley et al., 1973), Levy and Healzer (1980) developed an entrainment parameter β. G_F and β are evaluated by solving the following two equations simultaneously:

$$G(1 - X_{\text{eq}}) - G_F = G(1 - X_{\text{eq}})(\beta)^{-1/2} \tag{5-164}$$

where G is the total mass flow rate in an adiabatic flow, and X_{eq} = equilibrium quality, and

$$\beta = 1 + \left\{ \left[\left(\frac{\rho_L}{\rho_G} \right)^{1/\beta} - 1 \right] \left(\frac{\sigma}{k_r} \right) \frac{\rho_L}{(GX_{\text{eq}})^2} \right\}^{1/2} \tag{5-165}$$

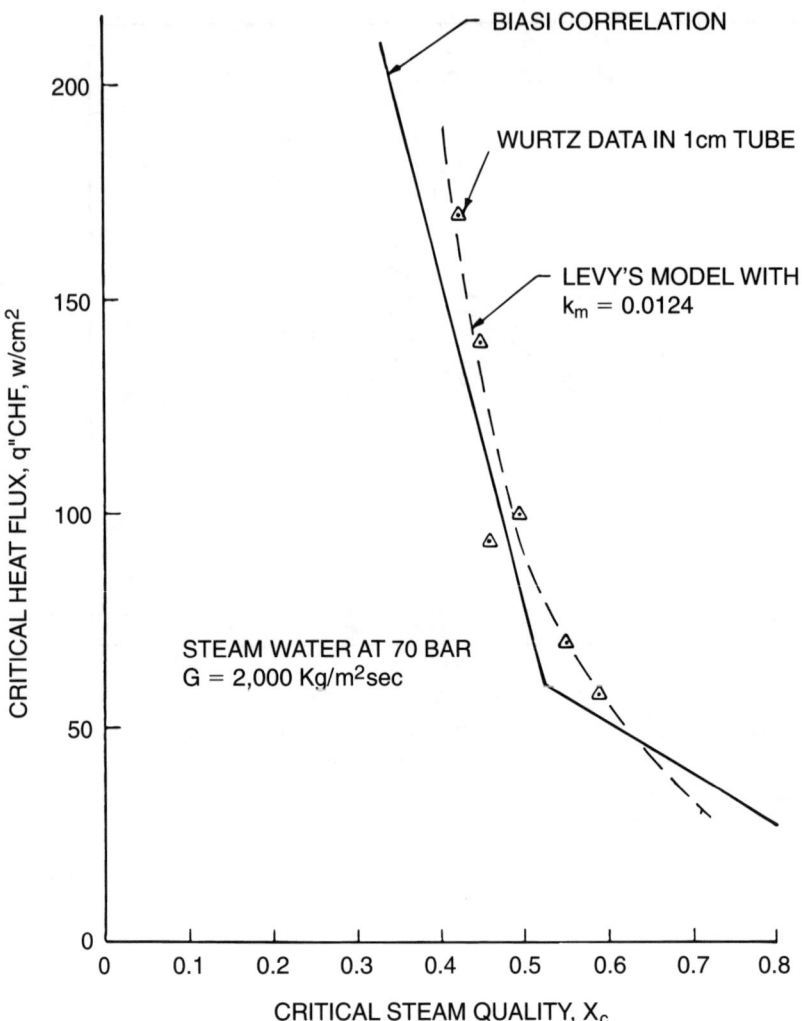

Figure 5.90 Comparison of model and critical heat flux data. (From Levy et al., 1980. Copyright © 1980 by Electric Power Research Institute, Palo Alto, CA. Reprinted with permission.)

For a very thin liquid film, the value of β cannot be evaluated, and it should be replaced by a new parameter β', using the generalized turbulent boundary-layer profile in an adiabatic flow as in Reference (Levy and Healzer, 1980). G_F can be solved stepwise along the pipe until the G value goes to zero, where dryout occurs. This analysis was performed to compare the calculated q''_{crit} with Wurtz data (Wurtz, 1978) and also to compare with the predictions by the well-known Biasi et al. correlation (1968), as shown in Figure 5.90. For the limited data points compared, the agreement was good.

5.5.12 Recommendations on Evaluation of CHF Margin in Reactor Design

1. The approach for developing a CHF correlation is usually based on phenomenological evidence obtained from visual observations or physical measurements. Most of the existing CHF correlations were conceived from visual observations as sketched in Figures 5-12 and 5-13, and developed from operating parameters in various flow regimes. Thus a CHF correlation is accurate only in the particular flow regime(s) within the ranges of the operating parameters in which it was developed. Consequently, its application should be limited to within these ranges of parameters.

2. The turbulence mixing in the coolant near the heating surface affects CHF strongly, and it is also very sensitive to the rod bundle geometry, especially the spacer grid. Reactor vendors have been diligently developing proprietary spacer grids to enhance reactor performance. The geometry of the grid is usually complex, and its effect on CHF is not amenable to analytical predictions. Therefore, the CHF correlation used for reactor design should be the one developed from the data tested in a prototype geometry.

3. The uncertainty in the predicted CHF of rod bundles depends on the combined performance of the subchannel code and the CHF correlation. Their sensitivities to various physical parameters or models, such as void fraction, turbulent mixing, etc., are complementary to each other. Therefore, in a comparison of the accuracy of the predictions from various rod bundle CHF correlations, they should be calculated by using their respective, "accompanied" computer codes. The word "accompanied" here means the particular code used in developing the particular CHF correlation of the rod bundle. To determine the individual uncertainties of the code or the correlation, both the subchannel code and the CHF correlation should be validated separately by experiments. For example, the subchannel code THINC II was validated in rod bundles (Weisman et al., 1968), while the W-3 CHF correlation was validated in round tubes (Tong, 1967a).

4. With the complicated geometry of subchannels and flux distributions in a rod bundle, the uncertainties in the CHF predictions may be expressed in the following three forms.

(a) A critical power ratio is defined as the ratio of the critical power (at which the minimum DNBR is unity) to the experimental critical power (Bowring, 1979; GE Report, 1973).

(b) A critical heat flux ratio is defined as the ratio of the calculated CHF at a predicted location on the predicted critical power profile to the heat flux at the predicted location on the measured critical power profile. The predicted critical power profile is identified in a subchannel code with the same test inlet conditions and the flux distribution, including the space factor, whenever CHF first occurs anywhere in the test section. The manner of predicting critical power is consistent with a critical power ratio concept (Rosal et al., 1974).

(c) The measured local heat flux at the location of the first CHF indication was compared with the predicted CHF, which was the calculated flux at the measured CHF location in a subchannel analysis for the measured critical bundle power and at the test inlet conditions (Reddy and Fighetti, 1983).

Since the uncertainty of the CHF predictions determines the safety margin of the protection systems and control systems for limiting the operating power of a reactor, the critical power ratio evaluated in (a) or (b) represents a realistic parameter for ensuring a proper safety margin. The simple CHF ratio as defined in (c) is rather too optimistic from a reactor safety point of view.

5.6 ADDITIONAL REFERENCES FOR FURTHER STUDY

Additional references are given here on recent research work on the subject of flow boiling crisis and are recommended for further study.

In spite of the large number of articles published to date, an exact theory of CHF has not yet been obtained. Katto (1994b), in his recent review of studies of CHF that were carried out during the last decade, sketched the mainstream investigations in a coherent style to determine the direction and subject of studies needed to further clarify the phenomenon of CHF. In a similar fashion, we are going to discuss seven different areas: CHF in high-quality annular flow; in subcooled flow boiling with low pressure and high flow rates; enhancement of CHF subcooled water flow boiling; in subcooled flow boiling with low quality; CHF in vertical upward and downward flow; on a cylinder in crossflow; and high flux boiling in low-flow-rates, low-pressure-drop, minichannel heat sinks.

The CHF in high-quality annular flow uses a liquid film dryout model, i.e., the concept of liquid film dryout (zero film flow rate) in annular flow. Sugawara (1990) reported an analytical prediction of CHF using FIDAS computer code based on a three-fluid and film dryout model.

The CHF in subcooled flow boiling at low pressures and high flow rates, typically applied to the design of high-heat-flux components of thermonuclear fusion reactors, was also predicted based on the liquid sublayer dryout mechanisms (Celata, 1991). Katto (1992) extended the prediction model to the low-pressure range. A data set of CHF in water subcooled flow boiling was presented by Celata and Mariani (1993). Celata et al. (1994a) rationalized the existing models based on basic mechanisms for CHF occurrence under flow boiling conditions (as discussed in Sec. 5.3.2). They proposed a model, similar to that of Lee and Mudawar (1988) and of Katto (1992). It is based on the observation that, during fully developed boiling, a vapor blanket forms in the vicinity of the heated wall by the coalescence of small bubbles, leaving a thin liquid sublayer in contact with the heated wall beneath the blanket. The CHF is assumed to occur when the liquid sublayer is extinguished by evaporation during the passage of the vapor blanket.

This rationalization process allowed the elimination of empirical constants needed in the two previous models mentioned above. Thus, this model is characterized by the very absence of empirical constants, which were always present in earlier models. The predicted CHF were compared with data of 1,888 data points for water and showed good agreement. Vandevort et al. (1994) reported an experimental study of forced-convection, subcooled boiling heat transfer to water at heat fluxes ranging from 10^7 to above 10^8 W/m^2 (3.2×10^6 to above 3×10^7 B/hr ft^2). To obtain predictive ability for the CHF at such high heat fluxes, experiments were performed with tubes of 0.3 to 2.7 mm (0.012 to 0.11 in.) in diameter, mass fluxes ranging from 5,000 to 40,000 kg/m^2 s (3.7×10^6 to 3×10^7 lb/hr ft^2), exit subcooling from 40 to 135°C (72 to 243°F), and exit pressures from 0.2 to 2.2 MPa (29 to 320 psia). A new empirical correlation specifically for the high-flux region was developed using a statistical approach to apply the CHF database to an assumed model and minimize the squares of the residuals of the dependent variables.

Enhancement of CHF subcooled water flow boiling was sought to improve the thermal hydraulic design of thermonuclear fusion reactor components. Experimental study was carried out by Celata et al. (1994b), who used two SS-304 test sections of inside diameters 0.6 and 0.8 cm (0.24 and 0.31 in.). Compared with smooth channels, an increase of the CHF up to 50% was reported. Weisman et al. (1994) suggested a phenomenological model for CHF in tubes containing twisted tapes.

A CHF model for subcooled flow boiling with low quality, based on a critical bubbly layer mechanism, was presented by Weisman and Pei (1983), Ying and Weisman (1986), Lin and Weisman (1990), and Lin et al. [1989]. These have been described in Section 5.3.2.2.

The CHF in vertical upward and downward, countercurrent flow was recently studied by Sudo et al. (1991) in a vertical rectangular channel. Sudo and Kaminaga (1993) later presented a new CHF correlation scheme for vertical rectangular channels heated from both sides in a nuclear research reactor.

For the CHF condition for two-phase crossflow on the shell side of horizontal tube bundles, few investigations have been conducted. Katto et al. (1987) reported CHF data on a uniformly heated cylinder in a crossflow of saturated liquid over a wide range of vapor-to-liquid density ratios. Recently, Dykas and Jensen (1992) and Leroux and Jensen (1992) obtained the CHF condition on individual tubes in a 5×27 bundle with known mass flux and quality. At qualities greater than zero, they found that the CHF data are a complex function of mass flux, local quality, pressure level, and bundle geometry.

High flux boiling in low-flow-rate, low-pressure-drop minichannel heat sinks becomes important because a need for new cooling technologies has been created by the increased demands for dissipating high heat fluxes from electronic, power, and laser devices. Bowers and Mudawar (1994) studied such heat sinks as a miniature heat sink of roughly 1 cm^2 (0.16 in.^{2}) in a heat surface area containing small channels for flow of cooling fluid.

CHAPTER
SIX

INSTABILITY OF TWO-PHASE FLOW

6.1 INTRODUCTION

In Chapter 3 the steady-state hydrodynamic aspects of two-phase flow were discussed and reference was made to their potential for instabilities. The instability of a system may be either static or dynamic. A flow is subject to a static instability if, when the flow conditions change by a small step from the original steady-state ones, another steady state is not possible in the vicinity of the original state. The cause of the phenomenon lies in the steady-state laws; hence, the threshold of the instability can be predicted only by using steady-state laws. A static instability can lead either to a different steady-state condition or to a periodic behavior (Bouré et al., 1973). A flow is subject to a dynamic instability when the inertia and other feedback effects have an essential part in the process. The system behaves like a servomechanism, and knowledge of the steady-state laws is not sufficient even for the threshold prediction. The steady-state may be a solution of the equations of the system, but is not the only solution. The above-mentioned fluctuations in a steady flow may be sufficient to start the instability. Three conditions are required for a system to possess a potential for oscillating instabilities:

1. Given certain external parameters, the system can exist in more than one state.
2. An external energy source is necessary to account for frictional dissipation.
3. Disturbances that can initiate the oscillations must be present.

All these conditions are satisfied for a two-phase flow with heat addition. When the conditions are favorable, sustained oscillations have also been found to occur. The analogy to oscillation of a mechanical system is clear when the mass flow rate, pressure drop, and voids are considered equivalent to the mass, exciting force, and spring of a mechanical system. In this connection, the relationship between flow rate and pressure drop plays an important role. A purely hydrodynamic instability may thus occur merely because a change in flow pattern (and flow rate) is possible without a change in pressure drop. The situation is aggravated when there is thermohydrodynamic coupling among heat transfer, void, flow pattern, and flow rate.

Flow instabilities are undesirable in boiling, condensing, and other two-phase flow processes for several reasons. Sustained flow oscillations may cause forced mechanical vibration of components or system control problems. Flow oscillations affect the local heat transfer characteristics and may induce boiling crisis (see Sec. 5.4.8). Flow stability becomes of particular importance in water-cooled and water-moderated nuclear reactors and steam generators. It can disturb control systems, or cause mechanical damage. Thus, the designer of such equipment must be able to predict the threshold of flow instability in order to design around it or compensate for it.

6.1.1 Classification of Flow Instabilities

The various flow instabilities are classified in Table 6.1. An instability is *compound* when several elementary mechanisms interact in the process and cannot be studied separately. It is *simple* (or fundamental) in the opposite sense. A *secondary phenomenon* is a phenomenon that occurs after the primary one. The term *secondary phenomenon* is used only in the very important particular case when the occurrence of the primary phenomenon is a necessary condition for the occurrence of the secondary one.

6.2 PHYSICAL MECHANISMS AND OBSERVATIONS OF FLOW INSTABILITIES

This section describes the physical mechanisms and summarizes the experimentally observed phenomena of flow instabilities, as classified in Table 6.1, and given by Bouré et al. (1973). The parameters considered are

Geometry—channel length, size, inlet and exit restrictions, single or multiple channels
Operation conditions—pressure, inlet cooling, mass flux, power input, forced or natural convection
Boundary conditions—axial heat flux distribution, pressure drop across channels

The effects of geometry and boundary conditions are usually interrelated, such as in flow redistribution among parallel channels. With common headers connected to the parallel channels, the flow distribution among channels is determined

Table 6.1 Two-phase flow instabilities

Category	Subcategory	Type	Mechanism	Characteristics
Static	Simple (fundamental)	Ledinegg (flow excursion) instability	$\left(\dfrac{\partial \Delta p}{\partial G}\right)_{\text{int}} \leq \left(\dfrac{\partial \Delta p}{\partial G}\right)_{\text{ext}}$	Flow undergoes a sudden, large-amplitude excursion, to a new, stable operating condition
		Thermal (boiling) crisis	Substantial decrease of heat transfer coefficient	Wall excursion with possible flow oscillation
		Flow-regime transition instability (relaxation instability)	Bubbly flow has less void but higher Δp than annular flow; condensation rate depends on flow regime	Cyclic flow regime transitions and flow rate variations
		Nonequilibrium-state instability	Transformation wave propagates along the system	Recoverable work disturbances and heating disturbances waves
	Compound	Unstable vapor formation (bumping, geysering, vapor burst) (relaxation instabilities)	Periodic adjustment of metastable condition, usually due to lack of nucleation sites	Occasional or periodic process of liquid superheat and violent vaporization with possible expulsion and refilling
		Condensation chugging	Bubble growth and condensation followed by surge of liquid (in steam discharge pipes)	Periodic interruption of vent steam flow due to condensation and surge of water up to downcomer
Dynamic	Simple (fundamental)	Acoustic oscillations	Resonance of pressure waves	High-frequency pressure oscillations (10–100 Hz) related to time required for pressure wave propagation in system
		Density wave oscillation	Delay and feedback effects in relationships among flow rate, density, pressure drops	Low-frequency oscillations (~1 Hz) related to transit time of a mass-continuity wave
	Compound	Thermal oscillations	Interaction of variable heat transfer coefficient with flow dynamics	Occurs close to film boiling
		Boiling-water reactor instability	Interaction of void/reactivity coupling with flow dynamics and heat transfer	Relevant only for a small fuel time constant and under low pressure
		Parallel channel instability	Interaction among a small number of parallel channels	Various modes of dynamic flow redistribution
		Condensation oscillation	Interaction of direct contact condensation interface with pool convection	Occurs with steam injection into vapor suppression pools
	Compound as a secondary phenomenon	Pressure drop oscillation	A flow excursion initiates a dynamic interaction between a channel and a compressible volume	Very-low-frequency periodic process (~0.1 Hz)

Source: Bouré et al., 1973. Copyright © 1973 by Elsevier Science SA, Lausanne, Switzerland. Reprinted with permission.

by the dynamic pressure variations in individual channels, caused by flow instability in each channel. Thus the boundary conditions of a heated channel separate the channel instability from the system instability.

6.2.1 Static Instabilities

6.2.1.1 Simple static instability. *Flow excursion (Ledinegg instability)* involves a sudden change in the flow rate to a lower value. It occurs when the slope of the channel demand pressure drop-versus-flow rate curve (internal characteristic of the channel) becomes algebraically smaller than the loop supply pressure drop-versus-flow rate curve (external characteristic of the channel). The criterion for this first-order instability is

$$\left(\frac{\partial \Delta p}{\partial G}\right)_{int} < \left(\frac{\partial \Delta p}{\partial G}\right)_{ext} \tag{6-1}$$

This behavior requires that the channel characteristics exhibit a region where the pressure drop decreases with increasing flow. In two-phase flow this situation does exist where the sum of the component terms (friction, momentum, and gravity terms) increases with decreasing flow. A low-pressure, subcooled boiling system was tested for excursive instability by paralleling the heated channel with a large bypass (Maulbetsch and Griffith, 1965). With a constant-pressure-drop boundary condition, excursions leading to critical heat flux were always observed near the minima in the pressure drop-versus-flow rate curves. It is shown by Figure 6.1 that the premature CHF is well below the actual CHF limit for the channel. The Ledinegg instability represents the limiting condition for a large bank of parallel tubes between common headers, since any individual tube sees an essentially constant pressure drop. Stable operation beyond the minimum and up to the CHF can be achieved by throttling individual channels at their inlets; however, the required increase in supply pressure may be considerable. Parallel-channel systems in downward flow are subject to a somewhat different type of excursion: that of flow reversal in some tubes (Bonilla, 1957). In heated downcomers, minima in pressure drop-versus-flow rate curves frequently occur due to interaction of momentum and gravitational pressure drop terms. The flow reversal reported by Giphshman and Levinzon (1966) in a pendant superheater can be explained in terms of the hydraulic characteristic.

Boiling crisis simultaneous with flow oscillations is caused by a change of heat transfer mechanism and is characterized by a sudden rise of wall temperature. The hydrodynamic and heat transfer relationships near the wall in a subcooled or low-quality boiling have been postulated as a boundary-layer separation during the boiling crisis by Kuteladze and Leont'ev (1966) and by Tong (1965, 1968b), although conclusive experimental evidence is still lacking. Mathisen (1967) observed that boiling crisis occurred simultaneously with flow oscillations in a boiling water

INSTABILITY OF TWO-PHASE FLOW **461**

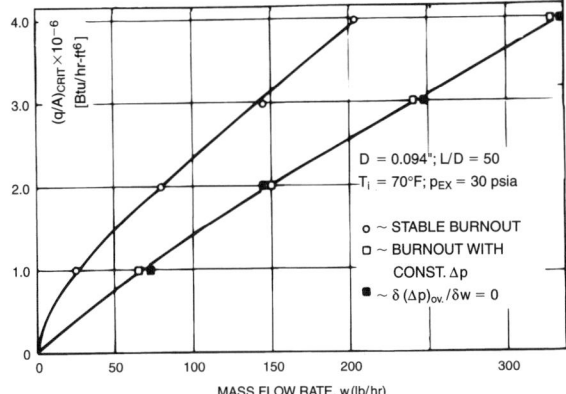

Figure 6.1 Critical heat flux versus mass flow rate for constant pressure drop. (From Maulbetsch and Griffith, 1965. Reprinted with permission of Massachusetts Institute of Technology, Cambridge, MA.)

channel at pressures higher than 870 psia (6.0 MPa), as did Dean et al. (1971) in a boiling Freon-113 flow with electrical heating on a stainless steel porous wall with vapor injection through the wall.

6.2.1.2 Simple (fundamental) relaxation instability. *Flow pattern (regimes) transition instabilities* have been postulated as occuring when the flow conditions are close to the point of transition between bubbly flow and annular flow. A temporary increase in bubble population in bubbly slug flow, arising from a temporary reduction in flow rate, may change the flow pattern to annular flow with its characteristically lower pressure drop. Thus the excess available driving pressure drop will speed up the flow rate momentarily. As the flow rate increases, however, the vapor generated may become insufficient to maintain the annular flow, and the flow pattern then reverts to that of bubbly slug flow. The cycle can be repeated, and this oscillatory behavior is partly due to the delay in acceleration and deceleration of the flow. In essence, each of the hydrodynamically compatible sets of conditions induces the transition toward the other; thus, typically, a relaxation mechanism sets in, resulting in a periodic behavior. In general, relaxation processes are characterized by finite amplitudes at the threshold. Bergles et al. (1967b) have suggested that low-pressure-water CHF data may be strongly influenced by the presence of an unstable flow pattern within the heated section. The complex effects of length, inlet temperature, mass flux, and pressure on the CHF appear to be related to the large-scale fluctuations characteristic of the slug flow regime. Since the slug flow regime may be viewed as a transition from bubbly to annular flow (Chap. 3), particularly for low-pressure diabatic flow, this phenomenon might be considered a flow pattern transition instability. The CHF would then be described as a secondary phenomenon. Cyclic flow pattern transitions have been observed in connection with oscillatory behavior (Fabrega, 1964; Jeglic and Grace, 1965), but it is not clear whether the flow pattern transition was the cause (through the above mechanism) or the consequence of a density wave or pressure drop oscillation. Grant (1971) also reported that shell-side slug flow was responsible for large-scale pressure fluctuation and exchanger vibration in a model of a segmentally baffled shell-and-tube heat exchanger.

Nonequilibrium state instability is similar to flow-regime transition instability. Nonequilibrium state instability is caused by transformation wave propagation along the system. This is characterized by recoverable work disturbances and heating disturbances waves.

6.2.1.3 Compound relaxation instability. *Bumping, geysering, and vapor burst* involve static phenomena that are coupled so as to produce a repetitive behavior which is not necessarily periodic. When the cycles are irregular, each flow excursion can be considered hydraulically independent of others. Bumping is exhibited in boiling of alkali metals at low pressure. As indicated in Figure 2.26 (Deane and Rohsenow, 1969), an erratic boiling region exists where the surface temperature moves between boiling and natural convection in rather irregular cycles. It has been postulated that this is due to the presence of gas in certain cavities, as the effect disappears at higher heat fluxes and higher pressures.

Geysering has been observed in a variety of closed-end, vertical columns of liquid that are heated at the base. When the heat flux is sufficiently high, boiling is initiated at the base. In low-pressure systems this results in a suddenly increased vapor generation due to the reduction in hydrostatic head, and usually an expulsion of vapor from the channel. The liquid then returns, the subcooled nonboiling condition is restored, and the cycle starts over again. An alternative mechanism for expulsion is vapor burst. Vapor burst instability is characterized by the sudden appearance and rapid growth of the vapor phase in a liquid where high values of superheat wave have been achieved. It occurs most frequently with alkali liquid metals and with fluorcarbons, both classes of fluids having near-zero contact angles on engineering surfaces. With good wettability, all of the larger cavities may be flooded out, with the result that a high superheat is required for nucleation (Chap. 2).

Condensation chugging refers to the cyclic phenomenon characterized by the periodic expulsion of coolant from a flow channel. The expulsion may range from simple transitory variations of the inlet and outlet flow rates to a violent ejection of large amounts of coolant, usually through both ends of the channel. The cycle, like the other phenomena described above, consists of incubation, nucleation, expulsion, and reentry of the liquid. The primary interest in these instabilities is in connection with fast reactor safety (Ford et al., 1971a, 1971b). Chexal and Bergles (1972) reported observations of instabilities in a small-scale natural-circulation loop resembling a thermosiphon reboiler. A typical flow regime map is shown in Figure 6.2. Regime II, periodic exit large bubble formation, was observed, which

INSTABILITY OF TWO-PHASE FLOW **463**

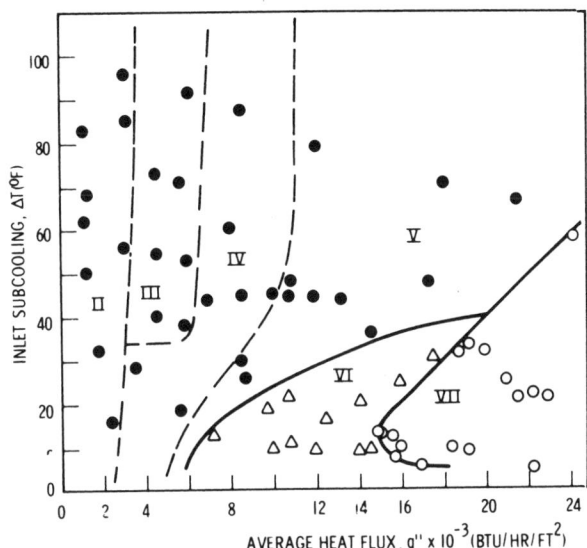

Figure 6.2 Typical flow regime boundaries for a small-scale thermosiphon reboiler. (From Chexal and Bergles, 1972. Copyright © 1972 by American Institute of Chemical Engineers, New York. Reprinted with permission.)

resembles chugging; regime IV, periodic exit small bubble formation, and regime V, periodic extensive small bubble formation, exhibit some characteristics of geysering. These instabilities would be expected during start-up of a thermosiphon reboiler. Condensation chugging occurs in steam discharge pipes as periodic interruption of vent steam flow due to condensation and surge of water up to the downcomer. This can be observed in a nuclear power plant when provisions are made to vent air and steam into a water pool to limit the rise in pressure during a reactor malfunction.

6.2.2 Dynamic Instabilities

6.2.2.1 Simple dynamic instability. Single dynamic instability involves the propagation of disturbances, which in two phase flow is itself a very complicated phenomenon. Disturbances are transported by two kinds of waves: pressure (or acoustic) waves, and void (or density) waves. In any real system, both kinds of waves are present and interact; but their velocities differ in general by one or two orders of magnitude, thus allowing the distinction between these two kinds of fundamental, primary dynamic instabilities.

Acoustic (or pressure wave) oscillations are characterized by a high frequency (10–100 Hz), the period being of the same order of magnitude as the time required for a pressure wave to travel through the system. Acoustic oscillations have been observed in subcooled boiling, bulk boiling, and film boiling. Bergles et al. (1967b) demonstrated that pressure drop amplitudes can be very large compared to the steady-state values, and inlet pressure fluctuations can be a significant fraction of the pressure level. Audible frequency oscillations of 1,000–10,000 Hz (whistle) were detected by a microphone in water flow at supercritical pressures of 3,200–3,500 psia (22–24 MPa) by Bishop et al. (1964). The minimum heat flux at which a whistle occurred was 0.84×10^6 Btu/hr ft² (2,640 kW/m²). In general, whenever a whistle was heard, the temperature of the fluid at the test section outlet was in the temperature range 690–750°F (366–399°C), and the wall temperature was above the pseudo-critical temperature. It is of interest to note that these oscillations have also been observed during blowdown experiments. Höppner (1971) found that subcooled decompression of a pressurized column of hot water was characterized by dampened cyclical pressure changes resulting from multiple wave reflections.

Density wave instability (oscillations) are low-frequency oscillations in which the period is approximately one to two times the time required for a fluid particle to travel through the channel (Saha et al., 1976). A temporary reduction of inlet flow in a heated channel increases the rate of enthalpy rise, thereby reducing the average density. This disturbance affects the pressure drop as well as the heat transfer behavior. For certain combinations of geometric arrangements, operating conditions, and boundary conditions, the perturbations can acquire a 180° out-of-phase pressure fluctuation at the exit, which is immediately transmitted to the inlet flow rate and becomes self-sustaining (Stenning and Veziroglu, 1965; Veziroglu et al., 1976). For boiling systems, the oscillations are due to multiple regenerative feedbacks among the flow rate, vapor generation rate, and pressure drop; hence the name "flow-void feedback instabilities" has been used (Neal et al., 1967). Since transportation delays are of paramount importance for the stability of the system, the alternative phrase "time-delay oscillations" has also been used (Bouré, 1966).

For fixed geometry, system pressure, inlet flow, and inlet subcooling, density wave oscillation can be started by increasing the test section power (heat flux). The fluctuation of the flow increases with increasing power (Saha et al., 1976), which means that increased heat flux always results in a smaller stability margin or in flow instability. In general, any increase in the frictional pressure drop in the liquid region has a stabilizing effect, as the pressure drop is in phase with the inlet flow, and it acts to damp the flow fluctuation. On the other hand, an increase in the two-phase-region pressure drop (such as an exit flow restriction) has a destabilizing effect, since the pressure drop is out of phase with the inlet flow, owing to the finite wave propagation time (Hetsroni, 1982). When the channel geometry is fixed, an increase in the inlet velocity has a stabilizing effect in terms of the heat flux, as the extent of two-phase-flow region and the density change due to boiling are significantly reduced by the increasing velocity. Further discussion of the parametric effects is given in Section 6.3.

6.2.2.2 Compound dynamic instability. *Thermal oscillations,* as identified by Stenning and Veziroglu (1965), appear to be associated with the thermal response of the heating wall after dryout. It was suggested that the flow could oscillate between film boiling and transition boiling at a given point, thus producing large-amplitude temperature oscillations in the channel wall subject to constant heat flux. An interaction with density wave oscillation is apparently required, with the higher-frequency density wave acting as a disturbance to destabilize the film boiling. Using a Freon-11 system with a Nichrome heater of 0.08-in. (2-mm) wall, Stenning and Veziroglu (1965) found the thermal oscillations to have a period of approximately 80 sec. Thermal oscillations are considered as a regular feature of dryout of steam–water mixtures at high pressures. Gandiosi (1965) and Quinn (1966) recorded wall temperature oscillations of several hundred degrees downstream of the dryout, with periods ranging from 2 to 20 sec. This was attributed to movement of the dryout point due to instabilities or even by small variations in pressure imposed by the loop control system. The fluctuation of the dryout point also causes the large-amplitude temperature oscillations in a channel wall subject to constant heat flux.

Boiling water reactor instability is complicated due to the feedback through a void–reactivity–power link. The feedback effect can be dominant when the time constant of a hydraulic oscillation is close to the magnitude of the time constant of the fuel element. Strong nuclear-coupled thermohydrodynamic instabilities therefore occurred in the early SPERT reactor cores, where a metallic fuel (small time constant) was operated in a low-pressure boiling water flow. The effect of pressure on the nuclear-coupled flow instability can be indicated by the magnitude of a pressure reactivity coefficient. Modern BWRs are operated at 1,000 psia (6.9 MPa) with uranium oxide fuel, which has a high time constant of 10 sec; the problem of flow instability is thus alleviated, and concern about density wave oscillations practically vanished for a couple of decades. Hydrodynamic instability was observed in the Experimental Boiling Water Reactor (EBWR) and was analyzed by Zivi and Jones (1966). Their results, using the FABLE code (Jones and Yarbrough, 1964–1965), were in excellent agreement with the experimental observations on the EBWR, and revealed that the principal mechanism of EBWR instability was a resonant hydrodynamic oscillation between the rod-bundle fuel assemblies and the other plate-type fuel assemblies.

Safety concerns about such thermohydrodynamic instabilities were raised after a rather unexpected occurrence of oscillations in the core of the LaSalle County Nuclear Station in 1988 (Phillips, 1990), following recirculation pump trips. Quite a large relative amplitude of oscillations was reached during that event, and the reactor was finally tripped from a high flux signal (Yadigaroglu, 1993). The safety concern is the potential fuel damage rather than the boiling crisis resulting from the critical bundle power being exceeded. The period of oscillation is dictated by the transit time of the coolant through the core. For a number of nonlinear dynamic studies of BWR instability margins, the reader is referred to (March-Leuba, 1990).

Parallel channel instability was reported by Gouse and Andrysiak (1963) for a two-channel Freon-113 system. The flow oscillations, as observed through electrically heated glass tubing, were generally 180° out of phase. Well within the stable region, the test sections began to oscillate in phase with very-large-amplitude flow oscillations and with the observed frequencies in the vicinity of the natural periods for the system of two or three tubes. This behavior was not observed when three heated channels were operated in parallel with a large bypass so as to maintain a constant pressure drop across the heated channels (Crowley et al., 1967). The stability boundaries and periods of oscillation were essentially identical for one, two, or three channels.

Koshelov et al. (1970) also reported tests results on a bank of three heated tubes in parallel. The phase shifts of flow oscillations were quite different for various tubes. Sometimes the flow oscillations in two tubes were in phase, while the flow oscillation in the third tube was in a phase shift of 120° or 180°. The amplitudes of the in-phase oscillations were different, that is, high in one tube and negligible in the other. Sometimes phase shift between individual tubes took place without apparent reason, but there were always tube in which the flow oscillations were 120° or 180° out of phase.

The effect of parallel channels is generally stabilizing, as compared with an identical single channel (Lee et al., 1976). This may be due to the damping effect of one channel with respect to the others, unless they are oscillating completely in phase. In other words, parallel channels have a tendency to equalize the pressure drop or pressure gradient if they are interconnected (Hetsroni, 1982).

Condensation oscillation can also occur, although experimental observations of density wave instabilities have been made mostly on boiling systems, the general behavior of condensation instability seems to indicate density wave oscillations. In connection with direct contact condensation, low-frequency pressure and interface oscillations were observed by Westendorf and Brown (1966). The oscillations were characterized by the annular condensing length alternatively growing to some maximum, then diminishing or collapsing (Fig. 6.3).

6.2.2.3 Compound dynamic instabilities as secondary phenomena. *Pressure-drop oscillations* are triggered by a static instability phenomenon. They occur in systems that have a compressible volume upstream of, or within, the heated section. Maulbetsch and Griffith (1965, 1967), in their study of instabilities in subcooled boiling water, found that the instability was associated with operation on the negative-sloping portion of the pressure drop-versus-flow curve. Pressure drop oscillations were predicted by an analysis (discussed in the next section), but because of the

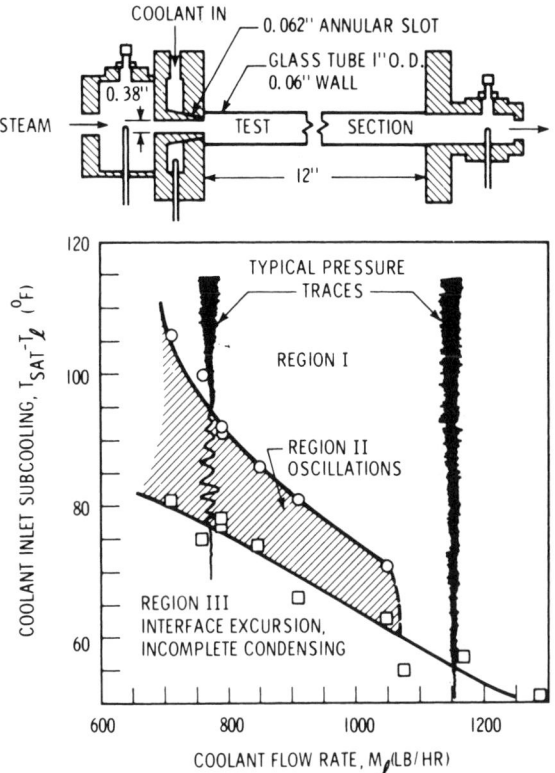

Figure 6.3 Instability regions for direct, constant condensation. (From Westendorf and Brown, 1966. Reprinted with permission of NASA Scientific & Technical Information, Linthicum Heights, MD.)

high heat fluxes, CHF always occurred during the excursion that initiated the first cycle.

A companion investigation (Daleas and Bergles, 1965) demonstrated that the amount of upstream compressibility required for unstable behavior is surprisingly small, and that relatively small volumes of cold water can produce large reductions in the CHF for small-diameter, thin-walled tubes. The reduction in CHF is less as subcooling, velocity, and tube size are increased. With tubes that have high thermal capacity, oscillations can be sustained without a boiling crisis (Aladyev et al., 1961); however, the amplitudes are usually large enough so that preventive measures are required. Stenning and Veziroglu (1967) encountered sustained oscillations when operating their bulk-boiling Freon-11 system with an upstream gas-loaded surge tank. Large-amplitude cycles began when the flow rate was reduced below the value corresponding to the minimum in the test section characteristic. A typical pressure-drop oscillation cycle with the indicated 40-sec period is shown

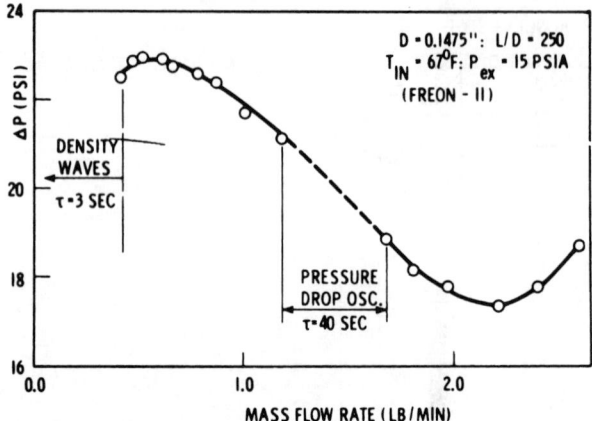

Figure 6.4 Density wave and pressure drop oscillation. (From Stenning and Verizoglu, 1965. Reprinted with permission of Stanford University Press, Stanford, CA.)

superimposed on the steady-state pressure drop-versus-flow curve in Figure 6.4. As mentioned before, the oscillations could be eliminated by throttling between the surge tank and heated section. This figure also illustrates that density wave oscillations occur at flow rates below those at which pressure drop oscillations are encountered. The frequencies of the oscillations are quite different, and it is easy to distinguish the mechanism by casual observation. However, the frequency of the pressure drop oscillations is determined to a large extent by the compressibility of the surge tank. With a relatively stiff system it might be possible to raise the frequency to the point where it becomes comparable to the density wave frequency, thus making it harder to distinguish the governing mechanism. Maulbetsch and Griffith (1965, 1967) also suggested that very long test sections of $(L/D) > 150$ may have sufficient internal compressibility to initiate pressure drop oscillations. In this case, no amount of inlet throttling will improve the situation.

6.3 OBSERVED PARAMETRIC EFFECTS ON FLOW INSTABILITY

The following parametric effects on density wave instability are summarized, as these effects have been often observed in the most common type of two-phase flow instability (Bouré et al., 1973):

Pressure effect
Inlet/exit restriction
Inlet subcooling
Channel length

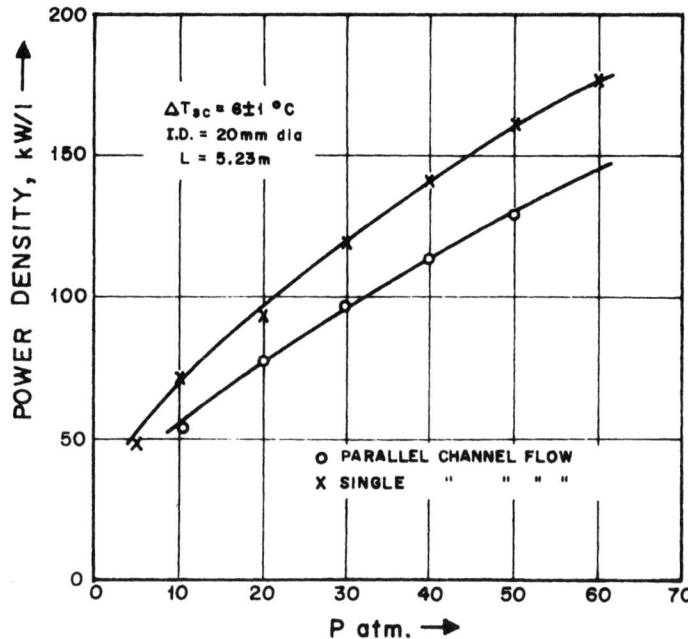

Figure 6.5 Effect of pressure on the power density at hydrodynamic instability. (From Mathisen, 1967. Copyright © 1967 by Office for Official Publications of the European Community, Luxembourg. Reprinted with permission.)

Bypass ratio of parallel channels
Mass flux and power
Nonuniform heat flux

6.3.1. Effect of Pressure on Flow Instability

The increase of system pressure at a given power input reduces the void fraction and thus the two-phase flow friction and momentum pressure drops. These effects are similar to that of a decrease of power input or an increase of flow rate, and thus stabilize the system. The increase of pressure decreases the amplitude of the void response to disturbances. However, it does not affect the frequency of oscillation significantly.

The effect of pressure on flow stability during natural circulation is given by Mathisen (1967) as shown in Figures 6.5 and 6.6. In Figure 6.6, the boiling crisis is shown to occur simultaneously with flow oscillation at pressures higher than 870 psia (6.0 MPa), as indicated in Section 6.2.1.1. Also noticeable in these figures is that an increase of power input reduces the flow stability. An increase of system pressure is a possible remedy to stabilize the flow at a high power input.

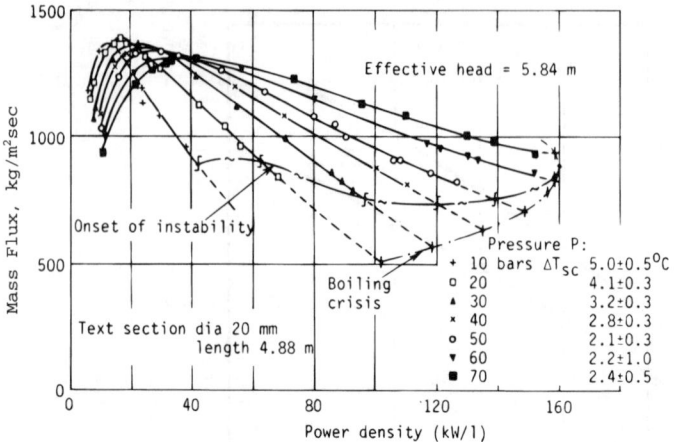

Figure 6.6 Effect of pressure and mass flux on the flow in a single boiling channel. (From Mathisen, 1967. Copyright © 1967 by Office for Official Publications of the European Community, Luxembourg. Reprinted with permission.)

6.3.2 Effect of Inlet and Exit Restrictions on Flow Instability

An inlet restriction increases single-phase friction, which provides a damping effect on the increasing flow and thereby increases flow stability. A restriction at the exit of a boiling channel increases two-phase friction, which is out of phase with the change of inlet flow. A low inlet (single-phase) flow increases void generation and exit pressure drop. It further slows down the flow. Thus, as observed by several investigators (Wallis and Hearsley, 1961; Maulbetsch and Griffith, 1965), an exit restriction reduces the flow stability. In a study of steam generator instabilities, McDonald and Johnson (1970) indicated that the laboratory version of the B&W Once-Through Steam Generator was subject to oscillations under certain combinations of operating conditions. The observed periods of 4–5 sec suggested a density wave instability. It was found that flow resistance in the feedwater heating chamber (equivalent to an inlet resistance) stabilized the unit.

6.3.3 Effect of Inlet Subcooling on Flow Instability

An increase in inlet subcooling decreases the void fraction and increases the non-boiling length and its transit time. Thus an increase of inlet subcooling stabilizes two-phase boiling flow at medium or high subcoolings. At small subcoolings, an incremental change of transit time is significant in the response delay of void generation from the inlet flow, and an increase of inlet subcooling destabilizes the flow. These stabilizing and destabilizing effects are competing. Thus, the effect of inlet subcooling on flow instability exhibits a minimum as the degree of subcooling is increased. Similar effects of inlet subcooling were observed by Crowley et al. (1967)

in a forced circulation. Moreover, they noticed that starting a system with high subcooling could lead to a large-amplitude oscillation.

6.3.4 Effect of Channel Length on Flow Instability

By cutting out a section of heated length of a Freon loop at the inlet and restoring the original flow rate, Crowley et al. (1967) found that the reduction of the heated length increased the flow stability in forced circulation with a constant power density. A similar effect was found in a natural-circulation loop (Mathisen, 1967). Crowley et al. (1967) further noticed that the change of heated length did not affect the period of oscillation, since the flow rate was kept constant.

6.3.5 Effects of Bypass Ratio of Parallel Channels

Studying the bypass ratio effect on the parallel channel flow instability, Collins and Gacesa (1969) tested the power at the threshold of flow oscillation in a 19-rod bundle with a length of 195 in. (4.95 m) by changing the bypass ratio from 2 to 18. The results show that a high bypass ratio destabilizes the flow in parallel channels. Their results agree with the analysis of Carver (1970). Veziroglu and Lee (1971) studied density wave instabilities in a cross-connected, parallel-channel system (Fig. 6.7). They found that the system was more stable than either a single channel or parallel channels without cross-connections. This work confirmed the common speculation that rod bundles are more stable than simple parallel-channel test sections.

6.3.6 Effects of Mass Flux and Power

Collins and Gacesa (1969) tested the effects of mass flux and power on flow oscillation frequency in a 19-rod bundle with steam–water flow at 800 psia. (5.5 MPa) They found that the oscillation frequency increases with mass flux as well as power input to the channel. Within the mass flux range 0.14–1.0×10^6 lb/hr ft² (189–1,350 kg/m² s), the frequency f can be expressed as

$$f = 0.27(P_{DO})^{0.73} - 0.1 \tag{6-2}$$

where the frequency f is in hertz and the power input, P_{DO}, is in megawatts. The mass flux effect on the threshold power of flow instability is also shown in Figure 6.6 (Mathisen, 1967).

6.3.7 Effect of Nonuniform Heat Flux

The effect of cosine heat flux distribution was tested by Dijkman (1969, 1971). He found that the cosine heat flux distribution stabilized the flow, which may be due

472 BOILING HEAT TRANSFER AND TWO-PHASE FLOW

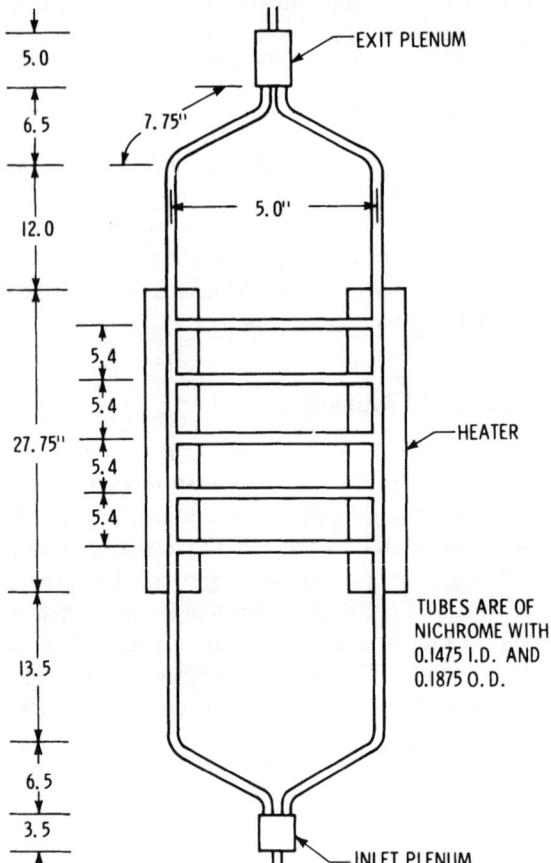

Figure 6.7 Cross-connected parallel channels. (From Verizoglu and Lee, 1971. Copyright © 1971 by American Society of Mechanical Engineers, New York. Reprinted with permission.)

to the decrease in local Δp at the exit, where the heat flux is lower than average. However, Yadigaroglu and Bergles (1969, 1972) found that the cosine distribution was generally *destabilizing,* which is also in agreement with data reported by Biancone et al. (1965). It should be noted that most of the flow instability tests were conducted with electric heaters of very small thermal capacitances, and the effects were negligible in evaluating the density wave oscillations. The thermal capacitance effect should be taken into consideration in cases where the time constant of the channel wall is comparable to that of the period of oscillation (Yadigaroglu and Bergles, 1969).

6.4 THEORETICAL ANALYSIS

6.4.1 Analysis of Static Instabilities

As indicated previously, static instabilities, being induced mostly by primary phenomena, can be predicted by using steady-state criteria or correlations. Therefore, the threshold of static instability can be predicted by using steady-state evaluations.

6.4.1.1 Analysis of simple (fundamental) static instabilities.

Analysis of flow excursion The threshold of flow excursion can be predicted by evaluating the Ledinegg instability criterion in a flow system or a loop, Eq. (6-1),

$$\left(\frac{\partial \Delta p}{\partial G}\right)_{int} \leq \left(\frac{\partial \Delta p}{\partial G}\right)_{ext}$$

where the Δp of the external head is supplied by the pump or by the natural-circulation head. The variation of mass flux, G, is the hot channel mass flux in a multichannel system or the loop mass flux in a single-channel system.

Analysis of boiling crisis instability The threshold of a boiling crisis instability can be predicted by the occurrence of a boiling crisis. Such predictions are given in Chapter 5.

6.4.1.2 Analysis of simple relaxation instabilities.

Analysis of flow-pattern transition instability The boundary of flow pattern transitions is not sharply defined, but is usually an operational band. As discussed in Chapter 3, analytical methods for predicting the stability of flow patterns are quite limited and require further development. The same is true for analyses of nonequilibrium state instability.

6.4.1.3 Analysis of compound relaxation instabilities.

Analysis of bumping, geysering, or chugging The primary phenomenon of this type of flow instability is a vapor burst at the maximum degree of superheat in the near-wall fluid. The value of maximum superheat varies with the heated wall surface conditions and the impurity contents of the fluid (see Sec. 2.2.1.2). The prediction of active nucleation site distributions for engineering surfaces must rely on experimental results, and nucleation instabilities cannot be predicted analytically. Considerable effort had been devoted to prediction of incipient boiling and subsequent voiding for LMFBR hypothetical loss-of-flow accidents (Fauske and Grolmes, 1971; Ford et al., 1971a, 1971b). Computer codes have been developed, based on a pure slug flow model, to calculate the thermal-hydraulic instability in a fast-reactor core (Schlechtendahl, 1970; Cronenberg et al., 1971).

6.4.2 Analysis of Dynamic Instabilities

The mathematical model describing the two-phase dynamic system consists of modeling of the flow and description of its boundary conditions. The description of the flow is based on the conservation equations as well as constitutive laws. The latter define the properties of the system with a certain degree of idealization, simplification, or empiricism, such as equation of state, steam table, friction, and heat transfer correlations (see Sec. 3.4). A typical set of six conservation equations is discussed by Bouré (1975), together with the number and nature of the necessary constitutive laws. With only a few general assumptions, these equations can be written, for a one-dimensional (z) flow of constant cross section, without injection or suction at the wall, as follows.

The mass conservation equation for each phase is

$$\partial \frac{(A_k \rho_k)}{\partial t} + \frac{\partial (A_k \rho_k u_k)}{\partial z} = M_k \tag{6-3}$$

with interface relationship

$$\sum M_k = 0 \tag{6-4}$$

where the subscript $k = L$ for liquid and $k = G$ for vapor, and M_k is mass transfer per unit time and volume to phase k.

The momentum equation for each phase is

$$A_k \left(\frac{\partial P}{\partial z}\right) + \frac{\partial (A_k \rho_k u_k)}{\partial t} + \frac{\partial (A_k \rho_k u_k^2)}{\partial z} = -A_k \rho_k g \cos \phi - F_{ki} - F_{kw} \tag{6-5}$$

where F_{ki} is the momentum loss for phase k to the interface, F_{kw} is the momentum loss from phase k to the wall, and ϕ is the angle from the vertical. If F_{TP} is the two-phase friction pressure drop, which is known, then

$$\sum F_{ki} = 0 \tag{6-6}$$

$$\sum F_{kw} = F_{TP} \tag{6-7}$$

The two equations (6-5) for liquid and vapor phases may be replaced by their sum [Eq. (6-13) below] and the "slip equation" may be written as

$$\rho_L \left[\left(\frac{\partial u_L}{\partial t}\right) + u_L \left(\frac{\partial u_L}{\partial z}\right)\right] - \rho_G \left[\left(\frac{\partial u_G}{\partial t}\right) + u_G \left(\frac{\partial u_G}{\partial z}\right)\right] = -(\rho_L - \rho_G) g \cos \phi$$
$$- \left[\left(\frac{F_{Li} + M_L u_L}{1 - \alpha}\right) - \left(\frac{F_{Gi} + M_G u_G}{\alpha}\right)\right] - \left(\frac{F_{Lw}}{1 - \alpha} - \frac{F_{Gw}}{\alpha}\right) \tag{6-8}$$

The energy equation for each phase is

$$\frac{\partial(A_k\rho_k H_k)}{\partial t} + \frac{\partial(A_k\rho_k u_k H_k)}{\partial z} = A_k\left(\frac{\partial P}{\partial t}\right) - A_k u_k\left(\frac{\partial P}{\partial z}\right) + Q_{ki} + Q_{kw} \quad (6\text{-}9)$$

with interface relationship

$$\sum Q_{ki} = 0 \quad (6\text{-}10)$$

$$\sum Q_{kw} = q''' \quad (6\text{-}11)$$

where Q_{ki} and Q_{kw} are the heat transferred per unit time and volume to phase k from the interface and wall, respectively. In the energy equation, the kinetic energy terms have been eliminated in combining the energy conservation with the other two conservation equations. Further simplification is obtained through the following assumptions (Bouré et al., 1973):

1. The $\partial P/\partial t$ and $\partial P/\partial z$ terms can be neglected in the energy equations.
2. The physical properties are independent of pressure. (This assumption is valid only for density wave instability at high pressures; it is not valid for acoustic instability.)
3. $H_k = (H_k)_{\text{sat}}$ for one (often the vapor) or both phases.
4. A finite correlation may replace the slip equations (6-8). With assumptions (1) and (2) and use of Eqs. (6-8)–(6-13), instead of two Eqs. (6-5), the pressure term is present only in Eq. (6-13), which may be solved separately. With assumptions (1) and (3), the phase energy equation (6-9) becomes equivalent to the phase mass conservation equation (6-3), thus reducing the order of the set.

It should be noted that the above assumptions are questionable when fast transient or large-pressure-drop flows are involved. By using these assumptions, the set is reduced to a set of three partial differential equations.

Mass conservation:
$$\frac{\partial[\alpha\rho_G - (1-\alpha)\rho_L]}{\partial t} + \frac{\partial[\alpha\rho_G u_G - (1-\alpha)\rho_L u_L]}{\partial z} = 0 \quad (6\text{-}12)$$

Momentum conservation:
$$\frac{\partial[\rho_L(1-\alpha)u_L + \rho_G\alpha u_G]}{\partial t}$$
$$+ \frac{\partial[\rho_L(1-\alpha)u_L^2 + \rho_G\alpha u_G^2]}{\partial z} + \frac{\partial P}{\partial z} + F_{wL}(1-\alpha) + F_{wG}\alpha$$
$$+ [\rho_L(1-\alpha) + \rho_G\alpha]g\cos\phi = 0 \quad (6\text{-}13)$$

Energy equation:
$$\frac{\partial[\rho_L(1-\alpha)H_L + \rho_G\alpha H_G]}{\partial t}$$
$$+ \frac{\partial[\rho_L(1-\alpha)u_L H_L + \rho_G\alpha u_G H_G]}{\partial z} = q''' \quad (6\text{-}14)$$

Note that frictional dissipation and pressure energy are neglected.

To analyze dynamic instabilities, the above equations have been programmed as computer codes such as STABLE (Jones and Dight, 1961–1964), DYNAM (Efferding, 1968), HYDNA (Currin et al., 1961), RAMONA (Solverg and Bakstad, 1967), and FLASH (Margolis and Redfield, 1965).

6.4.2.1 Analysis of simple dynamic instabilities.

Acoustic instability Bergles, Goldberg, and Maulbetsch (1967a) reported an analysis of acoustic oscillations because of its observed high frequencies. The test section was idealized as an inlet restriction, a line containing a homogeneous fluid, and a second restriction representing the test section pressure drop lumped at the exit. The pressure drop across the test section was presumed constant. The basic equations were solved by standard perturbation techniques and the method of characteristics. The condition for marginal dynamic stability was obtained as

$$\frac{\partial \Delta p_{ts}}{\partial u} = \frac{-(\beta/\alpha_o)^2}{2K_1 u} \tag{6-15}$$

with the associated frequency for threshold instability being given by

$$f = \frac{\alpha_o}{4L} \tag{6-16}$$

where β = volumetric compressibility of the two-phase flow
α_o = steady-state sonic velocity

K_1 = inlet orifice pressure coefficient = $\dfrac{\Delta P_{in}}{u_{in}^2 \rho}$

L = test-section heated length

The analytical predictions are in reasonable agreement with the observed frequency level and the variation of frequency with exit quality, X (Fig. 6.8). The analytical prediction of a threshold in the negative-sloping region of the pressure drop-versus-flow curve was corroborated by experimental data. However, the predicted slopes are somewhat steeper than those actually required for instability. A distributed parameter model such as FLASH (Margolis and Redfield, 1965) would be expected to improve both the threshold and the frequency predictions. In FLASH code, the one-dimensional space-dependent transient conservative equations are solved for compressible fluids by numerical integration. The results agree very well with experimental data obtained from various LOFT (loss-of-flow transient) testing programs.

Analysis of density wave instability Analyses have been made by using either simplified models or comprehensive computer codes.

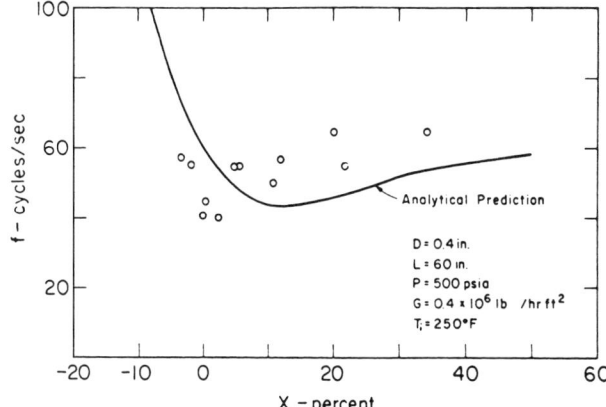

Figure 6.8 Comparison of predicted and observed frequencies for 500-psi run. (From Bergles et al., 1967a. Copyright © 1967 by Office for Official Publications of the European Community, Luxembourg. Reprinted with permission.)

Simplified models Simplified models have been used primarily to gain some physical understanding of the phenomena, although some models also give results that are in good agreement with experimental results (Bouré and Ihaila, 1967; Ishii and Zuber, 1970). Other simplified models can be found in Bouré (1966), Wallis and Hearsley (1961), and Yadigaroglu and Bergles (1969).

Dimensionless groups are useful to reduce the number of independent parameters and to serve as basis for developing scaling laws. For dimensional analysis it is important that dynamic similitude can be achieved only for the same type of flow instability, with similar geometry and similar boundary conditions. As would be expected, the dimensionless groups of several authors are more or less similar. For a given geometry with a uniform power density along the two-phase section, the main effects of fluid properties (including slip), gravity, pressure, inlet enthalpy, and heat flux are taken into account by three dimensionless groups:

A reduced velocity (Bouré, 1966; Ishii and Zuber, 1970) involving the flow rate-versus-heating power ratio or its reciprocal [phase change number in Ishii and Zuber (1970)]. This group includes the specific volume ratio in Bouré (1966) and in Ishii and Zuber (1970).
A subcooling number that also includes the specific volume ratio
A Froude number (Ishii and Zuber, 1970), or the equivalent reduced gravity (Bouré, 1966).

Some other dimensionless groups are also used individually to account for other minor effects. The basic concepts of the principal dimensionless groups used by Bouré (1966) and by Ishii and Zuber (1970) were compared by Bouré et al.

(1973). Blumenkrantz and Taborek (1971) applied the density effect model of Bouré to predict instability in natural-circulation systems in thermosiphon reboilers used in the petrochemical industry. An important conclusion of their work was that similarity analysis in terms of the model's dimensionless groups can be used to extrapolate threshold stability data from one fluid to another.

Computer codes Because of the computer's ability to handle the complicated mathematics, most of the compounded and feedback effects are built into computer codes for analyzing dynamic instabilities. Most of these codes can analyze one or more of the following instabilities: density wave instability, compound dynamic instabilities such as BWR instability and parallel-channel instability, and pressure drop oscillations.

Redfield and Murphy (1971) reported a comparison of the FLASH-4 (compressible, sectionalized) numerical solution of channel hydrodynamics with a simpler momentum integral solution. In the latter approach, neglect of density variations due to pressure changes and use of a single momentum equation for each section resulted in great savings in computer time. At typical LWR conditions (1,200 psia or 8.3 MPa), the threshold power levels for density wave instability agreed to within 2%. Frequency (~1 Hz) was also essentially identical. While both methods are acceptable for predicting the onset of oscillations, only the more complex FLASH-4 code is able to predict the limit cycles for higher power levels.

6.4.2.2 Analysis of compound dynamic instabilities. As mentioned in the previous section, most of the compounded and feedback effects are built into computer codes for analyzing dynamic instabilities. These computer codes can be used to analyze compound dynamic instabilities such as BWR instability and parallel-channel instability. However, thermal instability between transition boiling and film boiling cannot be analyzed, because of the lack of phenomenological correlation of transition boiling.

6.4.2.3 Analysis of compound dynamic instabilities as secondary phenomena (pressure drop oscillations). Maulbetsch and Griffith (1965, 1967) performed a stability analysis of an idealized model of the system in which pressure drop instabilities had been observed. The test section and upstream compressible volume were treated as lumped parameters. The governing equations for flows in the heated section and into the compressible volume were treated by a perturbation technique and standard stability criteria. At marginal stability, a critical slope of the heated-section pressure drop-versus-flow curve was obtained. Although the general expression is cumbersome, the limiting cases are of considerable interest. There is no oscillatory solution for a constant-pressure-drop supply system, since the pressure in the compressible volume is then constrained to be constant. For a constant-flow-rate delivery system and for a marginal flow stability,

$$\frac{\partial \Delta P_{ts}}{\partial Q} = 0 \qquad \omega^2 = -\frac{(dP/dV)_o}{I_1 + I_2} \qquad (6\text{-}17)$$

where Q = volumetric flow rate = Au

$\left(\dfrac{dP}{dV}\right)_o$ = measure of the system compressibility at the initial state

ω = frequency of oscillation
I_1 = flow inertia ($\rho L_1/A$) of the heated section after the compressible volume
I_2 = flow inertia ($\rho L_2/A$) of the section before the compressible volume

6.5. FLOW INSTABILITY PREDICTIONS AND ADDITIONAL REFERENCES FOR FURTHER STUDY

6.5.1 Recommended Steps for Instability Predictions

The following steps for predicting flow instability in boiling equipment are recommended (Bouré et al., 1973).

1. Check the system (or loop) instability by using the Ledinegg criterion with an average lumped channel pressure drop. If it does not satisfy the Ledinegg stability criterion, one or more of the three remedies can be taken: orifice the inlet, increase the steepness of the pump head-versus-flow curve; or increase the resistance of the downcomer of a natural-circulation loop.
2. Check the static instabilities by steady-state correlations, to avoid or alleviate the primary phenomenon of a potential static instability, namely, boiling crisis, vapor burst, flow pattern transition, and the physical conditions that extend the static instability into repetitive oscillations.
3. Check the onset of dynamic flow instability in the heated channel, if the flow is under supercritical pressure or in film boiling, by using Eqs. (6-1) and (6-18).

$$\frac{q''}{GH_{fg}} = 0.005\left(\frac{v_f}{v_{fg}}\right) \qquad (6\text{-}18)$$

where v_f and v_{fg} are the specific volume of saturated liquid and the difference in specific volumes of two phases, respectively.
4. Check the onset of density wave instability in a heated channel by using a simplified model or a nondimensional plot, if the geometry and boundary conditions of the equipment agree with that of the nondimensional plot analysis (e.g., Bouré, Zuber, etc.).

5. Finally, check the onset of density wave instability in a heated channel with specific boundary conditions by using a computer code, such as STABLE-5 (Jones and Dight, 1961–1964), RAMONA (Solverg and Bakstad, 1967), HYDNA (Currin et al., 1961) and SAT (Roy et al., 1988).

6.5.2 Additional References for Further Study

In the operation of BWRs, especially when operating near the threshold of instability, the "stability margin" of the stable system and the amplitude of the limit cycle under unstable condition become of importance. A number of nonlinear dynamic studies of BWRs have been reported, notably in an International Workshop on Boiling Water Reactor Stability (1990). The following references are mentioned for further study.

March-Leuba (1990) presented radial nodalization effects on the stability calculations. March-Leuba and Blakeman (1991) reported on out-of-phase power instabilities in BWRs. BWR stability analyses were reported by Anegawa et al. (1990) and by Haga et al. (1990). The experience and safety significance of BWR core-thermal-hydraulic stability was presented by Pfefferlen et al. (1990).

Lahey (1990) indicated the applications of fractal and chaos theory in the field of two-phase flow and heat transfer, especially during density wave oscillations in boiling flow.

APPENDIX

SUBCHANNEL ANALYSIS (TONG AND WEISMAN, 1979)

A.1 MATHEMATICAL REPRESENTATION

Adjacent subchannels are open to each other through the gap between two neighboring fuel rods; flow in one channel mixes with that in the other. In addition, as observed previously, there is crossflow between channels because of the pressure gradient. Local turbulent mixing reduces the enthalpy rise of the hot channel. On the other hand, flow leaving the hot channel increases its enthalpy rise. Calculation of the net result is complicated, although the equation describing enthalpy rise can easily be written.

For simplicity, in a segment of length ΔZ we consider the case of two adjacent subchannels that are linked only with each other. Such a situation is represented in Figure A.1. Quantities H, V, ρ, and P represent coolant enthalpy, velocity, density, and static pressure, respectively; A is the channel flow area; W_{mn} is the crossflow between channels; and w' is the flow exchange of diffusion mixing. The numbered subscripts refer to two different axial elevations in the core. Writing mass, energy, and momentum balance equations for channel m between elevations 1 and 2 results in the following relationships, where W_{mn} is considered positive in the direction from channel n into channel m.

Conservation of mass:

$$A_m V_{m1} \rho_{m1} + W_{mn} = A_m V_{m2} \rho_{m2} \qquad (A\text{-}1)$$

Conservation of heat:

$$A_m V_{m1} \rho_{m1} H_{m1} + Q_{mz} + W_{mn} \overline{H}_n + w'(H_n - H_m) \Delta Z = A_m V_{m2} \rho_{m2} H_{m2} \qquad (A\text{-}2)$$

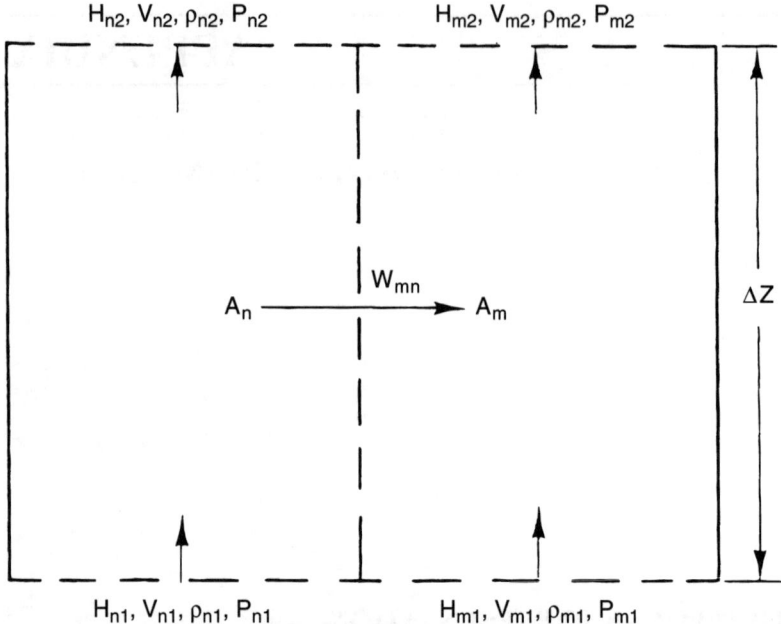

Figure A.1 Mathematical representation of flow redistribution. (From Tong and Weisman, 1979. Copyright © 1979 by American Nuclear Society, LaGrange Park, IL. Reprinted with permission.)

Conservation of axial momentum:

$$\frac{A_m P_{m1} + A_m \rho_{m1}(V_{m1})^2}{g_c} + \frac{\overline{W_{mn} V_n}}{g_c} = A_m P_{m2} + \frac{A_m \rho_{m2}(V_{m2})^2}{g_c} + \frac{K_{mz} A_m \rho_m (V_m)^2}{2 g_c}$$

$$+ \frac{\overline{\rho}_m \Delta Z g}{g_c} \qquad \text{(A-3)}$$

where Q_{m2} = heat input into channel m in interval ΔZ
$\overline{W}_{mn}$ = crossflow from channel n to channel m
$\overline{H}, \overline{\rho}, \overline{V}$ = mean values in ΔZ of enthalpy, density, and velocity, respectively
w' = flow exchange rate per unit length by mixing
K_{mz} = pressure loss coefficient for channel m in interval Z
P = pressure

Evaluation of ρ and K_{mz} requires determination of the void fraction and the two-phase pressure drop. Crossflow is determined from the appropriate lateral momentum balance equation. The interchange due to mixing, represented by w', is determined by the turbulent transverse fluctuating flow rate per foot of axial length (lb/hr ft), where

$$w' = \rho_e \left(\frac{\ell}{D_e} \right) \qquad \text{(A-4)}$$

and

ρ = fluid density, lb/ft³
e = eddy diffusivity, ft²/hr
ℓ = Prandtl mixing length, ft
D_e = equivalent diameter of channel, ft

Rowe and Angle (1967) correlated their data by

$$w' = cGD_e(\text{Re})^{-0.1} \qquad (A\text{-}5)$$

where G = fluid mass velocity, lb/hr ft²
Re = Reynolds number
c = constant = 0.0062

Rogers and Rosehart (1969) analyzed the data of a number of experimenters and found that they could use Eq. (A-5) to represent the data if $c = 0.004$. Rogers and Rosehart (1972) later modified their correlation such that for square-pitch rods (adjacent channels with the same D_e) the flow exchange rate could be represented by

$$w' = 0.005bG\left(\frac{D_e}{D}\right)(\text{Re})^{-0.1}\left(\frac{b}{D}\right)^{-0.894} \qquad (A\text{-}6)$$

where G = mass flux = m/A, in lb/hr ft²
b = gap between rods, ft
D = rod diameter, ft

The mixing data were correlated by defining a thermal diffusion coefficient, α, such that

$$\alpha = \frac{e}{Vb} \qquad (A\text{-}7)$$

where V is the superficial velocity, feet/second, and b is the gap between two rods. For their bundle (of rods of 0.422-in. diameter on a 0.535-in.² pitch), they found $\alpha \simeq 0.076$ and essentially independent of mass flux (for G up to 2.75×10^6 lb/hr ft², or 3,720 kg/m² s) and quality. Since their spacer grids contained small mixing vanes, it is expected that α for bundles without these vanes would be lower.

Some investigators preferred to correlate their data on the basis of modified Peclet number, Pe, where

$$Pe = \frac{VD_e}{e} \qquad (A\text{-}8)$$

Observing that the Peclet number equals $(D_e/\alpha b)$, Bell and LeTourneau (1960) reported, for bundles of bare rods,

$$\frac{1}{Pe} = \frac{e}{VD_E} = 0.003 \qquad 10^4 < Re < 5 \times 10^4 \tag{A-9}$$

They believed the data indicated that the correlation would be valid up to $Re = 2 \times 10^5$. For rectangular channels, Sembler (1960) reported

$$\frac{e}{VD_e} = 0.0015 \qquad Re > 10^5 \tag{A-10}$$

Waters (1963) reported on the mixing obtained in wire-wrapped bundles. He found that α increased from 0.1 to 1 as the wrap pitch changed from 0.7 to 3 wraps/ft.

A.2 COMPUTER SOLUTIONS

A similar set of equations can be written for every other channel in the region studied. A simultaneous solution of all of these equations is required to determine fluid conditions at the exit of the length interval. The complexity of the calculational procedure requires a computer solution; a number of computer codes have been written for this purpose. In all of these codes, the cross section of the region of interest is divided into a series of subchannels, while the length is divided into a series of axial segments. Thus, a set of connecting control volumes is provided. All of the calculational procedures assume that conditions are uniform within a given control volume. Gradients exist only across control-volume boundaries.

Earlier computer codes, such as THINC II (Weisman et al., 1968; Chelemer et al., 1972), COBRA II (Rowe, 1970), and HAMBO (Bowring, 1968), use "marching procedures." If conditions at the inlet of a given axial segment are known, the governing equations can be solved iteratively to give the conditions at the exit of the segment. By proceeding stepwise along the channel length, a marching solution for the whole channel is obtained. The iterative-interval calculation methods in COBRA II and HAMBO are quite similar. A set of crossflows between subchannels is "guessed." With this set, the energy equation is solved by forward differencing in COBRA II and central differencing in HAMBO. From the momentum equation, pressure drop in each subchannel is calculated. From this calculation, a new set of crossflow guesses that gives a pressure balance is obtained by backward differencing. The iteration is continued until an acceptable pressure balance is obtained.

The boundary conditions to be satisfied are that the lateral pressure difference between subchannels should be zero at the channel inlet and exit. Having passed once along the channel, this implies that iteration over the channel length may be necessary by using improved guesses of flow division between subchannels at the inlet. In practice, only one pass may be necessary, particularly for hydraulic model, in which lateral momentum transfer is neglected or only notionally included. Rowe (1969) has shown that for a single-pass solution to be stable and acceptable, the calculational increment of length must be greater than a critical value, $2C\,|\,w\,|\,g(mH)$, where

C = crossflow resistance coefficient
w = crossflow
$g(mH)$ = difference in axial pressure gradient caused by the crossflow
m = subchannel flow rate
H = enthalpy

For long enough increments, calculated exit conditions for an interval tend to compensate for errors in the assumed inlet conditions. This provides a self-correcting mechanism in the calculation and, conversely, means that large changes in assumed channel inlet conditions are required to affect calculated conditions and pressure balance at the channel exit. The acceptability of a single-pass marching solution depends on coupling between subchannels. If this coupling is weak (i.e., if the crossflows are small), a single-pass marching solution technique is adequate—following Rowe's criterion shown above. Upstream effects are confined entirely to the preceding interval. From the same criterion, stronger coupling (i.e., large crossflows) could mean that the intervals need to be impracticably large.

A multipass marching solution is used in COBRA IIIC (Rowe, 1973). The inlet flow division between subchannels is fixed as a boundary condition, and an iterated solution is obtained to satisfy the other boundary solution of zero pressure differential at the channel exit. The procedure is to guess a pattern of subchannel boundary pressure differentials for all mesh points simultaneously, and from this to compute, without further iteration, the corresponding pattern of crossflows using a marching technique up the channel. The pressure differentials are updated during each pass, and the overall channel iteration is completed when the fractional change in subchannel flows is less than a preset amount.

Pressure differentials at the exit of the length steps are used to calculate crossflows. Crossflows are then used to calculate pressure differentials at the outlet of the previous (upstream) length steps. These pressure differentials are saved for use during the next iteration. At the exit of the last length step, the boundary condition of zero pressure differential is imposed and crossflows at the exit are calculated on this basis. The iterative procedure forces agreement with the assumed boundary condition.

The TORC code (1975) is a modified version of COBRA IIIC. The basic numerics are those for COBRA IIIC, but TORC contains some additional features useful in overall core design.

The THINC II code handles the problem of zero lateral pressure gradient at the assembly exit by assuming that the lateral pressure gradient is zero everywhere. Chelemer et al. (1972) argue that because of close coupling between subchannels, there can be only a very low pressure gradient in an assembly. Under these conditions, they show that changes in the lateral pressure drop cause very small differences in axial flow. Hence the change in axial flow in a given channel is determined by the requirement that pressure drop be the same across each control volume at a given elevation. Since pressure gradients are not given, enthalpy and axial velocity associated with crossflow are not directly calculable. Therefore, these are taken as a weighted average of the values of the surrounding channels. Weighting factors depend on control volume net gain or loss in flow over the length step.

Use of a marching solution to determine the behavior of individual subchannels in an assembly requires that the inlet flow to that assembly be known. The assumption that all assemblies in a core have the same inlet flow can be appreciably in error. Flow must be divided so that the core pressure drop remains essentially constant. Therefore, higher pressure loss coefficients in high-power assemblies, due to the presence of significant exit quality, lead to lower flows in these assemblies.

In solving open channel flow equations, the THINC I code (Zernick et al., 1962) was the first calculational technique capable of satisfactorily assigning inlet flows to the assemblies within a semiopen core. In the THINC I approach, it was recognized that the total pressure distribution at the top of the core region is a function of inlet pressure, density, and velocity distributions. This functional dependence can be expressed as,

$$P_j^o = P_j^o(P_1^i, P_2^i, \ldots, P_n^i, \rho_1^i, \rho_2^i, \ldots, \rho_n^i, V_1^i, V_2^i, \ldots, V_n^i) \quad \text{(A-11)}$$

$$j = 1, 2, \ldots, n$$

where superscripts i and o represent inlet and outlet values, respectively, and j is the number of the channel. Since the flow leaving the core enters a large plenum across which significant pressure differentials do not exist, the basic criteria to be satisfied are that outlet total pressure is uniform and total mass flow rate must be constant; i.e.,

$$P_1^o = P_2^o = P_3^o = \ldots = P_n^o \quad \text{(A-12)}$$

and

$$\sum_{j=1}^{n} A_j \rho_j^i V_j^i = \text{a constant} \quad \text{(A-13)}$$

As flow enters the core from a large plenum with only small pressure differences across it, these basic criteria can be achieved only by adjusting the inlet velocity distribution, V_j^i.

Let us consider a two-channel core, for example. If inlet pressures and densities are fixed,

$$\begin{aligned} P_1^o &= \phi(V_1, V_2) \\ P_2^o &= \psi(V_1, V_2) \end{aligned} \quad \text{(A-14)}$$

then

$$dP_1^o = \phi_1 dV_1 + \phi_2 dV_2$$
$$dP_2^o = \psi_1 dV_1 + \psi_2 dV_2 \quad \text{(A-15)}$$

where

$$\phi_1 = \frac{\partial \phi}{\partial V_1} \quad \phi_2 = \frac{\partial \phi}{\partial V_2}$$
$$\psi_1 = \frac{\partial \psi}{\partial V_1} \quad \psi_2 = \frac{\partial \psi}{\partial V_2} \quad \text{(A-16)}$$

and

$$[dP^o] = \begin{bmatrix} \phi_1 & \phi_2 \\ \psi_1 & \psi_2 \end{bmatrix} [dV^i] \quad \text{(A-17)}$$

Let

$$[T] = \begin{bmatrix} \phi_1 & \phi_2 \\ \psi_1 & \psi_2 \end{bmatrix} \quad \text{(A-18)}$$

$$[T^{-1}] = \begin{bmatrix} \dfrac{\psi_2}{\phi_1\psi_2 - \phi_2\psi_1} & \dfrac{-\phi_2}{\phi_1\psi_2 - \phi_2\psi_1} \\ \dfrac{\psi_1}{\phi_1\psi_2 - \phi_2\psi_1} & \dfrac{\phi_1}{\phi_1\psi_2 - \phi_2\psi_1} \end{bmatrix} \quad \text{(A-19)}$$

Thus, for a given inlet pressure distribution and inlet density distribution, the following applies:

$$[dP^o] = [T][dV^i] \quad \text{(A-20)}$$

where $[dP^o]$ = column matrix with elements dP_j^o

$[T]$ = n by n matrix with elements $\dfrac{\partial P_j^o}{\partial V_k^i}$

$[dV^i]$ = column matrix with elements dV_k^i

For square matrices, since $[T][T^{-1}] = 1$, an identity matrix, premultiplication of Eq. (A-20) by $[T^{-1}]$ gives

$$[dV^i] = [T^{-1}][dP^o] \quad \text{(A-21)}$$

provided $[T]$ is not singular.

Consider the pressure residues at the region outlet to be given by the relation

$$\Delta P_j^o = \overline{P}^o - P_j^o \qquad (A\text{-}22)$$

where $\overline{P}^o$ is the average outlet total pressure. Let the changes in velocities ΔV_k^i associated with pressure residues ΔP_j^o be given by a relation analogous to Eq. (A-21),

$$[\Delta V^i] = [T^{-1}]\{[\Delta P^o] - \lambda[1]\} \qquad (A\text{-}23)$$

where $[\Delta V^i]$ = column matrix with elements ΔV_k^i
$[\Delta P^o]$ = column matrix with elements ΔP_j^o
$[1]$ = unit column matrix
λ = scalar quantity to be determined

Equation (A-23) allows us to determine λ. If the initial choice of velocities satisfies Eq. (A-13) and all subsequent choices satisfy this constraint, then

$$[A][\rho][\Delta V^i] = 0 \qquad (A\text{-}24)$$

where $[A]$ is a row matrix with elements A_j, $[\rho]$ is a square matrix with leading diagonal ρ_j^i, and all other elements are zero. Combining Eqs. (A-23) and (A-24) and solving for λ,

$$\lambda = \frac{[A][\rho][T^{-1}][\Delta P^o]}{[A][\rho][T^{-1}][1]} \qquad (A\text{-}25)$$

Thus, if the $[T]$ matrix is known and is nonsingular, λ can be calculated from Eq. (A-25) and substituted in Eq. (A-23) to give the changes in inlet velocities associated with outlet pressure distribution P_j^o. To obtain $[T]$, the inlet velocity to one channel is changed by a preassigned small fraction, ε, and the resulting changes in outlet pressures for all the channels are determined by the stepwise procedure described above. The elements of the first column of the $[T]$ matrix are determined from this calculation. The original velocity distribution is restored, and the procedure is repeated for each channel. Thus the inlet velocity to each channel is changed by the same small fraction, ε, and in turn, the elements of each column of the $[T]$ matrix are thereby determined. The inverse of the $[T]$ matrix is then determined and λ is calculated using Eq. (A-25). Substituting these values in Eq. (A-23) gives matrix $[\Delta V^i]$, which is then used to determine a new inlet velocity distribution using the relation

$$[V'^i] = [V^i] + [\Delta V^i] \qquad (A\text{-}26)$$

where $[V^i]$ and $[V'^i]$ are the column matrices for the original and new inlet velocities, respectively. This entire procedure is repeated until the error in the outlet distribution is less than a preassigned value.

Sha et al. (1976) observe that under some conditions the preceding procedure may not converge. Experience shows that the procedure always converges when there is no crossflow between channels. Advantage can be taken of this fact, and a solution can be obtained for an infinite lateral resistance (or zero lateral flow area) in the core. The velocity distribution as obtained is then used as input to a second problem with a finite lateral flow resistance (or finite lateral flow area). This velocity distribution is then used to solve a further problem in which the lateral resistance or flow area more closely approaches the actual value. The process continues until convergence is obtained at the actual lateral resistance or flow area. Such convergence difficulty, encountered when a marching solution is used to examine full core behavior, is one of the motivating factors that led to the development of other solution procedures. An additional motivation was the desire to be able to handle recirculating flows that could develop under severe blockage conditions. Marching solutions cannot handle reverse flows. Reverse flows can, however, be treated successfully by numerical procedures that solve the conservation equations for the control volumes at all axial levels simultaneously. This approach is followed in THINC IV code (Chu et al., 1972, 1973). Here, lateral velocity components are regarded as perturbed quantities much smaller than axial flow velocity. The original governing equations are split into a perturbed and an unperturbed system of equations. Perturbed momentum and continuity equations are then combined to form a field equation that is solved for the entire velocity field. Inlet velocities are determined such that a uniform outlet pressure is obtained. The initial solution obtained is used to update the properties and conditions assumed to exist in various control volumes. The iteration continues until assumed and calculated properties are in satisfactory agreement. The THINC IV code is written so that it can also be used for determining behavior in a core region where the subchannels are defined by four neighboring fuel rods.

REFERENCES*

Achener, P. Y., 1964, The Determination of the Latent Heat of Vaporization, Vapor Pressure, Enthalpy, and Density of Liquid Rubidium and Cesium up to 1,800°F, *Proc. 1963 High Temperature Liquid Metal Heat Transfer Technology Meeting,* Vol. 1, pp. 3–25 USAEC Rep. ORNL-3605. (2)*

Achener, P. Y., 1965, The Determination of the Latent Heat of Vaporization, Vapor Pressure of Potassium from 1,000–1,900°F, Aerojet-General Nucleonics Rep. AGN-8141. (2)

Adams, J. M., 1962. A Study of the Critical Heat Flux in an Accelerating Pool Boiling System, Report NSF G-19697, University of Washington, Seattle, WA. (2)

Addoms, J. N., 1948, Heat Transfer at High Rates to Water Boiling Outside Cylinders, D.Sc. thesis, Massachusetts Institute of Technology, Cambridge, MA. (2)

Agee, L. J., 1978, Power Series Solutions of the Thermal-Hydraulic Conservation Equations, in *Transient Two-Phase Flow,* Proc. 2nd Specialists Meeting, OECD Committee for the Safety of Nuclear Installations, Paris, Vol. 1, pp. 385–410. (3)

Agee, L. J., S. Banerjie, R. B. Duffey, and E. D. Hughes, 1978, Some Aspects of Two-Phase Models for Two-Phase Flow and Their Numerical Solutions, in *Transient Two-Phase Flow,* Proc. 2nd Specialists Meeting, OECD Committee for the Safety of Nuclear Installations, Paris, Vol. 1, pp. 27–58. (3)

Ahmadi, G., and D. Ma, 1990, A Thermodynamical Formulation for Disposed Multiphase Turbulent Flows I. Basic Theory, *Int. J. Multiphase Flow 16:*323. (3)

Akoski, J., R. D. Watson, P. L. Goranson, A. Hassanian, and J. Salmanson, 1991, Thermal Hydraulic Design Issues and Analysis for ITER Diverters, *Fusion Technol. 19:*1729–1735. (4)

Alad'yev, J. T., Z. L. Miropolsky, V. E. Doroshchuk, and M. A. Styrikovich, 1961, Boiling Crisis in Tubes, in *International Developments in Heat Transfer,* Part 2, pp. 237–243, ASME, New York. (5)

Alad'yev, I. T., N. D. Gavrilova, and L. D. Dodonov, 1969, Hydrodynamics of a Two-Phase Flow of Potassium in Tubes, *ASME Heat Transfer—Sov. Res. 1*(4):1–13. (3)

Alessandrini, A., S. Bertoletti, G. P. Gaspari, C. Lombardi, G. Soldrini, and R. Zavattarelli, 1963, Critical Heat Flux Data for Fully Developed Flow of Steam-Water Mixtures in Round Vertical Tubes with an Intermediate Nonheated Section, Centro Informationi Studi Esperienzi Rept. CISE-R-69, Milan, Italy. (5)

Alleman, R. T., A. J. McElfresh, A. S. Neuls, W. C. Townsend, N. P. Wilburn, and M. E. Witherspoon,

* The number in parentheses at the end of each reference refers to the chapter in which this reference is first cited.

1970, Experimental High Enthalpy Blowdown from a Single Vessel through a Bottom Outlet, BNWL-1411, Battelle Northwest Lab., Richland, WA. (3)

Amblard, A., J. M. Delhaye, and C. Favrean, 1983, Proc. et dispositif de determination de laire interfaciale dans un melange diphasique comprenant une phase gazeuse en écoulement sous forme de bulles, Brevet 83, 06473, CEA, Paris. (3)

Analytis, G. Th., and G. Yadigaroglu, 1987, Analytical Modeling of Inverted Annular Film Boiling, *Nuclear Eng. Des. 99*:201–212. (4)

Anderson, D. L. J., R. L. Judd, and H. Merte, Jr., 1970, Site Activation Phenomena in Saturated Pool Nucleate Boiling, ASME Paper 70-HT-14, Fluids Engineering, Heat Transfer, and Lubrication Conf., Detroit, MI. (2)

Anderson, R. P., and D. R. Armstrong, 1973, Comparison between Vapor Explosion Models and Recent Experimental Results, AIChE Preprint 16, 14th Natl. Heat Transfer Conf., Atlanta, GA. (2)

Andreani, M., and G. Yadigaroglu, 1992, Difficulties in Modeling Dispersed Flow Boiling, *Warme und Stoffubertragung 27*:37–49. (4)

Andreychek, T. S., L. E. Hochreiter, and R. E. Newton, 1989, Void Fraction Measurement in Two-Phase Downflow in Irregular Channels, *ANS Proc. 26th Natl. Heat Transfer Conf.*, HTC-Vol. 4, p. 123. (3)

Anegawa, T., S. Abata, O. Yakomizo, and Y. Yoshimoto, 1990, Space-Dependent Stability Analysis of Corewide and Regional Modes in BWRs, *Proc. Int. Workshop on BWR Stability*, Holtsville, NY, pp. 337–353, CSNI Rep. 178, OECD-NEA, Paris. (6)

Anklam, T. M., 1981a, ORNL Small-Break LOCA Heat Transfer Test Series 1: Rod Bundle Heat Transfer Analysis, NUREG/CR 2052, ORNL/NUREG/TM-445, Oak Ridge Natl. Lab., Oak Ridge, TN. (4)

Anklam, T. M., 1981b, ORNL Small-Break LOCA Heat Transfer Test Series 1: High Pressure Reflood Analysis, NUREG/CR 2114, ORNL/NUREG/TM-446, Oak Ridge Natl. Lab., Oak Ridge, TN. (4)

Anode, Y., Y. Kukita, H. Nakamura, and K. Tasaka, 1989, Flow Regime Transition in High-Pressure Large-Diameter Horizontal Two-Phase Flow, *ANS Proc. 26th Natl. Heat Transfer Conf.*, Philadelphia, PA. (3)

ANS/ASME/NRC, 1980, *Proc. of Int. Topical Meeting on Nuclear Reactor Thermal-Hydraulics*, Saratoga, FL, Hemisphere, New York. (3)

Armand, A. A., 1959, The Resistance during the Movement of a Two-Phase System in Horizontal Pipes, *AERE Trans. from Russian 828*, U.K. Atomic Energy Research Establishment, Harwell, England. (3)

Asali, J. C., T. J. Hanratty, and P. Andreussi, 1985, Interfacial Drag and Film Height for Vertical Annular Flow, *AIChE J. 31*(6):895. (3)

Avksentyuk, B. P., and N. N. Mamontova, 1973, Characteristics of Heat Transfer Crisis during Boiling of Alkali Metals and Organic Fluids under Forced Convection Conditions at Reduced Pressures, in *Progress in Heat and Mass Transfer*, Vol. 7, p. 355, O. E. Dwyer, Ed., Pergamon Press, New York. (2)

Babcock, D. F., 1964, Heavy Water Moderated Power Reactors, Progress Rep., Jan.–Feb. 1964, USAEC Rep. DP-895. (5)

Baker, O., 1954, Simultaneous Flow of Oil and Gas, *Oil Gas J. 53*:185–190. (3)

Baker, O., 1960, Designing Pipelines for Simultaneous Flow of Oil and Gas, *Handbook Section, Pipeline Engineering*, PH 69. (3)

Bakstad, P., and K. O. Solberg, 1967, A Model for the Dynamics of Nuclear Reactors with Boiling Coolant with a New Approach to the Vapor Generation Process, KR-121, U.K. Atomic Energy Research Establishment, Harwell, England. (3)

Bakstad, P., and K. O. Solberg, 1968, RAMONAI: A Fortran Code for Transient Analyses of Boiling Water Reactors and Boiling Loops, KR-135, U.K. Atomic Energy Research Establishment, Harwell, England. (3)

Bankoff, S. G., 1960, A Variable-Density, Single-Fluid Model for Two-Phase Flow with Particular Reference to Steam-Water Flow, *Trans. ASME, J. Heat Transfer 82*:265–272. (3)

Bankoff, S. G., 1961, On the Mechanism of Subcooled Nucleate Boiling, Parts I and II, Chem. Eng. Prog. Symp. Ser. 57(32):156–172. (5)
Bankoff, S. G., 1994, Significant Questions in Thin Liquid Film Heat Transfer, *Trans. ASME, J. Heat Transfer 116*:10–16. (2)
Bankoff, S. G., and S. C. Lee, 1983, A Critical Review of the Flooding Literature, NUREG/CR 3060. (3)
Bankoff, S. G., and J. P. Mason, 1962, Heat Transfer from the Surface of a Steam Bubble in a Turbulent Subcooled Liquid Stream, *AIChE J. 8*(1):30–33. (2)
Bankoff, S. G., and R. D. Mikesell, 1959, Bubble Growth Rates in Highly Subcooled Nucleate Boiling, *Chem. Eng. Prog. Symp. Ser.* 55(29):95–102. (2)
Barclay, F. J., T. J. Ledwidge, and G. C. Cornfield, 1969, Some Experiments on Sonic Velocity in Two-Phase Critical Flow, Symp. on Fluid Mechanics and Measurements in Two-Phase Flow Systems, *Proc. Inst. Mech. Eng. 184*(Part 3C):185–194. (3)
Bar-Cohen, A., 1992, Hysteresis Phenomena at the Onset of Nucleate Boiling, *Engineering Foundation Conf. on Pool and External Flow Boiling,* Santa Barbara, CA, pp. 1–14. (4)
Barnea, D., 1987, A Unified Model for Predicting Flow Pattern Transitions for the Whole Range of Pipe Inclinations, *Int. J. Multiphase Flow 13:*1. (3)
Barnea, D., 1990, Effect of Bubble Shape on Pressure-Drop Calculations in Vertical Slug Flow, *Int. J. Multiphase Flow, 16:*79–89. (3)
Barnea, D., and N. Brauner, 1985, Holdup of the Liquid Slug in Two-Phase Intermittent Flow, *Int. J. Multiphase Flow 11:*43–49. (3)
Barnea, D., Y. Luninski, and Y. Taitel, 1983, Flow Pattern in Horizontal and Vertical Two-Phase Flow in Small Diameter Pipes, *Can. J. Chem. Eng. 61:*617–620. (3)
Barnea, D., O. Shoham, and Y. Taitel, 1982a, Flow Pattern Transition for Downward Inclined Two-Phase Flow, Horizontal to Vertical, *Chem. Eng. Sci. 37*(5):735–740. (3)
Barnea, D., O. Shoham, and Y. Taitel, 1982b, Flow Pattern Transition for Vertical Downward Two-Phase Flow, *Chem. Eng. Sci. 37*(5):741–744. (3)
Barnea, D., O. Shoham, Y. Taitel, and A. E. Dukler, 1985, Gas-Liquid Flow in Inclined Tubes: Flow Pattern Transitions for Upward Flow, *Chem. Eng. Sci. 40*(1):131. (3)
Barnett, P. G., 1963, An Investigation into the Validity of Certain Hypotheses Implied by Various Burn-out Correlations, UK Rep. AEEW-R-214. (5)
Barnett, P. G., 1966, A Correlation of Burnout Data for Uniformly Heated Annuli and Its Use for Predicting Burnout in Uniformly Heated Rod Bundles, UK Rep. AEEW-R-463, Winfrith, England. (5)
Baroczy, C. J., 1966, A Systematic Correlation for Two-Phase Pressure Drop, NAA-SR-Memo-11858, North American Aviation Co., Canoga Park, CA. (3)
Baroczy, C. J., 1968, Pressure Drops for Two-Phase Potassium Flowing through a Circular Tube and an Orifice, *Chem. Eng. Prog. Symp. Ser.* 64(82):12. (3)
Bauer, E. G., G. R. Houdayer, and H. M. Sureau, 1978, A Nonequilibrium Axial Flow Model and Application to Loss-of-Coolant Accident, in Proc. Transient Two-Phase Flow CSNI Specialists Meeting, 1976, *Atomic Energy of Canada 1:*429–437. (3)
Baumeister, K. J., and G. J. Schoessow, 1968, Creeping Flow Solution of Leidenfrost Boiling with a Moving Surface, NASA TMX 52443. (2)
Beasant, W. R., and H. W. Jones, 1963, The Critical Heat Flux in Pool Boiling under Combined Effects of High Acceleration and Pressure, UK Rep. AEEW-R 275, Winfrith, England. (2)
Beattie, D. R. M., 1975, Friction Factors and Regime Transitions in High Pressure Steam-Water Flows, ASME Paper 75-WA/HT-4. (3)
Beattie, D. R. M., 1983, An Extension of Single Phase Flow Turbulent Pipe Flow Concepts to Two-Phase Flow, Ph.D. thesis, University of New South Wales, Sydney, Australia. (3)
Beattie, D. R. M., and P. B. Whalley, 1982, A Simple Two Phase Frictional Pressure Drop Calculation Methods, *Int. J. Multiphase Flow 8*(1):83–87. (3)
Becker, K. M., 1963, Burnout Conditions for Flow of Boiling Water in Vertical Rod Clusters, *AIChE J. 9:*216–222. (5)

REFERENCES

Becker, K. M., 1971, Measurement of Burnout Conditions for Flow of Boiling Water in Horizontal Round Tubes, Atomenergia-Aktieb Rep. AERL-1262, 25, Nyköping, Sweden. (3)

Becker, K. M., J. Bager, and D. Djursing, 1971, Round Tube Burnout Data for Flow of Boiling Water at Pressures between 30 and 200 Bar, Rep. KTH-NEL-14, A B Atomenergi, Nyköping, Sweden. (5)

Becker, K. M., D. Djursing, K. Lindberg, O. Eklind, and C. Osterdahl, 1973, Burnout Conditions for Round Tubes at Elevated Pressures, in *Progress in Heat and Mass Transfer, Vol. 6*, Pergamon Press, New York. (5)

Becker, K. M., J. Engstrom, O. Nylund, B. Schalin, and B. Soderquist, 1990, Analysis of the Dryout Incident in the Oskarshamn 2 BWR, *Int. J. Multiphase Flow 16:*959–974. (5)

Becker, K. M., G. Hernborg, and M. Bode, 1962, An Experimental Study of Pressure Gradients for Flow of Boiling Water in Vertical Round Ducts (Part 4), Rep. AE-86, A B Atomenergi, Nyköping, Sweden. (3)

Belda, W., 1975, Dryoutverzing bei Kuhlmittelverlust in Kernreaktoren, Ph.D. thesis, from Technische Universität Hannover, Hannover, FRG. (5)

Bell, W. H., and B. W. LeTourneau, 1960, Experimental Measurements of Mixing in Parallel Flow Rod Bundles, WAPD-TH-381, Bettis Atomic Power Lab., Pittsburgh, PA. (App.)

Bellman, R., and R. H. Pennington, 1954, Effects of Surface Tension and Viscosity on Taylor Instability, *Quarter. Appl. Math., 12:*151. (6)

Bennett, A. W., G. F. Hewitt, H. A. Kearsey, and R. K. F. Keeys, 1965a, Measurements of Burnout Heat Flux in Uniformly Heated Round Tubes at 1000 psia, UK Rep. AERE-R-5055, Harwell, England. (5)

Bennett, A. W., G. F. Hewitt, H. A. Kearsey, R. K. F. Keeys, and P. M. C. Lacey, 1965b, Flow Visualization Studies of Boiling at High Pressures, UK Rep. AERE R-4874, Harwell, England. (3)

Bennett, A. W., G. F. Hewitt, H. A. Kearsey, R. K. F. Keeys, and D. J. Pulling, 1966, Studies of Burnout in Boiling Heat Transfer to Water in Round Tubes with Non-Uniform Heating, UK Rep. AEEW-R-5076, Winfrith, England. (5)

Bennett, A. W., G. F. Hewitt, H. A. Kearsey, R. K. F. Keeys, and D. J. Pulling, 1967a, Studies of Burnout in Boiling Heat Transfer to Water in Round Tubes with Non-Uniform Heating, *Trans. Inst. Chem. Eng.* (London), *45:*319. (5)

Bennett, A. W., G. F. Hewitt, H. A. Kearsey, and R. K. F. Keeys, 1967b, Heat Transfer to Steam Water Mixture in Uniformly Heated Tubes in Which the CHF Has Been Exceeded, UK Rep. AERE-R-5373, Harwell, England. (4)

Bennett, A. W., J. G. Collier, and P. M. C. Lacey, 1963, Heat Transfer to Mixtures of High Pressure Steam and Water in an Annulus, Part III. The Effect of System Pressure on the Burnout Heat Flux for an Internally Heated Unit, UK Rep. AERE-R-3934, Harwell, England. (5)

Bennett, D. L., and J. C. Chen, 1980, Forced Convective Boiling in Vertical Tubes for Saturated Pure Components, and Binary Mixtures, *AIChE J. 26:*454. (4)

Bennett, D. L., M. W. Davis, and B. L. Hertzler, 1980, The Suppression of Saturated Nucleate Boiling by Forced Convective Flow, *AIChE Symp. Ser. 76*(199):91. (4)

Berenson, P. J., 1960, Transition Boiling Heat Transfer, 4th Natl. Heat Transfer Conf., AIChE Preprint 18, Buffalo, NY. (2)

Berenson, P. J., 1961, Film-Boiling Heat Transfer from a Horizontal Surface, *Trans. ASME, Ser. C, J. Heat Transfer 83:*351. (2)

Berenson, P. J., 1962, Experiments on Pool Boiling Heat Transfer, *Int. J. Heat Mass Transfer 5:*985. (2)

Bergelin, O. P., and C. Gazley, 1949, *Proc. Heat Transfer Fluid Mech. Inst.*, May. (3)

Bergles, A. E., 1992, What is the Real Mechanism of CHF in Pool Boiling, in *Pool and External Flow Boiling*, V. K. Dhir and A. E. Bergles (Eds.), ASME, New York. (2)

Bergles, A. E., P. Goldberg, and J. S. Maulbetsch, 1967a, Acoustic Oscillations in a High Pressure Single Channel Boiling System, EURATOM Rep., Proc. Symp. on Two-Phase Flow Dynamics, Eindhoven, pp. 525–550. (6)

Bergles, A. E., R. F. Lopina, and M. P. Fiori, 1967b, Critical Heat Flux and Flow Pattern Observations for Low Pressure Water Flowing in Tubes, *Trans. ASME, J. Heat Transfer 89:*69–74. (6)

REFERENCES 495

Bergles, A. E., and W. M. Rohsenow, 1964, The Determination of Forced Convection Surface Boiling Heat Transfer, *Trans. ASME, J. Heat Transfer* 86:365–372. (4)
Bergles, A. E., J. P. Ross, and J. G. Bourne, 1968, Investigation of Boiling Flow Regimes and Critical Heat Flux, NYO-3304-13, Dynatech Corp., Cambridge, MA. (3)
Bergles, A. E., and W. G. Thompson, Jr., 1970, The Relationship of Quench Data to Steady State Pool Boiling Data, *Int. J. Heat Mass Transfer* 13:55–68. (4)
Bernath, L., 1960, A Theory of Local Boiling Burnout and Its Application to Existing Data, *Chem. Eng. Prog., Symp. Ser.* 56(30):95–116. (2)
Bernier, R. N., and C. E. Brennen, 1983, Use of the Electromagnetic Flowmeter in a Two-Phase Flow, *Int. J. Multiphase Flow* 9:251. (3)
Bertoletti, S., G. P. Gaspari, C. Lombardi, G. Peterlongo, M. Silvestri, and F. A. Tacconi, 1965, Heat Transfer Crisis with Steam-Water Mixtures, *Energia Nucleare* 12(3):121–172. (5)
Bertoletti, S., G. P. Gaspari, C. Lombardi, G. Peterlongo, and F. A. Tacconi, 1964b, Heat Transfer Crisis with Steam-Water Mixtures, Centrol Inf. Studi Esperimenze Rep. CISE-R-99, Milan, Italy. (5)
Bertoletti, S., G. P. Gaspari, C. Lombardi, G. Soldaini, and R. Zavattarelli, 1964a, Heat Transfer Crisis in Steam-Mixtures, Experimental Data in Round Tubes and Vertical Upflow Obtained during the CAN 2 Program, Centrol Inf. Studi Esperimenze Rep. CISE-R-90, Milan, Italy. (5)
Bertoletti, S., J. Lesage, C. Lombardi, G. Peterlongo, M. Silvestri, and F. Weckermann, 1961, Heat Transfer and Pressure Drop with Steam Water Spray, Centrol Inf. Studi Esperimenze Rep. CISE-R-36, Milan, Italy. (4)
Besnard, D. C., and F. H. Harlow, 1988, Turbulence in Multiphase Flow, *Int. J. Multiphase Flow* 14:679. (3)
Bhat, A. M., R. Prakash, and J. S. Saini, 1983, Heat Transfer in Nucleate Pool Boiling at High Heat Flux, *Int. J. Heat Mass Transfer* 26:833–840. (2)
Biagioli, F., and A. Pramoli, 1967, A Measurement Technique for Voiding Transients in Two-Phase Flow, EURATOM Symp. on Two-Phase Flow Dynamics, Vol. II, p. 1743, Commission of the European Communities, Center for Information and Documentation, Brussels. (3)
Biancone, F., A. Campanile, G. Galimi, and M. Goffi, 1965, Forced Convection Burnout and Hydrodynamic Instability. Experiments for Water at High Pressure. I. Presentation of Data for Round Tubes with Uniform and Non-Uniform Power Distribution, Italian Rep. EUR-2490 e, European Atomic Energy Community, Brussels, Belgium. (5)
Biasi, L., G. C. Clerici, R. Sala, and A. Tozzi, 1967, Studies on Burnout: Pt. 3, *Energia Nucleare* 14:530. (5)
Biasi, L., G. C. Clerici, R. Sala, and A. Tozzi, 1968, Extention of A.R.S. Correlation to Burnout Prediction with Non-Uniform Heating, *J. Nuclear Energy* 22:705–716. (5)
Bibean, E. L., and M. Salcudean, 1994, A Study of Bubble Ebullition in Forced Convective Subcooled Nucleate Boiling at Low Pressures, *Int. J. Heat Mass Transfer* 37(15):2245–2259. (4)
Binder, J. L., and T. J. Hanratty, 1992, Use of Lagrangian Methods to Describe Drop Deposition and Distribution in Horizontal Gas-Liquid Annular Flows, *Int. J. Multiphase Flow* 18(6):803–821. (3)
Bishop, A. A., R. O. Sandberg, and L. S. Tong, 1965, Forced Convection Heat Transfer to Water at Near-Critical Temperature and Super-Critical Pressures, USAEC Rep. WCAP-2056-Part IIIB, and also Paper 2-7, AIChE/IChE Joint Meeting, London, June, 1965. (4)
Blumenkrantz, A., and J. Taborek, 1971, Application of Stability Analysis for Design of Natural Circulation Boiling Systems and Comparison with Experimental Data, AIChE Paper 13, Natl. Heat Transfer Conf., Tulsa, OK. (6)
Board, S. J., and R. B. Duffey, 1971, Spherical Vapor Bubble Growth in Superheated Liquids, *Chem. Eng. Sci.* 26:263. (2)
Bobrovish, G. I., B. P. Avksentyuk, and N. N. Mamontova, 1967, On the Mechanism of Boiling of Liquid Metals, Proc. Semi-Int. Symp. on Heat Transfer, Tokyo, Jpn. Soc. Mech Eng. p. 171. (2)
Bonilla, C. F., 1957, *Nuclear Energy*, chp. 9, Heat Removal, McGraw-Hill Co., New York. (6)

Bonilla, C. F., J. J. Grady, and G. W. Avery, 1965, Pool Boiling Heat Transfer from Scored Surface, *Chem. Eng. Prog. Symp. Ser.* 61(57):280. (2)

Bonilla, C. F., D. L. Sawhuey, and N. M. Makansi, 1962, Vapor Pressure of Alkali Metals III, Rubidium, Cesium, and Sodium-Potassium alloy up to 100 psia, *Proc. 1962 High Temperature Liquid Metal Heat Transfer Tech. Meeting*, BNL-756, Brookhaven, NY. (3)

Borgartz, B. O., T. P. O'Brien, N. J. M. Rees, and A. V. Smith, 1969, Experimental Studies of Water Decompression through Simple Pipe Systems, CREST Specialist Meeting on Depressurization Effects in Water-Cooled Reactors, Bettelle Institute, Frankfurt, FRG. (3)

Borishansky, V. M., 1953, Heat Transfer to Liquid Freely Flowing over a Surface Heated to a Temperature above the Boiling Point, in *Problems of Heat Transfer during a Change of State: A Collection of Articles*, S. S. Kutateladze, ed., Rep. AEC-tr-3405, p. 109. (2)

Borishansky, V. M., 1961, Allowing for the Influence of Pressure on the Heat Transfer and Critical Thermal Loads during Boiling in Accordance with the Theory of Thermodynamic Similarity, *Voprosy Teplootdachi i gidravliki Dvukhfaznykh Sred*, Gosenergoizdet, Moscow. (2)

Borishansky, V. M., and B. S. Fokin, 1969, Onset of Heat Transfer Crisis with Unsteady Increase in Heat Flux, *Heat Transfer Sov. Res.* 1(5):27–55. (5)

Borishansky, V. M., and I. I. Paleev, Eds., 1964, *Convective Heat Transfer in Two-Phase and One-Phase Flows*, transl. from Russian–Israeli Program for Scientific Translation, U.S. Dept. of Commerce, Washington, DC, 1969. (4)

Borishansky, V. M., I. I. Novikov, and S. S. Kutateladze, 1961, Use of Thermodynamic Similarity in Generalizing Experimental Data of Heat Transfer, *Int. Dev. Heat Transfer, Part 2*, pp. 475–482, ASME, New York. (5)

Borishansky, V. M., K. A. Zhokhov, A. A. Andreevsky, M. A. Putilin, A. P. Kozirev, and L. L. Shneiderman, 1965, Heat Transfer to Boiling Alkali Metals, *Atomic Energy 19:*191. (2)

Bouré, J. A., 1966, The Oscillator Behavior of Heated Channels, Part I & II, French Rep. CEA-R 3049, Grenoble, France. (6)

Bouré, J. A., 1975, Mathematical Modeling and the Two-Phase Constitutive Equations, European Two-Phase Flow Group Meeting, Haifa, Israel. (3)

Bouré, J. A., 1976, Mathematical Modeling of Two-Phase Flows, Its Bases—Problems: A Review, Services des Transferts Thermiques, Centre d'Etudes Nucleaires de Grenoble, France; see also Proc. OECD Specialists Meeting on Transient Two-Phase Flow, Toronto, Canada. (3)

Bouré, J. A., A. E. Bergles, and L. S. Tong, 1973, Review of Two-Phase Flow Instability, *Nuclear Eng. Design* 25:165–192. (1)

Bouré, J. A., and A. Ihaila, 1967, The Oscillator Behavior of Heated Channels, EURATOM Rep., Proc. Symp. on Two-Phase Flow Dynamics, Eindhoven. (6)

Bowers, M. B., and I. Mudawar, 1994, High Flux Boiling in Low Flow Rate, Low Pressure-Drop Minichannel Heat Sinks, *Int. J. Heat Mass Transfer* 37(2):321–332. (5)

Bowring, R. W., 1962, Physical Model, Based on Bubble Detachment and Calculation of Steam Voidage in the Subcooled Region of a Heated Channel, Inst. for Atomenergie Rep. HRP-10, Oslo, Norway. (3)

Bowring, R. W., 1967a, HAMBO: A Computer Programme for the Sub-Channel Analysis of the Hydraulic and Burnout Characteristics of Rod Clusters, European Two-Phase Heat Transfer Meeting, U.K., Durey Hall, Bournemouth, England. (3)

Bowring, R. W., 1967b, HAMBO: A Computer Programme for the Sub-Channel Analysis of the Hydraulic and Burnout Characteristics of Rod Cluster, Part I. General Description, Rep. AEEW-R-524, UK Atomic Energy Authority, Winfrith, England; and 1968, Part II. Equations, Rep. AEEW-R-582, UK Atomic Energy Authority, Winfrith, England. (3)

Bowring, R. W., 1972, A Simple but Accurate Round Tube, Uniform Heat Flux, Dryout Correlation over Pressure Range 0.7–17 MN/m² (100–2500 psia), Rep. AEEW-R-789, UK Atomic Energy Authority, Winfrith, England. (5)

Bowring, R. W., 1979, WSC-2: A Subchannel Dryout Correlation for Water Cooled Clusters over the

Pressure Range 3.4–15.9 MPa (500–2300 psia), Rep. AEEW-R-983, UK Atomic Energy Authority, Winfrith, England. (5)
Bowring, R., and P. Moreno, 1976, COBRA-IIIC/MIT Computer Code Manual, MIT Department of Nuclear Engineering, Cambridge, MA. (5)
Bradfield, W. S., 1967, On the Effect of Subcooling on Wall Superheat in Pool Boiling, *Trans. ASME, J. Heat Transfer 89:*269–270. (4)
Braver, H., and F. Mayinger, 1992, Onset of Nucleate Boiling and Hysteresis Effects under Convective and Pool Boiling, Engineering Foundation Conf. on Pool and External Flow Boiling, Santa Barbara, CA, pp. 1–14. (4)
Breen, B. P., and J. W. Westwater, 1962, Effect of Diameter of Horizontal Tubes on Film Boiling Heat Transfer, *Chem. Eng. Prog. 58:*67. (2)
Bromley, L. A., 1950, Heat Transfer in Stable Film Boiling, *Chem. Eng. Prog. 46:*221. (2)
Bromley, L. A., N. LeRoy, and J. A. Robbers, 1953, Heat Transfer in Forced Convection Film Boiling, *Ind. Eng. Chem. 45:*2639–2646. (4)
Brown, R. C., P. Andreussi, and S. Zanelli, 1978, *Can. J. Chem. Eng. 56:*754. (3)
Brown, W. T., Jr., 1967, A Study of Flow Surface Boiling, D.Sci. thesis, Massachusetts Institute of Technology, Cambridge, MA. (4)
Burnell, J. G., 1947, Flow of Boiling Water through Nozzles, Orifices and Pipes, *Engineering 164:*572–576. (3)
Butterworth, D., 1972, A Visual Study of Mechanism, in Horizontal Air-Water Systems, UK AERE Rep. M2556, Harwell, England. (3)
Butterworth, D., and A. W. Shock, 1982, Flow Boiling, *Proc. 7th Int. Heat Transfer Conf.*, Munich, Germany, Vol. 1, p. 15, Hemisphere, New York. (4)
Caira, M., E. Cipollone, M. Cumo, and A. Naviglio, 1985, Heat Transfer in Forced Convective Boiling in a Tube Bundle, 3rd Int. Topic Meeting on Reactor Thermohydraulics, ANS, Saratoga, New York. (4)
Campanile, F., G. Galimi, M. Goffi, and G. Passavanti, 1970, Forced Convection Burnout and Hydrodynamic Instability Experiments for Water at High Pressures: Part 6. Burnout Heat Flux Measurements on 9-Rod Bundles with Longitudinally and Transversally Uniform Heat Flux Generation, EUR-4468e. SORIN, Saluggia, Italy. (5)
Campolunghi, F., M. Cumo, G. Palazzi, and G. C. Urbani, 1977a. Subcooled and Bulk Boiling Correlations for Thermal Design of Steam Generators, *CNEN RT/ING* 77:10, Comitato Nazionale Per L'Energía Nucleare, Milan. Italy. (4)
Campolunghi, F., M. Cumo, G. Furari, and G. Palazzi, 1977b, Full Scale Tests and Thermal Design Correlations for Coiled, Once-Through Steam Generators, *CNEN RT/ING* 77:11, Comitato Nazionale Per L'Energía Nucleare, Milan, Italy. (4)
Caraher, D. L., and T. L. DeYoung, 1975, Interim Report on the Evaluation of Critical Flow Models, Aerojet Nuclear Company, San Diego, CA. (3)
Carey, Van P., 1992, Convective Boiling in Tubes and Channels, in *Liquid-Vapor Phase Change Phenomena,* chap. 12, Hemisphere, Washington, DC. (5)
Carne, M., and D. H. Charlesworth, 1966, Thermal Conduction Effects on the Critical Heat Flux in Pool Boiling, *Chem. Eng. Prog. Symp. Ser. 62*(64):24–34. (5)
Carver, M. B., 1970, Effect of By-Pass Characteristics on Parallel Channel Flow Instabilities, *Proc. Inst. Mech. Eng. 184,* Part 3C. (6)
C-E Report, 1975, C-E Critical Heat Flux Correlation for C-E Fuel Assemblies with Standard Spacer Grids, Part I, Uniform Axial Power Distribution, CENPD-162, Combustion Engineering Co., Winsor, CT. (5)
C-E Report, 1976, C-E Critical Heat Flux Correlation for C-E Fuel Assemblies with Standard Spacer Grids, Part II, Non-Uniform Axial Power Distribution, CENPD-207, Combustion Engineering Co., Winsor, CT. (5)
Celata, G. P., 1991, A Review of Recent Experiments and Predictive Aspects of Burnout at Very High Fluxes, Proc. Int. Conf. on Multiphase Flow, Tsukuba, Japan, September. (5)

Celata, G. P., M. Cumo, F. D'Annibale, G. E. Farello, and T. Setaro, 1985, Critical Heat Flux in Flow Transients, in *Advances in Heat Transfer,* Vol. IV, ENEA Publicatione, Rome. (5)

Celata, G. P., M. Cumo, A. Mariani, M. Simoncini, and G. Zummo, 1994a, Rationalization of Existing Mechanistic Models for the Prediction of Water Subcooled Flow Boiling CHF, *Int. J. Heat Mass Transfer 37*(Suppl.):347–360. (5)

Celata, G. P., M. Cumo, and A. Mariani, 1994b, Enhancement of CHF for Water Subcooled Flow Boiling in Tubes Using Helically Coiled Wires, *Int. J. Heat Mass Transfer 37*(1):53–67. (5)

Celata, G. P., M. Cumo, and A. Mariani, 1994c, Assessment of Correlations and Models for the Prediction of CHF in Subcooled Flow Boiling, *Int. J. Heat and Mass Transfer 37*(2):237–255. (5)

Celata, G. P., and A. Mariani, 1993, A Data Set of CHF in Water Subcooled Flow Boiling, Specialists' Workshop on the Thermal-Hydraulics of Hight Heat Flux Components in Fusion Reactors, Coordinated by G. P. Celata/A. Mariani, ENEA-C.R.E. Casaccia, Italy. (5)

Cermak, J. O., R. F. Farman, L. S. Tong, J. E. Casterline, S. Kokolis, and B. Matzner, 1970, The Departure from Nucleate Boiling in Rod Bundles during Pressure Blowdown, *Trans. ASME, Ser. C, J. Heat Transfer 92*(4):621–627. (4)

Cermak, J. O., R. Rosal, L. S. Tong, J. E. Casterline, S. Kokolis, and B. Matzner, 1971, High Pressure Rod-Bundle DNB Data with Axial Non-Uniform Heat Fluxes, Rep. WCAP-5727, Westinghouse Electric Corporation, Pittsburgh, PA. (5)

Chambré, P., and E. Elias, 1977, Rewetting Model Using a Generalized Boiling Curve, EPRI NP-571, Topical Rep. (University of California, Berkeley), Electric Power Research Inst., Palo Alto, CA. (4)

Chang, S. H., and K. W. Lee, 1988, A Derivation of Critical Heat Flux Model in Flow Boiling at Low Qualities Based on Mass, Energy, and Momentum Balance, Korea Advanced Institute of Science and Technology, Taejeon, Korea.

Chang, S. H., and K. W. Lee, 1989, A CHF Model Based on Mass, Energy, and Momentum Balance for Upflow Boiling at Low Qualities, *Nuclear Eng. Design 113:*35–50. (5)

Chang, Y. P., 1957, A Theoretical Analysis of Heat Transfer in Natural Convection and in Boiling, *Trans. ASME, J. Heat Transfer 79:*1501–1513. (2)

Chang, Y. P., 1959, Wave Theory of Heat Transfer in Film Boiling, *Trans. ASME, J. Heat Transfer 81:*1. (2)

Chang, Y. P., 1961, An Analysis of the Critical Conditions and Burnout in Boiling Heat Transfer, USAEC Rep. TID-14004, Washington, D.C. (2)

Chelemer, H., J. Weisman and L. S. Tong, 1972, Subchannel Thermal Analysis of Rod Bundle Cores, *Nuclear Engineering and Design, 21,* 35–45. (5)

Chen, J. C., 1966, A Correlation for Boiling Heat Transfer to Saturated Fluids in Convective Flow, *Ind. & Eng. Chem., Process Design and Dev. 5:*322. (4)

Chen, J. C., 1963b, A Proposed Mechanism and Method of Correlation for Convective Boiling Heat Transfer with Liquid Metals, BNL-7319, Brookhaven Natl. Lab., Brookhaven, NY. (4)

Chen, J. C., 1965, Non-equilibrium Inverse Temperature Profile in Boiling Liquid Metal Two-Phase Flow, *AIChE J. 11*(6):1145–1148. (4)

Chen, J. C., 1968, Incipient Boiling Superheats in Liquid Metals, *Trans. ASME, J. Heat Transfer 90:*303–312. (2)

Chen, J. C., 1970, An Experimental Investigation of Incipient Vaporization of Potassium in Convective Flow, in *Liquid-Metal Heat Transfer and Fluid Dynamics,* J. C. Chen and A. A. Bishop, Eds., Winter Annual Meeting, p. 129, ASME, New York. (4)

Chen, J. C., and S. Kalish, 1970, An Experimental Investigation of Two-Phase Pressure Drop for Potassium with and without Net Vaporization, 4th Int. Heat Transfer Conf., Paris-Versailles. (3)

Chen, J. C., et al., 1966, Heat Transfer Studies with Boiling Potassium, Nuclear Eng. Dept., Brookhaven Natl. Lab. Annual Report, BNL-50023 (S-69), pp. 52–54, Brookhaven, NY. (4)

Chen, J. C., F. T. Ozkaynak, and R. K. Sundaram, 1979, Vapor Heat Transfer in Post CHF Region Including the Effect of Thermal Nonequilibrium, *Nuclear Eng. Design 51:*143–155. (4)

Cheng, L. Y., D. A. Drew, and R. T. Lahey, Jr., 1983, An Analysis of Wave Dispersion, *Sonic Velocity and Critical Flow in Two-Phase Mixtures,* NUREG-CR-3372, US NRC, Washington, DC. (3)
Cheng, L. Y., D. A. Drew, and R. T. Lahey, Jr., 1985, An Analysis of Wave Propagation in Bubbly, Two-Component, Two Phase Flow, *Trans. ASME J. of Heat Transfer, 107:*402–408. (3)
Cheng, L. Y., and P. R. Tichler, 1991, CHF for Free Convection Boiling in Their Rectangular Channels, ANS Proc. Natl. Heat Transfer Conf., Minneapolis, MN, pp. 83–90. (2)
Cheremisinoff, N. P., and E. J. Davis, 1979, Stratified Turbulent-Turbulent Gas-Liquid Flow, *AIChE J. 25:*48–56. (3)
Chexal, V. K., and A. E. Bergles, 1973, Two-Phase Instabilities in a Low Pressure Natural Circulation Loop, *AIChE Symp. Ser. 69* (131):37–45. (3)
Chexal, G., J. Horowitz, and G. Lellouche, 1987, An Assessment of Eight Void Fraction Models, in *Heat Transfer—Pittsburgh, AIChE Symp. Ser. 83* (257):249–254; Also 1986, EPRI, Nuclear Safety Analysis Center, Rep. NSAC-107. (3)
Chisolm, D., 1967, A Theoretical Bases for the Lockhart/Martinelli Correlation for Two-Phase Flow, *Int. J. Heat Mass Transfer 10:*1767–1778. (3)
Chisolm, D., 1971, Prediction of Pressure Drop at Pipe Fittings during Two-Phase Flow, *Proc. 13th Int. Inst. of Refrigeration Congress,* Washington, DC, vol. 2, pp. 781–789. (3)
Chisolm, D., 1991, A Form of Correlation for Forced Convective Boiling in Tubes, in *Heat Transfer—Minneapolis, AIChE Symp. Ser. 87* (283):391–398. (4)
Chisolm, D., and L. A. Sutherland, 1969, Prediction of Pressure Gradients in Pipeline Systems during Two-Phase Flow, *Symp. on Fluid Mechanics and Measurements in Two-Phase Flow Systems,* Leeds, pp. 24–25. (3)
Chu, K. J., and A. E. Dukler, 1974, Statistical Characteristics of Thin Wavy Films, *AIChE J. 20:*695. (3)
Chu, P. T., H. Chelemer, and L. E. Hochreiter, 1973, THINC IV: An Improved Program for Thermal Hydraulic Analysis of Rod Bundle Cores, WCAP-7965, Westinghouse Electric Corporation, Pittsburgh, PA. (App.)
Chu, P. T., L. E. Hochreiter, H. Chelemer, H. Bowman, and L. S. Tong, 1972, THINC IV: A New Thermal Hydraulic Code for PWR Design, *Trans. Am. Nuclear Soc. 25:*876. (App.)
Chyu, M. C., 1987, Evaporation of Macrolayer in Nucleate Boiling Near Burnout, *Int. J. Heat Mass Transfer 30:*1531–1538. (2)
Chyu, M. C., 1989, Formation and Heat Transfer Mechanism of Vapor Mass during Nucleate Boiling, in *Thermal, Non-equilibrium in Two-Phase Flow,* pp. 157–181, ENEA, Rome. (2)
Cichelli, M. T., and C. F. Bonilla, 1945, Heat Transfer to Liquid Boiling under Pressure, *Trans. AIChE 41:*755. (2)
CISE, 1963, *A Research Program in Two-Phase Flow,* Centro Information Studi Esperienzi, Milan, Italy. (3)
Clark, H. B., P. S. Strenge, and J. W. Westwater, 1959, Active Sites for Nucleation, *Chem. Eng. Prog. Symp. Ser. 55 (29):*103–110. (2)
Clark, L. T., and M. F. Parkman, 1964, Effect of Additives on Wetting during Mercury Pool Boiling Heat Transfer, Rep. AGN-TP-55, Aerojet-General Nucleonics, San Diego, CA. (2)
Clements, L. D., and C. P. Colver, 1964, Generalized Correlation for Film Boiling, *Trans. ASME, J. Heat Transfer 86:*213. (2)
Clerici, G. C., S. Garribba, R. Sala, and A. Tozzi, 1966, Catalogue of Burnout Correlations for Forced Convection in the Quality Region, USAEC Rep. EURAEC-1729, *United States-Euratom Joint Research and Development Program,* Washington, DC. (5)
Clerici, G. C., S. Garribba, R. Sala, and A. Tozzi (ARS), 1967, Studies on Burnout Part 1—Alternative Forms for Some Burnout Correlations, *Energia Nucleare 14*(5)480. (5)
Clinch, J. M., and H. B. Karplus, 1964, An Analytical Study of the Propagation of Pressure Waves in Liquid Hydrogen-Vapor Mixtures, Report IITRI-N-6054-6, IIT Research Inst., NASA-CR-54015. (3)
Cobb, C. B., and E. L. Park, Jr., 1969, Nucleate Boiling—A Maximum Heat Flux Correlation for Corresponding States Liquids, *Chem. Eng. Prog. Symp. Ser. 65*(92):188–193. (4)
Coffield, R. J., Jr., W. M. Rohrer, Jr., and L. S. Tong, 1967, An Investigation of Departure from Nucle-

ate Boiling (DNB) in a Crossed Rod Matrix with Normal Flow of Freon 113 Coolant, *Nuclear Eng. Design* 6:147–154. (5)

Cohen, L. S., and T. J. Hanratty, 1968, Effects of Waves at a Gas-Liquid Interface on a Turbulent Air Flow, *J. Fluid Mech. 31:*467. (3)

Cohen, P., 1960, *Water Coolant Technology of Power Reactors,* AEC Monograph, Gorden & Breach, New York. (4)

Cole, R., 1960, A Photographic Study of Pool Boiling in the Region of the Critical Heat Flux, *AIChE J. 6:*533. (2)

Cole, R., 1967, Bubble Frequencies and Departure Volumes at Subatmospheric Pressures, *AIChE J. 13*(4):779–783. (2)

Cole, R., 1970, Boiling Nucleation, in *Advances in Heat Transfer* J. P. Hartnett and T. F. Irvine, Eds., vol. 10, pp. 86–166, Academic Press, New York. (2)

Cole, R., and W. R. Rohsenow, 1969, Correlation of Bubble Departure Diameters for Boiling of Saturated Liquids, *AIChE Chem. Eng. Prog. Symp. Ser.* 65(92):211. (2)

Cole, R., and H. L. Shulman, 1966, Bubble Departure Diameters at Subatmospheric Pressures, *AIChE Symp. Ser.* 62(64):6–16. (2)

Collier, J. G., and J. R. Thome, 1994, *Convective Boiling and Condensation,* 3rd ed., pp. 170, 175, McGraw-Oxford Univ. Press, London & New York. (4)

Collier, J. G., and D. J. Pulling, 1962, Heat Transfer to Two-Phase Gas-Liquid System, Part II, Further Data on Steam-Water Mixtures in the Liquid Dispersed Region in an Annulus, UK Rep. AERE-R-3809, Harwell, England. (4)

Collins, D. B., and M. Gacesa, 1969, Hydrodynamic Instability in a Full-Scale Simulated Reactor Channel, *Proc. Inst. Mech. Eng. 184.* (6)

Collins, D. B., M. Gacesa, and C. B. Parsons, 1971, Study of the Onset of Premature Heat Transfer Crisis during Hydrodynamic Instability in a Full Scale Reactor Channel, ASME Paper 71-HT-11, Natl. Heat Transfer Conf., Tulsa, OK. (5)

Colver, C. P., and R. E. Balzhiser, 1964, A Study of Saturated Pool Boiling of Potassium up to Burnout Heat Flux, AIChE Paper presented at 87th Nat. Heat Transfer Conf., Cleveland, OH. (2)

Cook, W. H., 1960, Fuel Cycle Program—A Boiling Water Research and Development Program, USAEC Rep. GEAP-3558, p. 35, General Electric, San Jose, CA. (5)

Cooper, M. G., 1984, Heat Flow Rates in Saturated Nucleate Pool Boiling—A Wide-Ranging Examination Using Reduced Properties, *Advances in Heat Transfer* 16:157–239. (2)

Cooper, M. G., and A. J. P. Lloyd, 1966, Transient Local Heat Flux in Nucleate Boiling, *Proc. 3rd Int. Heat Transfer Conf.,* vol. 3, pp. 193–203, AIChE, New York. (2)

Cooper, M. G., and A. J. P. Lloyd, 1969, The Microlayer in Nucleate Pool Boiling, *Int. J. Heat Mass Transfer* 12:895. (2)

Core, T. C., and K. Sato, 1958, Determination of Burnout Limits of Polyphenyl Coolants, USAEC Rep. IDO-28007, Washington, DC. (5)

Costello, C. P., and J. M. Adams, 1961, Burnout Heat Fluxes in Pool Boiling at High Acceleration, in *International Developments in Heat Transfer, Part II,* ASME, New York. (2)

Costello, C. P., C. O. Bock, and C. C. Nichols, 1965, A Study of Induced Convective Effects on Pool Boiling Burnout, *AIChE Chem. Eng. Prog. Symp. Ser.* 61:271–280. (2)

Costello, C. P., and W. J. Frea, 1965, A Salient Non-Hydrodynamic Effect on Pool Boiling Burnout of Small Semi Cylindrical Heaters, *AIChE Chem. Eng. Prog. Symp. Ser.* 61(57):258–268. (2)

Cousins, L. B., W. H. Denton, and G. F. Hewitt, 1965, Liquid Mass Transfer in Annular Two-Phase Flow, UK Rep. AERE-R-4926, Harwell, England. (5)

Cousins, L. B., and G. F. Hewitt, 1968, Liquid Phase Mass Transfer in Annular Two-Phase Flow, UK Rep. AERE-R-5657, Harwell, England; also Symp. on Two-Phase Flow, Paper C4, Exeter, England.

Cozzuol, J. M., O. M. Hanner, and G. G. Loomis, 1978, Inv. of Influence of Simulated Steam Generator Tube Ruptures during L-O-C Experiments in the Semiscale MOD-1 System, NUREG/CR-0175, EGG Idaho Inc. Idaho Nat. Eng. Lab., Idaho Falls, ID. (4)

Cravarolo, L., and A. Hassid, 1965, Liquid Volume Fraction in Two-Phase Adiabatic Systems, Exeter Symposium, England. (3)

Cronenberg, A. W., H. K. Fauske, S. G. Bankoff, and D. T. Eggen, 1971, A Single-Bubble Model for Sodium Expulsion from a Heated Channel, *Nuclear Eng. Design 16:*285–293. (4)

Crowley, J. D., C. Deane, and S. W. Gouse, Jr., 1967, Two-Phase Flow Oscillations in Vertical, Parallel Heated Channels, EURATOM Rep., Proc. Symp. on Two-Phase Flow Dynamics, Eindhoven, p. 1131. (6)

Crowley, C. J., G. B. Wallis, and J. J. Barry, 1992, Validation of One-Dimensional Wave Model for the Stratified-to-Slug Regime Transition, *Int. J. Multiphase Flow 18:*249–271. (3)

Crowley, C. J., G. B. Wallis, and J. J. Barry, 1993, Dimensionless Form of a One-Dimensional Wave Model for the Stratified Flow Regime Transition, *Int. J. Multiphase Flow 19*(2):369–376. (3)

Cumo, M., 1972, Personal communication, Comitato Nazionale Per L'energia Nucleare, Milan, Italy. (5)

Cumo, M., G. E. Farello, and G. Ferrari, 1969, Bubble Flow up to the Critical Pressure, ASME Paper 69-HT-30, Natl. Heat Transfer Conf., Minneapolis, MN. (5)

Cumo, M., G. E. Farello, and G. Ferrari, 1971, The Influence of Curvature in Post Dryout Heat Transfer, Rep. of XXVI Natl. ATI Annual Meeting, L'Aquilia, Italy. (4)

Cumo, M., and A. Naviglio, 1988, *Thermal Hydraulics,* vol. 1, pp. 49, 118, CRC Press, Boca Raton, FL. (3)

Cumo, M., and A. Palmieri, 1967, The Influence of Geometry on Critical Heat Flux in Subcooled Boiling, AIChE Preprint 18, 9th Natl. Heat Transfer Conf., Seattle, WA. (5)

Currin, H. B., C. M. Hunin, L. Rivlin, and L. S. Tong, 1961, HYDNA—Digital Computer Program for Hydrodynamic Transients in a Pressure Tube Reactor or a Closed Channel Core, USAEC Rep. CYNA-77, Washington, DC. (6)

Daleas, R. S., and A. E. Bergles, 1965, Effects of Upstream Compressibility on Subcooled Critical Heat Flux, ASME Paper 65-HT-67, Natl. Heat Transfer Conf., Los Angeles. (6)

Davis, E. J., and G. H. Anderson, 1966, The Incipience of Nucleate Boiling in Forced Convection Flow, *AIChE J. 12:*774–780. (4)

Dean, R. A., R. S. Dougall, and L. S. Tong, 1971, Effect of Vapor Injection on Critical Heat Flux in a Subcooled R-113 (Freon) Flow, Proc. Int. Symp. on Two-Phase Flow Systems, Haifa, Israel. (6)

Deane, C. W., and W. M. Rohsenow, 1969, Mechanism and Behavior of Nucleate Boiling Heat Transfer to the Alkali Liquid Metals, USAEC Rep. DSR 76303-65, Massachusetts Institute of Technology, Cambridge, MA; Also in 1970, *Liquid Metal Heat Transfer and Fluid Dynamics,* J. C. Chen and A. A. Bishop, Eds., ASME Winter Annual Meeting, New York. (4)

DeBortoli, R. A., S. J. Green, B. W. LeTourneau, M. Troy, and A. Weiss, 1958, Forced Convection Heat Transfer Burnout Studies for Water in Rectangular Channels and Round Tubes at Pressures above 500 psia, USAEC Rep. WAPD-188, Pittsburgh, PA. (5)

DeBortoli, R. A., and R. Masnovi, 1957, Effect of Dissolved Hydrogen on Burnout for Water Flowing Vertically Upward in Round Tubes at 2000 psia, USAEC Rep. WAPD-TH-318, Pittsburgh, PA. (5)

Ded, J., and J. H. Lienhard, 1972, The Peak Pool Boiling Heat Transfer from a Sphere, *AIChE J. 18*(2):337–342. (2)

Deev, V. I., Yu V. Gordeev, A. I. Pridantsev, et al., 1978, Pressure Drop in a Two-Phase Flow of Helium under Adiabatic Conditions and with Heat Supply, *Proc. 6th Int. Heat Transfer Conf.,* vol. 1, pp. 311–314, Toronto, Canada. (3)

Deissler, R. G., 1955, Analysis of Turbulent Heat Transfer, Mass Transfer, and Friction in Smooth Tubes at High Prandtl and Schmidt Numbers, NACA Rep. 1210, Lewis Res. Ctr., Cleveland, OH. (5)

DeJesus, J. M., and M. Kawaji, 1989, Measurement of Interfacial Area and Void Fraction in Upward, Cocurrent Gas-Liquid Flow, *ANS Proc. 1989 NHTC,* HTC-4, p. 137; see also 1990, Investigation of Interfacial Area and Void Fraction in Upward Cocurrent Gas-Liquid Flow, *Can. J. Chem. Eng. 68:*904–912, Dec. (3)

Delhaye, J. M., 1969a, General Equations for 2-Phase Systems and Their Applications to Air-Water

Bubble Flow and to Steam-Water Flushing Flow, ASME 69-HT-63, Natl. Heat Transfer Conf., ASME, New York. (3)

Delhaye, J. M., 1969b, Hot-Film Anemometry in Two-Phase Flow, ASME Symp. on Two-Phase Flow Instrumentation, Chicago, pp. 58–69. (3)

Delhaye, J. M., 1986, Recent Advances in Two-Phase Flow Instrumentation, *Proc. 8th Int. Heat Transfer Conf.*, San Francisco, vol. 1, p. 215, Hemisphere, New York. (3)

Delhaye, J. M., J. P. Galaup, M. Reocreux, and R. Ricque, 1973, Metrologie des Ecoulements Diphasiques—Quelques Procedes, CEA-R-4457, Comm. a L'Energie Atom., Grenoble, France. (3)

Dengler, C. E., and J. N. Addoms, 1956, Heat Transfer Mechanism for Vaporization of Water in a Vertical Tube, *AIChE Chem. Eng. Prog. Symp.* 52(18):95–103. (4)

Dergarabedian, P., 1960, Observations on Bubble Growth in Various Superheated Liquids, *J. Fluid Mech.* 9:39–48. (6)

Dhir, V. K., 1990, Nucleate and Transition Boiling Heat Transfer under Pool and External Flow Conditions, *Proc. 9th Int. Heat Transfer Conf.*, vol. 1, pp. 129–155; see also *Int. J. Heat Fluid Flow* 12(4):290. (2)

Dhir, V. K., 1992, Some Observations from Maximum Heat Flux Data Obtained on Surfaces Having Different Degrees of Wettability, in *Pool and External Flow Boiling*, V. K. Dhir and A. E. Bergles, Eds., pp. 185–192, ASME, New York. (2)

Dhir, V. K., and S. P. Liaw, 1989, Framework for a Unified Model for Nucleate and Transition Pool Boiling, *Trans. ASME J. Heat Transfer* 111:739–746. (2)

Dijkman, F. J. M., 1969, Some Hydrodynamic Aspects of a Boiling Water Channel. Thesis Technische Hogeschool to Eindhover, the Netherlands.

Dijkman, F. J. M., 1971. Some Hydrodynamic Aspects of a Boiling Water Channel, *Nuclear Eng. Design* 16:237–248. (6)

Dolgov, V. V., and O. A. Sudnitsyn, 1965, On Hydrodynamic Instability in Boiling Water Reactors, *Thermal Eng. (USSR)* (Eng. transl.) 12(3):51–55. (5)

Donaldson, M. R., and R. E. Pulfrey, 1979, Imaging Optical Probe for Pressurized Steam-Water Environments, *Proc. Review Group Conf. on Advanced Instrument for Reactor Safety Research*, USNRC, NUREG/CP-0007, III.17-1-27. (3)

Dougall, R. S., and T. E. Lippert, 1967, An Investigation into the Role of Thermal Fluctuations on Bubble Nucleation in Pool Boiling, Paper 67-WA/HT 31, ASME Winter Annual Meeting. (2)

Dougall, R. S., and T. E. Lippert, 1973, Net Vapor Generation Point in Boiling Flow of Trichlorotrifluoroethane at High Pressures, NASA, Contractor Report NASA-CR-2241, NASA Lewis Res. Ctr., Cleveland, OH (5)

Dougall, R. S., and W. M. Rohsenow, 1963, Film Boiling on the Inside of Vertical Tubes with Upward Flow of the Fluid at Low Qualities, MIT Heat Transfer Lab Rep. 9079-26, Cambridge, MA. (4)

Drew, D., S. Sim, and R. T. Lahey, Jr., 1978, Radial Phase-Distribution Mechanisms in Two-Phase Flow, Proc. OECD Committee for Safety in Nuclear Installations 2nd Specialists Meeting on Transient Two-Phase Flow, Paris. (3)

Drew, T. B., and A. C. Mueller, 1937, Boiling, *AIChE Trans.* 33:449–471. (4)

Dukler, A. E., 1960, Fluid Mechanics and Heat Transfer in Vertical Falling Film Systems, *AIChE Chem. Eng. Prog. Sym. Ser.* 56(30):1–10. (3)

Dukler, A. E., 1977, The Role of Waves in Two-Phase Flow: Some New Understanding, *Chem. Eng. Educ.* 11:108–118. (3)

Dukler, A. E., 1978, Modelling Two-Phase Flow and Heat Transfer, Keynote Paper KS-11, Proc. 6th Int. Heat Transfer Conf., Toronto, Canada. (3)

Dukler, A. E., and M. G. Hubbard, 1975, A Model for Gas-Liquid Slug Flow in Horizontal and Near Horizontal Tubes, *Ind. Eng. Chem. Fundam.* 14(4):337. (3)

Dukler, A. E., and Y. Taitel, 1991a, Modern Measuring Methods, in *Two-Phase Gas-Liquid Flow: A Short Course on Principles of Modelling Gas-Liquid Flow and on Modern Measuring Methods*, p. 115, University of Houston, Houston, TX. (3)

Dukler, A. E., and Y. Taitel, 1991b, Flow Pattern Transitions in Gas–Liquid Systems, chap. 3; Modeling Two-Phase Annular Flow, chap. 5, and Modeling Upward Gas-Liquid Flow, chap. 7, in *Two-Phase Gas-Liquid Flow: A Short Course on Principles of Modelling Gas-Liquid Flow and on Modern Measuring Methods,* University of Houston, Houston, TX. (3)

Dukler, A. E., M. Wicks, and R. E. Cleveland, 1964, Frictional Pressure Drop in Two Phase Flow, B. An Approach through Similarity Analysis, *AIChE J. 10:*44. (3)

Duncan, J. D., and J. E. Leonard, 1971, BWR Standby Cooling Heat Transfer Performance under Simulated LOCA Conditions Between 15 and 30 psia, GEAP 13190, General Electric, San Jose, CA. (4)

Duns, H., Jr., and N. C. J. Ros, 1963, Vertical Flow of Gas and Liquid Mixtures from Boreholes, Proc. 6th World Petroleum Congress, Frankfurt, FRG. (3)

Durant, W. S., and S. Mirshak, 1959, Roughening of Heat Transfer Surfaces as a Method of Increasing the Heat Flux at Burnout, Progress Rep. 1, USAEC Rep. DP-380, Savannah River Lab. (5)

Durant, W. S., and S. Mirshak, 1960, Roughening of Heat Transfer Surfaces as a Method of Increasing Heat Flux at Burnout, USAEC Rep. DPST-60-284, Savannah River Lab., Aiken, SC. (5)

Dwyer, O. E., 1969, On Incipient-Boiling Wall Superheats in Liquid Metals, *Int. J. Heat Mass Transfer 12:*1403. (2)

Dwyer, O. E., 1976, *Boiling Liquid-Metal Heat Transfer,* American Nuclear Society, LaGrange Park, IL. (3)

Dwyer, O. E., and C. J. Hsu, 1975, Liquid Microlayer Thickness in Nucleate Boiling on a Heated Surface, *Lett. Heat Mass Transfer 2:*179. (2)

Dwyer, O. E., and C. J. Hsu, 1976, Evaporation of the Microlayer in Hemispherical Bubble Growth in Nucleate Boiling of Liquid Metals, *Int. J. Heat Mass Transfer 19:*185. (2)

Dwyer, O. E., G. Strickland, S. Kalish, and P. J. Schoen, 1973a, Incipient-Boiling Superheat for Sodium in Turbulent Channel Flow: Effects of Heat Flux and Flow Rate, *Int. Heat Mass Transfer 16:*971–984. (4)

Dwyer, O. E., G. Strickland, S. Kalish, and P. J. Schoen, 1973b, Incipient-Boiling Superheats for Sodium in Turbulent Channel Flow: Effect of rate of Temperature Rise, *Trans. ASME, J. Heat Transfer 95:*159–165. (4)

Dykas, S., and M. K. Jensen, 1992, CHF on a Tube in a Horizontal Tube Bundle, *Expt. Thermal Fluid Sci. 5:*34–39. (5)

Dzakowic, G. S., 1967, An Analytical Study of Microlayer Evaporation and Related Bubble Growth Effects in Nucleate Boiling, Ph.D. thesis, University of Tennessee, Knoxville, TN. (2)

Edwards, A. R., 1968, Conduction Controlled Flashing of a Fluid and the Production of Critical Flow Rate in One-Dimensional System, AHSB (S)R 147, UK Atomic Energy Authority, Risley, England. (3)

Edwards, A. R., and D. J. Mather, 1973, Some U.K. Studies Related to the Loss of Coolant Accident, ANS Topical Meeting on Water Reactor Safety, Salt Lake City, UT. (3)

Edwards, P. A., J. D. Obertelli, 1966, 6 MW Rig Burnout and Pressure Drop Data on 37-Rod Cluster in High Pressure Water, UKAEA, Presentation at Ispra Symp. (5)

Edwards, A. R., and T. P. O'Brien, 1970, Studies of Phenomena Connected with the Depressurization of Water Reactors, *J. Br. Nuclear Energy Soc.* 9(2):125–135. (3)

Efferding, L. E., 1968, DYNAM—A Digital Computer Program for the Dynamic Stability of Once-Through Boiling Flow and Steam Superheat, USAEC Rep. GAMD-8656, Westinghouse Elec. Corp., Pittsburgh, Pa. (6)

Egen, R. A., D. A. Dingee, and J. W. Chatain, 1957, Vapor Formation and Behavior in Boiling Heat Transfer, USAEC Rep. BMI-1163, Batelle Memorial Inst., Columbus, OH. (3)

Elkassabgi, Y., and J. H. Lienhard, 1988, The Peak Pool Boiling Heat Fluxes from Horizontal Cylinders in Subcooled Liquids, *Trans. ASME, J. Heat Transfer 110:*479–496. (2)

Ellias, E., and G. S. Lellouche, 1994, Two-Phase Critical Flow, *Int. J. Multiphase Flow 20*(suppl.):91–168. (3)

Ellion, M. E., 1954, A Study of the Mechanism of Boiling Heat Transfer, Jet Propulsion Lab. Memo. 20–88, California Institute of Technology, Pasadena, CA. (2)

Elrod, E. C., J. A. Clark, E. R. Lady, and H. Merte, 1967, Boiling Heat Transfer Data at Low Heat Flux, *Trans. ASME, J. Heat Transfer 89:*235–241. (4)

Era, A., G. P. Gaspari, A. Hassid, A. Milani, and R. Lavattarelli, 1966, Heat Transfer Data in the Liquid Deficient Region for Steam-Water Mixtures at 70 kg/cm^2 Flowing in Tubular and Annular Conduits, Rep. CISE-R-184, Milan, Italy. (4)

EURAEC, 1966, Transition Boiling Heat Transfer Program, 12th Quarterly Progress Rep., Oct.–Dec. 1965, Rep. GEAP-5081, Joint US-EURATOM Research and Development Program. (5)

Evangelisti, R., G. P. Gaspari, L. Rubiera, and G. Vanoli, 1972, Heat Transfer Crisis Data with Steam Water Mixture in a Sixteen Rod Bundle, *Int. J. Heat Mass Transfer 55:*387–402. (5)

Ewing, C. T., J. P. Stone, J. R. Spam, and R. R. Miller, 1967, Molecular Association in Sodium, Potassium, and Cesium Vapors at High Temperatures, *J. Phys. Chem. 71:*473. (2)

Fabrega, S., 1964, Etude experimentale des instabilites hydrodynamiques survenant dans les reacteurs nucleaires a ebullition, CEA-R-2884, Commissariat á l'Energie Atomique, Cedex, France. (6)

Fajean, M., 1969, Programme FLICA, etude thermodynamicue d'un reacteur ou d'une bouch d'essai, CEA-R-3716, Cedex, France. (3)

Farber, E. A., and R. L. Scorah, 1948, Heat Transfer to Water Boiling under Pressure, *Trans. ASME 70:*369–384. (1)

Farmer, F. R., and E. V. Gilby, 1971, Reactor Safety Philosophy and Experimental Verification, Paper A/Conf. 49, p. 477, *Fourth Int. Conf. on the Peaceful Uses of Atomic Energy*, Geneva. (5)

Farrest, C. F., W. T. Hancox, and W. B. Nicoll, 1971, Axial Void Distributions, in Forced Convection Boiling; A Survey of Prediction Techniques and Their Efficiency, *Int. J. Heat Mass Transfer 14:*1377–1394. (3)

Fath, H. S., and R. L. Judd, 1978, Influence of System Pressure on Microlayer Evaporation Heat Transfer, *Trans. ASME, J. Heat Transfer 100:*49–55. (2)

Fauske, H. K., 1962, Contribution to the Theory of Two-Phase One-Component Critical Flow, USAEC TID-4500, 18th ed., Argonne National Lab. Rep. ANL-6633, Argonne, IL. (3)

Fauske, H. K., 1965, The Discharge of Saturated Water through Tubes, *AIChE Chem. Eng. Prog. Symp. Ser. 61*(59):210–216. (3)

Fauske, H. K., 1966, What's New in Two-Phase Flow? *Power Reactor Technol.* 9(1):35. (3)

Fauske, H. K., 1971, Transient Liquid-Metal Boiling and Two-Phase Flow, *Proc. Int. Seminar: Heat Transfer in Liquid Metals*, Trogir, Yugoslavia, September. (3)

Fauske, H. K., and M. A. Grolmes, 1970, Pressure Drop for Forced Convection Flashing Sodium, in *Liquid Metal Heat Transfer and Fluid Mechanics*, J. C. Chen and A. A. Bishop, Eds., pp. 135–143, ASME, New York. (3)

Fauske, H. K., and M. A. Grolmes, 1971, Analysis of Transient Forced Liquid Metal Boiling in Narrow Tubes, *AIChE Chem. Eng. Prog. Symp. Ser. 67*(119):64–71. (6)

Fighetti, C. F., and D. G. Reddy, 1982, Parametric Study of CHF Data, Vol. 1: Compilation of Rod Bundle CHF Data Available at the Columbia University Heat Transfer Research Facility; and 1983, Parametric Study of CHF Data, Vol 3: CHF Data (1983), EPRI Rep. NP-2609, EPRI, Palo Alto, CA. (5)

Fiori, M. P., and A. E. Bergles, 1968, Model of Critical Heat Flux in Subcooled Flow Boiling, *Heat Transfer 1970*, U. Grigull and E. Hahne, Eds., Vol. 6, Paper B 6.6, Elsevier, Amsterdam. (5)

Fisher, C. R., and P. Y. Achener, 1964, Alkali Metals Evaluation Program—Quarterly Progress Report for the period 1 July through 30 Sept. 1964, Rep. AGN-8121, Aerojet General—Nuclear Division, General Atomic, San Diego, CA. (5)

Fisher, C. R., and P. Y. Achener, 1965, Alkali Metals Evaluation Program—Swirl Flow Boiling of Alkali Metals Heat Transfer and Pressure Drop, Rep. AGN-8127, Aerojet General—Nuclear Division, General Atomic, San Diego, CA. (5)

Fisher, S. A., and D. L. Pearce, 1993, An Annular Flow Model for Predicting Liquid Carryover into Austenitic Superheaters, *Int. J. Multiphase Flow 19:*295–307. (3)

Flanagan, S., and A. R. Edwards, 1978, Some Nonequilibrium Effects in De-Pressurization Experiments of Water-Filled Systems, in *Transient Two-Phase Flow,* Proc. 2nd CSNI Specialists Meeting, Vol. 2, 487–516, M. Reocreux and G. Katz Eds., Commissariat a l'Energie Atomnizue, Fontenay aux Roses, France. (3)

Fletcher, C. D., and M. A. Bolander, 1986, Analysis of Instrument Tube Ruptures in Westinghouse 4-Loop PWRs, Rep. NUREG/CR 4672, EGG-2461, Idaho Natl. Eng. Lab., Idaho Falls, ID. (4)

Ford, W. D., H. K. Fauske, and S. G. Bankoff, 1971a, Slug Expulsion of Freon-113 by Rapid Depressurization of a Vertical Tube, *Int. J. Heat Mass Transfer 14:*133–140. (4)

Ford, W. D., S. G. Bankoff, and H. K. Fauske, 1971b, Slug Ejection of Freon-113 from a Vertical Channel with Nonuniform Initial Temperature Profiles, Paper 7-11, Int. Symp. on Two-Phase Systems, Haifa, Israel. (6)

Forslund, R. P., and W. M. Rohsenow, 1966, Thermal Non-Equilibrium in Dispersed Flow Film Boiling in a Vertical Tube, MIT Heat Transfer Lab. Rep. 75312-44, Massachusetts Institute of Technology, Cambridge, MA. (4)

Forster, H. K., and R. Greif, 1959, Heat Transfer to Boiling Liquid, Mechanism and Correlations, *Trans. ASME, J. Heat Transfer 81:*43–53. (2)

Forster, H. K., and N. Zuber, 1954, Growth of Vapor Bubble in a Superheated Fluid, *J. Appl. Phys. 25:*474–488. (2)

Forster, H. K., and N. Zuber, 1955, Dynamics of Vapor Bubbles and Boiling Heat Transfer, *AIChE J. 1*(4):531–535. (4)

France, D. M., R. D. Carlson, R. R. Rhode, and G. T. Charmoli, 1974, Experimental Determination of Sodium Superheat Employing LMFBR Simulation Parameters, *Trans. ASME, J. Heat Transfer 96:*359. (4)

Frederking, T. H. K., Y. C. Wu, and B. W. Clement, 1966, Effects of Interfacial Instability of Film Boiling of Saturated Liquid He above a Horizontal Surface, *AIChE J. 12:*238. (2)

Frenkel, J., 1955, *Kinetic Theory of Liquids,* Dover, New York, p. 376. (1)

Fritz, W., 1935, Maximum Volume of Vapor Bubbles, *Phys. Z. 36:*379–384. (2)

Fritz, W., and W. Eude, 1936, The Vaporization Process According to Cinematographic Pictures of Vapor Bubbles, *Phys. Z. 37:*391–401. (2)

Gaertner, R. F., 1961, Distribution of Active Sites in the Nucleate Boiling of Liquids, G.E. Rep. 61-RL-2826C, San Jose, CA. (2)

Gaertner, R. F., 1963a, Effect of Surface Chemistry on the Level of Burnout Heat Flux in Pool Boiling, G.E. Rep. 63-RL-3449C, San Jose, CA. (2)

Gaertner, R. F., 1963b, Distribution of Active Sites in the Nucleate Boiling of Liquids, *AIChE Chem. Eng. Prog. Symp. Ser. 59*(41):52. (2)

Gaertner, R. F., 1965, Photographic Study of Nucleate Pool Boiling on a Horizontal Surface, *Trans. ASME, J. Heat Transfer 87:*17–29. (2)

Gallagher, E. V., 1970, Water Decompression Experiments and Analysis for Blowdown of Nuclear Reactors, ITTRI-578-P-21-39, I.T.T. Res. Inst., Chicago, IL. (3)

Gandiosi, G., 1965, Experimental Results on the Dependence of Transition Boiling Heat Transfer on Loop Flow Disturbances, GEAP-4725, General Electric, San Jose, CA. (6)

Gaspari, G. P., A. Hassid, and G. Vanoli, 1968, Critical Heat Flux (CHF) Prediction in Complex Geometries (Annuli and Clusters) from a Correlation Developed for Circular Conduits, Rep. CISE-R-276, CISE, Milan, Italy. (5)

Gaspari, G. P., A. Hassid, and G. Vanoli (CISE), 1970, Some Consideration on Critical Heat Flux in Rod Clusters in Annular Dispersed Vertical Upward Two-Phase Flow, Proc. 1970 Intl. Heat Transfer Conference, Vol 6, Paper B 6.4, Paris, Hemisphere, Washington, DC. (5)

GE Report, 1962, Alkali Metals Boiling and Condensing Investigations, First Quarterly Rep., Contract NAS 3-2528, San Jose, CA. (4)

GE Report, 1973, BWR Analysis Basis GETAB: Data Correlation and Design Application, NEDO-10958. (5)

GE Report, 1975, BWR Blowdown Heat Transfer, Tenth Quarterly Progress Rep., Oct. 1–Dec. 31, 1974, GEAP-13319-10, San Jose, CA. (5)
Geiger, G. E., 1964, Sudden Contraction Losses in Single and Two-Phase Flow, Ph.D. thesis, University of Pittsburgh, Pittsburgh, PA. (3)
Gellerstedt, J. S., R. A. Lee, W. J. Oberjohn, R. H. Wilson, and L. J. Stanek, 1969, Correlation of Critical Heat Flux in a Bundle Cooled by Pressurized Water, in *Two Phase Flow and Heat Transfer in Rod Bundles,* pp. 63–71, ASME, New York. (5)
George, C. M., and D. M. France, 1991, Post-CHF Two-phase Flow and Low Wall-superheat, *Nuclear Eng. Design 125*:97–109. (4)
Gill, L. E., G. F. Hewitt, and P. M. C. Lacey, 1963, Sampling Probe Studies of the Gas Core in Annular Two Phase Flow, Part II: Studies of Flow Rates on Phase and Velocity Distributions, UK Rep. AERE-R-3955, Harwell, England. (3)
Giphshman, I. N., and V. M. Levinson, 1966, Disturbances in Flow Stability in the Pendent Superheater of the TPP-110 Boiler, *Teploenergetika 13*(5):45–49 (Translated by the Heating and Ventilating Research Association, Pergamon Press, Oxford, England). (6)
Golan, L. P., and A. H. Stenning, 1969–1970, Two-Phase Vertical Flow Maps, *Proc. Inst. Mech. Eng. 184*(Part 3C):108. (3)
Goldmann, K., 1953, Boiling Songs, US AEC Rep. NDA, pp. 10–68. (2)
Gorenflo, D., 1984, *Behaltersieden, VDI-Warmeatlas,* 4 Aufl., Abschnitt Ha VDI-Verlag, Dusseldorf. (2)
Gorenflo, D., V. Knabe, and V. Bieling, 1986, Bubble Density on Surfaces with Nucleate Boiling—Its Influence on Heat Transfer and Burnout Heat Flux at Elevated Saturation Pressure, *8th Int. Heat Transfer Conf.,* Vol. 4, San Francisco. (2)
Gouse, S. W., Jr., and C. D. Andrysiak, 1963, Fluid Oscillations in a Closed Looped with Transparent, Parallel, Vertical, Heated Channels, MIT Eng. Projects Lab. Rep. 8973-2, Massachusetts Institute of Technology, Cambridge, MA. (6)
Govan, A. H., G. F. Hewitt, H. J. Richter, and A. Scott, 1991, Flooding and Churn Flow in Vertical Pipes, *Int. J. Multiphase Flow 17*:27–44. (3)
Govier, G. W., and K. Aziz, 1972, *The Flow of Complex Mixtures in Pipes,* Van Nostrand Reinhold, New York. (3)
Graham, J., 1971, *Fast Reactor Safety,* p. 27, Academic Press, New York. (5)
Graham, R. W., and R. C. Hendricks, 1967, Assessment of Convection, Conduction and Evaporation in Nucleate Boiling, NASA-TN D-3943, Lewis Res. Ctr., Cleveland, OH. (2)
Graham, R. W., R. C. Hendricks, and R. C. Ehlers, 1965, Analytical and Experimental Study of Pool Heating of Liquid Hydrogen over a Range of Accelerations, NASA TN-D1883, Lewis Res. Ctr., Cleveland, OH. (2)
Grant, I. D. R., 1971, Shell-and-Tube Heat Exchangers for the Processing Industries, *Chem. Processing,* December. (6)
Granziera, R., and M. S. Kazimi, 1980, A Two-Dimensional Two-Fluid Model for Sodium Boiling in LMFBR Fuel Assemblies, Energy Laboratory Rep. No. MIT-EL-80-011, Massachusetts Institute of Technology, Cambridge, MA. (4)
Green, S. J., G. W. Mauer, and A. Weiss, 1962, Burnout and Pressure Drop Studies for Forced Convection Flow of Water Parallel to Rod Bundles, ASME Paper 62-HT-43, Natl. Heat Transfer Conf., Houston, TX. (3)
Griffel, J., and C. F. Bonilla, 1965, Forced Convection Boiling Burnout for Water in Uniformly Heated Tubular Test Sections, *Nuclear Structural Eng. 2*:1–35. (5)
Griffith, P., 1958, Bubble Growth Rates in Boiling, in *Trans. ASME, J. Heat Transfer, 80,* 721–727. (2)
Griffith, P., 1964, Two-Phase Flow in Pipes, in *Developments in Heat Transfer,* MIT Press, Cambridge, MA. (3)
Griffith, P., 1968, Flow Regimes, Lecture 2, Seminar in Boiling Heat Transfer and Two-Phase Flow, Massachusetts Institute of Technology, Cambridge, MA. (3)
Griffith, P., J. A. Clark, and W. M. Rohsenow, 1958, Void Volumes in Subcooled Boiling Systems, ASME Paper 58-HT-19, ASME, New York. (3)

Griffith, P., and G. B. Wallis, 1961, Two-Phase Slug Flow, *Trans. ASME, J. Heat Transfer 83:*307. (3)
Griffith, P., and J. D. Wallis, 1960, The Role of Surface Conditions in Nucleate Boiling, *AIChE Chem. Eng. Prog. Symp. Ser. 56*(30):49–63. (2)
Groeneveld, D. C., 1972, The Thermal Behavior of a Heated Surface at and Beyond Dryout, AECL-4309, Chalk River, Canada. (4)
Groeneveld, D. C., S. C. Cheng, and T. Doan, 1986, AECL—UO Critical Heat Flux Look up Table, *Heat Transfer Eng. 7:*46–64. (5)
Groeneveld, D. C., G. G. J. Delorme, 1976, Prediction of Thermal Non-equilibrium in the Post Dryout Regime, *Nuclear Eng. Design 36:*17–26. (4)
Groeneveld, D. C., and C. W. Snoek, 1986, A Comprehensive Examination of Heat Transfer Correlations Suitable for Reactor Safety Analysis, in *Multiphase Science & Technology* 2, G. F. Hewett, J. M. Delhaye, and N. Zuber, Eds., Hemisphere, Washington, DC. (5)
Grolmes, M. A., and H. K. Fauske, 1969, Propagation Characteristics of Compression and Rarefaction Pressure Pulses in One-Component Vapor-Liquid Mixtures, *Nuclear Eng. Design 11:*137–142. (3)
Grolmes, M. A., and H. K. Fauske, 1970, Modeling of Sodium Expulsion of Freon-11, ASME Paper 70-HT-24, Fluids Engineering Heat Transfer and Lubrication Conf., Detroit, MI. (4)
Guerneri, S. A., and R. D. Tatty, 1956, A Study of Heat Transfer to Organic Liquids in Single Tube Natural Circulation Vertical Tube Boilers, *AIChE Chem. Eng. Prog. Symp. Ser. 52*(18):69–77. (4)
Gungor, K. E., and R. H. S. Winterton, 1986, A General Correlation for Flow Boiling in Tubes and Annuli, *Int. J. Heat Mass Transfer 29*(3):351–358. (4)
Gunther, F. C., 1951, Photographic Study of Surface Boiling Heat Transfer to Water with Forced Convection, *Trans. ASME, J. Heat Transfer 73:*115–123. (3)
Gunther, F. C., and F. Kreith, 1950, Photographic Study of Bubble Formation in Heat Transfer to Subcooled Water, JPL Progress Rep. 4-120, pp. 1–29, JPL, Pasadena, CA. (2)
Hacker, D. S., 1963, Comment on Velocity Defect Law for a Transpired Turbulent Boundary Layer, *AIAA J. 1*(11):2676. (5)
Haga, T., T. Saitoh, and J. Tangji, 1990, Three-Dimensional Analysis of BWR Stability in Low Flow Conditions, *Proc. Int. Workshop on BWR Stability*, Hotsville, NY, pp. 163–174, CSNI Rep. 178, OECD-NEA, Paris. (6)
Hahne, E., K. Spindler, and H. Skok, 1993, A New Pressure Drop Correlation for Flow Boiling in Tubes and Annuli, *Int. J. Heat Mass Transfer 36*(17):4267–4274. (3)
Hall, W. B., 1958, Heat Transfer in Channels Composed of Rough and Smooth Surfaces, IGR-TN/W-832, UKAEA, Harwell, England. (5)
Hall, W. B., and W. G. Harrison, 1966, Transient Boiling of Water at Atmospheric Pressure, *Proc. Third Int. Heat Transfer Conf.*, vol. 3, p. 186, AIChE, New York. (5)
Hamill, T. D., and K. J. Baumeister, 1967, Effect of Subcooling and Radiation on Film Boiling Heat Transfer from a Flat Plate, NASA TND-3925, Lewis Res. Ctr., Cleveland, OH. (2)
Han, C. Y., and P. Griffith, 1965a, The Mechanism of Heat Transfer in Nucleate Pool Boiling, Part I, Bubble Initiation, Growth, and Departure, *Int. J. Heat Mass Transfer 8*(6):887–904. (2)
Han, C. Y., and P. Griffith, 1965b, The Mechanism of Heat Transfer in Nucleate Pool Boiling, Part II. The Heat Flux-Temperature Relation, *Int. J. Heat Mass Transfer 8*(6):905. (2)
Hancox, W. T., and W. B. Nicoll, 1971, A General Technique for the Prediction of Void Distributions in Non-Steady Two-Phase Forced Convection, *Int. J. Heat Mass Transfer 14:*1377–1394. (3)
Haramura, Y., and Y. Katto, 1983, A New Hydrodynamic Model of CHF Applicable Widely to Both Pool and Forced Convection Boiling on Submerged Bodies in Saturated Liquids, *Int. J. Heat Mass Transfer 26*(3):387–399. (2)
Harmathy, T., 1960, Velocity of Large Drops and Bubbles in Media of Infinite and Restricted Extent, *AIChE J. 6:*281. (3)
Hasson, P. T., 1965, HYDRO—A Digital Model for One-Dimensional Time-Dependent Two-Phase Hydrodynamics, Part 2, RFR-492/RFN-210, A B Atomenergi, Nyköping, Sweden. (3)
Hatton, A. P., and I. S. Hall, 1966, Photographic Study of Boiling on Prepared Surfaces, *Proc. Third Int. Heat Transfer Conf.*, vol. 4, pp. 24–37, AIChE, New York. (2)

REFERENCES

Heineman, J. B., J. F. Marchaterre, and S. Metha, 1963, Electromagnetic Flowmeters for Void Fraction Measurement in Two-Phase Metal Flow, *Rev. Sci. Instrum.* 34(4):319. (3)

Hench, J. E., and R. F. Boehm, 1966, Nine Rod Critical Heat Flux Investigation at 1000 psia, GEAP-4929, San Jose, CA. (5)

Hendricks, R. C., and R. R. Sharp, 1964, Initiation of Cooling due to Bubble Growth on a Heating Surface, NASA TN D-2290, Lewis Res. Ctr., Cleveland, OH. (2)

Henry, R. E., 1968, A Study of One- and Two-Component, Two-Phase Critical Flows at Low Qualities, Rep. ANL-7430, Argonne Natl. Lab., Argonne, IL. (3)

Henry, R. E., 1970, Pressure Wave Propagation in Two-Phase Mixtures, *AIChE Chem. Eng. Prog. Symp. Ser.* 66(102):1–10. (3)

Henry, R. E., 1971, Pressure Wave Propagation through Annular and Mist Flows, *AIChE Chem. Eng. Prog. Symp. Ser.* 67(113):38–47. (3)

Henry, R. E., and H. K. Fauske, 1971, The Two-Phase Critical Flow of One-Component Mixtures in Nozzles, Orifices and Short Tubes, *Trans. ASME, J. Heat Transfer* 93:179. (3)

Henry, R. E., M. A. Grolmes, and H. K. Fauske, 1968, Propagation Velocity of Pressure Waves in Gas-Liquid Mixtures, Gas-Liquid Flow Symp., Waterloo University, Waterloo, Canada. (3)

Henry, R. E., M. A. Grolmes, and H. K. Fauske, 1969, Propagation Velocity of Pressure Waves in Gas-Liquid Mixtures, in *Co-current Gas-Liquid Flow*, Edward Rhodes and D. S. Scott, Eds., pp. 1–18, Plenum Press, New York. (3)

Henry, R. E., M. A. Grolmes, and H. K. Fauske, 1971, Pressure Pulse Propagation in Two-Phase One and Two-Component Mixtures, Rep. ANL-7792, Argonne National Lab., Argonne, IL. (3)

Herd, K. G., W. P. Goss, and J. W. Connell, 1983, Correlation of Forced Flow Evaporation Heat Transfer Coefficient in Refrigerant Systems, Paper B2, in *Heat Exchangers for Two-Phase Applications*, Vol. 27, National Heat Transfer Conf., Seattle, WA. (4)

Hesson, G. M., D. E. Fitzsimmons, and J. M. Batch, 1964, Comparison of Boiling Burnout Data for 19-Rod Bundles in Horizontal and Vertical Positions, HQ-83443, Rev. 1., USAEC, Washington, DC. (3)

Hesson, G. M., D. E. Fitzsimmons, and J. M. Batch, 1965, Experimental Boiling Burnout Heat Fluxes with an Electrically Heated 19-Rod Bundle Test Section, BNWL-206, Battelle NorthWestern Lab., Richland, WA. (3)

Hetsroni, G., 1982, Two-Phase Flow Instabilities, in *Handbook of Multiphase Systems*, chap. 2, Hemisphere and McGraw-Hill, New York. (6)

Hetsroni, G., 1993, Two-Phase Heat Transfer, Workshop on Multiphase Flow and Heat Transfer, University of California, Santa Barbara, CA, September. (5)

Hewitt, G. F., 1964, A Method of Representing Burnout Data in Two-Phase Heat Transfer for Uniformly Heated Round Tubes, Rep. AEEW-R-4613, UK Atomic Energy Authority, Winfrith, England. (5)

Hewitt, G. F., 1970, Experimental Studies on the Mechanisms of Burnout in Heat Transfer to Steam Water Mixtures, Paper B6.6, in *Heat Transfer 1970*, Vol. 6, U. Grigull and E. Hahne, Eds., Elsevier, Amsterdam. (5)

Hewitt, G. F., 1969–1970, *Proc. Inst. Mech. Eng.* 184(Part 3C):142. (3)

Hewitt, G. F., 1978, *Measurement of Two-Phase Flow Parameters*, Academic Press, London. (3)

Hewitt, G. F., 1982, Burnout, in *Handbook of Multiphase Systems*, G. Hetsroni Ed., chap. 6, Hemisphere and McGraw-Hill, New York. (5)

Hewitt, G. F., 1991, Phenomena in Horizontal Two-Phase, *Int. Conf. Multiphase Flows. Proc.*, vol 3, pp. 1–10. (3)

Hewitt, G. F., and A. H. Govan, 1990, Phenomena & Prediction in Annular Two-Phase Flows, *Symp. Advances in Gas-Liquid Flows*, ASME FED-Vol. 99/HTD-Vol. 155, pp. 41–46, ASME Winter Annual Meeting, ASME, New York. (3)

Hewitt, G. F., H. A. Kearsey, P. M. C. Lacey, and D. J. Pulling, 1965, Burnout and Nucleation in Climbing Film Flow, *Int. J. Heat Mass Transfer* 8:793–814. (5)

Hewitt, G. F., H. A. Kearsey, P. M. C. Lacey, and D. J. Pulling, 1965–1966, Burnout and Film Flow in the Evaluation of Boiling Water in Tubes, *Proc. Inst. Mech. Eng. 180* (Part 3C):206–215. (5)

Hewitt, G. F., R. D. King, and P. C. Lovegrove, 1962, Techniques for Liquid Film and Pressure Drop Studies in Annular Two-Phase Flow, Rep. AERE-R-3921, UK AERE, Harwell, England. (3)

Hewitt, G. F., and N. S. Hall Taylor, 1970, *Annular Two-Phase Flow*, Pergamon Press, Oxford. (5)

Hill, K. W., F. E. Motley, F. F. Cadek, and J. E. Casterdine, 1974, Effects on Critical Heat Flux of Local Heat Flux Spikes or Local Flow Blockage in Pressurized Water Reactor Rod Bundles, ASME Paper 74-WA/HT-54, ASME, New York. (5)

Hill, K. W., F. E. Motley, F. F. Cadek, and J. E. Casterdine, 1975, Effect of a Rod Bowed to Contact on CHF in PWR Rod Bundles, *ASME Winter Annual Meeting* Paper 75-WA/HT-77, ASME, New York. (5)

Hill, W. S., and W. M. Rohsenow, 1982, Dryout Drop Distribution and Dispersed Flow Film Boiling, MIT Heat Transfer Lab. Rep. 85694-105, Massachusetts Institute of Technology, Cambridge, MA. (4)

Hodges, M. W., and D. H. Knoebel, 1973, Subcooled Burnout Phenomenon Adjacent to a Spacer Rib, ASME Paper 73-HT-40, ASME, New York. (5)

Hoffman, H. W., 1964, Recent Experimental Results in ORNL Studies with Boiling Potassium, Third Annual Conf. on High Temperature Liquid Metal Heat Transfer Technology, Oak Ridge Natl. Lab., Oak Ridge, TN, ORNL-3605, vol. 1, pp. 334–350. (4)

Hoffman, H. W., and J. J. Keyes, Jr., 1965, Studies on Heat Transfer and Fluid Mechanics Progress Report for Period Oct. 1, 1963–June 30, 1964, ORNL-TM-1148, Oak Ridge Natl. Lab, Oak Ridge, TN. (5)

Hoffman, H. W., and A. I. Kralsoviak, 1964, Convective Boiling with Liquid Potassium, Proc. 1964 Heat Transfer and Fluid Mechanics Inst., Berkeley, CA. (5)

Holland, P. K., and R. H. S. Winterton, 1973, The Radii of Surface Nucleation Sites Which Initiate Sodium Boiling, *Nuclear Eng. Design 24*:388. (3)

Holtz, R. E., 1966, The Effect of Pressure-Temperature History upon Incipient Boiling Superheats in Liquid Metals, ANL-7184, Argonne Natl. Lab., Argonne, IL. (2)

Holtz, R. E., 1971, On the Incipient Boiling of Sodium and Its Application to Reactor Systems, USAEC Rep. ANL-7884, Argonne Natl. Lab., Argonne, IL. (2)

Holtz, R. E., H. K. Fauske, and D. T. Eggen, 1971, The Prediction of Incipient Boiling Superheats in L/M Cooled Reactor Systems, *Nuclear Eng. Design 16*:253–265. (4)

Holtz, R. E., and R. M. Singer, 1967, On the Superheating of Sodium at Low Heat Fluxes, ANL-7383, Argonne Natl. Lab., Argonne, IL. (3)

Holtz, R. E., and R. M. Singer, 1968, Incipient Pool Boiling of Sodium, *AIChE J. 14*:654. (2)

Holtz, R. E., and R. M. Singer, 1969, On the Initiation of Pool Boiling in Sodium, *AIChE Chem. Eng. Prog. Symp. Ser. 65*(92):121. (2)

Höppner, G., 1971, Experimental Study of Phenomena Affecting the Loss of Coolant Accident, Ph.D. thesis, University of California, Berkeley, CA. (6)

Hosler, E. R., 1965, Visual Study of Boiling at High Pressure, AIChE Chem. Eng. Prog. Symp. Ser. 65 (57):269–279. (5)

Hosler, E. R., 1968, Flow Patterns in High Pressure Two-Phase (Steam-Water) Flow with Heat Addition, *AIChE Chem. Eng. Prog. Symp. Ser. 64*(82):54, AIChE, New York. (3)

Howard, C. L., 1963, Methods of Improving the Critical Heat Flux for BWR's, GE Rep. GEAP 4203, General Electric, San Jose, CA. (5)

Hsia, E. S., 1970, Forced Convective Annular-Flow Boiling with Liquid Mercury under Wetted and Swirl Flow Conditions, in *Liquid-Metal Heat Transfer and Fluid Mechanics*, J. C. Chen and A. A. Bishop, Eds., ASME, New York. (3)

Hsieh, Din-Yu, and M. S. Plesset, 1961, On the Propagation of Sound in a Liquid Containing Gas Bubbles, *Phys. Fluids 4*(8):970–975. (3)

Hsu, Y. Y., 1962, On the Size Range of Active Nucleation Cavities in a Heating Surface, *Trans. ASME, J. Heat Transfer 84*(3):207–216. (2)

Hsu, Y. Y., 1972, Review of Critical Flow Rate, Propagation of Pressure Pulse, and Sonic Velocity in Two-Phase Media, NASA TN D-6814, NASA Lewis Res. Ctr., Cleveland, OH. (3)

Hsu, Y. Y., and R. W. Graham, 1961, An Analytical and Experimental Study of the Thermal Boundary Layer & Ebullition Cycle in Nucleate Boiling, NASA TND-594, Lewis Res. Ctr., Cleveland, OH. (1)

Hsu, Y. Y., and R. W. Graham, 1976, *Transport Processes in Boiling and Two-Phase Systems,* chaps. 5 and 6, Hemisphere, New York. (2)

Hsu, Y. Y., F. F. Simon, and R. W. Graham, 1963, Application of Hot-Wire Anemometry for Two-Phase Flow Measurements Such as Void-Fraction and Slip Velocity, pp. 26–34, *ASME Symp. on Multiphase Flow,* Philadelphia, PA. (3)

Hsu, Y. Y., and J. W. Westwater, 1960, Approx. Theory for Film Boiling on Vertical Surfaces, *AIChE Chem. Eng. Prog. Symp. Ser.* 56(30):15, AIChE, New York. (3)

Hughmark, G. A., 1962, Holdup in Gas-Liquid Flow, *Chem. Eng. Prog.* 58(4):62. (3)

Hull, L. M., and W. M. Rohsenow, 1982, Thermal Boundary Layer Development in Dispersed Flow Film Boiling, MIT Heat Transfer Lab. Rep. 85694–104, Massachusetts Institute of Technology, Cambridge, MA. (4)

Huntley, W. R., 1969, Oak Ridge Natl. Lab., Oak Ridge, TN, Personal Communication to J. K. Jones, November. (5)

Hutchinson, P., and P. B. Whalley, 1973, A Possible Characterization of Entrainment in Annular Flow, *Chem. Eng. Sci.* 28:974–975. (5)

Hynek, S. J., 1969, Forced Convection Dispersed Flow Film Boiling, Ph.D. thesis, Massachusetts Institute of Technology, Cambridge, MA. (4)

Ibrahim, E. A., and R. L. Judd, 1985, An Experimental Investigation of the Effect of Subcooling on Bubble Growth and Waiting Time in Nucleate Boiling, *Trans. ASME, J. Heat Transfer* 107(1):168–174. (2)

Iloege, O. C., D. N. Plummer, W. M. Rohsenow, and P. Griffith, 1974, A Study of Wall Rewet and Heat Transfer in Dispersed Vertical Flow, M.I.T. Heat Transfer Lab. Rep. 72718–92, Massachusetts Institute of Technology, Cambridge, MA. (4)

Isbin, H. S., R. Vanderwater, H. K. Fauske, and S. Singher, 1961, A Model for Correlating Two-Phase, Steam-Water, Burnout Heat Transfer Fluxes, *Trans. ASME, J. Heat Transfer* 83:149–157. (5)

Ishii, M., 1975, *Thermo Fluid Dynamic Theory of Two-Phase Flow,* chaps. IX and X, Eyralles Press, Paris (Sci. and Medical Pub. of France, New York). (3)

Ishii, M., and T. C. Chawla, 1978, Multichannel Drift-Flux Model and Constitutive Relation for Transverse Drift-Velocity, Rep. ANL/RAS/LWR 78–2, Argonne Natl. Lab., Argonne, IL. (3)

Ishii, M., and G. Kocamustafaogullari, 1983, Two-Phase Flow Models and Their Limitations, p. 1, NATO Adv. Sci. Inst. Ser., *Advances in Two-Phase Flow and Heat Transfer,* S. Kakac and M. Ishii, Eds., Marinus Nijkoff, Brussil. (3)

Ishii, M., and N. Zuber, 1970, Thermally Induced Flow Instability in Two-Phase Mixtures, 4th Int. Heat Transfer Conf., Paris. (6)

Israel, S., J. F. Casterline, and B. Matzner, 1969, Critical Heat Flux Measurements in a 16-Rod Simulation of BWR Fuel Assembly, *J. Heat Transfer* 91:355, ASME, New York. (5)

Ivashkevitch, A. A., 1961, Critical Heat Flux and Heat Transfer Coefficient for Boiling Liquids in Forced Convection, *Teploenergetika,* October, Vol. 8. (5)

Ivey, H. J., 1967, Relationships between Bubble Frequency, Departure Diameter, and Rise Velocity in Nucleate Boiling, *Int. J. Heat Mass Transfer* 10:1023. (2)

Ivey, H. J., and D. J. Morris, 1962, On the Relevance of the Vapor Liquid Exchange Mechanism for Subcooled Boiling Heat Transfer at Higher Pressure, Rep. AEEW-R-137, UK Atomic Energy Authority, Winfrith, England. (2)

Ivey, H. J., and D. J. Morris, 1965, The Effect of Test Section Parameters on Saturation Pool Boiling Burnout at Atmospheric Pressure, *AIChE Chem. Eng. Prog. Symp. Ser.* 61(60):157–166. (2)

Jacket, H. S., J. D. Roarty, and J. E. Zerbe, 1958, Investigation of Burnout Heat Flux in Rectangular Channels at 2000 psia, *Trans. ASME, J. Heat Transfer* 80:391. (5)

Jairajkuri, A. M., and J. S. Saini, 1991, A New Model for Heat Flow through Macrolayer in Pool Boiling at High Heat Flux, *Int. J. Heat Mass Transfer* 34:1579–1591. (2)
Jakob, M., 1949, *Heat Transfer*, vol. 1, chap. 29, John Wiley, New York. (4)
Jakob, M., and W. Linke, 1933, Heat Transfer from a Horizontal Plate, *Forsch. Gebiete Ing.* 4(2):75–81. (2)
Jallouk, P. A., 1974, Two-Phase Flow, Pressure Drop and Heat Transfer Characteristics of Refrigerants in Vertical Tubes, Ph.D. thesis, University of Tennessee, Knoxville, TN. (4)
Janssen, E., 1962, Multirod Burnout at Low Pressure, ASME Paper 62-HT-26, Natl. Heat Transfer Conf. (3)
Janssen, E., 1967, Two Phase Flow Structure in a Nine-Rod Channel, Steam-Water at 1000 psia, Final Summary Rep. GEAP-5480, General Electric, San Jose, CA. (5)
Janssen, E., and J. A. Kervinen, 1963, Burnout Conditions for Single Rod in Annular Geometry, Water at 600 to 1400 psia, US AEC Rep. GEAP-3899, San Jose, CA. (5)
Janssen, E., and S. Levy, 1962, Burnout Limit Curves for Boiling Water Reactors, General Electric Co. Rep. APED-3892. (5)
Janssen, E., S. Levy, and J. A. Kervinen, 1963, Investigation of Burnout—Internally Heated Annulus Cooled by Water at 600 to 1450 psia, ASME Paper 63-WA-149, ASME Annual Meeting. (5)
Janssen, E., F. A. Schraub, R. B. Nixon, B. Matzner, and J. F. Casterline, 1969, Sixteen-Rod Heat Flux Investigation, Steam-Water at 600 to 1250 psia, ASME Paper, Winter Annual Meeting, Los Angeles, CA. (5)
Jawurek, H. H., 1969, Simultaneous Determination of Microlayer Geometry and Bubble Growth in Nucleate Boiling, *Int. J. Heat Mass Transfer* 12:843. (2)
Jayanti, S., and G. F. Hewitt, 1992, On the Prediction of the Slug-to-Churn Flow Transition in Vertical Two-Phase Flows, *Int. J. Multiphase Flow* 18:847–860. (3)
Jeglic, F. A., and T. M. Grace, 1965, Onset of Flow Oscillations in Forced Flow Subcooled Boiling, NASA-TN-D 2821, NASA Lewis Res. Ctr., Cleveland, OH. (6)
Jens, W. H., and P. A. Lottes, 1951, Analysis of Heat Transfer, Burnout, Pressure Drop, and Density Data for High Pressure Water, USAEC Rep. ANL-4627, Argonne Natl. Lab., Argonne, IL. (4)
Jepson, W. P., and R. E. Taylor, 1993, Slug Flow and Its Transitions in Large-Diameter Horizontal Pipes, *Int. J. Multiphase Flow* 19(3):410–420. (3)
Jiji, L. M., and J. A. Clark, 1964, Bubble Boundary Layer and Temperature Profiles for Forced Convection Boiling in Channel Flow, *Trans. ASME, J. Heat Transfer* 86:50–58. (4)
Johannessen, T., 1972, A Theoretical Solution of the Lockhart and Marinelli Flow Model for Calculating Two-Phase Flow Pressure Drop and Hold-up, *Int. J. Heat Mass Transfer* 15:1443. (3)
Jones, S., A. Amblard, and C. Favreau, 1986, Interaction of an Ultrasonic Wave with a Bubbly Mixture, *Exp. Fluids* 4:341–349. (3)
Jones, A. B., and A. G. Dight, 1961–1964, Hydrodynamic Stability of a Boiling Channel, USAEC Rep. KAPL-2170 (1961); KAPL-2208 (1962); KAPL-2290 (1963); KAPL-3070 (1964), Schenectady, New York. (6)
Jones, A. B., and W. M. Yarbrough, 1964–1965, Reactivity Stability of a Boiling Reactor, Part 1, US AEC Rep. KAPL-3072 (1964); Part 2, KAPL-3093 (1965), Albany, New York. (6)
Jones, J. K., 1969, An Experimental Study of the Critical Heat Flux for Low-Pressure Boiling Water in Forced Convection in a Vertical Seven-Rod Bundle, USAEC Rep. ORNL-TM-2122, Oak Ridge Natl. Lab., Oak Ridge, TN. (5)
Jones, J. K., and H. W. Hoffman, 1970, Critical Heat Flux for Boiling Water in a Rod Bundle as a Prelude to Boiling Potassium, ASME Paper 70-HT-22, ASME, New York. (5)
Jones, O. C., Jr., N. Abuof, G. A. Zimmer, and T. Feierabuend, 1981, Void Fluctuation Dynamics and Measurement Techniques, in *Two-Phase Flow Dynamics*, A. E. Bergles and S. Ishgai, Eds., Hemisphere, New York. (3)
Jones, O. C., Jr., and S. G. Bankoff, 1977, Two-Phase Flow in Light Water Reactors, *Proc. Symp. on Thermal and Hydraulic Aspects of Nuclear Reactors*, ASME, Atlanta GA. (3)
Judd, D. F., R. H. Wilson, C. P. Welch, R. A. Lee, and J. W. Ackerman, 1965, Nonuniform Heat Gener-

Convection Flow of Water, in *International Developments in Heat Transfer*, vol. 2, pp. 262–269, ASME, New York. (5)

Kirby, D. B., and J. W. Westwater, 1965, Bubble and Vapor Behavior on a Heated Horizontal Plate during Pool Boiling Near Burnout, *AIChE Chem. Eng. Prog. Symp. Ser.* 61(57):238–248. (2)

Kirby, G. J., 1966, A Model for Correlating Burnout in Round Tubes, UK Rep. AEEW-R-511, UK AEEW, Winfrith, England. (5)

Kirby, G. J., R. Staniforth, and J. H. Kinneir, 1965, A Visual Study of Forced Convection Boiling Part I, Results for a Flat Vertical Heater, UK Rep. AEEW-R-281, UK AEEW, Winfrith, England. (5)

Kirillov, P. L., 1968, A Generalized Functional Relationship between the CHF and Pressure in the Boiling of Metals in Large Quantities, *Transl. Atomic Energy* 24:143. (2)

Kirillov, P. L., L. P. Smogalev, M. Ya. Suvorow, R. V. Shumsky, Yu. Yu. Stein, 1978, Investigation of Steam-Water Flow Characteristics at High Pressures, 6th Int. Heat Transfer Conf., vol. 1, p. 315, Toronto, Canada. (3)

Kjellstrom, B., and A. E. Larson, 1967, Improvement of Reactor Fuel Element Heat Transfer by Surface Roughness, Rep. AE-R-271, A B Atomenergi, Nyköting, Sweden. (5)

Klausner, J. F., R. Mei, D. M. Bernhard, and L. Z. Zeng, 1993, Vapor Bubble Departure in Forced Convection Boiling, *Int. J. Heat Mass Transfer* 36(3):651–662. (4)

Knoebel, D. H., S. D. Harris, B. Craink, Jr., and R. M. Biderman, 1973, Forced Convection Subcooled Critical Heat Flux, USAEC Rep. DP-1306 Savannah River Lab., Aiken, SC. (5)

Knoll, K. E., 1991, Investigation of an Electromagnetic Flowmeter for Gas-Liquid Two-Phase Flow Measurement, *ANS Trans. TANSAO* 64:720. (3)

Kocamustafaogullari, G., and M. Ishii, 1987, Scaling of Two-Phase Flow Transients Using Reduced Pressure System and Simulant Field, *Nuclear Eng. Design* 104:121–132. (5)

Koestel, A., M. Gutstein, and R. T. Wainwright, 1963, Fog-Flow Mercury Condensing Pressure Drop Correlation, *Proc. 3rd Annual High Temperature Liquid Metal Heat Transfer Tech. Meeting,* ORNL-3605, Vol. 2, 198, ORNL, Oak Ridge, TN. (3)

Koffman, L. D., and M. S. Plesset, 1983, Experimental Observation of the Microlayer in Vapor Bubble Growth on a Heated Surface, *Trans. ASME, J. Heat Transfer* 105:625–632. (2)

Kopalinsky, E. M., and R. A. A. Bryant, 1976, Friction Coefficient for Bubbly Two-Phase Flow in Horizontal Pipes, *AIChE J.* 22(1):82. (3)

Kor'kev, A. A., and Y. D. Barulin, 1966, Heat Transfer in Boiling Subcooled Water, *Teploenergetika* 13:50. (4)

Korneev, M. I., 1955, Pool Boiling Heat Transfer with Mercury and Magnesium Amalgams, *Teploenergetika* 2:44. (2)

Koshelev, I. I., A. V. Surnov, and L. V. Nikitina, 1970, Inception of Pulsations Using a Model of Vertical Water-Wall Tubes, *Heat Transfer Sov. Res.* 2:3. (6)

Kosterin, S. I., 1949, Study of Influence of Tube Diameter and Position upon Hydraulic Resistance and Flow Structure of Gas-Liquid Mixtures, Izvestiza Akademii Nauk SSSR, Otdelema Tekhnicheskikh No. 12, 1824, Translation 3085, Henry Brutcher Tech. Translation, Altadena, CA. (3)

Kottowski, H., and G. Grass, 1970, Influence on Superheating by Suppression of Nucleation Cavities and Effect of Surface Microstructure on Nucleation Sites, *Proc. Symp. L/M Heat Transfer and Fluid Dynamics,* p. 108, ASME, New York. (2)

Koumoutsos, N., R. Moissis, and A. Spyridonos, 1967, A Study of Bubble Departure in Forced Convection Boiling, ASME Paper 67-HT-13, National Heat Transfer Conf. (3)

Kozlov, B. K., 1954, Forms of Flow of Gas-Liquid Mixtures and Their Stability Limits in Vertical Tubes, *Zh Tekh. Fiz.* 24(12):2285–2288. Transl. RJ-418, Assn. Tech. Service, East Orange, NJ. (3)

Kramer, A. M., 1958, *Boiling Water Reactors,* Addison-Wesley Pub. Co., New York. (4)

Kried, D. K., J. M. Creer, J. M. Bates, M. S. Quigley, A. M. Sutey, and D. S. Rowe, 1979, Fluid Flow Measurements in Rod Bundles Using Laser Doppler Anemometry Techniques, *Fluid Flow and Heat Transfer over Rod or Tube Bundles,* p. 13, ASME Winter Annual Meeting, ASME, New York. (3)

Kudryavtsev, A. P., D. M. Ovechkin, D. N. Sorokin, V. I. Subbotin, and A. A. Tsyganok, 1967, Transfer of Heat from Sodium Boiling in a Large Vessel (transl.), p. 268, Liquid Metals Atomidzdat, Moscow. (2)

Kumamaru, H., Y. Koizumi, and K. Tasake, 1987, Investigation of Pre- and Post-Dryout Heat Transfer of Steam-Water Two-Phase Flow in Rod Bundles, *Nuclear Eng. Design 102:*71–84. (4)

Kutateladze, S. S., 1952, Heat Transfer in Condensation and Boiling, USAEC Rep. AEC-tr-3770 (Translated from *Mashgiz,* 2d ed., pp. 76–107, State Sci. & Tech. Pub. House of Literature on Machinery, Moscow-Leningrad). (2)

Kutateladze, S. S., 1959, Critical Heat Flux in Subcooled Liquid Flow, *Energetika 7:*229–239. (2)

Kutateladze, S. S., 1963, *Fundamentals of Heat Transfer,* Academic Press, New York. (2)

Kutateladze, S. S., 1965, Criteria of Stability in Two-Phase Flow, *Symposium on Two Phase Flow,* vol. 1, pp. 83–92, University of Exeter, Devon, England. (5)

Kutateladze, S. S., 1961, Boiling Heat Transfer, *Int. J. Heat and Mass Transfer 4,* 31–45. (4)

Kutateladze, S. S., V. M. Borishansky, I. I. Novikov, and O. S. Fedyaskii, 1958, Liquid Metal Heat Transfer Media, *Atomnaya Energiva* (Moscow) Supplement No. 2, Translated by Consultants Bureau, Inc., New York (1959).

Kutateladze, S. S., and V. M. Borishansky, 1966, *A Concise Encyclopedia of Heat Transfer,* J. B. Arthur, trans., p. 212, Pergamon Press, New York. (4)

Kutateladze, S. S., and A. I. Leont'ev, 1964, Turbulent Boundary Layers in Compressible Gases, D. B. Spalding, trans., Academic Press, New York. (5)

Kutateladze, S. S., and A. I. Leont'ev, 1966, Some Applications of the Asymptotic Theory of the Turbulent Boundary Layer, *Proc. 3rd Int. Heat Transfer Conf.,* vol. 3, pp. 1–6, AIChE, New York. (5)

Kutateladze, S. S., and L. G. Malenkov, 1974, Heat Transfer at Boiling and Barbotage Similarity and Dissimilarity, *Proc. 5th Int.* Heat Transfer Conf., Tokyo, vol. IV, p. 1. (2)

Kutateladze, S. S., V. N. Moskvicheva, G. I. Bobrovich, N. N. Mamontova, and B. P. Avksentyuk, 1973, Some Peculiarities of Heat Transfer Crisis in Alkali Metals Boiling under Free Convection, *Int. J. Heat Mass Transfer 16:*705. (2)

Kutateladze, S. S., and L. L. Schneiderman, 1953, Experimental Study of Influence of Temperature of Liquid on the Change in the Rate of Boiling, USAEC Rep. tr-3405, 95–100, Washington, DC. (2)

Labuntsov, D. A., 1961, Critical Thermal Loads in Forced Motion of Water Which is Heated to a Temperature below the Saturation Temperature, *Sov. J. Atomic Energy* (English transl.) *10:*516–518. (5)

Lacy, C. E., A. E. Dukler, 1994, Flooding in Vertical Tubes—I: Experimental Studies of the Entrance Region, *Int. J. Multiphase Flow 20:*219–233; Flooding in Vertical Tubes—II: A Film Model for Entrance Region Flooding, *Int. J. Multiphase Flow 20:*235–247. (3)

Lahey, R. T., Jr., 1988, Turbulence and Phase Distribution Phenomena, in *Transient Phenomena in Multiphase Flow,* ICHMT Int. Seminar, N. H. Afgan, Ed., Hemisphere, New York. (3)

Lahey, R. T., Jr., 1990, Appl. of Fractal and Chaos Theory in the Field of Two-Phase Flow and Heat Transfer, *Advances in Gas-Liquid Flow,* ASME Winter Annual Meeting, FED-Vol. 99/HTD-Vol. 155, pp. 413–425. (6)

Lahey, R. T., Jr., and J. M. Gonzalez Santalo, 1977, The Effect of Non-Uniform Axial Heat Flux on Critical Power, Paper C29/77, Inst. Mech. Eng., London. (5)

Lahey, R. T., Jr., and F. J. Moody, 1977, *The Thermal Hydraulics of a Boiling Water Nuclear Reactor,* American Nuclear Society, LaGrange Park, IL. (3)

Lahey, R. T., Jr., and F. A. Schraub, 1969, Mixing, Flow Regimes and Void Fraction for Two-Phase Flow in Rod Bundles, in *Two-Phase Flow and Heat Transfer in Rod Bundles,* ASME, New York. (5)

Lahey, R. T., Jr., B. S. Shiralkar, and D. W. Radeliffe, 1971, Mass Flux and Enthalpy Distribution in a Rod-Bundle for Single and Two-Phase Flow Conditions, *Trans. ASME, J. Heat Transfer, 93:*197–209. (5)

Lahey, R. T., Jr., B. S. Shiralkar, D. W. Radeliffe, and E. E. Polomik, 1972, Out-of-Pile Subchannel Measurements in a Nine-Rod Bundle for Water at 1000 psia, in *Progress in Heat and Mass Transfer,* vol. VI, Pergamon Press, New York. (5)

Lahey, R. T., Jr., and G. Yadigaoglu, 1973, NUFREQ—A Computer Program to Investigate T/H Stability, NEDO-13344, General Electric Co., San Jose, CA. (5)

Lamb, H., 1945, *Hydrodynamics,* pp. 122, 374, 462, Dover, New York. (2)

Lang, C., 1888, *Trans. Inst. Mech. Eng. 32:*279–295, Shipbuilders, Scotland. (1)

Larson, P. S., and L. S. Tong, 1969, Void Fraction in Subcooled Flow Boiling, *Trans. ASME, J. Heat Transfer 91:*471–476. (3)

Latrobe, A., 1978, A Comparison of Some Implicit Finite Difference Schemes Used in Flow Boiling Analysis, in *Transient Two-Phase Flow* Proc. 2nd Specialists Meeting, OECD Comm. for Safety of Nuclear Installations, Paris, Vol. 1, 439–495. (3)

Laverty, W. F., and W. M. Rohsenow, 1964, Film Boiling of Saturated Liquid Flowing Upward through a Heated Tube: High Vapor Quality Range, MIT Heat Transfer Lab. Rep. 9857-32, Massachusetts Institute of Technology, Cambridge, MA. (4)

Le Coq, G., J. Lewi, and P. Raymond, 1978, Comments on the Formulation of the One-Dimension, Six Equations Two-Phase Flow Models, *Transient Two-Phase Flow,* Proc. 2nd Specialists Meeting, OECD Committee for Safety of Nuclear Installations, Paris, vol. 1, pp. 83–98. (3)

Ledinegg, M., 1938, Instability of Flow during Natural and Forced Circulation, *Die Warme 61:*8, AEC-tr-1861 (1954). (2)

Lee, C. H., and I. A. Mudawwar, 1988, A Mechanistic CHF Model for Subcooled Flow Boiling Based on Local Bulk Flow Conditions, *Int. J. Multiphase Flow 14:*711. (5)

Lee, C. S., D. L. Lee, and C. F. Bonilla, 1969, Calculation of the Thermodynamics and Transport Properties of Na, K, Rb, Cs vapors to 3000°K, *Nuclear Eng. Design 10:*83. (2)

Lee, D. H., 1965, An Experimental Investigation of Forced Convection Burnout in High Pressure Water, Part 3. Long Tubes with Uniform and Non-Uniform Axial Heating, UK Rep. AEEW-R-355, UK AEEW, Winfrith, England. (5)

Lee, D. H., 1966a, Burnout in a Channel with Non-Uniform Circumferential Heat Flux, UK Rep. AEEW-R-477, UK AEEW, Winfrith, England. (5)

Lee, D. H., 1966b, An Experimental Investigation of Forced Convection Burnout in High Pressure Water, Part 4. Large Diameter Tubes at about 1600 psia, UK Rep. AEEW-R-479, UK AEEW, Winfrith, England. (5)

Lee, D. H., 1970, Studies of Heat Transfer and Pressure Drop Relevant to Subcritical Once-through Evaporator, Paper IAEA-SM-130/56, *Symp. on Progress in Sodium-Cooled Fast Reactor Engineering,* Monte Carlo, Monaco. (4)

Lee, D. H., and R. B. Little, 1962, Experimental Studies into the Effect of Rod Spacing on Burnout in a Simulated Rod Bundle, UK Rep. AEEW-R-178, UK AEEW, Winfrith, England. (5)

Lee, D. H., and J. D. Obertelli, 1963, An Experimental Investigation of Forced Convection Burnout in High Pressure Water, Part 2. Preliminary Results for Round Tubes with Non-Uniform Axial Heat Flux Distribution, UK Rep. AEEW-R-309, UK AEEW, Winfrith, England. (5)

Lee, R. C., and J. E. Nydahl, 1989, Numerical Calculation of Bubble Growth in Nucleate Boiling from Inception through Departure, *Trans. ASME, J. Heat Transfer 111:*474–479. (2)

Lee, S. S., T. N. Veziroglu, and S. Kakac, 1976, Sustained and Transient Boiling Flow Instabilities in Two Parallel Channel Systems, *Proc. NATO Adv. Study Inst.* 1:467–510. (6)

Lee, Y., 1968, Pool-Boiling Heat Transfer with Mercury and Mercury Containing Dissolved Sodium, *Int. J. Heat Mass Transfer 11:*1807. (2)

Leidenfrost, J. G., 1756, *De Aguae Communis Nonnullis Qualitatibus Tractatus,* Duisburg. (1)

Leiner, W., 1994, Heat Transfer by Nucleate Pool Boiling—General Correlation Based on Thermodynamic Similarity, *Int. J. Heat Mass Transfer* 37(5):763–769. (2)

Lemmon, A. W., Jr., H. W. Deem, E. H. Hall and J. F. Walling, 1964, The Thermodynamic and Transport Properties of Potassium, *Proc. of High Temperature Liquid Metal Technology Meeting,* Vol. 1, 88–114, USAEC Rep. ORNL-3605. (2)

Leont'eva, L. A., and V. Ya Gal'tsov, 1968, Investigation of Heat Transfer during the Boiling of Solutions in a Vertical Tube under Conditions of Forced Motion, *Int. Chem. Eng.* 8(2):329–331. (4)

Leroux, K. M., and M. K. Jensen, 1992, CHF in Horizontal Tube Bundles in Vertical Crossflow of R-113, *Trans. ASME, J. Heat Transfer 114:*179–184. (5)

Levy, S., 1960, Steam Slip—Theoretical Prediction from Momentum Model, *Trans. ASME, J. Heat Transfer 82:*113. (3)

Levy, S., 1966, Forced Convection Subcooled Boiling—Prediction of Vapor Volumetric Fraction, Rep. GEAP 5157, General Electric Co., San Jose, CA. (5)

Levy, S., 1967, Forced Convection Subcooled Boiling—Prediction of Vapor Volumetric Fraction, *Int. J. Heat Mass Transfer 10:*951. (3)

Levy, S., and J. M. Healzer, 1980, Prediction of Annular Liquid-Gas Flow with Entrainment, Co-current Vertical Pipe Flow without Gravity, Rep. EPRI NP-1409, Electric Power Research Institute, Palo Alto, CA. (5)

Levy, S., and J. M. Healzer, 1980, Co-current Vertical Pipe Flow with Gravity, Rep. EPRI-NP-1521, Electric Power Research Institute, Palo Alto, CA. (5)

Levy, S., J. M. Healzer, and D. Abdollahian, 1980, Prediction of Critical Heat Flux for Annular Flow in Vertical Pipes, EPRI NP-1619, EPRI, Palo Alto, CA. (5)

Lewis, J. P., and D. E. Graesbeck, 1969, Tests of Sodium Boiling in a Single Tube-in-Shell Heat Exchanger over the Range 1720° to 1980°F, NASA TN D-5323, Lewis Res. Ctr., Cleveland, OH. (3)

Liaw, S. P., and V. K. Dhir, 1986, Effect of Surface Wettability on Transition Boiling Heat Transfer from a Vertical Surface, *Int. Heat Transfer Conf.*, San Francisco, CA, vol. 4. (2)

Lienhard, J. H., 1985, On the Two Regimes of Nucleate Boiling, *Trans. ASME, J. Heat Transfer 107:*262–264. (2)

Lienhard, J. H., 1988, Burnout on Cylinders, *Trans. ASME, J. Heat Transfer 110:*1271–1286. (2)

Lienhard, J. H., and V. K. Dhir, 1973a, Hydrodynamic Prediction of Peak Pool-Boiling Heat Fluxes from Finite Bodies, *Trans. ASME, J. Heat Transfer 95:*152. (2)

Lienhard, J. H., and V. K. Dhir, 1973b, Extended Hydrodynamic Theory of the Peak and Minimum Pool-Boiling Heat Fluxes, NASA CR-2270, UCLA, Los Angeles, CA. (2)

Lienhard, J. H., V. K. Dhir, and D. M. Riherd, 1973c, Peak Pool Boiling Heat Flux Measurement on Finite Horizontal Flat Plates, *Trans. ASME, J. Heat Transfer 95*(4):477–482. (2)

Lienhard, J. H., and M. M. Hasan, 1979, On Predicting Boiling Burnout with the Mechanical Energy Stability Criterion, *Trans. ASME, J. Heat Transfer 101:*276 (2)

Lienhard, J. H., and K. B. Keeling, Jr., 1970, An Induced Convection Effect upon the Peak Boiling Heat Flux, *Trans. ASME, J. Heat Transfer 92.* (5)

Lienhard, J. H., and V. E. Schrock, 1962, The Effect of Pressure, Geometry and the Equation of State upon the Peak and Min. Boiling Heat Flux, ASME Paper 62-HT-3, National Heat Transfer Conf., ASME, New York. (5)

Lin, J. C., and J. Weisman, 1990, A Phenomenologically Based Prediction of the CHF in Channels Containing an Unheated Wall, *Int. J. Heat Mass Transfer 33:*203–205. (5)

Lin, J. T., L. Ovacik, O. C. Jones, J. C. Newell, M. Cheney, and H. Suzuki, 1991, Use of Electrical Impedance Imaging in Two-Phase, Gas-Liquid Flows, *ANS, Proc. 27th Natl. Heat Transfer Conf.,* Minneapolis, MN, HTC-Vol. 5:190. (3)

Lin, P. Y., and T. J. Hanratty, 1987, Detection of Slug Flow from Pressure Measurements, *Int. J. Multiphase Flow 13*(1):13. (3)

Lin, Wen-Shan, Chien-Hsiung Lee, and Bao-Shei Pei, 1989, An Improved Theoretical CHF Model for Low-Quality Flow, *Nuclear Technol. 88:*294–306. (5)

Liu, Z., and R. H. S. Winterton, 1991, A General Correlation for Saturated and Subcooled Flow Boiling in Tubes and Annuli Based on a Nucleate Pool Boiling Equation, *Int. J. Heat Mass Transfer 34*(11):2759–2766. (4)

Lockhart, R. W., and R. C. Martinelli, 1949, Proposed Correlation of Data for Isothermal Two-Phase, Two-Component Flow in Pipes, *Chem. Eng. Prog. 45:*39. (3)

Logan, D., C. T. Baroczy, J. A. Landoni, and H. A. Morewitz, 1970, Effects of Velocity, Oxide Level, and Flow Transients on Boiling Initiation in Sodium, AI-AEC-12939; Rockwell Int., Canoga

ation Experimental Program, Quarterly Progress Rep. 7, Jan.–Mar. 1965, USAEC BAW-3238-7, Babcock & Wilcox Co., Lynchburg, VA. (5)

Judd, R. L., 1989, The Influence of Subcooling on the Frequency of Bubble Emission in Nucleate Boiling, *Trans. ASME, J. Heat Transfer 111:*747–751. (2)

Judd, R. L., and K. S. Hwang, 1976, A Comprehensive Model for Nucleate Pool Boiling Heat Transfer Including Microlayer Evaporation, *Trans. ASME, J. Heat Transfer* 98(4):623–624. (2)

Kandlikar, S. G., 1983, An Improved Correlation for Predicting Two-Phase Flow Boiling Heat Transfer Coefficient in Horizontal and Vertical Tubes, ASME HTD *Heat Exchangers for Two-Phase Flow Applications,* 21st Natl. Heat Transfer Conf., Seattle, WA. (4)

Kandlikar, S. G., 1989a, A General Correlation for Saturated Two-Phase Flow Boiling Heat Transfer inside Horizontal and Vertical Tubes, *Trans. ASME, J. Heat Transfer 112:*219–228. (4)

Kandlikar, S. G., 1989b, Development of a Flow Boiling Map for Subcooled and Saturated Flow Boiling of Different Fluids Inside Circular Tubes, *Heat Transfer with Phase Change,* ASME HTD Vol. 114, pp. 51–62, Winter Annual Meeting, San Francisco, CA. (4)

Kapitza, P. L., 1964, Wave Flow of Thin Layers of a Viscous Fluid, in *Collected Papers of P. L. Kapitza,* vol. II, Macmillan, New York. (3)

Karplus, H. B., 1958, The Velocity of Sound in a Liquid Containing Gas Bubbles, Rep. COO-248, Illinois Institute of Technology, Chicago, IL. (3)

Karplus, H. B., 1961, Propagation of Pressure Waves in a Mixture of Water and Steam, Rep. ARF-4132-12, Illinois Institute of Technology. (3)

Kast, W., 1964, Significance of Nucleating and Non-stationary Heat Transfer in the Heat Exchanger during Bubble Vaporization and Droplet Condensation, *Chem. Eng. Tech.* 36(9):933–940. (2)

Katto, Y., 1981, General Features of CHF of Forced Convection Boiling in Uniformly Heated Rectangular Channels, *Int. J. Heat Mass Transfer 24:*1413–1419. (5)

Katto, Y., 1983, Critical Heat Flux in Forced Convection, *Proc. ASME-JSME Thermal Eng. Joint Conf.,* Honolulu, HI, vol. 3, pp. 1–10, ASME, New York. (2)

Katto, Y., 1992, A Prediction Model of Subcooled Water Flow Boil-CHF for Pressures in the Range of 0.1–20.0 MPa, *Int. J. Heat Mass Transfer 35:*1115–1123. (5)

Katto, Y., 1994a, Limiting Conditions of Steady State Countercurrent Annular Flow and the Onset of Flooding, with Reference to the CHF of Boiling in a Bottom-Closed Vertical Tube, *Int. J. Multiphase Flow 20:*45–61. (3)

Katto, Y., 1994b, Critical Heat Flux, *Int. J. Multiphase Flow 20*(suppl.):53–90. (2)

Katto, Y., and S. Yokoya, 1966, Experimental Study of Nucleate Pool Boiling in Case of Making Interference-Plate Approach to the Heating Surface, Paper 103, Proc. 3rd Int. Heat Transfer Conf., Chicago, IL, vol. 3, p. 219. (2)

Katto, Y., S. Yokoya, S. Miake, and M. Taniguchi, 1987, CHF on a Uniformly Heated Cylinder in a Crossflow of Saturated Liquid over a Very Wide Range of Vapor-to-Liquid Density Ratio, *Int. J. Heat Mass Transfer 30:*1971–1977. (5)

Kawaji, M., and S. Banerjee, 1987, Application of a Multifield Model to Reflooding of a Hot Vertical Tube, Part I—Model Structure and Interfacial Phenomena, *Trans. ASME, J. Heat Transfer 109:*204–211. (4)

Kays, W. M., and A. L. London, 1958, *Compact Heat Exchangers,* National Press, Palo Alto, CA. (3)

Keeys, R. K. F., J. C. Ralph, and D. N. Roberts, 1971, Post Burnout Heat Transfer in High Pressure Steam Water Mixtures in a Tube with Cosine Heat Flux Distribution, UK Rep. AERE-R-6411, AEA, Harwell, England. (5)

Kendall, G. E., and W. M. Rohsenow, 1978, Heat Transfer to Impacting Drops and Post CHF Dispersed Flow, MIT Heat Transfer Lab. Rep. 85694-100, Massachusetts Institute of Technology, Cambridge, MA. (4)

Keshgi, H. S., and L. E. Scriven, 1983, Measurement of Liquid Film Profiles by Moiré Topography, *Chem. Eng. Sci.* 38(4):525–534. (3)

Kezios, S. P., T. S. King, and F. M. Rafchiek, 1961, Burnout in Crossed Rod Matrices and Forced

Park, CA. Also in *Liquid Metal Heat Transfer & Fluid Dynamics*, ASME Symp., pp. 116–128, ASME, New York. (2)

Longo, Ed., 1963, Alkali Metals Boiling and Condensing Investigations, Quarterly Progress Reps 2 and 3, Space & Propulsion Sec. General Electric Co., Sunnyvale, CA. (4)

Lorentz, J. J., B. B. Mikic, and W. M. Rohsenow, 1974, The Effect of Surface Conditions on Boiling Characteristics, *Proc. 5th Int. Heat Transfer Conf.*, vol. 5, Hemisphere, New York. (2)

Lottes, P. A., 1961, Expansion Losses in Two-Phase Flow, *Nuclear Sci. Eng.* 9:26–31. (3)

Lottes, P. A., and W. S. Flinn, 1956, A Method of Analysis of Natural Circulation Boiling System, *Nuclear Sci. Eng. Design 1:*461. (3)

Lowdermilk, W. H., C. D. Lanzo, and B. L. Siegel, 1958, Investigation of Boiling Burnout and Flow Stability for Water Flowing in Tubes, NACA-TN-4382, NASA Lewis Res. Ctr., Cleveland, OH. (1)

Lurie, H., 1965, Sodium Boiling Heat Transfer and Hydrodynamics, *Proc. Conf. on Applied High Temperature Instrumentation to Liquid Metal Experiments*, ANL-7100, pp. 549–571, Argonne Natl. Lab., Argonne, IL. (3)

Lurie, H., 1966, Steady State Sodium Boiling and Hydrodynamics, NAA-SR-11586, North American Aviation, Inc., Atomic Int. Div., Canoga Park, CA. (5)

Lyon, R. E., 1955, Liquid Metal Heat Transfer Coefficient, *Trans. AIChE* 47:75–79. (4)

Lyon, R. E., A. S. Foust, and D. L. Katz, 1955, Boiling Heat Transfer with Liquid Metals, *Chem. Eng. Prog. 51:*41. (2)

Macbeth, R. V., 1963a, Burnout Analysis: Pt. 2, The Basic Burnout Curve, UK Rep. AEEW-R-167; Pt. 3, The Low Velocity Burnout Regime, AEEW-R-222; Pt. 4, Application of Local Conditions Hypothesis to World Data for Uniformly Heated Round Tubes and Rectangular Channels, AEEW-R-267, UK AEEW, Winfrith, England. (5)

Macbeth, R. V., 1963b, Forced Convection Burnout in Simple, Uniformly Heated Channels: A Detailed Analysis of World Data, European Atomic Energy Community Symp. on Two Phase Flow, Steady State Burnout and Hydrodynamic Instability, Stockholm, Sweden. (5)

Macbeth, R. V., 1964, Burnout Analysis, Pt. 5, Examination of World Data for Rod Bundles, UK Rep. AEEW-R-358, UK AEEW, Winfrith, England. (5)

Macbeth, R. V., 1970, Personal communication.

Madden, J. M., 1968, Two-Phase Air-Water Flow in a Slot Type Distributor, M.S. thesis, University of Windsor, Canada. (3)

Madsen, N., and C. F. Bonilla, 1959, Heat Transfer to Boiling Sodium and Potassium Alloy, *AIChE Chem. Eng. Prog. Symp. Ser.* 56(30):251–259. (2)

Malenkov, I. G., 1971, Detachment Frequency as a Function of Size of Vapor Bubbles (transl.), *Inzh. Fiz. Zh. 20:*988. (2)

Malnes, D., 1966, Slip Ratios and Friction Factors in the Bubble Flow Regime in Vertical Tubes, Norwegian Rep. KR-110, Inst. for Atomenergi, Oslo, Norway. (5)

Malnes, D., and H. Boen, 1970, Dynamic Behavior of Hydraulic Channels, Meeting of European Two-Phase Flow Group, Milan, Italy. (3)

Mandhane, J. M., G. A. Gregory, and K. Aziz, 1974, Flow Pattern Map for Gas-Liquid Flow in Horizontal Pipes, *Int. J. Multiphase Flow 1:*537–553. (3)

Mandrusiak, G. D., and V. P. Carey, 1989, Convective Boiling in Vertical Channels with Different Offset Strip Fin Geometries, *Trans. ASME, J. Heat Transfer 111:*156–165. (4)

Marchaterre, J. F., 1956, The Effect of Pressure on Boiling Density in Multiple Rectangular Channels, USAEC Rep. ANL-5522, Argonne National Lab., Argonne, IL. (5)

Marchaterre, J. F., and B. M. Hoglund, 1962, Correlation for Two-Phase Flow, *Nucleonics* 20(8):142. (3)

Marchaterre, J. F., and M. Petrick, 1960, The Prediction of Steam Volume Fraction in Boiling Systems, *Nuclear Sci. Eng. 7:*525. (3)

March-Leuba, J., 1990, Radial Nodalization Effects on BWR Stability Calculations, *Int. Workshop on BWR Stability, Proc.*, Hottsville, NY, CSNI Rep. 178, pp. 232–240, OECD-NEA, Committee for the Safety of Nuclear Installations, Paris. (6)

March-Leuba, J., and E. D. Blakeman, 1991, A Mechanism for Out-of-Phase Power Instabilities in BWRs, *Nuclear Sci. Eng. 107:*173–179. (6)

Marcus, B. D., 1963, Experiments on the mechanism of Saturated Pool Boiling Heat Transfer, Ph.D. thesis, Cornell University, Ithaca, NY. (2)

Marcus, B. D., and D. Dropkin, 1965, Measured Temperature Profiles within the Superheated Boundary Layer above a Horizontal Surface in Saturated Nucleate Pool Boiling of Water, *Trans. ASME, J. Heat Transfer 87:*333–341. (2)

Margolis, S. G., and J. A. Redfield, 1965, FLASH: A Program for Digital Simulation of the Loss of Coolant Accident, WAPD-TM-534, Bettis Atomic Power Lab., Pittsburgh, PA. (6)

Martenson, A. J., 1962, Transient Boiling in Small Rectangular Channels, Ph.D. thesis, University of Pittsburgh, Pittsburgh, PA. (5)

Martin, C. S., 1973, Transition from Bubbly to Slug Flow of Vertically Downward Air-Water Flow, *Proc. ASME Symp.,* Atlanta, GA. (3)

Martinelli, R. C., and D. B. Nelson, 1948, Prediction of Pressure Drop during Forced Circulation Boiling of Water, *Trans. ASME 70:*695. (3)

Marto, P. J., and W. M. Rohsenow, 1965, The Effect of Surface Conditions on Nucleate Pool Boiling Heat Transfer to Sodium, Ph.D. thesis, MIT Rep. 5219-33 (USAEC Rep. MIT-3357-1), Massachusetts Institute of Technology, Cambridge, MA. (2)

Marto, P. J., and W. M. Rohsenow, 1966, The Effect of Surface Conditions on Nucleate Pool Boiling Heat Transfer to Sodium, *Trans. ASME, J. Heat Transfer 88:*196. (2)

Mathisen, R. P., 1967, Out of Pile Channel Instability in the Loop Skalv, *Symp. on Two-Phase Dynamics,* Eindhoven, The Netherlands. (6)

Mattson, R. J., F. G. Hammitt, and L. S. Tong, 1973, A Photographic Study of the Subcooled Flow Boiling Crisis in Freon-113, ASME Paper 73-HT-39, Natl. Heat Transfer Conf., Atlanta, GA. (5)

Matzner, B., J. E. Casterline, E. O. Moech, and G. A. Wilkhammer, Experimental Critical Heat Flux Measurement Applied to a Boiling Reactor Channel, ASME Paper 66-WA/HT-46, Winter Annual Meeting, ASME, New York. (5)

Maulbetsch, J. S., and P. Griffith, 1965, A Study of System-Induced Instabilities in Forced Convection Flows with Subcooled Boiling, MIT Engineering Projects Lab. Rep. 5382-35, Massachusetts Institute of Technology, Cambridge, MA. (5)

Maulbetsch, J. S., and P. Griffith, 1967, Prediction of the Onset of System-Induced Instabilities in Subcooled Boiling, Euratom Report, *Proc. Symp. on Two-Phase Dynamics,* Eindhoven, The Netherlands, pp. 799–825. (6)

Maurer, G. W., 1960, A Method of Predicting Steady State Boiling Vapor Fraction in Reactor Coolant Channels, Bettis Technical Review, USAEC Rep. WARD-BT-19, pp. 59–70. (3)

Mayinger, F., 1981, Scaling and Modelling Laws in Two Phase Flow and Boiling Heat Transfer, in *Two-Phase Flow and Heat Transfer in the Power and Processing Industries,* Hemisphere, Washington, DC. (5)

McAdams, W. H., W. K. Woods, and R. L. Bryan, 1941, Vaporization inside Horizontal Tubes, *Trans. ASME 63:*545–552. (1)

McAdams, W. H., W. K. Woods, and R. L. Bryan, 1949, Heat Transfer at High Rates to Water with Surface Boiling, *Ind. Eng. Chem., 41:*1945–55. (4)

McDonald, B. N., and L. E. Johnson, 1970, Nuclear Once-through Steam Generator, Rep. BR-923, TPO-67, ANS Topical Meeting, Williamsburg, VA., September. (6)

McQuillan, K. W., and P. B. Whalley, 1985, Flow Patterns in Vertical Two-Phase Flow, *Int. J. Multiphase Flow 11:*161–175. (3)

McWilliams, D., and R. K. Duggins, 1969, Speed of Sound in Bubbly Liquids, *Symp. on Fluid Mechanics & Measurements in Two-Phase Flow Systems, Proc. Inst. Mech. Eng. 184*(Part 3C):102–107. (3)

Merte, H., and J. A. Clark, 1961, A Study of Pool Boiling in an Accelerating System, *Trans. ASME, J. Heat Transfer 83:*233–242. (2)

Michiyoshi, I., A. Serizawa, O. Takahashi, K. Gakuhart, and T. Ida, 1986, Heat Transfer and Hydraulics of Liquid Metal-Gas Two Phase Magnetohydraulic Flow, *Proc. 8th Int. Heat Transfer Conf.,* vol. 1, p. 2391. (3)

Mikic, B. B., and W. M. Rohsenow, 1969, A New Correlation of Pool Boiling Data Including the Effect of Heating Surface Characteristics, *Trans. ASME, J. Heat Transfer 91:*245. (2)

Mikic, B. B., W. M. Rohsenow, and P. Griffith, 1970, On Bubble Growth Rates, *Int. J. Heat Mass Transfer 13:*657. (2)

Minchenko, F. P., 1960, On Heat Transfer during Nucleate Boiling, *Energomashinstroenie,* no. 6, p. 17. (2)

Mirshak, S., and R. H. Towell, 1961, Heat Transfer Burnout of a Surface Contacted by a Spacer Rib, USAEC Rep. DP-262, Washington, DC. (5)

Mishima, K., and M. Ishii, 1984, Flow Regime Transition Criteria for Upward Two-Phase Flow in Vertical Tubes, *Int. J. Heat Mass Transfer 27:*723. (3)

Mishima, K., and H. Nishihara, 1985, The Effect of Flow Direction and Magnitude on CHF for Low Pressure Water in Thin Rectangular Channels, *Nuclear Eng. Design 86:*165–181. (5)

Mishima, K., and H. Nishihara, 1987, Effect of Channel Geometry on CHF for Low Pressure Water, *Int. J. Heat Mass Transfer 30*(6):1169. (5)

MIT, 1964, MIT Engineering Projects Lab. Rep. 9847-37, Massachusetts Institute of Technology, Cambridge, MA. (5)

Mizukami, K., K. Nishikawa, and F. Abe, 1992, Inception of Boiling of Water on S. S. Surface, ASME HTD-Vol. 212, pp. 123–130. (4)

Moissis, R., and P. J. Berenson, 1962, On the Hydrodynamic Transitions in Nucleate Boiling, ASME 62-HT-8, Natl. Heat Transfer Conf., Houston, TX. (2)

Monde, M., H. Kusuda, and H. Uehara, 1982, Critical Heat Flux during Natural Convective Boiling in Vertical Rectangular Channels Submerged in Saturated Liquid, *Trans. ASME, J. Heat Transfer 104:*300–303. (5)

Moody, F. J., 1965, Maximum Flow Rate of Single Component, Two-Phase Mixture, *Trans. ASME, J. Heat Transfer 87:*134. (3)

Moody, F. J. 1975, Maximum Discharge Rate of Liquid-Vapor Mixtures from Vessels, in *Nonequilibrium Two-Phase Flows,* R. T. Lahey, Jr., and G. B. Wallis, Eds., ASME, New York. (3)

Moore, F. D., and R. B. Mesler, 1961, The Measurement of Rapid Surface Temperature Fluctuations during Nucleate Boiling of Water, *AIChE J. 7:*620–624. (2)

Mori, Y., T. Harada, M. Uchida, and T. Hara, 1970, Convective Boiling of a Binary Liquid Metal, in *L/M Heat Transfer and Fluid Dynamics,* J. C. Chen and A. A. Bishop, Eds., Winter Annual Meeting, ASME, New York. (4)

Moujaes, S. F., and R. S. Dougall, 1987, Experimental Investigation of Concurrent Two-Phase Flow in a Vertical Rectangular Channel, *Can. J. Chem. Eng. 65:*705. (3)

Moujaes, S. F., and R. S. Dougall, 1990, Experimental Measurements of Local Axial Gas Velocity and Void Fraction in Simulated PWR Steam Generator Rod Bundles, *Can. J. Chem. Eng. 68:*211. (3)

Moxon, D., and P. A. Edwards, 1967, Dryout during Flow and Power Transients, UK Rep. AEEW-R-553, UK AEEW, Harwell, England. (5)

Mozharov, N. A., 1959, An Investigation into the Critical Velocity at Which a Moisture Film Breaks away from the Wall of a Steam Pipe, *Teploenergetika 6*(2):50–53, DSIR trans. RTS-1581. (5)

Murdock, J. W., 1962, Two-Phase Flow Measurement with Orifices, *Trans. ASME, J. Basic Eng. 84:*419–433. (3)

Nakamura, H., Y. Anoda, and Y. Kukita, 1991, Flow Regime Transition in High Pressure Steam-Water Horizontal Pipe Two-Phase Flow, *Proc. ANS Natl. Heat Transfer Conf.,* Minneapolis, MN, p. 269. (3)

Nakanishi, S., S. Ishigai, and S. Yamanchi, 1979, in *Two-Phase Momentum and Heat and Mass Transfer,* Durst, Tsiklauri, and Afgan, Eds., vol. 1, p. 315, Hemisphere, New York. (3)

Neal, L. G., S. M. Zivi, and R. W. Wright, 1967, The Mechanisms of Hydrodynamic Instabilities in Boiling Channel, Euratom Rep., *Proc. Symp. on Two-Phase Flow Dynamics,* Eindhoven, The Netherlands. (6)

Nelson, R., and C. Ünal, 1992, A Phenomenological Model of the Thermal Hydraulics of Convective Boiling during the Quenching of Hot Rod Bundles, Part I: Thermal Hydraulic Model, *Nuclear Eng. Design 136:*277–298. (4)

Nishikawa, K., and Y. Fujita, 1990, Nucleate Boiling Heat Transfer and Its Augmentation, *Adv. Heat Transfer 20:*1–82. (1)

Nishikawa, K., Y. Fujita, S. Uchida, and H. Ohta, 1983, Effect of Heating Surface Orientation on Nucleate Boiling Heat Transfer, *Proc. ASME-JSME Thermal Engineering Joint Conf.,* Honolulu, HI, vol. 1, pp. 129–136, ASME, New York. (2)

No, H. C., and M. S. Kazimi, 1982, Wall Heat Transfer Coefficients for Condensation & Boiling in Forced Convection of Sodium, *Nuclear Sci. Eng. 81:*319–324. (4)

Noyes, R. C., 1963, An Experimental Study of Sodium Pool Boiling Heat Transfer, *Trans. ASME, J. Heat Transfer 85:*125–131. (2)

Noyes, R. C., and H. Lurie, 1966, Boiling Sodium Heat Transfer, *Proc. 3rd Int. Heat Transfer Conf.,* Chicago, IL, vol. 5, p. 92. (2)

Nukiyama, S., 1934, Maximum and Minimum Values of Heat Transmitted from Metal to Boiling Water under Atmospheric Pressure, *J. Soc. Mech. Eng. Jpn. 37:*367. (1)

Nylund, O., et al., 1968, Full Scale Loop Studies of the Hydrodynamic Behavior of BHWR Fuel Elements, European Two-Phase Group Meeting, Oslo, Norway. (3)

Nylund, O., et al., 1969, The Influence of Non-Uniform Heat Flux Distribution on the Thermodynamic Behavior of a BHWR 36-Rod Cluster, European Two-Phase Group Meeting, Karlsruhe, Germany. (3)

Oberjohn, W. J., and R. H. Wilson, 1966, The Effect of Non-uniform Axial Flux Shape on the Critical Heat Flux, ASME Paper 66-WA/HT-60, Winter Annual Meeting, ASME, New York. (5)

Ogasawara, H. et al., 1973, Cooling Mechanism of the Low Pressure Coolant-Injection System of BWR and Other Studies on the Loss-of-Coolant-Accident Phenomena, ANS Topical Meeting Water Reactor Safety, p. 351, Salt Lake City, UT. (4)

Ohba, K., T. Origuchi, and Y. Shimanaka, 1984, Multi-Fiber Optic Liquid Film Sensor, *Proc. 4th Sensor Symp.,* pp. 33–37. (3)

Oshimowo, T., and M. E. Charles, 1974, Vertical Two-Phase Flow, I. Flow Pattern Correlation, *Can. J. Chem. Eng. 52:*25–35. (3)

Owens, W. L., Jr., 1961, Two-Phase Pressure Gradients, in International Developments in Heat Transfer, Part II, pp. 363–368, ASME, New York. (3)

Owens, W. L., Jr., and V. E. Schrock, 1960, Local Pressure Gradients for Subcooled Boiling of Water in Vertical Tubes, ASME Paper 60-WA-249, ASME, New York. (4)

Padilla, A., 1966, Film Boiling of Potassium on a Horizontal Plate, Ph.D. thesis, University of Michigan, Ann Arbor, MI. (2)

Palen, J. W., G. Breber, and J. Taborek, 1977, Prediction of Flow Regimes in Horizontal Tubeside Condensation, 17th Natl. Heat Transfer Conf., AIChE/ASME, Salt Lake City, UT. (3)

Palen, J. W., G. Breber, and J. Taborek, 1980, Prediction of Horizontal Tubeside Condensation of Pure Components Using Flow Regime Criteria, *Trans. ASME, J. Heat Transfer, 102:*471. (3)

Pan, C., and T. L. Lin, 1989, Marongoni Flow on Pool Boiling near CHF, *Int. Communun. Heat Mass Transfer 16:*475–486. (2)

Pan, C., and T. L. Lin, 1991, Prediction of Parametric Effects on Transition Boiling under Pool Boiling Conditions, *Int. J. Heat Mass Transfer* 34(6):1355–1370. (2)

Parker, J. D., and R. J. Grosh, 1961, Heat Transfer to a Mist Flow, USAEC Rep. ANL-6291, Argonne, IL. (4)

Patankar, S. V., 1980, *Numerical Heat Transfer and Fluid Flow,* Hemisphere, New York. (3)

Peebles, F. N., and H. J. Garber, 1953, Studies on the Motion of Gas Bubbles in Liquids, *Chem. Eng. Prog. 49:*88–97. (3)

Pei, B. S., 1981, Prediction of Critical Heat Flux in Flow Boiling at Low Qualities, Ph.D. thesis, University of Cincinnati, Cincinnati, OH. (5)

Peppler, W., E. G. Schlechtendhl, and G. F. Schultheiss, 1970, Investigation of the Dynamics of Boiling Events in Sodium-Cooled Reactors, *Nuclear Eng. Design 14:*23–42. (4)

Pfefferlen, H., R. Raush, and G. Watford, 1990, BWR Core Thermal-Hydraulic Stability Experience and Safety Significance, *Proc. Int. Workshop on BWR Instability,* Hottsville, NY, CSNI Rep. 178, OECD-NEA, Paris. (6)

Phillips, L. E., 1990, Resolution of U.S. Regulatory Issues Involving BWR Stability, *Proc. Int. Workshop on BWR Instability,* Hottsville, NY, CSNI Rep. 178, OECD-NEA, Paris. (6)

Plesset, M. S., and S. A. Zwick, 1954, The Growth of Vapor Bubbles in Superheated Liquids, *J. Appl. Phys. 25:*493. (2)

Plummer, D. N., O. C. Iloge, W. M. Rohsenow, P. Griffith, and E. Ganic, 1974, Post Critical Heat Flux to Flowing Liquid in a Vertical Tube, MIT Heat Lab. Rep. 72718-91, Massachusetts Institute of Technology, Cambridge, MA. (4)

Polomik, E. E., S. Levy, and S. G. Sawochika, 1960, Film Boiling of Steam-Water Mixture in Annular Flow at 800, 1100, 1400 psi, ASME Paper 62-WA-136, Winter Annual Meeting, ASME, New York. (4)

Pomerantz, M. L., 1964, Film Boiling on a Horizontal Tube in Increased Gravity Fields, *Trans. ASME, J. Heat Transfer 86:*213–219. (2)

Povarnin, P. I., and S. T. Semenov, 1960, An Investigation of Burnout during the Flow of Subcooled Water through Small Diameter Tubes at High Pressures, *Teploenergetika* 7(1):79–85. (5)

Pramuk, F. S., and J. W. Westwater, 1956, Effect of Agitation on the Critical Temperature Difference for a Boiling Liquid, *AICHE Chem. Eng. Prog. Symp. Ser.* 52(18):79–83. (2)

Proskuryakov, K. N., 1965, Self Oscillation in a Single Steam Generating Duct, *Thermal Eng.* (USSR) 12(3):96–100. (5)

Quandt, E., 1965, Measurement of Some Basic Parameters in Two-Phase Annular Flow, *AIChE J.* 11(12):311–318. (5)

Quandt, E., 1965, Analysis of Gas-Liquid Flow Patterns, *AIChE Chem. Eng. Prog. Symp. Ser.* 61(57):128–135. (3)

Quinn, E. P., 1963, Transition Boiling Heat Transfer Program, General Electric Corp. Quarterly Rep., Oct.–Dec. 1963, GEAP-4487. San Jose, CA. (4)

Quinn, E. P., 1966, Forced-Flow Heat Transfer to High-Pressure Water beyond the Critical Heat Flux, ASME Paper 66-WA/HT-36, Winter Annual Meeting, ASME, New York. (6)

Ramilison, J. M., and J. H. Lienhard, 1987, Transition Boiling Heat Transfer and the Film Transition Regime, *Trans. ASME, J. Heat Transfer 109:*746. (2)

Rayleigh, J. W. S., 1917, *Phil. Mag.* XXXIV: 94, cited in H. Lamb, *Hydrodynamics,* p. 122, Dover, New York, 1945. (2)

Reddy, D. G., and C. F. Fighetti, 1982, Subchannel Analysis of Multiple CHF Events, NUREG/CR 2855, Columbia University, New York. (5)

Reddy, D. G., and C. F. Fighetti, 1983, Parametric Study of CHF Data Vol. 2: A Generalized Subchannel CHF Correlation for PWR & BWR Fuel Element Assemblies, EPRI Rep. NP-2609, Electric Power Research Institute, Palo Alto, CA. (5)

Redfield, J. A., 1965, CHIC-KIN, A Fortran Program for Intermediate and Fast Transients in a Water Moderated Reactor, USAEC Rep. WAPD TM-479, Westinghouse Electric Corp., Pittsburgh, PA. (5)

Redfield, J. A., and J. H. Murphy, 1971, Sectionalized Compressible and Momentum Integral Models for Channel Hydrodynamics, ASME Paper 71-HT-14. (6)

Re'ocreux, M., 1977, Experimental Study of Steam-Water Choked Flow, *Proc. Transient Two-Phase Flow Specialists Meeting,* CSNI, Aug. 1976; *Atomic Energy of Canada 2:*637–669. (3)

Richter, H. J., 1983, Separated Two-Phase Flow Model, Application to Critical Two-Phase Flow, *Int. J. Multiphase Flow* 9(5):511–530. (5)

Riedle, K., H. P. Gaul, K. Ruthrof, and J. Sarkar, 1976, Reflood and Spray Cooling Heat Transfer in PWR and BWR Bundles, ASME Paper 76-HT-10, Natl. Heat Transfer Conf., St. Louis, MO, ASME, New York. (4)

Robertson, J. M., 1983, The Boiling Characteristics of Perforated Plate-Fin Channels with Liquid Nitrogen in Upflow, in *Heat Exchangers for Two-Phase Flow Applications,* ASME HTD-Vol. 27, pp. 35–40. (4)

Robinson, J. M., and H. Lurie, 1962, Critical Heat Flux of Some Polyphenyl Coolants, AIChE Preprint 156, 55th Annual Natl. Meeting, Chicago, IL. (5)

Rodgers, J. T., and R. G. Rosehart, 1969, Turbulent Interchange Mixing in Fuel Bundles, Trans. Conf. Applied Mechanics, University of Waterloo, Waterloo, Canada. (App.)

Rodgers, J. T., and R. G. Rosehart, 1972, Mixing by Turbulent Interchange Bundles, Correlation and Inference, Paper 72-HT-53, ASME-AIChE Natl. Heat Transfer Conf., ASME, New York. (App.)

Rohsenow, W. M., 1952, A Method of Correlating Heat Transfer Data for Surface Boiling of Liquids, *Trans. ASME* 74:969–976. (2)

Rohsenow, W. M., 1953, Heat Transfer with Evaporation, *Heat Transfer—A Symposium Held at the University of Michigan,* Summer 1952, pp. 101–150, University of Michigan Press, Ann Arbor. (4)

Rosal, E. E., J. O. Cermak, L. S. Tong, L. E. Casterline, S. Kokolis, and B. Matzner, 1974, High Pressure Rod Bundle DNB Data with Axially Non-uniform Heat Flux, *Nuclear Eng. Design* 31:1–20. (5)

Rose, W. J., H. L. Gilles, and V. W. Uhl, 1963, Subcooled Boiling Heat Transfer to Aqueous Binary Mixtures, *Chem. Eng. Prog. Symp. Ser.* 59(41):62–70. (4)

Rosenthal, M. W., and R. L. Miller, 1957, An Experimental Study of Transient Boiling, USAEC Rep. ORNL-2294, Oak Ridge Natl. Lab., Oak Ridge, TN. (5)

Ross, H., R. Radermacher, M. DiMarzo, and D. Dirdion, 1987, Horizontal Flow Boiling of Pure and Mixed Refrigerants, *Int. J. Heat Mass Transfer* 30(5):979. (3)

Rouhani, S. Zia, 1967, Calculation of Steam Volume Fraction in Subcooled Boiling, ASME Paper 67-HT-31, Natl. Heat Transfer Conf. (3)

Rouhani, S. Z., 1973, Axial and Transverse Momentum Balance in Subchannel Analysis, Topical Meeting Requirements and Status of the Prediction of the Physics Parameters for Thermal and Fast Reactors, Julich, Germany. (3)

Rowe, D. S., 1969, Initial and Boundary Value Flow Solutions during Boiling in Two Interconnected Parallel Channels, *Trans. Am. Nuclear Soc.* 12:834. (App.)

Rowe, D. S., 1970, COBRA II: Digital Computer Program for Thermal Hydraulic Subchannel Analysis of Rod Bundle Nuclear Fuel Elements, BNWL 1229, Battelle Northwest Laboratory, Richland, WA. (3)

Rowe, D. S., 1973, COBRA IIIC: A Digital Computer Program for Steady State and Transient Thermal Hydraulic Analysis of Rod Bundle Nuclear Fuel Elements, BNWL-1695, Battelle Northwest Laboratory, Richland, WA. (3)

Rowe, D. S., and C. W. Angle, 1967, Cross Flow Mixing between Parallel Flow Channels during Boiling Part II: Measurement of Flow Enthalpy in Two Parallel Channels, BNWL-371 Pt II, Battelle Northwest Lab., Richland, WA. (App.)

Roy, G. M., 1966, Getting More out of BWR's, *Nucleonics* 24(11):41. (5)

Roy, R. P., R. C. Dykuizen, M. G. Su, and P. Jain, 1988, The Stability Analysis Using Two-Fluid SAT™ Code for Boiling Flow Systems, Vol 1, Theory; Vol. 4, Experiments and Model Validation, EPRI NP-6103-CCM, Palo Alto, CA. (6)

Ruddick, M., 1953, An Experimental Investigation of the Heat Transfer at High Rates between a Tube and Water with Conditions at or near Boiling, Ph.D. thesis, University of London, England. (1)

Ruder, Z., and T. J. Hanratty, 1990, A Definition of Gas-Liquid Plug Flow in Horizontal Pipes, *Int. J. Multiphase Flow* 16(2):233. (3)

Ruggles, A. E., 1987, An Inv. of the Propagation of Pressure Perturbations in Two-Phase Flow, Ph.D. Thesis, Rensselaer Polytechnic Inst., Troy, NY. (3)

Ruggles, A. E., R. T. Lahey, Jr., D. A. Drew and H. A. Scartan, 1988, An Inv. of the Propagation of Pressure Perturbations in Bubbly Air-Water Flows, *Trans. ASME, J. Heat Transfer,* 110(2):494–499. (3)

Ruggles, A. E., R. T. Lahey, Jr., D. A. Drew and H. A. Scartan, 1989, The Relationship Between Standing Waves, Pressure Pulse Propagation, and Critical Flow Rate in Two-Phase Mixtures, *Trans. ASME, J. Heat Transfer,* 111(2):467–473. (3)

Ryzhov, V. A., and A. P. Arkhipov, 1985, Investigation of Relationship Governing the Forced Convection Boiling Crisis in Rod Bundles, *Heat Transfer Sov. Res.* 17(6). (5)

Sabersky, R. H., and H. E. Mulligan, 1955, On the Relationship between Fluid Friction and Heat Transfer in Nucleate Boiling, *Jet Propulsion* 25(9):12. (5)

Saha, P., M. Ishii, and N. Zuber, 1976, An Experimental Investigation of the Thermally Induced Flow Oscillations in Two-Phase Systems, *Trans. ASME, J. Heat Transfer* 98:616–622. (6)

Salcudean, M., J. H. Chun, and D. C. Groeneveld, 1983a, Effect of Flow Obstructions on the Flow Pattern Transitions in Horizontal Two-Phase Flow, *Int. J. Multiphase Flow,* 9(1):87. (3)

Salcudean, M., D. C. Groeneveld, and L. Leung, 1983b, Effects of Flow Obstruction Geometry on Pressure Drops in Horizontal Air-Water Flow, *Int. J. Multiphase Flow* 9(1):73–85. (3)

Sani, R. L., 1960, Downward Boiling and Non-boiling Heat Transfer in a Uniformly Heated Tube, UCRL-9023, University of California, Los Angeles, CA. (4)

Schlechtendahl, E. G., 1967, Die Ejektion von Natrium aus Reaktorkuhlkanalen, *Nukleonik* 10(5). (4)

Schlechtendahl, E. G., 1969, Sieden des Kuhlmittels in Natriumgekuhlten Schnellen Reaktoren, Gesellschaft für Kernforschung, Karlsruhe, Rep. KFK 1020; also Rep. EUR-4302d. (3)

Schlechtendahl, E. G., 1970, Theoretical Investigation on Sodium Boiling in Fast Reactors, *Nuclear Sci. Eng.* 41:99. (6)

Schmidt, K. R., 1959, Thermodynamic Investigation of Highly Loaded Boiling Heating Surfaces, USAEC Rep. AEC-tr-4033, transl. from *Mitt. Ver. Grosskesselbesitzer* 63:391. (4)

Schoneberg, R., 1968, Stabilitats untersuchungen für Siederwasser Reaktoren, AEGE 31/E-1201, Germany. (3)

Schraub, F. A., 1968, Isokinetic Sampling Probe Technique Applied to Two-Phase Flow, ASME 67-WA/FE-28, Annual Meeting, ASME, New York. (3)

Schraub, F. A., 1969, Spray Cooling Heat Transfer Effectiveness during Simulated Loss-of-Coolant Transients, ASME Paper 69WA/NE-8, ASME, New York. (4)

Schraub, F. A., R. L. Simpson, and E. Janssen, 1969, Two-Phase Flow and Heat Transfer in Multirod Geometries: Air-Water Flow Structure Data for a Round Tube, Concentric & Eccentric Annulus, and Nine-Rod Bundle, GEAP-5739, General Electric Co., San Jose, CA. (5)

Schrock, V. E., 1969, Radiation Attenuation Techniques in Two-Phase Flow Measurement, *ASME Symp. on Two-Phase Instrumentation,* Chicago, IL, pp. 24–35. (3)

Schrock, V. E., and L. M. Grossman, 1959, Forced Convection Boiling Studies, Forced Convection Vaporization Project—Final Rep. 73308 UCX 2182, University of California, Berkeley, CA. (4)

Schrock, V. E., and L. M. Grossman, 1962, Forced Convection Boiling in Tubes, *Nuclear Sci. Eng.* 12:474–480. (4)

Schrock, V. E., H. H. Johnson, A. Gopalakrishnan, K. E. Lavezzo, and S. M. Cho, 1966, Transient Boiling Phenomena, USAEC Rep. SAN-1013, University of California, Berkeley, CA. (5)

Schultheiss, G. F., 1970, Experimental Inv. of Incipient-Boiling Superheat in Wall Cavities, *Proc. Symp. Liquid-Metal Heat Transfer and Fluid Dynamics,* November p. 100, ASME, New York. (2)

Schultheiss, G. F., and D. Smidt, 1969, Discussion on J. C. Chen's paper [Chen, 1968], *Trans. ASME, J. Heat Transfer* 91:198. (2)

Sciance, C. T., C. P. Colver, and C. M. Sliepcevich, 1967, Film Boiling Measurements and Correlations for Liquified Hydrocarbon Gases, *Chem. Eng. Prog. Symp. Ser.* 63(77):115. (2)

Scott, D. S., 1963, Properties of Co-current Gas-Liquid Flow, in *Advances in Chemical Engineering,* Vol. 4, p. 199, Academic Press, New York. (3)

Scriven, L. E., 1959, On the Dynamics of Phase Growth, *Chem. Eng. Sci.* 10:1–13. (2)

Sekoguchi, K., H. Fukui, and Y. Sato, 1981, Flow Characteristics and Heat Transfer in Vertical Bubble Flow, in *Two-Phase Flow Dynamics Japan-US Seminar,* A. E. Bergles and S. Ishigai, Eds., Hemisphere, New York. (3)

Sekoguchi, K., M. Takeishi, T. Nishiure, H. Kano, and T. Nomura, 1985 Multiple Optical Fiber Probe

Technique for Measuring Profiles of Gas-Liquid Interface and its Velocity, *Int. Symp. on Laser Anemometry,* ASME Winter Annual Meeting, *FED 33,* 97–102. (3)

Sekoguchi, K., O. Tanaka, and S. Esaki, 1980, *Bull. Jpn. Soc. Mech. Eng. 23:*1475. (4)

Sekoguchi, K., O. Tanaka, T. Ueno, M. Yamashita, and S. Esaki, 1982, Heat Transfer Characteristics of Boiling Flow in Subcooled and Low Quality Regions, *7th Int. Heat Transfer Conf.* Paper FB12, Munich, Germany, Hemisphere, Washington, DC. (4)

Sembler, R.J., 1960, Mixing in Rectangular Nuclear Reactor Channels, WAPD-T-653, Westinghouse Bettis Atomic Power Lab, Pittsburgh, PA. (App.)

Semeria, R. L., 1962, An Experimental Study of the Characteristics of Vapor Bubbles, *Symp. on Two-Phase Fluid Flow,* pp. 57–65, Inst. Mech. Eng., London. (5)

Semiat, R., and A. E. Dukler, 1981, The Simultaneous Measurement of Size and Velocity of Bubbles or Drops, *AIChE J. 27:*148. (3)

Serizawa, A., I. Kataoka, and I. Michiyoshi, 1975, Turbulence Structure of Air-Water Bubbly Flow— II, Local Properties, *Int. J. Multiphase Flow 2:*235–246. (3)

Serizawa, A., I. Kataoka, and L. Van Wijngaarden, 1992, Dispersed Flow, *3rd Int. Workshop on Two-Phase Fundamentals,* London, June. (3)

Sha, W. T., R. C. Schmitt, and P. Huebotter, 1976, Boundary Value Thermal Hydraulic Analysis of a Reactor Fuel Rod Bundle, *Nuclear Sci. Eng. 59:*140. (App.)

Shah, M. M., 1976, A New Correlation for Heat Transfer during Boiling Flow through Pipes, *ASHRAE Trans.* 82(II):66. (4)

Shah, M. M., 1982, Chart Correlation for Saturated Boiling Heat Transfer: Equations and Further Study, *ASHRAE Trans. 88*(Part I). (4)

Shai, I., 1967, The Mechanism of Nucleate Pool Boiling Heat Transfer to Sodium and the Criterion for Stable Boiling, Ph.D. thesis, Massachusetts Institute of Technology, Cambridge, MA. (2)

Shapiro, A. H., 1953, *The Dynamics and Thermodynamics of Compressible Fluid Flow, Vol. 1,* Ronald Press, New York. (3)

Sharp, R. R., 1964, The Nature of Liquid Film Evaporation during Nucleate Boiling, NASA-TN-D-1997, Lewis Res. Ctr., Cleveland, OH. (2)

Shiralkar, B. S., and R. T. Lahey, Jr., 1973, The Effect of Obstacles on a Liquid Film, *Trans. ASME, J. Heat Transfer 95:*528–533. (5)

Shitsman, M. E., 1963, On the Methods for Calculating the Critical Heat Flux with Water in Forced Convection, *Teploenergetika,* August, *10*(8). (5)

Shitsman, M. E., 1966, The Effect of Duct Diameter on Critical Heat Flux, *Teploenergetika* 13(4):70–72. (5)

Siegel, R., 1967, Effect of Reduced Gravity on Heat Transfer, in *Advances in Heat Transfer, Vol. 4,* J. D. Harnett and T. F. Irvine, Eds., pp. 144–228, Academic Press, New York. (2)

Silvestri, M., 1966, On the Burnout Equation and on Location of Burnout Points, *Energia Nucleare 13*(9). (5)

Simon, F. F., and Y. Y. Hsu, 1970, Thermocapillary Induced Breakdown of a Falling Liquid Film, NASA-TN-D-5624, NASA Lewis Res. Ctr., Cleveland, OH. (5)

Simpson, H. C., D. H. Rooney, E. Gratton, and F. A. A. Al-Samarral, 1981, Two-Phase Flow in Large Diameter Horizontal Tubes, NEL Rep. 677, Nuclear Energy Lab., Washington, DC. (3)

Singer, R. M., and R. E. Holtz, 1970, Comparison of the Expulsion Dynamics of Sodium and Nonmetallic Fluids, ASME Paper 70-HT-23, ASME Fluids Engineering Heat Transfer and Lubrication Conf., Detroit, MI. (5)

Singh, A., B. B. Mikic, and W. M. Rohsenow, 1974, Effect of Surface Condition on Nucleation and Boiling Characteristics, Rep. DSR-7341-93, Massachusetts Institute of Technology, Cambridge, MA. (2)

Slifer, B. C., and J. E. Hench, 1971, Loss of Coolant Accident and Emergency Core Cooling Models for General Electric Boiling Water Reactors, NEDO-10329, General Electric Co., San Jose, CA. (5)

Smith, A. M., 1969, Oak Ridge National Lab., Personal communication to J. K. Jones, November. (5)

Smith, O. G., W. M. Rohrer, Jr., and L. S. Tong, 1965, Burnout in Steam-Water Flows with Axially Non-uniform Heat Flux, ASME Paper 65-WA/HT-33, ASME, New York. (5)
Smith, S. L., 1970, Void Fraction in Two-Phase Flow: A Correlation Based upon an Equal Velocity Head Model, *Proc. Inst. Mech. Engrs., 184*(Part I), (36):657. (3)
Snyder, N. R., and D. K. Edwards, 1956, Post-Conference Comments: Conference on Bubble Dynamics and Boiling Heat Transfer, Memo 20-137, p. 38, Jet Propulsion Lab., Pasadena, CA. (2)
Soliman, H. M., and N. Z. Azer, 1971, Flow Patterns during Condensation inside a Horizontal Tube, *ASHRAE Trans. 77*:210. (3)
Solverg, K. O., and P. Bakstad, 1967, A Model for the Dynamics of Nuclear Reactors with Boiling Coolant with a New Approach to the Vapor Generation, *Proc. Symp. on Two-Phase Flow Dynamics* at Eindhoven, EURATOM Rep. (6)
Sozzi, G. L., and W. A. Sutherland, 1975, Critical Flow of Saturated and Subcooled Water at High Pressure, in *Non-equilibrium Two-Phase Flows*, R. T. Lahey and G. B. Wallis, Eds., ASME, New York. (3)
Spedding, P. L., and V. T. Nguyen, 1980, Regime Maps for Air-Water Two-Phase Flow, *Chem. Eng. Sci. 35*:779. (3)
Spedding, P. L., and D. R. Spence, 1993, Flow Regimes in Two Phase Gas-Liquid Flow, *Int. J. Multiphase Flow 19*(2):245–280. (3)
Spiegler, P., J. Hopenfeld, M. Silverberg, and C. F. Bumpus, Jr., 1964, In-Pile Experimental Studies of Transient Boiling with Organic Reactor Coolant, USAEC Rep. NAA-SR-9010, North American Aviation, Rockwell Int., Inc., Canoga Park, CA. (5)
Spiller, K. H., Grass, and Perschke, 1967, Superheating and Single Bubble Ejection in the Vaporization of Stagnating Liquid Metals, *Atomkernenergie 12*(3/4):111–114. (4)
Spindler, K., 1994, Flow Boiling, Invited lecture, Proc. 10th Int. Heat Transfer Conf., *Heat Transfer 1994*, vol. 1, pp. 349–368, Brighton, England, Taylor & Francis, New York, Washington, DC. (4)
Staniszewski, B. E., 1959, Nucleate Boiling Bubble Growth and Departure, Tech. Rep. 7673-16, Massachusetts Institute of Technology, Cambridge, MA. (2)
Staub, F. W., 1967, The Void Fraction in Subcooled Boiling—Prediction of the Initial Point of Net Vapor Generation, ASME Paper 67-HT-36, National Heat Transfer Conf., ASME, New York. (3)
Staub, F. W., 1969, Two Phase Fluid Modeling, The Critical Heat Flux, *Nuclear Sci. Eng. 35*:190–199. (5)
Stefanovic, M., N. Afgan, V. Pislar, and L. Jovanovic, 1970, Experimental Investigation of the Superheated Boundary Layer in Forced Convection Boiling, *Proc. 4th Int. Heat Transfer Conf.*, Versailles, Paper B 4.12. (3)
Steiner, D., and E. U. Schlunder, 1977, Heat Transfer and Pressure Drop for Boiling Nitrogen Flowing in a Horizontal Tube, in *Heat Transfer in Boiling*, E. Hahne and U. Grigull, Eds., p. 263, Hemisphere, Washington, DC. (4)
Stenning, A. H., and T. N. Veziroglu, 1965, Flow Oscillations Modes in Forced Convection Boiling, *Proc. 1965 Heat Transfer and Fluid Mechanics Inst.*, pp. 301–316, Stanford University Press, Palo Alto, CA. (6)
Stenning, A. H., and T. N. Veziroglu, 1967, Oscillations in Two-Component, Two-Phase Flow, Vol. 1, NASA CR-72121; and Flow Oscillations in Forced Convective Boiling, Vol. 2, NASA CR-72122, University of Miami, Coral Gables, FL. (6)
Sternling, V. C., 1965, Two-Phase Flow Theory and Engineering Decision, Award Lecture presented at AIChE Annual Meeting. (3)
Stevens, G. F., and G. J. Kirby, 1964, A Quantitative Comparison between Burnout Data for Water at 1000 psia and Freon-12 at 155 psia Uniformly Heated Round Tubes, Vertical Flow, Report AEEW-R-327, UK Atomic Energy Authority, Winfrith, England. (5)
Stevens, G. F., and R. W. Wood, 1966, A Comparison between Burnout Data for 19 Rod Cluster Test Sections Cooled by Freon-12 at 155 psia and by Water at 1000 psia in Vertical Upflow, Rep. AEEW-R-468, UK Atomic Energy Authority, Winfrith, England. (5)

Stevens, J. W., R. L. Bullock, L. C. Witte, and J. E. Cox, 1970, The Vapor Explosion—Heat Transfer and Fragmentation II, Transition Boiling from Sphere to Water, Tech. Rep. ORD-3936-3, University of Houston, Houston, TX. (4)

Stock, B. J., 1960, Observations on Transition Boiling Heat Transfer Phenomena, Rep. ANL-6175, Argonne National Lab., Argonne, IL. (2)

St. Pierre, C. C., and S. G. Bankoff, 1967, Vapor Volume Profiles in Developing Two-Phase Flow, *Int. J. Heat Mass Transfer 10:*237. (3)

Stravs, A. A., and V. Van Stocker, 1985, Measurement of Interfacial areas in Gas-Liquid Dispersions by Ultrasound Pulse Transmission, *Chem. Eng. Sci. 40*(7):1169. (3)

Styrikovich, M. A., et al., 1960, Effect of Upstream Elements on Critical Boiling in a Vapor Generating Pipe, USAEC Rep. AEC-tr-4740, translated from *Teploenergetika 7*(5):81–87. (5)

Styrikovich, M. A., Z. L. Miropol'skii, and V. V. Eva, 1963, The Influence of Local Raised Heat Fluxes along the Length of a Channel on the Boiling Crisis, *Sov. Phys. Dokl. 7:*597–599. (5)

Styrikovich, M. A., E. I. Nevstrueva, I. M. Romanovsky, and V. S. Polonsky, 1970, Interconnection between Mass and Heat Transfer in Boiling, in *Heat Transfer 1970, Vol. 6,* Paper B-7.4. U. Grigull and E. Hahne Eds., Hemisphere, Washington, DC. (5)

Subbotin, V. I., D. M. Ovechkin, D. N. Sorokin, and A. P. Kudryavtsev, 1968, Critical Heat Flux during Pool Boiling of Cesium, transl. from *Teploenergetika 6:*58. (2)

Subbotin, V. I., D. N. Sorokin, and A. P. Kudryavtsev, 1970, Generalized Relationship for Calculating Heat Transfer in the Developed Boiling of Alkali Metals, transl. US AEC from *Atomnaya Energiya 29:*45. (2)

Sudo, Y., M. Kaminaga, 1993, A New CHF Correlation Scheme Proposed for Vertical Rectangular Channels Heated from Both Sides in Nuclear Reactors, *Trans. ASME, J. Heat Transfer 115:*426–434. (5)

Sudo, Y., T. Usui, and M. Kaminaga, 1991, Experimental Study of Falling Water Limitation under a Counter-current Flow in a Vertical Rectangular Channel, *JSME Int. J. Ser. II 34:*169–174. (5)

Sugawara, S., 1990, Analytical Prediction of CHF by FIDAS Code Based on Three-Fluid and Film-Dryout Model, *J. Nuclear Sci. Technol. 27:*12–29. (5)

Sun, K. H., and J. H. Lienhard, 1970, The Peak Pool Boiling Heat Flux on Horizontal Cylinders, *Int. J. Heat Mass Transfer 13:*1425–1439. (2)

Swenson, H. S., J. R. Carver, and C. R. Kakarala, 1962a, The Influence of Axial Heat Flux Distribution on the Departure from Nucleate Boiling in a Water Cooled Tube, ASME Paper 62-WA-297, Winter Annual Meeting, ASME, New York. (5)

Swenson, H. S., J. R. Carver, and G. Szoeke, 1962b, The Effects of Nucleate Boiling vs Film Boiling on Heat Transfer in Power Boiler Tubes, *Trans. ASME, J. Eng. Power 84:*365–371. (4)

Tachibana, F., M. Akiyama, and H. Kawamura, 1968, Heat Transfer and Critical Heat Flux in Transient Boiling, 1. An Experimental Study in Saturated Pool Boiling, *J. Nuclear Sci. Technol. of Japan 5*(3):117. (5)

Tagami, T., 1966, Interim Report on Safety Assessments and Facilities Establishment Project (unpublished document), private communication. (4)

Taitel, Y., and D. Barnea, 1990, Two-Phase Slug Flow, *Adv. Heat Transfer 20:*83. (3)

Taitel, Y., D. Barnea, and A. E. Dukler, 1980, Modelling Flow Pattern Transitions for Steady Upward Gas-Liquid Flow in Vertical Tubes, *AIChE J. 26*(3):345. (3)

Taitel, Y., and A. E. Dukler, 1976a, A Theoretical Approach to the Lockhart-Martinelli Correlation for Stratified Flow, *Int. J. Multiphase Flow 2:*591–595. (3)

Taitel, Y., and A. E. Dukler, 1976b, A Model for Predicting Flow Regime Transitions in Horizontal and Near Horizontal Gas-Liquid Flow, *AIChE J. 22:*47. (3)

Taitel, Y., N. Lee, and A. E. Dukler, 1978, Transient Gas-Liquid Flow in Horizontal Pipes Modeling the Flow Pattern Transitions, *AIChE J. 24:*920–934. (3)

Taitel, Y., S. Vierkandt, O. Shoham, and J. P. Brill, 1990, Severe Slugging in a Riser System: Experiments and Modeling, *Int. J. Multiphase Flow 16*(1):57–68. (3)

REFERENCES 527

Tang, Y. S., P. T. Rose, R. C. Nicholson, and C. R. Smith, 1964, Forced Convection Boiling of Potassium-Mercury Systems, *AIChE J. 10*(5):617–620. (4)

Tarasova, N. V., A. I. Leontiev, V. I. Hlopushin, and V. M. Orlov, 1966, Pressure Drop of Boiling Subcooled Water and Steam-Water Mixture Flow in Heated Channels, *Proc. 3rd Int. Heat Transfer Conf.*, vol. 4, pp. 178–183, ASME, New York. (3)

Tatterson, D. F., J. C. Dallman, and T. J. Hanratty, 1977, Drop Sizes in Annular Gas-Liquid Flows, *AIChE J. 23*(1):68–76. (4)

Taylor, C. E., and J. F. Steinhaus, 1958, High Flux Boiling Heat Transfer from a Flat Plate, US AEC Rep. UCRL-5414, Los Angeles, CA. (2)

Telles, A. S., and A. E. Dukler, 1970, Statistical Characteristics of Thin Vertical Wavy Films, *Ind. Eng. Chem. Fundam. 9:*412. (3)

Temkin, S., 1966, Attenuation and Dispersion of Sound by Particulate Relaxation Processes, Brown University (available from DDC as AD-630326). (3)

Tepper, F., A. Murchison, J. Zelenak, and F. Roehlich, 1964, Thermophysical Properties of Rubidium and Cesium, *Proc. 1963 High Temperature Liquid Metal Technology Meeting* Vol. 1, 26–65, US AEC Rep. ORNL-3605. (5)

Theofanous, T., L. Biasi, H. S. Isbin, and H. K. Fauske, 1969, A Theoretical Study on Bubble Growth in Constant and Time-Dependent Pressure Fields, *Chem. Eng. Sci. 24:*885. (2)

Thom, J. R. S., W. M. Walker, T. A. Fallon, and G. F. S. Reising, 1966, Boiling in Subcooled Water during Flow up Heated Tubes or Annuli, *Proc. Inst. Mech. Eng.* 180(Part 3C). (3)

Thompson, B., and R. V. Macbeth, 1964, Boiling Water Heat Transfer in Uniformly Heated Tubes: A Compilation of World Data with Accurate Correlations, Rep. AEEW-R-356, UK Atomic Energy Authority, Winfrith, England. (5)

Thorgerson, E. J., 1969, Hydrodynamic Aspects of the Critical Heat Flux in Subcooled Convection Boiling, Ph.D. thesis, University of South Carolina, Columbia, SC. (5)

Tippets, F. E., 1962, Critical Heat Fluxes and Flow Patterns in High Pressure Boiling Water Flows, ASME Paper 62-WA-162, Winter Annual Meeting, ASME, New York. (5)

Tippets, F. E., J. A. Bond, and J. R. Peterson, 1965, Heat Transfer and Pressure Drop Measurements for High Temperature Boiling Potassium in Forced Convection, Proc. Conf. on Applied Heat Transfer Instrumentation to Liquid Metal Experiments, ANL-7100, p. 53–95, Argonne National Lab., Argonne, IL. (3)

Todreas, N. E., and M. S. Kazimi, 1990, Two-Phase Flow Dynamics, in *Nuclear Systems I: Thermal Hydraulic Fundamentals,* pp. 476–479, Hemisphere, Washington, DC. (3)

Todreas, N. E., and W. M. Rohsenow, 1966, The Effect of Axial Distribution on Critical Heat Flux in Annular Two Phase Flow, *Proc. Third Int. Heat Transfer Conf.*, vol. 3, pp. 78–85, AIChE, New York. (5)

Tong, L. S., 1965, *Boiling Heat Transfer and Two Phase Flow,* John Wiley, New York. (2)

Tong, L. S., 1967a, Prediction of Departure from Nucleate Boiling for an Axially Non-uniform Heat Flux Distribution, *J. Nuclear Energy* 21:241–248. (3)

Tong, L. S., 1967b, Heat Transfer in Water-Cooled Nuclear Reactors, *Nuclear Eng. Design 6:*301. (3)

Tong, L. S., 1968a, An Evaluation of the Departure from Nucleate Boiling in Bundles of Reactor Fuel Rods, *Nuclear Sci. Eng. 33:*7–15. (5)

Tong, L. S., 1968b, Boundary Layer Analysis of the Flow Boiling Crisis, *Int. J. Heat Mass Transfer* 11:1208–1211. (5)

Tong, L. S., 1969, Critical Heat Fluxes in Rod Bundles, in *Two-Phase Flow and Heat Transfer in Rod Bundles,* pp. 31–41, ASME, New York. (5)

Tong, L. S., 1972, *Boiling Crisis and Critical Heat Flux,* AEC Critical Review Series, USAEC, Washington, DC. (1)

Tong, L. S., 1975, A Phenomenological Study of Critical Heat Flux, ASME Paper 75-HT-68, Natl. Heat Transfer Conf., San Francisco, CA. (5)

Tong, L. S., 1988, *Principles of Design Improvement for Light Water Reactors,* Hemisphere, New York. (5)

Tong, L. S., A. A. Bishop, J. E. Casterline, and B. Matzner, 1965, Transient DNB Test on CVTR Fuel Assembly, ASME Paper 65-WA/NE-3 Winter Annual Meeting, ASME, New York. (4)

Tong, L. S., H. Chelemer, J. E. Casterline, and B. Matzner, 1967a, Critical Heat Flux (DNB) in the Square and Triangular Array Rod Bundles, Symp. Proc. JSME Semi-International Symp., Tokyo, September, Jpn. Soc. Mech. Eng., Tokyo. (5)

Tong, L. S., and H. B. Currin, 1964, Interpreting DNB Data in Reactor Design, *Nucleonics 22*(11). (5)

Tong, L. S., H. B. Currin, and F. C. Engel, 1964, DNB (Burnout) Studies in an Open Lattice Core, USAEC Rep. WCAP-3736, Pittsburgh, PA. (5)

Tong, L. S., H. B. Currin, P. S. Larsen, and O. G. Smith, 1966a, Influence of Axially Non-uniform Heat Flux on DNB, *AIChE Chem. Eng. Prog. Symp. Ser. 62*(64):35–40. (5)

Tong, L. S., H. B. Currin, and A. G. Thorp, II, 1963, New Correlations Predict DNB Conditions, *Nucleonics 21*(5):43–47. (5)

Tong, L. S., L. E. Efferding, and A. A. Bishop, 1966b, A Photographic Study of Subcooled Boiling Flow and DNB of Freon-113 in a Vertical Channel, ASME Paper 66-WA/HT-39, Winter Annual Meeting, ASME, New York. (3)

Tong, L. S., and G. F. Hewitt, 1972, Overall Viewpoint of Flow Boiling CHF Mechanisms, ASME Paper 72-HT-54, Natl. Heat Transfer Conf., Denver, CO. (1)

Tong, L. S., A. S. Kitzes, J. Green, and T. D. Stromer, 1967c, Departure from Nucleate Boiling on a Finned Surface Heater Rod, *Nuclear Eng. Design 5:*386–390. (5)

Tong, L. S., F. E. Motley, and J. O. Cermak, 1970, Scaling Law of Flow Boiling Crisis, in *Heat Transfer, 1970, Vol. 6,* Paper B 6.12, U. Grigull and E. Hahne, Eds., Hemisphere, Washington, DC. (5)

Tong, L. S., G. Previti, and R. T. Berringer, 1960, Flow Redistribution in an Open Lattice Core, Westinghouse Rep. WAPD-1645, Nuclear Center, Westinghouse Electric Corp., Pittsburgh, PA. (5)

Tong, L. S., R. W. Steer, A. H. Wenzel, M. Bogaardt, and C. L. Spigt, 1967b, Critical Heat Flux on a Heater Rod in the Center of Smooth and Rough Square Sleeves, and in Line Contact with an Unheated Wall, ASME Paper 67-WA/HT-29, Winter Annual Meeting, ASME, New York. (5)

Tong, L. S., and J. Weisman, 1979, *Thermal Analysis of Pressurized Water Reactors,* 2d ed., American Nuclear Society, LaGrange Park, IL. (3)

Topper, L., 1963, A Diffusion Theory Analysis of Boiling Burnout in the Fog Flow Regime, *Trans. ASME, J. Heat Transfer 85:*284–285. (5)

TORC Code, 1975, TORC Code: A Computer Code for Determining the Thermal Margin of a Reactor Core, CENPD-161, Combustion Engineering Co., Winsor, CT. (App.)

Towell, R. H., 1965, Effect of Rod Spacing on Heat Transfer Burnout in Rod Bundles, USAEC Rep. DP-1013, E. I. duPont de Nemours and Co., Wilmington, DE. (5)

Townsend, A. A., 1956, *The Structure of Turbulent Shear Flow,* Cambridge University Press, Cambridge, England. (5)

Travis, D. P., and M. W. Rohsenow, 1973, Flow Regimes in Horizontal Two-Phase Flow with Condensation, *AIChE J. 24:*920. (3)

Trimble, G. D., and W. J. Turner, 1976, Report AAEC-E378, Australian Atomic Energy Commission Research Establishment, Sydney, New South Wales, Australia. (3)

Turner, J. B., and C. P. Colver, 1971, Heat Transfer to Pool-Boiling Mercury from Horizontal Cylindrical Heaters at Heat Fluxes up to Burnout, *Trans. ASME, J. Heat Transfer 93:*1. (2)

Turner, J. M., and G. B. Wallis, 1965, Analysis of the Liquid Film in Annular Flow, Dartmouth College Rep. NYO-3114-13, Hanover, NH. (5)

Turner, W. J., and G. D. Trimble, 1976, Calculation of Transient Two-Phase Flow, Specialists' Meeting on Transient Two-Phase Flow, Toronto, Canada. (3)

Uchida, H., and H. Nariai, 1966, Discharge of Saturated Water Through Pipes and Orifices, *Proc. 3rd Int. Heat Transfer Conf.* Vol 5, pp. 1–12, AIChE, New York. (3)

Ünal, C., K. Tuza, O. Bach, S. Neti, and J. C. Chen, 1991b, Convective Boiling in a Rod Bundle: Transverse Variation of Vapor Superheat Temperature under Stabilized Post-CHF Conditions, *Int. J. Heat Mass Transfer 34:*1695–1706. (4)

Ünal, C., K. Tuza, A. F. Cokmez-Tuzla, and J. C. Chen, 1991a, Vapor Generation Rate Model for Dispersed Drop Flow, *Nuclear Eng. Design 125:*161–173. (4)

Van Wijingaarden, L., 1966, Linear and Nonlinear Dispersion of Pressure Pulses in Liquid Bubble Mixtures, 6th Symp. on Naval Hydrodynamics, Washington, DC. (3)

Van Wijingaarden, L., 1968, On the Equations of Motion for Mixtures of Liquid and Gas Bubbles, *J. Fluid Mech. 33* (pt. 3): 465–474. (3)

Vandervort, C. L., A. E. Bergles, and M. K. Jensen, 1994, An Experimental Study of CHF in Very High Heat Flux Subcooled Boiling, *Int. J. Heat Mass Transfer 37*(suppl.):161–173. (5)

Vanderwater, R. G., 1956, An Analysis of Burnout in Two Phase, Liquid Vapor Flow, Ph.D. thesis, University of Minnesota, Minneapolis, MN. (5)

Varone, Jr. A. F., and W. M. Rohsenow, 1986, Post-dryout Heat Transfer Prediction, *Nuclear Eng. Design 95:*315–317. (4)

Varone, Jr. A. F., and W. M. Rohsenow, 1990, The Influence of the Dispersed Flow Film Boiling, MIT Rep. 71999-106, Heat Transfer Lab., Massachusetts Institute of Technology, Cambridge, MA. (3)

Venkateswararao, P., R. Semiat, and A. E. Dukler, 1982, Film Pattern Transition for Gas Liquid Flow in a Vertical Rod Bundle, *Int. J. Multiphase Flow 8*(5):509. (3)

Vernier, P., and J. M. Delhaye, 1968, General Two-Phase Flow Equations Applied to the Thermohydrodynamics of BWR's, Centre d'Etudes Nucleaires de Grenoble, Service des Transports Thermiques; Also see *Energie Primaire 4*(1). (3)

Veziroglu, T. N., and S. S. Lee, 1971, Boiling-Flow Instabilities in a Cross-Connected Parallel Channel Upflow System, Nat. Heat Transfer Conf., ASME Paper 71-HT-12, ASME, New York. (6)

Veziroglu, T. N., S. S. Lee, and S. Kakac, 1976, Fundamentals of Two-Phase Flow Oscillations and Experiments in Single Channel Systems, *NATO Adv. Study Inst.* 1:423–466, Hannover, Germany. (6)

Vliet, G. C., and G. Leppert, 1962, Critical Heat Flux for Subcooled Water Flowing Normal to a Cylinder, ASME Paper 62-WA-174. Winter Annual Meeting, ASME, New York. (5)

Vohr, J. H., 1970, Evaporative Processes in Superheated Forced Convective Boiling, Rep. MTI-70TR15, p. 3, Mechanical Technology, Inc., Washington, DC. (2)

Voutsinos, C. M., and R. L. Judd, 1975, Laser Interferometric Investigation of the Microlayer Evaporation Phenomenon, *Trans. ASME, J. Heat Transfer 97*(1):88–92. (2)

Wallis, G. B., 1969, Annular Flow in *One Dimensional Two Phase Flow,* chap. 11, McGraw-Hill, New York. (3)

Wallis, G. B., 1970, Annular Two Phase Flow: Part 2, Additional Effects, *Trans. ASME, J. Basic Eng. 92:*73–82. (5)

Wallis, G. B., 1980, Theoretical Model of Gas-Liquid Flow, *Proc. Annual Meeting, Society of Engineering and Science,* Atlanta, GA, p. 207. (3)

Wallis, G. B., and J. E. Dobson, 1973, The Onset of Slugging in Horizontal Stratified Air-Water Flow, *Int. J. Multiphase Flow 1:*173. (3)

Wallis, G. B., and J. H. Hearsley, 1961, Oscillations in Two-Phase Flow Systems, *Trans. ASME, J. Heat Transfer 83:*363–369. (5)

Wark, J. W., 1933, The Physical Chemistry of Flotation, I, *J. Phys. Chem. 37:*623–644. (2)

Wasden, F. K., and A. E. Dukler, 1989, Insights into the Hydraulic of Free Falling Wavy Films, *AIChE J. 35*(2):187–195. (3)

Waters, E. D., 1963, Fluid Mixing Experiments with a Wire-Wrapper 7-Rod Bundle Fuel Assembly, Rep. HW-70178 Rev., Hanford Lab., General Electric Co., Hanford, WA. (App.)

Waters, E. D., J. K. Anderson, W. L. Thorne, and J. M. Batch, 1965, Experimental Observations of Upstream Boiling Burnout, *AIChE Chem. Eng. Prog. Symp. Ser. 61*(57):230–237. (5)

Weatherhead, R. J., 1962, Hydrodynamic Instability and the Critical Heat Flux Occurrence in Forced Convection Vertical Boiling Channels, USAEC Rep. TID-16539, Washington, DC; USAEC Rep. ANL-6675, Argonne National Lab, Argonne, IL. (5)

Weisman, J., 1973, Review of Two-Phase Mixing and Division Cross Section in Subchannel Analysis, Rep. AEEW-R-928, UK Atomic Energy Authority, Winfrith, England. (3)

Weisman, J., 1985, Theoretically Based Predictions of Critical Heat Flux in Rod Bundles, Third Int. Conf. on Reactor Thermal Hydraulics, Newport, RI. (5)

Weisman, J., 1992, The Current Status of Theoretically Based Approaches to the Prediction of the Critical Heat Flux in Flow Boiling, *Nuclear Technol. 99*:1–21. (1)

Weisman, J., A. Husain, and B. Harshe, 1978, Two-Phase Pressure Drop Across Abrupt Changes and Restriction, in *Two-Phase Transport and Reactor Safety,* T. N. Vezetroglu and S. Kakac, Eds., Taylor & Francis, Inc., Washington, DC. (3)

Weisman, J., and B. S. Pei, 1983, Prediction of CHF in Flow Boiling at Low Qualities, *Int. J. Heat Mass Transfer 26*:1463. (5)

Weisman, J., A. H. Wenzel, L. S. Tong, D. Fitzsimmons, W. Thorne, and J. Batch, 1968, Experimental Determination of the Departure from Nucleate Boiling in Large Rod Bundles at High Pressures, *AIChE Chem. Eng. Prog. Symp. Ser. 64*(82):114–125. (5)

Weisman, J., J. Y. Yang, and S. Usman, 1994, A Phenomenological Model for Boiling Heat Transfer and the CHF in Tubes Containing Twisted Tapes, *Int. J. Heat Mass Transfer 37*(1):69–80. (4)

Weisman, J., and S. H. Ying, 1983, Theoretically Based CHF Prediction at Low Qualities and Intermediate Flows, *Trans. Am. Nuclear Soc. 45*:832. (5)

Weisman, J., and S. H. Ying, 1985, A Theoretically Based Critical Heat Flux Prediction for Rod Bundles at PWR Conditions, *Nuclear Eng. Design 85*:239–250. (5)

Westendorf, W. H., and W. F. Brown, 1966, Stability of Intermixing of High-Velocity Vapor with Its Subcooled Liquid in Cocurrent Streams, NASA TN D-3553, Lewis Res. Ctr., Cleveland, OH. (6)

Westinghouse Electric Corp., 1969, *Thermal Conductivity of Crud,* Rep. WAPD-TM-918, Pittsburgh, PA. (14)

Westwater, J. W., 1956, Boiling of Liquids, *Adv. Chem. Eng. 1*:2–76. (2)

Westwater, J. W., and D. B. Kirby, 1963, Bubble and Vapor Behavior on a Heated Horizontal Plate during Pool Boiling near Burnout, AIChE Preprint 14, 6th Natl. Heat Transfer Conf., Boston, MA. (5)

Westwater, J. W., A. J. Lowery, and F. S. Pramuk, 1955, Sound of Boiling, *Science 122*:332–333. (2)

Westwater, J. W., and J. G. Santangelo, 1955, Photographic Study of Boiling, *Ind. Eng. Chem. 47*:1605. (2)

Westwater, J. W., J. C. Zinn, and K. J. Brobeck, 1989, Correlation of Pool Boiling Curves for the Nomologous Group—Freons, *Trans. ASME, J. Heat Transfer 111*:204–207. (2)

Whalley, P. B., 1976, The Calculation of Dryout in a Rod Bundle, Rep. UK AERE-R-8319, UK AERE, Harwell, England. (5)

Whalley, P. B., P. Hutchinson, and G. F. Hewitt, 1973, The Calculation of Critical Heat Flux in Forced Convective Boiling, Rep. AERE-R-7520, European Two Phase Flow Group Meeting, Brussels. (5)

Whalley, P. B., P. Hutchinson, and G. F. Hewitt, 1974, The Calculation of Critical Heat Flux in Forced Convection Boiling, *Heat Transfer 1974,* vol. IV, pp. 290–294, Int. Heat Transfer Conf., Tokyo. (5)

Wichner, R. P., and H. W. Hoffman, 1965, Pressure Drop with Forced Convection Boiling of Potassium, *Proc. Conf. on Applications of Heat Transfer Instrumentation to Liquid Metals Experiments,* ANL-7100, p. 535, Argonne National Lab., Argonne, IL. (3)

Williams, C. L., and A. C. Peterson, Jr., 1978, Two Phase Flow Patterns with High Pressure Water in a Heated Four-Rod Bundle, *Nuclear Sci. Eng. 68*:155. (3)

Wilson, R. H., L. J. Stanek, J. S. Gellerstedt, and R. A. Lee, 1969, Critical Heat Flux in a Nonuniformly Heated Rod Bundle, in *Two Phase Flow and Heat Transfer in Rod Bundles,* pp. 56–62, ASME, New York. (5)

Witte, L. C., J. W. Stevens, and P. J. Hemingson, 1969, The Effect of Subcooling on the Onset of Transition Boiling, *Am. Nuclear Soc. Trans. 12*(2) 806. (4)

Wong, Y. L., D. C. Groeneveld, and S. C. Cheng, 1990, Semi-analytical CHF Predictions for Horizontal Tubes, *Int. J. Multiphase Flow 16*:123–138. (5)

Wright, R. M., 1961, Downflow Forced Convection Boiling of Water in Uniformly Heated Tubes, USAEC Rep. UCRL-9744, Los Angeles, CA. (4)

Wulff, W., 1978, Lump-Parameter Modeling of One-Dimensional Two-Phase Flow, in *Transient Two-Phase Flow,* Proc. 2nd Specialists Meeting, vol. 1, pp. 191–219, OECD Committee on Safety of Nuclear Installations, Paris. (3)
Wurtz, J., 1978, An Experimental and Theoretical Investigation of Annular Steam-Water Flow in Tubes and Annular Channels, Riso Natl. Lab., Oslo, Norway. (5)
Wyllie, G., 1965, Evaporation and Surface Structure of Liquid, *Proc. Roy. Soc. A197:*383. (2)
Yadigaroglu, G., 1993, Instabilities in Two-Phase Flow, in *Workshop on Multiphase Flow and Heat Transfer: Bases, Modeling and Applications,* chapter 12, University of California, Santa Barbara, CA. (4)
Yadigaroglu, G., and M. Andreani, 1989, Two Fluid Modeling of Thermal-Hydraulic Phenomena for Best Estimate LWR Safety Analysis, *Proc. 4th Int. Topical Meeting on Nuclear Reactors Thermal-Hydraulics,* Karlsruhe, U. Mueller, K. Rehnee, and K. Rust, Eds., Rep. NURETH-4, pp. 980–996. (3)
Yadigaroglu, G., and A. E. Bergles, 1969, An Experimental and Theoretical Study of Density-Wave Oscillation in Two-Phase Flow, M.I.T. Rep. DSR 74629-3 (HTL 74629–67), Massachusetts Institute of Technology, Cambridge, MA. (6)
Yadigaroglu, G., and A. E. Bergles, 1972, Fundamental and Higher-Mode Density-Wave Oscillation in Two-Phase Flow: The Importance of Single Phase Region, *Trans. ASME, J. Heat Transfer 94:*189–195. (6)
Yang, J., L. C. Chow, and M. R. Pais, 1993, Nucleate Boiling Heat Transfer in Spray Cooling, ASME Paper 93-HT-29, Natl. Heat Transfer Conf., Atlanta, GA, ASME, New York. (4)
Ying, S. H., and J. Weisman, 1986, Prediction of the CHF in Flow Boiling at Intermediate Qualities, *Int. J. Mass Transfer 29*(11):1639–1648. (5)
Yoder, G. L., Jr., and W. M. Rohsenow, 1980, Dispersed Flow Film Boiling, MIT Heat Transfer Lab. Rep. 85694-103, Massachusetts Institute of Technology, Cambridge, MA. (4)
Zaker, T. A., and A. H. Wiedermann, 1966, Water Depressurization Studies, IITRI-578-P, pp. 21–26, IIT Res. Inst., Chicago, Illinois. (3)
Zaloudek, F. R., 1963, The Critical Flow of Hot Water through Short Tubes, USAEC Rep. HW-77594, Hanford, WA. (3)
Zeigarnick, Y. A., and V. D. Litvinov, 1980, Heat Transfer and Pressure Drop in Sodium Boiling in Tubes, *Nuclear Sci. Eng. 73:*19–28. (4)
Zeng, L. Z., and J. F. Kausner, 1993, Nucleation Site Density in Forced Convection Boiling, *Trans. ASME, J. Heat Transfer 115:*215–221. (4)
Zenkevich, B. A., and V. I. Subbotin, 1959, Critical Heat Fluxes in Subcooled Water Forced Circulation, *J. Nuclear Energy, Part B, Reactor Technol. 1:*134–140. (5)
Zenkevich, B. A., et al., 1969, An Analysis and Correlation of the Experimental Data on Burnout in the Case of Forced Boiling Water in Pipes, Physico Energy Institute, Afomizdat, Moscow, HTFS Transl. 12022. (5)
Zernick, W., H. B. Curren, E. Elyash, and G. Prevette, 1962, THINC, A Thermal Hydraulic Interaction Code for Semi-open or Closed Channel Cores, Rep. WCAP-3704. Westinghouse Electric Corp., Pittsburgh, PA. (App.)
Zetzmann, K., 1981, Flow Pattern of Two Phase Flow inside Cooled Tubes, in *Two-Phase Flow and Heat Transfer in the Power and Processing Industries,* Hemisphere, Washington, DC. (3)
Zivi, S. M., and A. B. Jones, 1966, An Analysis of EBWR Instability by FABLE Program, *Trans. Am. Nuclear Soc.,* ANS 1966 Annual Meeting, American Nuclear Society, LaGrange Park, IL. (6)
Zuber, N., 1958, On Stability of Boiling Heat Transfer, *Trans. ASME, J. Heat Transfer 80:*711–720. (2)
Zuber, N., 1959, Hydrodynamic Aspects of Boiling Heat Transfer, USAEC Rep. AECU-4439, Ph.D. thesis, University of California, Los Angeles, CA. (2)
Zuber, N., 1961, The Dynamics of Vapor Bubbles in Non-uniform Temperature Fields, *Int. J. Heat Mass Transfer 2:*83–98. (2)
Zuber, N., and J. A. Findlay, 1965, Average Volumetric Concentration in Two-Phase Flow System, *Trans. ASME, J. Heat Transfer 87:*453. (3)

Zuber, N., M. Tribus, and J. W. Westwater, 1961, The Hydrodynamics Crisis in Pool Boiling of Saturated and Subcooled Liquids, in *International Developments in Heat Transfer, Part II*, pp. 230–236, ASME, New York. (2)

Zun, I., 1985, The Transverse Migration of Bubbles Influenced by Walls in Vertical Two-Phase Bubbly Flow, 2nd Int. Conf. on Multiphase Flow, London, pp. 127–139, BHRA, The Fluid Engineering Center, Cranfield, England. (3)

Zun, I., 1988, Transition from Wall Void Peaking to Core Void Peaking in Turbulent Bubbly Flow, in *Transient Phenomena in Multiphase Flow*, ICHMT Int. Seminar, N. H. Afgan, Ed., pp. 225–245, Hemisphere, Washington, DC. (3)

Zun, I., 1990, The Mechanism of Bubble Non-homogenous Distribution in 2-Phase Shear Flow, *Nuclear Eng. Design 118:*155–162. (3)

INDEX

Accommodation coefficient, 31, 34
Alkali metals, 17, 39, 71–73, 79, 101, 110–112, 140, 252, 266, 272, 364, 462
Analysis, one dimensional, 129
Anemometer, hot wire, 161
Anemometry:
 Doppler, 164
 optical, 164
 thermal, 164
Apex angle, 13, 14
Area:
 of influence, 60, 62
 local preferable, 137
Attenuation coefficient, 163

Bernoulli effect, 146
Blowdown, 225, 227–228, 230, 283, 286–288
 experiments, 219–220, 222
 heat transfer, 283
Boiling, flow, 1–4, 7, 44, 80, 86, 117, 245, 249, 251
 burnout, (*see* flow boiling crisis)
 crisis, 4, 303–310, 311–322, 328, 331, 333–334, 338, 342–343, 346–348, 352, 353, 357, 361–363, 366–367, 370, 378–380, 382–384, 388, 392, 399–401, 404–406, 428, 433, 440, 454, 458–460, 466, 469, 473, 479
 critical heat flux (CHF), 258, 272–274, 283, 286, 303–305, 309, 314, 317, 322–329, 332–334, 336, 338, 342, 344, 348–351, 357–363, 366–371, 373–399, 401–406, 413–417, 420–433, 436, 438, 441–443, 452–455, 460, 461, 467
 film, 274, 277, 279, 283, 306, 307, 311, 313, 459, 478, 479
  departure from, (DFB), 258, 287
  dispersed flow, 182, 277
  inverted annular, 301
  partial, 245, 248, 283, 286, 289
  stable, 245, 271, 274, 276, 283, 306, 307
  with liquid metals, 252, 258, 265, 268, 271, 274
  (*See also* Liquid metals)
 local, 2, 143, 144, 152, 246, 249, 258, 259, 301, 311, 331
 nucleate, 202, 245, 248, 251, 252, 257, 266, 283, 303, 306, 307, 313
  departure from, (DNB), 258, 287, 288, 303–307, 310–312, 314, 315, 318, 365, 401, 402, 408–415
  partial, 245, 248, 249, 250, 252
  subcooled, 143, 248, 249, 257, 258, 305, 338, 341, 454
  (*See also* local flow boiling)
 saturated, 143, 144, 155, 258, 266, 301, 304, 305, 312, 326
 transition, (*see* partial film boiling)
Boiling, pool, 1–4, 7, 15, 44, 46, 50, 54–55, 65, 71, 72, 78, 80–81, 83, 86, 99
 burnout, 44, 80, 87–88, 93, 97, 117, 259
 crisis, 5, 50, 56, 80–81, 99, 101
 (*See also* burnout)

533

Boiling, pool (*continued*)
  critical heat flux, 3–4, 54, 81, 83–86, 88, 90, 91, 95, 97, 99–102, 116–117, 393
    hydrodynamic prediction of, 88
  film, 44, 50, 52, 57, 87–88, 102, 104, 108–109, 112, 115, 277
    partial, 2–3, 44, 50, 52, 57, 81, 84–88, 102–104, 106, 112, 116–117
    stable, 2–3, 81, 86–87, 122
  incipient, 14, 16–19, 79, 80, 159–161, 248, 255, 298
  liquid metals, with, 16–17, 20, 43, 48, 72, 74, 75, 78, 100–102, 110–112, 114–115
    (*See also* Liquid metals)
  nucleate, 1–3, 17, 22, 39, 44–46, 50, 55, 60, 62, 65–67, 69–74, 79, 80, 84, 86–88, 95, 97, 101, 102, 116, 261, 299
    departure from, (DNB), 3, 80
    with liquid metals, 14, 39, 44, 48–49, 71, 77–80, 102, 248
  saturated, 16, 39, 81, 82, 98, 287
  stable, 17, 72, 74
  subcooled, 25, 44, 83–84, 93, 97, 98
  unstable, 72–74, 102
Boiling regimes, 1–3
Boiling sound, 37, 44, 45
Boiling superheat, 17, 79, 80, 252–254, 256
  incipient, 159–161, 252, 254–257, 300
Boiling surface, fouling, 268
  thermal conductivity of, 269
Boiling suppression, 18, 19, 260, 266
Boltzmann constant, 8, 38
Boltzmann distribution, 9
Boundary layer, 262, 271, 312
  separation, 320, 321
  technique, 173
Bubble:
  activation, 12
  agglomeration, 126
  agitation, 58, 59, 248
  blanket, 326, 328
    (*See also* vapor blanket)
  boundary layer, 143, 154, 173, 257, 271
  counts, 16
  departure, 1, 2, 7, 16, 19, 37, 38, 40, 41, 50, 58, 60, 62, 67, 154, 321, 340,
    diameter, 299, 340
  deposition theory, 174
  detached, 143, 153, 154, 331, 343
  diameter, 38, 41, 43, 56, 60, 62, 103, 122, 300, 369
  dispersed, 124, 135, 175
  dynamics, 23–25
  elongated, 175, 300
  embryo, 11, 19
  equilibrium equation, 17, 74
  frequencies, 24, 25, 40, 67, 81
  generation, 1, 17, 41, 81, 147, 245, 326
  growth, 1, 4, 7, 10, 12, 14, 15, 17, 20, 22–28, 30–34, 37–39, 49, 62, 68, 71, 81, 88, 142, 246, 250, 263, 264, 300
  interface, 8, 23, 25–27, 29
  isolated, 122
  layer, 143, 145, 148, 304, 307, 320, 326, 328–334, 336, 337, 340, 342, 343, 367, 405, 455
  life, 26, 300
  nucleation, 1, 6, 7, 15, 19, 20, 23, 116, 250
  population, 9, 40, 50, 67, 245, 246
  radius, 25, 35, 61, 237, 238, 300
  release, 103
  Reynolds number, 57, 189, 209
    (*See also* boiling Reynolds number)
  rise velocity, 40, 43, 157, 189
  segregation, 143
  shape, 119, 122, 300
  slip velocity, 157
  site, 8, 14, 81
  size, 3, 8, 24–26, 29, 34, 36, 37, 40, 41, 50, 54, 81, 237, 300, 323, 351, 352, 357, 369, 392
    equilibrium, 10, 17
  Taylor, 137, 190, 210
  unstable, 122
Bubbly flow (*see flow*)
Bumping, 72, 73, 77, 459, 462
Buoyancy, 1, 7, 37, 41, 42, 81, 151, 189
  force, 130, 132
  modulus, 55
Burnell correlation, 225, 227
Burnout, 314, 430–432
  fast, 314
  power, 309
  slow, 316
    (*See also* pool boiling burnout)

Capillary:
  effect, 10
  wave, 52, 105
Cavity, 8, 10–13, 17–19, 48, 122
  conical, 13
  cylindrical, 17, 72–74

INDEX 535

deactivation, 252–254
depth, 14
reentrant, 100, 271, 300
site, 60, 80
 active, 60, 62, 63, 67, 72–73, 77
 size, 11, 15, 17, 20, 38, 39, 64, 74, 77, 83, 117, 251
Channel voiding, 297
Chemical reactions, 110, 111
Cladding temperature, peak, 297
Clausiues-Clapeyron equation, 10, 13, 29, 31
Coolant depression, 230
Computer codes:
 COBRA, 183
 COBRA II, 484
 COBRA IIIC, 425, 485–6
 COBRA IIIC/MIT-1, 428
 DYNAM, 476
 FIDAS, 454
 FLICA, 209
 HAMBO, 417–418, 422, 484
 HYDRO, 209
 HYKAMO, 209
 NAIAD, 218
 NATOF, 271
 RELAP 4/MOD 7, 290
 RELAP-5, 183, 293–294
 RETRAN, 183
 ROMONA, 209
 STABLE, 476
 THINC I, 486
 THINC II, 403, 410, 412, 484, 486
 THINC IV, 486, 489
 TORC, 486
 TRAC, 183
Colburn factor, 120
Conduction-convection contribution, 97, 98, 102
Constitutive laws, intrinsic, 170
Contact angle, 13, 14, 37–38, 68, 86, 87, 96, 300
Convection number, 264
Critical discharge rate, 239
Critical distance, 73
Critical flow, 120, 209, 219, 220, 222, 224–228
 homogeneous choking, 219, 222, 226, 227
 mass flux, 222, 223, 227, 228
 model, 227, 244
 pattern, 226
 velocity, 241
Critical heat flux, (see boiling crisis)
 correlation, 342, 355, 357, 359, 360, 367, 373, 376–378, 401, 404, 453

A.R.S. correlation, 429–433
B & W 2 correlation, 415–416, 423, 426
CISE-1 correlation, 355, 367, 433–438
CISE-3 correlation, 433, 439–441
CISE-GE correlation, 443–445
Cincinnati correlation, 427
Cincinnati model, improved 428–430
CE-1 correlation, 416–417, 426
Columbia correlation, 423–424, 426, 429–430
GE lower-envelope correlation, 441–443
integral concept, 334
local condition concept, 334, 370–371, 379, 383
for liquid metals, 360
for organic fluids, 357–359
Russian correlation of Ryzhov & Ankhipow, 426–427
system parameter concept of, 370–372, 369, 384
W-3 DNB correlation, 401, 405–410, 423, 426, 429–430
WSC-2 correlation, 417–423, 426
mechanism:
 microscopic analysis of, 303–304, 406, 310, 317–318
 phenomenological analysis of, 343
effects of ribs and spacers on, 393–394
flow instability effect on, 398
flow obstruction effect on, 391
inverse mass flux effect on, 369, 370
nondimensional, 376
parameter effect on, 366–369, 377–381
surface roughness effect on, 391
transient, 399–402
wall thermal capacitance effect on, 392
Critical power, 334–335, 348, 384–387, 394, 428, 431–436, 438, 443, 453, 454, 466
 GEXL correlation, 443, 445
Critical quality, 354, 434, 437, 443, 444

Deissler expression on eddy viscosity, 178, 192
Density, apparent, 153, 173
Density, flow average, 151
Density ratio, vapor to liquid, 161
Diffusion mixing, 481
Diffusion velocity, 169
Dimensionless group, 4, 55, 57, 122, 128, 355
 for heat transfer parameter, 140
 for heater radius, 90

Dimensionless group (*continued*)
  for inverse viscosity, 180
  for two-phase parameter, 132, 182, 193
  for superficial velocity, 133
Discharge coefficient, 213
Discharge mass flux, 225
Discharge rate, 220
Drift flux models, 154, 168, 174, 184
  Chexal-Lellouche model, 154
  Dix model, 154
  Liao, Parlos and Griffith model, 154
  multichannel, 187
  Ohkawa-Lahey model, 154
  Wilson bubble rise model, 154
  Yeh-Hockreiter model, 154
Drift velocity, 169, 187
  effective, 174
Droplet:
  breakup, 282
  deposition velocity, 182, 280, 343, 348
  dryout diameter, 281, 282
  entrainment, 280, 405
  formation, 278
  liquid, 136, 138
    evaporation, 277
    suspension, 239, 278
    wall collisions, 277, 280
  vapor, 144
Dry patch, 20, 48, 291, 306
Dry spot, 68
Dryout, 247, 258, 277–279, 281, 282, 286, 288,
    303–305, 309, 314, 316–318, 338, 339,
    343, 346–351, 355, 395, 433, 447–449,
    451, 452, 454, 465
  model, 116, 117, 181
  quality, 279, 282, 347
  (*See also* boiling crisis)
Dynamic similitude, 351

Ebullition, 681
  cycle, 8, 17, 20, 22, 23, 47, 55, 67, 300
  site, 12
Electromagnetic flow meter, 162
Energy, activation, 9
Energy, free, 8, 9
Enthalpy, 230
  critical, 328–329
  stagnation, 218, 222
Emergency cooling, 286
  coolant injection, 220

Euler number, 55
Evaporation film, 272, 273, 309, 314
Expulsion, slug-type, 297
  model of, 297–298

Fiber optic probe, 161, 165
  fiber optic video probe, 161
Flooding, 180, 242, 287, 291, 373
  bottom, 288, 289
Flow, two-phase, 1, 3, 4, 55, 119, 123, 128, 136,
    139
  adiabatic, 119–120, 124, 128, 138, 142, 157,
    201
  boiling, 119, 138, 141, 142, 145, 147, 149, 159,
    195–202
  dynamic quality, 151
  gravity-controlled, 129
  hydrodynamics of, 119, 120
  instability, 4, 5, 122, 364–365, 383, 398, 457,
    459–460, 464–469, 471, 472, 476–479
  (*See also* Oscillations)
    compound, 4, 5, 458, 459
    condensation chugging, 459, 462, 463
    dynamic, 4, 5, 459, 463–468, 474, 476–479
    fundamental, 4, 5, 459
    flow-void feedback, 464
    hydrodynamic, 458, 465, 468
    Ledinegg, (flow excursion), 4, 459, 473
    nonequilibrium-state, 459, 462, 473
    oscillating, 398, 457
    parallel channel, 459, 466, 471, 478
    parametric effect on, 468
    relaxation, 459, 461–463, 473
    static, 4, 457, 459–463, 466, 473, 479
    system, 460
    thermal, (boiling crisis), 459, 473
    thermo-hydrodynamic, (BWR), 459, 465,
      466, 478, 480
    threshold of, 458, 471, 476, 480
  liquid metal-gas, 139, 140, 159, 195, 196
  (*See also* Liquid metals)
    flow regime map, 140, 141
  models, 120, 168, 356
    for annular flow, 177
    for bubbly flow, 173
    drift flux model, 184
    for flow pattern transition, 172
    homogeneous model, 168, 173, 184, 189,
      198, 203, 208, 212, 219, 221–222, 225,
      341

INDEX **537**

separate flow model, 125, 168, 182, 184, 191, 195, 222
   for slug flow, 174
   for stratified flow, 182
   unified model, 172
patterns (regimes), 3, 4, 7, 119–122, 124, 125, 128, 136, 138, 139, 142, 207, 214, 226, 245, 246, 258, 308, 352–353, 369, 370, 399
   annular flow, 121, 122, 129, 130, 135–140, 143, 146, 172, 222, 235, 242, 246–247, 258, 265, 277–279, 281, 304–309, 343, 344, 347, 351, 352, 372, 454
   annular, dispersed, 123, 127, 129, 132
   annular, inverse, 122, 180, 277–279, 281–282, 311
   annular, wispy, 122, 136
   bubbly flow, 121–123, 126–127, 129, 133, 135, 136, 143, 172, 173, 246, 247, 258, 265, 304, 307, 377, 405
   bubbly, dispersed, 124, 129, 132
   churn flow, 121, 122, 126–128, 135, 138, 140, 265
   dispersed flow, 180, 181, 227, 232, 234, 243, 277–279, 282, 288, 289, 302
   dispersed flow, dispersion coefficient of, 238
   drop flow (liquid deficient region), 247, 248, 259, 274
   effect of obstructions on, 138
   film flow, climbing, 146, 309, 372
   film flow, falling, 127–129, 177
   froth flow, 123, 136
   intermittent flow, 132, 139, 172
   mist flow, 121, 235
   plug flow, 121, 122, 132
   slug flow, 121–123, 125–127, 129–133, 135–138, 140, 143, 159, 172, 174, 246, 247, 258, 265
   slug flow, pseudo-, 132
   stratified flow, 120, 123–125, 129, 136, 172, 191
   stratified wavy flow, 121, 130, 139
pattern criteria, 125, 128, 129
pattern map, 120, 123, 128, 134, 135, 140, 288, 390
   for horizontal flow, 123, 124, 130, 142
   for vertical flow, 125–127, 144
pattern transition, 3, 4, 123, 129–133, 135–140, 146, 147, 172, 175, 185, 207, 219, 226, 227, 376, 458, 459
   pressure-gradient controlled, 129

steady state, 123, 207
transient, 120, 183, 210
Fluorescence method, 167
Forced convection boiling, (see flow boiling)
Fourier decomposition technique, 241
Froude number, 56, 128, 130, 352
   density modified, 133
   gas, 242
   liquid, 242, 264
   mixture, 129
Fuel-coolant interaction, 115

Gas-liquid relative velocity, 132
Geometric optics, principles of, 164
Geysering, 459, 462
Grashof number, 108

Heat diffusion effect, 24
Heat flux parameter, axial, 420
Heat imbalance factor, subchannel, 418, 421, 422
Heat transfer:
   coefficient, 36, 79, 80, 103, 173, 260, 295
      boiling, 47, 58, 62, 78, 88, 111, 275
      convective, 103, 107
      local, 105, 266
      nucleate boiling, 260
      nuclear boiling, forced convection factor in, 261–262
      nuclear boiling, suppression factor in, 261, 263
   convective, 259, 266, 277
   parameter, dimensionless, 140
   mist, 277
   vapor-drop, 277, 279, 280
   wall-drop, 277, 279
   wall-drop, effective, 182, 280–282
Helmholtz critical wave length, 54, 340
Helmholtz critical wave number, (see critical wave number)
Helmholtz instability, 50, 51, 93, 119, 281, 338
Helmholtz stability requirement, 81, 84
Hot channel factor, 403
Hydraulic diameter concept, 183
Hydrodynamic:
   equilibrium, 347, 348
   factors, 41
   instability model, 116, 117
   region, 41, 42
   similitude, 351

Induced-convection buoyancy parameter, 93
Induced-convection scale parameter, 93
Inert gas, 17–19, 80
Inertia effect, 232, 238
Inertial force, 39, 40, 48, 55–57, 81, 82, 93, 125
Instruments:
 conductance probe, 120
 hot wire probe, 120
 impedance imaging method, 161
 impedance probe, 140
 photographic definition, 120
Interface:
 area, 163, 235
 heat balance, 29
 jump condition (discontinuities), 170
 liquid-vapor, 48, 50, 62, 91, 111, 117, 132, 138, 141, 388
 (See also vapor-liquid interface)
 smooth, 234
 shear stress at, 192, 193, 350
 vapor-liquid, 9, 13, 27, 30, 48, 50, 62, 80, 84, 91, 105, 107, 111, 117, 175
 velocity, 23
 wavy, 235
Interfacial:
 friction factor ratio, 242
 heat transfer, 233, 302
 momentum transfer, 233
 pressure model, 241
 transport, 234, 239, 329

Jacob number, 28, 29, 39, 43, 44, 50, 56, 58, 323
Joule heating, 254

Kelvin-Helmholtz instability, 130
Kutateladze number, 56, 85, 87
Kutateladze correlation, 71, 83, 98, 101

Laplace equation, 9, 74
Laplace reference length, 104
Latent heat of vaporization, 45, 56, 101, 103–105, 115, 141
Latent heat transport, 45, 50, 55, 248
Law of corresponding states, 100
Leidenfrost point, 102, 245, 275–277, 288
Liquid:
 boundary layer, 202
 drops in vapor upflow (carryover), 136

entrainment, 138, 248, 295, 343, 345–348, 389
film, falling, 178, 179
film thickness, 161, 166, 167, 266, 316, 343–345, 349, 351
film thickness, dimensionless, 201
film, wavy, 166, 305, 307
flowrate, dimensionless, 193, 194
holdup, 125, 193, 194
layer, 142, 146, 346, 347
level, equilibrium, 185
Liquid metals, 71, 79, 80, 97–99, 105, 111, 115, 140, 243, 254, 255, 256
 boiling crisis, 98–101, 102
 critical heat flux, 98, 100–101, 361–364, 367
 fuel cell, 273
 two-phase flow, with gas, (see two-phase flow)
 Magnetohydrodynamic (MHD) power generator, 140, 273
Liquid sublayer, evaporation of, 454
Liquid-vapor exchange, 56, 58–61, 248, 327
Liquid-vapor interface mass flux, 124
Liquid volume fraction, 136, 140

Macrolayer, 116, 117
Martinelli parameter, 124, 125, 182, 264
Mass transfer coefficient, 350, 351, 451
Matrix, column, 488
Memory effect, upstream, 333–334, 395
Microconvection, 43, 47, 248
Microlayer, 20, 21, 45–48, 59–62, 68–71, 116, 117, 338–340
 evaporation, 59, 62, 68–71, 248
 evaporative mechanism, 58
Microscopic parameters, 4
Microstructure of surface, 79, 80, 101
Microthermocouples, 161, 162
Mikic-Rohsenow correlation, 62
Mishima and Ishii correlation, 140
Moiré fringes, 167
Molecular diffusion, 110, 111, 115
Momentum exchange model, 272

Natural convection, 2, 7, 64–65, 68–74, 100, 116
Needle-contact probe, 166
Nondimensional groups, (see Dimensionless group)
Nuclear instability, 5
Nuclear reactivity, 297
Nuclear reactors, 5, 119, 132, 138, 140, 152, 154,

163, 170, 183, 186, 187, 210, 230, 275, 276, 279, 283, 287, 295, 303, 328, 357, 366, 379, 391, 395, 401, 455
Boiling Water Reactor (BWR), 5, 154, 195, 243, 288, 304, 398, 399, 418, 422, 437, 445, 447, 465, 478, 480
Fusion reactor, 140, 301, 454, 455
Liquid Metal Cooled Reactor (LMR), 243, 254–256, 267, 363
  Liquid Metal Fast Breeder Reactor (LMFBR), loop-type, 255–256
  Liquid Metal Fast Breeder Reactor (LMFBR), pot-type, 255–256
  Pressurized Water Reactor (PWR), 132, 136, 139, 243, 257, 267, 286, 288, 290, 292, 295–297, 304, 394, 395, 410, 418, 422, 437
Nuclear reactor, accidents in, 230, 279, 285
  breaking instrument tubes, 288, 292
  cold-leg break, 285
  emergency coolant injection, 220
  emergency core cooling, 287, 293
  loss-of-coolant, (LOCA), 120, 136, 220, 279, 288, 274, 443, 473
  loss-of-coolant test, 294, 296
  MOD-1 system, 296
  transient cooling, 283, 286
  tube rupture, 230, 296, 297
Nuclear reactor safety, 136
  analysis, 182, 183, 297, 363, 473
  fast reactors, of, 462
  ratio of Departure from Nucleate Boiling (DNBR), 401–403, 412, 413, 429, 430
  safety injection system, 293
  safety margin in limiting the reactor power, 454
  seismic event, 294
Nucleate boiling, (see Flow boiling and pool boiling)
Nucleation, 8, 9, 11, 14, 19, 23, 71, 72, 77, 248, 251
  cavity, 252, 253
  sites, 2, 10, 16, 19, 40, 54, 67, 68, 79, 245, 257, 259, 268, 392
  size, 8, 10, 14
Nusselt number, 108, 268, 282, 283

Oscillations, flow, 460
  acoustic wave, 5, 459, 464, 475–476
  (See also pressure wave)

condensation, 459, 466
density wave, 399, 459, 462–464, 466, 468, 470, 475–480
dynamic, 399, 459
flow regime induced, 399, 459, 461–462, 473
pressure-drop, 399, 459, 466–468, 478
resonant hydrodynamic, 465
thermal, 459, 465
time-delay, 464
Oxide level, 80
Oxygen concentration, 79, 101, 252

Parametric distortion, 366
Peclet number, 95
Plesset-Zwick solution, 24, 37
Poisson distribution, 67
Prandtl number, 55–57, 61, 154, 261
Prandtl's mixing length theory, 120, 173, 483
Pressure drop in two-phase flow, 56, 119, 120, 180, 186, 187, 202, 206
  analytic models for, 188
  for annular flow, 191, 211
  for bubbly flow, 188
  for slug flow, 190
  for stratified flow, 191
  in critical flow, 209
  dimensionless, 125, 182, 191, 215, 243
  elevation, 187
  in flow restrictions, 120, 210
    through abrupt, 212
    through abrupt expansion, 210
    Borda-Carnot coefficient, 210
    through orifice, 213
    vena contracta area ratio, 213, 226, 227
  frictional loss, 3, 56, 187, 188, 190, 195, 464
    correlation of, 196, 202
    Fanning friction factor, 173
    Fanning friction loss equation, 188
    friction coefficient, average, 198, 199
    friction factor, 177, 188, 192, 195, 201
    friction factor, mixture, 173, 185
    friction factor, normalizing, 188
    Lockhart-Martinelli correlation, 158, 194, 195, 202
    Martinelli-Nelson correlation, 158, 203
    Martinelli-Nelson-Chisholm method, 203, 206
  momentum change in, 187
  pressure coefficient of, 130, 327

Pressure drop in two-phase flow (*continued*)
  relative, (frictional loss multiplier), 125, 191, 195–197, 202, 204, 205, 208, 214–217
  velocity, friction, 193, 194, 341
 obstructions, with, 214, 230
 pressure reactivity coefficient, 465
 pressure loss coefficient, overall, 216
 reduced, 243
 rod bundles, in, 194, 207
  grid spacer factor for, 327
 transient flow, in, 209, 217, 228
  energy density, average, 218
  energy flux, average, 218
  mass flux, average, 218
  momentum flux, average, 218
Pressure propagation rate, 239, 240
Pressure pulse, 230, 231, 236
 propagation of, 231, 231, 240
Pressure waves, 230, 236, 238
 propagation of, 232, 234, 239
 rarefaction, 232, 236
Probability distribution for uncertainties, 404

Quench front, 302
 velocity of, 295
Quench, saturated pool, 287

Radiation attenuation technique, 161
Rayleigh equation, 23, 29, 31
Rayleigh unstable wavelength, 97
Reflood, analysis of, 295, 297
Relaxation length, 73
Relaxation time, 231, 238, 240
 ratio of, 239
Rewetting model (EPRI), 295
Reynolds number, 188, 198, 199, 209, 261, 322, 323, 337, 341, 352, 483
 analogy, 173
 boiling, 57, 130
 gas phase, 192, 201
 liquid phase, 193, 199, 261
Rod bundle:
 critical heat flux data with, 363, 364, 385
  bowing effect on, 395
  spacer effect on, 395, 422
 film boiling in, 277
 pressure drop in, 207
 subcooled flow boiling in, 257
 transient flow in, 186
  pressure drop of, 257
  model for, 187

Scaling:
 criteria, 351, 355, 356
 factor, 352, 354–356
 number, dimensionless, 355
Scattering coefficient, 163
Shadowgraph, 8, 58
Shape fact for nonuniform flux, 332, 333, 366, 406, 416
Shock tube, 230
Slip equation, 170, 218
Slip flow, model, 222
Slip ratio, 150, 218, 219, 223, 224, 405
 critical, 225
Slip velocity, bubble, 157, 281
Slug flow:
 geometry, 175
 mixing zone, 191
 model for, 174
 pressure drop of, 190
 (*See also* Pressure drop of two-phase flow)
 transient, 186, 236
 slug unit in, 175
  frictional pressure gradient across, 190
  gas zone of, 190
  slug zone, 175, 190
Sonic wave, propagation, 236, 240
Sound, velocity of (acoustic velocity), 86, 209, 237, 239
Spheroidal modulus (S_o), 57
Spheroidal state, 245, 275
Spray, top, 287, 290
Sputtering, 291
Stanton number, 57
Static quality, 151, 158
Steam generator, 136, 257, 266, 267, 276, 279, 293, 296–297, 470
Stefan-Boltzmann constant, 107
Stokes law, 238
Stratified flow, pressure drop model of, 191
Streamline map, for a wave, 178, 179
Subchannel:
 analysis code, 403, 410–411, 425–428, 445, 481
 crossflow between, 481, 482, 484–486
 pressure difference between, 485
 pressure drop in, 484
Subcooling, degree of, 246, 254, 259, 305
Subcooling on flow instability, inlet, 470
Superficial velocity, 141, 189
 mixture average, 169
 non-dimensional, 134
Superposition, assumption of, 260
Surface conditions, 4, 88, 89, 393

Surface energy, 9
Surface factor, 36
Surface roughness, 1, 73, 74, 88, 90, 391, 392

Taylor series, 13
Taylor instability, 50, 52, 84, 91, 103, 374
Technology, new cooling, 455
Thermal contact, 79
Thermal:
  boundary layer, 19, 23, 29, 36, 152, 264
  diffusion coefficient, nondimensional, 411, 483
  Hydraulic Test Facility, 294
  hysteresis, 287, 288
  inertia, 23
  layer, 11, 15, 20, 37, 61, 68
Thermocapillarity effect, 62, 257, 258
Thermodynamic:
  conditions, 41
  critical property, 116
  equilibrium, 8, 9, 23, 31, 81, 119, 218, 221, 222, 227, 258
  region, 41–43
  similarity, 116
  similitude, 351
Thermohydrodynamic problem, coupled, 119
Transient cooling, 283, 286
Transient two-phase flow, 357, 358
Transition boiling (see partial film boiling)
Turbulent mixing at interface, 337
Turbulent mixing in subchannels, 447, 448

Ultrasonic techniques, 163

Vapor:
  blanket, 311, 312, 318, 340
  burst, 459, 462
  downflow in liquid stream (carryunder), 136
  film, 133
  slip, 151, 189
Vaporization (evaporation), 46–48, 50, 109
  forced convection, 265
    correlation for, 266
Velocity:
  boundary layer, 152
  profile, parabolic, 178
  terminal, 157
  weighted mean, 157
View factor, 107
Virtual mass effect, 232, 233, 235

Viscosity, apparent, 173
Void distribution, 3, 148, 150, 155, 161, 174
Void drift, 447
Void fraction, 122, 142, 147, 148, 152, 153, 155, 158, 162, 163, 173, 233, 236–238, 263, 265, 272, 277, 288, 290, 341, 343, 482
  average of a slug unit, 176
  channel average, 148
  correlation, (see pressure drop, frictional loss)
  critical value of, 135, 336, 360
  model, 154, 160
  space average, 189, 190
Void reactivity coefficient, 398
Volume factor, 36
Volumetric:
  coefficient, 96
  flow rate, 189
  flux density, 155
  fluxes, average, 139
  interfacial area, 163
  vapor quality, 128, 149
Von Karman's relationship, 178

Waiting period, 8, 10, 20, 22, 23, 37, 40, 48, 67, 68
Waiting time, 20, 71, 73
Wall:
  shear stress, 192
  superheat, 2, 10, 17, 67, 77, 86, 108, 111, 160, 245, 246, 250, 252, 260, 268, 272, 287, 368, 369
  voidage, 152, 153, 195
Wave:
  amplitude of, 51, 52
  analysis, numerical, 231
  angular velocity, 51
  compressional (shock), 230, 235
    propagation of, 232
  crest, 202, 203
  decompressional, 230
  form, 51
  front, 231, 239, 240
  frequency, 237
  interfacial, 84, 132, 136
  model, one dimensional, 242
  number, 51, 237, 238
    critical, 84
  period of, 231, 238
  pressure, 230, 231, 236
  propagation of pressure, 230
  roll, 202

Wave (*continued*)
  shape, 178, 232
  sonic (acoustic), 236
    propagation of, 237, 240
  velocity, 178
Wavelength, 52–54, 91, 237
  critical, 108, 340
  most dangerous, 54, 103
  unstable, 85

Weber number, 82, 130, 281, 282
  breakup criterion, 278
  critical, 81, 83, 282
  mixture, 129
Weber-Reynolds number, 57, 352
Wetting agent, 75, 78–80
Work-heat conversion factor, 61, 218
Work interaction between two phases, 185